Preface

The *Uniform Building Code* ™ is dedicated to the development of better building construction and greater safety to the public by uniformity in building laws. The code is founded on broad-based principles that make possible the use of new materials and new construction systems.

The *Uniform Building Code* was first enacted by the International Conference of Building Officials at the Sixth Annual Business Meeting held in Phoenix, Arizona, October 18-21, 1927. Revised editions of this code have been published since that time at approximate three-year intervals. New editions incorporate changes approved since the last edition.

The *Uniform Building Code* is designed to be compatible with related publications to provide a complete set of documents for regulatory use. See the publications list following this preface for a listing of the complete family of Uniform Codes and related publications.

Code Changes. The ICBO code development process has been suspended by the Board of Directors and, because of this action, changes to the *Uniform Building Code* will not be processed. For more information, write to the International Conference of Building Officials, 5360 Workman Mill Road, Whittier, California 90601-2298. An analysis of changes between editions is published in the *Analysis of Revisions to the Uniform Codes*.

Marginal Markings. Solid vertical lines in the margins within the body of the code indicate a change from the requirements of the 1994 edition except where an entire chapter was revised, a new chapter was added or a change was minor. Where an entire chapter was revised or a new chapter was added, a notation appears at the beginning of that chapter. The letter **F** repeating in line vertically in the margin indicates that the provision is maintained under the code change procedures of the International Fire Code Institute. Deletion indicators (♦) are provided in the margin where a paragraph or item listing has been deleted if the deletion resulted in a change of requirements.

Three-Volume Set. Provisions of the *Uniform Building Code* have been divided into a three-volume set. Volume 1 accommodates administrative, fire- and life-safety, and field inspection provisions. Chapters 1 through 15 and Chapters 24 through 35 are printed in Volume 1 in their entirety. Any appendix chapters associated with these chapters are printed in their entirety at the end of Volume 1. Excerpts of certain chapters from Volume 2 are reprinted in Volume 1 to provide greater usability.

Volume 2 accommodates structural engineering design provisions, and specifically contains Chapters 16 through 23 printed in their entirety. Included in this volume are design standards that have been added to their respective chapters as divisions of the chapters. Any appendix chapters associated with these chapters are printed in their entirety at the end of Volume 2. Excerpts of certain chapters from Volume 1 are reprinted in Volume 2 to provide greater usability.

Volume 3 contains material, testing and installation standards.

Metrication. The *Uniform Building Code* was metricated in the 1994 edition. The metric conversions are provided in parenthesis following the English units. Where industry has made metric conversions available, the conversions conform to current industry standards.

Formulas are also provided with metric equivalents. Metric equivalent formulas immediately follow the English formula and are denoted by "For **SI**:" preceding the metric equivalent. Some formulas do not use dimensions and, thus, are not provided with a metric equivalent. Multiplying conversion factors have been provided for formulas where metric forms were unavailable. Tables are provided with multiplying conversion factors in subheadings for each tabulated unit of measurement.

Reference Collection
Kalamazoo Public Library

R 348 U58 v.1 1997

Uniform building code.

1997

UNIFORM BUILDING CODE™

VOLUME 1

ADMINISTRATIVE, FIRE- AND LIFE-SAFETY, AND FIELD INSPECTION PROVISIONS

KALAMAZOO PUBLIC LIBRARY

Fourth Printing

Publication Date: April 1997

ISSN 0896-9655
ISBN 1-884590-87-X (soft cover edition)
ISBN 1-884590-88-8 (loose leaf edition)
ISBN 1-884590-93-4 (3-vol. set—soft cover)
ISBN 1-884590-94-2 (3-vol. set—loose leaf)

COPYRIGHT © 1994, 1995, 1996, 1997

by

International Conference of Building Officials

5360 WORKMAN MILL ROAD
WHITTIER, CALIFORNIA 90601-2298
(800) 284-4406 • (562) 699-0541

PRINTED IN THE U.S.A.

CODES AND RELATED PUBLICATIONS

The International Conference of Building Officials (ICBO) publishes a family of codes, each correlated with the *Uniform Building Code*™ to provide jurisdictions with a complete set of building-related regulations for adoption. Some of these codes are published in affiliation with other organizations such as the International Fire Code Institute (IFCI) and the International Code Council (ICC). Reference materials and related codes also are available to improve knowledge of code enforcement and administration of building inspection programs. Publications and products are continually being added, so inquiries should be directed to Conference headquarters for a listing of available products. Many codes and references are also available on CD-ROM or floppy disk. These are denoted by (*). The following publications and products are available from ICBO:

CODES

*Uniform Building Code, Volumes 1, 2 and 3. The most widely adopted model building code in the United States, the performance-based *Uniform Building Code* is a proven document, meeting the needs of government units charged with the enforcement of building regulations. Volume 1 contains administrative, fire- and life-safety and field inspection provisions; Volume 2 contains structural engineering design provisions; and Volume 3 contains material, testing and installation standards.

*Uniform Mechanical Code™. Provides a complete set of requirements for the design, construction, installation and maintenance of heating, ventilating, cooling and refrigeration systems; incinerators and other heat-producing appliances.

International Plumbing Code™. Provides consistent and technically advanced requirements that can be used across the country to provide comprehensive regulations of modern plumbing systems. Setting minimum regulations for plumbing facilities in terms of performance objectives, the IPC provides for the acceptance of new and innovative products, materials and systems.

International Private Sewage Disposal Code™. Provides flexibility in the development of safety and sanitary individual sewage disposal systems and includes detailed provisions for all aspects of design, installation and inspection of private sewage disposal systems.

International Mechanical Code™. Establishes minimum regulations for mechanical systems using prescriptive and performance-related provisions. It is founded on broad-based principles that make possible the use of new materials and new mechanical designs.

Uniform Zoning Code™. This code is dedicated to intelligent community development and to the benefit of the public welfare by providing a means of promoting uniformity in zoning laws and enforcement.

*Uniform Fire Code™, Volumes 1 and 2. The premier model fire code in the United States, the *Uniform Fire Code* sets forth provisions necessary for fire prevention and fire protection. Published by the International Fire Code Institute, the *Uniform Fire Code* is endorsed by the Western Fire Chiefs Association, the International Association of Fire Chiefs and ICBO. Volume 1 contains code provisions compatible with the *Uniform Building Code,* and Volume 2 contains standards referenced from the code provisions.

*Urban-Wildland Interface Code™. Promulgated by IFCI, this code regulates both land use and the built environment in designated urban-wildland interface areas. This newly developed code is the only model code that bases construction requirements on the fire-hazard severity exposed to the structure. Developed under a grant from the Federal Emergency Management Agency, this code is the direct result of hazard mitigation meetings held after devastating wildfires.

Uniform Housing Code™. Provides complete requirements affecting conservation and rehabilitation of housing. Its regulations are compatible with the *Uniform Building Code.*

Uniform Code for the Abatement of Dangerous Buildings™. A code compatible with the *Uniform Building Code* and the *Uniform Housing Code* which provides equitable remedies consistent with other laws for the repair, vacation or demolition of dangerous buildings.

Uniform Sign Code™. Dedicated to the development of better sign regulation, its requirements pertain to all signs and sign construction attached to buildings.

Uniform Administrative Code™. This code covers administrative areas in connection with adoption of the *Uniform Building Code,* *Uniform Mechanical Code* and related codes. It contains provisions which relate to site preparation, construction, alteration, moving, repair and use and occupancies of buildings or structures and building service equipment, including plumbing, electrical and mechanical regulations. The code is compatible with the administrative provisions of all codes published by the Conference.

Uniform Building Security Code™. This code establishes minimum standards to make dwelling units resistant to unlawful entry. It regulates swinging doors, sliding doors, windows and hardware in connection with dwelling units of apartment houses or one- and two-family dwellings. The code gives consideration to the concerns of police, fire and building officials in establishing requirements for resistance to burglary which are compatible with fire and life safety.

Uniform Code for Building Conservation™. A building conservation guideline presented in code format which will provide a community with the means to preserve its existing buildings while achieving appropriate levels of safety. It is formatted in the same manner as the *Uniform Building Code,* is compatible with other Uniform Codes, and may be adopted as a code or used as a guideline.

Dwelling Construction under the Uniform Building Code™. Designed primarily for use in home building and apprentice training, this book contains requirements applicable to the construction of one- and two-story dwellings based on the requirements of the *Uniform Building Code.* Available in English or Spanish.

Dwelling Construction under the Uniform Mechanical Code™. This publication is for the convenience of the homeowner or contractor interested in installing mechanical equipment in a one- or two-family dwelling in conformance with the *Uniform Mechanical Code.*

Supplements to UBC and related codes. Published in the years between editions, the Supplements contain all approved changes, plus an analysis of those changes.

Uniform Building Code—1927 Edition. A special 60th anniversary printing of the first published *Uniform Building Code.*

One and Two Family Dwelling Code. Promulgated by ICC, this code eliminates conflicts and duplications among the model codes to achieve national uniformity. Covers mechanical and plumbing requirements as well as construction and occupancy.

Application and Commentary on the One and Two Family Dwelling Code. An interpretative commentary on the *One and Two Family Dwelling Code* intended to enhance uniformity of interpretation and application of the code nationwide. Developed by the three model code organizations, this document includes numerous illustrations of code requirements and the rationale for individual provisions.

Model Energy Code. This code includes minimum requirements for effective use of energy in the design of new buildings and structures and additions to existing buildings. It is based on American Society of Heating, Refrigeration and Air-conditioning Engineers Standard 90A-1980 and was originally developed jointly by ICBO, BOCA, SBCCI and the National Conference of States on Building Codes and Standards under a contract funded by the United States Department of Energy. The code is now maintained by ICC and is adopted by reference in the *Uniform Building Code.*

National Electrical Code®. The electrical code used throughout the United States. Published by the National Fire Protection Association, it is an indispensable aid to every electrician, contractor, architect, builder, inspector and anyone who must specify or certify electrical installations.

TECHNICAL REFERENCES AND EDUCATIONAL MATERIALS

Analysis of Revisions to the Uniform Codes™. An analysis of changes between the previous and new editions of the Uniform Codes is provided. Changes between code editions are noted either at the beginning of chapters or in the margins of the code text.

*Handbook to the Uniform Building Code. The handbook is a completely detailed and illustrated commentary on the *Uniform Building Code,* tracing historical background and rationale of the codes through the current edition. Also included are numerous drawings and figures clarifying the application and intent of the code provisions. Also available in electronic format.

*Handbook to the Uniform Mechanical Code. An indispensable tool for understanding the provisions of the current UMC, the handbook traces the historical background and rationale behind the UMC provisions, includes 160 figures which clarify the intent and application of the code, and provides a chapter-by-chapter analysis of the UMC.

*Uniform Building Code Application Manual. This manual discusses sections of the *Uniform Building Code* with a question-and-answer format, providing a comprehensive analysis of the intent of the code sections. Most sections include illustrative examples. The manual is in loose-leaf format so that code applications published in *Building Standards* magazine may be inserted. Also available in electronic format.

*Uniform Mechanical Code Application Manual. As a companion document to the *Uniform Mechanical Code,* this manual provides a comprehensive analysis of the intent of a number of code sections in an easy-to-use question-and-answer format. The manual is available in a loose-leaf format and includes illustrative examples for many code sections.

*Uniform Fire Code Applications Manual. This newly developed manual provides questions and answers regarding UFC provisions. A comprehensive analysis of the intent of numerous code sections, the manual is in a loose-leaf format for easy insertion of code applications published in IFCI's *Fire Code Journal.*

Quick-Reference Guide to the Occupancy Requirements of the 1997 UBC. Code requirements are compiled in this publication by occupancy groups for quick access. These tabulations assemble requirements for each occupancy classification in the code. Provisions, such as fire-resistive ratings for occupancy separations in Table 3-B, exterior wall and opening protection requirements in Table 5-A-1, and fire-resistive ratings for types of construction in Table 6-A, are tabulated for quick reference and comparison.

Plan Review Manual. A practical text that will assist and guide both the field inspector and plan reviewer in applying the code requirements. This manual covers the nonstructural and basic structural aspects of plan review.

Field Inspection Manual. An important fundamental text for courses of study at the community college and trade or technical school level. It is an effective text for those studying building construction or architecture and includes sample forms and checklists for use in the field.

Building Department Administration. An excellent guide for improvement of skills in departmental management and in the enforcement and application of the Building Code and other regulations administered by a building inspection department. This textbook will also be a valuable aid to instructors, students and those in related professional fields.

Building Department Guide to Disaster Mitigation. This new, expanded guide is designed to assist building departments in developing or updating disaster mitigation plans. Subjects covered include guidelines for damage mitigation, disaster-response management, immediate response, mutual aid and inspections, working with the media, repair and recovery policies, and public information bulletins. This publication is a must for those involved in preparing for and responding to disaster.

Building Official Management Manual. This manual addresses the unique nature of code administration and the managerial duties of the building official. A supplementary insert addresses the budgetary and financial aspects of a building department. It is also an ideal resource for those preparing for the management module of the CABO Building Official Certification Examination.

Legal Aspects of Code Administration. A manual developed by the three model code organizations to inform the building official on the legal aspects of the profession. The text is written in a logical sequence with explanation of legal terminology. It is designed to serve as a refresher for those preparing to take the legal module of the CABO Building Official Certification Examination.

Illustrated Guide to Conventional Construction Provisions of the UBC. This comprehensive guide and commentary provides detailed explanations of the conventional construction provisions in the UBC, including descriptive discussions and illustrated drawings to convey the prescriptive provisions related to wood-frame construction.

Introduction to the Uniform Building Code. A workbook that provides an overview of the basics of the UBC.

Uniform Building Code Update Workbook. This manual addresses many of the changes to the administrative, fire- and life-safety, and inspection provisions appearing in the UBC.

UMC Workbook. Designed for independent study or use with instructor-led programs based on the *Uniform Mechanical Code,* this comprehensive study guide consists of 16 learning sessions, with the first two sessions reviewing the purpose, scope, definitions and administrative provisions and the remaining 14 sessions progressively exploring the requirements for installing, inspecting and maintaining heating, ventilating, cooling and refrigeration systems.

UBC Field Inspection Workbook. A comprehensive workbook for studying the provisions of the UBC. Divided into 12 sessions, this workbook focuses on the UBC combustible construction requirements for the inspection of wood-framed construction.

Concrete Manual. A publication for individuals seeking an understanding of the fundamentals of concrete field technology and inspection practices. Of particular interest to concrete construction inspectors, it will also benefit employees of concrete producers, contractors, testing and inspection laboratories and material suppliers.

Reinforced Concrete Masonry Construction Inspector's Handbook. A comprehensive information source written especially for masonry inspection covering terminology, technology, materials, quality control, inspection and standards. Published jointly by ICBO and the Masonry Institute of America.

You Can Build It! Sponsored by ICBO in cooperation with CABO, this booklet contains information and advice to aid "do-it-yourselfers" with building projects. Provides guidance in necessary procedures such as permit requirements, codes, plans, cost estimation, etc.

Guidelines for Manufactured Housing Installations. A guideline in code form implementing the *Uniform Building Code* and its companion code documents to regulate the permanent installation of a manufactured home on a privately owned, nonrental site. A commentary is included to explain specific provisions, and codes applying to each component part are defined.

Accessibility Reference Guide. This guide is a valuable resource for architects, interior designers, plan reviewers and others who design and enforce accessibility provisions. Features include accessibility requirements, along with detailed commentary and graphics to clarify the provisions; cross-references to other applicable sections of the UBC and the Americans with Disabilities Act Accessibility Guidelines; a checklist of UBC provisions on access and usability requirements; and many other useful references.

Educational and Technical Reference Materials. The Conference has been a leader in the development of texts and course material to assist in the educational process. These materials include vital information necessary for the building official and subordinates in carrying out their responsibilities and have proven to be excellent references in connection with community college curricula and higher-level courses in the field of building construction technology and inspection and in the administration of building departments. Included are plan review checklists for structural, nonstructural, mechanical and fire-safety provisions and a full line of videotapes and automated products.

Table of Contents—Volume 1
Administrative, Fire- and Life-Safety, and Field Inspection Provisions

Table of Contents—Volume 2
Structural Engineering Design Provisions

Table of Contents—Volume 3
Material, Testing and Installation Standards

EFFECTIVE USE OF THE
UNIFORM BUILDING CODE

The following procedure may be helpful in using the *Uniform Building Code:*

1. Classify the building:
    A. **OCCUPANCY CLASSIFICATION:** Compute the floor area and occupant load of the building or portion thereof. See Sections 207 and 1002 and Table 10-A. Determine the occupancy group which the use of the building or portion thereof most nearly resembles. See Sections 301, 303.1.1, 304.1, 305.1, 306.1, 307.1, 308.1, 309.1, 310.1, 311.1 and 312.1. See Section 302 for buildings with mixed occupancies.
    B. **TYPE OF CONSTRUCTION:** Determine the type of construction of the building by the building materials used and the fire resistance of the parts of the building. See Chapter 6.
    C. **LOCATION ON PROPERTY:** Determine the location of the building on the site and clearances to property lines and other buildings from the plot plan. See Table 5-A and Sections 602.3, 603.3, 604.3, 605.3 and 606.3 for fire resistance of exterior walls and wall opening requirements based on proximity to property lines. See Section 503.
    D. **ALLOWABLE FLOOR AREA:** Determine the allowable floor area of the building. See Table 5-B for basic allowable floor area based on occupancy group and type of construction. See Section 505 for allowable increases based on location on property and installation of an approved automatic fire sprinkler system. See Section 504.2 for allowable floor area of multistory buildings.
    E. **HEIGHT AND NUMBER OF STORIES:** Compute the height of the building, Section 209, and determine the number of stories, Section 220. See Table 5-B for the maximum height and number of stories permitted based on occupancy group and type of construction. See Section 506 for allowable story increase based on the installation of an approved automatic fire-sprinkler system.

2. Review the building for conformity with the occupancy requirements in Sections 303 through 312.

3. Review the building for conformity with the type of construction requirements in Chapter 6.

4. Review the building for conformity with the exiting requirements in Chapter 10.

5. Review the building for other detailed code regulations in Chapters 4, 7 through 11, 14, 15, 24 through 26, and 30 through 33, and the appendix.

6. Review the building for conformity with structural engineering regulations and requirements for materials of construction. See Chapters 16 through 23.

SAMPLE ORDINANCE FOR ADOPTION OF THE
UNIFORM BUILDING CODE,
VOLUMES 1, 2 AND 3
ORDINANCE NO. _____

An ordinance of the _____(jurisdiction)_____ adopting the 1997 edition of the *Uniform Building Code,* Volumes 1, 2 and 3, regulating the erection, construction, enlargement, alteration, repair, moving, removal, demolition, conversion, occupancy, equipment, use, height, area and maintenance of all buildings or structures in the _____(jurisdiction)_____; providing for the issuance of permits and collection of fees therefor; providing for penalties for the violation thereof, repealing Ordinance No. _____ of the _____(jurisdiction)_____ and all other ordinances and parts of the ordinances in conflict therewith.

The _____(governing body)_____ of the _____(jurisdiction)_____ does ordain as follows:

Section 1. That certain documents, three (3) copies of which are on file and are open for inspection of the public in the office of the ___(jurisdiction's keeper of records)___ of the _____(jurisdiction)_____ , being marked and designated as:

Uniform Building Code, 1997 Edition, published by the International Conference of Building Officials, including the generic fire-resistive assemblies listed in the *Fire Resistance Design Manual,* Fourteenth Edition, dated April 1994, published by the Gypsum Association as referenced in Tables 7-A, 7-B and 7-C (also reference Appendix Chapter 12, Division II, if adopted) of the specified *Uniform Building Code,* including Appendix Chapters _____. *[Fill in the applicable appendix chapters (see Uniform Building Code Section 101.3, last paragraph). If reference is made to Appendix Chapter 30, an additional reference to ANSI/ASME A17.1, 1987, Safety Code for Elevators and Escalators, including Supplements A17.1a-1988, A17.1b-1989, and to ANSI/ASME A17.3a-1986, Safety Code for Existing Elevators and Escalators, including Supplements A17.3a-1989, published by the American Society of Mechanical Engineers, should be added and three (3) copies of this code should also be on file (see Appendix Sections 3010 and 3012)], and*

Structural Welding Code—Reinforcing Steel, AWS D1.4-92 (UBC Standard 19-1); American National Standard for Accessible and Useable Buildings and Facilities, A117.1-1992 (see *Uniform Building Code* Section 1101.2), published by the Council of American Building Officials; *Load and Resistance Factor Design Specifications for Structural Steel Buildings,* December 1, 1993 (Chapter 22, Division II); *Specification for Structural Steel Buildings Allowable Stress Design and Plastic Design,* June 1, 1989 (Chapter 22, Division III); *Load and Resistance Factor Design Specification for Cold Formed Steel Structural Members,* 1986 (with December, 1989 Addendum) (Chapter 22, Division VI); *Specification for Design of Cold-Formed Steel Structural Members,* 1986 (Chapter 22, Division VII); *Standard Specification for Steel Joists, K-Series, LH-Series, DLH-Series and Joist Girders,* 1994 (Chapter 22, Division IX); *Structural Applications of Steel Cables for Buildings,* ASCE 17-95 (Chapter 22, Division XI); and *National Design Specification for Wood Construction,* Revised 1991 Edition (Chapter 22, Division III, Part I), as modified or amended in the *Uniform Building Code* referenced herein:

be and the same are hereby adopted as the code of the _____(jurisdiction)_____ for regulating the erection, construction, enlargement, alteration, repair, moving, removal, demolition, conversion, occupancy, equipment, use, height, area and maintenance of all buildings or structures in the _____(jurisdiction)_____ providing for issuance of permits and collection of fees therefor; and each and all of the regulations, provisions, conditions and terms of such *Uniform Building Code, 1997 Edition,* Volumes 1, 2 and 3, published by the International Conference of Building Officials, and the secondary publications referenced above, all of which are on file in the office of the _____(jurisdiction)_____ are hereby referred to, adopted and made a part hereof as if fully set out in this ordinance.

Section 2. (Incorporate penalties for violations. See Section 103.)

Section 3. That Ordinance No. _____ of _____(jurisdiction)_____ entitled *(fill in the title of building ordinance or ordinances in effect at the present time)* and all other ordinances or parts of ordinances in conflict herewith are hereby repealed.

Section 4. That if any section, sentence, clause or phrase of this ordinance is, for any reason, held to be invalid or unconstitutional, such decision shall not affect the validity or constitutionality of the remaining portions of this ordinance. The _____(governing body)_____ hereby declares that it would have passed this ordinance, and each section, clause or phrase hereof, irrespective of the fact that any one or more sections, sentences, clauses and phrases be declared unconstitutional.

Section 5. That the ___(jurisdiction's keeper of records)___ is hereby ordered and directed to cause this ordinance to be published. *(An additional provision may be required to direct the number of times the ordinance is to be published and to specify that it is to be in a newspaper in general circulation. Posting may also be required.)*

Section 6. That this ordinance and the rules, regulations, provisions, requirements, orders and matters established and adopted hereby shall take effect and be in full force and effect _(time period)_ from and after the date of its final passage and adoption.

Volume 1

Chapter 1
ADMINISTRATION

SECTION 101 — TITLE, PURPOSE AND SCOPE

101.1 Title. These regulations shall be known as the *Uniform Building Code,* may be cited as such and will be referred to herein as "this code."

101.2 Purpose. The purpose of this code is to provide minimum standards to safeguard life or limb, health, property and public welfare by regulating and controlling the design, construction, quality of materials, use and occupancy, location and maintenance of all buildings and structures within this jurisdiction and certain equipment specifically regulated herein.

The purpose of this code is not to create or otherwise establish or designate any particular class or group of persons who will or should be especially protected or benefited by the terms of this code.

101.3 Scope. The provisions of this code shall apply to the construction, alteration, moving, demolition, repair, maintenance and use of any building or structure within this jurisdiction, except work located primarily in a public way, public utility towers and poles, mechanical equipment not specifically regulated in this code, and hydraulic flood control structures.

For additions, alterations, moving and maintenance of buildings and structures, see Chapter 34. For temporary buildings and structures see Section 3103 and Appendix Chapter 31.

Where, in any specific case, different sections of this code specify different materials, methods of construction or other requirements, the most restrictive shall govern. Where there is a conflict between a general requirement and a specific requirement, the specific requirement shall be applicable.

Wherever in this code reference is made to the appendix, the provisions in the appendix shall not apply unless specifically adopted.

SECTION 102 — UNSAFE BUILDINGS OR STRUCTURES

All buildings or structures regulated by this code that are structurally unsafe or not provided with adequate egress, or that constitute a fire hazard, or are otherwise dangerous to human life are, for the purpose of this section, unsafe. Any use of buildings or structures constituting a hazard to safety, health or public welfare by reason of inadequate maintenance, dilapidation, obsolescence, fire hazard, disaster, damage or abandonment is, for the purpose of this section, an unsafe use. Parapet walls, cornices, spires, towers, tanks, statuary and other appendages or structural members that are supported by, attached to, or a part of a building and that are in deteriorated condition or otherwise unable to sustain the design loads that are specified in this code are hereby designated as unsafe building appendages.

All such unsafe buildings, structures or appendages are hereby declared to be public nuisances and shall be abated by repair, rehabilitation, demolition or removal in accordance with the procedures set forth in the Dangerous Buildings Code or such alternate procedures as may have been or as may be adopted by this jurisdiction. As an alternative, the building official, or other employee or official of this jurisdiction as designated by the governing body,

may institute any other appropriate action to prevent, restrain, correct or abate the violation.

SECTION 103 — VIOLATIONS

It shall be unlawful for any person, firm or corporation to erect, construct, enlarge, alter, repair, move, improve, remove, convert or demolish, equip, use, occupy or maintain any building or structure or cause or permit the same to be done in violation of this code.

SECTION 104 — ORGANIZATION AND ENFORCEMENT

104.1 Creation of Enforcement Agency. There is hereby established in this jurisdiction a code enforcement agency which shall be under the administrative and operational control of the building official.

104.2 Powers and Duties of Building Official.

104.2.1 General. The building official is hereby authorized and directed to enforce all the provisions of this code. For such purposes, the building official shall have the powers of a law enforcement officer.

The building official shall have the power to render interpretations of this code and to adopt and enforce rules and supplemental regulations to clarify the application of its provisions. Such interpretations, rules and regulations shall be in conformance with the intent and purpose of this code.

104.2.2 Deputies. In accordance with prescribed procedures and with the approval of the appointing authority, the building official may appoint such number of technical officers and inspectors and other employees as shall be authorized from time to time. The building official may deputize such inspectors or employees as may be necessary to carry out the functions of the code enforcement agency.

104.2.3 Right of entry. When it is necessary to make an inspection to enforce the provisions of this code, or when the building official has reasonable cause to believe that there exists in a building or upon a premises a condition that is contrary to or in violation of this code that makes the building or premises unsafe, dangerous or hazardous, the building official may enter the building or premises at reasonable times to inspect or to perform the duties imposed by this code, provided that if such building or premises be occupied that credentials be presented to the occupant and entry requested. If such building or premises be unoccupied, the building official shall first make a reasonable effort to locate the owner or other person having charge or control of the building or premises and request entry. If entry is refused, the building official shall have recourse to the remedies provided by law to secure entry.

104.2.4 Stop orders. Whenever any work is being done contrary to the provisions of this code, or other pertinent laws or ordinances implemented through the enforcement of this code, the building official may order the work stopped by notice in writing served on any persons engaged in the doing or causing such work to be done, and any such persons shall forthwith stop such work until authorized by the building official to proceed with the work.

104.2.5 Occupancy violations. Whenever any building or structure or equipment therein regulated by this code is being used contrary to the provisions of this code, the building official may order such use discontinued and the structure, or portion thereof, vacated by notice served on any person causing such use to be continued. Such person shall discontinue the use within the time prescribed by the building official after receipt of such notice to make the structure, or portion thereof, comply with the requirements of this code.

104.2.6 Liability. The building official charged with the enforcement of this code, acting in good faith and without malice in the discharge of the duties required by this code or other pertinent law or ordinance shall not thereby be rendered personally liable for damages that may accrue to persons or property as a result of an act or by reason of an act or omission in the discharge of such duties. A suit brought against the building official or employee because of such act or omission performed by the building official or employee in the enforcement of any provision of such codes or other pertinent laws or ordinances implemented through the enforcement of this code or enforced by the code enforcement agency shall be defended by this jurisdiction until final termination of such proceedings, and any judgment resulting therefrom shall be assumed by this jurisdiction.

This code shall not be construed to relieve from or lessen the responsibility of any person owning, operating or controlling any building or structure for any damages to persons or property caused by defects, nor shall the code enforcement agency or its parent jurisdiction be held as assuming any such liability by reason of the inspections authorized by this code or any permits or certificates issued under this code.

104.2.7 Modifications. When there are practical difficulties involved in carrying out the provisions of this code, the building official may grant modifications for individual cases. The building official shall first find that a special individual reason makes the strict letter of this code impractical and that the modification is in conformance with the intent and purpose of this code and that such modification does not lessen any fire-protection requirements or any degree of structural integrity. The details of any action granting modifications shall be recorded and entered in the files of the code enforcement agency.

104.2.8 Alternate materials, alternate design and methods of construction. The provisions of this code are not intended to prevent the use of any material, alternate design or method of construction not specifically prescribed by this code, provided any alternate has been approved and its use authorized by the building official.

The building official may approve any such alternate, provided the building official finds that the proposed design is satisfactory and complies with the provisions of this code and that the material, method or work offered is, for the purpose intended, at least the equivalent of that prescribed in this code in suitability, strength, effectiveness, fire resistance, durability, safety and sanitation.

The building official shall require that sufficient evidence or proof be submitted to substantiate any claims that may be made regarding its use. The details of any action granting approval of an alternate shall be recorded and entered in the files of the code enforcement agency.

104.2.9 Tests. Whenever there is insufficient evidence of compliance with any of the provisions of this code or evidence that any material or construction does not conform to the requirements of this code, the building official may require tests as proof of compliance to be made at no expense to this jurisdiction.

Test methods shall be as specified by this code or by other recognized test standards. If there are no recognized and accepted test methods for the proposed alternate, the building official shall determine test procedures.

All tests shall be made by an approved agency. Reports of such tests shall be retained by the building official for the period required for the retention of public records.

104.2.10 Cooperation of other officials and officers. The building official may request, and shall receive, the assistance and cooperation of other officials of this jurisdiction so far as is required in the discharge of the duties required by this code or other pertinent law or ordinance.

SECTION 105 — BOARD OF APPEALS

105.1 General. In order to hear and decide appeals of orders, decisions or determinations made by the building official relative to the application and interpretation of this code, there shall be and is hereby created a board of appeals consisting of members who are qualified by experience and training to pass on matters pertaining to building construction and who are not employees of the jurisdiction. The building official shall be an ex officio member of and shall act as secretary to said board but shall have no vote on any matter before the board. The board of appeals shall be appointed by the governing body and shall hold office at its pleasure. The board shall adopt rules of procedure for conducting its business, and shall render all decisions and findings in writing to the appellant with a duplicate copy to the building official.

105.2 Limitations of Authority. The board of appeals shall have no authority relative to interpretation of the administrative provisions of this code nor shall the board be empowered to waive requirements of this code.

SECTION 106 — PERMITS

106.1 Permits Required. Except as specified in Section 106.2, no building or structure regulated by this code shall be erected, constructed, enlarged, altered, repaired, moved, improved, removed, converted or demolished unless a separate permit for each building or structure has first been obtained from the building official.

106.2 Work Exempt from Permit. A building permit shall not be required for the following:

1. One-story detached accessory buildings used as tool and storage sheds, playhouses, and similar uses, provided the floor area does not exceed 120 square feet (11.15 m²).

2 Fences not over 6 feet (1829 mm) high.

3. Oil derricks.

4. Movable cases, counters and partitions not over 5 feet 9 inches (1753 mm) high.

5. Retaining walls that are not over 4 feet (1219 mm) in height measured from the bottom of the footing to the top of the wall, unless supporting a surcharge or impounding Class I, II or III-A liquids.

6. Water tanks supported directly upon grade if the capacity does not exceed 5,000 gallons (18 927 L) and the ratio of height to diameter or width does not exceed 2:1.

7. Platforms, walks and driveways not more than 30 inches (762 mm) above grade and not over any basement or story below.

8. Painting, papering and similar finish work.

9. Temporary motion picture, television and theater stage sets and scenery.

10. Window awnings supported by an exterior wall of Group R, Division 3, and Group U Occupancies when projecting not more than 54 inches (1372 mm).

11. Prefabricated swimming pools accessory to a Group R, Division 3 Occupancy in which the pool walls are entirely above the adjacent grade and if the capacity does not exceed 5,000 gallons (18 927 L).

Unless otherwise exempted, separate plumbing, electrical and mechanical permits will be required for the above-exempted items.

Exemption from the permit requirements of this code shall not be deemed to grant authorization for any work to be done in any manner in violation of the provisions of this code or any other laws or ordinances of this jurisdiction.

106.3 Application for Permit.

106.3.1 Application. To obtain a permit, the applicant shall first file an application therefor in writing on a form furnished by the code enforcement agency for that purpose. Every such application shall:

1. Identify and describe the work to be covered by the permit for which application is made.

2. Describe the land on which the proposed work is to be done by legal description, street address or similar description that will readily identify and definitely locate the proposed building or work.

3. Indicate the use or occupancy for which the proposed work is intended.

4. Be accompanied by plans, diagrams, computations and specifications and other data as required in Section 106.3.2.

5. State the valuation of any new building or structure or any addition, remodeling or alteration to an existing building.

6. Be signed by the applicant, or the applicant's authorized agent.

7. Give such other data and information as may be required by the building official.

106.3.2 Submittal documents. Plans, specifications, engineering calculations, diagrams, soil investigation reports, special inspection and structural observation programs and other data shall constitute the submittal documents and shall be submitted in one or more sets with each application for a permit. When such plans are not prepared by an architect or engineer, the building official may require the applicant submitting such plans or other data to demonstrate that state law does not require that the plans be prepared by a licensed architect or engineer. The building official may require plans, computations and specifications to be prepared and designed by an engineer or architect licensed by the state to practice as such even if not required by state law.

> **EXCEPTION:** The building official may waive the submission of plans, calculations, construction inspection requirements and other data if it is found that the nature of the work applied for is such that reviewing of plans is not necessary to obtain compliance with this code.

106.3.3 Information on plans and specifications. Plans and specifications shall be drawn to scale upon substantial paper or cloth and shall be of sufficient clarity to indicate the location, nature and extent of the work proposed and show in detail that it will conform to the provisions of this code and all relevant laws, ordinances, rules and regulations.

Plans for buildings of other than Group R, Division 3 and Group U Occupancies shall indicate how required structural and fire-resistive integrity will be maintained where penetrations will be made for electrical, mechanical, plumbing and communication conduits, pipes and similar systems.

106.3.4 Architect or engineer of record.

106.3.4.1 General. When it is required that documents be prepared by an architect or engineer, the building official may require the owner to engage and designate on the building permit application an architect or engineer who shall act as the architect or engineer of record. If the circumstances require, the owner may designate a substitute architect or engineer of record who shall perform all of the duties required of the original architect or engineer of record. The building official shall be notified in writing by the owner if the architect or engineer of record is changed or is unable to continue to perform the duties.

The architect or engineer of record shall be responsible for reviewing and coordinating all submittal documents prepared by others, including deferred submittal items, for compatibility with the design of the building.

106.3.4.2 Deferred submittals. For the purposes of this section, deferred submittals are defined as those portions of the design that are not submitted at the time of the application and that are to be submitted to the building official within a specified period.

Deferral of any submittal items shall have prior approval of the building official. The architect or engineer of record shall list the deferred submittals on the plans and shall submit the deferred submittal documents for review by the building official.

Submittal documents for deferred submittal items shall be submitted to the architect or engineer of record who shall review them and forward them to the building official with a notation indicating that the deferred submittal documents have been reviewed and that they have been found to be in general conformance with the design of the building. The deferred submittal items shall not be installed until their design and submittal documents have been approved by the building official.

106.3.5 Inspection and observation program. When special inspection is required by Section 1701, the architect or engineer of record shall prepare an inspection program that shall be submitted to the building official for approval prior to issuance of the building permit. The inspection program shall designate the portions of the work that require special inspection and the name or names of the individuals or firms who are to perform the special inspections, and indicate the duties of the special inspectors.

The special inspector shall be employed by the owner, the engineer or architect of record, or an agent of the owner, but not the contractor or any other person responsible for the work.

When structural observation is required by Section 1702, the inspection program shall name the individuals or firms who are to perform structural observation and describe the stages of construction at which structural observation is to occur.

The inspection program shall include samples of inspection reports and provide time limits for submission of reports.

106.4 Permits Issuance.

106.4.1 Issuance. The application, plans, specifications, computations and other data filed by an applicant for a permit shall be reviewed by the building official. Such plans may be reviewed by other departments of this jurisdiction to verify compliance with any applicable laws under their jurisdiction. If the building official finds that the work described in an application for a permit and the plans, specifications and other data filed therewith conform to the requirements of this code and other pertinent laws and ordinances, and that the fees specified in Section 107 have been paid, the building official shall issue a permit therefor to the applicant.

When the building official issues the permit where plans are required, the building official shall endorse in writing or stamp the plans and specifications APPROVED. Such approved plans and specifications shall not be changed, modified or altered without authorizations from the building official, and all work regulated by this code shall be done in accordance with the approved plans.

The building official may issue a permit for the construction of part of a building or structure before the entire plans and specifications for the whole building or structure have been submitted or approved, provided adequate information and detailed statements have been filed complying with all pertinent requirements of this code. The holder of a partial permit shall proceed without assurance that the permit for the entire building or structure will be granted.

106.4.2 Retention of plans. One set of approved plans, specifications and computations shall be retained by the building official for a period of not less than 90 days from date of completion of the work covered therein; and one set of approved plans and specifications shall be returned to the applicant, and said set shall be kept on the site of the building or work at all times during which the work authorized thereby is in progress.

106.4.3 Validity of permit. The issuance or granting of a permit or approval of plans, specifications and computations shall not be construed to be a permit for, or an approval of, any violation of any of the provisions of this code or of any other ordinance of the jurisdiction. Permits presuming to give authority to violate or cancel the provisions of this code or other ordinances of the jurisdiction shall not be valid.

The issuance of a permit based on plans, specifications and other data shall not prevent the building official from thereafter requiring the correction of errors in said plans, specifications and other data, or from preventing building operations being carried on thereunder when in violation of this code or of any other ordinances of this jurisdiction.

106.4.4 Expiration. Every permit issued by the building official under the provisions of this code shall expire by limitation and become null and void if the building or work authorized by such permit is not commenced within 180 days from the date of such permit, or if the building or work authorized by such permit is suspended or abandoned at any time after the work is commenced for a period of 180 days. Before such work can be recommenced, a new permit shall be first obtained to do so, and the fee therefor shall be one half the amount required for a new permit for such work, provided no changes have been made or will be made in the original plans and specifications for such work, and provided further that such suspension or abandonment has not exceeded one year. In order to renew action on a permit after expiration, the permittee shall pay a new full permit fee.

Any permittee holding an unexpired permit may apply for an extension of the time within which work may commence under that permit when the permittee is unable to commence work within the time required by this section for good and satisfactory reasons. The building official may extend the time for action by the permittee for a period not exceeding 180 days on written request by the permittee showing that circumstances beyond the control of the permittee have prevented action from being taken. No permit shall be extended more than once.

106.4.5 Suspension or revocation. The building official may, in writing, suspend or revoke a permit issued under the provisions of this code whenever the permit is issued in error or on the basis of

incorrect information supplied, or in violation of any ordinance or regulation or any of the provisions of this code.

SECTION 107 — FEES

107.1 General. Fees shall be assessed in accordance with the provisions of this section or shall be as set forth in the fee schedule adopted by the jurisdiction.

107.2 Permit Fees. The fee for each permit shall be as set forth in Table 1-A.

The determination of value or valuation under any of the provisions of this code shall be made by the building official. The value to be used in computing the building permit and building plan review fees shall be the total value of all construction work for which the permit is issued, as well as all finish work, painting, roofing, electrical, plumbing, heating, air conditioning, elevators, fire-extinguishing systems and any other permanent equipment.

107.3 Plan Review Fees. When submittal documents are required by Section 106.3.2, a plan review fee shall be paid at the time of submitting the submittal documents for plan review. Said plan review fee shall be 65 percent of the building permit fee as shown in Table 1-A.

The plan review fees specified in this section are separate fees from the permit fees specified in Section 107.2 and are in addition to the permit fees.

When submittal documents are incomplete or changed so as to require additional plan review or when the project involves deferred submittal items as defined in Section 106.3.4.2, an additional plan review fee shall be charged at the rate shown in Table 1-A.

107.4 Expiration of Plan Review. Applications for which no permit is issued within 180 days following the date of application shall expire by limitation, and plans and other data submitted for review may thereafter be returned to the applicant or destroyed by the building official. The building official may extend the time for action by the applicant for a period not exceeding 180 days on request by the applicant showing that circumstances beyond the control of the applicant have prevented action from being taken. No application shall be extended more than once. In order to renew action on an application after expiration, the applicant shall resubmit plans and pay a new plan review fee.

107.5 Investigation Fees: Work without a Permit.

107.5.1 Investigation. Whenever any work for which a permit is required by this code has been commenced without first obtaining said permit, a special investigation shall be made before a permit may be issued for such work.

107.5.2 Fee. An investigation fee, in addition to the permit fee, shall be collected whether or not a permit is then or subsequently issued. The investigation fee shall be equal to the amount of the permit fee required by this code. The minimum investigation fee shall be the same as the minimum fee set forth in Table 1-A. The payment of such investigation fee shall not exempt any person from compliance with all other provisions of this code nor from any penalty prescribed by law.

107.6 Fee Refunds. The building official may authorize refunding of any fee paid hereunder which was erroneously paid or collected.

The building official may authorize refunding of not more than 80 percent of the permit fee paid when no work has been done under a permit issued in accordance with this code.

The building official may authorize refunding of not more than 80 percent of the plan review fee paid when an application for a permit for which a plan review fee has been paid is withdrawn or canceled before any plan reviewing is done.

The building official shall not authorize refunding of any fee paid except on written application filed by the original permittee not later than 180 days after the date of fee payment.

SECTION 108 — INSPECTIONS

108.1 General. All construction or work for which a permit is required shall be subject to inspection by the building official and all such construction or work shall remain accessible and exposed for inspection purposes until approved by the building official. In addition, certain types of construction shall have continuous inspection, as specified in Section 1701.5.

Approval as a result of an inspection shall not be construed to be an approval of a violation of the provisions of this code or of other ordinances of the jurisdiction. Inspections presuming to give authority to violate or cancel the provisions of this code or of other ordinances of the jurisdiction shall not be valid.

It shall be the duty of the permit applicant to cause the work to remain accessible and exposed for inspection purposes. Neither the building official nor the jurisdiction shall be liable for expense entailed in the removal or replacement of any material required to allow inspection.

A survey of the lot may be required by the building official to verify that the structure is located in accordance with the approved plans.

108.2 Inspection Record Card. Work requiring a permit shall not be commenced until the permit holder or an agent of the permit holder shall have posted or otherwise made available an inspection record card such as to allow the building official to conveniently make the required entries thereon regarding inspection of the work. This card shall be maintained available by the permit holder until final approval has been granted by the building official.

108.3 Inspection Requests. It shall be the duty of the person doing the work authorized by a permit to notify the building official that such work is ready for inspection. The building official may require that every request for inspection be filed at least one working day before such inspection is desired. Such request may be in writing or by telephone at the option of the building official.

It shall be the duty of the person requesting any inspections required by this code to provide access to and means for inspection of such work.

108.4 Approval Required. Work shall not be done beyond the point indicated in each successive inspection without first obtaining the approval of the building official. The building official, upon notification, shall make the requested inspections and shall either indicate that portion of the construction is satisfactory as completed, or shall notify the permit holder or an agent of the permit holder wherein the same fails to comply with this code. Any portions that do not comply shall be corrected and such portion shall not be covered or concealed until authorized by the building official.

There shall be a final inspection and approval of all buildings and structures when completed and ready for occupancy and use.

108.5 Required Inspections.

108.5.1 General. Reinforcing steel or structural framework of any part of any building or structure shall not be covered or concealed without first obtaining the approval of the building official.

Protection of joints and penetrations in fire-resistive assemblies shall not be concealed from view until inspected and approved.

The building official, upon notification, shall make the inspections set forth in the following sections.

108.5.2 Foundation inspection. To be made after excavations for footings are complete and any required reinforcing steel is in place. For concrete foundations, any required forms shall be in place prior to inspection. All materials for the foundation shall be on the job, except where concrete is ready mixed in accordance with approved nationally recognized standards, the concrete need not be on the job. Where the foundation is to be constructed of approved treated wood, additional inspections may be required by the building official.

108.5.3 Concrete slab or under-floor inspection. To be made after all in-slab or under-floor building service equipment, conduit, piping accessories and other ancillary equipment items are in place, but before any concrete is placed or floor sheathing installed, including the subfloor.

108.5.4 Frame inspection. To be made after the roof, all framing, fire blocking and bracing are in place and all pipes, chimneys and vents are complete and the rough electrical, plumbing, and heating wires, pipes and ducts are approved.

108.5.5 Lath or gypsum board inspection. To be made after all lathing and gypsum board, interior and exterior, is in place, but before any plastering is applied or before gypsum board joints and fasteners are taped and finished.

108.5.6 Final inspection. To be made after finish grading and the building is completed and ready for occupancy.

108.6 Special Inspections. For special inspections, see Chapter 17.

108.7 Other Inspections. In addition to the called inspections specified above, the building official may make or require other inspections of any construction work to ascertain compliance with the provisions of this code and other laws which are enforced by the code enforcement agency.

108.8 Reinspections. A reinspection fee may be assessed for each inspection or reinspection when such portion of work for which inspection is called is not complete or when corrections called for are not made.

This section is not to be interpreted as requiring reinspection fees the first time a job is rejected for failure to comply with the requirements of this code, but as controlling the practice of calling for inspections before the job is ready for such inspection or reinspection.

Reinspection fees may be assessed when the inspection record card is not posted or otherwise available on the work site, the approved plans are not readily available to the inspector, for failure to provide access on the date for which inspection is requested, or for deviating from plans requiring the approval of the building official.

To obtain a reinspection, the applicant shall file an application therefor in writing on a form furnished for that purpose and pay the reinspection fee in accordance with Table 1-A or as set forth in the fee schedule adopted by the jurisdiction.

In instances where reinspection fees have been assessed, no additional inspection of the work will be performed until the required fees have been paid.

SECTION 109 — CERTIFICATE OF OCCUPANCY

109.1 Use and Occupancy. No building or structure shall be used or occupied, and no change in the existing occupancy classification of a building or structure or portion thereof shall be made until the building official has issued a certificate of occupancy therefor as provided herein.

> **EXCEPTION:** Group R, Division 3 and Group U Occupancies.

Issuance of a certificate of occupancy shall not be construed as an approval of a violation of the provisions of this code or of other ordinances of the jurisdiction. Certificates presuming to give authority to violate or cancel the provisions of this code or other ordinances of the jurisdiction shall not be valid.

109.2 Change in Use. Changes in the character or use of a building shall not be made except as specified in Section 3405 of this code.

109.3 Certificate Issued. After the building official inspects the building or structure and finds no violations of the provisions of this code or other laws that are enforced by the code enforcement agency, the building official shall issue a certificate of occupancy that shall contain the following:

1. The building permit number.

2. The address of the building.

3. The name and address of the owner.

4. A description of that portion of the building for which the certificate is issued.

5. A statement that the described portion of the building has been inspected for compliance with the requirements of this code for the group and division of occupancy and the use for which the proposed occupancy is classified.

6. The name of the building official.

109.4 Temporary Certificate. If the building official finds that no substantial hazard will result from occupancy of any building or portion thereof before the same is completed, a temporary certificate of occupancy may be issued for the use of a portion or portions of a building or structure prior to the completion of the entire building or structure.

109.5 Posting. The certificate of occupancy shall be posted in a conspicuous place on the premises and shall not be removed except by the building official.

109.6 Revocation. The building official may, in writing, suspend or revoke a certificate of occupancy issued under the provisions of this code whenever the certificate is issued in error, or on the basis of incorrect information supplied, or when it is determined that the building or structure or portion thereof is in violation of any ordinance or regulation or any of the provisions of this code.

TABLE 1-A—BUILDING PERMIT FEES

TOTAL VALUATION	FEE
$1.00 to $500.00	$23.50
$501.00 to $2,000.00	$23.50 for the first $500.00 plus $3.05 for each additional $100.00, or fraction thereof, to and including $2,000.00
$2,001.00 to $25,000.00	$69.25 for the first $2,000.00 plus $14.00 for each additional $1,000.00, or fraction thereof, to and including $25,000.00
$25,001.00 to $50,000.00	$391.25 for the first $25,000.00 plus $10.10 for each additional $1,000.00, or fraction thereof, to and including $50,000.00
$50,001.00 to $100,000.00	$643.75 for the first $50,000.00 plus $7.00 for each additional $1,000.00, or fraction thereof, to and including $100,000.00
$100,001.00 to $500,000.00	$993.75 for the first $100,000.00 plus $5.60 for each additional $1,000.00, or fraction thereof, to and including $500,000.00
$500,001.00 to $1,000,000.00	$3,233.75 for the first $500,000.00 plus $4.75 for each additional $1,000.00, or fraction thereof, to and including $1,000,000.00
$1,000,001.00 and up	$5,608.75 for the first $1,000,000.00 plus $3.15 for each additional $1,000.00, or fraction thereof

Other Inspections and Fees:
1. Inspections outside of normal business hours .... $47.00 per hour[1]
   (minimum charge—two hours)
2. Reinspection fees assessed under provisions of Section 305.8 .... $47.00 per hour[1]
3. Inspections for which no fee is specifically indicated .... $47.00 per hour[1]
   (minimum charge—one-half hour)
4. Additional plan review required by changes, additions or revisions to plans .... $47.00 per hour[1]
   (minimum charge—one-half hour)
5. For use of outside consultants for plan checking and inspections, or both .... Actual costs[2]

[1]Or the total hourly cost to the jurisdiction, whichever is the greatest. This cost shall include supervision, overhead, equipment, hourly wages and fringe benefits of the employees involved.
[2]Actual costs include administrative and overhead costs.

Chapter 2
DEFINITIONS AND ABBREVIATIONS

SECTION 201 — DEFINITIONS

201.1 General. For the purpose of this code, certain terms, phrases, words and their derivatives shall be construed as specified in this chapter and elsewhere in this code where specific definitions are provided. Terms, phrases and words used in the singular include the plural and the plural the singular. Terms, phrases and words used in the masculine gender include the feminine and the feminine the masculine.

Where terms, phrases and words are not defined, they shall have their ordinary accepted meanings within the context with which they are used. *Webster's Third New International Dictionary of the English Language, Unabridged,* copyright 1986, shall be considered as providing ordinarily accepted meanings.

201.2 Standards of Quality.

201.2.1 General. The standards listed below labeled a "UBC Standard" are also listed in Chapter 35, Part II, and are part of this code. The other standards listed below are recognized standards (see Sections 3503 and 3504).

201.2.2 Noncombustible material.

UBC Standard 2-1, Noncombustible Material Test

201.2.3 Burning characteristics of building materials.

1. UBC Standard 8-1, Test Method for Surface-burning Characteristics of Building Materials

2. UBC Standard 23-4, Fire-retardant-treated Wood Tests on Durability and Hygroscopic Properties

3. UBC Standard 26-5, Chamber Method of Test for Measuring the Density of Smoke from the Burning or Decomposition of Plastic Materials

4. UBC Standard 26-6, Ignition Properties of Plastics

201.2.4 Corrosives and irritants.

1. 49 C.F.R. 173, Appendix A, Testing for Corrosiveness

2. 16 C.F.R. 1500.41 and 1500.42, Methods of Testing Primary Irritant Substances and Test for Eye Irritants

201.2.5 Ranking of hazardous materials.

UFC Standard 79-3, Identification of the Health, Flammability and Reactivity of Hazardous Materials

201.2.6 Classification of plastics.

UBC Standard 26-7, Method of Test for Determining Classification of Approved Light-transmitting Plastics

SECTION 202 — A

ACCESS FLOOR SYSTEM is an assembly consisting of panels mounted on pedestals to provide an under-floor space for the installations of mechanical, electrical, communication or similar systems or to serve as an air-supply or return-air plenum.

ACCREDITATION BODY is an approved, third-party organization that initially accredits and subsequently monitors, on a continuing basis, the competency and performance of a grading or inspection agency related to carrying out specific tasks.

ACI is the American Concrete Institute, P.O. Box 9094, Farmington Hills, Michigan 48333.

ADDITION is an extension or increase in floor area or height of a building or structure.

AEROSOL is a product that is dispensed by a propellant from a metal can up to a maximum size of 33.8 fluid ounces (1000 mL) or a glass or plastic bottle up to a size of 4 fluid ounces (118.3 mL), other than a rim-vented container.

AGRICULTURAL BUILDING is a structure designed and constructed to house farm implements, hay, grain, poultry, livestock or other horticultural products. This structure shall not be a place of human habitation or a place of employment where agricultural products are processed, treated or packaged, nor shall it be a place used by the public.

AISC is the American Institute of Steel Construction, Inc., One East Wacker Drive, Suite 3100, Chicago, Illinois 60601-2001.

ALLEY is any public way or thoroughfare less than 16 feet (4877 mm) but not less than 10 feet (3048 mm) in width that has been dedicated or deeded to the public for public use.

ALTER or **ALTERATION** is any change, addition or modification in construction or occupancy.

AMUSEMENT BUILDING. See Section 408.2.

ANSI is the American National Standards Institute, 1430 Broadway, New York, New York 10018.

APARTMENT HOUSE is any building or portion thereof that contains three or more dwelling units and, for the purpose of this code, includes residential condominiums.

APPROVED, as to materials and types of construction, refers to approval by the building official as the result of investigation and tests conducted by the building official, or by reason of accepted principles or tests by recognized authorities, technical or scientific organizations.

APPROVED AGENCY is an established and recognized agency regularly engaged in conducting tests or furnishing inspection services, when such agency has been approved.

APPROVED FABRICATOR is an established and qualified person, firm or corporation approved by the building official pursuant to Section 1701.7 of this code.

AREA. See "floor area."

ASSEMBLY BUILDING is a building or portion of a building used for the gathering together of 50 or more persons for such purposes as deliberation, education, instruction, worship, entertainment, amusement, drinking or dining, or awaiting transportation.

ASTM is the American Society for Testing and Materials, 100 Barr Harbor Drive, West Conshohocken, Pennsylvania 19428.

ATRIUM is an opening through two or more floor levels other than enclosed stairways, elevators, hoistways, escalators, plumbing, electrical, air-conditioning or other equipment, which is closed at the top and not defined as a mall. Floor levels, as used in this definition, do not include balconies within assembly occupancies or mezzanines that comply with Section 507.

AUTOMATIC, as applied to fire-protection devices, is a device or system providing an emergency function without the necessity of human intervention and activated as a result of a predetermined temperature rise, rate of rise of temperature or increase in the level of combustion products.

SECTION 203 — B

BALCONY is that portion of the seating space of an assembly room, the lowest part of which is raised 4 feet (1219 mm) or more above the level of the main floor and shall include the area providing access to the seating area or serving only as a foyer.

BALCONY, EXTERIOR EXIT. See Section 1006.3.

BASEMENT is any floor level below the first story in a building, except that a floor level in a building having only one floor level shall be classified as a basement unless such floor level qualifies as a first story as defined herein.

BOILER, HIGH-PRESSURE, is a boiler furnishing steam at pressures in excess of 15 pounds per square inch (psi) (103.4 kPa) or hot water at temperatures in excess of 250°F (121°C), or at pressures in excess of 160 psi (1103.2 kPa).

BOILER ROOM is any room containing a steam or hot-water boiler.

BUILDING is any structure used or intended for supporting or sheltering any use or occupancy.

BUILDING, EXISTING, is a building erected prior to the adoption of this code, or one for which a legal building permit has been issued.

BUILDING OFFICIAL is the officer or other designated authority charged with the administration and enforcement of this code, or the building official's duly authorized representative.

BULK HANDLING is the transferring of flammable or combustible liquids from tanks or drums into smaller containers for distribution.

SECTION 204 — C

CAST STONE is a precast building stone manufactured from portland cement concrete and used as a trim, veneer or facing on or in buildings or structures.

CENTRAL HEATING PLANT is environmental heating equipment that directly utilizes fuel to generate heat in a medium for distribution by means of ducts or pipes to areas other than the room or space in which the equipment is located.

C.F.R. is the Code of Federal Regulations, a regulation of the United States of America available from the Superintendent of Documents, United States Government Printing Office, Washington, DC 20402.

CHIEF OF THE FIRE DEPARTMENT is the head of the fire department or a regularly authorized deputy.

COMBUSTIBLE LIQUID. See the Fire Code.

CONDOMINIUM, RESIDENTIAL. See "apartment house."

CONGREGATE RESIDENCE is any building or portion thereof that contains facilities for living, sleeping and sanitation, as required by this code, and may include facilities for eating and cooking, for occupancy by other than a family. A congregate residence may be a shelter, convent, monastery, dormitory, fraternity or sorority house, but does not include jails, hospitals, nursing homes, hotels or lodging houses.

CONTROL AREA is a building or portion of a building within which the exempted amounts of hazardous materials may be stored, dispensed, handled or used.

CORROSIVE is a chemical that causes visible destruction of, or irreversible alterations in, living tissue by chemical action at the site of contact. A chemical is considered to be corrosive if, when tested on the intact skin of albino rabbits by the method described in the United States Department of Transportation in Appendix A to 49 C.F.R. 173, it destroys or changes irreversibly the structure of the tissue at the site of contact following an exposure period of four hours. This term shall not refer to action on inanimate surfaces.

COURT is a space, open and unobstructed to the sky, located at or above grade level on a lot and bounded on three or more sides by walls of a building.

SECTION 205 — D

DANGEROUS BUILDINGS CODE is the *Uniform Code for the Abatement of Dangerous Buildings* promulgated by the International Conference of Building Officials, as adopted by this jurisdiction.

DISPENSING is the pouring or transferring of any material from a container, tank or similar vessel, whereby vapors, dusts, fumes, mists or gases may be liberated to the atmosphere.

DISPERSAL AREA, SAFE. See Section 1008.2.

DRAFT STOP is a material, device or construction installed to restrict the movement of air within open spaces of concealed areas of building components such as crawl spaces, floor-ceiling assemblies, roof-ceiling assemblies and attics.

DWELLING is any building or portion thereof that contains not more than two dwelling units.

DWELLING UNIT is any building or portion thereof that contains living facilities, including provisions for sleeping, eating, cooking and sanitation, as required by this code, for not more than one family, or a congregate residence for 10 or less persons.

SECTION 206 — E

EFFICIENCY DWELLING UNIT is a dwelling unit containing only one habitable room.

ELECTRICAL CODE is the *National Electrical Code* promulgated by the National Fire Protection Association, as adopted by this jurisdiction.

ELEVATOR CODE is the safety code for elevators, dumbwaiters, escalators and moving walks as adopted by this jurisdiction (see Appendix Chapter 30).

EMERGENCY CONTROL STATION is an approved location on the premises of a Group H, Division 6 Occupancy where signals from emergency equipment are received and that is continually staffed by trained personnel.

EXISTING BUILDINGS. See "building, existing."

EXIT. See Section 1005.1.

EXIT COURT. See Section 1006.3.5.1.

SECTION 207 — F

FABRICATION AREA (fab area) is an area within a semiconductor fabrication facility and related research and development areas in which there are processes using hazardous production materials. Such areas are allowed to include ancillary rooms or areas such as dressing rooms and offices that are directly related to the fab area processes.

FAMILY is an individual or two or more persons related by blood or marriage or a group of not more than five persons (excluding servants) who need not be related by blood or marriage living together in a dwelling unit.

FIRE ASSEMBLY. See Section 713.2.

FIRE CODE is the *Uniform Fire Code* promulgated by the International Fire Code Institute, as adopted by this jurisdiction.

FIRE RESISTANCE or **FIRE-RESISTIVE CONSTRUCTION** is construction to resist the spread of fire, details of which are specified in this code.

FIRE-RETARDANT-TREATED WOOD is any wood product impregnated with chemicals by a pressure process or other means during manufacture, and which, when tested in accordance with UBC Standard 8-1 for a period of 30 minutes, shall have a flame spread of not over 25 and show no evidence of progressive combustion. In addition, the flame front shall not progress more than $10^{1}/_{2}$ feet (3200 mm) beyond the center line of the burner at any time during the test. Materials that may be exposed to the weather shall pass the accelerated weathering test and be identified as Exterior type, in accordance with UBC Standard 23-4. Where material is not directly exposed to rainfall but exposed to high humidity conditions, it shall be subjected to the hygroscopic test and identified as Interior Type A in accordance with UBC Standard 23-4.

All materials shall bear identification showing the fire performance rating thereof. Such identifications shall be issued by an approved agency having a service for inspection of materials at the factory.

FLAMMABLE LIQUID. See the Fire Code.

FLOOR AREA is the area included within the surrounding exterior walls of a building or portion thereof, exclusive of vent shafts and courts. The floor area of a building, or portion thereof, not provided with surrounding exterior walls shall be the usable area under the horizontal projection of the roof or floor above.

FM is Factory Mutual Engineering and Research, 1151 Boston-Providence Turnpike, Norwood, Massachusetts 02062.

FOAM PLASTIC INSULATION is a plastic that is intentionally expanded by the use of a foaming agent to produce a reduced-density plastic containing voids consisting of hollow spheres or interconnected cells distributed throughout the plastic for thermal insulating or acoustical purposes and that has a density less than 20 pounds per cubic foot (320 kg/m^3).

FOOTING is that portion of the foundation of a structure that spreads and transmits loads directly to the soil or the piles.

FRONT OF LOT is the front boundary line of a lot bordering on the street and, in the case of a corner lot, may be either frontage.

SECTION 208 — G

GARAGE is a building or portion thereof in which a motor vehicle containing flammable or combustible liquids or gas in its tank is stored, repaired or kept.

GARAGE, PRIVATE, is a building or a portion of a building, not more than 1,000 square feet (93 m^2) in area, in which only motor vehicles used by the tenants of the building or buildings on the premises are stored or kept. (See Section 312.)

GARAGE, PUBLIC, is any garage other than a private garage.

GAS ROOM is a separately ventilated, fully enclosed room in which only toxic and highly toxic compressed gases and associated equipment and supplies are stored or used.

GRADE (Adjacent Ground Elevation) is the lowest point of elevation of the finished surface of the ground, paving or sidewalk within the area between the building and the property line or, when the property line is more than 5 feet (1524 mm) from the building, between the building and a line 5 feet (1524 mm) from the building.

GRADE (Lumber) is the classification of lumber in regard to strength and utility.

GUARDRAIL is a system of building components located near the open sides of elevated walking surfaces for the purpose of minimizing the possibility of an accidental fall from the walking surface to the lower level.

GUEST is any person hiring or occupying a room for living or sleeping purposes.

GUEST ROOM is any room or rooms used or intended to be used by a guest for sleeping purposes. Every 100 square feet (9.3 m^2) of superficial floor area in a dormitory shall be considered to be a guest room.

SECTION 209 — H

HABITABLE SPACE (ROOM) is space in a structure for living, sleeping, eating or cooking. Bathrooms, toilet compartments, closets, halls, storage or utility space, and similar areas, are not considered habitable space.

HANDLING is the deliberate movement of material by any means to a point of storage or use.

HANDRAIL is a railing provided for grasping with the hand for support. See also "guardrail."

HAZARDOUS PRODUCTION MATERIAL (HPM) is a solid, liquid or gas that has a degree of hazard rating in health, flammability or reactivity of 3 or 4 and that is used directly in research, laboratory or production processes that have, as their end product, materials that are not hazardous.

HEALTH HAZARD is a classification of a chemical for which there is statistically significant evidence based on at least one study conducted in accordance with established scientific principles that acute or chronic health effects may occur in exposed persons. The term "health hazard" includes chemicals that are carcinogens, toxic or highly toxic agents, reproductive toxins, irritants, corrosives, sensitizers, hepatotoxins, nephrotoxins, neurotoxins, agents that act on the hematopoietic system, and agents that damage the lungs, skin, eyes or mucous membranes.

HEIGHT OF BUILDING is the vertical distance above a reference datum measured to the highest point of the coping of a flat roof or to the deck line of a mansard roof or to the average height of the highest gable of a pitched or hipped roof. The reference datum shall be selected by either of the following, whichever yields a greater height of building:

1. The elevation of the highest adjoining sidewalk or ground surface within a 5-foot (1524 mm) horizontal distance of the exterior wall of the building when such sidewalk or ground surface is not more than 10 feet (3048 mm) above lowest grade.

2. An elevation 10 feet (3048 mm) higher than the lowest grade when the sidewalk or ground surface described in Item 1 is more than 10 feet (3048 mm) above lowest grade.

The height of a stepped or terraced building is the maximum height of any segment of the building.

HELIPORT is an area of land or water or a structural surface that is used, or intended for use, for the landing and take-off of helicopters, and any appurtenant areas that are used, or intended for use, for heliport buildings and other heliport facilities.

HELISTOP is the same as a heliport, except that no refueling, maintenance, repairs or storage of helicopters is permitted.

HIGHLY TOXIC MATERIAL is a material that produces a lethal dose or a lethal concentration that falls within any of the following categories:

1. A chemical that has a median lethal dose (LD_{50}) of 50 milligrams or less per kilogram of body weight when administered orally to albino rats weighing between 200 and 300 grams each.

2. A chemical that has a median lethal dose (LD_{50}) of 200 milligrams or less per kilogram of body weight when administered by continuous contact for 24 hours (or less if death occurs within 24 hours) with the bare skin of albino rabbits weighing between 2 and 3 kilograms each.

3. A chemical that has a median lethal concentration (LC_{50}) in air of 200 parts per million by volume or less of gas or vapor, or 2 milligrams per liter or less of mist, fume or dust, when administered by continuous inhalation for one hour (or less if death occurs within one hour) to albino rats weighing between 200 and 300 grams each.

Mixtures of these materials with ordinary materials, such as water, may not warrant a classification of highly toxic. While this system is basically simple in application, any hazard evaluation that is required for the precise categorization of this type of material shall be performed by experienced, technically competent persons.

HORIZONTAL EXIT. See Section 1005.3.5.

HOTEL is any building containing six or more guest rooms intended or designed to be used, or that are used, rented or hired out to be occupied, or that are occupied for sleeping purposes by guests.

HOT-WATER-HEATING BOILER is a boiler having a volume exceeding 120 gallons (454.2 L), or a heat input exceeding 200,000 Btu/h (149 540 kW), or an operating temperature exceeding 210°F (99°C) that provides hot water to be used externally to itself.

HPM ROOM is a room used in conjunction with or serving a Group H, Division 6 Occupancy that hazardous production materials (HPM) are stored or used and that is classified as a Group H, Division 2, 3 or 7 Occupancy.

SECTION 210 — I

IRRITANT is a chemical that is not corrosive but that causes a reversible inflammatory effect on living tissue by chemical action at the site of contact. A chemical is a skin irritant if, when tested on the intact skin of albino rabbits by the methods of 16 C.F.R. 1500.41 for four hours' exposure or by other appropriate techniques, it results in an empirical score of 5 or more. A chemical is an eye irritant if so determined under the procedure listed in 16 C.F.R. 1500.42 or other appropriate techniques.

SECTION 211 — J

JURISDICTION, as used in this code, is any political subdivision that adopts this code for administrative regulations within its sphere of authority.

SECTION 212 — K

No definitions.

SECTION 213 — L

LINTEL is a structural member placed over an opening or a recess in a wall and supporting construction above.

LIQUID is any material that has a fluidity greater than that of 300 penetration asphalt when tested in accordance with the *Uniform Fire Code* standards. When not otherwise identified, the term "liquid" is both flammable and combustible liquids.

LIQUID STORAGE ROOM is a room classified as a Group H, Division 3 Occupancy used only for the storage of flammable or combustible liquids in a closed condition. The quantities of flammable or combustible liquids in storage shall not exceed the limits set forth in the Fire Code.

LIQUID STORAGE WAREHOUSE is a Group H, Division 3 Occupancy used only for the storage of flammable or combustible liquids in an unopened condition. The quantities of flammable or combustible liquids stored are not limited.

LISTED and **LISTING** are terms referring to equipment or materials included in a list published by an approved testing laboratory, inspection agency or other organization concerned with product evaluation that maintains periodic inspection of current productions of listed equipment or materials. The published list shall state that the material or equipment complies with approved nationally recognized codes, standards or tests and has been tested or evaluated and found suitable for use in a specified manner.

LOADS. See Chapter 16.

LODGING HOUSE is any building or portion thereof containing not more than five guest rooms where rent is paid in money, goods, labor or otherwise.

LOW-PRESSURE HOT-WATER-HEATING BOILER is a boiler furnishing hot water at pressures not exceeding 160 psi (1103.2 kPa) and at temperatures not exceeding 250°F (121°C).

LOW-PRESSURE STEAM-HEATING BOILER is a boiler furnishing steam at pressures not exceeding 15 psi (103.4 kPa).

SECTION 214 — M

MARQUEE is a permanent roofed structure attached to and supported by the building and projecting over public property. Marquees are regulated in Chapter 32.

MASONRY is that form of construction composed of stone, brick, concrete, gypsum, hollow-clay tile, concrete block or tile, glass block or other similar building units or materials or combination of these materials laid up unit by unit and set in mortar.

MASONRY, SOLID, is masonry of solid units built without hollow spaces.

MECHANICAL CODE is the *Uniform Mechanical Code* promulgated by the International Conference of Building Officials, as adopted by this jurisdiction.

MEZZANINE or **MEZZANINE FLOOR** is an intermediate floor placed within a room.

MOTEL shall mean hotel as defined in this code.

MOTOR VEHICLE FUEL-DISPENSING STATION is that portion of a building where flammable or combustible liquids or gases used as motor fuels are stored and dispensed from fixed equipment into the fuel tanks of motor vehicles.

SECTION 215 — N

NONCOMBUSTIBLE, as applied to building construction material, means a material that, in the form in which it is used, is either one of the following:

1. Material of which no part will ignite and burn when subjected to fire. Any material conforming to UBC Standard 2-1 shall be considered noncombustible within the meaning of this section.

2. Material having a structural base of noncombustible material as defined in Item 1, with a surfacing material not over $1/8$ inch (3.2 mm) thick which has a flame-spread rating of 50 or less.

"Noncombustible" does not apply to surface finish materials. Material required to be noncombustible for reduced clearances to flues, heating appliances or other sources of high temperature shall refer to material conforming to Item 1. No material shall be classed as noncombustible, which is subject to increase in combustibility or flame-spread rating, beyond the limits herein established, through the effects of age, moisture or other atmospheric condition.

Flame-spread rating as used herein refers to rating obtained according to tests conducted as specified in UBC Standard 8-1.

SECTION 216 — O

OCCUPANCY is the purpose for that a building, or part thereof, is used or intended to be used.

ORIEL WINDOW is a window that projects from the main line of an enclosing wall of a building and is carried on brackets or corbels.

OWNER is any person, agent, firm or corporation having a legal or equitable interest in the property.

SECTION 217 — P

PANIC HARDWARE. See Section 1002.

PEDESTRIAN WALKWAY is a walkway used exclusively as a pedestrian trafficway.

PERMIT is an official document or certificate issued by the building official authorizing performance of a specified activity.

PERSON is a natural person, heirs, executors, administrators or assigns, and includes a firm, partnership or corporation, its or their successors or assigns, or the agent of any of the aforesaid.

PHOTOLUMINESCENT is the property of emitting light as the result of absorption of visible or invisible light, which continues for a length of time after excitation.

PLASTIC MATERIALS, APPROVED, other than foam plastics regulated under Sections 601.5.5 and 2602, are those plastic materials having a self-ignition temperature of 650°F (343°C) or greater as determined in accordance with UBC Standard 26-6, and a smoke-density rating not greater than 450 when tested in accordance with UBC Standard 8-1, in the way intended for use, or a smoke-density rating not greater than 75 when tested in accordance with UBC Standard 26-5 in the thickness intended for use. Approved plastics shall be classified as either CC1 or CC2 in accordance with UBC Standard 26-7. See also "foam plastic insulation."

PLATFORM. See Section 405.1.2.

PLUMBING CODE is the *Plumbing Code,* as adopted by this jurisdiction.

PROTECTIVE MEMBRANE is a surface material that forms the required outer layer or layers of a fire-resistive assembly containing concealed spaces.

PUBLIC WAY. See Section 1002.

SECTION 218 — Q

No definitions.

SECTION 219 — R

REPAIR is the reconstruction or renewal of any part of an existing building for the purpose of its maintenance.

SECTION 220 — S

SELF-LUMINOUS means powered continuously by a self-contained power source other than a battery or batteries, such as radioactive tritium gas. A self-luminous sign is independent of external power supplies or other energy for its operation.

SENSITIZER is a chemical that causes a substantial proportion of exposed people or animals to develop an allergic reaction in normal tissue after repeated exposure to the chemical.

SERVICE CORRIDOR is a fully enclosed passage used for transporting hazardous production materials and for purposes other than required exiting.

SHAFT is an interior space, enclosed by walls or construction, extending through one or more stories or basements that connects openings in successive floors, or floors and roof, to accommodate elevators, dumbwaiters, mechanical equipment or similar devices or to transmit light or ventilation air.

SHAFT ENCLOSURE is the walls or construction forming the boundaries of a shaft.

SHALL, as used in this code, is mandatory.

SMOKE DETECTOR is an approved, listed device that senses visible or invisible particles of combustion.

STAGE. See Chapter 4.

STORY is that portion of a building included between the upper surface of any floor and the upper surface of the floor next above, except that the topmost story shall be that portion of a building included between the upper surface of the topmost floor and the ceiling or roof above. If the finished floor level directly above a usable or unused under-floor space is more than 6 feet (1829 mm) above grade, as defined herein, for more than 50 percent of the total perimeter or is more than 12 feet (3658 mm) above grade, as defined herein, at any point, such usable or unused under-floor space shall be considered as a story.

STORY, FIRST, is the lowest story in a building that qualifies as a story, as defined herein, except that a floor level in a building having only one floor level shall be classified as a first story, provided such floor level is not more than 4 feet (1219 mm) below grade, as defined herein, for more than 50 percent of the total perimeter, or not more than 8 feet (2438 mm) below grade, as defined herein, at any point.

STREET is any thoroughfare or public way not less than 16 feet (4877 mm) in width that has been dedicated or deeded to the public for public use.

STRUCTURAL OBSERVATION means the visual observation of the structural system, for general conformance to the approved plans and specifications, at significant construction stages and at completion of the structural system. Structural observation does not include or waive the responsibility for the inspections required by Section 108, 1701 or other sections of this code.

STRUCTURE is that which is built or constructed, an edifice or building of any kind, or any piece of work artificially built up or composed of parts joined together in some definite manner.

SURGICAL AREA is the preoperating, operating, recovery and similar rooms within an outpatient health-care center.

SECTION 221 — T

TRAVEL DISTANCE. See Section 1004.2.5.

SECTION 222 — U

UBC STANDARDS are those standards published in Volume 3 of the *Uniform Building Code* promulgated by the International Conference of Building Officials, as adopted by this jurisdiction. (See Chapter 35.)

UL is the Underwriters Laboratories Inc., 333 Pfingsten Road, Northbrook, Illinois 60062.

USE, with reference to flammable or combustible liquids, is the placing in action or service of flammable or combustible liquids whereby flammable vapors may be liberated to the atmosphere.

USE, with reference to hazardous materials other than flammable or combustible liquids, is the placing in action or making available for service by opening or connecting any container utilized for confinement of material whether a solid, liquid or gas.

USE, CLOSED SYSTEM, is use of a solid or liquid hazardous material in a closed vessel or system that remains closed during normal operations where vapors emitted by the product are not liberated outside of the vessel or system and the product is not exposed to the atmosphere during normal operations, and all uses of compressed gases. Examples of closed systems for solids and liquids include product conveyed through a piping system into a closed vessel, system or piece of equipment, and reaction process operations.

USE, OPEN SYSTEM, is use of a solid or liquid hazardous material in a vessel or system that is continuously open to the atmosphere during normal operations and where vapors are liberated, or the product is exposed to the atmosphere during normal operations. Examples of open systems for solids and liquids include dispensing from or into open beakers or containers, dip tank and plating tank operations.

SECTION 223 — V

VALUE or **VALUATION** of a building shall be the estimated cost to replace the building and structure in kind, based on current replacement costs, as determined in Section 107.2.

VENEER. See Section 1403.2.

SECTION 224 — W

WALLS shall be defined as follows:

Bearing Wall is any wall meeting either of the following classifications:

1. Any metal or wood stud wall that supports more than 100 pounds per lineal foot (1.459 kN per lineal meter) of superimposed load.

2. Any masonry or concrete wall that supports more than 200 pounds per lineal foot (2.918 kN per lineal meter) superimposed load, or any such wall supporting its own weight for more than one story.

Exterior Wall is any wall or element of a wall, or any member or group of members, that defines the exterior boundaries or courts of a building and that has a slope of 60 degrees or greater with the horizontal plane.

Faced Wall is a wall in which the masonry facing and backing are so bonded as to exert a common action under load.

Nonbearing Wall is any wall that is not a bearing wall.

Parapet Wall is that part of any wall entirely above the roof line.

Retaining Wall is a wall designed to resist the lateral displacement of soil or other materials.

WATER HEATER is an appliance designed primarily to supply hot water and is equipped with automatic controls limiting water temperature to a maximum of 210°F (99°C).

WEATHER-EXPOSED SURFACES are all surfaces of walls, ceilings, floors, roofs, soffits and similar surfaces exposed to the weather, excepting the following:

1. Ceilings and roof soffits enclosed by walls or by beams, which extend a minimum of 12 inches (305 mm) below such ceiling or roof soffits.

2. Walls or portions of walls within an unenclosed roof area, when located a horizontal distance from an exterior opening equal to twice the height of the opening.

3. Ceiling and roof soffits beyond a horizontal distance of 10 feet (3048 mm) from the outer edge of the ceiling or roof soffits.

WINDOW WELL is a soil-retaining structure at a window having a sill height lower than the adjacent ground elevation.

SECTION 225 — X

No definitions.

SECTION 226 — Y

YARD is an open space, other than a court, unobstructed from the ground to the sky, except where specifically provided by this code, on the lot on which a building is situated.

SECTION 227 — Z

No definitions.

Chapter 3
USE OR OCCUPANCY

SECTION 301 — OCCUPANCY CLASSIFIED

Every building, whether existing or hereafter erected, shall be classified by the building official according to its use or the character of its occupancy, as set forth in Table 3-A, as a building of one of the following occupancy groups:

Group A—Assembly (see Section 303.1.1)

Group B—Business (see Section 304.1)

Group E—Educational (see Section 305.1)

Group F—Factory and Industrial (see Section 306.1)

Group H—Hazardous (see Section 307.1)

Group I—Institutional (see Section 308.1)

Group M—Mercantile (see Section 309.1)

Group R—Residential (see Section 310.1)

Group S—Storage (see Section 311.1)

Group U—Utility (see Section 312.1)

Any occupancy not mentioned specifically or about which there is any question shall be classified by the building official and included in the group that its use most nearly resembles, based on the existing or proposed fire and life hazard.

For changes in use, see Section 3405.

SECTION 302 — MIXED USE OR OCCUPANCY

302.1 General. When a building is used for more than one occupancy purpose, each part of the building comprising a distinct "occupancy," as described in Section 301, shall be separated from any other occupancy as specified in Section 302.4.

> **EXCEPTIONS: 1.** When an approved spray booth constructed in accordance with the Fire Code is installed, such booth need not be separated from Group B, F, H, M or S Occupancies.
>
> 2. The following occupancies need not be separated from the uses to which they are accessory:
>
> 2.1 Assembly rooms having a floor area of not over 750 square feet (69.7 m²).
>
> 2.2 Administrative and clerical offices and similar rooms that do not exceed 25 percent of the floor area of the major use when not related to Group H, Division 2 and Group H, Division 3 Occupancies.
>
> 2.3 Gift shops, administrative offices and similar rooms in Group R, Division 1 Occupancies not exceeding 10 percent of the floor area of the major use.
>
> 2.4 The kitchen serving the dining area of which it is a part.
>
> 2.5 Customer waiting rooms not exceeding 450 square feet (41.8 m²) when not related to Group H Occupancies and when such waiting rooms have an exit directly to the exterior.
>
> 3. An occupancy separation need not be provided between a Group R, Division 3 Occupancy and a carport having no enclosed uses above, provided the carport is entirely open on two or more sides.
>
> 4. A Group S, Division 3 Occupancy used exclusively for the parking or storage of private or pleasure-type motor vehicles need not be separated from a Group S, Division 4 Occupancy open parking garage as defined in Section 311.1.

When a building houses more than one occupancy, each portion of the building shall conform to the requirements for the occupancy housed therein.

An occupancy shall not be located above the story or height set forth in Table 5-B, except as provided in Section 506. When a mixed occupancy building contains a Group H, Division 6 Occupancy, the portion containing the Group H, Division 6 Occupancy shall not exceed three stories or 55 feet (16 764 mm) in height.

302.2 Forms of Occupancy Separations. Occupancy separations shall be vertical or horizontal or both or, when necessary, of such other form as may be required to afford a complete separation between the various occupancy divisions in the building.

Where the occupancy separation is horizontal, structural members supporting the separation shall be protected by equivalent fire-resistive construction.

302.3 Types of Occupancy Separations. Occupancy separations shall be classed as "four-hour fire-resistive," "three-hour fire-resistive," "two-hour fire-resistive" and "one-hour fire-resistive."

1. A four-hour fire-resistive occupancy separation shall have no openings therein and shall not be of less than four-hour fire-resistive construction.

2. A three-hour fire-resistive occupancy separation shall not be of less than three-hour fire-resistive construction. All openings in walls forming such separation shall be protected by a fire assembly having a three-hour fire-protection rating. The total width of all openings in any three-hour fire-resistive occupancy separation wall in any one story shall not exceed 25 percent of the length of the wall in that story and no single opening shall have an area greater than 120 square feet (11 m²).

All openings in floors forming a three-hour fire-resistive occupancy separation shall be protected by shaft, stairway, ramp or escalator enclosures extending above and below such openings. The walls of such enclosures shall not be of less than two-hour fire-resistive construction and all openings therein shall be protected by a fire assembly having a one- and one-half-hour fire-protection rating.

> **EXCEPTION:** When the walls of such enclosure extending below the three-hour fire-resistive occupancy separation to the foundation are provided with a fire-resistive rating of not less than three hours with openings therein protected as required for walls forming three-hour occupancy separations, the enclosure walls extending above such floor used as the three-hour fire-resistive occupancy separation may have a one-hour fire-resistive rating, provided:
>
> 1. The occupancy above is not required to be of Type I or II fire-resistive construction, and
>
> 2. The enclosure walls do not enclose an exit stairway, a ramp or an escalator required to have enclosure walls of not less than two-hour fire-resistive construction.

3. A two-hour fire-resistive occupancy separation shall not be of less than two-hour fire-resistive construction. All openings in such separation shall be protected by a fire assembly having a one- and one-half-hour fire-protection rating.

4. A one-hour fire-resistive occupancy separation shall not be of less than one-hour fire-resistive construction. All openings in such separation shall be protected by a fire assembly having a one-hour fire-protection rating.

302.4 Fire Ratings for Occupancy Separations. Occupancy separations shall be provided between the various groups and divisions of occupancies as set forth in Table 3-B. For required separation of specific uses in Group I, Division 1 hospitals and nursing homes, see Table 3-C. See also Section 504.6.1.

EXCEPTIONS: 1. A three-hour occupancy separation may be used between a Group A, Division 1 and a Group S, Division 3 Occupancy used exclusively for the parking or storage of private or pleasure-type motor vehicles provided no repair or fueling is done. A two-hour occupancy separation may be used between a Group A, Division 2, 2.1, 3 or 4 or E or I Occupancy and a Group S, Division 3 Occupancy used exclusively for the parking or storage of private or pleasure-type motor vehicles provided no repair or fueling is done.

2. Unless required by Section 311.2.2, the three-hour occupancy separation between a Group R, Division 1 Occupancy and a Group S, Division 3 Occupancy used only for the parking or storage of private or pleasure-type motor vehicles with no repair or fueling may be reduced to two hours. Such occupancy separation may be further reduced to one hour where the area of such Group S, Division 3 Occupancy does not exceed 3,000 square feet (279 m²).

3. In the one-hour occupancy separation between Group R, Division 3 and Group U Occupancies, the separation may be limited to the installation of materials approved for one-hour fire-resistive construction on the garage side and a self-closing, tightfitting solid-wood door 1³/₈ inches (35 mm) in thickness, or a self-closing, tightfitting door having a fire-protection rating of not less than 20 minutes when tested in accordance with Part II of UBC Standard 7-2, which is a part of this code, is permitted in lieu of a one-hour fire assembly. Fire dampers need not be installed in air ducts passing through the wall, floor or ceiling separating a Group R, Division 3 Occupancy from a Group U Occupancy, provided such ducts within the Group U Occupancy are constructed of steel having a thickness not less than 0.019 inch (0.48 mm) (No. 26 galvanized sheet gage) and have no openings into the Group U Occupancy.

4. Group H, Division 2 and Group H, Division 3 Occupancies need not be separated from Group H, Division 7 Occupancies when such occupancies also comply with the requirements for a Group H, Division 7 Occupancy.

302.5 Heating Equipment Room Occupancy Separation. In Groups A; B; E; F; I; M; R, Division 1; and S Occupancies, rooms containing a boiler, central heating plant or hot-water supply boiler shall be separated from the rest of the building by not less than a one-hour occupancy separation.

EXCEPTIONS: 1. In Groups A, B, E, F, M and S Occupancies, boilers, central heating plants or hot-water supply boilers where the largest piece of fuel equipment does not exceed 400,000 Btu per hour (117.2 kW) input.

2. In Group R, Division 1 Occupancies, a separation need not be provided for such rooms with equipment serving only one dwelling unit.

In Group E Occupancies, when the opening for a heater or equipment room is protected by a pair of fire doors, the inactive leaf shall be normally secured in the closed position and shall be openable only by the use of a tool. An astragal shall be provided and the active leaf shall be self-closing.

In Group H Occupancies, rooms containing a boiler, central heating plant or hot-water supply boiler shall be separated from the rest of the building by not less than a two-hour occupancy separation. In Divisions 1 and 2, there shall be no openings in such occupancy separation except for necessary ducts and piping.

For opening in exterior walls of equipment rooms in Group A, E or I Occupancies, see Sections 303.8, 305.8 and 308.8.

302.6 Water Closet Room Separation. A room in which a water closet is located shall be separated from food preparation or storage rooms by a tightfitting door.

SECTION 303 — REQUIREMENTS FOR GROUP A OCCUPANCIES

303.1 General.

303.1.1 Group A Occupancies defined. Group A Occupancies include the use of a building or structure, or a portion thereof, for the gathering together of 50 or more persons for purposes such as civic, social or religious functions, recreation, education or instruction, food or drink consumption, or awaiting transportation. A room or space used for assembly purposes by less than 50 persons and accessory to another occupancy shall be included as a part of that major occupancy. Assembly occupancies shall include the following:

Division 1. A building or portion of a building having an assembly room with an occupant load of 1,000 or more and a legitimate stage.

Division 2. A building or portion of a building having an assembly room with an occupant load of less than 1,000 and a legitimate stage.

Division 2.1. A building or portion of a building having an assembly room with an occupant load of 300 or more without a legitimate stage, including such buildings used for educational purposes and not classed as Group B or E Occupancies.

Division 3. A building or portion of a building having an assembly room with an occupant load of less than 300 without a legitimate stage, including such buildings used for educational purposes and not classed as Group B or E Occupancies.

Division 4. Stadiums, reviewing stands and amusement park structures not included within other Group A Occupancies. Specific and general requirements for grandstands, bleachers and reviewing stands are to be found in Chapter 10.

303.1.2 Occupancy separations. For occupancy separations, see Table 3-B.

303.1.3 Amusement buildings. Amusement buildings shall conform with the requirements of this code for their occupancy classification in addition to the provisions set forth in Sections 408, 904.2.3 and 1007.2.7.

EXCEPTION: Amusement buildings or portions thereof that are without walls or a roof and constructed to prevent the accumulation of smoke in assembly areas.

303.2 Construction, Height and Allowable Area.

303.2.1 General. Unless otherwise specified in this section, buildings or portions of buildings classed in Group A Occupancy, because of the use or character of the occupancy, shall be limited to the types of construction set forth in Table 5-B, and shall not exceed in area or height the limits specified in Sections 504, 505 and 506.

303.2.2 Special provisions.

The roof-framing system for the roof-ceiling assembly in one-story portions of buildings of Type II One-hour, Type III One-hour or Type V One-hour construction may be of unprotected construction when such roof-framing system is open to the assembly area and does not contain concealed spaces.

Stages and platforms shall be constructed in accordance with the provisions of Section 405.

The slope of the main floor of an assembly room shall not exceed the slopes permitted in Section 1003.3.4.

Group A assembly rooms having an aggregate occupant load of 1,000 or more shall not be located in a basement, except basements in buildings of Type I or Type II-F.R. construction.

Gymnasiums and similar occupancies may have floor surfaces constructed of wood or unprotected steel or iron.

In gymnasiums having an area not greater than 3,200 square feet (297 m²), 1-inch (25 mm) nominal thickness tight tongue-and-grooved boards or ³/₄-inch (19 mm) plywood wall covering may be used on the inner side in lieu of fire-resistive plaster.

For attic space partitions and draft stops, see Section 708.

303.2.2.1 Division 2.1 provisions. Division 2.1 Occupancies with an occupant load of 1,000 or more shall be of Type I, Type II-F.R., Type II One-hour, Type III One-hour or Type IV construction, except that the roof-framing system for one-story portions of buildings of Type II One-hour or Type III One-hour construction may be of unprotected construction when such roof-framing system is open to the assembly area and does not contain concealed spaces.

303.2.2.2 Division 3 provisions. Division 3 Occupancies located in a basement or above the first floor shall not be of less than one-hour fire-resistive construction.

Division 3 Occupancies with an occupant load of 50 or more which are located over usable space shall be separated from such space by not less than one-hour fire-resistive construction.

For Division 3 Occupancies with a Group S, Division 3 parking garage in the basement or first floor, see Section 311.2.2.

303.2.2.3 Division 4 provisions. Grandstands, bleachers or reviewing stands of Type III One-hour, Type IV or Type V One-hour construction shall not exceed 40 feet (12 192 mm) to the highest level of seat boards; 20 feet (6096 mm) in cases where construction is Type III-N or Type V-N; and 12 feet (3658 mm) in cases where construction is with combustible members in the structural frame and located indoors.

Division 4 structures other than Type III-N and Type V-N grandstands, bleachers, reviewing stands and folding and telescoping seating of open skeleton-frametype without roof, cover or enclosed usable space are not limited in area or height.

Erection and structural maintenance shall conform to these special requirements as well as with other applicable provisions of this code.

When the space under a Division 4 Occupancy is used for any purpose, including means of egress, it shall be separated from all parts of such Division 4 Occupancy, including means of egress, by walls, floor and ceiling of not less than one-hour fire-resistive construction.

> **EXCEPTIONS:** 1. A means of egress under temporary grandstands need not be separated.
>
> 2. The underside of continuous steel deck grandstands when erected outdoors need not be fire protected when occupied for public toilets.

The building official may cause Division 4 structures to be reinspected at least once every six months.

Grandstands, bleachers or folding and telescoping seating may have seat boards, toeboards, bearing or base pads and footboards of combustible materials regardless of construction type.

Seating and exiting requirements for reviewing stands, grandstands, bleachers, and folding and telescoping seating are provided under Section 1008.

303.3 Location on Property. Buildings housing Group A Occupancies shall front directly on or discharge to a public street not less than 20 feet (6096 mm) in width. The exit discharge to the public street shall be a minimum 20-foot-wide (6096 mm) right-of-way, unobstructed and maintained only as exit discharge to the public street. The main entrance to the building shall be located on a public street or on the exit discharge.

For fire-resistive protection of exterior walls and openings, as determined by location on property, see Section 503 and Chapter 6.

303.4 Access and Exit Facilities. Exits shall be provided as specified in Chapter 10. (For special exiting requirements, see Section 1007.2.) Access to, and egress from, buildings required to be accessible shall be provided as specified in Chapter 11.

For amusement buildings, see Section 408.

303.5 Light, Ventilation and Sanitation. Light and ventilation shall be in accordance with Chapter 12. The number of plumbing fixtures shall not be less than specified in Section 2902.2.

303.6 Shaft and Exit Enclosures. Exits shall be enclosed as specified in Chapter 10.

Elevator shafts, vent shafts and other vertical openings shall be enclosed and the enclosure shall be as specified in Section 711.

303.7 Sprinkler and Standpipe Systems. When required by Section 904.2.1 or other provisions of this code, automatic sprinkler systems and standpipes shall be designed and installed as specified in Chapter 9.

303.8 Special Hazards. Stages shall be equipped with automatic ventilators as required in Section 405.3.3.

Chimneys and heating apparatus shall conform to the requirements of Chapter 31 of this code and the Mechanical Code.

Motion picture machine booths shall conform to the requirements of Section 406.

Proscenium curtains shall conform to the requirements set forth in UBC Standard 4-1, which is a part of this code. (See Chapter 35, Part II.)

Class I, II or III-A liquids shall not be placed or stored in any Group A Occupancy.

When heating equipment rooms are required to be separated in accordance with Section 302.5, exterior openings in a boiler room or room containing central heating equipment if located below openings in another story or if less than 10 feet (3048 mm) from other doors or windows of the same building shall be protected by a fire assembly having a three-fourths-hour fire-protection rating. Such fire assemblies shall be fixed, automatic or self-closing. For heating equipment occupancy separation, see Section 302.5.

303.9 Fire Alarm Systems. An approved fire alarm system shall be installed as set forth in the Fire Code in Group A, Divisions 1, 2 and 2.1 Occupancies.

For amusement building alarm systems, see Section 408.5.1.

SECTION 304 — REQUIREMENTS FOR GROUP B OCCUPANCIES

304.1 Group B Occupancies Defined.

Group B Occupancies shall include buildings, structures, or portions thereof, for office, professional or service-type transactions, which are not classified as Group H Occupancies. Such occupancies include occupancies for the storage of records and accounts, and eating and drinking establishments with an occupant load of less than 50. Business occupancies shall include, but not be limited to, the following:

1. Animal hospitals, kennels, pounds.

2. Automobile and other motor vehicle showrooms.

3. Banks.

4. Barber shops.

5. Beauty shops.

6. Car washes.

7. Civic administration.

8. Outpatient clinic and medical offices (where five or less patients in a tenant space are incapable of unassisted self-preservation).

9. Dry cleaning pick-up and delivery stations and self-service.

10. Educational occupancies above the 12th grade.

11. Electronic data processing.

12. Fire stations.

13. Florists and nurseries.

14. Laboratories—testing and research.

15. Laundry pick-up and delivery stations and self-service.

16. Police stations.

17. Post offices.

18. Print shops.

19. Professional services such as attorney, dentist, physician, engineer.

20. Radio and television stations.

21. Telephone exchanges.

For occupancy separations, see Table 3-B.

304.2 Construction, Height and Allowable Area.

304.2.1 General. Buildings or parts of buildings classed as Group B Occupancies because of the use or character of the occupancy shall be limited to the types of construction set forth in Table 5-B. Such occupancies shall not exceed, in area or height, the limits specified in Sections 504, 505 and 506 and shall comply with the provisions of this section.

304.2.2 Special provisions.

304.2.2.1 Laboratories and vocational shops. Laboratories and vocational shops in buildings used for educational purposes, and similar areas containing hazardous materials, shall be separated from each other and other portions of the building by not less than a one-hour fire-resistive occupancy separation. When the quantities of hazardous materials in such uses do not exceed those listed in Table 3-D or 3-E, the requirements of Sections 307.5 and 307.8 shall apply. When the quantities of hazardous materials in such uses exceed those listed in Table 3-D or 3-E, the use shall be classified as the appropriate Group H Occupancy.

Occupants in laboratories having an area in excess of 200 square feet (18.6 m^2) shall have access to at least two exits or exit-access doors from the room and all portions of the room shall be within 75 feet (22 860 mm) of an exit or exit-access door.

304.2.2.2 Amusement buildings. Amusement buildings with an occupant load of less than 50 shall comply with Section 408.

304.3 Location on Property. For fire-resistive protection of exterior walls and openings, as determined by location on property, see Section 503 and Chapter 6.

304.4 Access and Means of Egress Facilities. Means of egress shall be provided as specified in Chapter 10. See also Section 304.2.2.1 for means of egress from laboratories.

Access to, and egress from, buildings required to be accessible shall be provided as specified in Chapter 11.

304.5 Light, Ventilation and Sanitation. Light, ventilation and sanitation shall be in accordance with Chapters 12 and 29 and this section.

304.5.1 Ventilation of flammable vapors. See Section 1202.2.2 for ventilation of flammable vapors.

304.5.2 Sanitation. The number of plumbing fixtures shall not be less than specified in Section 2902.3.

304.6 Shaft and Exit Enclosures. Exits shall be enclosed as specified in Chapter 10.

Elevator shafts, vent shafts and other openings through floors shall be enclosed, and the enclosure shall be as specified in Section 711.

In buildings housing Group B Occupancies equipped with automatic sprinkler systems throughout, enclosures need not be provided for escalators where the top of the escalator opening at each story is provided with a draft curtain and automatic fire sprinklers are installed around the perimeter of the opening within 2 feet (610 mm) of the draft curtain. The draft curtain shall enclose the perimeter of the unenclosed opening and extend from the ceiling downward at least 12 inches (305 mm) on all sides. The spacing between sprinklers shall not exceed 6 feet (1829 mm).

304.7 Sprinkler and Standpipe Systems. When required by Section 904.2.1 or other provisions of this code, automatic sprinkler systems and standpipes shall be installed as specified in Chapter 9.

304.8 Special Hazards. Chimneys and heating apparatus shall conform to the requirements of Chapter 31 of this code and the Mechanical Code.

Storage and use of flammable and combustible liquids shall be in accordance with the Fire Code.

Devices generating a glow, spark or flame capable of igniting flammable vapors shall be installed such that sources of ignition are at least 18 inches (457 mm) above the floor of any room in which Class I flammable liquids or flammable gases are used or stored.

Stationary lead-acid battery systems used for facility standby, emergency power or uninterrupted power supplies shall be installed and maintained in accordance with the Fire Code.

SECTION 305 — REQUIREMENTS FOR GROUP E OCCUPANCIES

305.1 Group E Occupancies Defined. Group E Occupancies shall be:

Division 1. Any building used for educational purposes through the 12th grade by 50 or more persons for more than 12 hours per week or four hours in any one day.

Division 2. Any building used for educational purposes through the 12th grade by less than 50 persons for more than 12 hours per week or four hours in any one day.

Division 3. Any building or portion thereof used for day-care purposes for more than six persons.

For occupancy separations, see Table 3-B.

305.2 Construction, Height and Allowable Area.

305.2.1 General. Buildings or parts of buildings classed in Group E because of the use or character of the occupancy shall be limited to the types of construction set forth in Table 5-B and shall not exceed, in area or height, the limits specified in Sections 504, 505 and 506, except that the area may be increased by 50 percent when the maximum travel distance specified in Section 1004.2.5 is reduced by 50 percent.

305.2.2 Atmospheric separation requirements.

305.2.2.1 Definitions. For the purpose of this chapter and Section 1007.3, the following definitions are applicable:

COMMON ATMOSPHERE exists between rooms, spaces or areas within a building that are not separated by an approved smoke- and draft-stop barrier.

SEPARATE ATMOSPHERE exists between rooms, spaces or areas that are separated by an approved smoke barrier.

SMOKE BARRIER consists of walls, partitions, floors and openings therein as will prevent the transmission of smoke or gases through the construction. See Section 905.

305.2.2.2 General provisions. The provisions of this section apply when a separate exit system is required in accordance with Section 1007.3.

Walls, partitions and floors forming all or part of an atmospheric separation shall be as required by Section 905.2.3. Glass lights of approved wired glass set in steel frames may be installed in such walls or partitions.

All automatic-closing fire assemblies installed in the atmospheric separation shall be activated by approved smoke detectors.

The specific requirements of this section are not intended to prevent the design or use of other systems, equipment or techniques that will effectively prevent the products of combustion from breaching the atmospheric separation.

305.2.3 Special provisions. Rooms in Divisions 1 and 2 Occupancies used for kindergarten, first- or second-grade pupils, and Division 3 Occupancies shall not be located above or below the first story.

> **EXCEPTIONS:** 1. Basements or stories having floor levels located within 4 feet (1219 mm), measured vertically, from adjacent ground level at the level of exit discharge, provided the basement or story has exterior exit doors at that level.
>
> 2. In buildings equipped with an automatic sprinkler system throughout, rooms used for kindergarten, first- and second-grade children or for day-care purposes may be located on the second story, provided there are at least two exterior exit doors for the exclusive use of such occupants.
>
> 3. Division 3 Occupancies may be located above the first story in buildings of Type I construction and in Types II-F.R., II One-hour and III One-hour construction, subject to the limitation of Section 506 when:
>
> > 3.1 Division 3 Occupancies with children under the age of seven or containing more than 12 children per story shall not be located above the fourth floor; and
> >
> > 3.2 The entire story. in which the day-care facility is located is equipped with an approved manual fire alarm and smoke-detection system. (See the Fire Code.) Actuation of an initiating device shall sound an audible alarm throughout the entire story. When a building fire alarm system is required by other provisions of this code or the Fire Code, the alarm system shall be connected to the building alarm system.
> >
> > An approved alarm signal shall sound at an approved location in the day-care occupancy to indicate a fire alarm or sprinkler flow condition in other portions of the building; and
> >
> > 3.3 The day-care facility, if more than 1,000 square feet (92.9 m²) in area, is divided into at least two compartments of approximately the same size by a smoke barrier with door openings protected by smoke- and draft-control assemblies having a fire-protection rating of not less than 20 minutes. Smoke barriers shall have a fire-resistive rating of not less than one hour. In addition to the requirements of Section 302, occupancy separations between Division 3 Occupancies and other occupancies shall be constructed as smoke barriers. Door openings in the smoke barrier shall be tightfitting, with gaskets installed as required by Section 1005, and shall be automatic closing by actuation of the automatic sprinklers, fire alarm or smoke-detection system. Openings for ducts and other heating, ventilating and air-conditioning openings shall be equipped with a minimum Class I, 250°F (121°C) smoke damper as defined and tested in accordance with approved recognized standards. See Chapter 35, Part IV. The damper shall close upon detection of smoke by an approved smoke detector located within the duct, or upon the activation of the fire alarm system; and

> > 3.4 Each compartment formed by the smoke barrier has not less than two exits or exit-access doors, one of which is permitted to pass through the adjoining compartment; and
> >
> > 3.5 At least one exit or exit-access door from the Division 3 Occupancy shall be into a separate means of egress as defined in Section 1007.3; and
> >
> > 3.6 The building is equipped with an automatic sprinkler system throughout.

Stages and platforms shall be constructed in accordance with Chapter 4. For attic space partitions and draft stops, see Section 708.

305.2.4 Special hazards. Laboratories, vocational shops and similar areas containing hazardous materials shall be separated from each other and from other portions of the building by not less than a one-hour fire-resistive occupancy separation. When the quantities of hazardous materials in such uses do not exceed those listed in Table 3-D or 3-E, the requirements of Sections 307.5.2 and 307.8 shall apply. When the quantities of hazardous materials in such uses exceed those listed in Table 3-D or 3-E, the use shall be classified as the appropriate Group H Occupancy.

See Section 1007.3 for means of egress from laboratories in Group E Occupancies.

Equipment in rooms or groups of rooms sharing a common atmosphere where flammable liquids, combustible dust or hazardous materials are used, stored, developed or handled shall conform to the requirements of the Fire Code.

305.3 Location on Property. All buildings housing Group E Occupancies shall front directly on a public street or an exit discharge not less than 20 feet (6096 mm) in width. The exit discharge to the public street shall be a minimum 20-foot-wide (6096 mm) right-of-way, unobstructed and maintained only as access to the public street. At least one required exit shall be located on the public street or on the exit discharge.

For fire-resistive protection of exterior walls and openings, as determined by location on property, see Section 503 and Chapter 6.

305.4 Access and Means of Egress Facilities. Means of egress shall be provided as specified in Chapter 10. (For special provisions, see Section 1007.3.)

Access to, and egress from, buildings required to be accessible shall be provided as specified in Chapter 11.

305.5 Light, Ventilation and Sanitation. All portions of Group E Occupancies customarily occupied by human beings shall be provided with light and ventilation, either natural or artificial, as specified in Chapter 12. See Section 1003.2.9 for required means of egress illumination.

The number of urinals and drinking fountains shall be as specified in Section 2902.4.

305.6 Shaft and Exit Enclosures. Exits shall be enclosed as specified in Chapter 10. Elevator shafts, vent shafts and other vertical openings shall be enclosed, and the enclosure shall be as specified in Section 711.

305.7 Sprinkler and Standpipe Systems. When required by Section 904.2.1 or other provisions of this code, automatic sprinkler systems and standpipes shall be designed and installed as specified in Chapter 9.

305.8 Special Hazards. Chimneys and heating apparatus shall conform to the requirements of Chapter 31 of this code and the Mechanical Code.

Motion picture machine rooms shall conform to the requirements of Chapter 4.

All exterior openings in a boiler room or rooms containing central heating equipment, if located below openings in another story or if less than 10 feet (3048 mm) from other doors or windows of the same building, shall be protected by a fire assembly having a three-fourths-hour fire-protection rating. Such fire assemblies shall be fixed, automatic closing or self-closing.

Class I, II or III-A liquids shall not be placed, stored or used in Group E Occupancies, except in approved quantities as necessary in laboratories and classrooms and for operation and maintenance as set forth in the Fire Code.

305.9 Fire Alarm Systems. An approved fire alarm system shall be provided for Group E Occupancies with an occupant load of 50 or more persons. In Group E Occupancies provided with an automatic sprinkler or detection system, the operation of such system shall automatically activate the school fire alarm system, which shall include an alarm mounted on the exterior of the building.

See Chapter 10 for smoke-detection requirements.

For installation requirements, see the Fire Code.

SECTION 306 — REQUIREMENTS FOR GROUP F OCCUPANCIES

306.1 Group F Occupancies Defined. Group F Occupancies shall include the use of a building or structure, or a portion thereof, for assembling, disassembling, fabricating, finishing, manufacturing, packaging, repair or processing operations that are not classified as Group H Occupancies. Factory and industrial occupancies shall include the following:

Division 1. Moderate-hazard factory and industrial occupancies shall include factory and industrial uses that are not classified as Group F, Division 2 Occupancies, but are not limited to facilities producing the following:

1. Aircraft.
2. Appliances.
3. Athletic equipment.
4. Automobiles and other motor vehicles.
5. Bakeries.
6. Alcoholic beverages.
7. Bicycles.
8. Boats.
9. Brooms and brushes.
10. Business machines.
11. Canvas or similar fabric.
12. Cameras and photo equipment.
13. Carpets and rugs, including cleaning.
14. Clothing.
15. Construction and agricultural machinery.
16. Dry cleaning and dyeing.
17. Electronics assembly.
18. Engines, including rebuilding.
19. Photographic film.
20. Food processing.
21. Furniture.
22. Hemp products.
23. Jute products.
24. Laundries.
25. Leather products.
26. Machinery.
27. Metal.
28. Motion pictures and television filming and videotaping.
29. Musical instruments.
30. Optical goods.
31. Paper mills or products.
32. Plastic products.
33. Printing or publishing.
34. Recreational vehicles.
35. Refuse incineration.
36. Shoes.
37. Soaps and detergents.
38. Tobacco.
39. Trailers.
40. Wood, distillation.
41. Millwork (sash and door).
42. Woodworking, cabinet.

Division 2. Low-hazard factory and industrial occupancies shall include facilities producing noncombustible or nonexplosive materials which, during finishing, packing or processing, do not involve a significant fire hazard, including, but not limited to, the following:

1. Nonalcoholic beverages.
2. Brick and masonry.
3. Ceramic products.
4. Foundries.
5. Glass products.
6. Gypsum.
7. Steel products—fabrication and assembly.

For occupancy separations, see Table 3-B.

306.2 Construction, Height and Allowable Area.

306.2.1 General. Buildings or parts of buildings classed as Group F Occupancies because of the use or character of the occupancies shall be limited to the types of construction set forth in Table 5-B and shall not exceed, in area or height, the limits specified in Sections 504, 505 and 506.

306.2.2 Special provisions, Group F, Division 2 roof framing. In Division 2 Occupancies, the roof-framing system may be of unprotected construction.

306.3 Location on Property. For fire-resistive protection of exterior walls and openings, as determined by location on property, see Section 503.

306.4 Access and Means of Egress Facilities. Means of egress shall be provided as specified in Chapter 10.

Access to, and egress from, buildings required to be accessible shall be provided as specified in Chapter 11.

306.5 Light, Ventilation and Sanitation. In Group F Occupancies, light, ventilation and sanitation shall be as specified in Chapters 12 and 29.

306.6 Shaft and Exit Enclosures. Exits shall be enclosed as specified in Chapter 10.

Elevator shafts, vent shafts and other openings through floors shall be enclosed, and the enclosure shall be as specified in Section 711.

> **EXCEPTION:** In Group F, Division 2 Occupancies, exits shall be enclosed as specified in Chapter 10, but other through-floor openings need not be enclosed.

In buildings housing Group F Occupancies equipped with automatic sprinkler systems throughout, enclosures need not be provided for escalators where the top of the escalator opening at each story is provided with a draft curtain and automatic fire sprinklers are installed around the perimeter of the opening within 2 feet (610 mm) of the draft curtain. The draft curtain shall enclose the perimeter of the unenclosed opening and extend from the ceiling downward at least 12 inches (305 mm) on all sides. The spacing between sprinklers shall not exceed 6 feet (1829 mm).

306.7 Sprinkler and Standpipe Systems. When required by Section 904.2 or other provisions of this code, automatic sprinkler systems and standpipes shall be installed as specified in Chapter 9.

306.8 Special Hazards. For special hazards of Group F Occupancies, see Section 304.8.

Storage and use of flammable and combustible liquids shall be in accordance with the Fire Code.

Buildings erected or converted to house high-piled combustible stock or aerosols shall comply with the Fire Code.

Equipment, machinery or appliances that generate finely divided combustible waste or that use finely divided combustible material shall be equipped with an approved method of collection and removal.

SECTION 307 — REQUIREMENTS FOR GROUP H OCCUPANCIES

307.1 Group H Occupancies Defined.

307.1.1 General. Group H Occupancies shall include buildings or structures, or portions thereof, that involve the manufacturing, processing, generation or storage of materials that constitute a high fire, explosion or health hazard. For definitions, identification and control of hazardous materials and pesticides, and the display of nonflammable solid and nonflammable and noncombustible liquid hazardous materials in Group B, F, M or S Occupancies, see the Fire Code. For hazardous materials used as refrigerants or lubricants within closed cycle refrigeration systems and the areas served by them, see Chapter 28 of this code, the Mechanical Code and the Fire Code. For the application and use of control areas, see Footnote 1 of Tables 3-D and 3-E. Group H Occupancies shall be:

Division 1. Occupancies with a quantity of material in the building in excess of those listed in Table 3-D, which present a high explosion hazard, including, but not limited to:

1. Explosives, blasting agents, Class 1.3G (Class B, Special) fireworks and black powder.

> **EXCEPTIONS:** 1. Storage and use of pyrotechnic special effect materials in motion picture, television, theatrical and group entertainment production when under permit as required in the Fire Code. The time period for storage shall not exceed 90 days.

> 2. Indoor storage and display of smokeless powder, black sporting powder, and primers or percussion caps exceeding the exempt amounts for Group M retail sales need not be classified as a Group H, Division 1 Occupancy where stored and displayed in accordance with the Fire Code.

2. Manufacturing of Class 1.4G (Class C, Common) fireworks.

3. Unclassified detonatable organic peroxides.

4. Class 4 oxidizers.

5. Class 4 or Class 3 detonatable unstable (reactive) materials.

Division 2. Occupancies where combustible dust is manufactured, used or generated in such a manner that concentrations and conditions create a fire or explosion potential; occupancies with a quantity of material in the building in excess of those listed in Table 3-D, which present a moderate explosion hazard or a hazard from accelerated burning, including, but not limited to:

1. Class I organic peroxides.

2. Class 3 nondetonatable unstable (reactive) materials.

3. Pyrophoric gases.

4. Flammable or oxidizing gases.

5. Class I, II or III-A flammable or combustible liquids which are used or stored in normally open containers or systems, or in closed containers or systems pressurized at more than 15-pounds-per-square-inch (psi) (103.4 kPa) gage.

> **EXCEPTION:** Aerosols.

6. Class 3 oxidizers.

7. Class 3 water-reactive materials.

Division 3. Occupancies where flammable solids, other than combustible dust, are manufactured, used or generated.

Division 3 Occupancies also include uses in which the quantity of material in the building in excess of those listed in Table 3-D presents a high physical hazard, including, but not limited to:

1. Class II, III or IV organic peroxides.

2. Class 1 or 2 oxidizers.

3. Class I, II or III-A flammable or combustible liquids that are used or stored in normally closed containers or systems and containers or systems pressurized at 15 psi (103.4 kPa) gage or less, and aerosols.

4. Class III-B combustible liquids.

5. Pyrophoric liquids or solids.

6. Class 1 or 2 water-reactive materials.

7. Flammable solids in storage.

8. Flammable or oxidizing cryogenic fluids (other than inert).

9. Class 1 unstable (reactive) gas or Class 2 unstable (reactive) materials.

10. Storage of Class 1.4G (Class C, Common) fireworks.

Division 4. Repair garages not classified as Group S, Division 3 Occupancies.

Division 5. Aircraft repair hangars not classified as Group S, Division 5 Occupancies and heliports.

Division 6. Semiconductor fabrication facilities and comparable research and development areas in which hazardous production materials (HPM) are used and the aggregate quantity of materials are in excess of those listed in Table 3-D or 3-E. Such facilities and areas shall be designed and constructed in accordance with Section 307.11.

Division 7. Occupancies having quantities of materials in excess of those listed in Table 3-E that are health hazards, including:

1. Corrosives.

> EXCEPTION: Stationary lead-acid battery systems.

2. Toxic and highly toxic materials.

3. Irritants.

4. Sensitizers.

5. Other health hazards.

307.1.2 Multiple hazards. When a hazardous material has multiple hazards, all hazards shall be addressed and controlled in accordance with the provisions of this chapter.

307.1.3 Liquid use, dispensing and mixing rooms. Rooms in which Class I, Class II and Class III-A flammable or combustible liquids are used, dispensed or mixed in open containers shall be constructed in accordance with the requirements for a Group H, Division 2 Occupancy and the following:

1. Rooms in excess of 500 square feet (46.5 m^2) shall have at least one exterior exit door approved for fire department access.

2. Rooms shall be separated from other areas by an occupancy separation having a fire-resistive rating of not less than one hour for rooms up to 150 square feet (13.9 m^2) in area and not less than two hours where the room is more than 150 square feet (13.9 m^2) in area. Separations from other occupancies shall not be less than required by Section 302 and Table 3-B.

3. Shelving, racks and wainscoting in such areas shall be of noncombustible construction or wood not less than 1-inch (25 mm) nominal thickness.

4. Liquid use, dispensing and mixing rooms shall not be located in basements.

307.1.4 Liquid storage rooms. Rooms in which Class I, Class II and Class III-A flammable or combustible liquids are stored in closed containers shall be constructed in accordance with the requirements for a Group H, Division 3 Occupancy and to the following:

1. Rooms in excess of 500 square feet (46.5 m^2) shall have at least one exterior exit door approved for fire department access.

2. Rooms shall be separated from other areas by an occupancy separation having a fire-resistive rating of not less than one hour for rooms up to 150 square feet (13.9 m^2) in area and not less than two hours where the room is more than 150 square feet (13.9 m^2) in area. Separations from other occupancies shall not be less than required by Section 302 and Table 3-B.

3. Shelving, racks and wainscoting in such areas shall be of noncombustible construction or wood of not less than 1-inch (25 mm) nominal thickness.

4. Rooms used for the storage of Class I flammable liquids shall not be located in a basement.

307.1.5 Flammable or combustible liquid storage warehouses. Liquid storage warehouses in which Class I, Class II and Class III-A flammable or combustible liquids are stored in closed containers shall be constructed in accordance with the requirements for a Group H, Division 3 Occupancy and the following:

1. Liquid storage warehouses shall be separated from all other uses by a four-hour area separation wall.

2. Shelving, racks and wainscoting in such warehouses shall be of noncombustible construction or wood not less than 1-inch (25 mm) nominal thickness.

3. Rooms used for the storage of Class I flammable liquids shall not be located in a basement.

307.1.6 Requirement for report. The building official may require a technical opinion and report to identify and develop methods of protection from the hazards presented by the hazardous material. The opinion and report shall be prepared by a qualified person, firm or corporation approved by the building official and shall be provided without charge to the enforcing agency.

The opinion and report may include, but is not limited to, the preparation of a hazardous material management plan (HMMP); chemical analysis; recommendations for methods of isolation, separation, containment or protection of hazardous materials or processes, including appropriate engineering controls to be applied; the extent of changes in the hazardous behavior to be anticipated under conditions of exposure to fire or from hazard control procedures; and the limitations or conditions of use necessary to achieve and maintain control of the hazardous materials or operations. The report shall be entered into the files of the code enforcement agencies. Proprietary and trade secret information shall be protected under the laws of the state or jurisdiction having authority.

307.2 Construction, Height and Allowable Area.

307.2.1 General. Buildings or parts of buildings classed in Group H because of the use or character of the occupancy shall be limited to the types of construction set forth in Table 5-B and shall not exceed, in area or height, the limits specified in Sections 504, 505 and 506.

307.2.2 Floors. Except for surfacing, floors in areas containing hazardous materials and in areas where motor vehicles, boats, helicopters or airplanes are stored, repaired or operated shall be of noncombustible, liquid-tight construction.

> EXCEPTION: In Group H, Divisions 4 and 5 Occupancies, floors may be surfaced or waterproofed with asphaltic paving materials in that portion of the facility where no repair work is done.

307.2.3 Spill control and secondary containment for the storage of hazardous materials liquids and solids.

307.2.3.1 Applicability. When required by the Fire Code, rooms, buildings or areas used for the storage of liquid or solid hazardous materials shall be provided with spill control and secondary containment in accordance with Section 307.2.3.

See the Fire Code for outdoor storage provisions.

307.2.3.2 Spill control for hazardous materials liquids. Rooms, buildings or areas used for the storage of hazardous materials liquids in individual vessels having a capacity of more than 55 gallons (208.2 L) or when the aggregate capacity of multiple vessels exceeds 1,000 gallons (3785 L) shall be provided with spill control to prevent the flow of liquids to adjoining areas. Floors shall be constructed to contain a spill from the largest single vessel by one of the following methods:

1. Liquid-tight sloped or recessed floors,

2. Liquid-tight floors provided with liquid-tight raised or recessed sills or dikes, or

3. Sumps and collection systems.

Except for surfacing, the floors, sills, dikes, sumps and collection systems shall be constructed of noncombustible material, and the liquid-tight seal shall be compatible with the material stored. When liquid-tight sills or dikes are provided, they are not required at perimeter openings, which are provided with an open-grate trench across the opening that connects to an approved collection system.

307.2.3.3 Secondary containment for hazardous materials liquids and solids. When required by the Fire Code, buildings, rooms or areas used for the storage of hazardous materials liquids or solids shall be provided with secondary containment in accordance with this section when the capacity of an individual vessel or the aggregate capacity of multiple vessels exceeds the following:

Liquids: Capacity of an individual vessel exceeds 55 gallons (208.2 L) or the aggregate capacity of multiple vessels exceeds 1,000 gallons (3785 L).

Solids: Capacity of an individual vessel exceeds 550 pounds (248.8 kg) or the aggregate capacity of multiple vessels exceeds 10,000 pounds (4524.8 kg).

The building, room or area shall contain or drain the hazardous materials and fire-protection water through the use of one of the following methods:

1. Liquid-tight sloped or recessed floors,

2. Liquid-tight floors provided with liquid-tight raised or recessed sills or dikes,

3. Sumps and collection systems, or

4. Drainage systems leading to an approved location.

Incompatible materials shall be separated from each other in the secondary containment system.

Secondary containment for indoor storage areas shall be designed to contain a spill from the largest vessel, plus the design flow volume of fire-protection water calculated to discharge from the fire-extinguishing system over the minimum required system design area or area of the room or area in which the storage is located, whichever is smaller, for a period of 20 minutes.

A monitoring method shall be provided to detect hazardous materials in the secondary containment system. The monitoring method is allowed to be visual inspection of the primary or secondary containment, or other approved means. Where secondary containment is subject to the intrusion of water, a monitoring method for detecting water shall be provided. When monitoring devices are provided, they shall be connected to distinct visual or audible alarms.

Drainage systems shall be in accordance with the Plumbing Code and the following:

1. The slope of floors to drains shall not be less than 1 percent,

2. Drains shall be sized to carry the volume of the fire-protection water as determined by the design density discharged from the automatic fire-extinguishing system over the minimum required system design area or area of the room or area in which the storage is located, whichever is smaller,

3. Materials of construction for drainage systems shall be compatible with the materials stored,

4. Incompatible materials shall be separated from each other in the drainage system, and

5. Drains shall terminate in an approved location away from buildings, valves, means of egress, fire-access roadways, adjoining property and storm drains.

307.2.4 Spill control and secondary containment for use of hazardous materials liquids.

307.2.4.1 Open containers and systems.

307.2.4.1.1 Spill control for hazardous materials liquids. When required by the Fire Code, buildings, rooms or areas where hazardous materials liquids are dispensed into vessels exceeding a 1.1-gallon (4 L) capacity or used in open systems exceeding a 5.3-gallon (20 L) capacity shall be provided with spill control in accordance with Section 307.2.3.2.

307.2.4.1.2 Secondary containment for hazardous materials liquids. When required by the Fire Code, buildings, rooms or areas where hazardous materials liquids are dispensed or used in open systems shall be provided with secondary containment in accordance with Section 307.2.3.3 when the capacity of an individual vessel or system or the capacity of multiple vessels or systems exceeds the following:

Individual vessel or system: Greater than 1.1 gallons (4 L)

Multiple vessels or systems: Greater than 5.3 gallons (20 L)

307.2.4.2 Closed containers and systems.

307.2.4.2.1 Spill control for hazardous materials liquids. When required by the Fire Code, buildings, rooms or areas where hazardous materials liquids are used in individual vessels exceeding a 55-gallon (208.2 L) capacity shall be provided with spill control in accordance with Section 307.2.3.2.

307.2.4.2.2 Secondary containment for hazardous materials liquids. When required by the Fire Code, buildings, rooms or areas where hazardous materials liquids are used in vessels or systems shall be provided with secondary containment in accordance with Section 307.2.3.3 when the capacity of an individual vessel or system or the capacity of multiple vessels or systems exceeds the following:

Individual vessel or system: Greater than 55 gallons (208.2 L)

Multiple vessels or systems: Greater than 1,000 gallons (3785 L)

307.2.5 Smoke and heat vents. Smoke and heat venting shall be provided in areas containing hazardous materials as set forth in the Fire Code in addition to the provisions of this code.

307.2.6 Standby power. Standby power shall be provided in Group H, Divisions 1 and 2 Occupancies and in Group H, Division 3 Occupancies in which Class I or II organic peroxides are stored. The standby power system shall be designed and installed in accordance with the Electrical Code to automatically supply power to all required electrical equipment when the normal electrical supply system is interrupted.

307.2.7 Emergency power. An emergency power system shall be provided in Group H, Divisions 6 and 7 Occupancies. The emergency power system shall be designed and installed in accordance with the Electrical Code to automatically supply power to all required electrical equipment when the normal electrical supply system is interrupted.

The exhaust system may be designed to operate at not less than one half the normal fan speed on the emergency power system when it is demonstrated that the level of exhaust will maintain a safe atmosphere.

307.2.8 Special provisions for Group H, Division 1 Occupancies. Group H, Division 1 Occupancies shall be in buildings used for no other purpose, without basements, crawl spaces or other under-floor spaces. Roofs shall be of lightweight construction with suitable thermal insulation to prevent sensitive material from reaching its decomposition temperature.

Group H, Division 1 Occupancies containing materials, which are in themselves both physical and health hazards in quantities exceeding the exempt amounts in Table 3-E, shall comply with requirements for both Group H, Division 1 and Group H, Division 7 Occupancies.

307.2.9 Special provisions for Group H, Divisions 2 and 3 Occupancies. Group H, Divisions 2 and 3 Occupancies containing quantities of hazardous materials in excess of those set forth in

Table 3-G shall be in buildings used for no other purpose, shall not exceed one story in height and shall be without basements, crawl spaces or other under-floor spaces.

Group H, Divisions 2 and 3 Occupancies containing water-reactive materials shall be resistant to water penetration. Piping for conveying liquids shall not be over or through areas containing water reactives, unless isolated by approved liquid-tight construction.

EXCEPTION: Fire-protection piping may be installed over reactives without isolation.

307.2.10 Special provisions for Group H, Division 4 Occupancies. Group H, Division 4 Occupancies having a floor area not exceeding 2,500 square feet (232 m^2) may have exterior walls of not less than two-hour fire-resistive construction when less than 5 feet (1524 mm) from a property line and not less than one-hour fire-resistive construction when less than 20 feet (6096 mm) from a property line.

307.2.11 Special provisions for Group H, Division 6 Occupancies. See Section 307.11.

307.3 Location on Property. Group H Occupancies shall be located on property in accordance with Section 503, Table 3-F and other provisions of this chapter. In Group H, Division 2 or 3 Occupancies, not less than 25 percent of the perimeter wall of the occupancy shall be an exterior wall.

EXCEPTIONS: 1. Liquid use, dispensing and mixing rooms having a floor area of not more than 500 square feet (46.5 m^2) need not be located on the outer perimeter of the building when they are in accordance with Section 307.1.3.

2. Liquid storage rooms having a floor area of not more than 1,000 square feet (93 m^2) need not be located on the outer perimeter when they are in accordance with Section 307.1.4.

3. Spray paint booths that comply with the Fire Code need not be located on the outer perimeter.

307.4 Access and Means of Egress Facilities. Means of egress shall be provided as specified in Chapter 10. (For special provisions, see Section 1007.4.)

Access to, and egress from, buildings required to be accessible shall be provided as specified in Chapter 11.

307.5 Light, Ventilation and Sanitation.

307.5.1 General. Light, ventilation and sanitation in Group H Occupancies shall comply with requirements in this section and Chapters 12 and 29.

307.5.2 Ventilation in hazardous locations. See Section 1202.2.3 for ventilation requirements in hazadous locations.

307.5.3 Ventilation in Group H, Division 4 Occupancies. See Section 1202.2.4 for ventilation requirements in Group H, Division 4 Occupancies.

307.5.4 Sanitation. The number of plumbing fixtures shall not be less than specified in Section 2902.3.

307.6 Shaft and Exit Enclosures. Exits shall be enclosed as specified in Chapter 10.

Elevator shafts, vent shafts and other openings through floors shall be enclosed, and the enclosure shall be as specified in Section 711.

Doors which are a part of an automobile ramp enclosure shall be equipped with automatic-closing devices.

For Group H, Division 6 Occupancies, see Section 307.11.2.3.

307.7 Sprinkler and Standpipe Systems. When required by Section 904.2.1 or other provisions of this code, automatic fire-extinguishing systems and standpipes shall be designed and installed as specified in Chapter 9.

307.8 Special Hazards. Chimneys and heating apparatus shall conform to the requirements of Chapter 31 of this code and the Mechanical Code.

In Divisions 4 and 5 Occupancies, devices that generate a glow, spark or flame capable of igniting flammable vapors shall be installed with sources of ignition at least 18 inches (457 mm) above the floor. See the Mechanical Code for additional restrictions.

Equipment or machinery that generates or emits combustible or explosive dust or fibers shall be provided with an adequate dust-collecting and exhaust system installed in conformance with the Mechanical Code. Equipment or systems that are used to collect, process or convey combustible dusts or fibers shall be provided with an approved explosion venting or containment system.

Combustible fiber storage rooms with a fiber storage capacity not exceeding 500 cubic feet (14.2 m^3) shall be separated from the remainder of the building by a one-hour fire-resistive occupancy separation. Combustible fiber storage vaults having a fiber storage capacity of more than 500 cubic feet (14.2 m^3) shall be separated from the remainder of the building by a two-hour fire-resistive occupancy separation.

Cellulose nitrate film storage and handling shall be in accordance with Section 407.

307.9 Fire Alarm Systems. An approved manual fire alarm system shall be provided in Group H Occupancies used for the manufacturing of organic coatings. Approved automatic smoke detection shall be provided for rooms used for the storage, dispensing, use and handling of hazardous materials when required by the Fire Code.

For Group H, Division 6 Occupancies, see Section 307.11.

For installation requirements, see the Fire Code.

For aerosol storage warehouses, see the Fire Code.

307.10 Explosion Control. Explosion control, equivalent protective devices or suppression systems; or barricades shall be provided to control or vent the gases resulting from deflagrations of dusts, gases or mists in rooms, buildings or other enclosures as required by the Fire Code so as to minimize structural or mechanical damage. If detonation rather than deflagration is considered likely, protective devices or systems such as fully contained barricades shall be provided, except that explosion venting to minimize damage from less than 2.0 grams of trinitrotoluene (TNT) (equivalence) is permitted. Walls, floors and roofs separating a use from an explosion exposure shall be designed to resist a minimum internal pressure of 100 pounds per square foot (psf) (4.79 kPa) in addition to the loads required by Chapter 16.

Explosion venting shall be provided in exterior walls or roof only. The venting shall be designed to prevent serious structural damage and production of lethal projectiles. The aggregate clear vent relief area shall be regulated by the pressure resistance of the nonrelieving portions of the building and be designed by persons competent in such design. The design shall recognize the nature of the material and its behavior in an explosion. Vents shall consist of any one or any combination of the following to relieve at a maximum internal pressure of 20 psf (958 Pa), but not less than the loads required by Chapter 16:

1. Walls of lightweight material.

2. Lightly fastened hatch covers.

3. Lightly fastened, outward-opening swinging doors in exterior walls.

4. Lightly fastened walls or roof.

Venting devices shall discharge vertically or directly to an unoccupied yard not less than 50 feet (15 240 mm) in width on the same lot. Releasing devices shall be so located that the discharge end shall not be less than 10 feet (3048 mm) vertically and 20 feet (6096 mm) horizontally from window openings or exits in the same or adjoining buildings or structures. The exhaust shall always be in the direction of least exposure and never into the interior of the building unless a suitably designed shaft is provided that discharges to the exterior. See Footnote 12 of Table 3-D.

307.11 Group H, Division 6 Occupancies.

307.11.1 General. In addition to the requirements set forth elsewhere in this code, Group H, Division 6 Occupancies shall comply with the provisions of this section and the Fire Code.

307.11.2 Fabrication area.

307.11.2.1 Separation and location. Fabrication areas, whose sizes are limited by the quantity of hazardous production materials (HPM) permitted by the Fire Code, shall be separated from each other, from corridors, and from other parts of the building by not less than one-hour fire-resistive occupancy separations. Occupied levels of fabrication areas shall be located at or above the first story.

> **EXCEPTIONS:** 1. Doors within such occupancy separation, including doors to corridors, shall be only self-closing fire assemblies having a fire-protection rating of not less than three-fourths hour.
>
> 2. Windows between fabrication areas and corridors may be in accordance with Section 1004.3.4.3.2.2.

307.11.2.2 Floors. Except for surfacing, floors within fabrication areas shall be of noncombustible construction. Openings through floors of fabrication areas may be unprotected when the interconnected levels are used solely for mechanical equipment directly related to such fabrication area. See also Section 307.11.2.3. When forming a part of an occupancy separation, floors shall be liquid tight.

307.11.2.3 Shaft and exit enclosures. Exits shall be enclosed as specified in Chapter 10.

Elevator shafts, vent shafts and other openings through floors shall be enclosed and the enclosure shall be as specified in Section 711. A fabrication area may have mechanical, duct and piping penetrations that extend through not more than two floors within that fabrication area. The annular space around penetrations for cables, cable trays, tubing, piping, conduit or ducts shall be sealed at the floor level to restrict the movement of air. The fabrication area, including the areas through which the ductwork and piping extend, shall be considered a single conditioned environment.

307.11.2.4 Ventilation. See Section 1202.2.5 for ventilation requirements.

307.11.2.5 Transporting hazardous production materials. Hazardous production materials shall be transported to fabrication areas through enclosed piping or tubing systems that comply with Section 307.11.6, through service corridors or in corridors as permitted in the exception to Section 307.11.3. The handling or transporting of hazardous production materials within service corridors shall comply with the Fire Code.

307.11.2.6 Electrical. Electrical equipment and devices within the fabrication area shall comply with the Electrical Code. The requirements for hazardous locations need not be applied when the average air change is at least four times that set forth in Section 307.11.2.4 and when the number of air changes at any location is

not less than three times that required by Section 307.11.2.4 and the Fire Code.

Electrical equipment and devices within 5 feet (1524 mm) of work stations in which flammable or pyrophoric gases or flammable liquids are used shall be in accordance with the Electrical Code for Class I, Division 2 hazardous locations. Work stations shall not be energized without adequate exhaust ventilation. See Section 1202.2.5 for work station exhaust ventilation requirements.

> **EXCEPTION:** Class I, Division 2 hazardous electrical is not required when the air removal from the work station or dilution will provide nonflammable atmospheres on a continuous basis.

307.11.3 Corridors. Corridors shall comply with Section 1004.3.4 and shall be separated from fabrication areas as specified in Section 307.11.2.1. Corridors shall not be used for transporting hazardous production materials except as provided in Section 307.11.6.2.

> **EXCEPTION:** In existing Group H, Division 6 Occupancies when there are alterations or modifications to existing fabrication areas, the building official may permit the transportation of hazardous production materials in corridors subject to the requirements of the Fire Code and as follows:
>
> 1. Exit-access corridors adjacent to the fabrication area where the alteration work is to be done shall comply with Section 1004.3.4 for a length determined as follows:
>
> > 1.1 The length of the common wall of the corridor and the fabrication area, and
> >
> > 1.2 For the distance along the corridor to the point of entry of HPM into the corridor serving that fabrication area.
>
> 2. There shall be an emergency telephone system or a local alarm manual pull station or approved signal device within corridors at not more than 150-foot (45 720 mm) intervals or fraction thereof and at each stair doorway. The signal shall be relayed to the emergency control station and a local signaling device shall be provided.
>
> 3. Sprinkler protection shall be designed in accordance with UBC Standard 9-1 for Ordinary Hazard Group 3, except that when one row of sprinklers is used in the corridor protection, the maximum number of sprinklers that need be calculated is 13. UBC Standard 9-1 is a part of this code. (See Chapter 35, Part II.)
>
> 4. Self-closing doors having a fire-protection rating of not less than one hour shall separate pass-throughs from existing corridors. Pass-throughs shall be constructed as required for the corridors. Pass-throughs shall be protected by an approved automatic fire sprinkler system.

307.11.4 Service corridors. Service corridors shall be classified as Group H, Division 6 Occupancies. Service corridors shall be separated from corridors as required by Section 307.11.2.1.

Service corridors shall be mechanically ventilated as required by Section 307.11.2.4 or at not less than six air changes per hour, whichever is greater.

The maximum travel distance from any point in a service corridor to an exit or door into a fabrication area shall not exceed 75 feet (22 860 mm). Dead ends shall not exceed 4 feet (1219 mm) in length. There shall not be less than two means of egress, and not more than one half of the required means of egress shall be into the fabrication area. Doors from service corridors shall swing in the direction of egress and shall be self-closing.

The minimum clear width of a service corridor shall be 5 feet (1524 mm), or 33 inches (838 mm) wider than the widest cart or truck used in the corridor, whichever is greater.

307.11.5 Storage of hazardous production materials.

307.11.5.1 Construction. The storage of hazardous production materials in quantities greater than those listed in Table 3-D or 3-E shall be in liquid storage rooms, HPM rooms or gas rooms, as required by the Fire Code. HPM rooms and gas rooms shall be separated from all other areas by not less than a two-hour fire-resistive occupancy separation when the area is 300 square feet (27.9 m²) or more and not less than one-hour fire-resistive occupancy separation when the area is less than 300 square feet (27.9 m²). The provisions of Section 302.1 shall apply.

Except for surfacing, floors of storage and HPM rooms shall be of noncombustible liquid-tight construction. Raised grating over floors shall be of noncombustible materials. See Section 307.2.3 for sill requirements for liquid storage rooms.

307.11.5.2 Location within building. When HPM rooms are provided, they shall have at least one exterior wall and such wall shall not be less than 30 feet (9144 mm) from property lines, including property lines adjacent to public ways. Explosion control shall be provided when required by Section 307.10.

307.11.5.3 Means of egress. When two means of egress are required from HPM rooms, one shall be directly to the outside of the building. See Section 307.11.2.1, Exception 1.

307.11.5.4 Ventilation. Mechanical exhaust ventilation shall be provided in liquid storage rooms, HPM rooms and gas rooms at the rate of not less than 1 cubic foot per minute per square foot (0.044 L/s/m²) of floor area or six air changes per hour, whichever is greater, for all categories of material.

Exhaust ventilation for gas rooms shall be designed to operate at a negative pressure in relation to the surrounding areas and direct the exhaust ventilation to an exhaust system.

307.11.5.5 Fire and emergency alarm. An approved manual fire alarm system shall be provided throughout buildings containing Group H, Division 6 Occupancies.

An approved emergency alarm system shall be provided for HPM rooms, liquid storage rooms and gas rooms. Emergency alarm-initiating devices shall be installed outside of each interior door of such rooms. Activation of an emergency alarm-initiating device shall sound a local alarm and transmit a signal to the emergency control station.

For installation requirements, see the Fire Code.

307.11.5.6 Electrical. Electrical wiring and equipment in HPM rooms, gas rooms and liquid storage rooms shall comply with the Electrical Code.

307.11.6 Piping and tubing.

307.11.6.1 General. Hazardous production materials piping and tubing shall comply with this section and shall be installed in accordance with nationally recognized standards. Piping and tubing systems shall be metallic unless the material being transported is incompatible with such system. Systems supplying gaseous HPM having a health hazard ranking of 3 or 4 shall be welded throughout, except for connections, valves and fittings, to the systems which are within a ventilated enclosure. Hazardous production materials supply piping or tubing in service corridors shall be exposed to view.

307.11.6.2 Installations in corridors and above other occupancies. Hazardous production materials shall not be located within corridors or above areas not classified as Group H, Division 6 Occupancies except as permitted by this section.

Hazardous production material piping and tubing may be installed within the space defined by the walls of corridors and the floor or roof above or in concealed spaces above other occupancies under the following conditions:

1. Automatic sprinklers shall be installed within the space unless the space is less than 6 inches (152 mm) in least dimension.

2. Ventilation at not less than six air changes per hour shall be provided. The space shall not be used to convey air from any other area.

3. When the piping or tubing is used to transport HPM liquids, a receptor shall be installed below such piping or tubing. The receptor shall be designed to collect any discharge or leakage and drain it to an approved location. The one-hour enclosure shall not be used as part of the receptor.

4. All HPM supply piping and tubing and HPM nonmetallic waste lines shall be separated from the corridor and from any occupancy other than Group H, Division 6 by construction as required for walls or partitions that have a fire-protection rating of not less than one hour. When gypsum wallboard is used, joints on the piping side of the enclosure need not be taped, provided the joints occur over framing members. Access openings into the enclosure shall be protected by approved fire assemblies.

5. Readily accessible manual or automatic remotely activated fail-safe emergency shutoff valves shall be installed on piping and tubing other than waste lines at the following locations:

    5.1 At branch connections into the fabrication area.

    5.2 At entries into corridors.

Excess flow valves shall be installed as required by the Fire Code.

> **EXCEPTION:** Occasional transverse crossings of the corridors by supply piping which is enclosed within a ferrous pipe or tube for the width of the corridor need not comply with Items 1 through 5.

307.11.6.3 Identification. Piping, tubing and HPM waste lines shall be identified in accordance with nationally recognized standards to indicate the material being transported.

307.12 Heliports. Heliports may be erected on buildings or other locations if they are constructed in accordance with this chapter and with Section 311.10.

SECTION 308 — REQUIREMENTS FOR GROUP I OCCUPANCIES

308.1 Group I Occupancies Defined. Group I Occupancies shall be:

Division 1.1. Nurseries for the full-time care of children under the age of six (each accommodating more than five children).

Hospitals, sanitariums, nursing homes with nonambulatory patients and similar buildings (each accommodating more than five patients).

Division 1.2. Health-care centers for ambulatory patients receiving outpatient medical care that may render the patient incapable of unassisted self-preservation (each tenant space accommodating more than five such patients).

Division 2. Nursing homes for ambulatory patients, homes for children six years of age or over (each accommodating more than five patients or children).

Division 3. Mental hospitals, mental sanitariums, jails, prisons, reformatories and buildings where personal liberties of inmates are similarly restrained.

For occupancy separations, see Table 3-B.

> **EXCEPTION:** Group I Occupancies shall not include buildings used only for private residential purposes for a family group.

308.2 Construction, Height and Allowable Area.

308.2.1 General. Buildings or parts of buildings classed in Group I because of the use or character of the occupancy shall be limited to the types of construction set forth in Table 5-B and shall not exceed, in area or height, the limits specified in Sections 504, 505 and 506.

> **EXCEPTIONS: 1.** Hospitals and nursing homes classified as Group I, Division 1.1 Occupancies, and health-care centers for ambu-

latory patients classified as Group I, Division 1.2 Occupancies that are equipped with an automatic sprinkler system throughout shall not exceed one story in height when in Type III One-hour, Type IV or Type V One-hour construction.

2. Hospitals and nursing homes classified as Group I, Division 1.1 Occupancies, and health-care centers for ambulatory patients classified as Group I, Division 1.2 Occupancies that are equipped with automatic sprinkler systems throughout may be five stories when of Type II-F.R. construction and three stories when of Type II One-hour construction. The allowable area increase specified in Section 505.3 applies only when the number of stories in the building is one less than set forth above.

3. Hospitals and nursing homes classified as Group I, Division 1.1 Occupancies, and health-care centers for ambulatory patients classified as Group I, Division 1.2 Occupancies that are equipped with automatic sprinkler systems throughout may be housed within one-story buildings of Type II-N construction. The area of such building shall not exceed 13,500 square feet (1254 m^2) plus the allowable area increase for separation by public space or yards as set forth in Section 505.1.

308.2.2 Specific-use provisions.

308.2.2.1 Group I, Division 1.1 smoke barriers. Floor levels of Group I, Division 1.1 Occupancies used by inpatients for sleeping or treatment, or having an occupant load of 50 or more, shall be divided into at least two compartments by smoke barriers of not less than one-hour fire resistance meeting the requirements of Section 905.2.3. The area within a smoke-control zone shall not exceed 22,500 square feet (2090 m^2) and its width or length shall not exceed 150 feet (45 720 mm). The area of a smoke zone shall not be less than that required to accommodate the occupants of the zone plus the occupants from any adjoining zone. Not less than 30 square feet (2.8 m^2) net clear floor area for bed and litter patients and 6 square feet (0.6 m^2) net clear floor area for other occupants shall be used to compute the required areas.

Doors in smoke barriers shall be tightfitting smoke- and draft-control assemblies having a fire-protection rating of not less than 20 minutes and shall comply with Section 1007.5.1. When doors are installed across corridors, a pair of opposite-swinging doors without a center mullion or horizontal sliding doors that comply with UBC Standard 7-8, which is part of this code (see Chapter 35, Part II), shall be installed. Smoke-barrier doors shall:

1. When installed across corridors, have vision panels. The area of the vision panels shall not exceed that tested.

2. Be closefitting with only the clearance necessary for proper operation and shall be without undercuts, louvers or grilles.

3. Have stops at the head and jambs. Opposite-swinging corridor doors shall have rabbets or astragals at the meeting edges.

4. Have positive latching devices, except on doors installed across corridors.

5. Be automatic closing. Doors installed across corridors shall comply with Section 713.6.1, Item 3, and doors on the floor or in the affected zone shall automatically close if the fire alarm or sprinkler system is activated.

At least two means of egress shall be provided from each smoke zone. Means of egress may pass through adjacent zones, provided the means of egress does not return through the compartment zone from which exiting originated. Exit or exit-access doors at zone boundaries shall be equipped with approved vision panels.

308.2.2.2 Group I, Division 3 Occupancies. Group I, Division 3 Occupancies shall be housed in buildings of Type I or Type II-F.R. construction.

> **EXCEPTION:** Such occupancies may be housed in one-story buildings of Type II One-hour, Type III One-hour or Type V One-hour construction provided the floor area does not exceed 3,900 square feet

(362 m^2) between separation walls of two-hour fire-resistive construction with openings protected by fire assemblies having one- and one-half-hour fire-protection rating.

Rooms occupied by inmates or patients whose personal liberties are restrained shall have noncombustible floor surfaces.

308.3 Location on Property. For fire-resistive protection of exterior walls and openings, as determined by location on property, see Section 503 and Chapter 6.

308.4 Access and Means of Egress Facilities. Means of egress shall be provided as specified in Chapter 10. (For special provisions, see Section 1007.5.)

Access to, and egress from, buildings required to be accessible shall be provided as specified in Chapter 11.

308.5 Light, Ventilation and Sanitation.

308.5.1 Light and ventilation. All portions of enclosed Group I Occupancies customarily occupied by human beings shall be provided with light and ventilation, either natural or artificial, as specified in Section 1202. See Section 1003.2.9 for required exit illumination.

308.5.2 Sanitation. The number of plumbing fixtures shall not be less than specified in Section 2902.5.

308.6 Shaft and Exit Enclosures. Exits shall be enclosed as specified in Chapter 10.

Elevator shafts, vent shafts and other vertical openings shall be enclosed, and the enclosure shall be as specified in Section 711.

308.7 Sprinkler and Standpipe Systems. When required by Section 904.2.1 or other provisions of this code, automatic sprinkler systems and standpipes shall be designed and installed as specified in Chapter 9.

308.8 Special Hazards. Chimneys and heating apparatus shall conform to the requirements of Chapter 31 of this code and the Mechanical Code.

Motion picture projection rooms shall conform to the requirements of Section 406.

Specific use areas shall be separated from Group I, Division 1.1 Occupancies used for hospitals or nursing homes in accordance with Table 3-C. Doors shall be maintained self-closing or shall be automatic closing by actuation of a smoke detector.

Storage and handling of flammable and combustible liquids shall be in accordance with the Fire Code.

All exterior openings in a boiler room or room containing central heating equipment if located below openings in another story, or if less than 10 feet (3048 mm) from the other doors or windows of the same building, shall be protected by a fire assembly having a three-fourths-hour fire-protection rating.

308.9 Fire Alarm Systems. An approved manual and automatic fire alarm system shall be provided for Group I Occupancies. Audible alarm devices shall be used in nonpatient areas. Visible alarm devices may be used in lieu of audible devices in patient-occupied areas. For installation requirements, see the Fire Code.

308.10 Smoke Detectors. Smoke detectors that receive their primary power from the building wiring shall be installed in patient sleeping rooms of hospital and nursing homes. Actuation of such detectors shall cause a visual display on the corridor side of the room in which the detector is located and shall cause an audible and visual alarm at the respective nurses' station. When single-station detectors and related devices are combined with a nursing call system, the nursing call system shall be listed for the intended combined use.

EXCEPTION: In rooms equipped with automatic door closers having integral smoke detectors on the room side, the integral detector may substitute for the room smoke detector, provided it performs the required alerting functions.

SECTION 309 — REQUIREMENTS FOR GROUP M OCCUPANCIES

309.1 Group M Occupancies Defined. Group M Occupancies shall include buildings, structures, or portions thereof, used for the display and sale of merchandise, and involving stocks of goods, wares or merchandise incidental to such purposes and accessible to the public. Mercantile occupancies shall include, but are not limited to, the following:

1. Department stores.

2. Drug stores.

3. Markets.

4. Paint stores without bulk handling.

5. Shopping centers.

6. Sales rooms.

7. Wholesale and retail stores.

For occupancy separations, see Table 3-B.

309.2 Construction, Height and Allowable Area.

309.2.1 General. Buildings or parts of buildings classed in Group M Occupancy because of the use or character of the occupancy shall be limited to the types of construction set forth in Table 5-B and shall not exceed, in area or height, the limits specified in Sections 504, 505 and 506.

309.2.2 Special provisions. Storage areas in connection with wholesale or retail sales shall be separated from the public area by a one-hour fire-resistive occupancy separation.

> **EXCEPTION:** Occupancy separations need not be provided when any one of the following conditions exist:
>
> 1. The storage area does not exceed 1,000 square feet (93 m²),
> 2. The storage area is sprinklered and does not exceed 3,000 square feet (279 m²), or
> 3. The building is provided with an approved automatic sprinkler system throughout.

309.3 Location on Property. For fire-resistive protection of exterior walls and openings, as determined by location on property, see Section 503.

309.4 Access and Means of Egress Facilities. Means of egress shall be provided as specified in Chapter 10.

Access to, and egress from, buildings required to be accessible shall be provided as specified in Chapter 11.

309.5 Light, Ventilation and Sanitation. In Group M Occupancies, light, ventilation and sanitation shall be as specified in Chapters 12 and 29.

309.6 Shaft and Exit Enclosures. Exits shall be enclosed as specified in Chapter 10.

Elevator shafts, vent shafts and other openings through floors shall be enclosed, and the enclosure shall be as specified in Section 711.

In buildings housing Group M Occupancies equipped with automatic sprinkler systems throughout, enclosures need not be provided for escalators where the top of the escalator opening at each story is provided with a draft curtain and automatic fire sprinklers are installed around the perimeter of the opening within 2 feet (610 mm) of the draft curtain. The draft curtain shall enclose the perimeter of the unenclosed opening and extend from the ceiling downward at least 12 inches (305 mm) on all sides. The spacing between sprinklers shall not exceed 6 feet (1829 mm).

309.7 Sprinkler and Standpipe Systems. When required by other provisions of this code, automatic sprinkler systems and standpipes shall be installed as specified in Chapter 9.

309.8 Special Hazards. For special hazards of Group M Occupancies, see Section 304.8.

Storage and use of flammable and combustible liquids shall be in accordance with the Fire Code.

Buildings erected or converted to house high-piled combustible stock or aerosols shall comply with the Fire Code.

SECTION 310 — REQUIREMENTS FOR GROUP R OCCUPANCIES

310.1 Group R Occupancies Defined. Group R Occupancies shall be:

Division 1. Hotels and apartment houses.

Congregate residences (each accommodating more than 10 persons).

Division 2. Not used.

Division 3. Dwellings and lodging houses.

Congregate residences (each accommodating 10 persons or less).

For occupancy separations, see Table 3-B.

A complete code for construction of detached one- and two-family dwellings is in Appendix Chapter 3, Division III, of this code. When adopted, as set forth in Section 101.3, it will take precedence over the other requirements set forth in this code.

310.2 Construction, Height and Allowable Area.

310.2.1 General. Buildings or parts of buildings classed in Group R because of the use or character of the occupancy shall be limited to the types of construction set forth in Table 5-B and shall not exceed, in area or height, the limits specified in Sections 504, 505 and 506.

310.2.2 Special provisions. Walls and floors separating dwelling units in the same building, or guest rooms in Group R, Division 1 hotel occupancies, shall not be of less than one-hour fire-resistive construction.

Group R, Division 1 Occupancies more than two stories in height or having more than 3,000 square feet (279 m²) of floor area above the first story shall not be of less than one-hour fire-resistive construction throughout, except as provided in Section 601.5.2.2.

Storage or laundry rooms that are within Group R, Division 1 Occupancies that are used in common by tenants shall be separated from the rest of the building by not less than one-hour fire-resistive occupancy separation.

For Group R, Division 1 Occupancies with a Group S, Division 3 parking garage in the basement or first story, see Section 311.2.2.

For attic space partitions and draft stops, see Section 708.

310.3 Location on Property. For fire-resistive protection of exterior walls and openings, as determined by location on property, see Section 503 and Chapter 6.

310.4 Access and Means of Egress Facilities and Emergency Escapes. Means of egress shall be provided as specified in Chapter 10. (See also Section 1007.6.2 for exit markings.)

Access to, and egress from, buildings required to be accessible shall be provided as specified in Chapter 11.

Basements in dwelling units and every sleeping room below the fourth story shall have at least one operable window or door approved for emergency escape or rescue that shall open directly into a public street, public alley, yard or exit court. The emergency door or window shall be operable from the inside to provide a full, clear opening without the use of separate tools.

> **EXCEPTION:** The window or door may open into an atrium complying with Section 402 provided the window or door opens onto an exit-access balcony and the dwelling unit or guest room has an exit or exit-access doorway that does not open into the atrium.

Escape or rescue windows shall have a minimum net clear openable area of 5.7 square feet (0.53 m²). The minimum net clear openable height dimension shall be 24 inches (610 mm). The minimum net clear openable width dimension shall be 20 inches (508 mm). When windows are provided as a means of escape or rescue, they shall have a finished sill height not more than 44 inches (1118 mm) above the floor.

Escape and rescue windows with a finished sill height below the adjacent ground elevation shall have a window well. Window wells at escape or rescue windows shall comply with the following:

1. The clear horizontal dimensions shall allow the window to be fully opened and provide a minimum accessible net clear opening of 9 square feet (0.84 m²), with a minimum dimension of 36 inches (914 mm).

2. Window wells with a vertical depth of more than 44 inches (1118 mm) shall be equipped with an approved permanently affixed ladder or stairs that are accessible with the window in the fully open position. The ladder or stairs shall not encroach into the required dimensions of the window well by more than 6 inches (152 mm).

Bars, grilles, grates or similar devices may be installed on emergency escape or rescue windows, doors or window wells, provided:

1. The devices are equipped with approved release mechanisms that are openable from the inside without the use of a key or special knowledge or effort; and

2. The building is equipped with smoke detectors installed in accordance with Section 310.9.

310.5 Light, Ventilation and Sanitation. Light and ventilation shall be as specified in Chapter 12. The number of plumbing fixtures shall not be less than specified in Section 2902.6.

310.6 Room Dimensions.

310.6.1 Ceiling heights. Habitable space shall have a ceiling height of not less than 7 feet 6 inches (2286 mm) except as otherwise permitted in this section. Kitchens, halls, bathrooms and toilet compartments may have a ceiling height of not less than 7 feet (2134 mm) measured to the lowest projection from the ceiling. Where exposed beam ceiling members are spaced at less than 48 inches (1219 mm) on center, ceiling height shall be measured to the bottom of these members. Where exposed beam ceiling members are spaced at 48 inches (1219 mm) or more on center, ceiling height shall be measured to the bottom of the deck supported by these members, provided that the bottom of the members is not less than 7 feet (2134 mm) above the floor.

If any room in a building has a sloping ceiling, the prescribed ceiling height for the room is required in only one half the area thereof. No portion of the room measuring less than 5 feet (1524 mm) from the finished floor to the finished ceiling shall be included in any computation of the minimum area thereof.

If any room has a furred ceiling, the prescribed ceiling height is required in two thirds the area thereof, but in no case shall the height of the furred ceiling be less than 7 feet (2134 mm).

310.6.2 Floor area. Dwelling units and congregate residences shall have at least one room that shall have not less than 120 square feet (11.2 m²) of floor area. Other habitable rooms except kitchens shall have an area of not less than 70 square feet (6.5 m²). Efficiency dwelling units shall comply with the requirements of Section 310.7.

310.6.3 Width. Habitable rooms other than a kitchen shall not be less than 7 feet (2134 mm) in any dimension.

310.7 Efficiency Dwelling Units. An efficiency dwelling unit shall conform to the requirements of the code except as herein provided:

1. The unit shall have a living room of not less than 220 square feet (20.4 m²) of superficial floor area. An additional 100 square feet (9.3 m²) of superficial floor area shall be provided for each occupant of such unit in excess of two.

2. The unit shall be provided with a separate closet.

3. The unit shall be provided with a kitchen sink, cooking appliance and refrigeration facilities, each having a clear working space of not less than 30 inches (762 mm) in front. Light and ventilation conforming to this code shall be provided.

4. The unit shall be provided with a separate bathroom containing a water closet, lavatory and bathtub or shower.

310.8 Shaft and Exit Enclosures. Exits shall be enclosed as specified in Chapter 10.

Elevator shafts, vent shafts, dumbwaiter shafts, clothes chutes and other vertical openings shall be enclosed and the enclosure shall be as specified in Section 711.

In nonsprinklered Group R, Division 1 Occupancies, corridors serving an occupant load of 10 or more shall be separated from corridors and other areas on adjacent floors by not less than approved fixed wired glass set in steel frames or by 20-minute smoke- and draft-control assemblies, which are automatic closing by smoke detection.

310.9 Smoke Detectors and Sprinkler Systems.

310.9.1 Smoke detectors.

310.9.1.1 General. Dwelling units, congregate residences and hotel or lodging house guest rooms that are used for sleeping purposes shall be provided with smoke detectors. Detectors shall be installed in accordance with the approved manufacturer's instructions.

310.9.1.2 Additions, alterations or repairs to Group R Occupancies. When the valuation of an addition, alteration or repair to a Group R Occupancy exceeds $1,000 and a permit is required, or when one or more sleeping rooms are added or created in existing Group R Occupancies, smoke detectors shall be installed in accordance with Sections 310.9.1.3, 310.9.1.4 and 310.9.1.5 of this section.

> **EXCEPTION:** Repairs to the exterior surfaces of a Group R Occupancy are exempt from the requirements of this section.

310.9.1.3 Power source. In new construction, required smoke detectors shall receive their primary power from the building wiring when such wiring is served from a commercial source and shall be equipped with a battery backup. The detector shall emit a signal when the batteries are low. Wiring shall be permanent and without a disconnecting switch other than those required for overcurrent protection. Smoke detectors may be solely battery operated when installed in existing buildings; or in buildings without commercial power; or in buildings which undergo alterations, repairs or additions regulated by Section 310.9.1.2.

310.9.1.4 Location within dwelling units.
In dwelling units, a detector shall be installed in each sleeping room and at a point centrally located in the corridor or area giving access to each separate sleeping area. When the dwelling unit has more than one story and in dwellings with basements, a detector shall be installed on each story and in the basement. In dwelling units where a story or basement is split into two or more levels, the smoke detector shall be installed on the upper level, except that when the lower level contains a sleeping area, a detector shall be installed on each level. When sleeping rooms are on an upper level, the detector shall be placed at the ceiling of the upper level in close proximity to the stairway. In dwelling units where the ceiling height of a room open to the hallway serving the bedrooms exceeds that of the hallway by 24 inches (610 mm) or more, smoke detectors shall be installed in the hallway and in the adjacent room. Detectors shall sound an alarm audible in all sleeping areas of the dwelling unit in which they are located.

310.9.1.5 Location in efficiency dwelling units, congregate residences and hotels.
In efficiency dwelling units, hotel suites and in hotel and congregate residence sleeping rooms, detectors shall be located on the ceiling or wall of the main room or each sleeping room. When sleeping rooms within an efficiency dwelling unit or hotel suite are on an upper level, the detector shall be placed at the ceiling of the upper level in close proximity to the stairway. When actuated, the detector shall sound an alarm audible within the sleeping area of the dwelling unit or congregate residence, hotel suite, or sleeping room in which it is located.

310.9.2 Sprinkler and standpipe systems.
When required by Section 904.2.1 or other provisions of this code, automatic sprinkler systems and standpipes shall be designed and installed as specified in Chapter 9.

310.10 Fire Alarm Systems.
Group R, Division 1 Occupancies shall be provided with a manual and automatic fire alarm system in apartment houses three or more stories in height or containing 16 or more dwelling units, in hotels three or more stories in height or containing 20 or more guest rooms and in congregate residences three or more stories in height or having an occupant load of 20 or more. A fire alarm and communication system shall be provided in Group R, Division 1 Occupancies located in a high-rise building.

> **EXCEPTIONS:** 1. A manual fire alarm system need not be provided in buildings not over two stories in height when all individual dwelling units and contiguous attic and crawl spaces are separated from each other and public or common areas by at least one-hour fire-resistive occupancy separations and each individual dwelling unit or guest room has an exit directly to a public way, exit court or yard.
>
> 2. A separate fire alarm system need not be provided in buildings that are protected throughout by an approved supervised fire sprinkler system having a local alarm to notify all occupants.

The alarm signal shall be a distinctive sound that is not used for any other purpose other than the fire alarm. Alarm-signaling devices shall produce a sound that exceeds the prevailing equivalent sound level in the room or space by 15 decibels minimum, or exceeds any maximum sound level with a duration of 30 seconds minimum by 5 decibels minimum, whichever is louder. Sound levels for alarm signals shall be 120 decibels maximum.

For the purposes of this section, area separation walls shall not define separate buildings.

310.11 Heating.
Dwelling units, guest rooms and congregate residences shall be provided with heating facilities capable of maintaining a room temperature of 70°F (21°C) at a point 3 feet (914 mm) above the floor in all habitable rooms.

310.12 Special Hazards.
Chimneys and heating apparatus shall conform to the requirements of Chapter 31 and the Mechanical Code.

The storage, use and handling of flammable and combustible liquids in Division 1 Occupancies shall be in accordance with the Fire Code.

In Division 1 Occupancies, doors leading into rooms in which Class I flammable liquids are stored or used shall be protected by a fire assembly having a one-hour fire-protection rating. Such fire assembly shall be self-closing and shall be posted with a sign on each side of the door in 1-inch (25.4 mm) block letters stating: FIRE DOOR—KEEP CLOSED.

SECTION 311 — REQUIREMENTS FOR GROUP S OCCUPANCIES

311.1 Group S Occupancies Defined.
Group S Occupancies shall include the use of a building or structure, or a portion thereof, for storage not classified as a hazardous occupancy. Storage occupancies shall include the following:

Division 1. Moderate hazard storage occupancies shall include buildings or portions of buildings used for storage of combustible materials that are not classified as a Group S, Division 2 or as a Group H Occupancy.

Division 2. Low-hazard storage occupancies shall include buildings, structures, or portions thereof, used for storage of noncombustible materials, such as products on wood pallets or in paper cartons with or without single-thickness divisions, or in paper wrappings and shall include ice plants, power plants and pumping plants. Such products may have a negligible amount of plastic trim such as knobs, handles or film wrapping. Low-hazard storage occupancies shall include, but are not limited to, storage of the following items:

1. Beer or wine (in metal, glass or ceramic containers).
2. Cement in bags.
3. Cold storage and creameries.
4. Dairy products in nonwax-coated paper containers.
5. Dry-cell batteries.
6. Dryers.
7. Dry pesticides in a building not classed as a Group H Occupancy.
8. Electrical coils.
9. Electrical insulators.
10. Electrical motors.
11. Empty cans.
12. Foods in noncombustible containers.
13. Fresh fruits in nonplastic trays or containers.
14. Frozen foods.
15. Glass bottles (empty or filled with nonflammable liquids).
16. Gypsum board.
17. Inert pigments.
18. Meats.
19. Metal cabinets.
20. Metal furniture.
21. Oil-filled distribution transformers.
22. Stoves.
23. Washers.

Division 3. Division 3 Occupancies shall include repair garages where work is limited to exchange of parts and maintenance requiring no open flame or welding, motor vehicle fuel-dispensing stations, and parking garages not classed as Group S, Division 4 open parking garages or Group U private garages.

For the use of flammable and combustible liquids, see Section 307 and the Fire Code.

Division 4. Open parking garages per Section 311.9.

Division 5. Aircraft hangars where work is limited to exchange of parts and maintenance requiring no open flame or welding and helistops.

For occupancy separations, see Table 3-B.

311.2 Construction, Height and Allowable Area.

311.2.1 General. Buildings or parts of buildings classed in Group S Occupancy because of the use or character of the occupancy shall be limited to the types of construction set forth in Table 5-B and shall not exceed, in area or height, the limits specified in Sections 504, 505 and 506.

311.2.2 Special provisions.

311.2.2.1 Group S, Division 3 with Group A, Division 3; Group B; Group M or R, Division 1 Occupancy above. Other provisions of this code notwithstanding, a basement or first story of a building may be considered as a separate and distinct building for the purpose of area limitations, limitation of number of stories and type of construction, when all of the following conditions are met:

1. The basement or first story is of Type I construction and is separated from the building above with a three-hour occupancy separation. See Section 302.3.

2. The building above the three-hour occupancy separation contains only Group A, Division 3; Group B; or Group M or R, Division 1 Occupancies.

3. The building below the three-hour occupancy separation is a Group S, Division 3 Occupancy used exclusively for the parking and storage of private or pleasure-type motor vehicles.

> **EXCEPTIONS:** 1. Entry lobbies, mechanical rooms and similar uses incidental to the operation of the building.
> 2. Group A, Division 3 and Group B office, drinking and dining establishments and Group M retail occupancies in addition to those uses incidental to the operation of the building (including storage areas), provided that the entire structure below the three-hour occupancy separation is protected throughout by an automatic sprinkler system.

4. The maximum building height in feet shall not exceed the limits set forth in Table 5-B for the least type of construction involved.

311.2.2.2 Group S, Division 3 Occupancy with Group S, Division 4 Occupancy above. Other provisions of this code notwithstanding, a Group S, Division 3 Occupancy, located in the basement or first story below a Group S, Division 4 Occupancy, as defined in Section 311.9, may be classified as a separate and distinct building for the purpose of determining the type of construction when all of the following conditions are met:

1. The allowable area of the structure shall be such that the sum of the ratios of the actual area divided by the allowable area for each separate occupancy shall not exceed one.

2. The Group S, Division 3 Occupancy is of Type I or II construction and is at least equal to the fire resistance of the Group S, Division 4 Occupancy.

3. The height and the number of the tiers above the basement shall be limited as specified in Table 3-H or Section 311.9.5.

4. The floor-ceiling assembly separating the Group S, Division 3 and Group S, Division 4 Occupancy shall be protected as required for the floor-ceiling assembly of the Group S, Division 3 Occupancy. Openings between the Group S, Division 3 and Group S, Division 4 Occupancy, except exit openings, need not be protected.

5. The Group S, Division 3 Occupancy is used exclusively for the parking or storage of private or pleasure-type motor vehicles, but may contain (i) mechanical equipment rooms incidental to the operation of the building and (ii) an office, and waiting and toilet rooms having a total area of not more than 1,000 square feet (93 m^2).

311.2.3 Specific use provisions.

311.2.3.1 Group S, Divisions 3 and 5 Occupancies. In areas where motor vehicles, boats or aircraft are stored, and in motor vehicle fuel-dispensing stations and repair garages, floor surfaces shall be of noncombustible, nonabsorbent materials. Floors shall drain to an approved oil separator or trap discharging to sewers in accordance with the Plumbing Code.

> **EXCEPTION:** Floors may be surfaced or waterproofed with asphaltic paving materials in areas where motor vehicles or airplanes are stored or operated.

311.2.3.2 Marine or motor vehicle fuel-dispensing stations. Marine or motor vehicle fuel dispensing stations, including canopies and supports over fuel dispensers, shall be of noncombustible, fire-retardant-treated wood or of one-hour fire-resistive construction.

> **EXCEPTIONS:** 1. Roofs of one-story fuel-dispensing stations may be of heavy-timber construction.
> 2. Canopies conforming to Section 2603.13 may be erected over pumps.

Canopies under which fuels are dispensed shall have a clear, unobstructed height of not less than 13 feet 6 inches (4114 mm) to the lowest projecting element in the vehicle drive-through area.

A one-hour occupancy separation need not be provided between fuel dispensers covered with a canopy that is open on three or more sides, and a Group M Occupancy retail store having an area of less than 2,500 square feet (232 m^2) when the following conditions exist:

1. The Group M Occupancy is provided with two exits or exit-access doorways separated as required by Section 1004.2.4 and not located in the same exterior wall.

2. Fuel-dispenser islands are not located within 20 feet (6096 mm) of the Group M Occupancy retail store.

311.2.3.3 Parking garage headroom. Parking garages shall have an unobstructed headroom clearance of not less than 7 feet (2134 mm) above the finish floor to any ceiling, beam, pipe or similar obstruction, except for wall-mounted shelves, storage surfaces, racks or cabinets.

311.2.3.4 Group S, Division 2 Occupancy roof framing. In Division 2 Occupancies, the roof-framing system may be of unprotected construction.

311.2.3.5 Vehicle barriers. In parking garages where any parking area is located more than 5 feet (1524 mm) above the adjacent grade, vehicle barriers shall be provided.

> **EXCEPTION:** Parking garages of Group U, Division 1 Occupancies.

Vehicle barriers shall have a minimum vertical dimension of 12 inches (305 mm) and shall be centered at 18 inches (457 mm) above the parking surface. See Table 16-B for load criterion.

311.3 Location on Property. For fire-resistive protection of exterior walls and openings, as determined by location on property, see Section 503.

311.4 Access and Means of Egress Facilities. Means of egress shall be provided as specified in Chapter 10.

Access to, and egress from, buildings required to be accessible shall be provided as specified in Chapter 11.

311.5 Light, Ventilation and Sanitation. In Group S Occupancies, light, ventilation and sanitation shall be as contained in Chapters 12 and 29, except as noted below.

311.5.1 Repair and storage garages, aircraft hangars. See Section 1202.2.6 for ventilation requirements for Group S, Division 3 repair garages, storage garages and Group S, Division 5 aircraft hangars.

311.5.2 Parking garages. See Section 1202.2.7 for ventilation requirements for parking garages.

311.6 Shaft and Exit Enclosures. Exits shall be enclosed as specified in Chapter 10.

Elevator shafts, vent shafts and other openings through floors shall be enclosed, and the enclosure shall be as specified in Section 711.

> **EXCEPTION:** In Group S, Division 2 Occupancies, exits shall be enclosed as specified in Chapter 10, but other through-floor openings need not be enclosed.

In buildings housing Group S Occupancies equipped with automatic sprinkler systems throughout, enclosures need not be provided for escalators where the top of the escalator opening at each story is provided with a draft curtain and automatic fire sprinklers are installed around the perimeter of the opening within 2 feet (610 mm) of the draft curtain. The draft curtain shall enclose the perimeter of the unenclosed opening and extend from the ceiling downward at least 12 inches (305 mm) on all sides. The spacing between sprinklers shall not exceed 6 feet (1829 mm).

311.7 Sprinkler and Standpipe Systems. When required by Section 904.2 or other provisions of this code, automatic sprinkler systems and standpipes shall be installed as specified in Chapter 9.

311.8 Special Hazards. Storage and use of flammable and combustible liquids shall be in accordance with the Fire Code.

Devices generating a glow, spark or flame capable of igniting flammable vapors shall be installed such that sources of ignition are at least 18 inches (457 mm) above the floor of any room in which Class I flammable liquids or flammable gases are used or stored.

Buildings erected or converted to house high-piled combustible stock or aerosols shall comply with the Fire Code.

311.9 Group S, Division 4 Open Parking Garages.

311.9.1 Scope. Except where specific provisions are made in the following sections, other requirements of this code shall apply.

311.9.2 Definitions.

311.9.2.1 General. For the purpose of this section, certain terms are defined as follows:

MECHANICAL-ACCESS OPEN PARKING GARAGES are open parking garages employing parking machines, lifts, elevators or other mechanical devices for vehicles moving from and to street level and in that public occupancy is prohibited above the street level.

OPEN PARKING GARAGE is a structure of Type I or II construction with the openings as described in Section 311.9.2.2 on two or more sides and that is used exclusively for the parking or storage of private or pleasure-type motor vehicles.

> **EXCEPTION:** The grade-level tier may contain an office, and waiting and toilet rooms having a total area of not more than 1,000 square feet (93 m²). Such area need not be separated from the open parking garage.

RAMP-ACCESS OPEN PARKING GARAGES are open parking garages employing a series of continuously rising floors or a series of interconnecting ramps between floors permitting the movement of vehicles under their own power from and to the street level.

311.9.2.2 Openings. For natural ventilation purposes, the exterior side of the structure shall have uniformly distributed openings on two or more sides. The area of such openings in exterior walls on a tier must be at least 20 percent of the total perimeter wall area of each tier. The aggregate length of the openings considered to be providing natural ventilation shall constitute a minimum of 40 percent of the perimeter of the tier. Interior wall lines and column lines shall be at least 20 percent open with uniformly distributed openings.

311.9.3 Construction. Construction shall be of noncombustible materials. Open parking garages shall meet the design requirements of Chapter 16. For vehicle barriers, see Section 311.2.3.5.

311.9.4 Area and height. Area and height of open parking garages shall be limited as set forth in Table 3-H, except for increases allowed by Section 311.9.5.

In structures having a spiral or sloping floor, the horizontal projection of the structure at any cross section shall not exceed the allowable area per parking tier. In the case of a structure having a continuous spiral floor, each 9 feet 6 inches (2896 mm) of height, or portion thereof, shall be considered a tier.

The clear height of a parking tier shall not be less than 7 feet (2134 mm), except that a lower clear height may be permitted in mechanical-access open parking garages when approved by the building official.

311.9.5 Area and height increases. The area and height of structures with cross ventilation throughout may be increased in accordance with provisions of this section. Structures with sides open on three fourths of the building perimeter may be increased by 25 percent in area and one tier in height. Structures with sides open around the entire building perimeter may be increased 50 percent in area and one tier in height. For a side to be considered open under the above provisions, the total area of openings along the side shall not be less than 50 percent of the interior area of the side at each tier, and such openings shall be equally distributed along the length of the tier.

Open parking garages constructed to heights less than the maximums established by Table 3-H may have individual tier areas exceeding those otherwise permitted, provided the gross tier area of the structure does not exceed that permitted for the higher structure. At least three sides of each such larger tier shall have continuous horizontal openings not less than 30 inches (762 mm) in clear height extending for at least 80 percent of the length of the sides, and no part of such larger tier shall be more than 200 feet (60 960 mm) horizontally from such an opening. In addition, each such opening shall face a street or yard accessible to a street with a width of at least 30 feet (9144 mm) for the full length of the opening, and standpipes shall be provided in each such tier.

Structures of Type II-F.R., Type II One-hour or Type II-N construction, with all sides open, may be unlimited in area when the height does not exceed 75 feet (22 860 mm). For a side to be considered open, the total area of openings along the side shall not be less than 50 percent of the interior area of the side at each tier, and such openings shall be equally distributed along the length of the tier. All portions of tiers shall be within 200 feet (60 960 mm) horizontally from such openings.

311.9.6 Location on property. Exterior walls and openings in exterior walls shall comply with Table 5-A. The distance from an adjacent property line shall be determined in accordance with Section 503.

311.9.7 Stairs and means of egress. Where persons other than parking attendants are permitted, the means of egress shall meet the requirements of Chapter 10, based on an occupant load of 200 square feet (18.6 m^2) per occupant. Where no persons other than parking attendants are permitted, there shall not be less than two 3-foot-wide (914 mm) stairs. Lifts may be installed for use of employees only, provided they are completely enclosed by noncombustible materials.

311.9.8 Standpipes. Standpipes shall be installed when required by the provisions of Chapter 9.

311.9.9 Sprinkler systems. When required by other provisions of this code, automatic sprinkler systems and standpipes shall be installed in accordance with the provisions of Chapter 9.

311.9.10 Enclosure of vertical openings. Enclosure shall not be required for vertical openings except as specified in Section 311.9.7 for lifts.

311.9.11 Ventilation. Ventilation, other than the percentage of openings specified in Section 311.9.2.2, shall not be required.

311.9.12 Prohibitions. The following uses and alterations are not permitted:

1. Automobile repair work.

2. Parking of buses, trucks and similar vehicles.

3. Partial or complete closing of required openings in exterior walls by tarpaulins or any other means.

4. Dispensing of fuel.

311.10 Helistops.

311.10.1 General. Helistops may be erected on buildings or other locations if they are constructed in accordance with this section.

311.10.2 Size. The touchdown or landing area for helicopters of less than 3,500 pounds (1589 kg) shall be a minimum of 20 feet by 20 feet (6096 mm by 6096 mm) in size. The touchdown area shall be surrounded on all sides by a clear area having a minimum average width at roof level of 15 feet (4572 mm) but with no width less than 5 feet (1524 mm).

311.10.3 Design. Helicopter landing areas and supports therefor on the roof of a building shall be of noncombustible construction. Landing areas shall be designed to confine any Class I, II or III-A liquid spillage to the landing area itself and provision shall be made to drain such spillage away from any exit or stairway serving the helicopter landing area or from a structure housing such exit or stairway.

311.10.4 Means of egress. Means of egress from helistops shall comply with the provisions of Chapter 10 of this code, except that all landing areas located on buildings or structures shall have two or more means of egress. For landing platforms or roof areas less than 60 feet (18 288 mm) in length, or less than 2,000 square feet (186 m^2) in area, the second means of egress may be a fire escape or ladder leading to the floor below.

311.10.5 Federal Aviation Administration approval. Before operating helicopters from helistops, approval must be obtained from the Federal Aviation Administration.

SECTION 312 — REQUIREMENTS FOR GROUP U OCCUPANCIES

312.1 Group U Occupancies Defined. Group U Occupancies shall include buildings or structures, or portions thereof, and shall be:

Division 1. Private garages, carports, sheds and agricultural buildings.

> **EXCEPTION:** Where applicable (see Section 101.3) for agricultural buildings, see Appendix Chapter 3.

Division 2. Fences over 6 feet (1829 mm) high, tanks and towers.

For occupancy separations, see Table 3-B.

312.2 Construction, Height and Allowable Area.

312.2.1 General. Buildings or parts of buildings classed as Group U, Division 1 Occupancies because of the use or character of the occupancy shall not exceed 1,000 square feet (92.9 m^2) in area or one story in height except as provided in Section 312.2.2. Any building or portion thereof that exceeds the limitations specified in this chapter shall be classed in the occupancy group other than Group U, Division 1 that it most nearly resembles.

312.2.2 Special area provisions. The total area of a private garage used only as a parking garage for private or pleasure-type motor vehicles where no repair work is done or fuel dispensed may be 3,000 square feet (279 m^2), provided the provisions set forth in Item 1 or 2 are satisfied. More than one 3,000-square-foot (279 m^2) Group U, Division 1 Occupancy may be within the same building, provided each 3,000-square-foot (279 m^2) area is separated by area separation walls complying with Section 504.6.

1. For a mixed-occupancy building, the exterior wall and opening protection for the Group U, Division 1 portion of the building shall be as required for the major occupancy of the building. For such mixed-occupancy building, the allowable floor area of the building shall be as permitted for the major occupancy contained therein.

2. For a building containing only a Group U, Division 1 Occupancy, the exterior wall and opening protection shall be as required for a building classified as a Group R, Division 1 Occupancy.

312.2.3 Headroom clearance. Garages in connection with Group R, Division 1 Occupancies shall have an unobstructed headroom clearance of not less than 7 feet (2134 mm) above the finish floor to any ceiling, beam, pipe or similar construction except for wall-mounted shelves, storage surfaces, racks or cabinets.

312.3 Location on Property. For fire-resistive protection of exterior walls and openings, as determined by location on property, see Section 503 and Chapter 6.

312.4 Special Hazards. Chimneys and heating apparatus shall conform to the requirements of Chapter 31 and the Mechanical Code.

Under no circumstances shall a private garage have any opening into a room used for sleeping purposes.

Class I, II or III-A liquids shall not be stored, handled or used in Group U Occupancies unless such storage or handling shall comply with the Fire Code.

312.5 Garage Floor Surfaces. In areas where motor vehicles are stored or operated, floor surfaces shall be of noncombustible materials or asphaltic paving materials.

312.6 Agricultural Buildings. Where applicable (see Section 101.3) for agricultural buildings, see Appendix Chapter 3.

TABLE 3-A—DESCRIPTION OF OCCUPANCIES BY GROUP AND DIVISION[1]

GROUP AND DIVISION	SECTION	DESCRIPTION OF OCCUPANCY
A-1		A building or portion of a building having an assembly room with an occupant load of 1,000 or more and a legitimate stage.
A-2		A building or portion of a building having an assembly room with an occupant load of less than 1,000 and a legitimate stage.
A-2.1	303.1.1	A building or portion of a building having an assembly room with an occupant load of 300 or more without a legitimate stage, including such buildings used for educational purposes and not classed as a Group E or Group B Occupancy.
A-3		Any building or portion of a building having an assembly room with an occupant load of less than 300 without a legitimate stage, including such buildings used for educational purposes and not classed as a Group E or Group B Occupancy.
A-4		Stadiums, reviewing stands and amusement park structures not included within other Group A Occupancies.
B	304.1	A building or structure, or a portion thereof, for office, professional or service-type transactions, including storage of records and accounts; eating and drinking establishments with an occupant load of less than 50.
E-1		Any building used for educational purposes through the 12th grade by 50 or more persons for more than 12 hours per week or four hours in any one day.
E-2	305.1	Any building used for educational purposes through the 12th grade by less than 50 persons for more than 12 hours per week or four hours in any one day.
E-3		Any building or portion thereof used for day-care purposes for more than six persons.
F-1	306.1	Moderate-hazard factory and industrial occupancies include factory and industrial uses not classified as Group F, Division 2 Occupancies.
F-2		Low-hazard factory and industrial occupancies include facilities producing noncombustible or nonexplosive materials that during finishing, packing or processing do not involve a significant fire hazard.
H-1		Occupancies with a quantity of material in the building in excess of those listed in Table 3-D that present a high explosion hazard as listed in Section 307.1.1.
H-2	307.1	Occupancies with a quantity of material in the building in excess of those listed in Table 3-D that present a moderate explosion hazard or a hazard from accelerated burning as listed in Section 307.1.1.
H-3		Occupancies with a quantity of material in the building in excess of those listed in Table 3-D that present a high fire or physical hazard as listed in Section 307.1.1.
H-4		Repair garages not classified as Group S, Division 3 Occupancies.
H-5		Aircraft repair hangars not classified as Group S, Division 5 Occupancies and heliports.
H-6	307.1 and 307.11	Semiconductor fabrication facilities and comparable research and development areas when the facilities in which hazardous production materials are used, and the aggregate quantity of material is in excess of those listed in Table 3-D or 3-E.
H-7	307.1	Occupancies having quantities of materials in excess of those listed in Table 3-E that are health hazards as listed in Section 307.1.1.
I-1.1		Nurseries for the full-time care of children under the age of six (each accommodating more than five children), hospitals, sanitariums, nursing homes with nonambulatory patients and similar buildings (each accommodating more than five patients).
I-1.2	308.1	Health-care centers for ambulatory patients receiving outpatient medical care which may render the patient incapable of unassisted self-preservation (each tenant space accommodating more than five such patients).
I-2		Nursing homes for ambulatory patients, homes for children six years of age or over (each accommodating more than five persons).
I-3		Mental hospitals, mental sanitariums, jails, prisons, reformatories and buildings where personal liberties of inmates are similarly restrained.
M	309.1	A building or structure, or a portion thereof, for the display and sale of merchandise, and involving stocks of goods, wares or merchandise, incidental to such purposes and accessible to the public.
R-1	310.1	Hotels and apartment houses, congregate residences (each accommodating more than 10 persons).
R-3		Dwellings, lodging houses, congregate residences (each accommodating 10 or fewer persons).
S-1		Moderate hazard storage occupancies including buildings or portions of buildings used for storage of combustible materials not classified as Group S, Division 2 or Group H Occupancies.
S-2		Low-hazard storage occupancies including buildings or portions of buildings used for storage of noncombustible materials.
S-3	311.1	Repair garages where work is limited to exchange of parts and maintenance not requiring open flame or welding, and parking garages not classified as Group S, Division 4 Occupancies.
S-4		Open parking garages.
S-5		Aircraft hangars and helistops.
U-1	312.1	Private garages, carports, sheds and agricultural buildings.
U-2		Fences over 6 feet (1829 mm) high, tanks and towers.

[1]For detailed descriptions, see the occupancy definitions in the noted sections.

TABLE 3-B
TABLE 3-C

1997 UNIFORM BUILDING CODE

TABLE 3-B—REQUIRED SEPARATION IN BUILDINGS OF MIXED OCCUPANCY[1] (HOURS)

	A-1	A-2	A-2.1	A-3	A-4	B	E	F-1	F-2	H-2	H-3	H-4,5	H-6,7[2]	I	M	R-1	R-3	S-1	S-2	S-3	S-5	U-1[3]
A-1		N	N	N	N	3	N	3	3	4	4	4	4	3	3	1	1	3	3	4	3	1
A-2			N	N	N	1	N	1	1	4	4	4	4	3	1	1	1	1	1	3	1	1
A-2.1				N	N	1	N	1	1	4	4	4	4	3	1	1	1	1	1	3	1	1
A-3					N	N	N	N	N	4	4	4	3	2	N	1	1	N	N	3	1	1
A-4						1	N	1	1	4	4	4	4	3	1	1	1	1	1	3	1	1
B							1	N[5]	N	2	1	1	1	2	N	1	1	N	N	1	1	1
E								1	1	4	4	4	3	1	1	1	1	1	1	3	1	1
F-1									1	2	1	1	1	3	N[5]	1	1	N	N	1	1	1
F-2										2	1	1	1	2	1	1	1	N	N	1	1	1
H-1									NOT PERMITTED IN MIXED OCCUPANCIES. SEE SECTION 307.2.8													
H-2											1	1	2	4	2	4	4	2	2	2	2	1
H-3												1	1	4	1	3	3	1	1	1	1	1
H-4,5													1	4	1	3	3	1	1	1	1	1
H-6,7[2]														4	1	4	4	1	1	1	1	3
I															2	1	1	2	2	4	3	1
M																1	1	1[4]	1[4]	1	1	1
R-1																	N	3	1	3	1	1
R-3																		1	1	1	1	1
S-1																			1	1	1	1
S-2																				1	1	N
S-3																					1	1
S-4									OPEN PARKING GARAGES ARE EXCLUDED EXCEPT AS PROVIDED IN SECTION 311.2													
S-5																						N

N—No requirements for fire resistance.

[1]For detailed requirements and exceptions, see Section 302.4.

[2]For special provisions on highly toxic materials, see the Fire Code.

[3]For agricultural buildings, see also Appendix Chapter 3.

[4]See Section 309.2.2 for exception.

[5]For Group F, Division 1 woodworking establishments with more than 2,500 square feet (232.3 m²), the occupancy separation shall be one hour.

TABLE 3-C—REQUIRED SEPARATION OF SPECIFIC-USE AREAS IN GROUP I, DIVISION 1.1 HOSPITAL AND NURSING HOMES

DESCRIPTION	OCCUPANCY SEPARATION
1. Employee locker rooms	None
2. Gift/retail shops	None
3. Handicraft shops	None
4. Kitchens	None
5. Laboratories that employ hazardous materials in quantities less than that which would cause classification as a Group H Occupancy	One hour
6. Laundries greater than 100 square feet (9.3 m²)[1]	One hour
7. Paint shops employing hazardous substances and materials in quantities less than that which would cause classification as a Group H Occupancy	One hour
8. Physical plant maintenance shop	One hour
9. Soiled linen room[1]	One hour
10. Storage rooms 100 square feet (9.3 m²) or less in area storing combustible material	None
11. Storage rooms more than 100 square feet (9.3 m²) storing combustible material	One hour
12. Trash-collection rooms[1]	One hour

[1]For rubbish and linen chute termination rooms, see Section 711.5.

TABLE 3-D—EXEMPT AMOUNTS OF HAZARDOUS MATERIALS PRESENTING A PHYSICAL HAZARD
MAXIMUM QUANTITIES PER CONTROL AREA[1]
When two units are given, values within parentheses are in cubic feet (cu. ft.) or pounds (lbs.)

CONDITION		STORAGE[2]			USE[2]—CLOSED SYSTEMS			USE[2]—OPEN SYSTEMS	
		Solid Lbs.[3] (Cu. Ft.)	Liquid Gallons[3] (Lbs.)	Gas Cu. Ft.	Solid Lbs. (Cu. Ft.)	Liquid Gallons (Lbs.)	Gas Cu. Ft.	Solid Lbs. (Cu. Ft.)	Liquid Gallons (Lbs.)
Material	Class	$\times$ 0.4536 for kg $\times$ 0.0283 for m³	$\times$ 3.785 for L $\times$ 0.4536 for kg	$\times$ 0.0283 for m³	$\times$ 0.4536 for kg $\times$ 0.0283 for m³	$\times$ 3.785 for L $\times$ 0.4536 for kg	$\times$ 0.0283 for m³	$\times$ 0.4536 for kg $\times$ 0.0283 for m³	$\times$ 3.785 for L $\times$ 0.4536 for kg
1.1 Combustible liquid[4,5,6,7,8,9]	II	N.A.	120[10]	N.A.	N.A.	120	N.A.	N.A.	30
	III-A	N.A.	330[10]	N.A.	N.A.	330	N.A.	N.A.	80
	III-B	N.A.	13,200[10,11]	N.A.	N.A.	13,200[11]	N.A.	N.A.	3,300[11]
1.2 Combustible fiber (loose) (baled)		(100) (1,000)	N.A. N.A.	N.A. N.A.	(100) (1,000)	N.A. N.A.	N.A. N.A.	(20) (200)	N.A. N.A.
1.3 Cryogenic, flammable or oxidizing		N.A.	45	N.A.	N.A.	45	N.A.	N.A.	10
2.1 Explosives		1[10,13]	(1)[10,13]	N.A.	$^{1}/_{4}$[12]	$(^{1}/_{4})$[12]	N.A.	$^{1}/_{4}$[12]	$(^{1}/_{4})$[12]
3.1 Flammable solid		125[6,10]	N.A.	N.A.	14	N.A.	N.A.	14	N.A.
3.2 Flammable gas (gaseous) (liquefied)		N.A. N.A.	N.A. 15[6,10]	750[6,10] N.A.	N.A. N.A.	N.A. 15[6,10]	750[6,10] N.A.	N.A. N.A.	N.A. N.A.
3.3 Flammable liquid[4,5,6,7,8,9]	I-A	N.A.	30[10]	N.A.	N.A.	30	N.A.	N.A.	10
	I-B	N.A.	60[10]	N.A.	N.A.	60	N.A.	N.A.	15
	I-C	N.A.	90[10]	N.A.	N.A.	90	N.A.	N.A.	20
Combination I-A, I-B, I-C[15]		N.A.	120[10]	N.A.	N.A.	120	N.A.	N.A.	30
4.1 Organic peroxide, unclassified detonatable		1[10,12]	(1)[10,12]	N.A.	$^{1}/_{4}$[12]	$(^{1}/_{4})$[12]	N.A.	$^{1}/_{4}$[12]	$(^{1}/_{4})$[12]
4.2 Organic peroxide	I	5[6,10]	(5)[6,10]	N.A.	1[6]	(1)[6]	N.A.	1[6]	(1)[6]
	II	50[6,10]	(50)[6,10]	N.A.	50[6]	(50)[6]	N.A.	10[6]	(10)[6]
	III	125[6,10]	(125)[6,10]	N.A.	125[6]	(125)[6]	N.A.	25[6]	(25)[6]
	IV	500[6,10]	(500)[6,10]	N.A.	500[6]	(500)[6]	N..A.	100[6]	(100)[6]
	V	N.L.	N.L.	N.A.	N.L.	N.L.	N.A.	N.L.	N.L.
4.3 Oxidizer	4	1[10,12]	(1)[10,12]	N.A.	$^{1}/_{4}$[12]	$(^{1}/_{4})$[12]	N.A.	$^{1}/_{4}$[12]	$(^{1}/_{4})$[12]
	3[16]	10[6,10]	(10)[6,10]	N.A.	2[6]	(2)[6]	N.A.	2[6]	(2)[6]
	2	250[6,10]	(250)[6,10]	N.A.	250[6]	(250)[6]	N.A.	50[6]	(50)[6]
	1	4,000[6,10]	(4,000)[6,10]	N.A.	4,000[6]	(4,000)[6]	N.A.	1,000[6]	(1,000)[6]
4.4 Oxidizer—gas (gaseous)[6,10] (liquefied)[6,10]		N.A. N.A.	N.A. 15	1,500 N.A.	N.A. N.A.	N.A. 15	1,500 N.A.	N.A. N.A.	N.A. N.A.
5.1 Pyrophoric		4[10,12]	(4)[10,12]	50[10,12]	1[12]	(1)[12]	10[10,12]	0	0
6.1 Unstable (reactive)	4	1[10,12]	(1)[10,12]	10[10,12]	$^{1}/_{4}$[12]	$(^{1}/_{4})$[12]	2[10,12]	$^{1}/_{4}$[12]	$(^{1}/_{4})$[12]
	3	5[6,10]	(5)[6,10]	50[6,10]	1[6]	(1)[6]	10[6,10]	1[6]	(1)[6]
	2	50[6,10]	(50)[6,10]	250[6,10]	50[6]	(50)[6]	250[6,10]	10[6]	(10)[6]
	1	N.L.	N.L.	750[6,10]	N.L.	N.L.	N.L.	N.L.	N.L.
7.1 Water reactive	3	5[6,10]	(5)[6,10]	N.A.	5[6]	(5)[6]	N.A.	1[6]	(1)[6]
	2	50[6,10]	(50)[6,10]	N.A.	50[6]	(50)[6]	N.A.	10[6]	(10)[6]
	1	125[10,11]	(125)[10,11]	N.A.	125[11]	(125)[11]	N.A.	25[11]	(25)[11]

N.A.—Not applicable.   N.L.—Not limited.

[1]Control areas shall be separated from each other by not less than a one-hour fire-resistive occupancy separation. The number of control areas within a building used for retail or wholesale sales shall not exceed two. The number of control areas in buildings with other uses shall not exceed four. See Section 204.

[2]The aggregate quantity in use and storage shall not exceed the quantity listed for storage.

[3]The aggregate quantity of nonflammable solid and nonflammable or noncombustible liquid hazardous materials within a single control area of Group M Occupancies used for retail sales may exceed the exempt amounts when such areas are in compliance with the Fire Code.

[4]The quantities of alcoholic beverages in retail sales uses are unlimited provided the liquids are packaged in individual containers not exceeding 4 liters.
The quantities of medicines, foodstuffs and cosmetics containing not more than 50 percent of volume of water-miscible liquids and with the remainder of the solutions not being flammable in retail sales or storage occupancies are unlimited when packaged in individual containers not exceeding 4 liters.

[5]For aerosols, see the Fire Code.

[6]Quantities may be increased 100 percent in sprinklered buildings. When Footnote 10 also applies, the increase for both footnotes may be applied.

[7]For storage and use of flammable and combustible liquids in Groups A, B, E, F, H, I, M, R, S and U Occupancies, see Sections 303.8, 304.8, 305.8, 306.8, 307.1.3 through 307.1.5, 308.8, 309.8, 310.12, 311.8 and 312.4.

[8]For wholesale and retail sales use, also see the Fire Code.

[9]Spray application of any quantity of flammable or combustible liquids shall be conducted as set forth in the Fire Code.

[10]Quantities may be increased 100 percent when stored in approved storage cabinets, gas cabinets or exhausted enclosures as specified in the Fire Code. When Footnote 6 also applies, the increase for both footnotes may be applied.

[11]The quantities permitted in a sprinklered building are not limited.

[12]Permitted in sprinklered buildings only. None is allowed in unsprinklered buildings.

[13]One pound of black sporting powder and 20 pounds (9 kg) of smokeless powder are permitted in sprinklered or unsprinklered buildings.

[14]See definitions of Divisions 2 and 3 in Section 307.1.

TABLE 3-D
TABLE 3-E

1997 UNIFORM BUILDING CODE

FOOTNOTES TO TABLE 3-D—(Continued)

[15]Containing not more than the exempt amounts of Class I-A, Class I-B or Class I-C flammable liquids.

[16]A maximum quantity of 200 pounds (90.7 kg) of solid or 20 gallons (75.7 L) of liquid Class 3 oxidizers may be permitted when such materials are necessary for maintenance purposes or operation of equipment as set forth in the Fire Code.

TABLE 3-E—EXEMPT AMOUNTS OF HAZARDOUS MATERIALS PRESENTING A HEALTH HAZARD MAXIMUM QUANTITIES PER CONTROL AREA[1,2]
When two units are given, values within parentheses are in pounds (lbs.)

MATERIAL	STORAGE[3]			USE[3]—CLOSED SYSTEMS			USE[3]—OPEN SYSTEMS	
	Solid Lbs.[4,5,6]	Liquid Gallons[4,5,6] (Lbs.)	Gas Cu. Ft.[5]	Solid Lbs.[4,5]	Liquid Gallons[4,5] (Lbs.)	Gas Cu. Ft.[5]	Solid Lbs.[4,5]	Liquid Gallons[4,5] (Lbs.)
	$\times$ 0.4536 for kg	$\times$ 3.785 for L $\times$ 0.4536 for kg	$\times$ 0.028 for m^3	$\times$ 0.4536 for kg	$\times$ 3.785 for L $\times$ 0.4536 for kg	$\times$ 0.028 for m^3	$\times$ 0.4536 for kg	$\times$ 3.785 for L $\times$ 0.4536 for kg
1. Corrosives[10]	5,000	500	810[6]	5,000	500	810[6]	1,000	100
2. Highly toxics[7]	10	(10)	20[8]	10	(10)	20[8]	3	(3)
3. Irritants[9]	N.L.	N.L.	810[6,11]	N.L.	N.L.	810[6,11]	5,000[11]	500[11]
4. Sensitizers[9]	N.L.	N.L.	810[6,11]	N.L.	N.L.	810[6,11]	5,000[11]	500[11]
5. Other health hazards[9]	N.L.	N.L.	810[6,11]	N.L.	N.L.	810[6,11]	5,000[11]	500[11]
6. Toxics[7]	500	(500)	810[6]	500	(500)	810[8]	125	(125)

N.L. = Not limited.

[1]Control areas shall be separated from each other by not less than a one-hour fire-resistive occupancy separation. The number of control areas within a building used for retail or wholesale sales shall not exceed two. The number of control areas in buildings with other uses shall not exceed four. See Section 204.

[2]The quantities of medicines, foodstuffs and cosmetics, containing not more than 50 percent by volume of water-miscible liquids and with the remainder of the solutions not being flammable, in retail sales uses are unlimited when packaged in individual containers not exceeding 4 liters.

[3]The aggregate quantity in use and storage shall not exceed the quantity listed for storage.

[4]The aggregate quantity of nonflammable solid and nonflammable or noncombustible liquid health hazard materials within a single control area of Group M Occupancies used for retail sales may exceed the exempt amounts when such areas are in compliance with the Fire Code.

[5]Quantities may be increased 100 percent in sprinklered buildings. When Footnote 6 also applies, the increase for both footnotes may be applied.

[6]Quantities may be increased 100 percent when stored in approved storage cabinets, gas cabinets or exhausted enclosures as specified in the Fire Code. When Footnote 5 also applies, the increase for both footnotes may be applied.

[7]For special provisions, see the Fire Code.

[8]Permitted only when stored in approved exhausted gas cabinets, exhausted enclosures or fume hoods.

[9]Irritants, sensitizers and other health hazards do not include commonly used building materials and consumer products that are not otherwise regulated by this code.

[10]For stationary lead-acid battery systems, see the Fire Code.

[11]The quantities allowed in a sprinklered building are not limited when exhaust ventilation is provided in accordance with the Fire Code. See Table 8001.15-B, Footnote 12.

TABLE 3-F—MINIMUM DISTANCES FOR BUILDINGS CONTAINING EXPLOSIVE MATERIALS

QUANTITY OF EXPLOSIVE MATERIAL[1]		MINIMUM DISTANCE (feet)		
		× 304.8 for mm		
Pounds Over	Pounds Not Over	Property Lines[2] and Inhabited Buildings[3]		
× 0.4536 for kg		Barricaded[4]	Unbarricaded	Separation of Magazines[4,5,6]
2	5	70	140	12
5	10	90	180	16
10	20	110	220	20
20	30	125	250	22
30	40	140	280	24
40	50	150	300	28
50	75	170	340	30
75	100	190	380	32
100	125	200	400	36
125	150	215	430	38
150	200	235	470	42
200	250	255	510	46
250	300	270	540	48
300	400	295	590	54
400	500	320	640	58
500	600	340	680	62
600	700	355	710	64
700	800	375	750	66
800	900	390	780	70
900	1,000	400	800	72
1,000	1,200	425	850	78
1,200	1,400	450	900	82
1,400	1,600	470	940	86
1,600	1,800	490	980	88
1,800	2,000	505	1,010	90
2,000	2,500	545	1,090	98
2,500	3,000	580	1,160	104
3,000	4,000	635	1,270	116
4,000	5,000	685	1,370	122
5,000	6,000	730	1,460	130
6,000	7,000	770	1,540	136
7,000	8,000	800	1,600	144
8,000	9,000	835	1,670	150
9,000	10,000	865	1,730	156
10,000	12,000	875	1,750	164
12,000	14,000	885	1,770	174
14,000	16,000	900	1,800	180
16,000	18,000	940	1,880	188
18,000	20,000	975	1,950	196
20,000	25,000	1,055	2,000	210
25,000	30,000	1,130	2,000	224
30,000	35,000	1,205	2,000	238
35,000	40,000	1,275	2,000	248
40,000	45,000	1,340	2,000	258
45,000	50,000	1,400	2,000	270
50,000	55,000	1,460	2,000	280
55,000	60,000	1,515	2,000	290
60,000	65,000	1,565	2,000	300
65,000	70,000	1,610	2,000	310
70,000	75,000	1,655	2,000	320
75,000	80,000	1,695	2,000	330
80,000	85,000	1,730	2,000	340
85,000	90,000	1,760	2,000	350
90,000	95,000	1,790	2,000	360
95,000	100,000	1,815	2,000	370

(Continued)

TABLE 3-F
TABLE 3-G

1997 UNIFORM BUILDING CODE

TABLE 3-F—MINIMUM DISTANCES FOR BUILDINGS CONTAINING EXPLOSIVE MATERIALS—(Continued)

QUANTITY OF EXPLOSIVE MATERIAL[1]		MINIMUM DISTANCE (feet)		
		× 304.8 for mm		
Pounds Over	Pounds Not Over	Property Lines[2] and Inhabited Buildings[3]		Separation of Magazines[4,5,6]
× 0.4536 for kg		Barricaded[4]	Unbarricaded	
100,000	110,000	1,835	2,000	390
110,000	120,000	1,855	2,000	410
120,000	130,000	1,875	2,000	430
130,000	140,000	1,890	2,000	450
140,000	150,000	1,900	2,000	470
150,000	160,000	1,935	2,000	490
160,000	170,000	1,965	2,000	510
170,000	180,000	1,990	2,000	530
180,000	190,000	2,010	2,010	550
190,000	200,000	2,030	2,030	570
200,000	210,000	2,055	2,055	590
210,000	230,000	2,100	2,100	630
230,000	250,000	2,155	2,155	670
250,000	275,000	2,215	2,215	720
275,000	300,000	2,275	2,275	770

[1]The number of pounds (kg) of explosives listed is the number of pounds of trinitrotoluene (TNT) or the equivalent pounds (kg) of other explosive.

[2]The distance listed is the distance to property line, including property lines at public ways.

[3]Inhabited building is any building on the same property that is regularly occupied by human beings. When two or more buildings containing explosives or magazines are located on the same property, each building or magazine shall comply with the minimum distances specified from inhabited buildings, and, in addition, they shall be separated from each other by not less than the distances shown for "Separation of Magazines," except that the quantity of explosive materials contained in detonator buildings or magazines shall govern in regard to the spacing of said detonator buildings or magazines from buildings or magazines containing other explosive materials. If any two or more buildings or magazines are separated from each other by less than the specified "Separation of Magazines" distances, then such two or more buildings or magazines, as a group, shall be considered as one building or magazine, and the total quantity of explosive materials stored in such group shall be treated as if the explosive were in a single building or magazine located on the site of any building or magazine of the group, and shall comply with the minimum distance specified from other magazines or inhabited buildings.

[4]Barricades shall effectively screen the building containing explosives from other buildings, public ways or magazines. When mounds or revetted walls of earth are used for barricades, they shall not be less than 3 feet (914 mm) in thickness. A straight line from the top of any side wall of the building containing explosive materials to the eave line of any other building, magazine or a point 12 feet (3658 mm) above the center line of a public way shall pass through the barricades.

[5]Magazine is a building or structure approved for storage of explosive materials. In addition to the requirements of this code, magazines shall comply with the Fire Code.

[6]The distance listed may be reduced by 50 percent when approved natural or artificial barriers are provided in accordance with the requirements in Footnote 4.

TABLE 3-G—REQUIRED DETACHED STORAGE

DETACHED STORAGE IS REQUIRED WHEN THE QUANTITY OF MATERIAL EXCEEDS THAT LISTED			
Material		Solids and Liquids (tons)[1,2]	Gases (cubic feet)[1,2]
		× 907.2 for kg	× 0.0283 for m³
1. Explosives, blasting agents, black powder, fireworks, detonatable organic peroxides 2. Class 4 oxidizers 3. Class 4 or Class 3 detonatable unstable (reactives)		Over exempt amounts	Over exempt amounts
4. Oxidizers, liquids and solids	Class 3 Class 2	1,200 2,000	N.A. N.A.
5. Organic peroxides	Class I Class II Class III	Over exempt amounts 25 50	N.A. N.A. N.A.
6. Unstable (reactives)	Class 4 Class 3 Class 2	1/1,000 1 25	20 2,000 10,000
7. Water reactives	Class 3 Class 2	1 25	N.A. N.A.
8. Pyrophoric gases		N.A.	2,000

N.A.—Not applicable.

[1]For materials that are detonable, the distance to other buildings or property lines shall be as specified in Table 3-F based on trinitrotoluene (TNT) equivalence of the material. For all other materials, the distance shall be as indicated in Table 5-A.

[2]Over exempt amounts mean over the quantities listed in Table 3-D.

TABLE 3-H—OPEN PARKING GARAGES AREA AND HEIGHT

TYPE OF CONSTRUCTION	AREA PER TIER (square feet) × 0.0929 for m²	HEIGHT (in tiers)		
			Mechanical Access	
			Automatic Fire-extinguishing System	
		Ramp Access	No	Yes
I	Unlimited	Unlimited	Unlimited	Unlimited
II-F.R.	125,000	12 tiers	12 tiers	18 tiers
II One-hour	50,000	10 tiers	10 tiers	15 tiers
II-N	30,000	8 tiers	8 tiers	12 tiers

Chapter 4
SPECIAL USE AND OCCUPANCY

SECTION 401 — SCOPE

In addition to the occupancy and construction requirements in this code, the provisions of this chapter apply to the special uses described herein.

SECTION 402 — ATRIA

402.1 General. Buildings, of other than Group H Occupancy, with automatic sprinkler protection throughout may have atria complying with the provisions of this section. Such atria shall have a minimum opening area and dimension as set forth in Table 4-A.

402.2 Smoke-control System. A smoke-control system meeting the requirements of Section 905 shall be provided within the atrium and areas open to the atrium. The smoke-control system shall operate automatically upon actuation of the automatic sprinkler system within the atrium or areas open to the atrium and as required by Section 905.9.

402.3 Enclosure of Atria. Atria shall be separated from adjacent spaces by not less than one-hour fire-resistive construction.

> **EXCEPTIONS:** 1. The separation between atria and tenant spaces that are not guest rooms, congregate residences or dwelling units may be omitted at three floor levels.
>
> 2. Open exit-access balconies are permitted within the atrium.

Openings in the atrium enclosure other than fixed glazing shall be protected by smoke- and draft-control assemblies conforming to Section 1004.3.4.3.2.

> **EXCEPTION:** Other tightfitting doors that are maintained automatic closing, in accordance with Section 713.2, by actuation of a smoke detector, or self-closing may be used when protected as required for glazed openings in Exception 2.

Fixed glazed openings in the atrium enclosure shall be equipped with fire windows having a fire-resistive rating of not less than three-fourths hour, and the total area of such openings shall not exceed 25 percent of the area of the common wall between the atrium and the room into which the opening is provided.

> **EXCEPTIONS:** 1. In Group R, Division 1 Occupancies, openings may be unprotected when the floor area of each guest room, congregate residence or dwelling unit does not exceed 1,000 square feet (92.9 m²) and each room or unit has an approved means of egress not entering the atrium.
>
> 2. Guest rooms, dwelling units, congregate residences and tenant spaces may be separated from the atrium by approved fixed wired glass set in steel frames. In lieu thereof, tempered or laminated glass or listed glass block may be used, subject to the following:
>
> 2.1 The glass shall be protected by a sprinkler system equipped with listed quick-response sprinklers. The sprinkler system shall completely wet the entire surface of the glass wall when actuated. Where there are walking surfaces on both sides of the glass, both sides of the glass shall be so protected.
>
> 2.2 The tempered or laminated glass shall be in a gasketed frame so installed that the glazing system may deflect without breaking (loading) the glass before the sprinkler system operates.
>
> 2.3 The glass block wall assembly shall be installed in accordance with its listing for a three-fourths-hour fire-resistive rating and Section 2110.
>
> 2.4 Obstructions such as curtain rods, drapery traverse rods, curtains, drapes or similar materials shall not be installed between the sprinkler and the glass.

402.4 Escalators and Elevators. Escalators and elevators located entirely within the atrium enclosure need not be enclosed unless required by Chapter 30.

402.5 Means of Egress.

402.5.1 Travel distance. Not more than 100 feet (30 480 mm) of the travel distance allowed by Section 1004.2.5 may be on an open exit-access balcony within the atrium.

402.5.2 Group I Occupancy means of egress. Required means of egress from sleeping rooms in Group I Occupancies other than jails, prisons and reformatories shall not pass through the atrium.

402.5.3 Stairs and ramps. Stairways and ramps in the atrium space shall be enclosed.

> **EXCEPTIONS:** 1. Stairs and ramps not required for egress need not be enclosed.
>
> 2. Stairs and ramps connecting only the lowest two floors in the atrium space need not be enclosed.
>
> 3. Stairs and ramps connecting floor levels within a story need not be enclosed.

402.6 Occupancy Separation Exceptions. The vertical portion of the occupancy separation that is adjacent to the atrium may be omitted between a Group B Occupancy office, Group M Occupancy sales area or Group A, Division 3 Occupancy and Group R, Division 1 apartment, congregate residence or guest room located on another level.

402.7 Standby Power. Smoke control for the atrium and the smoke-control system for the tenant space shall be provided with standby power as required in Section 905.8.

402.8 Interior Finish. The interior finish of walls and ceilings of the atrium and all unseparated tenant spaces allowed under Exception 1 to the first paragraph of Section 402.3 shall be Class I with no reduction in class for sprinkler protection.

402.9 Acceptance of the Smoke-control System. Acceptance shall be as required by Section 905.15.

402.10 Combustible Furnishings in Atria. The quantity of combustible furnishings in atria shall not exceed that specified in the Fire Code.

SECTION 403 — SPECIAL PROVISIONS FOR GROUP B OFFICE BUILDINGS AND GROUP R, DIVISION 1 OCCUPANCIES

403.1 Scope. This section applies to all Group B office buildings and Group R, Division 1 Occupancies, each having floors used for human occupancy located more than 75 feet (22 860 mm) above the lowest level of fire department vehicle access. Such buildings shall be of Type I or II-F.R. construction and shall be provided with an approved automatic sprinkler system in accordance with Section 403.2.

403.2 Automatic Sprinkler System.

403.2.1 System design. The automatic sprinkler system shall be provided throughout the building as specified by UBC Standard 9-1, and shall be designed in accordance with that standard and the following:

1. Shutoff valves and a water-flow device shall be provided for each floor. The sprinkler riser may be combined with the standpipe riser.

2. In Seismic Zones 2, 3 and 4, in addition to the main water supply, a secondary on-site supply of water equal to the hydraulically calculated sprinkler design demand plus 100 gallons per minute (378.5 L/m) additional for the total standpipe system shall be provided. This supply shall be automatically available if the principal supply fails and shall have a duration of 30 minutes.

403.2.2 Modifications. The following modifications of code requirements are permitted:

1. In buildings of Type I construction, the fire-resistive time periods set forth in Table 6-A may be reduced by one hour for interior-bearing walls, exterior-bearing and nonbearing walls, roofs and the beams supporting roofs, provided they do not frame into columns. In buildings of Type II-F.R. construction, the fire-resistive time period set forth in Table 6-A may be reduced by one hour for interior-bearing walls, exterior-bearing and nonbearing walls, but no reduction is allowed for roofs. The fire-resistive time period reduction as specified herein shall not apply to exterior-bearing and nonbearing walls whose fire-resistive rating is less than four hours.

Shafts other than stairway enclosures and elevator shafts may be reduced to one hour when sprinklers are installed within the shafts at alternate floors.

2. Except for corridors in Group B offices and Group R, Division 1 Occupancies, and partitions separating dwelling units or guest rooms, all interior-nonbearing partitions required to be one-hour fire-resistive construction by Table 6-A may be of noncombustible construction without a fire-resistive time period.

3. Fire dampers, other than those needed to protect floor-ceiling assemblies to maintain the fire resistance of the assembly, are not required.

4. Emergency windows required by Section 310.4 are not required.

403.3 Smoke Detection. Smoke detectors shall be provided in accordance with this section. Smoke detectors shall be connected to an automatic fire alarm system installed in accordance with the Fire Code. The actuation of any detector required by this section shall operate the emergency voice alarm signaling system and shall place into operation all equipment necessary to prevent the recirculation of smoke.

Smoke detectors shall be located as follows:

1. In every mechanical equipment, electrical, transformer, telephone equipment, elevator machine or similar room and in elevator lobbies. Elevator lobby detectors shall be connected to an alarm verification zone or be listed as releasing devices.

2. In the main return-air and exhaust-air plenum of each air-conditioning system. Such detector shall be located in a serviceable area downstream of the last duct inlet.

3. At each connection to a vertical duct or riser serving two or more stories from a return-air duct or plenum of an air-conditioning system. In Group R, Division 1 Occupancies, an approved smoke detector may be used in each return-air riser carrying not more than 5,000 cubic feet per minute (2360 L/s) and serving not more than 10 air inlet openings.

4. For Group R, Division 1 Occupancies in all interior corridors serving as a means of egress for an occupant load of 10 or more.

403.4 Smoke Control. A smoke-control system meeting the requirements of Chapter 9 shall be provided.

403.5 Fire Alarm and Communication Systems.

403.5.1 General. The fire alarm, emergency voice/alarm signaling system and fire department communication systems shall be designed and installed as set forth in this code and the Fire Code.

403.5.2 Emergency voice alarm signaling system. The operation of any automatic fire detector, sprinkler or water-flow device shall automatically sound an alert tone followed by voice instructions giving appropriate information and direction on a general or selective basis to the following terminal areas:

1. Elevators.

2. Elevator lobbies.

3. Corridors.

4. Exit stairways.

5. Rooms and tenant spaces exceeding 1,000 square feet (93 m^2) in area.

6. Dwelling units in apartment houses.

7. Hotel guest rooms or suites.

8. Areas of refuge (as defined in Section 1102).

A manual override for emergency voice communication shall be provided for all paging zones.

403.5.3 Fire department communication system. A two-way, approved fire department communication system shall be provided for fire department use. It shall operate between the central control station and elevators, elevator lobbies, emergency and standby power rooms and at entries into enclosed stairways.

403.6 Central Control Station.

403.6.1 General. A central control station room for fire department operations shall be provided. The location and accessibility of the central control station room shall be approved by the fire department. The central control station room shall be separated from the remainder of the building by not less than a one-hour fire-resistive occupancy separation. The room shall be a minimum of 96 square feet (9 m^2) with a minimum dimension of 8 feet (2438 mm). It shall contain the following as a minimum:

1. The voice alarm and public address system panels.

2. The fire department communications panel.

3. Fire-detection and alarm system annunciator panels.

4. Annunciator visually indicating the location of the elevators and whether they are operational.

5. Status indicators and controls for air-handling systems.

6. Controls for unlocking all stairway doors simultaneously.

7. Sprinkler valve and water-flow detector display panels.

8. Emergency and standby power status indicators.

9. A telephone for fire department use with controlled access to the public telephone system.

10. Fire pump status indicators.

11. Schematic building plans indicating the typical floor plan and detailing the building core, means of egress, fire-protection systems, firefighting equipment and fire department access.

12. Work table.

403.6.2 Annunciation identification. Control panels in the central control station shall be permanently identified as to function.

Alarm, supervisory and trouble signals as required by Items 3 and 7 above shall be annunciated in compliance with the Fire Code in the central control station by means of an audible and visual indicator. For purposes of annunciation, zoning shall be in accordance with the following:

1. When the system serves more than one building, each building shall be considered separately.

2. Each floor shall be considered a separate zone. When one or more sprinkler risers serve the same floor, each riser shall be considered a separate zone.

> **EXCEPTION:** When more than one riser serves the same system on the floor.

403.7 Elevators. Elevators and elevator lobbies shall comply with the provisions of Chapter 30 and the following:

> **NOTE:** A bank of elevators is a group of elevators or a single elevator controlled by a common operating system; that is, all those elevators that respond to a single call button constitute a bank of elevators. There is no limit on the number of cars that may be in a bank or group, but there may not be more than four cars within a common hoistway.

1. Elevators on all floors shall open into elevator lobbies that are separated from the remainder of the building, including corridors and other means of egress, by walls extending from the floor to the underside of the fire-resistive floor or roof above. Such walls shall not be of less than one-hour fire-resistive construction. Openings through such walls shall conform to Section 1004.3.4.3.2.

> **EXCEPTIONS:** 1. The main entrance-level elevator lobby in office buildings.
> 2. Elevator lobbies located within an atrium complying with the provisions of Section 402.
> 3. In fully sprinklered office buildings, corridors may lead through enclosed elevator lobbies if all areas of the building have access to at least one required means of egress without passing through the elevator lobby.

2. Each elevator lobby shall be provided with approved smoke detector(s) installed in accordance with their listings. When the detector is activated, elevator doors shall not open and all cars serving that lobby are to return to the main floor and be under manual control only. If the main floor detector or a transfer floor detector is activated, all cars serving the main floor or transfer floor shall return to a location approved by the fire department and building official and be under manual control only. The detector may serve to close the lobby doors, additional doors at the hoistway opening allowed in Section 3007 and smoke dampers serving the lobby.

3. Elevator hoistways shall not be vented through an elevator machine room. Each elevator machine room shall be treated as a separate smoke-control zone.

403.8 Standby Power, Light and Emergency Systems.

403.8.1 Standby power. A standby power-generator set conforming to the Electrical Code shall be provided on the premises. The set shall supply all functions required by this section at full power. Set supervisions with manual start and transfer override features shall be provided at the central control station.

An on-premises fuel supply sufficient for not less than two hours' full-demand operation of the system shall be provided.

The standby system shall have a capacity and rating that would supply all equipment required to be operational at the same time. The generating capacity need not be sized to operate all the connected electrical equipment simultaneously.

All power, lighting, signal and communication facilities specified in Sections 403.3, 403.4, 403.5, 403.6, 403.7 and 403.8, as applicable; fire pumps required to maintain pressure, standby lighting and normal circuits supplying exit signs and means of egress illumination shall be transferable to the standby source.

403.8.2 Standby lighting. Standby lighting shall be provided as follows:

1. Separate lighting circuits and fixtures sufficient to provide light with an intensity of not less than 1 footcandle (10.76 lx) measured at floor level in all corridors, stairways, pressurized enclosures, elevator cars and lobbies and other areas that are clearly a part of the escape route.

2. All circuits supply lighting for the central control station and mechanical equipment room.

403.8.3 Emergency systems. The following are classified as emergency systems and shall operate within 10 seconds of failure of the normal power supply:

1. Exit sign and means of egress illumination as required by Sections 1003.2.8 and 1003.2.9.

2. Elevator car lighting.

403.9 Means of Egress. Means of egress shall comply with other requirements of this code and the following:

1. All stairway doors that are locked from the stairway side shall have the capability of being unlocked simultaneously without unlatching upon a signal from the central control station.

2. A telephone or other two-way communications system connected to an approved emergency service that operates continuously shall be provided at not less than every fifth floor in each required stairway where other provisions of this code permit the doors to be locked.

403.10 Seismic Considerations. In Seismic Zones 2, 3 and 4, the anchorage of mechanical and electrical equipment required for life-safety systems, including fire pumps and elevator drive and suspension systems, shall be designed in accordance with the requirements of Section 1626.

SECTION 404 — COVERED MALL BUILDINGS

404.1 General.

404.1.1 Purpose. The purpose of this section is to establish minimum standards of safety for the construction and use of covered mall buildings having not more than three levels

404.1.2 Scope. The provisions of Section 404 shall apply to buildings or structures defined herein as covered mall buildings and shall supersede other similar requirements in other chapters of the code.

> **EXCEPTIONS:** 1. Covered mall buildings conforming with all other applicable provisions of this code.
> 2. Terminals for transportation facilities and lobbies of hotels, apartments and office buildings.

404.1.3 Definitions. For the purpose of this chapter, certain terms are defined as follows:

ANCHOR BUILDING is an exterior perimeter department store, major merchandising center or Group R, Division 1 Occupancy having direct access to a covered mall building but having all required means of egress independent of the mall.

COVERED MALL BUILDING is a single building enclosing a number of tenants and occupancies such as retail stores, drinking and dining establishments, entertainment and amusement facilities, offices and other similar uses wherein two or more tenants have a main entrance into the mall.

FOOD COURT is a public seating area located in the mall that serves adjacent food preparation tenant spaces.

GROSS LEASABLE AREA is the total floor area designed for tenant occupancy and exclusive use. The area of tenant occupancy is measured from the center lines of joint partitions to the outside of the tenant walls. All tenant areas, including areas used for storage, shall be included in calculating gross leasable area.

MALL is a roofed or covered common pedestrian area within a covered mall building that serves as access for two or more tenants and may have three levels that are open to one another.

404.1.4 Applicability of other provisions. Except as specifically required by this chapter, covered mall buildings shall meet all applicable provisions of this code.

404.1.5 Standards of quality. The standard listed below is a recognized standard. (See Sections 3503 and 3504.)

1. UL 1975, Standard for Fire Tests for Foamed Plastics Used for Decorative Purposes

404.2 Types of Construction and Required Yards for Unlimited Area.

404.2.1 Type of construction. One- and two-level covered mall buildings may be of any type of construction permitted by this code. Three-level covered mall buildings shall be at least Type II One-hour construction.

Anchor buildings and parking garages shall be limited in height and area in accordance with Sections 504, 505 and 506.

404.2.2 Required yards for unlimited area. Covered mall buildings may be of unlimited area, provided the covered mall building, attached anchor buildings and parking garages are adjoined by public ways, streets or yards not less than 60 feet (18 288 mm) in width along all exterior walls.

404.3 Special Provisions.

404.3.1 Automatic sprinkler systems. The covered mall building shall be provided with an automatic sprinkler system conforming to the provisions of UBC Standard 9-1, which is a part of this code. See Chapter 35. In addition to these standards, the automatic sprinkler system shall comply with the following:

1. All automatic sprinkler system control valves shall be electrically supervised by an approved central, proprietary or remote station or a local alarm service that will give an audible signal at a constantly attended location.

2. The automatic sprinkler system shall be complete and operative throughout the covered mall building prior to occupancy of any of the tenant spaces. The separation between an unoccupied tenant space and the covered mall building shall be subject to the approval of the building official and the fire department.

3. Sprinkler protection for the mall shall be independent from that provided for tenant spaces. However, tenant spaces may be supplied by the same system if they can be independently controlled.

The respective increases for area and height for covered mall buildings, including anchor buildings, specified in Sections 311.9, 505 and 506, shall be permitted.

404.3.2 Standpipes. There shall be a combined Class I standpipe outlet connected to a system sized to deliver 250 gallons per minute (946.4 L/m) at the most hydraulically remote outlet. The outlet shall be supplied from the mall zone sprinkler system and shall be hydraulically calculated. Standpipe outlets shall be provided at each of the following locations:

1. Within the mall at the entrance to each exit passage or corridor.

2. At each floor-level landing within enclosed stairways opening directly onto the mall.

3. At exterior public entrances to the mall.

404.3.3 Smoke-control system. A smoke-control system meeting the requirements of Section 905 shall be provided.

> **EXCEPTION:** A smoke-control system need not be provided when both of the following conditions exist:
>
> 1. The mall does not exceed one story, and
>
> 2. The gross leasable area does not exceed 24,000 square feet (2230 m^2).

404.3.4 Fire department access to equipment. Rooms or areas containing controls for air-conditioning systems, automatic fire-extinguishing systems or other detection, suppression or control elements shall be identified for use by the fire department.

404.3.5 Tenant separation. Each tenant space shall be separated from other tenant spaces by a wall having a fire-resistive rating of not less than one hour. The separation wall shall extend from the floor to the underside of the ceiling above. Except as required by other provisions of this code, the ceiling need not be a fire-resistive assembly. A separation is not required between any tenant space and a mall except for occupancy separations required by Section 404.5 or for smoke-control purposes.

404.3.6 Public address system. Covered mall buildings exceeding 50,000 square feet (4645 m^2) in total floor area shall be provided with a public address system accessible for use by the fire department. Covered mall buildings of 50,000 square feet (4645 m^2) or less in total floor area, when provided with a public address system, shall have such system accessible for use by the fire department.

404.3.7 Plastic panels and plastic signs. Within every story or level and from side wall to side wall of each tenant space or mall, plastic panels and plastic signs shall comply with the following:

1. Plastics other than foam plastics shall be approved plastic materials as defined in Section 217.

2. Foam plastics shall have a maximum heat-release rate of 150 kilowatts when tested in accordance with approved recognized standards (see Chapter 35, Part IV) and shall have the following physical characteristics:

> 2.1 A density not less than 20 pounds per cubic foot (320.4 kg/m^3) and
>
> 2.2 A thickness not greater than $^1/_2$ inch (12.7 mm).

3. They shall not exceed 20 percent of the wall area facing the mall.

4. They shall not exceed a height of 36 inches (914 mm) except that if the sign is vertical, then the height shall not exceed 96 inches (2438 mm) and the width shall not exceed 36 inches (914 mm).

5. They shall be located a minimum distance of 18 inches (457 mm) from adjacent tenants.

404.3.8 Lease plan. Each covered mall building owner shall provide both the building and fire departments with a lease plan showing the location of each occupancy and its means of egress after the certificate of occupancy has been issued. Such plans shall be kept current. No modifications or changes in occupancy or use shall be made from that shown on the lease plan without prior approval of the building official.

404.3.9 Openings between anchor building and mall. Except for the occupancy separation between Group R, Division 1 sleeping rooms and the mall, openings between anchor buildings of

Type I, Type II-F.R., Type II One-hour or Type II-N construction and the mall need not be protected.

404.3.10 Standby power. Covered mall buildings exceeding 50,000 square feet (4645 m^2) shall be provided with standby power systems that are capable of operating the public address system, the smoke-control activation system and the smoke-control equipment as required by Section 905.

404.4 Means of Egress.

404.4.1 General. Each tenant space and the covered mall building shall be provided with means of egress as required by this section and Chapter 10. Where there is a conflict between the requirements of Chapter 10 and the requirements of this section, the requirements of this section shall apply.

404.4.2 Determination of occupant load. The occupant load permitted in any individual tenant space in a covered mall building shall be determined as required by Section 1002. Means of egress requirements for individual tenant spaces shall be based on the occupant load thus determined.

The occupant load permitted for the covered mall building, assuming all portions, including individual tenant spaces and the mall to be occupied at the same time, shall be determined by dividing the gross leasable area by 30 for covered mall buildings containing up to 150,000 square feet (13 935 m^2) of gross leasable area, by 40 for covered mall buildings containing between 150,001 and 350,000 square feet (13 935 m^2 and 32 515 m^2) of gross leasable area, and by 50 for covered mall buildings containing more than 350,000 square feet (32 515 m^2) of gross leasable area. Means of egress requirements for the mall shall be based on the occupant load thus determined.

The occupant load of a food court shall be determined in accordance with Section 1003.2.2. For purposes of determining the means of egress requirements for the mall, the food court occupant load shall be added to the occupant load of the covered mall building as calculated above.

The occupant load of anchor buildings opening into the mall shall not be included in determining means of egress requirements for the mall.

404.4.3 Number of means of egress. When the distance of travel to the mall exceeds 75 feet (22 860 mm) within the public area of a tenant space or when the occupant load served by the means of egress to the mall exceeds 50, not less than two means of egress shall be provided. The occupant load of a public sales area shall be computed at 30 square feet (2.8 m^2) per occupant. Occupant loads for other areas shall be computed in accordance with Table 10-A.

404.4.4 Arrangement of means of egress. Group A, Divisions 1, 2 and 2.1 Occupancies, other than drinking and dining establishments, shall be so located in the covered mall building that their entrance will be immediately adjacent to a principal entrance to the mall and shall have not less than one half of their required means of egress opening directly to the exterior of the covered mall building.

Required means of egress for anchor buildings shall be provided independently from the mall means of egress system.

Malls shall not egress through anchor buildings. Malls terminating at an anchor building where no other means of egress has been provided shall be considered a dead-end mall.

404.4.5 Travel distance. Within each individual tenant space in a covered mall building the maximum travel distance shall not exceed 200 feet (60 960 mm).

The maximum travel distance from any point within a mall shall not exceed 200 feet (60 960 mm).

404.4.6 Exit access. The means of egress shall be so arranged that it is possible to go in either direction from any point in a mall to a separate exit, except for dead ends not exceeding a length equal to twice the width of the mall measured at the narrowest location within the dead-end portion of the mall.

The minimum width of the means of egress from a mall shall be 66 inches (1676 mm).

Storage is prohibited in exit passageways, which are also used for service to the tenants. Such exit passageways shall be posted with conspicuous signs so stating.

404.4.7 Malls. For the purpose of providing required egress, malls may be considered as corridors but need not comply with the requirements of Sections 1004.3.4.3.1 and 1004.3.4.3.2 when the width of mall is as specified in this section.

The minimum aggregate clear width of the mall shall be 20 feet (6096 mm). There shall be a minimum of 10 feet (3048 mm) clear width to a height of 8 feet (2438 mm) on each side of the mall between any projection from a tenant space bordering the mall and the nearest kiosk, vending machine, bench, display, food court or other obstruction to egress. Kiosks, vending machines and similar uses shall be spaced at least 20 feet (6096 mm) from one another and shall not be more than 300 square feet (28 m^2) in area.

Malls that do not conform to the requirements of this section shall comply with the requirements of Sections 1004.3.4.3.1 and 1004.3.4.3.2.

404.4.8 Security grilles and doors. Horizontal sliding or vertical security grilles or doors that are a part of a required means of egress shall conform to the following:

1. They must remain secured in the full open position during the period of occupancy by the general public.

2. Doors or grilles shall not be brought to the closed position when there are more than 10 persons occupying spaces served by a single exit or 50 persons occupying spaces served by more than one exit.

3. The doors or grilles shall be openable from within without the use of any special knowledge or effort when the space is occupied.

4. When two or more exits are required, not more than one half of the exits may be equipped with horizontal sliding or vertical rolling grilles or doors.

404.5 Occupancy.

404.5.1 General. Covered mall buildings shall be classified as a Group B or M Occupancy and may contain accessory uses consisting of Group A, E or R, Division 1 Occupancies. The area of individual accessory uses within a covered mall building shall not exceed three times the basic area permitted by Table 5-B for the type of construction and the occupancy involved. The aggregate area of all accessory uses within a covered mall building shall not exceed 25 percent of the gross leasable area.

An attached garage for the parking or storage of private or pleasure-type motor vehicles having a capacity of not more than nine persons and open parking garages may be considered as separate buildings when they are separated from the covered mall building by an occupancy separation having a fire-endurance time period of at least two hours.

404.5.2 Mixed occupancy. Individual tenant spaces within a covered mall building that comprise a distinct "occupancy," as described in Chapter 3, shall be separated from any other occupancy as specified in Section 302.4.

EXCEPTIONS: 1. A main entrance that opens onto a mall need not be separated.

2. An occupancy separation is not required between a food court and adjacent tenant spaces or mall.

SECTION 405 — STAGES AND PLATFORMS

405.1 Scope.

405.1.1 Standards of quality. Stages, platforms and accessory spaces in assembly occupancies shall conform with the requirements of Section 405.

The standards listed below labeled a "UBC Standard" are also listed in Chapter 35, Part II, and are part of this code.

1. UBC Standard 4-1, Proscenium Firesafety Curtains

2. UBC Standard 7-1, Fire Tests of Building Construction and Materials

3. UBC Standard 8-1, Test Method for Surface-burning Characteristics of Building Materials

4. UBC Standard 9-1, Installation of Sprinkler Systems

405.1.2 Definitions. For the purpose of this chapter, certain terms are defined as follows:

BATTEN is a flown metal pipe or shape on which lights or scenery are fastened.

DROP is a large piece of scenic canvas or cloth that hangs vertically, usually across the stage area.

FLY is the space over the stage of a theater where scenery and equipment can be hung out of view. Also called lofts and rigging lofts.

FLY GALLERY is a raised area above a stage from which the movement of scenery and operation of other stage effects are controlled.

GRIDIRON is the structural framing over a stage supporting equipment for hanging or flying scenery and other stage effects. A gridiron grating shall not be considered a floor.

LEG DROP is a long narrow strip of fabric used for masking. When used on either or both sides of the acting area, it is provided to designate an entry onto the stage by the actors. It is also used to mask the side stage area. They may also be called "wings."

PINRAIL is a rail on or above a stage that has belaying pins to which lines are fastened.

PLATFORM is that raised area within a building used for the presentation of music, plays or other entertainment; the head table for special guests; the raised area for lectures and speakers; boxing and wrestling rings; theater in the round; and similar purposes wherein there are not overhead hanging curtains, drops, scenery or stage effects other than lighting.

PLATFORM, PERMANENT, is a platform used within an area for more than 30 days.

PLATFORM, TEMPORARY, is a platform used within an area for not more than 30 days.

PROSCENIUM WALL is the wall that separates the stage from the auditorium or house.

STAGE is a space within a building used for entertainment or presentations, with a stage height of 50 feet (15 240 mm) or less. Curtains, drops, scenery, lighting devices and other stage effects are hung and not retractable except for a single lighting bank; single main curtain, border and legs; and single backdrop.

STAGE AREAS are the entire performance area and adjacent backstage and support areas not separated from the performance area by fire-resistive construction.

STAGE HEIGHT is the dimension between the lowest point on the stage floor and the highest point of the underside of the roof or floor deck above the stage.

STAGE, LEGITIMATE, is a stage wherein curtains, drops, leg drops, scenery, lighting devices or other stage effects are retractable horizontally or suspended overhead and the stage height is greater than 50 feet (15 240 mm).

THEATER-IN-THE-ROUND is an acting area in the middle of a room with the audience sitting all around it.

405.1.3 Materials and design. Materials used in the construction of platforms and stages shall conform to the applicable materials and design requirements as set forth in this code. All assumed design live loads shall be indicated on the construction documents submitted for approval.

405.2 Platforms. Temporary platforms may be constructed of any materials. The space between the floor and the platform above shall not be used for any purpose other than electrical wiring or plumbing to platform equipment.

Platforms shall be constructed of materials as required for the type of construction of the building in which the platform is located. When the space beneath a raised platform is used for storage or any purpose other than equipment wiring or plumbing, the floor construction shall not be less than one-hour fire-resistive construction. When the space beneath the platform is not used for any purpose other than equipment wiring or plumbing, the underside of the platform shall be fireblocked and may be constructed of any type of materials permitted by this code. The floor finish may be of wood in all types of construction.

405.3 Stages.

405.3.1 Construction. The minimum type of construction for stages shall be as required for the building except that the finish floor, in all types of construction, may be of wood.

Stages having a stage height exceeding 50 feet (15 240 mm) shall be separated from the balance of the building by not less than a two-hour occupancy separation.

> **EXCEPTION:** The opening in the proscenium wall used for viewing performances may be protected by a proscenium firesafety curtain conforming to UBC Standard 4-1.

Where permitted by the building construction type or where the stage is separated from all other areas as required in the paragraph above, the stage floor may be of unprotected noncombustible or heavy-timber framing members with a minimum $1^1/_2$-inch-thick (38 mm) wood deck.

Where a stage floor is required to be of one-hour fire-resistive-rated construction, the stage floor may be unprotected when the space below the stage is sprinklered throughout.

Where the stage height is 50 feet (15 240 mm) or less, the stage area shall be separated from accessory spaces by a one-hour fire-resistive occupancy separation.

> **EXCEPTION:** Control rooms and follow spot rooms may be open to the audience.

405.3.2 Accessory rooms. Dressing rooms, workshops, storerooms and other accessory spaces contiguous to stages shall be separated from one another and other building areas by a one-hour fire-resistive occupancy separation.

> **EXCEPTION:** A separation is not required for stages having a floor area not exceeding 500 square feet (46.5 m²).

405.3.3 Ventilation. Emergency ventilation shall be provided for all stage areas greater than 1,000 square feet (93 m²) or with a stage height of greater than 50 feet (15 240 mm) to provide a means of removing smoke and combustion gases directly to the outside in the event of a fire. Ventilation shall be by one or a combination of the following methods in Section 405.3.3.1 and 405.3.3.2.

405.3.3.1 Smoke control. A means shall be provided to maintain the smoke level not less than 6 feet (1829 mm) above the highest level of assembly seating or above the top of the proscenium opening where proscenium wall and opening protection is provided. The system shall be activated independently by each of the following: (1) activation of the sprinkler system in the stage area and (2) by a manually operated switch at an approved location. The emergency ventilation system shall be connected to both normal and standby power. The fan(s) power wiring and ducts shall be located and properly protected to ensure a minimum 20 minutes of operation in the event of activation.

405.3.3.2 Roof vents. Two or more vents shall be located near the center of and above the highest part of the stage area. They shall be raised above the roof and provide a net free vent area equal to 5 percent of the stage area. Vents shall be constructed to open automatically by approved heat-activated devices. Supplemental means shall be provided for manual operation of the ventilator from the stage floor. Vents shall be of an approved type.

405.3.4 Proscenium walls. The proscenium opening shall be protected by an approved fire curtain or an approved water curtain complying with UBC Standard 4-1. The fire curtain shall be designed to close automatically upon automatic detection of a fire and upon manual activation and shall resist the passage of flame and smoke for 20 minutes between the stage area and the audience area.

405.3.5 Gridirons, fly galleries and pinrails. Beams designed only for the attachment of portable or fixed theater equipment, gridirons, galleries and catwalks shall be constructed of materials consistent with the building type of construction. A fire-resistance rating is not required.

> **EXCEPTION:** Combustible materials shall be permitted for use as the floors of galleries and catwalks of all types of construction.

405.3.6 Flame-retardant requirements. Combustible scenery of cloth, film, dry vegetation and similar materials shall meet the requirements of the Fire Code. Foam plastics shall have a maximum heat release rate of 100 kilowatts.

SECTION 406 — MOTION PICTURE PROJECTION ROOMS

406.1 General.

406.1.1 Scope. The provisions of this section shall apply where ribbon-type cellulose acetate or other safety film is used in conjunction with electric arc, xenon or other light-source projection equipment, which develops hazardous gases, dust or radiation. Where cellulose nitrate film is used, projection rooms shall comply with the Fire Code.

406.1.2 Projection room required. Every motion picture machine projecting film as mentioned within the scope of this chapter shall be enclosed in a projection room. Appurtenant electrical equipment, such as rheostats, transformers and generators, may be within the projection room or in an adjacent room of equivalent construction.

There shall be posted on the outside of each projection room door and within the projection room itself a conspicuous sign with 1-inch (25.4 mm) block letters stating: SAFETY FILM ONLY PERMITTED IN THIS ROOM.

406.2 Construction. Every projection room shall be of permanent construction consistent with the construction requirements for the type of building in which the projection room is located. Openings need not be protected.

The room shall have a floor area of not less than 80 square feet (7.4 m^2) for a single machine and at least 40 square feet (3.7 m^2) for each additional machine. Each motion picture projector, floodlight, spotlight or similar piece of equipment shall not be used unless approved and shall have a clear working space not less than 30 inches by 30 inches (762 mm by 762 mm) on each side and at the rear thereof, but only one such space shall be required between two adjacent projectors.

The projection room and the rooms appurtenant thereto shall have a ceiling height of not less than 7 feet 6 inches (2286 mm).

406.3 Means of Egress. Means of egress shall be provided as required in Chapter 10. Motion picture projection rooms used for projection of safety film only are required to have only one exit or exit-access door.

406.4 Projection Ports and Openings. The aggregate of openings for projection equipment shall not exceed 25 percent of the area of the wall between the projection room and the auditorium.

All openings shall be provided with glass or other approved material so as to completely close the opening.

406.5 Ventilation.

406.5.1 General. Ventilation shall be provided in accordance with the provisions of this section.

406.5.2 Projection booth.

406.5.2.1 Supply air. Each projection room shall be provided with adequate air-supply inlets so arranged as to provide well-distributed air throughout the room. Air-inlet ducts shall provide an amount of air equivalent to the amount of air being exhausted by projection equipment. Air may be taken from the outside; from adjacent spaces within the building, provided the volume and infiltration rate is sufficient; or from the building air-conditioning system, provided it is so arranged as to provide sufficient air when other systems are not in operation.

406.5.2.2 Exhaust air. Projection booths may be exhausted through the lamp exhaust system. The lamp exhaust system shall be positively interconnected with the lamp so that the lamp will not operate unless there is the air flow required for the lamp. Exhaust air ducts shall terminate at the exterior of the building in such a location that the exhaust air cannot be readily recirculated into any air-supply system. The projection room ventilation system may also serve appurtenant rooms such as the generator room and the rewind room.

Each projection machine shall be provided with an exhaust duct that will draw air from each lamp and exhaust it directly to the outside of the building. The lamp exhaust may serve to exhaust air from the projection room to provide room air circulation. Such ducts shall be of rigid materials, except for a flexible connector approved for the purpose. The projection lamp or projection room exhaust system or both may be combined but shall not be interconnected with any other exhaust or return system, or both, within the building.

406.5.3 Projection equipment ventilation.

406.5.3.1 General. Each projection machine shall be provided with an exhaust duct that will draw air from each lamp and exhaust it directly to the outside of the building in such a fashion that it will not be picked up by supply inlets. Such a duct shall be of rigid materials, except for a continuous flexible connector approved for the purpose. The lamp exhaust system shall not be interconnected with any other system.

406.5.3.2 Electric arc projection equipment. The exhaust capacity shall be 200 cubic feet per minute (94.4 L/s) for each lamp connected to the lamp exhaust system, or as recommended by the equipment manufacturer. Auxiliary air may be introduced into the system through a screened opening to stabilize the arc.

406.5.3.3 Xenon projection equipment. The lamp exhaust system shall exhaust not less than 300 cubic feet per minute (142 L/s) per lamp or not less than that exhaust volume required or recommended by the equipment manufacturer, whichever is the greater. The external temperature of the lamp housing shall not exceed 130°F (54.4°C) when operating.

406.6 Miscellaneous Equipment. Each projection room shall be provided with rewind and film storage facilities.

A maximum of four containers for flammable liquids not greater than 16-ounce (473.2 mL) capacity and of a nonbreakable type may be permitted in each projection booth.

406.7 Sanitary Facilities. Every projection room shall be provided with a lavatory and a water closet.

> **EXCEPTION:** A projection room where completely automated projection equipment is installed that does not require a projectionist in attendance for projection or rewinding film.

SECTION 407 — CELLULOSE NITRATE FILM

The handling and storage of cellulose nitrate film shall be in accordance with the Fire Code. For exits, see Section 1007.7.4.

SECTION 408 — AMUSEMENT BUILDINGS

408.1 General. Amusement buildings having an occupant load of 50 or more shall comply with the requirements for the appropriate Group A Occupancy and this section. Amusement buildings having an occupant load of less than 50 shall comply with the requirements for a Group B Occupancy and this section.

> **EXCEPTION:** Amusement buildings or portions thereof that are without walls or a roof and constructed to prevent the accumulation of smoke in assembly areas.

For flammable decorative materials, see the Fire Code.

408.2 Definition. For the purposes of this code, the following definition applies:

AMUSEMENT BUILDING is a building or portion thereof, temporary or permanent, used for entertainment or educational purposes and that contains a system that transports passengers or provides a walkway through a course so arranged that the means of egress are not apparent due to theatrical distractions, are disguised or are not readily available due to the method of transportation through the building or structure.

408.3 Means of Egress and Exit Signs. Means of egress and exit signs for amusement buildings shall be approved by the building official and, where practical, shall comply with the requirements specified in Chapter 10. For exit marking, see Section 1007.2.7 for all amusement buildings.

408.4 Automatic Fire-extinguishing Systems. An automatic fire-extinguishing system shall be installed in amusement buildings as set forth in Section 904.2.3.6.

408.5 Alarm Systems.

408.5.1 General. An approved smoke-detection system installed in accordance with the Fire Code shall be provided in amusement buildings.

> **EXCEPTION:** In areas where ambient conditions will cause a smoke-detector system to alarm, an approved alternate type of automatic detector shall be installed.

408.5.2 Alarm system. Activation of any single smoke detector, the automatic sprinkler system or other automatic fire-detection device shall immediately sound an alarm in the building at a constantly supervised location from which the manual operation of systems noted in Section 408.5.3, Items 1, 2 and 3, may be initiated.

408.5.3 System response. The activation of two or more smoke detectors, a single smoke detector monitored by an alarm verification zone, the automatic sprinkler system or other approved automatic fire-detection device shall automatically:

1. Stop confusing sounds and visual effects,

2. Activate an approved directional exit marking, and

3. Cause illumination of the means of egress with light of not less than 1 footcandle (10.76 lx) at the walking surface.

408.5.4 Public address system. A public address system that is audible throughout the amusement building shall be provided. The public address system may also serve as an alarm system.

SECTION 409 — PEDESTRIAN WALKWAYS

409.1 General. A pedestrian walkway shall be considered a building when determining the roof covering permitted by Table 15-A. Pedestrian walkways connecting separate buildings need not be considered as buildings and need not be considered in the determination of the allowable floor area of the connected buildings when the pedestrian walkway complies with the provisions of this section.

409.2 Construction. Pedestrian walkways shall be constructed of noncombustible materials.

> **EXCEPTIONS:** 1. Pedestrian walkways connecting buildings of Type III, IV or V construction may be constructed of one-hour fire-resistive construction or of heavy-timber construction in accordance with Section 605.6.
>
> 2. Pedestrian walkways located on grade having both sides open by at least 50 percent and connecting buildings of Type III, IV or V construction may be constructed with any materials allowed by this code.

409.3 Openings between Pedestrian Walkways and Buildings. Openings from buildings to pedestrian walkways shall conform to the requirements of Table 5-A and Sections 503.3, 602.3, 603.3, 604.3, 605.3 and 606.3. In addition, pedestrian walkways connecting buildings shall be either provided with opening protection at connections to buildings in accordance with Section 1004.3.4.3.2 or constructed with both sides of the pedestrian walkway at least 50 percent open with the open area distributed so as to prevent the accumulation of smoke and toxic gases.

409.4 Width. The unobstructed width of pedestrian walkways shall not be less than 44 inches (1118 mm). The total width of a pedestrian walkway shall not exceed 30 feet (9144 mm).

409.5 Maximum Length. The length of a pedestrian walkway shall not exceed 300 feet (91 440 mm).

> **EXCEPTIONS:** 1. Pedestrian walkways that are fully sprinklered may be 400 feet (121 920 mm) in length.
>
> 2. Unenclosed walkways at grade.

409.6 Multiple Pedestrian Walkways. The distance between any two pedestrian walkways on the same horizontal plane shall not be less than 40 feet (12 192 mm).

409.7 Required Means of Egress. Pedestrian walkways at other than grade shall not be used as a means of egress. Pedestrian walkways at grade level used as required means of egress shall be unobstructed and shall have a minimum width in accordance with Chapter 10.

> **EXCEPTION:** Pedestrian walkways conforming to the requirements of a horizontal exit may be used as a required means of egress.

409.8 Pedestrian Walkways over Public Streets. Pedestrian walkways over public streets shall be subject to the approval of local jurisdictions.

SECTION 410 — MEDICAL GAS SYSTEMS IN GROUPS B AND I OCCUPANCIES

Medical gas systems in Groups B and I Occupancies shall be installed and maintained in accordance with this section and the Fire Code. When nonflammable gas cylinders for such systems are located inside buildings, they shall be in a separate room or enclosure separated from the rest of the building by not less than one-hour fire-resistive construction. Doors to the room or enclosure shall be self-closing smoke- and draft-control assemblies having a fire-protection rating of not less than one hour. Rooms shall have at least one exterior wall in which there are not less than two vents of not less than 36 square inches (0.023 m^2) in area per vent. One vent shall be within 6 inches (152 mm) of the floor and one shall be within 6 inches (152 mm) of the ceiling.

> **EXCEPTION:** When an exterior wall cannot be provided for the room, automatic sprinklers shall be installed within the room and the room shall be vented to the exterior through ducting contained within a one-hour-rated shaft enclosure. Approved mechanical ventilation shall provide six air changes per hour for the room.

SECTION 411 — COMPRESSED GASES

The storage and handling of compressed gases shall comply with the Fire Code.

SECTION 412 — AVIATION CONTROL TOWERS

Where applicable (see Section 101.3) for aviation control towers, see Appendix Chapter 4, Division II.

SECTION 413 — DETENTION AND CORRECTION FACILITIES

Where applicable (see Section 101.3) for detention and correction facilities, see Appendix Chapter 3, Division I.

SECTION 414 — AGRICULTURAL BUILDINGS

Where applicable (see Section 101.3) for agricultural buildings, see Appendix Chapter 3, Division II.

SECTION 415 — GROUP R, DIVISION 3 OCCUPANCIES

Where applicable (see Section 101.3) for Group R, Division 3 Occupancies, see Appendix Chapter 3, Division III.

SECTION 416 — GROUP R, DIVISION 4 OCCUPANCIES

Where applicable (see Section 101.3) for Group R, Division 4 Occupancies, see Appendix Chapter 3, Division IV.

SECTION 417 — BARRIERS FOR SWIMMING POOLS

Where applicable (see Section 101.3) for barriers for swimming pools, see Appendix Chapter 4, Division I.

SECTION 418 — RESERVED

TABLE 4-A—ATRIUM OPENING AND AREA

HEIGHT IN STORIES	MINIMUM CLEAR OPENING[1] (feet)	MINIMUM AREA (square feet)
	× 304.8 for mm	× 0.0929 for m²
3-4	20	400
5-7	30	900
8 or more	40	1,600

[1]The specified dimensions are the diameters of inscribed circles whose centers fall on a common axis for the full height of the atrium.

Chapter 5
GENERAL BUILDING LIMITATIONS

SECTION 501 — SCOPE

Buildings and structures shall comply with the location on property, area, height and other provisions of this chapter.

For additional limitations or allowances for special uses or occupancies, see the following:

SECTION	SUBJECT
402	Atria
403	High-rise office buildings and Group R, Division 1 Occupancies
404	Malls
311.9	Open parking structures
307	Group H, Division 6 Occupancies
412	Aviation control towers
414	Agricultural buildings
3111	Membrane structures

SECTION 502 — PREMISES IDENTIFICATION

Approved numbers or addresses shall be provided for all new buildings in such a position as to be plainly visible and legible from the street or road fronting the property.

SECTION 503 — LOCATION ON PROPERTY

503.1 General. Buildings shall adjoin or have access to a public way or yard on not less than one side. Required yards shall be permanently maintained.

For the purpose of this section, the center line of an adjoining public way shall be considered an adjacent property line. (See also Section 1203.4.)

503.2 Fire Resistance of Walls.

503.2.1 General. Exterior walls shall have fire resistance and opening protection as set forth in Table 5-A and in accordance with such additional provisions as are set forth in Chapter 6. Distance shall be measured at right angles from the property line. The above provisions shall not apply to walls at right angles to the property line.

Projections beyond the exterior wall shall comply with Section 705 and shall not extend beyond:

1. A point one third the distance to the property line from an assumed vertical plane located where fire-resistive protection of openings is first required due to location on property; or

2. More than 12 inches (305 mm) into areas where openings are prohibited.

503.2.2 Area of openings. When openings in exterior walls are required to be protected due to distance from property line, the sum of the area of such openings shall not exceed 50 percent of the total area of the wall in each story.

503.3 Buildings on Same Property and Buildings Containing Courts. For the purposes of determining the required wall and opening protection and roof-covering requirements, buildings on the same property and court walls of buildings over one story in height shall be assumed to have a property line between them.

EXCEPTION: In court walls where opening protection is required, such protection may be omitted, provided (1) not more than two levels open into the court, (2) the aggregate area of the building including the court is within the allowable area and (3) the building is not classified as a Group I Occupancy.

When a new building is to be erected on the same property as an existing building, the location of the assumed property line with relation to the existing building shall be such that the exterior wall and opening protection of the existing building meet the criteria as set forth in Table 5-A and Chapter 6.

EXCEPTION: Two or more buildings on the same property may be considered as portions of one building if the aggregate area of such buildings is within the limits specified in Section 504 for a single building.

When the buildings so considered house different occupancies or are of different types of construction, the area shall be that allowed for the most restricted occupancy or construction.

503.4 Special Provisions and Exceptions to Table 5-A.

503.4.1 General. The provisions of this section are exceptions to, or special provisions of, the construction requirements of Table 5-A, Chapters 3 and 6.

503.4.2 One-story Groups B, F, M and S Occupancies. In Groups B, F, M and S Occupancies, a fire-resistive time period will not be required for an exterior wall of a one-story, Type II-N building, provided the floor area of the building does not exceed 1,000 square feet (93 m^2) and such wall is located not less than 5 feet (1524 mm) from a property line.

503.4.3 Fire-retardant-treated wood framing. In Types III and IV construction, approved fire-retardant-treated wood framing may be used within the assembly of exterior walls when Table 5-A allows a fire-resistive rating of two hours or less, provided the required fire resistance is maintained and the exposed outer and inner faces of such walls are noncombustible.

503.4.4 Wood columns and arches. In Types III and IV construction, wood columns and arches conforming to heavy-timber sizes may be used externally when exterior walls are permitted to be unprotected, noncombustible construction or when one-hour fire-resistive noncombustible exterior walls are permitted.

503.4.5 Group H Occupancies—minimum distance to property lines. Regardless of any other provisions, Group H Occupancies shall be set back a minimum distance from property lines as set forth in Items 1 through 4. Distances shall be measured from the walls enclosing the occupancy to all property lines, including those on a public way.

1. Group H, Division 1 Occupancies. Not less than 75 feet (22 860 mm) and not less than required by Table 3-F.

2. Group H, Division 2 Occupancies. Not less than 30 feet (9144 mm) when the area of the occupancy exceeds 1,000 square feet (93 m^2) and it is not required to be located in a detached building.

3. Group H, Divisions 2 and 3 Occupancies. Not less than 50 feet (15 240 mm) when a detached building is required. See Table 3-G.

4. Group H, Divisions 2 and 3 Occupancies containing materials with explosive characteristics. Not less than the distances required by Table 3-F.

503.4.6 Group H, Division 1, 2 or 3 Occupancies—detached buildings. When a detached building is required by Table 3-G, there are no requirements for wall and opening protection based on location on property.

503.4.7 Group H, Division 4 Occupancies. Group H, Division 4 Occupancies having a floor area not exceeding 2,500 square feet (232 m^2) may have exterior-bearing walls of not less than two-hour fire-resistive construction when less than 5 feet (1524 mm) from a property line, and not less than one hour when less than 20 feet (6096 mm) from a property line.

503.4.8 Group U, Division 1 Occupancies. In Group U, Division 1 Occupancies, exterior walls that are required to be of one-hour fire-resistive construction due to location on property may be protected only on the exterior side with materials approved for one-hour fire-resistive construction.

When work is exempt from a permit as listed in Section 106.2, Item 1, there are no requirements for wall and opening protection based on location on property when accessory to a Group R, Division 3 Occupancy.

503.4.9 Exterior wall assemblies. Exterior wall assemblies complying with Section 2602.5.2 may be used in all types of construction.

SECTION 504 — ALLOWABLE FLOOR AREAS

504.1 One-story Areas. The area of a one-story building shall not exceed the limits set forth in Table 5-B, except as provided in Section 505.

504.2 Areas of Buildings over One Story. The total combined floor area for multistory buildings may be twice that permitted by Table 5-B for one-story buildings, and the floor area of any single story shall not exceed that permitted for a one-story building.

504.3 Allowable Floor Area of Mixed Occupancies. When a building houses more than one occupancy, the area of the building shall be such that the sum of the ratios of the actual area for each separate occupancy divided by the total allowable area for each separate occupancy shall not exceed one.

> **EXCEPTIONS:** 1. The major occupancy classification of a building may be used to determine the allowable area of such building when the major use occupies not less than 90 percent of the area of any floor of the building and provided that other minor accessory uses shall not exceed the basic area permitted by Table 5-B for such minor uses and that various uses are separated as specified in Section 302.4.
>
> 2. Groups B, F, M and S and Group H, Division 5 Occupancies complying with the provisions of Section 505.2 may contain other occupancies provided that such occupancies do not occupy more than 10 percent of the area of any floor of a building, nor more than the basic area permitted in the occupancy by Table 5-B for such occupancy, and further provided that such occupancies are separated as specified in Section 302.4.

504.4 Mezzanines. Unless considered as a separate story, the floor area of all mezzanines shall be included in calculating the allowable floor area of the stories in which the mezzanines are located.

504.5 Basements. A basement need not be included in the total allowable area, provided such basement does not exceed the area permitted for a one-story building.

504.6 Area Separation Walls.

504.6.1 General. Each portion of a building separated by one or more area separation walls that comply with the provisions of this section may be considered a separate building. The extent and location of such area separation walls shall provide a complete separation.

When an area separation wall also separates occupancies that are required to be separated by an occupancy separation, the most restrictive requirements of each separation shall apply.

504.6.2 Fire resistance and openings. Area separation walls shall not be less than four-hour fire-resistive construction in Types I, II-F.R., III and IV buildings and two-hour fire-resistive construction in Type II One-hour, Type II-N or Type V buildings. The total width of all openings in such walls shall not exceed 25 percent of the length of the wall in each story. All openings shall be protected by a fire assembly having a three-hour fire-protection rating in four-hour fire-resistive walls and one- and one-half-hour fire-protection rating in two-hour fire-resistive walls.

504.6.3 Extensions beyond exterior walls. Area separation walls shall extend horizontally to the outer edges of horizontal projecting elements such as balconies, roof overhangs, canopies, marquees or architectural projections extending beyond the floor area as defined in Section 207.

> **EXCEPTIONS:** 1. When horizontal projecting elements do not contain concealed spaces, the area separation wall may terminate at the exterior wall.
>
> 2. When the horizontal projecting elements contain concealed spaces, the area separation wall need only extend through the concealed space to the outer edges of the projecting elements.
>
> In either Exception 1 or 2, the exterior walls and the projecting elements above shall not be of less than one-hour fire-resistive construction for a distance not less than the depth of the projecting elements on both sides of the area separation wall. Openings within such widths shall be protected by fire assemblies having a fire-protection rating of not less than three-fourths hour.

504.6.4 Terminating. Area separation walls shall extend vertically from the foundation to a point at least 30 inches (762 mm) above the roof.

> **EXCEPTIONS:** 1. Any area separation wall may terminate at the underside of the roof sheathing, deck or slab, provided the roof-ceiling assembly is of at least two-hour fire-resistive construction.
>
> 2. Two-hour area separation walls may terminate at the underside of the roof sheathing, deck or slab, provided:
>
> 2.1 When the roof-ceiling framing elements are parallel to the walls, such framing and elements supporting such framing shall not be of less than one-hour fire-resistive construction for a width of not less than 5 feet (1524 mm) on each side of the wall.
>
> 2.2 When roof-ceiling framing elements are not parallel to the wall, the entire span of such framing and elements supporting such framing shall not be of less than one-hour fire-resistive construction.
>
> 2.3 Openings in the roof shall not be located within 5 feet (1524 mm) of the area separation wall.
>
> 2.4 The entire building shall be provided with not less than a Class B roof assembly as specified in Table 15-A.
>
> 3. Two-hour area separation walls may terminate at the underside of noncombustible roof sheathing, deck or slabs of roofs of noncombustible construction, provided:
>
> 3.1 Openings in the roof are not located within 5 feet (1524 mm) of the area separation wall.
>
> 3.2 The entire building is provided with not less than a Class B roofing assembly as specified in Table 15-A.

504.6.5 Parapet faces. Parapets of area separation walls shall have noncombustible faces for the uppermost 18 inches (457 mm), including counterflashing and coping materials.

504.6.6 Building of different heights. Where an area separation wall separates portions of a building having different heights, such wall may terminate at a point 30 inches (762 mm) above the lower roof level, provided the exterior wall for a height of 10 feet (3048 mm) above the lower roof is of one-hour fire-resistive construction with openings protected by assemblies having a three-fourths-hour fire-protection rating.

EXCEPTION: Two-hour area separation walls may terminate at the underside of the roof sheathing, deck or slab of the lower roof, provided:

1. When the roof-ceiling framing elements are parallel to the wall, such framing and elements supporting such framing shall not be of less than one-hour fire-resistive construction for a width of 10 feet (3048 mm) along the wall at the lower roof.

2. When the lower roof-ceiling framing elements are not parallel to the wall, the entire span of such framing and elements supporting such framing shall not be of less than one-hour fire-resistive construction.

3. Openings in the lower roof shall not be located within 10 feet (3048 mm) of the area separation wall.

See Chapters 3 and 4 for special occupancy provisions.

504.6.7 Combustible framing in area separation walls. Adjacent combustible members entering into a masonry area separation wall from opposite sides shall not have less than a 4-inch (102 mm) distance between embedded ends. Where combustible members frame into hollow walls or walls of hollow units, all hollow spaces shall be solidly filled for the full thickness of the wall and for a distance not less than 4 inches (102 mm) above, below and between the structural members, with noncombustible materials approved for fireblocking.

SECTION 505 — ALLOWABLE AREA INCREASES

505.1 General. The floor areas specified in Section 504 may be increased by employing one of the provisions of this section.

505.1.1 Separation on two sides. Where public ways or yards more than 20 feet (6096 mm) in width extend along and adjoin two sides of the building, floor areas may be increased at a rate of $1^1/_4$ percent for each foot (305 mm) by which the minimum width exceeds 20 feet (6096 mm), but the increase shall not exceed 50 percent.

505.1.2 Separation on three sides. Where public ways or yards more than 20 feet (6096 mm) in width extend along and adjoin three sides of the building, floor areas may be increased at a rate of $2^1/_2$ percent for each foot (305 mm) by which the minimum width exceeds 20 feet (6096 mm), but the increase shall not exceed 100 percent.

505.1.3 Separation on all sides. Where public ways or yards more than 20 feet (6096 mm) in width extend on all sides of a building and adjoin the entire perimeter, floor areas may be increased at a rate of 5 percent for each foot (305 mm) by which the minimum width exceeds 20 feet (6096 mm). Such increases shall not exceed 100 percent, except that greater increases shall be permitted for the following occupancies:

1. Group S, Division 5 aircraft storage hangars not exceeding one story in height.

2. Group S, Division 2 or Group F, Division 2 Occupancies not exceeding two stories in height.

3. Group H, Division 5 aircraft repair hangars not exceeding one story in height. Area increases shall not exceed 500 percent for aircraft repair hangars except as provided in Section 505.2.

505.2 Unlimited Area. The area of any one- or two-story building of Groups B; F, Division 1 or 2; M; S, Division 1, 2, 3, 4 or 5; and H, Division 5 Occupancies shall not be limited if the building is provided with an approved automatic sprinkler system throughout as specified in Chapter 9, and entirely surrounded and adjoined by public ways or yards not less than 60 feet (18 288 mm) in width.

The area of a Group S, Division 2 or Group F, Division 2 Occupancy in a one-story Type II, Type III One-hour or Type IV build-

ing shall not be limited if the building is entirely surrounded and adjoined by public ways or yards not less than 60 feet (18 288 mm) in width.

505.3 Automatic Sprinkler Systems. The areas specified in Table 5-B and Section 504.2 may be tripled in one-story buildings and doubled in buildings of more than one story if the building is provided with an approved automatic sprinkler system throughout. The area increases permitted in this section may be compounded with that specified in Section 505.1.1, 505.1.2 or 505.1.3. The increases permitted in this section shall not apply when automatic sprinkler systems are installed under the following provisions:

1. Section 506 for an increase in allowable number of stories.

2. Section 904.2.6.1 for Group H, Divisions 1 and 2 Occupancies.

3. Substitution for one-hour fire-resistive construction pursuant to Section 508.

4. Section 402, Atria.

SECTION 506 — MAXIMUM HEIGHT OF BUILDINGS AND INCREASES

The maximum height and number of stories of buildings shall be dependent on the character of the occupancy and the type of construction and shall not exceed the limits set forth in Table 5-B, except as provided in this section and as specified in Section 302.1 for mixed occupancy buildings.

EXCEPTIONS: 1. Towers, spires and steeples erected as a part of a building and not used for habitation or storage are limited as to height only by structural design if completely of noncombustible materials, or may extend not to exceed 20 feet (6096 mm) above the height limit in Table 5-B if of combustible materials.

2. The height of one-story aircraft hangars and buildings used for manufacture of aircraft shall not be limited if the building is provided with automatic sprinkler systems throughout as specified in Chapter 9 and is entirely surrounded by public ways or yards not less in width than one- and one-half times the height of the building.

The story limits set forth in Table 5-B may be increased by one story if the building is provided with an approved automatic sprinkler system throughout. The increase in the number of stories for automatic sprinkler systems shall not apply when the automatic sprinkler systems throughout are installed under the following provisions:

1. Section 904.2.6 for Group H, Divisions 1, 2, 3, 6 and 7 Occupancies.

2. Section 505 for an increase in allowable area.

3. Substitution for one-hour fire-resistive construction pursuant to Section 508.

4. Section 402, Atria.

5. Section 904.2.7 for Group I, Divisions 1.1 and 1.2 Occupancies used as hospitals, nursing homes or health-care centers in Type II One-hour, Type III One-hour, Type IV or Type V One-hour construction.

See Chapters 3 and 4 for special occupancy provisions.

SECTION 507 — MEZZANINES

A mezzanine need not be counted as a story for determining the allowable number of stories when constructed in accordance with the following:

1. The construction of a mezzanine shall be consistent with the requirements for the type of construction in which the mezzanine is located, but the fire-resistive time period need not exceed one

hour for unenclosed mezzanines. The clear height above and below the mezzanine floor construction shall not be less than 7 feet (2134 mm).

2. There shall not be more than two levels of mezzanines in a room. However, there is no limitation on the number of mezzanines within a room.

3. The aggregate area of mezzanines within a room shall not exceed one third of the area of the room in which they are located.

4. All portions of a mezzanine shall be open and unobstructed to the room in which they are located, except for columns and posts and protective walls or railings not more than 44 inches (1118 mm) in height.

EXCEPTIONS: 1. Partitioning may be installed if either of the following conditions exist:

1.1 The aggregate floor area of the enclosed space does not exceed 10 percent of the mezzanine area, or

1.2 The occupant load of the enclosed area of the mezzanine does not exceed 10.

2. A mezzanine having two or more means of egress need not be open into the room in which it is located, provided at least one of the means of egress gives direct access to a protected corridor, exit court or exit.

3. In industry facilities, mezzanines used for control equipment may be glazed on all sides.

5. Two means of egress shall be provided from a mezzanine when two are required by Table 10-A.

6. If any required means of egress enters the room below, the occupant load of the mezzanine shall be added to the occupant load of the room in which it is located.

SECTION 508 — FIRE-RESISTIVE SUBSTITUTION

When an approved automatic sprinkler system is not required throughout a building by other sections of this code, it may be used in a building of Type II One-hour, Type III One-hour and Type V One-hour construction to substitute for the one-hour fire-resistive construction. Such substitution shall not waive or reduce the required fire-resistive construction for:

1. Occupancy separations (Section 302.3).

2. Exterior wall protection due to proximity of property lines (Section 503.2).

3. Area separations (Section 504.6).

4. Dwelling unit separations (Section 310.2.2).

5. Shaft enclosures (Section 711).

6. Corridors (Sections 1004.3.4.3.1 and 1004.3.4.3.2).

7. Stair enclosures (Section 1005.3.3).

8. Exit passageways (Section 1005.3.4).

9. Type of construction separation (Section 601.1).

10. Boiler, central heating plant or hot-water supply boiler room enclosures (Section 302.5).

SECTION 509 — GUARDRAILS

509.1 Where Required. Unenclosed floor and roof openings, open and glazed sides of stairways, aisles, landings and ramps, balconies or porches, which are more than 30 inches (762 mm) above grade or floor below, and roofs used for other than service of the building shall be protected by a guardrail. Guardrails shall be provided at the ends of aisles where they terminate at a fascia of boxes, balconies and galleries.

EXCEPTION: Guardrails need not be provided at the following locations:

1. On the loading side of loading docks.

2. On the auditorium side of a stage, raised platforms and other raised floor areas such as runways, ramps and side stages used for entertainment or presentation. Along the side of an elevated walking surface when used for the normal functioning of special lighting or for access and use of other special equipment. At vertical openings in the performance area of stages.

3. Along vehicle service pits not accessible to the public.

509.2 Height. The top of guardrails shall not be less than 42 inches (1067 mm) in height.

EXCEPTIONS: 1. The top of guardrails for Group R, Division 3 and Group U, Division 1 Occupancies and interior guardrails within individual dwelling units, Group R, Division 3 congregate residences and guest rooms of Group R, Division 1 Occupancies may be 36 inches (914 mm) in height.

2. The top of guardrails on a balcony immediately in front of the first row of fixed seats and that are not at the end of an aisle may be 26 inches (660 mm) in height.

3. The top of guardrails for stairways, exclusive of their landings, may have a height as specified in Section 1003.3.3.6 for handrails.

Where an elevation change of 30 inches (762 mm) or less occurs between an aisle parallel to the seats (cross aisle) and the adjacent floor or grade below, guardrails not less than 26 inches (660 mm) above the aisle floor shall be provided.

EXCEPTION: Where the backs of seats on the front of the cross aisle project 24 inches (610 mm) or more above the adjacent floor of the aisle, a guardrail need not be provided.

The top of guardrails at the ends of aisles terminating at the fascia of boxes, balconies and galleries shall extend for the width of the aisle and be no closer than 42 inches (1067 mm) to the closest surface of the aisle where there are steps and 36 inches (914 mm) otherwise.

509.3 Openings. Open guardrails shall have intermediate rails or an ornamental pattern such that a sphere 4 inches (102 mm) in diameter cannot pass through.

EXCEPTIONS: 1. The open space between the intermediate rails or ornamental pattern of guardrails in areas of commercial and industrial-type occupancies which are not accessible to the public may be such that a sphere 12 inches (305 mm) in diameter cannot pass through.

2. The triangular openings formed by the riser, tread and bottom element of a guardrail at the open side of a stairway may be of such size that a sphere 6 inches (152 mm) in diameter cannot pass through.

For guardrail requirements at grandstands, bleachers or other elevated seating facilities, see Section 1008.5.7.

TABLE 5-A—EXTERIOR WALL AND OPENING PROTECTION BASED ON LOCATION ON PROPERTY FOR ALL CONSTRUCTION TYPES[1,2,3]

For exceptions, see Section 503.4.

OCCUPANCY GROUP[4]	CONSTRUCTION TYPE	EXTERIOR WALLS — Bearing	EXTERIOR WALLS — Nonbearing	OPENINGS[5]
		Distances are measured to property lines (see Section 503).		
		× 304.8 for mm		
A-1	I-F.R. II-F.R.	Four-hour N/C	Four-hour N/C less than 5 feet Two-hour N/C less than 20 feet One-hour N/C less than 40 feet NR, N/C elsewhere	Not permitted less than 5 feet Protected less than 20 feet
A-1	II One-hour II-N III One-hour III-N IV-H.T. V One-hour V-N	Group A, Division 1 Occupancies are not allowed in these construction types.		
A-2 A-2.1 A-3 A-4	I-F.R. II-F.R. III One-hour IV-H.T.	Four-hour N/C	Four-hour N/C less than 5 feet Two-hour N/C less than 20 feet One-hour N/C less than 40 feet NR, N/C elsewhere	Not permitted less than 5 feet Protected less than 20 feet
A-2 A-2.1[2]	II One-hour	Two-hour N/C less than 10 feet One-hour N/C elsewhere	Same as bearing except NR, N/C 40 feet or greater	Not permitted less than 5 feet Protected less than 10 feet
A-2 A-2.1[2]	II-N III-N V-N	Group A, Divisions 2 and 2.1 Occupancies are not allowed in these construction types.		
A-2 A-2.1[2]	V One-hour	Two-hour less than 10 feet One-hour elsewhere	Same as bearing	Not permitted less than 5 feet Protected less than 10 feet
A-3	II One-hour	Two-hour N/C less than 5 feet One-hour N/C elsewhere	Same as bearing except NR, N/C 40 feet or greater	Not permitted less than 5 feet Protected less than 10 feet
A-3	II-N	Two-hour N/C less than 5 feet One-hour N/C less than 20 feet NR, N/C elsewhere	Same as bearing	Not permitted less than 5 feet Protected less than 10 feet
A-3	III-N	Four-hour N/C	Four-hour N/C less than 5 feet Two-hour N/C less than 20 feet One-hour N/C less than 40 feet NR, N/C elsewhere	Not permitted less than 5 feet Protected less than 20 feet
A-3	V One-hour	Two-hour less than 5 feet One-hour elsewhere	Same as bearing	Not permitted less than 5 feet Protected less than 10 feet
A-3	V-N	Two-hour less than 5 feet One-hour less than 20 feet NR elsewhere	Same as bearing	Not permitted less than 5 feet Protected less than 10 feet
A-4	II One-hour	One-hour N/C	Same as bearing except NR, N/C 40 feet or greater	Protected less than 10 feet
A-4	II-N	One-hour N/C less than 10 feet NR, N/C elsewhere	Same as bearing	Protected less than 10 feet
A-4	III-N	Four-hour N/C	Four-hour N/C less than 5 feet Two-hour N/C less than 20 feet One-hour N/C less than 40 feet NR, N/C elsewhere	Not permitted less than 5 feet Protected less than 10 feet
A-4	V One-hour	One-hour	Same as bearing	Protected less than 10 feet
A-4	V-N	One-hour less than 10 feet NR elsewhere	Same as bearing	Protected less than 10 feet
B, F-1, M, S-1, S-3	I-F.R. II-F.R. III One-hour III-N IV-H.T.	Four-hour N/C less than 5 feet Two-hour N/C elsewhere	Four-hour N/C less than 5 feet Two-hour N/C less than 20 feet One-hour N/C less than 40 feet NR, N/C elsewhere	Not permitted less than 5 feet Protected less than 20 feet
B F-1 M S-1, S-3	II One-hour	One-hour N/C	Same as bearing except NR, N/C 40 feet or greater	Not permitted less than 5 feet Protected less than 10 feet
B F-1 M S-1, S-3	II-N[3]	One-hour N/C less than 20 feet NR, N/C elsewhere	Same as bearing	Not permitted less than 5 feet Protected less than 10 feet
B F-1 M S-1, S-3	V One-hour	One-hour	Same as bearing	Not permitted less than 5 feet Protected less than 10 feet
B F-1 M S-1, S-3	V-N	One-hour less than 20 feet NR elsewhere	Same as bearing	Not permitted less than 5 feet Protected less than 10 feet

(Continued)

TABLE 5-A

1997 UNIFORM BUILDING CODE

TABLE 5-A—EXTERIOR WALL AND OPENING PROTECTION BASED ON LOCATION ON PROPERTY FOR ALL CONSTRUCTION TYPES[1,2,3]—(Continued)

OCCUPANCY GROUP[4]	CONSTRUCTION TYPE	EXTERIOR WALLS		OPENINGS[5]
		Bearing	Nonbearing	
		Distances are measured to property lines (see Section 503).		
		× 304.8 for mm		
E-1 E-2[6] E-3[6]	I-F.R. II-F.R. III One-hour III-N IV-H.T.	Four-hour N/C	Four-hour N/C less than 5 feet Two-hour N/C less than 20 feet One-hour N/C less than 40 feet NR, N/C elsewhere	Not permitted less than 5 feet Protected less than 20 feet
	II One-hour	Two-hour N/C less than 5 feet One-hour N/C elsewhere	Same as bearing except NR, N/C 40 feet or greater	Not permitted less than 5 feet Protected less than 10 feet
	II-N	Two-hour N/C less than 5 feet One-hour N/C less than 10 feet NR, N/C elsewhere	Same as bearing	Not permitted less than 5 feet Protected less than 10 feet
	V One-hour	Two-hour less than 5 feet One-hour elsewhere	Same as bearing	Not permitted less than 5 feet Protected less than 10 feet
	V-N	Two-hour less than 5 feet One-hour less than 10 feet NR elsewhere	Same as bearing	Not permitted less than 5 feet Protected less than 10 feet
F-2 S-2	I-F.R. II-F.R. III One-hour III-N IV-H.T.	Four-hour N/C less than 5 feet Two-hour N/C elsewhere	Four-hour N/C less than 5 feet Two-hour N/C less than 20 feet One-hour N/C less than 40 feet NR, N/C elsewhere	Not permitted less than 3 feet Protected less than 20 feet
	II One-hour	One-hour N/C	Same as bearing NR, N/C 40 feet or greater	Not permitted less than 5 feet Protected less than 10 feet
	II-N[3]	One-hour N/C less than 5 feet NR, N/C elsewhere	Same as bearing	Not permitted less than 5 feet Protected less than 10 feet
	V One-hour	One-hour	Same as bearing	Not permitted less than 5 feet Protected less than 10 feet
	V-N	One-hour less than 5 feet NR elsewhere	Same as bearing	Not permitted less than 5 feet Protected less than 10 feet
H-1[2,3]	I-F.R. II-F.R.	Four-hour N/C	NR N/C	Not restricted[3]
	II One-hour	One-hour N/C	NR N/C	Not restricted[3]
	II-N	NR N/C	Same as bearing	Not restricted[3]
	III One-hour III-N IV-H.T. V One-hour V-N	Group H, Division 1 Occupancies are not allowed in buildings of these construction types.		
H-2[2,3] H-3[2,3] H-4[3] H-6 H-7	I-F.R. II-F.R. III One-hour III-N IV-H.T.	Four-hour N/C	Four-hour N/C less than 5 feet Two-hour N/C less than 10 feet One-hour N/C less than 40 feet NR, N/C elsewhere	Not permitted less than 5 feet Protected less than 20 feet
	II One-hour	Four-hour N/C less than 5 feet Two-hour N/C less than 10 feet One-hour N/C elsewhere	Four-hour N/C less than 5 feet Two-hour N/C less than 10 feet One-hour N/C less than 20 feet NR, N/C elsewhere	Not permitted less than 5 feet Protected less than 20 feet
	II-N	Four-hour N/C less than 5 feet Two-hour N/C less than 10 feet One-hour N/C less than 20 feet NR, N/C elsewhere	Same as bearing	Not permitted less than 5 feet Protected less than 20 feet
	V One-hour	Four-hour less than 5 feet Two-hour less than 10 feet One-hour elsewhere	Same as bearing	Not permitted less than 5 feet Protected less than 20 feet
	V-N	Four-hour less than 5 feet Two-hour less than 10 feet One-hour less than 20 feet NR elsewhere	Same as bearing	Not permitted less than 5 feet Protected less than 20 feet

(Continued)

TABLE 5-A—EXTERIOR WALL AND OPENING PROTECTION BASED ON LOCATION ON PROPERTY FOR ALL CONSTRUCTION TYPES[1,2,3]—(Continued)

OCCUPANCY GROUP[4]	CONSTRUCTION TYPE	EXTERIOR WALLS		OPENINGS[5]
		Bearing	Nonbearing	
		Distances are measured to property lines (see Section 503).		
		× 304.8 for mm		
H-5[2]	I-F.R. II-F.R. III One-hour III-N IV-H.T.	Four-hour N/C	Four-hour N/C less than 40 feet One-hour N/C less than 60 feet NR, N/C elsewhere	Protected less than 60 feet
	II One-hour	One-hour N/C	Same as bearing, except NR, N/C 60 feet or greater	Protected less than 60 feet
	II-N	One-hour N/C less than 60 feet NR, N/C elsewhere	Same as bearing	Protected less than 60 feet
	V One-hour	One-hour	Same as bearing	Protected less than 60 feet
	V-N	One-hour less than 60 feet NR elsewhere	Same as bearing	Protected less than 60 feet
I-1.1 I-1.2 I-2 I-3	I-F.R. II-F.R.	Four-hour N/C	Four-hour N/C less than 5 feet Two-hour N/C less than 20 feet One-hour N/C less than 40 feet NR, N/C elsewhere	Not permitted less than 5 feet Protected less than 20 feet
I-1.1 I-1.2 I-3[2]	II One-hour	Two-hour N/C less than 5 feet One-hour N/C elsewhere	Same as bearing except NR, N/C 40 feet or greater	Not permitted less than 5 feet Protected less than 10 feet
	V One-hour	Two-hour less than 5 feet One-hour elsewhere	Same as bearing	Not permitted less than 5 feet Protected less than 10 feet
I-1.1 I-1.2 I-2 I-3	II-N III-N V-N	These occupancies are not allowed in buildings of these construction types.[7]		
I-3	IV-H.T.	Group I, Division 3 Occupancies are not allowed in buildings of this construction type.		
I-1.1 I-1.2 I-2 I-3	III One-hour	Four-hour N/C	Same as bearing except NR, N/C 40 feet or greater	Not permitted less than 5 feet Protected less than 20 feet
I-1.1 I-1.2 I-2	IV-H.T.	Four-hour N/C	Same as bearing except NR, N/C 40 feet or greater	Not permitted less than 5 feet Protected less than 20 feet
I-2	II One-hour	One-hour N/C	Same as bearing except NR, N/C 40 feet or greater	Not permitted less than 5 feet Protected less than 10 feet
	V One-hour	One-hour	Same as bearing	Not permitted less than 5 feet Protected less than 10 feet
R-1	I-F.R. II-F.R. III One-hour III-N IV-H.T.	Four-hour N/C less than 3 feet Two-hour N/C elsewhere	Four-hour N/C less than 3 feet Two-hour N/C less than 20 feet One-hour N/C less than 40 feet NR, N/C elsewhere	Not permitted less than 3 feet Protected less than 20 feet
	II One-hour	One-hour N/C	Same as bearing except NR, N/C 40 feet or greater	Not permitted less than 5 feet
	II-N	One-hour N/C less than 5 feet NR, N/C elsewhere	Same as bearing	Not permitted less than 5 feet
	V One-hour	One-hour	Same as bearing	Not permitted less than 5 feet
	V-N	One-hour less than 5 feet NR elsewhere	Same as bearing	Not permitted less than 5 feet
R-3	I-F.R. II-F.R. III One-hour III-N IV-H.T.	Four-hour N/C	Four-hour N/C less than 3 feet Two-hour N/C less than 20 feet One-hour N/C less than 40 feet NR, N/C elsewhere	Not permitted less than 3 feet Protected less than 20 feet
	II One-hour	One-hour N/C	Same as bearing except NR, N/C 40 feet or greater	Not permitted less than 3 feet
	II-N	One-hour N/C less than 3 feet NR, N/C elsewhere	Same as bearing	Not permitted less than 3 feet
	V One-hour	One-hour	Same as bearing	Not permitted less than 3 feet
	V-N	One-hour less than 3 feet NR elsewhere	Same as bearing	Not permitted less than 3 feet

(Continued)

TABLE 5-A   1997 UNIFORM BUILDING CODE

TABLE 5-A—EXTERIOR WALL AND OPENING PROTECTION BASED ON LOCATION ON PROPERTY FOR ALL CONSTRUCTION TYPES[1,2,3]—(Continued)

OCCUPANCY GROUP[4]	CONSTRUCTION TYPE	EXTERIOR WALLS		OPENINGS[5]
		Bearing	Nonbearing	
		Distances are measured to property lines (see Section 503).		
		× 304.8 for mm		
S-4	I-F.R. II-F.R. II One-hour II-N[3]	One-hour N/C less than 10 feet NR, N/C elsewhere	Same as bearing	Not permitted less than 5 feet Protected less than 10 feet
	III One-hour III-N IV-H.T. V One-hour V-N	Group S, Division 4 open parking garages are not permitted in these types of construction.		
S-5	I-F.R. II-F.R. III One-hour III-N IV-H.T.	Four-hour N/C less than 5 feet Two-hour N/C elsewhere	Four-hour N/C less than 5 feet Two-hour N/C less than 20 feet One-hour N/C less than 40 feet NR, N/C elsewhere	Not permitted less than 5 feet Protected less than 20 feet
	II One-hour	One-hour N/C	Same as bearing except NR, N/C 40 feet or greater	Not permitted less than 5 feet Protected less than 20 feet
	II-N[3]	One-hour N/C less than 20 feet NR, N/C elsewhere	Same as bearing	Not permitted less than 5 feet Protected less than 20 feet
	V One-hour	One-hour	Same as bearing	Not permitted less than 5 feet Protected less than 20 feet
	V-N[3]	One-hour less than 20 feet NR elsewhere	Same as bearing	Not permitted less than 5 feet Protected less than 20 feet
U-1[3]	I-F.R. II-F.R. III One-hour III-N IV-H.T.	Four-hour N/C	Four-hour N/C less than 3 feet Two-hour N/C less than 20 feet One-hour N/C less than 40 feet NR, N/C elsewhere	Not permitted less than 3 feet Protected less than 20 feet
	II One-hour	One-hour N/C	Same as bearing except NR, N/C 40 feet or greater	Not permitted less than 3 feet
	V One-hour	One-hour	Same as bearing	Not permitted less than 3 feet
	II-N[2]	One-hour N/C less than 3 feet[3] NR, N/C elsewhere	Same as bearing	Not permitted less than 3 feet
	V-N	One-hour less than 3 feet[3] NR elsewhere	Same as bearing	Not permitted less than 3 feet
U-2	All	Not regulated		

N/C— Noncombustible.
NR — Nonrated.
H.T.— Heavy timber.
F.R. — Fire resistive.

[1]See Section 503 for types of walls affected and requirements covering percentage of openings permitted in exterior walls. For walls facing streets, yards and public ways, see also Section 601.5.

[2]For additional restrictions, see Chapters 3 and 6.

[3]For special provisions and exceptions, see also Section 503.4.

[4]See Table 3-A for a description of each occupancy type.

[5]Openings requiring protection in exterior walls shall be protected by a fire assembly having at least a three-fourths-hour fire-protection rating.

[6]Group E, Divisions 2 and 3 Occupancies having an occupant load of not more than 20 may have exterior wall and opening protection as required for Group R, Division 3 Occupancies.

[7]See Section 308.2.1, Exception 3.

TABLE 5-B—BASIC ALLOWABLE BUILDING HEIGHTS AND BASIC ALLOWABLE FLOOR AREA FOR BUILDINGS ONE STORY IN HEIGHT[1]

Use Group	Height/Area	I F.R.	II F.R.	II One-hour	II N	III One-hour	III N	IV H.T.	V One-hour	V N
			Maximum Height (feet)							
		UL	160 (48 768 mm)	65 (19 812 mm)	55 (16 764 mm)	65 (19 812 mm)	55 (16 764 mm)	65 (19 812 mm)	50 (15 240 mm)	40 (12 192 mm)
		Maximum Height (stories) and Maximum Area (sq. ft.) ($\times$ 0.0929 for m²)								
A-1	H	UL	4	Not Permitted						
	A	UL	29,900							
A-2, 2.1[2]	H	UL	4	2	NP	2	NP	2	2	NP
	A	UL	29,900	13,500	NP	13,500	NP	13,500	10,500	NP
A-3, 4[2]	H	UL	12	2	1	2	1	2	2	1
	A	UL	29,900	13,500	9,100	13,500	9,100	13,500	10,500	6,000
B, F-1, M, S-1, S-3, S-5	H	UL	12	4	2	4	2	4	3	2
	A	UL	39,900	18,000	12,000	18,000	12,000	18,000	14,000	8,000
E-1, 2, 3[4]	H	UL	4	2	1	2	1	2	2	1
	A	UL	45,200	20,200	13,500	20,200	13,500	20,200	15,700	9,100
F-2, S-2	H	UL	12	4	2	4	2	4	3	2
	A	UL	59,900	27,000	18,000	27,000	18,000	27,000	21,000	12,000
H-1[5]	H	1	1	1	1	Not Permitted				
	A	15,000	12,400	5,600	3,700					
H-2[5]	H	UL	2	1	1	1	1	1	1	1
	A	15,000	12,400	5,600	3,700	5,600	3,700	5,600	4,400	2,500
H-3, 4, 5[5]	H	UL	5	2	1	2	1	2	2	1
	A	UL	24,800	11,200	7,500	11,200	7,500	11,200	8,800	5,100
H-6, 7	H	3	3	3	2	3	2	3	3	1
	A	UL	39,900	18,000	12,000	18,000	12,000	18,000	14,000	8,000
I-1.1, 1.2[6,10]	H	UL	3	1	NP	1	NP	1	1	NP
	A	UL	15,100	6,800	NP	6,800	NP	6,800	5,200	NP
I-2	H	UL	3	2	NP	2	NP	2	2	NP
	A	UL	15,100	6,800	NP	6,800	NP	6,800	5,200	NP
I-3	H	UL	2	Not Permitted[7]						
	A	UL	15,100							
R-1	H	UL	12	4	2[9]	4	2[9]	4	3	2[9]
	A	UL	29,900	13,500	9,100[9]	13,500	9,100[9]	13,500	10,500	6,000[9]
R-3	H	UL	3	3	3	3	3	3	3	3
	A	Unlimited								
S-4[3]	H	See Table 3-H								
	A									
U[8]	H	See Chapter 3								
	A									

A—Building area in square feet.
H—Building height in number of stories.
H.T.—Heavy timber.
NP—Not permitted.

N—No requirements for fire resistance.
F.R.—Fire resistive.
UL—Unlimited.

[1]For multistory buildings, see Section 504.2.

[2]For limitations and exceptions, see Section 303.2.

[3]For open parking garages, see Section 311.9.

[4]See Section 305.2.3.

[5]See Section 307.

[6]See Section 308.2.1 for exception to the allowable area and number of stories in hospitals, nursing homes and health-care centers.

[7]See Section 308.2.2.2.

[8]For agricultural buildings, see also Appendix Chapter 3.

[9]For limitations and exceptions, see Section 310.2.

[10]For Type II F.R., the maximum height of Group I, Division 1.1 Occupancies is limited to 75 feet (22 860 mm). For Type II, One-hour construction, the maximum height of Group I, Division 1.1 Occupancies is limited to 45 feet (13 716 mm).

Chapter 6
TYPES OF CONSTRUCTION

SECTION 601 — CLASSIFICATION OF ALL BUILDINGS BY TYPES OF CONSTRUCTION AND GENERAL REQUIREMENTS

601.1 General. The requirements of this chapter are for the various types of construction and represent varying degrees of public safety and resistance to fire. Every building shall be classified by the building official into one of the types of construction set forth in Table 6-A. Any building that does not entirely conform to a type of construction set forth in Table 6-A shall be classified by the building official into a type having an equal or lesser degree of fire resistance.

A building or portion thereof shall not be required to conform to the details of a type of construction higher than that type that meets the minimum requirements based on occupancy even though certain features of such building actually conform to a higher type of construction.

When specific materials, types of construction or fire-resistive protection are required, such requirements shall be the minimum requirements, and any materials, types of construction or fire-resistive protection that will afford equal or greater public safety or resistance to fire, as specified in this code, may be used.

For additional limitations or allowances for special uses or occupancies, see the following:

SECTION	SUBJECT
402	Atria
403	High-rise office buildings and Group R, Division 1 Occupancies
404	Malls
405	Open parking structures
307.11	Group H, Division 6 Occupancies
411	Aviation control structures
413	Agricultural buildings
3111	Membrane structures

601.2 Mixed Types of Construction. When a building contains more than one distinct type of construction, the area of the entire building shall not exceed the least area permitted for the types of construction involved.

> **EXCEPTION:** Each portion of a building separated by one or more area separation walls as specified in Section 504.6 may be considered a separate building for the purpose of classification of types of construction. The fire-resistive time period for such type of construction separation shall not be less than the most restrictive requirement in Section 504.6.2 based on the types of construction involved.

601.3 Standards of Quality. The standards listed below labeled a "UBC Standard" are also listed in Chapter 35, Part II, and are part of this code. The other standards listed below are recognized standards. (See Sections 3503 and 3504.)

1. **Building paper.**

    1.1 UBC Standard 14-1, Kraft Waterproof Building Paper

    1.2 Underwriters Laboratories Inc. Standard Specification 55A, Materials for Use in Construction of Built-up Roof Coverings

2. **Potential heat of building materials.**

    UBC Standard 26-1, Test Method to Determine Potential Heat of Building Materials

3. **Foam plastic tests.**

    3.1 UBC Standard 26-2, Test Method for the Evaluation of Thermal Barriers

    3.2 Factory Mutual Standard Fire Test Standard for Insulated Roof Deck Construction

    3.3 Underwriters Laboratories Inc. 1256, Fire Test Standard for Insulated Roof Deck Construction

    3.4 UBC Standard 26-3, Room Fire Test Standard for Interior of Foam Plastic Systems

    3.5 UBC Standard 26-4, Method of Test for the Evaluation of Flammability Characteristics of Exterior, Nonload-bearing Wall Panel Assemblies Using Foam Plastic Insulation

    3.6 UBC Standard 26-8, Room Fire Test Standard for Garage Doors Using Foam Plastic Insulation

    3.7 UBC Standard 26-9, Method of Test for the Evaluation of Flammability Characteristics of Exterior Nonload-Bearing Wall Assemblies Containing Combustible Components Using the Intermediate-Scale, Multistory Test Apparatus

4. **Roofing.**

    4.1 Underwriters Laboratories Inc. Standard Specification 55A, Materials for Use in Construction of Built-up Roof Coverings

    4.2 UBC Standard 15-2, Test Method for Determining the Fire Retardancy of Roofing Assemblies

5. **Surface-burning characteristics and fire resistance of building materials and assemblies.**

    5.1 UBC Standard 7-1, Fire Test of Building Construction and Materials

    5.2 UBC Standard 8-1, Test Method for Surface-burning Characteristics of Building Materials

6. **Self-ignition properties of plastics.**

    UBC Standard 26-6, Ignition Properties of Plastics

7. **Fire dampers.**

    UL 555, Fire Dampers

601.4 Structural Frame. The structural frame shall be considered to be the columns and the girders, beams, trusses, and spandrels having direct connections to the columns and bracing members designed to carry gravity loads. The members of floor or roof panels that have no connection to the columns shall be considered secondary members and not a part of the structural frame.

601.5 Exceptions to Table 6-A.

601.5.1 General. The provisions of this section are exceptions to the construction requirements of Table 6-A, Chapter 3 and Sections 602 through 606.

601.5.2 Fixed partitions.

601.5.2.1 Stores and offices. Interior nonload-bearing partitions dividing portions of stores, offices or similar places occupied by one tenant only and that do not establish a corridor that is required to be of fire-resistive construction under the provisions of Section 1004.3.4.3.1 may be constructed of:

1. Noncombustible materials.

2. Fire-retardant-treated wood.

3. One-hour fire-resistive construction.

4. Wood panels or similar light construction up to three fourths the height of the room in which placed; when more than three fourths the height of the room, such partitions shall not have less than the upper one fourth of the partition constructed of glass.

601.5.2.2 Hotels and apartments. Interior nonload-bearing partitions within individual dwelling units in apartment houses and guest rooms or suites in hotels when such dwelling units, guest rooms or suites are separated from each other and from corridors by not less than one-hour fire-resistive construction may be constructed of:

1. Noncombustible materials or fire-retardant-treated wood in buildings of any type of construction; or

2. Combustible framing with noncombustible materials applied to the framing in buildings of Type III or V construction.

Openings to such corridors shall be equipped with doors conforming to Section 1004.3.4.3.2 regardless of the occupant load served.

For use of plastics in partitions, see Section 2603.10.

601.5.3 Folding, portable or movable partitions. Approved folding, portable or movable partitions need not have a fire-resistive rating, provided:

1. They do not block required exits or exit-access doors (without providing alternative conforming exits or exit-access doors) and they do not establish a corridor.

2. Their location is restricted by means of permanent tracks, guides or other approved methods.

3. Flammability shall be limited to materials having a flame-spread classification as set forth in Table 8-B for rooms or areas.

601.5.4 Walls fronting on streets or yards. Regardless of fire-resistive requirements for exterior walls, certain elements of the walls fronting on streets or yards having a width of 40 feet (12 192 mm) may be constructed as follows:

1. Bulkheads below show windows, show-window frames, aprons and showcases may be of combustible materials, provided the height of such construction does not exceed 15 feet (4572 mm) above grade.

2. Wood veneer of boards not less than 1-inch (25 mm) nominal thickness or exterior-type panels not less than $^3/_8$-inch (9.5 mm) nominal thickness may be applied to walls, provided the veneer does not exceed 15 feet (4572 mm) above grade, and further provided such veneer shall be placed either directly against noncombustible surfaces or furred out from such surfaces not to exceed $1^5/_8$ inches (41 mm) with all concealed spaces fire-blocked as provided in Section 708. Where boards, panels and furring as described above comply with Section 207 as fire-retardant-treated wood suitable for exterior exposure, the height above grade may be increased to 35 feet (10 668 mm).

601.5.5 Trim. Trim, picture molds, chair rails, baseboards, handrails and show-window backing may be of wood. Unprotected wood doors and windows may be used except where openings are required to be fire protected.

Foam plastic trim covering not more than 10 percent of the wall or ceiling area may be used, provided such trim (1) has a density of no less than 20 pounds per cubic foot (320.4 kg/m^3), (2) has a maximum thickness of $^1/_2$ inch (12.7 mm) and a maximum width of 4 inches (102 mm), and (3) has a flame-spread rating no greater than 75.

Materials used for interior finish of walls and ceilings, including wainscoting, shall be as specified in Chapter 8.

601.5.6 Loading platforms. Exterior loading platforms may be of noncombustible construction or heavy-timber construction with wood floors not less than 2-inch (51 mm) nominal thickness. Such wood construction shall not be carried through the exterior walls.

601.5.7 Insulating boards. Combustible insulating boards may be used under finished flooring.

601.5.8 Walls within health-care suites. In health-care suites that comply with Section 1007.5, interior nonload-bearing partitions of noncombustible construction need not be of fire-resistive construction. In buildings of combustible construction, interior nonload-bearing partitions within suites may be of combustible framing covered with noncombustible materials having an approved thermal barrier with an index of 15 in accordance with UBC Standard 26-2.

SECTION 602 — TYPE I FIRE-RESISTIVE BUILDINGS

602.1 Definition. The structural elements in Type I fire-resistive buildings shall be of steel, iron, concrete or masonry.

Walls and permanent partitions shall be of noncombustible fire-resistive construction except that permanent nonbearing partitions of one-hour or two-hour fire-resistive construction, which are not part of a shaft enclosure, may have fire-retardant-treated wood (see Section 207) within the assembly.

Materials of construction and fire-resistive requirements shall be as specified in Section 601 and Chapter 7.

602.2 Structural Framework. Structural framework shall be of structural steel or iron as specified in Chapter 22, reinforced concrete as specified in Chapter 19, or reinforced masonry as specified in Chapter 21.

For additional requirements for Group H Occupancies, see Section 307.2.

602.3 Exterior Walls and Openings.

602.3.1 Exterior walls. Exterior walls and all structural members shall comply with the requirements specified in Section 503 and Table 5-A and the fire-resistive provisions set forth in Table 6-A.

602.3.2 Openings in walls. All openings in exterior walls shall conform to the requirements of Section 503.2 and Table 5-A.

602.4 Stairway Construction. Stairways shall be constructed of reinforced concrete, iron or steel with treads and risers of concrete, iron or steel. Brick, marble, tile or other hard noncombustible materials may be used for the finish of such treads and risers.

> **EXCEPTION:** On stairs not required to be enclosed by Section 1005.3.3, the finish material of treads and risers may be of any material permitted by the code.

Stairways shall comply with the requirements of Chapter 10.

602.5 Roofs. Except in retail sales and storage areas classified as Groups M and S, Division 1 Occupancies and in Group H Occupancies, roofs and their members, other than the structural frame, may be of unprotected noncombustible materials when every part of the roof framing, including the structural frame, is 25 feet (7620 mm) or more above the floor, balcony or gallery immediately below. Heavy-timber members in accordance with Section 605.6 may be used for such unprotected members in one-story buildings.

When every part of the structural framework of the roof of a Group A or E Occupancy or of an atrium is not less than 25 feet (7620 mm) above any floor, balcony or gallery, fire protection of all members of the roof construction, including those of the structural frame, may be omitted. Heavy-timber members in accordance with Section 605.6 may be used for such unprotected members in one-story buildings.

Roofs of unprotected noncombustible or heavy-timber construction conforming to Section 605.6.4 may be less than 25 feet (7620 mm) above any floor, balcony or gallery of a Group A, Division 2.1 Occupancy having an occupant load of 10,000 or more when all of the following conditions are met:

1. The building is not more than one story in height, except for multilevel areas located under the roof and used for locker rooms, exiting, concession stands, mechanical rooms and others accessory to the assembly room.

2. The area in which the roof clearance is less than 25 feet (7620 mm) does not exceed 35 percent of the area encompassed by the exterior walls.

3. An approved supervised automatic sprinkler system shall be installed throughout.

Where every part of the structural steel framework of the roof of a Group A or E Occupancy is more than 18 feet (5486 mm) and less than 25 feet (7620 mm) above any floor, balcony or gallery, the roof construction shall be protected by a ceiling of not less than one-hour fire-resistive construction.

Roof coverings shall be as specified in Chapter 15.

SECTION 603 — TYPE II BUILDINGS

603.1 Definition. The structural elements in Type II-F.R. buildings shall be of steel, iron, concrete or masonry.

The structural elements of Type II One-hour or Type II-N buildings shall be of noncombustible materials.

Floor construction of Type II One-hour and Type II-N buildings shall be of noncombustible material, provided, however, that a wood surface or finish may be applied over such noncombustible material.

Walls and permanent partitions of Type II-F.R. buildings shall be of noncombustible fire-resistive construction, except that permanent nonbearing partitions of one-hour or two-hour fire-resistive construction, which are not part of a shaft enclosure, may have fire-retardant-treated wood (see Section 207) within the assembly.

Type II One-hour buildings shall be of noncombustible construction and one-hour fire resistive throughout, except that permanent nonbearing partitions may use fire-retardant-treated wood (see Section 207) within the assembly, provided fire-resistive requirements are maintained.

Walls and permanent partitions of Type II-N buildings shall be of noncombustible materials.

Materials of construction and fire-resistive requirements shall be as specified in Section 601.

For requirements due to occupancy, see Chapter 3.

603.2 Structural Framework. Structural framework shall be as specified in Chapter 22 for iron and steel, Chapter 19 for concrete and Chapter 21 for masonry.

603.3 Exterior Walls and Openings.

603.3.1 Exterior walls. Exterior walls and all structural members shall comply with the requirements specified in Section 503 and Table 5-A and the fire-resistive provisions set forth in Table 6-A.

603.3.2 Openings in walls. All openings in exterior walls shall conform to the requirements of Section 503.2 and Table 5-A.

603.4 Stairway Construction. Stairways of Type II-F.R. buildings shall be constructed of reinforced concrete, iron or steel with treads and risers of concrete, iron or steel. Brick, marble, tile or other hard noncombustible materials may be used for the finish of such treads and risers. Stairways of Type II, One-hour and Type II-N buildings shall be of noncombustible construction.

> **EXCEPTION:** On stairs not required to be enclosed by Section 1005.3.3, the finish material of treads and risers may be of any material permitted by the code.

Stairways shall comply with the requirements of Chapter 10.

603.5 Roofs. Roofs shall be of noncombustible construction, except that in Type II-F.R. and Type II One-hour buildings, roofs may be as specified in Section 602.5.

Roof coverings shall be as specified in Chapter 15.

SECTION 604 — TYPE III BUILDINGS

604.1 Definition. Structural elements in Type III buildings may be of any materials permitted by this code.

Type III One-hour buildings shall be of one-hour fire-resistive construction throughout.

604.2 Structural Framework. Structural framework shall be of steel or iron as specified in Chapter 22, concrete as specified in Chapter 19, masonry as specified in Chapter 21, or wood as specified in Chapter 23 and this chapter.

604.3 Exterior Walls, Openings and Partitions.

604.3.1 Exterior walls. Exterior walls shall be constructed of noncombustible materials and shall comply with the fire-resistive requirements set forth in Section 503 and Tables 5-A and 6-A.

604.3.2 Openings in walls. Openings in exterior walls shall conform to the requirements of Section 503.2 and Table 5-A.

604.3.3 Partitions. Bearing partitions, when constructed of wood, shall comply with Section 2308.

604.4 Stairway Construction.

604.4.1 General. Stairways shall comply with the requirements of Chapter 10.

604.4.2 Interior. Interior stairways serving buildings not exceeding three stories in height may be constructed of any material permitted by this code.

In buildings more than three stories in height, interior stairways shall be constructed as required for Type I buildings.

604.4.3 Exterior. Exterior stairways shall be of noncombustible material except that on buildings not exceeding two stories in height, they may be of wood not less than 2 inches (51 mm) in nominal thickness.

604.5 Roofs. Roof coverings shall be as specified in Chapter 15.

Except in retail sales and storage areas classified as Group M or S, Division 1 Occupancies and in Group H Occupancies, roofs and their members other than the structural frame may be of unprotected noncombustible materials when every part of the roof framing, including the structural frame, is 25 feet (7620 mm) or more above the floor, balcony or gallery immediately below. Heavy-timber members in accordance with Section 605.6 may be used for such unprotected members in one-story buildings.

SECTION 605 — TYPE IV BUILDINGS

605.1 Definition. Structural elements of Type IV buildings may be of any materials permitted by this code.

Type IV construction shall conform to Section 605.6, except that permanent partitions and members of the structural frame may be of other materials, provided they have a fire resistance of not less than one hour.

605.2 Structural Framework. Structural framework shall be of steel or iron as specified in Chapter 22, concrete as specified in Chapter 19, masonry as specified in Chapter 21, or wood as specified in Chapter 23 and this chapter.

605.3 Exterior Walls, Openings and Partitions.

605.3.1 Exterior walls. Exterior walls shall be constructed of noncombustible materials and shall comply with the fire-resistive requirements set forth in Section 503 and Tables 5-A and 6-A.

605.3.2 Openings in walls. Openings in exterior walls shall conform to the requirements of Section 503.2 and Table 5-A.

605.3.3 Partitions. Bearing partitions, when constructed of wood, shall comply with Section 2308.

605.4 Stairway Construction.

605.4.1 General. Stairways shall comply with the requirements of Chapter 10.

605.4.2 Interior. Interior stairways serving buildings not exceeding three stories in height may be constructed of wood or as required for Type I buildings. If constructed of wood, treads and risers shall not be less than 2 inches (51 mm) in thickness, except where built on laminated or plank inclines as required for floors, where they may be of 1-inch (25 mm) thickness. Wood stair stringers shall be a minimum of 3 inches (76 mm) in thickness and not less than 10 inches (254 mm) in depth.

In buildings more than three stories in height, interior stairways shall be constructed as required for Type I buildings.

605.4.3 Exterior. Exterior stairways shall be of noncombustible material except that on buildings not exceeding two stories in height they may be of wood not less than 2 inches (51 mm) in nominal thickness.

605.5 Roofs. Roof coverings shall be as specified in Chapter 15.

605.6 Heavy-timber Construction.

605.6.1 General. Details of heavy-timber construction shall be in accordance with the provisions of this section. Unless otherwise specified, all dimensions are nominal as defined in Section 2302.

605.6.2 Columns. Wood columns may be of sawn timber or structural glued-laminated timber not less than 8 inches (203 mm) in any dimension when supporting roof or floor loads except as specified in Section 605.6.4.

Columns shall be continuous or superimposed and connected in an approved manner.

605.6.3 Floor framing. Beams and girders may be of sawn timber or structural glued-laminated timber and shall not be less than 6 inches (152 mm) in width and not less than 10 inches (254 mm) in depth.

Framed sawn timber or structural glued-laminated timber arches, which spring from the floor line and support floor loads, shall not be less than 8 inches (203 mm) in any dimension.

Framed lumber or structural glued-laminated timber trusses supporting floor loads shall have members of not less than 8 inches (203 mm) in any dimension.

605.6.4 Roof framing. Framed sawn timber arches or structural glued-laminated timber arches for roof construction, which spring from the floor line and do not support floor loads, shall have members not less than 6 inches (152 mm) in width and not less than 8 inches (203 mm) in depth for the lower half of the height and not less than 6 inches (152 mm) in depth for the upper half.

Framed sawn timber or structural glued-laminated timber arches for roof construction, which spring from the top of walls or wall abutments, framed lumber or structural glued-laminated timber trusses, and other roof framing that does not support floor loads, shall have members not less than 4 inches (102 mm) in width and not less than 6 inches (152 mm) in depth. Spaced members may be composed of two or more pieces not less than 3 inches (76 mm) in thickness, when blocked solidly throughout their intervening spaces, or when such spaces are tightly closed by a continuous wood cover plate of not less than 2 inches (51 mm) in thickness, secured to the underside of the members. Splice plates shall not be less than 3 inches (76 mm) in thickness. When protected by an approved automatic sprinkler system under the roof deck, framing members shall not be less than 3 inches (76 mm) in thickness.

605.6.5 Floors. Floors shall be without concealed spaces. Floors shall be of planks, splined or tongue and groove, of not less than 3 inches (76 mm) in thickness covered with 1-inch (25 mm) tongue-and-groove flooring laid crosswise or diagonally, or $^{15}/_{32}$-inch (12 mm) wood structural panels, or of plank not less than 4 inches (102 mm) in width set on edge close together and well spiked, and covered with 1-inch (25 mm) flooring or $^{15}/_{32}$-inch (12 mm) wood structural panels. The lumber shall be laid so that no continuous line of joints will occur except at points of support. Floors shall not extend closer than $^1/_2$ inch (12.7 mm) to walls. Such $^1/_2$-inch (12.7 mm) space shall be covered by a molding fastened to the wall and arranged so that it will not obstruct the swelling or shrinkage movements of the floor. Corbeling of masonry walls under floors may be used in place of such molding.

605.6.6 Roof decks. Roofs shall be without concealed spaces and roof decks shall be of planks, splined or tongue and groove, of not less than 2-inch (51 mm) thickness, or $1^1/_8$-inch (29 mm) tongue-and-groove wood structural panels with exterior glue, or of a double thickness of 1-inch (25 mm) boards with tongue-and-groove joints, or with staggered joints, of lumber not less than 3 inches (76 mm) nominal in width, set on edge close together and laid as required for floors.

605.6.7 Construction details. Approved wall plate boxes or hangers shall be provided where wood beams, girders or trusses rest on masonry or concrete walls.

Girders and beams shall be closely fitted around columns, and adjoining ends shall be cross tied to each other, or intertied by caps or ties, to transfer horizontal loads across the joints. Wood bolsters may be placed on top of columns which support roof loads only.

Where intermediate beams are used to support a floor, they shall rest on top of the girders, or shall be supported by ledgers or blocks securely fastened to the sides of the girders, or they may be supported by approved metal hangers into which the ends of the beams shall be closely fitted.

In heavy-timber roof construction, every roof girder and at least every alternate roof beam shall be anchored to its supporting member; roof decks, where supported by a wall, shall be anchored to such wall at intervals not exceeding 20 feet (6096 mm); every monitor and every sawtooth construction shall be anchored to the main roof construction. Such anchors shall consist of steel or iron bolts of sufficient strength to resist vertical uplift of the roof.

605.6.8 Mechanically laminated floors and roof decks. Mechanically laminated floors and roof decks conforming to Section 2313 may be used as heavy-timber floors or roof decks, provided the minimum thickness and other applicable requirements of the section are followed.

605.6.9 Partitions. Partitions shall be of solid wood construction formed by not less than two layers of 1-inch (25 mm) matched boards or laminated construction of 4-inch (102 mm) thickness, or of one-hour fire-resistive construction.

SECTION 606 — TYPE V BUILDINGS

606.1 Definition. Type V buildings may be of any materials allowed by this code.

Type V One-hour buildings shall be of one-hour fire-resistive construction throughout.

Materials of construction and fire-resistive requirements shall be as specified in Section 601.

For requirements due to occupancy, see Chapter 3.

606.2 Structural Framework. Structural framework shall be of steel or iron as specified in Chapter 22, concrete as specified in Chapter 19, masonry as specified in Chapter 21, or wood as specified in Chapter 23 and this chapter.

606.3 Exterior Walls and Openings. Exterior walls shall comply with fire-resistive requirements set forth in Section 503 and Tables 5-A and 6-A. Openings in exterior walls shall conform to requirements of Section 503.2 and Table 5-A.

606.4 Stairway Construction.

606.4.1 General. Stairways shall comply with the requirements of Chapter 10.

606.4.2 Interior. Interior stairways may be constructed of any materials permitted by this code.

606.4.3 Exterior. Exterior stairways shall be constructed of wood not less than 2 inches (51 mm) in nominal thickness, or may be of noncombustible materials.

606.5 Roofs. Roof coverings shall be as specified in Chapter 15.

Except in retail sales and storage areas classified as Group M or S, Division 1 Occupancies and in Group H Occupancies, roofs and their members other than the structural frame may be of unprotected noncombustible materials when every part of the roof framing, including the structural frame, is 25 feet (7620 mm) or more above the floor, balcony or gallery immediately below. Heavy-timber members in accordance with Section 605.6 may be used for such unprotected members in one-story buildings.

TABLE 6-A

1997 UNIFORM BUILDING CODE

TABLE 6-A—TYPES OF CONSTRUCTION—FIRE-RESISTIVE REQUIREMENTS (In Hours)
For details, see occupancy section in Chapter 3, type of construction sections in this chapter and sections referenced in this table.

BUILDING ELEMENT	TYPE I	TYPE II			TYPE III		TYPE IV	TYPE V	
	Noncombustible				Combustible				
	Fire-resistive	Fire-resistive	1-Hr.	N	1-Hr.	N	H.T.	1-Hr.	N
1. Bearing walls—exterior	4 Sec. 602.3.1	4 Sec. 603.3.1	1	N	4 Sec. 604.3.1	4 Sec. 604.3.1	4 Sec. 605.3.1	1	N
2. Bearing walls—interior	3	2	1	N	1	N	1	1	N
3. Nonbearing walls—exterior	4 Sec. 602.3.1	4 Sec. 603.3.1	1 Sec. 603.3.1	N	4 Sec. 604.3.1	4 Sec. 604.3.1	4 Sec. 605.3.1	1	N
4. Structural frame[1]	3	2	1	N	1	N	1 or H.T.	1	N
5. Partitions—permanent	1[2]	1[2]	1[2]	N	1	N	1 or H.T.	1	N
6. Shaft enclosures[3]	2	2	1	1	1	1	1	1	1
7. Floors and floor-ceilings	2	2	1	N	1	N	H.T.	1	N
8. Roofs and roof-ceilings	2 Sec. 602.5	1 Sec. 603.5	1 Sec. 603.5	N	1	N	H.T.	1	N
9. Exterior doors and windows	Sec. 602.3.2	Sec. 603.3.2	Sec. 603.3.2	Sec. 603.3.2	Sec. 604.3.2	Sec. 604.3.2	Sec. 605.3.2	Sec. 606.3	Sec. 606.3
10. Stairway construction	Sec. 602.4	Sec. 603.4	Sec. 603.4	Sec. 603.4	Sec. 604.4	Sec. 604.4	Sec. 605.4	Sec. 606.4	Sec. 606.4

N—No general requirements for fire resistance.
H.T.—Heavy timber.

[1]Structural frame elements in an exterior wall that is located where openings are not permitted, or where protection of openings is required, shall be protected against external fire exposure as required for exterior-bearing walls or the structural frame, whichever is greater.
[2]Fire-retardant-treated wood (see Section 207) may be used in the assembly, provided fire-resistance requirements are maintained. See Sections 602 and 603.
[3]For special provisions, see Sections 304.6, 306.6 and 711.

Chapter 7
FIRE-RESISTANT MATERIALS AND CONSTRUCTION

SECTION 701 — SCOPE

This chapter applies to materials and systems used in the design and construction of a building to safeguard against the spread of fire and smoke within a building and the spread of fire to or from buildings.

SECTION 702 — DEFINITIONS

For the purposes of this chapter, the terms, phrases and words listed in this section and their derivatives shall have the indicated meanings.

ANNULAR SPACE is the opening around the penetrating item.

CONCRETE, CARBONATE AGGREGATE, is concrete made with aggregates consisting mainly of calcium or magnesium carbonate, e.g., limestone or dolomite, and containing 40 percent or less quartz, chert or flint.

CONCRETE, LIGHTWEIGHT AGGREGATE, is concrete made with aggregates of expanded clay, shale, slag or slate or sintered fly ash or any natural lightweight aggregate meeting ASTM C 330 and possessing equivalent fire-resistive properties and weighing 85 to 115 pounds per cubic foot (pcf) (1360 to 1840 kg/m^3).

CONCRETE, SAND-LIGHTWEIGHT, is concrete made with a combination of expanded clay, shale, slag or slate or sintered fly ash or any natural lightweight aggregate meeting ASTM C 330 and possessing equivalent fire-resistive properties and natural sand. Its unit weight is generally between 105 and 120 pcf (1680 and 1920 kg/m^3).

CONCRETE, SILICEOUS AGGREGATE, is concrete made with normal-weight aggregates consisting mainly of silica or compounds other than calcium or magnesium carbonate, and may contain more than 40 percent quartz, chert or flint.

F RATING is the time period the penetration firestop system limits the passage of fire through the penetration when tested in accordance with UBC Standard 7-5.

FIREBLOCKING is building material installed to resist the free passage of flame and gases to other areas of the building through small concealed spaces.

FIRE-RESISTIVE JOINT SYSTEM is an assemblage of specific materials or products that are designed, tested and fire resistive in accordance with UBC Standard 7-1 to resist, for a prescribed period of time, the passage of fire through joints.

JOINT is the linear opening between adjacent fire-resistive assemblies. A joint is a division of a building that allows independent movement of the building, in any plane, which may be caused by thermal, seismic, wind loading or any other loading.

MEMBRANE PENETRATION is an opening made through one side (wall, floor or ceiling membrane) of an assembly.

PENETRATION is an opening created in a membrane or assembly to accommodate penetrating items for electrical, mechanical, plumbing, environmental and communication systems.

EXCEPTION: Ducts.

PENETRATION FIRESTOP SYSTEM is an assemblage of specific materials or products that are designed, tested and fire re-

sistive in accordance with UBC Standard 7-5 to resist, for a prescribed period of time, the passage of fire through penetrations.

SPLICE is the result of a factory or field method of joining or connecting two or more lengths of a fire-resistive joint system into a continuous entity.

T RATING is the time period that the penetration firestop system including the penetrating item, limits the maximum temperature rise to 325°F (163°C) above its initial temperature through the penetration on the nonfire side, when tested in accordance with UBC Standard 7-5.

THROUGH-PENETRATION is an opening that passes through both sides of an assembly.

SECTION 703 — FIRE-RESISTIVE MATERIALS AND SYSTEMS

703.1 General. Materials and systems used for fire-resistive purposes shall be limited to those specified in this chapter, unless accepted under the procedure given in Section 703.2 or 703.3.

The materials and details of construction for the fire-resistive systems described in this chapter shall be in accordance with all other provisions of this code except as modified herein.

For the purpose of determining the degree of fire resistance afforded, the materials of construction listed in this chapter shall be assumed to have the fire-resistance rating indicated in Table 7-A, 7-B or 7-C.

As an alternate to Table 7-A, 7-B or 7-C, fire-resistive construction may be approved by the building official on the basis of evidence submitted showing that the construction meets the required fire-resistive classification.

703.2 Qualification by Testing. Material or assembly of materials of construction tested in accordance with the requirements set forth in UBC Standard 7-1 shall be rated for fire resistance in accordance with the results and conditions of such tests.

EXCEPTION: The acceptance criteria of UBC Standard 7-1 for exterior-bearing walls shall not be required to be greater with respect to heat transmission and passage of flame or hot gases than would be required of a nonbearing wall in the same building with the same distance to the property line. The fire exposure time period, water pressure and duration of application for the hose stream test shall be based on the fire-resistive rating determined by this exception.

Fire-resistive assemblies tested under UBC Standard 7-1 shall not be considered to be restrained unless evidence satisfactory to the building official is furnished by the person responsible for the structural design showing that the construction qualifies for a restrained classification in accordance with UBC Standard 7-1. Restrained construction shall be identified on the plans.

703.3 Calculating Fire Resistance. The fire-resistive rating of a material or assembly may be established by calculations. The procedures used for such calculations shall be in accordance with UBC Standard 7-7.

703.4 Standards of Quality. In addition to all the other requirements of this code, fire-resistive materials shall meet the requirements for fire-resistive construction given in this chapter.

The standards listed below labeled a "UBC standard" are also listed in Chapter 35, Part II, and are part of this code. The standards listed below labeled an "Adopted Standard" are also listed in Chapter 35, Part III, and are part of this code. The other standards listed below are recognized standards. (See Sections 3503 and 3504.)

1. UBC Standard 7-1, Fire Tests of Building Construction and Materials

2. UBC Standard 7-2, Fire Tests of Door Assemblies

3. UBC Standard 7-3, Tinclad Fire Doors

4. UBC Standard 7-4, Fire Tests of Window Assemblies

5. UBC Standard 7-5, Fire Tests of Through-penetration Fire Stops

6. UBC Standard 7-6, Thickness, Density Determination and Cohesion/Adhesion for Spray-applied Fire-resistive Fireproofing

7. UBC Standard 7-7, Methods for Calculating Fire Resistance of Steel, Concrete, Wood, Concrete Masonry and Clay Masonry Construction

8. ASTM C 516, Vermiculite Loose-fill Insulation

9. ASTM C 549, Perlite Loose-fill Insulation

10. ANSI/NFPA 80, Standard for Fire Doors and Fire Windows

11. ASTM C 587 and C 588, Gypsum Base for Veneer Plaster and Gypsum Veneer

12. ASTM C 332, Lightweight Aggregates for Insulating Concrete

13. ASTM C 331, Lightweight Aggregates for Concrete Masonry Units

14. UL 555, Fire Dampers

15. UL 555C, Ceiling Dampers

16. UL 555S, Leakage Rated Dampers for Use in Smoke Control Systems

17. UL 33, Heat Response Links for Fire Protection Service

18. UL 353, Limit Controls

19. ASTM E 1399, Cyclic Movement and Measuring the Minimum and Maximum Joint Widths of Architectural Joint Systems

20. Adopted standard—Fire-Resistance Design Manual, Fourteenth Edition

21. Adopted standard—ASTM C 330, Lightweight Aggregates for Structural Concrete

22. Adopted standard—CPSC 16 CFR, Part 1209 Interim Safety Standard for Cellulose Insulation and Part 1404 Cellulose Insulation

SECTION 704 — PROTECTION OF STRUCTURAL MEMBERS

704.1 General. Structural members having the fire-resistive protection set forth in Table 7-A shall be assumed to have the fire-resistance ratings set forth therein.

704.2 Protective Coverings.

704.2.1 Thickness of protection. The thickness of fire-resistive materials required for protection of structural members shall be not less than set forth in Table 7-A, except as modified in this section. The figures shown shall be the net thickness of the protecting materials and shall not include any hollow space back of the protection.

704.2.2 Unit masonry protection. Where required, metal ties shall be embedded in transverse joints of unit masonry for protection of steel columns. Such ties shall be as set forth in Table 7-A or be equivalent thereto.

704.2.3 Reinforcement for cast-in-place concrete column protection. Cast-in-place concrete protection for steel columns shall be reinforced at the edges of such members with wire ties of not less than 0.18 inch (4.6 mm) in diameter wound spirally around the columns on a pitch of not more than 8 inches (203 mm) or by equivalent reinforcement.

704.2.4 Embedment of pipes. Conduits and pipes shall not be embedded in required fire protection of structural members.

704.2.5 Column jacketing. Where the fire-resistive covering on columns is exposed to injury from moving vehicles, the handling of merchandise or other means, it shall be protected in an approved manner.

704.2.6 Ceiling membrane protection. When a ceiling forms the protective membrane for fire-resistive assemblies, the assemblies and their supporting horizontal structural members need not be individually fire protected except where such members support directly applied loads from a floor and roof or more than one floor. The required fire resistance shall not be less than that required for individual protection of members.

704.2.7 Plaster application. Plaster protective coatings may be applied with the finish coat omitted when they comply with the design mix and thickness requirements of Tables 7-A, 7-B and 7-C.

704.2.8 Truss protection. Where trusses are used as all or part of the structural frame and protection is required by Table 6-A, such protection may be provided by fire-resistive materials enclosing the entire truss assembly on all sides for its entire length and height. The required thickness and construction of fire-resistive assemblies enclosing trusses shall be based on the results of full-scale tests or combinations of tests on truss components or on approved calculations based on such tests that satisfactorily demonstrate that the assembly has the required fire resistance.

704.3 Protected Members.

704.3.1 Attached metal members. The edges of lugs, brackets, rivets and bolt heads attached to structural members may extend to within 1 inch (25 mm) of the surface of the fire protection.

704.3.2 Reinforcing. Thickness of protection for concrete or masonry reinforcement shall be measured to the outside of the reinforcement except that stirrups and spiral reinforcement ties may project not more than $1/2$ inch (12.7 mm) into the protection.

704.3.3 Bonded prestressed concrete tendons. For members having a single tendon or more than one tendon installed with equal concrete cover measured from the nearest surface, the cover shall not be less than that set forth in Table 7-A.

For members having multiple tendons installed with variable concrete cover, the average tendon cover shall not be less than that set forth in Table 7-A, provided:

1. The clearance from each tendon to the nearest exposed surface is used to determine the average cover.

2. In no case can the clear cover for individual tendons be less than one half of that set forth in Table 7-A. A minimum cover of $3/4$ inch (19.1 mm) for slabs and 1 inch (25.4 mm) for beams is required for any aggregate concrete.

3. For the purpose of establishing a fire-resistive rating, tendons having a clear covering less than that set forth in Table 7-A shall not contribute more than 50 percent of the required ultimate moment capacity for members less than 350 square inches (0.226 m^2) in cross-sectional area and 65 percent for larger members. For structural design purposes, however, tendons having a reduced cover are assumed to be fully effective.

704.4 Members Carrying Masonry or Concrete. All members carrying masonry or concrete walls in buildings over one story in height shall be fire protected with one-hour fire protection or the fire-resistive requirement of the wall, whichever is greater.

704.5 Fire-resistive Material Omitted. Fire-resistive material may be omitted from the bottom flange of lintels spanning not over 6 feet (1829 mm), shelf angles or plates that are not a part of the structural frame.

704.6 Spray-applied Fire-resistive Materials. The density and thickness of spray-applied fire-resistive materials shall be determined following the procedures set forth in UBC Standard 7-6.

SECTION 705 — PROJECTIONS

Cornices, eave overhangs, exterior balconies and similar architectural appendages extending beyond the floor area as defined in Section 207 shall conform to the requirements of this section. (See Section 1006 for additional requirements applicable to exterior exit balconies and stairways.)

Projections from walls of Type I or II construction shall be of noncombustible materials.

Projections from walls of Type III, IV or V construction may be of noncombustible or combustible materials.

Combustible projections located where openings are not permitted or where protection of openings is required shall be of one-hour fire-resistive or heavy-timber construction conforming to Section 605.6.

For projections extending over public property, see Chapter 32.

For combustible ornamentation, see Section 601.5.4.

For limitations on projection distances, see Sections 503.2 and 1204.

SECTION 706 — FIRE-RESISTIVE JOINT SYSTEMS

706.1 General. Joints installed in or between fire-resistive walls, fire-resistive floor or floor-ceiling assemblies and fire-resistive roof or roof-ceiling assemblies shall be protected by an approved fire-resistive joint system designed to resist the passage of fire for a time period not less than the required fire-resistance rating of the floor, roof or wall in or between which it is installed. Fire-resistive joint systems shall be tested in accordance with Section 706.2.

> **EXCEPTION:** Fire-resistive joint systems are not required for joints in the following locations:
> 1. Floors within a single dwelling unit.
> 2. Floors where the joint is protected by a shaft enclosure in accordance with Section 711.
> 3. Floors with atriums where the space adjacent to the atrium is included in the volume of the atrium for smoke-control purposes.
> 4. Floors within malls.
> 5. Floors within open parking structures.
> 6. Mezzanine floors.
> 7. Walls that are permitted to have unprotected openings.
> 8. Roofs where openings are permitted.

Such material or construction assembly shall be securely installed in or on the joint for its entire length so as not to dislodge, loosen or otherwise impair its ability to accommodate expected building movements and to resist the passage of fire and hot gases.

706.2 Fire-resistive Joint Systems. Fire-resistive joint systems shall be tested in accordance with UBC Standard 7-1 under the following conditions:

1. Joint systems shall be installed full height in wall assemblies and full length in floor and roof assemblies.

2. Floor and roof assemblies shall be tested with a minimum positive pressure differential of 0.01 inch of water column (2.5 Pa).

3. Wall assemblies shall be tested with a minimum positive pressure differential of 0.01 inch of water column (2.5 Pa) measured at the mid-height of the wall assembly.

4. Joint systems shall contain a splice. For wall assemblies, the splice shall be located above the mid-height of the wall assembly.

5. Joint systems shall be tested at the maximum joint width for which they are designed. Joint systems designed to accommodate movement shall be expanded to the maximum joint opening width for which they are intended to function.

6. Joint systems designed to be load-bearing shall be loaded to the maximum design load in accordance with their intended application.

7. Joint systems designed to accommodate movement shall be preconditioned by cycling between the minimum and the maximum joint opening width for which they are intended to function for the number of cycles specified in Table 7-D.

8. Nonsymmetrical wall joint systems shall be tested in accordance with Sections 706 and 709.5.

SECTION 707 — INSULATION

707.1 General. Thermal and acoustical insulation located on or within floor-ceiling and roof-ceiling assemblies, crawl spaces, walls, partitions and insulation on pipes and tubing shall comply with this section. Duct insulation and insulation in plenums shall conform to the requirements of the Mechanical Code.

> **EXCEPTION:** Roof insulation shall comply with Section 1510.

707.2 Insulation and Covering on Pipe and Tubing. Insulation and covering on pipe and tubing shall have a flame-spread rating not to exceed 25 and a smoke density not to exceed 450 when tested in accordance with UBC Standard 8-1.

> **EXCEPTION:** Foam plastic insulation shall comply with Section 2602.

707.3 Insulation. Cellulose loose-fill insulation shall comply with CPSC 16 CFR, Parts 1209 and 1404. All other insulation materials, including facings, such as vapor barriers or breather papers installed within floor-ceiling assemblies, roof-ceiling assemblies, walls, crawl spaces or attics, shall have a flame-spread rating not to exceed 25 and a smoke density not to exceed 450 when tested in accordance with UBC Standard 8-1.

> **EXCEPTIONS:** 1. Foam plastic insulation shall comply with Section 2602.
>
> 2. When such materials are installed in concealed spaces of Types III, IV and V construction, the flame-spread and smoke-developed limitations do not apply to facings, provided that the facing is installed in substantial contact with the unexposed surface of the ceiling, floor or wall finish.

SECTION 708 — FIRE BLOCKS AND DRAFT STOPS

708.1 General. In combustible construction, fireblocking and draftstopping shall be installed to cut off all concealed draft openings (both vertical and horizontal) and shall form an effective barrier between floors, between a top story and a roof or attic space, and shall subdivide attic spaces, concealed roof spaces and floor-ceiling assemblies. The integrity of all fire blocks and draft stops shall be maintained.

708.2 Fire Blocks.

708.2.1 Where required. Fireblocking shall be provided in the following locations:

1. In concealed spaces of stud walls and partitions, including furred spaces, at the ceiling and floor levels and at 10-foot (3048 mm) intervals both vertical and horizontal. See also Section 803, Item 1.

EXCEPTION: Fire blocks may be omitted at floor and ceiling levels when approved smoke-actuated fire dampers are installed at these levels.

2. At all interconnections between concealed vertical and horizontal spaces such as occur at soffits, drop ceilings and cove ceilings.

3. In concealed spaces between stair stringers at the top and bottom of the run and between studs along and in line with the run of stairs if the walls under the stairs are unfinished.

4. In openings around vents, pipes, ducts, chimneys, fireplaces and similar openings that afford a passage for fire at ceiling and floor levels, with noncombustible materials.

5. At openings between attic spaces and chimney chases for factory-built chimneys.

6. Where wood sleepers are used for laying wood flooring on masonry or concrete fire-resistive floors, the space between the floor slab and the underside of the wood flooring shall be filled with noncombustible material or fire blocked in such a manner that there will be no open spaces under the flooring that will exceed 100 square feet (9.3 m^2) in area and such space shall be filled solidly under all permanent partitions so that there is no communication under the flooring between adjoining rooms.

EXCEPTIONS: 1. Fire blocking need not be provided in such floors when at or below grade level in gymnasiums.

2. Fire blocking need be provided only at the juncture of each alternate lane and at the ends of each lane in a bowling facility.

708.2.2 Fire block construction. Except as provided in Item 4 above, fireblocking shall consist of 2 inches (51 mm) nominal lumber or two thicknesses of 1-inch (25 mm) nominal lumber with broken lap joints or one thickness of $^{23}/_{32}$-inch (18.3 mm) wood structural panel with joints backed by $^{23}/_{32}$-inch (18.3 mm) wood structural panel or one thickness of $^3/_4$-inch (19.1 mm) Type 2-M particleboard with joints backed by $^3/_4$-inch (19.1 mm) Type 2-M particleboard.

Fire blocks may also be of gypsum board, cement fiber board, batts or blankets of mineral or glass fiber, or other approved materials installed in such a manner as to be securely retained in place. Loose-fill insulation material shall not be used as a fire block unless specifically tested in the form and manner intended for use to demonstrate its ability to remain in place and to retard the spread of fire and hot gases.

Walls having parallel or staggered studs for sound-transmission control shall have fire blocks of batts or blankets of mineral or glass fiber or other approved flexible materials.

708.3 Draft Stops.

708.3.1 Where required. Draftstopping shall be provided in the locations set forth in this section.

708.3.1.1 Floor-ceiling assemblies.

708.3.1.1.1 Single-family dwellings. When there is usable space above and below the concealed space of a floor-ceiling assembly in a single-family dwelling, draft stops shall be installed so that the area of the concealed space does not exceed 1,000 square feet (93 m^2). Draftstopping shall divide the concealed space into approximately equal areas.

708.3.1.1.2 Two or more dwelling units and hotels. Draft stops shall be installed in floor-ceiling assemblies of buildings having more than one dwelling unit and in hotels. Such draft stops shall be in line with walls separating individual dwelling units and guest rooms from each other and from other areas.

708.3.1.1.3 Other uses. Draft stops shall be installed in floor-ceiling assemblies of buildings or portions of buildings used for other than dwelling or hotel occupancies so that the area of the concealed space does not exceed 1,000 square feet (93 m^2) and so that the horizontal dimension between stops does not exceed 60 feet (18 288 mm).

EXCEPTION: Where approved automatic sprinklers are installed within the concealed space, the area between draft stops may be 3,000 square feet (279 m^2) and the horizontal dimension may be 100 feet (30 480 mm).

708.3.1.2 Attics.

708.3.1.2.1 Two or more dwelling units and hotels. Draft stops shall be installed in the attics, mansards, overhangs, false fronts set out from walls and similar concealed spaces of buildings containing more than one dwelling unit and in hotels. Such draft stops shall be above and in line with the walls separating individual dwelling units and guest rooms from each other and from other uses.

EXCEPTIONS: 1. Draft stops may be omitted along one of the corridor walls, provided draft stops at walls separating individual dwelling units and guest rooms from each other and from other uses, extend to the remaining corridor draft stop.

2. Where approved sprinklers are installed, draftstopping may be as specified in the exception to Section 708.3.1.2.2.

708.3.1.2.2 Other uses. Draft stops shall be installed in attics, mansards, overhangs, false fronts set out from walls and similar concealed spaces of buildings having uses other than dwellings or hotels so that the area between draft stops does not exceed 3,000 square feet (279 m^2) and the greatest horizontal dimension does not exceed 60 feet (18 288 mm).

EXCEPTION: Where approved automatic sprinklers are installed, the area between draft stops may be 9,000 square feet (836 m^2) and the greatest horizontal dimension may be 100 feet (30 480 mm).

708.3.1.3 Draft stop construction. Draftstopping materials shall not be less than $^1/_2$-inch (12.7 mm) gypsum board, $^3/_8$-inch (9.5 mm) wood structural panel, $^3/_8$-inch (9.5 mm) Type 2-M particleboard or other approved materials adequately supported.

Openings in the partitions shall be protected by self-closing doors with automatic latches constructed as required for the partitions.

Ventilation of concealed roof spaces shall be maintained in accordance with Section 1505.

708.4 Draft Stops or Fire Blocks in Other Locations. Fireblocking of veneer on noncombustible walls shall be in accordance with Section 708.2.1, Item 1.

For fireblocking ceilings applied against noncombustible construction, see Section 803, Item 1.

SECTION 709 — WALLS AND PARTITIONS

709.1 General. Fire-resistive walls and partitions shall be assumed to have the fire-resistance ratings set forth in Table 7-B.

Where materials, systems or devices are incorporated into the assembly that have not been tested as part of the assembly, sufficient data shall be made available to the building official to show that the required fire-resistive rating is not reduced. Materials and methods of construction used to protect joints and penetrations in fire-resistive, fire-rated building assemblies shall not reduce the required fire-resistive rating.

709.2 Combustible Members. Combustible members framed into a wall shall be protected at their ends by not less than one half the required fire-resistive thickness of such wall.

709.3 Exterior Walls.

709.3.1 Extension through attics and concealed spaces. In fire-resistive exterior wall construction, the fire-resistive rating shall be maintained for such walls passing through attic areas or other areas containing concealed spaces.

709.3.2 Vertical fire spread at exterior walls.

709.3.2.1 General. The provisions of this section are intended to restrict the passage of smoke, flame and hot gases from one floor to another at exterior walls. See Section 710 for floor penetrations.

709.3.2.2 Interior. When fire-resistive floor or floor-ceiling assemblies are required, voids created at the intersection of the exterior wall assemblies and such floor assemblies shall be sealed with an approved material. Such material shall be securely installed and capable of preventing the passage of flame and hot gases sufficient to ignite cotton waste when subjected to UBC Standard 7-1 time-temperature fire conditions under a minimum positive pressure differential of 0.01 inch of water column (2.5 Pa) for the time period at least equal to the fire-resistance rating of the floor assembly.

709.3.2.3 Exterior. When openings in an exterior wall are above and within 5 feet (1524 mm) laterally of an opening in the story below, such openings shall be separated by an approved flame barrier extending 30 inches (762 mm) beyond the exterior wall in the plane of the floor or by approved vertical flame barriers not less than 3 feet (914 mm) high measured vertically above the top of the lower opening. Flame barriers shall have a fire resistance of not less than three-fourths hour.

> **EXCEPTIONS: 1.** Flame barriers are not required in buildings equipped with an approved automatic sprinkler system throughout.
>
> 2. This section shall not apply to buildings of three stories or less in height.
>
> 3. Flame barriers are not required on Group S, Division 4 Occupancies.

709.4 Parapets.

709.4.1 General. Parapets shall be provided on all exterior walls of buildings.

> **EXCEPTION:** A parapet need not be provided on an exterior wall when any of the following conditions exist:
>
> 1. The wall is not required to be of fire-resistive construction.
>
> 2. The wall, due to location on property line, may have unprotected openings.
>
> 3. The building has an area of not more than 1,000 square feet (93 m²) on any floor.
>
> 4. Walls that terminate at roofs of not less than two-hour fire-resistive construction or roofs constructed entirely of noncombustible materials.
>
> 5. One-hour fire-resistive exterior walls may terminate at the underside of the roof sheathing, deck or slab, provided:
>
> > 5.1 Where the roof-ceiling framing elements are parallel to the walls, such framing and elements supporting such framing shall not be of less than one-hour fire-resistive construction for a width of 5 feet (1524 mm) measured from the interior side of the wall for Groups R and U Occupancies and 10 feet (3048 mm) for all other occupancies.
> >
> > 5.2 Where roof-ceiling framing elements are not parallel to the wall, the entire span of such framing and elements supporting such framing shall not be of less than one-hour fire-resistive construction.

> > 5.3 Openings in the roof shall not be located within 5 feet (1524 mm) of the one-hour fire-resistive exterior wall for Groups R and U Occupancies and 10 feet (3048 mm) for all other occupancies.
> >
> > 5.4 The entire building shall be provided with not less than a Class B roofing assembly.

709.4.2 Construction. Parapets shall have the same degree of fire resistance required for the wall upon which they are erected, and on any side adjacent to a roof surface, shall have noncombustible faces for the uppermost 18 inches (457 mm), including counterflashing and coping materials. The height of the parapet shall not be less than 30 inches (762 mm) above the point where the roof surface and the wall intersect. Where the roof slopes toward a parapet at slopes greater than 2 units vertical in 12 units horizontal (16.7% slope), the parapet shall extend to the same height as any portion of the roof that is within the distance where protection of wall openings would be required, but in no case shall the height be less than 30 inches (762 mm).

709.5 Nonsymmetrical Wall Construction. Walls and partitions of nonsymmetrical construction shall be tested with both faces exposed to the furnace, and the assigned fire-resistive rating will be the shortest duration obtained from the two tests conducted in conformance with UBC Standard 7-1. When evidence is furnished to show that the wall was tested with the least fire-resistive side exposed to the furnace, subject to acceptance of the building official, the wall need not be subjected to tests from the opposite side.

709.6 Through Penetrations.

709.6.1 General. Through penetrations of the fire-resistive walls shall comply with Section 709.6.2 or 709.6.3.

> **EXCEPTION:** Where the penetrating items are steel, ferrous or copper pipes or steel conduits, the annular space shall be permitted to be protected as follows:
>
> 1. In concrete or masonry walls where the penetrating items are a maximum 6-inch (152 mm) nominal diameter and the opening is a maximum 144 square inches (92 903 mm²) concrete, grout or mortar shall be permitted when installed the full thickness of the wall or the thickness required to maintain the fire rating, or
>
> 2. The material used to fill the annular space shall prevent the passage of flame and hot gases sufficient to ignite cotton waste when subjected to UBC Standard 7-1 time-temperature fire conditions under a minimum positive pressure differential of 0.01 inch of water column (2.5 Pa) at the location of the penetration for the time period equivalent to the fire rating of the construction penetrated.

709.6.2 Fire-rated assembly. Penetrations shall be installed as tested in the approved UBC Standard 7-1 rated assembly.

709.6.3 Penetration firestop system. Penetrations shall be protected by an approved penetration firestop system installed as tested in accordance with UBC Standard 7-5 and shall have an F rating of not less than the required rating of the wall penetrated.

709.7 Membrane Penetrations. Membrane penetrations of the fire-resistive walls shall comply with Section 709.6.

> **EXCEPTIONS: 1.** Steel electrical boxes that do not exceed 16 square inches (10 323 mm²) in area, provided that the area of such openings does not exceed 100 square inches for any 100 square feet (694 mm²/m²) of wall area. Outlet boxes on opposite sides of the wall shall be separated by a horizontal distance of not less than 24 inches (610 mm). Membrane penetrations for electrical outlet boxes of any material are permitted, provided that such boxes are tested for use in fire-resistive assemblies and installed in accordance with the tested assembly.
>
> 2. The annular space created by the penetration of a fire sprinkler shall be permitted to be unprotected, provided such space is covered by a metal escutcheon plate.

Noncombustible penetrating items shall not be connected to combustible materials on both sides of the membrane unless it can be confirmed that the fire-resistive integrity of the wall is maintained in accordance with UBC Standard 7-1.

709.8 Joints. The protection of joints shall comply with the requirements of Section 706.

SECTION 710 — FLOOR CEILINGS OR ROOF CEILINGS

710.1 General. Fire-resistive floors, floor-ceiling or roof-ceiling assemblies shall be assumed to have the fire-resistance ratings set forth in Table 7-C. When materials are incorporated into an otherwise fire-resistive assembly that may change the capacity for heat dissipation, fire test results or other substantiating data shall be made available to the building official to show that the required fire-resistive time period is not reduced.

Where the weight of lay-in ceiling panels used as part of fire-resistive floor-ceiling or roof-ceiling assemblies is not adequate to resist an upward force of 1 pound per square foot (0.048 kN/m^2), wire holddowns or other approved devices shall be installed above the panels to prevent vertical displacement under such upward force.

710.2 Through Penetrations.

710.2.1 General. Through penetrations of fire-resistive horizontal assemblies shall be enclosed in fire-resistive shaft enclosures in accordance with Section 711.1 or shall comply with Section 710.2.2 or 710.2.3.

> **EXCEPTIONS: 1.** Steel, ferrous or copper conduits, pipes, tubes, vents, concrete, or masonry penetrating items that penetrate a single fire-rated floor assembly where the annular space is protected with materials that prevent the passage of flame and hot gases sufficient to ignite cotton waste when subjected to UBC Standard 7-1 time-temperature fire conditions under a minimum positive pressure differential of 0.01 inch of water column (2.5 Pa) at the location of the penetration for the time period equivalent to the fire-resistive rating of the construction penetrated. Penetrating items with a maximum 6-inch (152 mm) nominal diameter shall not be limited to the penetration of a single-fire-resistive floor assembly, provided that the area of the penetration does not exceed 144 square inches in any 100 square feet (100 000 mm^2 in 10 m^2) of floor area.
>
> 2. Penetrations in a single concrete floor by steel, ferrous or copper conduits, pipes, tubes and vents with a maximum 6-inch (152 mm) nominal diameter provided concrete, grout or mortar is installed the full thickness of the floor or the thickness required to maintain the fire-resistive rating. The penetrating items with a maximum 6-inch (152 mm) nominal diameter shall not be limited to the penetration of a single concrete floor, provided that the area of the penetration does not exceed 144 square inches (92 903 mm^2).
>
> 3. Electrical outlet boxes of any material are permitted provided that such boxes are tested for use in fire-resistive assemblies and installed in accordance with the tested assembly.

710.2.2 Fire-rated assemblies. Penetrations shall be installed as tested in the approved UBC Standard 7-1.

710.2.3 Penetration firestop system. Penetration shall be protected by an approved penetration firestop system installed as tested in accordance with UBC Standard 7-5. The system shall have an F rating and a T rating of not less than one hour but not less than the required rating of the floor penetrated.

> **EXCEPTION:** Floor penetrations contained and located within the cavity of a wall do not require a T rating.

710.3 Membrane Penetrations. Penetrations of membranes that are part of a fire-resistive horizontal assembly shall comply with Section 710.2.

> **EXCEPTIONS: 1.** Membrane penetrations of steel, ferrous or copper conduits, electrical outlet boxes, pipes, tubes, vents, concrete, or masonry penetrating items where the annular space is protected in

accordance with Section 709.6 or 710.2 or is protected to prevent the free passage of flame and the products of combustion. Such penetrations shall not exceed an aggregate area of 100 square inches in any 100 square feet (694 mm^2/m^2) of ceiling area in assemblies tested without penetrations.

> 2. Membrane penetrations for electrical outlet boxes of any material are permitted, provided that such boxes are tested for use in fire-resistive assemblies and installed in accordance with the tested assembly.
>
> 3. The annular space created by the penetration of a fire sprinkler shall be permitted to be unprotected, provided such space is covered by a metal escutcheon plate.

710.4 Roofs. Fire-resistive roofs may have unprotected openings. See Chapter 24 for skylight construction.

710.5 Wiring in Plenums. Wiring in plenums shall comply with the Mechanical Code.

710.6 Joints. The protection of joints in fire-resistive floors and roofs shall comply with the requirements of Section 706.

SECTION 711 — SHAFT ENCLOSURES

711.1 General. Openings through floors shall be enclosed in a shaft enclosure of fire-resistive construction having the time period set forth in Table 6-A for "shaft enclosures" except as permitted in Sections 711.3, 711.5 and 711.6. See also Section 304.6 for shafts in Group B Occupancies, Section 306.6 for shafts in Group F Occupancies, Sections 307.6 and 307.11.2.3 for shafts in Group H Occupancies, Section 309.6 for shafts in Group M Occupancies and Section 311.6 for shafts in Group S Occupancies.

711.2 Extent of Enclosures. Shaft enclosures shall extend from the lowest floor opening through successive floor openings and shall be enclosed at the top and bottom.

> **EXCEPTIONS: 1.** Shafts extending through or to the underside of the roof sheathing, deck or slab need not be enclosed at the top.
>
> 2. Shafts need not be enclosed at the bottom when protected by fire dampers conforming to approved recognized standards, installed at the lowest floor level within the shaft enclosure.

Shaft enclosures shall be constructed to continuously maintain the required fire-resistive integrity.

711.3 Special Provision. In other than Group I Occupancies, openings that penetrate only one floor and are not connected with openings communicating with other stories or basements and that are not concealed within building construction assemblies need not be enclosed.

Exit enclosures shall conform to the applicable provisions of Section 1005.3.3.

In one- and two-story buildings other than Group I Occupancies, gas vents, ducts, piping and factory-built chimneys that extend through not more than two floors need not be enclosed, provided the openings around the penetrations are firestopped at each floor.

> **EXCEPTION:** BW gas vents installed in accordance with their listing.

Gas vents and factory-built chimneys shall be protected as required by the Mechanical Code.

Walls containing gas vents or noncombustible piping that pass through three floors or less need not provide the fire-resistance rating specified in Table 6-A for "shaft enclosures," provided the annular space around the vents or piping is filled at each floor or ceiling with noncombustible materials.

> **EXCEPTION:** BW gas vents installed in accordance with their listing.

Openings made through a floor for penetrations such as cables, cable trays, conduit, pipes or tubing that are protected with approved through-penetration fire stops to provide the same degree of fire resistance as the floor construction need not be enclosed. For floor-ceiling assemblies, see Section 710.

711.4 Protection of Openings. Openings into a shaft enclosure shall be protected by a self-closing or an automatic-closing fire assembly conforming to Section 713 and having a fire-protection rating of one hour for openings through one-hour fire-resistive walls and one and one-half hours for openings through two-hour fire-resistive walls.

> **EXCEPTIONS:** 1. Openings to the exterior may be unprotected when permitted by Table 5-A.
>
> 2. Openings protected by through-penetration fire stops to provide the same degree of fire resistance as the shaft enclosure. See Sections 709 and 710.
>
> 3. Noncombustible ducts, vents or chimneys used to convey vapors, dusts or combustion products may penetrate the enclosure at the bottom.

Openings in shaft enclosures penetrating smoke barriers shall be further protected by smoke dampers conforming with approved recognized standards. See Chapter 35, Part IV.

> **EXCEPTIONS:** 1. Exhaust-only openings serving continuously operating fans and protected using the provisions of Chapter 9.
>
> 2. Smoke dampers are not required when their operation would interfere with the function of a smoke-control system.

711.5 Rubbish and Linen Chute Termination Rooms. In other than Group R, Division 3 Occupancies, rubbish and linen chutes shall terminate in rooms separated from the remainder of the building by an occupancy separation having the same fire resistance as required for the shaft enclosure, but not less than one hour. Openings into chutes and chute termination rooms shall not be located in corridors or stairways. For sprinklers, see Section 904.2.2.

711.6 Chute and Dumbwaiter Shafts. In buildings of Type V construction, chutes and dumbwaiter shafts with a cross-sectional area of not more than 9 square feet (0.84 m^2) may be either of approved fire-resistive wall construction or may have the inside layers of the approved fire-resistive assembly replaced by a lining of not less than 0.019-inch (0.48 mm) No. 26 galvanized sheet gage metal with all joints locklapped. The outside layers of the wall shall be as required for the approved construction. All openings into any such enclosure shall be protected by not less than a self-closing solid-wood door 1^3/$_8$ inches (35 mm) thick or equivalent.

SECTION 712 — USABLE SPACE UNDER FLOORS

Usable space under the first story shall be enclosed, and such enclosure, when constructed of metal or wood, shall be protected on the side of the usable space as required for one-hour fire-resistive construction. Doors shall be self-closing, tightfitting of solid-wood construction 1^3/$_8$ inches (35 mm) in thickness or self-closing, tightfitting doors acceptable as a part of an assembly having a fire-protection rating of not less than 20 minutes when tested in accordance with Part II of UBC Standard 7-2.

> **EXCEPTIONS:** 1. Group R, Division 3 and Group U Occupancies.
>
> 2. Basements in single-story Group S, Division 3 repair garages where 10 percent or more of the area of the floor-ceiling is open to the first floor.
>
> 3. Underfloor spaces protected by an automatic sprinkler system.

SECTION 713 — FIRE-RESISTIVE ASSEMBLIES FOR PROTECTION OF OPENINGS

713.1 General. Where required by this code for the fire protection of openings, fire assemblies shall meet the requirements of this section.

713.2 Definitions.

FIRE ASSEMBLY is the assembly of a fire door, fire windows or fire damper, including all required hardware, anchorage, frames and sills.

FIRE ASSEMBLY, AUTOMATIC-CLOSING, is a fire assembly that may remain in an open position and that will close automatically when subjected to one or the other of the following:

1. An increase in temperature.

Unless otherwise specified, the closing device shall be one rated at a maximum temperature of 165°F (74°C).

2. Actuation of a smoke detector.

The closing device shall operate by the activation of an approved listed smoke detector. Smoke detectors shall be installed and maintained as set forth in approved nationally recognized standards.

FIRE ASSEMBLY, SELF-CLOSING, is a fire assembly that is kept in a normally closed position and is equipped with an approved device to ensure closing and latching after having been opened for use.

713.3 Identification of Fire Doors, Fire Windows and Fire Dampers. Fire doors, fire windows and fire dampers shall have an approved label or listing mark, indicating the fire-protection rating, which is permanently affixed at the factory where fabrication and assembly are done. Periodic inspections shall be made by an approved inspection agency during fabrication and assembly.

Labels for fire doors used to protect openings into exit enclosures shall indicate that the temperature rise on the unexposed surface does not exceed 450°F (232°C) above ambient at the end of 30 minutes of the fire exposure specified in UBC Standard 7-2 to show compliance with Section 1005.3.

Oversized fire doors may be installed when approved by the building official. The doors shall be labeled or be furnished with a certificate of inspection from an approved agency.

713.4 Installation of Fire Doors, Hardware and Frames, and Fire Dampers. Approved fire door hardware and fire door frames including the anchorage thereof shall be installed in accordance with their listing. Fire dampers shall be fabricated and installed in an approved manner.

713.5 Fire-resistive Tests. The fire-protection rating of all types of required fire assemblies shall be determined in accordance with the requirements specified in UBC Standards 7-2, 7-3 and 7-4. The fire-protection rating of fire dampers shall be determined in accordance with the requirements specified within approved recognized standards.

713.6 Hardware.

713.6.1 Closing devices. Every fire assembly shall be provided with a closing device as follows:

1. Fire assemblies required to have a three-hour fire-protection rating shall be automatic-closing fire assemblies. Automatic-closing fire assemblies to be activated by an increase in temperature shall have one heat-actuating device installed on each side of the wall at the top of the opening and one on each side of the wall at the ceiling height where the ceiling is more than 3 feet (914 mm) above the top of the opening.

2. Fire assemblies required to have a one- and one-half-hour, one-hour or three-fourths-hour fire-protection rating shall be either automatic- or self-closing fire assemblies. Automatic-closing fire assemblies to be activated by an increase in temperature shall have heat-actuating devices located as required in Item 1 or by a single fusible link in the opening incorporated in the closing device.

3. Fire door assemblies required to have fire-protection rating, which are installed across a corridor, shall be automatic-closing fire assemblies. Such fire assemblies shall be activated by a smoke detector. All hold-open devices shall be listed for the purpose and shall release or close the door in the event of a power failure at the device.

4. Fire assemblies required by provisions of Chapter 10 shall have closing devices as specified in Chapter 10.

5. Doors that are a part of an automobile ramp enclosure shall be equipped with automatic-closing devices.

Fire doors that are automatic closing by smoke detection shall not have a closing or reclosing delay of more than 10 seconds.

713.6.2 Hinges. Swinging fire doors shall not have less than two hinges, and when such door exceeds 60 inches (1524 mm) in height, an additional hinge shall be installed for each additional 30 inches (762 mm) of height or fraction thereof. Hinges, except for spring hinges, shall be of the ball-bearing or antifriction type. When spring hinges are used for door-closing purposes, not less than one half of the hinges shall be spring hinges.

713.6.3 Latch. Unless otherwise specifically permitted, all single doors and both leaves of pairs of side-hinged swinging doors shall be provided with an automatic latch that will secure the door when it is closed.

713.7 Glazed Openings in Fire Doors. Glazed openings in fire doors shall not be permitted in a fire assembly required to have a three-hour fire-resistive rating.

The area of glazed openings in a fire door required to have one- and one-half-hour or one-hour fire-resistive rating shall be limited to 100 square inches (64 500 mm^2) with a minimum dimension of 4 inches (102 mm). When both leaves of a pair of doors have observation panels, the total area of the glazed openings shall not exceed 100 square inches (64 500 mm^2) for each leaf.

Glazed openings shall be limited to 1,296 square inches (0.84 m^2) in wood and plastic-faced composite or hollow metal doors, per light, when fire-resistive assemblies are required to have a three-fourths-hour fire-resistive rating.

713.8 Fire Window Size. Fire windows required to have a three-fourths-hour fire-protection rating for protection of openings in exterior walls shall have an area not greater than 84 square feet (7.8 m^2) with neither width nor height exceeding 12 feet (3658 mm) and for protection of openings in interior walls shall be limited in area and size to that tested.

713.9 Glazing. Glazing materials and glass block assemblies shall be qualified by tests in accordance with UBC Standard 7-2 (for fire doors) or UBC Standard 7-4 (for fire windows) as appropriate for the use, and they shall be labeled for the required fire-protection rating and installed in accordance with their listing. Glazing in fire door assemblies and in fire window assemblies subject to human impact in hazardous locations as indicated in Section 2406.4 shall comply with Section 2406.3.

713.10 Smoke Dampers. Not less than Class II, 250°F (121°C) smoke dampers complying with approved recognized standards (see Chapter 35, Part IV) shall be installed and be accessible for

inspection and servicing in the following ducted or unducted air openings at:

1. Penetrations of area or occupancy separation walls.

2. Penetrations of the fire-resistive construction of horizontal exit walls or corridors serving as a means of egress.

   EXCEPTION: Openings for steel ducts penetrating the required fire-resistive construction of corridors are not required to have smoke dampers when such ducts are of not less than 0.019-inch (0.48 mm) thickness (No. 26 galvanized sheet steel gage) and have no openings serving the corridor.

3. Penetrations of shaft enclosures.

   EXCEPTION: Exhaust-only openings serving continuously operating fans and protected using the provisions of Chapter 9.

4. Penetrations of smoke barriers.

5. Penetrations of elevator lobbies required by Section 403.7 or 1004.3.4.5.

6. Penetrations of areas of refuge.

   EXCEPTION: Ventilation systems specifically designed and protected to supply outside air to these areas during an emergency.

A smoke damper need not be provided when it can be demonstrated that the smoke damper is not essential to limit the passage of smoke under passive conditions and the proper function of a smoke-control system complying with Chapter 9 does not depend on the operation of the damper. Smoke dampers may be omitted at openings that must be maintained open for proper operation of a mechanical smoke-control system, provided that adequate protection against smoke migration, in the event of system failure, has been provided.

Smoke dampers shall be closed by actuation of a smoke detector installed in accordance with the Fire Code and one of the following applicable methods:

1. Where a damper is installed within a duct, a smoke detector shall be installed in the duct within 5 feet (1524 mm) of the damper with no air outlets or inlets between the detector and the damper. The detector shall be listed for the air velocity, temperature and humidity anticipated at the point where it is installed.

2. Where a damper is installed within an unducted opening in a wall, a spot-type detector listed for releasing service shall be installed within 5 feet (1524 mm) horizontally of the damper.

3. Where a damper is installed in a ceiling, a spot-type detector listed for releasing service shall be installed on the ceiling within 5 feet (1524 mm) of the damper.

4. Where a damper is installed in a corridor wall or ceiling, the damper may be controlled by a smoke-detection system installed in the corridor.

5. When a total-coverage smoke-detection system is provided within all areas served by an HVAC system, dampers may be controlled by the smoke-detection system.

713.11 Fire Dampers. Fire dampers complying with the requirements of approved recognized standards (see Chapter 35, Part IV) shall be installed and be accessible for inspection and servicing in the following ducted and unducted air openings at:

1. Penetrations through area separation walls or occupancy separations.

2. Penetrations of the fire-resistive construction of horizontal exit walls or corridors serving as a means of egress.

   EXCEPTION: Openings for steel ducts penetrating the required fire-resistive construction of corridors are not required to have dampers when such ducts are of not less than 0.019-inch (0.48 mm) thickness (No. 26 galvanized sheet steel gage) and have no openings serving the corridor.

3. Penetrations of shaft enclosures.

EXCEPTIONS: 1. Duct penetrations by steel exhaust air subducts extending vertically upward at least 22 inches (559 mm) above the top of the opening in a vented shaft where the airflow is upward.

2. Penetrations of a fire-resistive floor forming the base of a shaft enclosure may be protected by fire dampers listed for installation in the horizontal position.

4. Penetrations of the ceiling of fire-resistive floor-ceiling or roof-ceiling assemblies.

5. Penetrations of an atrium enclosure element.

6. Penetrations of the building exterior required to have protected openings by Section 503.

7. Penetrations of areas of refuge.

EXCEPTION: Ventilation systems specifically designed and protected to supply outside air to these areas during an emergency.

A fire damper is not required where fire tests have demonstrated that fire dampers are not required to maintain the fire resistance of the construction.

The operating temperature of the fire-damper actuating device shall be approximately 50°F (27.8°C) above the normal temperature within the duct system, but not less than 160°F (71°C). The operating temperature of the actuating device may be increased to not more than 286°F (141°C) when located in a smoke-control system complying with Chapter 9.

713.12 Installation. Fire assemblies shall be installed in accordance with their listing. Only fire dampers labeled for use in dynamic systems shall be installed in heating, ventilation and air-conditioning systems intended to operate with fans on during a fire.

713.13 Signs. When required by the building official, a sign shall be displayed permanently near or on each required fire door in letters not less than 1 inch (25 mm) high to read as follows:

FIRE DOOR
DO NOT OBSTRUCT

SECTION 714 — THROUGH-PENETRATION FIRE STOPS

Through-penetration fire stops required by this code shall have an F or T rating as determined by tests conducted in accordance with UBC Standard 7-5.

Through-penetration fire stops may be used for membrane penetrations.

The F rating shall apply to all through penetrations and shall not be less than the required fire-resistance rating of the assembly penetrated.

The T rating shall apply to those through-penetration locations required to have T ratings as specified in Section 710.2.3 and shall not be less than the required fire-resistance rating of the assembly penetrated.

Where sleeves are used, the sleeves shall be securely fastened to the assembly penetrated. All space between the item contained in the sleeve and the sleeve itself and any space between the sleeve and the assembly penetrated shall be protected. Insulation and coverings on the penetrating item shall not penetrate the assembly unless the specific materials used have been tested as part of the assembly.

EXCEPTION: Fire damper or combination fire damper/smoke damper sleeves shall be installed in accordance with their listing.

TABLE 7-A   1997 UNIFORM BUILDING CODE

TABLE 7-A—MINIMUM PROTECTION OF STRUCTURAL PARTS BASED ON TIME PERIODS FOR VARIOUS NONCOMBUSTIBLE INSULATING MATERIALS[a]

STRUCTURAL PARTS TO BE PROTECTED	ITEM NUMBER	INSULATING MATERIAL USED	MINIMUM THICKNESS OF INSULATING MATERIAL FOR FOLLOWING FIRE-RESISTIVE PERIODS (inches) $\times$ 25.4 for mm			
			4 Hr.	3 Hr.	2 Hr.	1 Hr.
1. Steel columns and all members of primary trusses	1-1.1	Carbonate, lightweight and sand-lightweight aggregate concrete, members 6″ by 6″ (152 mm by 152 mm) or greater (not including sandstone, granite and siliceous gravel).[1]	$2^1/_2$	2	$1^1/_2$	1
	1-1.2	Carbonate, lightweight and sand-lightweight aggregate concrete, members 8″ by 8″ (203 mm by 203 mm) or greater (not including sandstone, granite and siliceous gravel).[1]	2	$1^1/_2$	1	1
	1-1.3	Carbonate, lightweight and sand-lightweight aggregate concrete, members 12″ by 12″ (305 mm by 305 mm) or greater (not including sandstone, granite and siliceous gravel).[1]	$1^1/_2$	1	1	1
	1-1.4	Siliceous aggregate concrete and concrete excluded in Item 1-1.1, members 6″ by 6″ (152 mm by 152 mm) or greater.[1]	3	2	$1^1/_2$	1
	1-1.5	Siliceous aggregate concrete and concrete excluded in Item 1-1.1, members 8″ by 8″ (203 mm by 203 mm) or greater.[1]	$2^1/_2$	2	1	1
	1-1.6	Siliceous aggregate concrete and concrete excluded in Item 1-1.1, members 12″ by 12″ (305 mm by 305 mm) or greater.[1]	2	1	1	1
	1-2.1	Clay or shale brick with brick and mortar fill.[1]	$3^3/_4$			$2^1/_4$
	1-3.1	4″ (102 mm) hollow clay tile in two 2″ (51 mm) layers; $^1/_2$″ (12.7 mm) mortar between tile and column; $^3/_8$″ (9.5 mm) metal mesh [0.046″ (1.2 mm) wire diameter] in horizontal joints; tile fill.[1]	4			
	1-3.2	2″ (51 mm) hollow clay tile; $^3/_4$″ (19 mm) mortar between tile and column; $^3/_8$″ (9.5 mm) metal mesh [0.046″ (1.2 mm) wire diameter] in horizontal joints; limestone concrete fill;[1] plastered with $^3/_4$″ (19 mm) gypsum plaster.	3			
	1-3.3	2″ (51 mm) hollow clay tile with outside wire ties [0.08″ (2 mm) diameter] at each course of tile or $^3/_8$″ (9.5 mm) metal mesh [0.046″ (1.2 mm) diameter wire] in horizontal joints; limestone or trap-rock concrete fill[1] extending 1″ (25 mm) outside column on all sides.			3	
	1-3.4	2″ (51 mm) hollow clay tile with outside wire ties [0.08″ (2 mm) diameter] at each course of tile with or without concrete fill; $^3/_4$″ (19 mm) mortar between tile and column.				2
	1-4.1	Cement plaster over metal lath wire tied to $^3/_4$″ (19 mm) cold-rolled vertical channels with 0.049 inch (1.24 mm) (No. 18 B.W. gage) wire ties spaced 3″ to 6″ (76 mm to 152 mm) on center. Plaster mixed 1:2$^1/_2$ by volume, cement to sand.			$2^1/_2{}^2$	$^7/_8$
	1-5.1	Vermiculite concrete, 1:4 mix by volume over paperbacked wire fabric lath wrapped directly around column with additional 2″ by 2″ (51 mm by 51 mm) 0.065 inch/0.065 inch (1.65 mm/1.65 mm) (No. 16/16 B.W. gage) wire fabric placed $^3/_4$″ (19 mm) from outer concrete surface. Wire fabric tied with 0.049 inch (1.24 mm) (No. 18 B.W. gage) wire spaced 6″ (152 mm) on center for inner layer and 2″ (51 mm) on center for outer layer.	2			
	1-6.1	Perlite or vermiculite gypsum plaster over metal lath wrapped around column and furred $1^1/_4$″ (32 mm) from column flanges. Sheets lapped at ends and tied at 6″ (152 mm) intervals with 0.049 inch (1.24 mm) (No. 18 B.W. gage) tie wire. Plaster pushed through to flanges.	$1^1/_2$	1		
	1-6.2	Perlite or vermiculite gypsum plaster over self-furring metal lath wrapped directly around column, lapped 1″ (25 mm) and tied at 6″ (152 mm) intervals with 0.049 inch (1.24 mm) (No. 18 B.W. gage) wire.	$1^3/_4$	$1^3/_8$	1	
	1-6.3	Perlite or vermiculite gypsum plaster on metal lath applied to $^3/_4$″ (19 mm) cold-rolled channels spaced 24 inches (610 mm) apart vertically and wrapped flatwise around column.	$1^1/_2$			
	1-6.4	Perlite or vermiculite gypsum plaster over two layers of $^1/_2$″ (12.7 mm) plain full-length gypsum lath applied tight to column flanges. Lath wrapped with 1″ (25.4 mm) hexagonal mesh of No. 20 gage wire and tied with doubled 0.035 inch diameter (0.89 mm) (No. 18 B.W. gage) wire ties spaced 23″ (584 mm) on center. For three-coat work the plaster mix for the second coat shall not exceed 100 pounds (45.4 kg) of gypsum to $2^1/_2$ cubic feet (0.07 m³) of aggregate for the three-hour system.	$2^1/_2$	2		
	1-6.5	Perlite or vermiculite gypsum plaster over one layer of $^1/_2$″ (12.7 mm) plain full-length gypsum lath applied tight to column flanges. Lath tied with doubled 0.049 inch (1.24 mm) (No. 18 B.W. gage) wire ties spaced 23″ (584 mm) on center and scratch coat wrapped with 1″ (25 mm) hexagonal mesh 0.035 inch (0.89 mm) (No. 20 B.W. gage) wire fabric. For three-coat work, the plaster mix for the second coat shall not exceed 100 pounds (45.4 kg) of gypsum to $2^1/_2$ cubic feet (0.07 m³) of aggregate.		2		
	1-7.1	Multiple layers of $^1/_2$″ (12.7 mm) gypsum wallboard[3] adhesively[4] secured to column flanges and successive layers. Wallboard applied without horizontal joints. Corner edges of each layer staggered. Wallboard layer below outer layer secured to column with doubled 0.049 inch (1.24 mm) (No. 18 B.W. gage) steel wire ties spaced 15″ (381 mm) on center. Exposed corners taped and treated.			2	1

(Continued)

TABLE 7-A—MINIMUM PROTECTION OF STRUCTURAL PARTS BASED ON TIME PERIODS
FOR VARIOUS NONCOMBUSTIBLE INSULATING MATERIALS[a]—(Continued)

STRUCTURAL PARTS TO BE PROTECTED	ITEM NUMBER	INSULATING MATERIAL USED	MINIMUM THICKNESS OF INSULATING MATERIAL FOR FOLLOWING FIRE-RESISTIVE PERIODS (inches)			
			× 25.4 for mm			
			4 Hr.	3 Hr.	2 Hr.	1 Hr.
1. Steel columns and all members of primary trusses (cont.)	1-7.2	Three layers of $5/8''$ (15.9 mm) Type X gypsum wallboard.[3] First and second layer held in place by $1/8''$ (3.2 mm) diameter by $1^3/8''$ (35 mm) long ring shank nails with $5/16''$ (7.9 mm) diameter heads spaced $24''$ (610 mm) on center at corners. Middle layer also secured with metal straps at mid-height and $18''$ (457 mm) from each end, and by metal corner bead at each corner held by the metal straps. Third layer attached to corner bead with $1''$ (25 mm) long gypsum wallboard screws spaced $12''$ (305 mm) on center.			$1^7/8$	
	1-7.3	Three layers of $5/8''$ (15.9 mm) Type X gypsum wallboard,[3] each layer screw attached to $1^5/8''$ (41 mm) steel studs 0.018 inch thick (0.46 mm) (No. 25 carbon sheet steel gage) at each corner of column. Middle layer also secured with 0.049 inch (0.12 mm) (No. 18 B.W. gage) double strand steel wire ties, $24''$ (610 mm) on center. Screws are No. 6 by $1''$ (25 mm) spaced $24''$ (610 mm) on center for inner layer, No. 6 by $1^5/8''$ (41 mm) spaced $12''$ (305 mm) on center for middle layer and No. 8 by $2^1/4''$ (57 mm) spaced $12''$ (305 mm) on center for outer layer.		$1^7/8$		
	1-8.1	Wood-fibered gypsum plaster mixed 1:1 by weight gypsum to sand aggregate applied over metal lath. Lath lapped $1''$ (25 mm) and tied $6''$ (152 mm) on center at all ends, edges and spacers with 0.049 inch (0.12 mm) (No. 18 B.W. gage) steel tie wires. Lath applied over $1/2''$ (12.7 mm) spacers made of $3/4''$ (19 mm) furring channel with $2''$ (51 mm) legs bent around each corner. Spacers located $1''$ (25 mm) from top and bottom of member and a maximum of $40''$ (1016 mm) on center and wire tied with a single strand of 0.049 inch (0.12 mm) (No. 18 B.W. gage) steel tie wires. Corner bead tied to the lath at $6''$ (152 mm) on center along each corner to provide plaster thickness.			$1^5/8$	
2. Webs or flanges of steel beams and girders	2-1.1	Carbonate, lightweight and sand-lightweight aggregate concrete (not including sandstone, granite and siliceous gravel) with $3''$ (76 mm) or finer metal mesh placed $1''$ (25 mm) from the finished surface anchored to the top flange and providing not less than 0.025 square inch of steel area per foot (53 mm² of steel area per meter) in each direction.	2	$1^1/2$	1	1
	2-1.2	Siliceous aggregate concrete and concrete excluded in Item 2-1.1 with $3''$ (76 mm) or finer metal mesh placed $1''$ (25 mm) from the finished surface anchored to the top flange and providing not less than 0.025 square inch of steel area per foot (53 mm² of steel area per meter) in each direction.	$2^1/2$	2	$1^1/2$	1
	2-2.1	Cement plaster on metal lath attached to $3/4''$ (19 mm) cold-rolled channels with 0.049 inch (1.24 mm) (No. 18 B.W. gage) wire ties spaced $3''$ to $6''$ (76 mm to 152 mm) on center. Plaster mixed $1:2^1/2$ by volume, cement to sand.			$2^1/2^2$	$7/8$
	2-3.1	Vermiculite gypsum plaster on a metal lath cage, wire tied to 0.165 inch (4.19 mm) diameter (No. 8 B.W. gage) steel wire hangers wrapped around beam and spaced $16''$ (406 mm) on center. Metal lath ties spaced approximately $5''$ (127 mm) on center at cage sides and bottom.		$7/8$		
	2-4.1	Two layers of $5/8''$ (15.9 mm) Type X gypsum wallboard[3] are attached to U-shaped brackets spaced $24''$ (610 mm) on center. 0.018 inch (0.46 mm) (No. 25 carbon sheet steel gage) $1^5/8''$ deep by $1''$ (41 mm deep by 25 mm) galvanized steel runner channels are first installed parallel to and on each side of the top beam flange to provide a $1/2''$ (12.7 mm) clearance to the flange. The channel runners are attached to steel deck or concrete floor construction with approved fasteners spaced $12''$ (305 mm) on center. U-shaped brackets are formed from members identical to the channel runners. At the bent portion of the U-shaped bracket, the flanges of the channel are cut out so that $1^5/8''$ (41 mm) deep corner channels can be inserted without attachment parallel to each side of the lower flange. As an alternate, 0.021 inch (0.41 mm) (No. 24 carbon sheet steel gage) $1''$ by $2''$ (25 mm by 51 mm) runner and corner angles may be used in lieu of channels, and the web cutouts in the U-shaped brackets may be omitted. Each angle is attached to the bracket with $1/2''$ (12.7 mm) long No. 8 self-drilling screws. The vertical legs of the U-shaped bracket are attached to the runners with one $1/2''$ (12.7 mm) long No. 8 self-drilling screw. The completed steel framing provides a $2^1/8''$ and $1^1/2''$ (54 mm and 38 mm) space between the inner layer of wallboard and the sides and bottom of the steel beam, respectively. The inner layer of wallboard is attached to the top runners and bottom corner channels or corner angles with $1^1/4''$ (52 mm) long No. 6 self-drilling screws spaced $16''$ (406 mm) on center. The outer layer of wallboard is applied with $1^3/4''$ (44.5 mm) long No. 6 self-drilling screws spaced $8''$ (203 mm) on center. The bottom corners are reinforced with metal corner beads.			$1^1/4$	

(Continued)

TABLE 7-A   1997 UNIFORM BUILDING CODE

TABLE 7-A—MINIMUM PROTECTION OF STRUCTURAL PARTS BASED ON TIME PERIODS FOR VARIOUS NONCOMBUSTIBLE INSULATING MATERIALS[a]—(Continued)

STRUCTURAL PARTS TO BE PROTECTED	ITEM NUMBER	INSULATING MATERIAL USED	MINIMUM THICKNESS OF INSULATING MATERIAL FOR FOLLOWING FIRE-RESISTIVE PERIODS (inches) × 25.4 for mm			
			4 Hr.	3 Hr.	2 Hr.	1 Hr.
2. Webs or flanges of steel beams and girders (cont.)	2-4.2	Three layers of $5/8''$ (15.9 mm) Type X gypsum wallboard[3] attached to a steel suspension system as described immediately above utilizing the 0.018 inch (0.46 mm) (No. 25 carbon sheet steel gage) 1″ by 2″ (25 mm by 51 mm) lower corner angles. The framing is located so that a $2^1/8''$ and 2″ (54 mm and 51 mm) space is provided between the inner layer of wallboard and the sides and bottom of the beam, respectively. The first two layers of wallboard are attached as described immediately above. A layer of 0.035 inch (0.89 mm) (No. 20 B.W. gage) 1″ (25 mm) hexagonal galvanized wire mesh is applied under the soffit of the middle layer and up the sides approximately 2″ (51 mm). The mesh is held in position with the No. 6 $1^5/8''$ (41 mm) long screws installed in the vertical leg of the bottom corner angles. The outer layer of wallboard is attached with No. 6 $2^1/4''$ (57 mm) long screws spaced 8″ (203 mm) on center. One screw is also installed at the mid-depth of the bracket in each layer. Bottom corners are finished as described above.		$1^7/8$		

3. Bonded pretensioned reinforcement in prestressed concrete[5]	3-1.1	Carbonate, lightweight, sand-lightweight and siliceous[6] aggregate concrete
		  Beams or girders
		  Solid slabs[8]
			4^7	3^7
				2
					$2^1/2$
					$1^1/2$
						$1^1/2$
						1

4. Bonded or unbonded posttensioned tendons in prestressed concrete[5,9]	4-1.1	Carbonate, lightweight, sand-lightweight and siliceous aggregate concrete
		Unrestrained members:
		  Solid slabs[8]
		  Beams and girders[10]
		  8 in. (203 mm) wide
		  > 12 in. (305 mm) wide
			3	2
				
				$4^1/2$
				$2^1/2$
					$1^1/2$
					
					$2^1/2$
					2
						
						$1^3/4$
						$1^1/2$

	4-1.2	Carbonate, lightweight, sand-lightweight and siliceous aggregate
		Restrained members:[11]
		  Solid slabs[8]
		  Beams and girders[10]
		  8 in. (203 mm) wide
		  > 12 in. (305 mm) wide
			$1^1/4$
			
			$2^1/2$
			2
				1
				
				2
				$1^3/4$
					$3/4$
					
					$1^3/4$
					$1^1/2$	

5. Reinforcing steel in reinforced concrete columns, beams, girders and trusses	5-1.1	Carbonate, lightweight and sand-lightweight aggregate concrete, members 12″ (305 mm) or larger, square or round. (Size limit does not apply to beams and girders monolithic with floors.)	$1^1/2$	$1^1/2$	$1^1/2$	$1^1/2$
	5-1.2	Siliceous aggregate concrete, members 12″ (305 mm) or larger, square or round. (Size limit does not apply to beams and girders monolithic with floors.)	2	$1^1/2$	$1^1/2$	$1^1/2$
6. Reinforcing steel in reinforced concrete joists[12]	6-1.1	Carbonate, lightweight and sand-lightweight aggregate concrete.	$1^1/4$	$1^1/4$	1	$3/4$
	6-1.2	Siliceous aggregate concrete.	$1^3/4$	$1^1/2$	1	$3/4$
7. Reinforcing and tie rods in floor and roof slabs[12]	7-1.1	Carbonate, lightweight and sand-lightweight aggregate concrete.	1	1	$3/4$	$3/4$
	7-1.2	Siliceous aggregate concrete.	$1^1/4$	1	1	$3/4$

[a]Generic fire-resistance ratings (those not designated as PROPRIETARY* in the listing) in the *Fire-Resistance Design Manual,* Fourteenth Edition, dated April 1994, as published by the Gypsum Association, may be accepted as if herein listed.

[1]Reentrant parts of protected members to be filled solidly.

[2]Two layers of equal thickness with a $3/4$-inch (19 mm) air space between.

[3]For all of the construction with gypsum wallboard described in Table 7-A, gypsum base for veneer plaster of the same size, thickness and core type may be substituted for gypsum wallboard, provided attachment is identical to that specified for the wallboard and the joints on the face layer are reinforced, and the entire surface is covered with a minimum of $1/16$-inch (1.6 mm) gypsum veneer plaster.

[4]An approved adhesive qualified under UBC Standard 7-1.

[5]Where lightweight or sand-lightweight concrete having an oven-dry weight of 110 pounds per cubic foot (1762 kg/m^3) or less is used, the tabulated minimum cover may be reduced 25 percent, except that in no case shall the cover be less than $3/4$ inch (19 mm) in slabs or $1^1/2$ inches (38 mm) in beams or girders.

[6]For solid slabs of siliceous aggregate concrete, increase tendon cover 20 percent.

[7]Adequate provisions against spalling shall be provided by U-shaped or hooped stirrups spaced not to exceed the depth of the member with a clear cover of 1 inch (25 mm).

[8]Prestressed slabs shall have a thickness not less than that required in Table 7-C for the respective fire-resistive time period.

[9]Fire coverage and end anchorages shall be as follows: Cover to the prestressing steel at the anchor shall be $1/2$ inch (12.7 mm) greater than that required away from the anchor. Minimum cover to steel-bearing plate shall be 1 inch (25 mm) in beams and $3/4$ inch (19 mm) in slabs.

[10]For beam widths between 8 inches and 12 inches (203 mm and 305 mm), cover thickness can be determined by interpolation.

[11]Interior spans of continuous slabs, beams and girders may be considered restrained.

[12]For use with concrete slabs having a comparable fire endurance where members are framed into the structure in such a manner as to provide equivalent performance to that of monolithic concrete construction.

TABLE 7-B—RATED FIRE-RESISTIVE PERIODS FOR VARIOUS WALLS AND PARTITIONS[a,1]

MATERIAL	ITEM NUMBER	CONSTRUCTION	MINIMUM FINISHED THICKNESS FACE-TO-FACE[2] (inches) × 25.4 for mm			
			4 Hr.	3 Hr.	2 Hr.	1 Hr.
1. Brick of clay or shale	1-1.1	Solid units (at least 75 percent solid).	8		6[3]	4
	1-2.1	Hollow brick units[4] (at least 71 percent solid).		8		
	1-2.2	Hollow brick units (at least 60 percent solid, cells filled with loose-fill insulation).	8			
	1-2.3	Hollow brick units at least 64 percent solid.	12			
	1-2.4	Hollow brick, not filled.	5.0	4.3	3.4	2.3
	1-2.5	Hollow brick unit wall, grout or filled with perlite vermiculite or expanded shale aggregate.	6.6	5.5	4.4	3.0
	1-3.1	Hollow (rowlock[5]).	12		8	
	1-4.1	Cavity wall consisting of two 3″ (76 mm) (actual) solid clay brick units separated by 2″ (51 mm) air space, joint reinforcement every 16″ (406 mm) on center vertically.		8		
	1-4.2	Cavity wall consisting of two 4″ (100 mm) nominal solid clay brick units separated by 2″ (51 mm) air space, 1/4″ (6.4 mm) metal ties for 3 square feet (0.28 m²) of wall area.	10			
	1-5.1	4″ (102 mm) nominal thick units at least 75 percent solid backed with a hat-shaped metal furring channel 3/4″ (19 mm) thick formed from 0.021″ (0.53 mm) sheet metal attached to the brick wall on 24″ (610 mm) centers with approved fasteners, and 1/2″ (12.7 mm) Type X gypsum wallboard[7] attached to the metal furring strips with 1″ (25 mm) long Type S screws spaced 8″ (203 mm) on center.			5[6]	
2. Hollow clay tile, nonload-bearing	2-1.1	Two cells in wall thickness, units at least 40 percent solid.				8
	2-1.2	Two cells in wall thickness, units at least 43 percent solid.				
	2-1.3	Two cells in wall thickness, units at least 46 percent solid.				8
	2-1.4	Two cells in wall thickness, units at least 49 percent solid.			8	
	2-1.5	Three or four cells in wall thickness, units at least 40 percent solid.				8
	2-1.6	Three or four cells in wall thickness, units at least 43 percent solid.			8	
	2-1.7	Three or four cells in wall thickness, units at least 48 percent solid.			8	
	2-1.8	Three or four cells in wall thickness, units at least 53 percent solid.		8		
	2-1.9	Three cells in wall thickness, units at least 40 percent solid.			12	
	2-1.10	Three cells in wall thickness, units at least 45 percent solid.		12		
	2-1.11	Three cells in wall thickness, units at least 49 percent solid.		12		
	2-1.12	Two units and three or four cells in wall thickness, units at least 40 percent solid.		12		
	2-1.13	Two units and three or four cells in wall thickness, units at least 45 percent solid.	12			
	2-1.14	Two units and three or four cells in wall thickness, units at least 53 percent solid.	12			
	2-1.15	Two or three units and four or five cells in wall thickness, units at least 40 percent solid.	16			
3. Structural clay tile, load-bearing	3-1.1	One cell in wall thickness, units at least 40 percent solid.[8,9]				4
	3-1.2	One cell in wall thickness, units at least 30 percent solid.[8,9]			6	
	3-1.3	Two cells in wall thickness, units at least 45 percent solid.[10]				6
	3-1.4	One cell in wall thickness, units at least 40 percent solid.[9,10]				4
	3-1.5	One cell in wall thickness, units at least 30 percent solid.[9,10]			6	
4. Hollow structural clay tile, load-bearing	4-1.1	Two cells in wall thickness, units at least 40 percent solid.				8
	4-1.2	Two cells in wall thickness, units at least 49 percent solid.			8	
	4-1.3	Three or four cells in wall thickness, units at least 53 percent solid.		8		
	4-1.4	Two cells in wall thickness, units at least 46 percent solid.				8
	4-1.5	Three cells in wall thickness, units at least 40 percent solid.			12	
	4-1.6	Two units and three cells in wall thickness, units at least 40 percent solid.		12		
	4-1.7	Two units and three or four cells in wall thickness, units at least 45 percent solid.	12			
	4-1.8	Three cells in wall thickness, units at least 45 percent solid.		12		
	4-1.9	Three cells in wall thickness, units at least 49 percent solid.		12		
	4-1.10	Two units and four cells in wall thickness, units at least 43 percent solid.	16			
	4-1.11	Two or three units and four or five cells in wall thickness, units at least 40 percent solid.	16			

(Continued)

TABLE 7-B

1997 UNIFORM BUILDING CODE

TABLE 7-B—RATED FIRE-RESISTIVE PERIODS FOR VARIOUS WALLS AND PARTITIONS[a,1]—(Continued)

MATERIAL	ITEM NUMBER	CONSTRUCTION	MINIMUM FINISHED THICKNESS FACE-TO-FACE[2] (inches) × 25.4 for mm			
			4 Hr.	3 Hr.	2 Hr.	1 Hr.
5. Combination of clay brick and load-bearing hollow clay tile	5.1.1	4″ (102 mm) solid brick and 4″ (102 mm) tile (at least 40 percent solid).		8		
	5.1.2	4″ (102 mm) solid brick and 8″ (203 mm) tile (at least 40 percent solid).	12			
6. Concrete masonry units	6-1.1[11, 12]	Expanded slag or pumice.	4.7	4.0	3.2	2.1
	6-1.2[11, 12]	Expanded clay, shale or slate.	5.1	4.4	3.6	2.6
	6-1.3[11]	Limestone, cinders or air-cooled slag.	5.9	5.0	4.0	2.7
	6-1.4[11, 12]	Calcareous or siliceous gravel.	6.2	5.3	4.2	2.8
7. Solid concrete[13, 14]	7-1.1	Siliceous aggregate concrete.	7.0	6.2	5.0	3.5
		Carbonate aggregate concrete.	6.6	5.7	4.6	3.2
		Sand-lightweight concrete.	5.4	4.6	3.8	2.7
		Lightweight concrete.	5.1	4.4	3.6	2.5
8. Glazed or unglazed facing tile, nonload-bearing	8-1.1	One 2″ (51 mm) unit cored 15 percent maximum and one 4″ (102 mm) unit cored 25 percent maximum with $^3/_4$″ (19 mm) mortar-filled collar joint. Unit positions reversed in alternate courses.		$6^3/_8$		
	8-1.2	One 2″ (51 mm) unit cored 15 percent maximum and one 4″ (102 mm) unit cored 40 percent maximum with $^3/_8$″ (9.5 mm) mortar-filled collar joint. Plastered one side with $^3/_4$″ (19 mm) gypsum plaster. Two wythes tied together every fourth course with No. 22 gage corrugated metal ties.		$6^3/_4$		
	8-1.3	One unit with three cells in wall thickness, cored 29 percent maximum.		6		
	8-1.4	One 2″ (51 mm) unit cored 22 percent maximum and one 4″ (102 mm) unit cored 41 percent maximum with $^1/_4$″ (6 mm) mortar-filled collar joint. Two wythes tied together every third course with 0.030 inch (0.76 mm) (No. 22 galvanized sheet steel gage) corrugated metal ties.		6		
	8-1.5	One 4″ (102 mm) unit cored 25 percent maximum with $^3/_4$″ (19 mm) gypsum plaster on one side.			$4^3/_4$	
	8-1.6	One 4″ (102 mm) unit with two cells in wall thickness, cored 22 percent maximum.				4
	8-1.7	One 4″ (102 mm) unit cored 30 percent maximum with $^3/_4$″ (19 mm) vermiculite gypsum plaster on one side.			$4^1/_2$	
	8-1.8	One 4″ (102 mm) unit cored 39 percent maximum with $^3/_4$″ (19 mm) gypsum plaster on one side.				$4^1/_2$
9. Solid gypsum plaster	9-1.1	$^3/_4$″ (19 mm) by 0.055 inch (1.4 mm) (No. 16 carbon sheet steel gage) vertical cold-rolled channels, 16″ (406 mm) on center with 2.5-pound (1.13 kg) flat metal lath applied to one face and tied with 0.049 inch (1.24 mm) (No. 18 B.W. gage) wire at 6″ (152 mm) spacing. Gypsum plaster each side mixed 1:2 by weight, gypsum to sand aggregate.				2^6
	9-1.2	$^3/_4$″ (19 mm) by 0.055 inch (1.4 mm) (No. 16 carbon sheet steel gage) cold-rolled channels 16″ (406 mm) on center with metal lath applied to one face and tied with 0.049 inch (1.24 mm) (No. 18 B.W. gage) wire at 6″ (152 mm) spacing. Perlite or vermiculite gypsum plaster each side. For three-coat work, the plaster mix for the second coat shall not exceed 100 pounds (45.4 kg) of gypsum to $2^1/_2$ cubic feet (0.071 m^3) of aggregate for the one-hour system.			$2^1/_2{}^6$	2^6
	9-1.3	$^3/_4$″ (19 mm) by 0.055 inch (1.4 mm) (No. 16 carbon sheet steel gage) vertical cold-rolled channels, 16″ (406 mm) on center, with $^3/_8$″ (9.5 mm) gypsum lath applied to one face and attached with sheet metal clips. Gypsum plaster each side mixed 1:2 by weight, gypsum to sand aggregate.				2^6
	9-2.1	Studless with $^1/_2$″ (12.7 mm) full-length plain gypsum lath and gypsum plaster each side. Plaster mixed 1:1 for scratch coat and 1:2 for brown coat, by weight, gypsum to sand aggregate.				2^6
	9-2.2	Studless with $^1/_2$″ (12.7 mm) full-length plain gypsum lath and perlite or vermiculite gypsum plaster each side.			$2^1/_2{}^6$	2^6
	9-2.3	Studless partition with $^3/_8$″ (9.5 mm) rib metal lath installed vertically, adjacent edges tied 6″ (152 mm) on center with No. 18 gage wire ties, gypsum plaster each side mixed 1:2 by weight, gypsum to sand aggregate.				2^6
10. Solid perlite and portland cement	10-1.1	Perlite mixed in the ratio of 3 cubic feet (0.085 m^3) to 100 pounds (45.4 kg) of portland cement and machine applied to stud side of $1^1/_2$″ (38 mm) mesh by 0.058 inch (1.47 mm) (No. 17 B.W. gage) paper-backed woven wire fabric lath wire-tied to 4″ (102 mm) deep steel trussed wire[15] studs 16″ (406 mm) on center. Wire ties of 0.049 inch (1.24 mm) (No. 18 B.W. gage) galvanized steel wire 6″ (152 mm) on center vertically.			$3^1/_8{}^6$	

(Continued)

TABLE 7-B—RATED FIRE-RESISTIVE PERIODS FOR VARIOUS WALLS AND PARTITIONS[a,1]—(Continued)

MATERIAL	ITEM NUMBER	CONSTRUCTION	MINIMUM FINISHED THICKNESS FACE-TO-FACE[2] (inches) × 25.4 for mm			
			4 Hr.	3 Hr.	2 Hr.	1 Hr.
11. Solid neat wood fibered gypsum plaster	11-1.1	$3/4''$ (19 mm) by 0.055 inch (1.4 mm) (No. 16 carbon sheet steel gage) cold-rolled channels, $12''$ (305 mm) on center with 2.5-pound (1.13 kg) flat metal lath applied to one face and tied with 0.049 inch (1.24 mm) (No. 18 B.W. gage) wire at $6''$ (152 mm) spacing. Neat gypsum plaster applied each side.			2^6	
12. Solid gypsum wallboard partition	12-1.1	One full-length layer $1/2''$ (12.7 mm) Type X gypsum wallboard[7] laminated to each side of $1''$ (25 mm) full-length V-edge gypsum coreboard with approved laminating compound. Vertical joints of face layer and coreboard staggered at least $3''$ (76 mm).			2^6	
13. Hollow (studless) gypsum wallboard partition	13-1.1	One full-length layer of $5/8''$ (15.9 mm) Type X gypsum wallboard[7] attached to both sides of wood or metal top and bottom runners laminated to each side of $1''$ by $6''$ (25 mm by 152 mm) full-length gypsum coreboard ribs spaced $24''$ (610 mm) on center with approved laminating compound. Ribs centered at vertical joints of face plies and joints staggered $24''$ (610 mm) in opposing faces. Ribs may be recessed $6''$ (152 mm) from the top and bottom.				$2^1/4^6$
	13-1.2	$1''$ (25 mm) regular gypsum V-edge full-length backing board attached to both sides of wood or metal top and bottom runners with nails or $1^5/8''$ (41 mm) drywall screws at $24''$ (610 mm) on center. Minimum width of runners $1^5/8''$ (41 mm). Face layer of $1/2''$ (12.7 mm) regular full-length gypsum wallboard laminated to outer faces of backing board with approved laminating compound.				$4^5/8^6$
14. Noncombustible studs—interior partition with plaster each side	14-1.1	$3^1/4''$ (82 mm) by 0.044 inch (1.12 mm) (No. 18 carbon sheet steel gage) steel studs spaced $24''$ (610 mm) on center. $5/8''$ (15.9 mm) gypsum plaster on metal lath each side mixed 1:2 by weight, gypsum to sand aggregate.				$4^3/4^6$
	14-1.2	$3^5/8''$ (92 mm) 0.055 inch (1.4 mm) (No. 16 carbon sheet steel gage) approved nailable[16] studs spaced $24''$ (610 mm) on center. $5/8''$ (15.9 mm) neat gypsum wood fibered plaster each side over $3/8''$ (9.5 mm) rib metal lath nailed to studs with 6d common nails, $8''$ (203 mm) on center. Nails driven $1^1/4''$ (32 mm) and bent over.			$5^5/8$	
	14-1.3	$4''$ (102 mm) 0.044 inch (1.12 mm) (No. 18 carbon sheet steel gage) channel-shaped steel studs at $16''$ (406 mm) on center. On each side approved resilient clips pressed onto stud flange at $16''$ (406 mm) vertical spacing, $1/4''$ (6.4 mm) pencil rods snapped into or wire-tied onto outer loop of clips, metal lath wire-tied to pencil rods at $6''$ (152 mm) intervals, $1''$ (25 mm) perlite gypsum plaster, each side.	$7^5/8^6$			
	14-1.4	$2^1/2''$ (63.5 mm) 0.044 inch (1.12 mm) (No. 18 carbon sheet steel gage) steel studs spaced $16''$ (406 mm) on center. Wood fibered gypsum plaster mixed 1:1 by weight gypsum to sand aggregate applied on 3.4-pound (1.54 kg) metal lath wire tied to studs, each side. $3/4''$ (19 mm) plaster applied over each face, including finish coat.				$4^1/4^6$
15. Wood studs interior partition with plaster each side	15-1.1[17, 18]	$2''$ by $4''$ (51 mm by 102 mm) wood studs $16''$ (406 mm) on center with $5/8''$ (15.9 mm) gypsum plaster on metal lath. Lath attached by 4d common nails bent over or No. 14 gage by $1^1/4''$ by $3/4''$ (31.7 mm by 19 mm) crown width staples spaced $6''$ (152 mm) on center. Plaster mixed $1:1^1/2$ for scratch coat and 1:3 for brown coat, by weight, gypsum to sand aggregate.				$5^1/8$
	15-1.2[17]	$2''$ by $4''$ (51 mm by 102 mm) wood studs $16''$ (406 mm) on center with metal lath and $7/8''$ (22 mm) neat wood fibered gypsum plaster each side. Lath attached by 6d common nails, $7''$ (178 mm) on center. Nails driven $1^1/4''$ (31.7 mm) and bent over.				$5^1/2^6$
	15-1.3[11, 17]	$2''$ by $4''$ (51 mm by 102 mm) wood studs $16''$ (406 mm) on center with $3/8''$ (9.5 mm) perforated or plain gypsum lath and $1/2''$ (12.7 mm) gypsum plaster each side. Lath nailed with $1^1/8''$ (28.6 mm) by No. 13 gage by $19/64''$ (7.5 mm) head plasterboard blued nails, $4''$ (102 mm) on center. Plaster mixed 1:2 by weight, gypsum to sand aggregate.				$5^1/4$
	15-1.4[11, 17]	$2''$ by $4''$ (51 mm by 102 mm) wood studs $16''$ (406 mm) on center with $3/8''$ (9.5 mm) Type X gypsum lath and $1/2''$ (12.7 mm) gypsum plaster each side. Lath nailed with $1^1/8''$ (28.6 mm) by No. 13 gage by $19/64''$ (7.5 mm) head plasterboard blued nails, $5''$ (127 mm) on center. Plaster mixed 1:2 by weight, gypsum to sand aggregate.				$5^1/4$
16. Noncombustible studs—interior partition with gypsum wallboard each side	16-1.1	0.018 inch (0.46 mm) (No. 25 carbon sheet steel gage) channel-shaped studs $24''$ (610 mm) on center with one full-length layer of $5/8''$ (15.9 mm) Type X gypsum wallboard[7] applied vertically attached with $1''$ (25 mm) long No. 6 drywall screws to each stud. Screws are $8''$ (203 mm) on center around the perimeter and $12''$ (305 mm) on center on the intermediate stud. The wallboard may be applied horizontally when attached to $3^5/8''$ (92 mm) studs and the horizontal joints are staggered with those on the opposite side. Screws for the horizontal application shall be $8''$ (203 mm) on center at vertical edges and $12''$ (305 mm) on center at intermediate studs.				$2^7/8^6$
	16-1.2	0.018 inch (0.46 mm) (No. 25 carbon sheet steel gage) channel-shaped studs $24''$ (610 mm) on center with two full-length layers of $1/2''$ (12.7 mm) Type X gypsum wallboard[7] applied vertically each side. First layer attached with $1''$ (25 mm) long, No. 6 drywall screws, $8''$ (203 mm) on center around the perimeter and $12''$ (305 mm) on center on the intermediate stud. Second layer applied with vertical joints offset one stud space from first layer using $1^5/8''$ (41.3 mm) long, No. 6 drywall screws spaced $9''$ (229 mm) on center along vertical joints, $12''$ (305 mm) on center at intermediate studs and $24''$ (610 mm) on center along top and bottom runners.			$3^5/8^6$	

(Continued)

TABLE 7-B—RATED FIRE-RESISTIVE PERIODS FOR VARIOUS WALLS AND PARTITIONS[a,1]—(Continued)

MATERIAL	ITEM NUMBER	CONSTRUCTION	MINIMUM FINISHED THICKNESS FACE-TO-FACE[2] (inches) × 25.4 for mm			
			4 Hr.	3 Hr.	2 Hr.	1 Hr.
16. Noncombustible studs—interior partition with gypsum wallboard each side (cont.)	16-1.3	0.055 inch (1.40 mm) (No. 16 carbon sheet steel gage) approved nailable metal studs[16] 24″ (610 mm) on center with full-length $^5/_8$″ (15.9 mm) Type X gypsum wallboard[7] applied vertically and nailed 7″ (178 mm) on center with 6d cement-coated common nails. Approved metal fastener grips used with nails at vertical butt joints along studs.				$4^7/_8$
17. Wood studs—interior partition with gypsum wallboard each side	17-1.1[13, 18]	2″ by 4″ (51 mm by 102 mm) wood studs 16″ (406 mm) on center with two layers of $^3/_8$″ (9.5 mm) regular gypsum wallboard[7] each side, 4d cooler[19] or wallboard[19] nails at 8″ (203 mm) on center first layer, 5d cooler[19] or wallboard[19] nails at 8″ (203 mm) on center second layer with laminating compound between layers. Joints staggered. First layer applied full length vertically, second layer applied horizontally or vertically.				5
	17-1.2[17, 18]	2″ by 4″ (51 mm by 102 mm) wood studs 16″ (406 mm) on center with two layers $^1/_2$″ (12.7 mm) regular gypsum wallboard[7] applied vertically or horizontally each side, joints staggered. Nail base layer with 5d cooler[19] or wallboard[15] nails at 8″ (203 mm) on center, face layer with 8d cooler[19] or wallboard[19] nails at 8″ (203 mm) on center.				$5^1/_2$
	17-1.3[17, 18]	2″ by 4″ (51 mm by 102 mm) wood studs 24″ (610 mm) on center with $^5/_8$″ (15.9 mm) Type X gypsum wallboard[7] applied vertically or horizontally nailed with 6d cooler[19] or wallboard[19] nails at 7″ (178 mm) on center with end joints on nailing members. Stagger joints each side.				$4^3/_4$
	17-1.4[17]	2″ by 4″ (51 mm by 102 mm) fire-retardant-treated wood studs spaced 24″ (610 mm) on center with one layer of $^5/_8$″ (15.9 mm) thick Type X gypsum wallboard[7] applied with face paper grain (long dimension) parallel to studs. Wallboard attached with 6d cooler[19] or wallboard[19] nails at 7″ (178 mm) on center.				$4^3/_4$[6]
	17-1.5[17, 18]	2″ by 4″ (51 mm by 102 mm) wood studs 16″ (406 mm) on center with two layers $^5/_8$″ (15.9 mm) Type X gypsum wallboard[7] each side. Base layers applied vertically and nailed with 6d cooler[19] or wallboard[19] nails at 9″ (229 mm) on center. Face layer applied vertically or horizontally and nailed with 8d cooler[19] or wallboard[19] nails at 7″ (178 mm) on center. For nail-adhesive application, base layers are nailed 6″ (152 mm) on center. Face layers applied with coating of approved wallboard adhesive and nailed 12″ (305 mm) on center.			6	
	17-1.6[17]	2″ by 3″ (51 mm by 76 mm) fire-retardant-treated wood studs spaced 24″ (610 mm) on center with one layer of $^5/_8$″ (15.9 mm) thick Type X gypsum wallboard[7] applied with face paper grain (long dimension) at right angles to studs. Wallboard attached with 6d cement-coated box nails spaced 7″ (178 mm) on center.				$3^5/_8$[6]
18. Exterior or interior walls	18-1.1[17, 18]	Exterior surface with $^3/_4$″ (19 mm) drop siding over $^1/_2$″ (12.7 mm) gypsum sheathing on 2″ by 4″ (51 mm by 102 mm) wood studs at 16″ (406 mm) on center; interior surface treatment as required for one-hour-rated exterior or interior 2″ by 4″ (51 mm by 102 mm) wood stud partitions. Gypsum sheathing nailed with $1^3/_4$″ (44.5 mm) by No. 11 gage by $^7/_{16}$″ (11.1 mm) head galvanized nails at 8″ (203 mm) on center. Siding nailed with 7d galvanized smooth box nails.				Varies
	18-1.2[17, 18]	2″ by 4″ (51 mm by 102 mm) wood studs 16″ (406 mm) on center with metal lath and $^3/_4$″ (19 mm) cement plaster on each side. Lath attached with 6d common nails 7″ (178 mm) on center driven to 1″ (25 mm) minimum penetration and bent over. Plaster mix 1:4 for scratch coat and 1:5 for brown coat, by volume, cement to sand.				$5^3/_8$
	18-1.3[17, 18]	2″ by 4″ (51 mm by 102 mm) wood studs 16″ (406 mm) on center with $^7/_8$″ (22 mm) cement plaster (measured from the face of studs) on the exterior surface with interior surface treatment as required for interior wood stud partitions in this table. Plaster mix 1:4 for scratch coat and 1:5 for brown coat, by volume, cement to sand.				Varies
	18-1.4	$3^5/_8$″ (92 mm) No. 16 gage noncombustible studs 16″ (406 mm) on center with $^7/_8$″ (22 mm) cement plaster (measured from the face of the studs) on the exterior surface with interior surface treatment as required for interior, nonbearing, noncombustible stud partitions in this table. Plaster mix 1:4 for scratch coat and 1:5 for brown coat, by volume, cement to sand.				Varies[6]
	18-1.5[18]	$2^1/_4$″ by $3^3$4″ (57 mm by 95 mm) clay face brick with cored holes over $^1/_2$″ (12.7 mm) gypsum sheathing on exterior surface of 2″ by 4″ (51 mm by 102 mm) wood studs at 16″ (406 mm) on center and two layers $^5/_8$″ (15.9 mm) Type X gypsum wallboard[7] on interior surface. Sheathing placed horizontally or vertically with vertical joints over studs nailed 6″ (152 mm) on center with $1^3/_4$″ (44.5 mm) by No. 11 gage by $^7/_{16}$″ (11.1 mm) head galvanized nails. Inner layer of wallboard placed horizontally or vertically and nailed 8″ (203 mm) on center with 6d cooler[19] or wallboard[19] nails. Outer layer of wallboard placed horizontally or vertically and nailed 8″ (203 mm) on center with 8d cooler[19] or wallboard[19] nails. All joints staggered with vertical joints over studs. Outer layer joints taped and finished with compound. Nail heads covered with joint compound. 0.035 inch (0.91 mm) (No. 20 galvanized sheet gage) corrugated galvanized steel wall ties $^3/_4$″ by $6^5/_8$″ (19 mm by 168 mm) attached to each stud with two 8d cooler[19] or wallboard[19] nails every sixth course of bricks.			10	

(Continued)

TABLE 7-B—RATED FIRE-RESISTIVE PERIODS FOR VARIOUS WALLS AND PARTITIONS[a,1]—(Continued)

MATERIAL	ITEM NUMBER	CONSTRUCTION	MINIMUM FINISHED THICKNESS FACE-TO-FACE[2] (inches) × 25.4 for mm			
			4 Hr.	3 Hr.	2 Hr.	1 Hr.
18. Exterior or interior walls (cont.)	18-1.6[17, 18]	2″ by 6″ (51 mm by 152 mm) fire-retardant-treated wood studs 16″ (406 mm) on center. Interior face has two layers of $^5/_8$″ (15.9 mm) Type X gypsum wallboard[7] with the base layer placed vertically and attached with 6d box nails 12″ (305 mm) on center. The face layer is placed horizontally and attached with 8d box nails 8″ (203 mm) on center at joints and 12″ (305 mm) on center elsewhere. The exterior face has a base layer of $^5/_8$″ (15.9 mm) Type X gypsum wallboard placed vertically with 6d box nails 8″ (203 mm) on center at joints and 12″ (305 mm) on center elsewhere. An approved building paper is next applied, followed by self-furred exterior lath attached with $2^1/_2$″ (63.5 mm), No. 12 gage galvanized roofing nails with a $^3/_8$″ (9.5 mm) diameter head and spaced 6″ (152 mm) on center along each stud. Cement plaster consisting of a $^1/_2$″ (12.7 mm) brown coat is then applied. The scratch coat is mixed in the proportion of 1:3 by weight, cement to sand with 10 pounds (4.54 kg) of hydrated lime and 3 pounds (1.36 kg) of approved additives or admixtures per sack of cement. The brown coat is mixed in the proportion of 1:4 by weight, cement to sand with the same amounts of hydrated lime and approved additives or admixtures used in the scratch coat.			$8^1/_4$	
	18-1.7[17, 18]	2″ by 6″ (51 mm by 152 mm) wood studs 16″ (406 mm) on center. The exterior face has a layer of $^5/_8$″ (15.9 mm) Type X gypsum wallboard[7] placed vertically with 6d box nails 8″ (203 mm) on center at joints and 12″ (305 mm) on center elsewhere. An approved building paper is next applied, followed by 1″ (25 mm) by No. 18 gage self-furred exterior lath attached with 8d by $2^1/_2$″ (63.5 mm) long galvanized roofing nails spaced 6″ (152 mm) on center along each stud. Cement plaster consisting of a $^1/_2$″ (12.7 mm) scratch coat, a bonding agent and a $^1/_2$″ (12.7 mm) brown coat and a finish coat is then applied. The scratch coat is mixed in the proportion of 1:3 by weight, cement to sand with 10 pounds (4.54 kg) of hydrated lime and 3 pounds (1.36 kg) of approved additives or admixtures per sack of cement. The brown coat is mixed in the proportion of 1:4 by weight, cement to sand with the same amounts of hydrated lime and approved additives or admixtures used in the scratch coat. The interior is covered with $^3/_8$″ (9.5 mm) gypsum lath with 1″ (25 mm) hexagonal mesh of 0.035 inch (0.89 mm) (No. 20 B.W. gage) woven wire lath furred out $^5/_{16}$″ (8 mm) and 1″ (25 mm) perlite or vermiculite gypsum plaster. Lath nailed with $1^1/_8$″ (28.6 mm) by No. 13 gage by $^{19}/_{64}$″ (7.5 mm) head plasterboard blued nails spaced 5″ (127 mm) on center. Mesh attached by $1^3/_4$″ (44.5 mm) by No. 12 gage by $^3/_8$″ (9.5 mm) head nails with $^3/_8$″ (9.5 mm) furrings, spaced 8″ (203 mm) on center. The plaster mix shall not exceed 100 pounds (45.4 kg) of gypsum to $2^1/_2$ cubic feet (0.071 m^3) of aggregate.			$8^3/_8$	
	18-1.8[17, 18]	2″ by 6″ (51 mm by 152 mm) wood studs 16″ (406 mm) on center. The exterior face has a layer of $^5/_8$″ (15.9 mm) Type X gypsum wallboard[7] placed vertically with 6d box nails 8″ (203 mm) on center at joints and 12″ (305 mm) on center elsewhere. An approved building paper is next applied, followed by $1^1/_2$″ (38 mm) by No. 17 gage self-furred exterior lath attached with 8d by $2^1/_2$″ (63.5 mm) long galvanized roofing nails spaced 6″ (153 mm) on center along each stud. Cement plaster consisting of a $^1/_2$″ (12.7 mm) scratch coat, and a $^1/_2$″ (12.7 mm) brown coat is then applied. The plaster may be placed by machine. The scratch coat is mixed in the proportion of 1:4 by weight, plastic cement to sand. The brown coat is mixed in the proportion of 1:5 by weight, plastic cement to sand. The interior is covered with $^3/_8$″ (9.5 mm) gypsum lath with 1″ (25 mm) hexagonal mesh of No. 20 gage woven wire lath furred out $^5/_{16}$″ (8 mm) and 1″ (25 mm) perlite or vermiculite gypsum plaster. Lath nailed with $1^1/_8$″ (28.6 mm) by No. 13 gage by $^{19}/_{64}$″ (7.5 mm) head plasterboard blued nails spaced 5″ (127 mm) on center. Mesh attached by $1^3/_4$″ (44.5 mm) by No. 12 gage by $^3/_8$″ (9.5 mm) head nails with $^3/_8$″ (9.5 mm) furrings, spaced 8″ (203 mm) on center. The plaster mix shall not exceed 100 pounds (45.4 kg) of gypsum to $2^1/_2$ cubic feet (0.071 m^3) of aggregate.			$8^3/_8$	
	18-1.9	4″ (102 mm) No. 18 gage, nonload-bearing metal studs, 16″ (406 mm) on center, with 1″ (25 mm) portland cement lime plaster [measured from the back side of the 3.4-pound (1.54 kg) expanded metal lath] on the exterior surface. Interior surface to be covered with 1″ (25 mm) of gypsum plaster on 3.4-pound (1.54 kg) expanded metal lath proportioned by weight—1:2 for scratch coat, 1:3 for brown, gypsum to sand. Lath on one side of the partition fastened to $^1/_4$″ (6.4 mm) diameter pencil rods supported by No. 20 gage metal clips, located 16″ (406 mm) on center vertically, on each stud. 3″ (76 mm) thick mineral fiber insulating batts friction fitted between the studs.			$6^1/_2$[6]	

(Continued)

TABLE 7-B   1997 UNIFORM BUILDING CODE

TABLE 7-B—RATED FIRE-RESISTIVE PERIODS FOR VARIOUS WALLS AND PARTITIONS[a,1]—(Continued)

MATERIAL	ITEM NUMBER	CONSTRUCTION	MINIMUM FINISHED THICKNESS FACE-TO-FACE[2] (inches) × 25.4 for mm			
			4 Hr.	3 Hr.	2 Hr.	1 Hr.
18. Exterior or interior walls (cont.)	18-1.10	Steel studs 0.060″ (1.52 mm) thick, 4″ deep (102 mm) or 6″ (152 mm) at 16″ (244 mm) or 24″ (610 mm) centers, with $1/2$″ (12.7 mm) Glass Fiber Reinforced Concrete (GFRC) on the exterior surface. GFRC is attached with flex anchors at 24″ (610 mm) on center, with 5″ (127 mm) leg welded to studs with two $1/2$″-long (12.7 mm) flare-bevel welds, and 4″ (102 mm) foot attached to the GFRC skin with $5/8$″ (16 mm) thick GFRC bonding pads that extend $2^1/2$″ (63.5 mm) beyond the flex anchor foot on both sides. Interior surface to have two layers of $1/2$″ (12.7 mm) Type X gypsum wallboard.[7] The first layer of wallboard to be attached with 1-inch-long (25 mm) Type S buglehead screws spaced 24″ (610 mm) on center and the second layer is attached with $1^5/8$-inch-long (40 mm) Type S screws spaced at 12″ (305 mm) on center. Cavity is to be filled with 5″ (127 mm) of 4 pcf (64 kg/m^3) (nominal) mineral fiber batts. GFRC has $1^1/2$″ (38 mm) returns packed with mineral fiber and caulked on the exterior.			$6^1/2$	
	18-1.11	Steel studs 0.060″ (1.52 mm) thick, 4″ deep (102 mm) or 6″ (152 mm) at 16″ (406 mm) or 24″ (610 mm) centers, with $1/2$″ (12.7 mm) Glass Fiber Reinforced Concrete (GFRC) on the exterior surface. GFRC is attached with flex anchors at 24″ (610 mm) on center, with 5″ (127 mm) leg welded to studs with two $1/2$″-long (12.7 mm) flare-bevel welds, and 4″ (102 mm) foot attached to the GFRC skin with $5/8$″ (16 mm) thick GFRC bonding pads that extend $2^1/2$″ (63.5 mm) beyond the flex anchor foot on both sides. Interior surface to have one layer of $5/8$″ (16 mm) Type X gypsum wallboard,[7] attached with $1^1/4$-inch-long (32 mm) Type S buglehead screws spaced 12″ (305 mm) on center. Cavity is to be filled with 5″ (127 mm) of 4 pcf (64 kg/m^3) (nominal) mineral fiber batts. GFRC has $1^1/2$″ (38 mm) returns packed with mineral fiber and caulked on the exterior.				$6^1/8$

[a]Generic fire-resistance ratings (those not designated as PROPRIETARY* in the listing) in the *Fire-Resistance Design Manual,* Fourteenth Edition, dated April 1994, as published by the Gypsum Association, may be accepted as if herein listed.

[1]Staples with equivalent holding power and penetration may be used as alternate fasteners to nails for attachment to wood framing.

[2]Thickness shown for brick and clay tile are nominal thicknesses unless plastered, in which case thicknesses are net. Thickness shown for concrete masonry and hollow clay or shale brick is equivalent thickness defined as the average thickness of solid material in the wall and is represented by the formula:

$$T_E = \frac{V}{L \times H}$$

WHERE:

H = height of block or brick using specified dimensions as defined in Chapter 21, in inches (mm).

L = length of block or brick using specified dimensions as defined in Chapter 21, in inches (mm).

T_E = equivalent thickness, in inches (mm).

V = net volume (gross volume less volume of voids), in cubic inches (mm^3).

When all cells are solid grouted or filled with silicone-treated perlite loose-fill insulation; vermiculite loose-fill insulation; or expanded clay, shale or slate light-weight aggregate, the equivalent thickness shall be the thickness of the block or brick using specified dimensions as defined in Chapter 21. Equivalent thickness may also include the thickness of applied plaster and lath or gypsum wallboard, where specified.

[3]Single-wythe brick.

[4]Hollow brick units 4-inch by 8-inch by 12-inch (102 mm by 203 mm by 305 mm) nominal with two interior cells having a $1^1/2$-inch (38 mm) web thickness between cells and $1^3/4$-inch-thick (44.5 mm) face shells.

[5]Rowlock design employs clay brick with all or part of bricks laid on edge with the bond broken vertically.

[6]Shall be used for nonbearing purposes only.

[7]For all of the construction with gypsum wallboard described in this table, gypsum base for veneer plaster of the same size, thickness and core type may be substituted for gypsum wallboard, provided attachment is identical to that specified for the wallboard, and the joints on the face layer are reinforced and the entire surface is covered with a minimum of $1/16$-inch (1.6 mm) gypsum veneer plaster.

[8]Ratings are for hard-burned clay or shale tile.

[9]Cells filled with tile, stone, slag, cinders or sand mixed with mortar.

[10]Ratings are for medium-burned clay tile.

[11]The fire-resistive time period for concrete masonry units meeting the equivalent thicknesses required for a two-hour fire-resistive rating in Item 6, and having a thickness of not less than $7^5/8$ inches (194 mm) is four hours when cores which are not grouted are filled with silicone-treated perlite loose-fill insulation; vermiculite loose-fill insulation; or expanded clay, shale or slate lightweight aggregate, sand or slag having a maximum particle size of $3/8$ inch (9.5 mm).

[12]For determining the fire-resistance rating of concrete masonry units composed of a combination of aggregate types or where plaster is applied directly to the concrete masonry, see UBC Standard 7-7, Part III. Lightweight aggregates shall have a maximum combined density of 65 pounds per cubic foot (1049 kg/m^3).

[13]See also Footnote 2. The equivalent thickness may include the thickness of cement plaster or 1.5 times the thickness of gypsum plaster applied in accordance with the requirements of Chapter 25.

[14]Concrete walls shall be reinforced with horizontal and vertical temperature reinforcement as required by Sections 1914.3.2 and 1914.3.3.

[15]Studs are welded truss wire studs with 0.18 inch (4.57 mm) (No. 7 B.W. gage) flange wire and 0.18 inch (4.57 mm) (No. 7 B.W. gage) truss wires.

[16]Nailable metal studs consist of two channel studs spot welded back to back with a crimped web forming a nailing groove.

[17]Wood structural panels may be installed between the fire protection and the wood studs on either the interior or exterior side of the wood-frame assemblies in this table, provided the length of the fasteners used to attach the fire protection are increased by an amount at least equal to the thickness of the wood structural panel.

[18]The design stress of studs shall be reduced to 78 percent of allowable F'_c with the maximum not greater than 78 percent of the calculated stress with studs having a slenderness ratio l_e/d of 33.

[19]For properties of cooler or wallboard nails, see approved nationally recognized standards.

TABLE 7-C—MINIMUM PROTECTION FOR FLOOR AND ROOF SYSTEMS[a,1]

FLOOR OR ROOF CONSTRUCTION	ITEM NUMBER	CEILING CONSTRUCTION	THICKNESS OF FLOOR OR ROOF SLAB (inches) $\times$ 25.4 for mm				MINIMUM THICKNESS OF CEILING (inches)			
			4 Hr.	3 Hr.	2 Hr.	1 Hr.	4 Hr.	3 Hr.	2 Hr.	1 Hr.
1. Siliceous aggregate concrete	1-1.1	Slab (no ceiling required). Minimum cover over nonprestressed reinforcement shall be not less than $^3/_4$ inch (19 mm).[2]	7.0	6.2	5.0	3.5				
2. Carbonate aggregate concrete	2-1.1		6.6	5.7	4.6	3.2				
3. Sand-lightweight concrete	3-1.1		5.4	4.6	3.8	2.7				
4. Lightweight concrete	4-1.1		5.1	4.4	3.6	2.5				
5. Reinforced concrete joists	5-1.1	Slab with suspended ceiling of vermiculite gypsum plaster over metal lath attached to $^3/_4''$ (19 mm) cold-rolled channels spaced 12″ (305 mm) on center. Ceiling located 6″ (152 mm) minimum below joists.	3	2			1	$^3/_4$		
	5-2.1	$^3/_8$ (9.5 mm) Type X gypsum wallboard[3] attached to 0.018 inch (0.53 mm) (No. 25 carbon sheet steel gage) by $^7/_8''$ deep by $2^5/_8''$ (22.2 mm deep by 66.7 mm) hat-shaped galvanized steel channels with 1″ (25 mm) long No. 6 screws. The channels are spaced 24″ (610 mm) on center, span 35″ (889 mm) and are supported along their length at 35″ (889 mm) intervals by 0.033 inch (0.84 mm) (No. 21 galvanized sheet gage) galvanized steel flat strap hangers having formed edges that engage the lips of the channel. The strap hangers are attached to the side of the concrete joists with $^5/_{32}''$ by $1^1/_4''$ (4 mm by 31.8 mm) long powder-driven fasteners. The wallboard is installed with the long dimension perpendicular to the channels. All end joints occur on channels and supplementary channels are installed parallel to the main channels, 12″ (305 mm) each side, at end joint occurrences. The finish ceiling is located approximately 12″ (305 mm) below the soffit of the floor slab.			$2^1/_2$				$^5/_8$	
6. Steel joists constructed with a poured reinforced concrete slab on metal lath forms or steel form units.[4,5]	6-1.1	Gypsum plaster on metal lath attached to the bottom cord with single No. 16 gage or doubled No. 18 gage wire ties spaced 6″ (152 mm) on center. Plaster mixed 1:2 for scratch coat, 1:3 for brown coat, by weight, gypsum to sand aggregate for two-hour system. For three-hour system plaster is neat.			$2^1/_2$	$2^1/_4$			$^3/_4$	$^5/_8$
	6-2.1	Vermiculite gypsum plaster on metal lath attached to the bottom chord with single No. 16 gage or doubled 0.049 inch (1.24 mm) (No. 18 B.W. gage) wire ties 6″ (152 mm) on center.			2				$^5/_8$	
	6-3.1	Cement plaster over metal lath attached to the bottom chord of joists with single No. 16 gage or doubled 0.049 inch (1.24 mm) (No. 18 B.W. gage) wire ties spaced 6″ (152 mm) on center. Plaster mixed 1:2 for scratch coat, 1:3 for brown coat for one-hour system and 1:1 for scratch coat, $1:1^1/_2$ for brown coat for two-hour system, by weight, cement to sand.				2				$^5/_8$[6]

(Continued)

TABLE 7-C 1997 UNIFORM BUILDING CODE

TABLE 7-C—MINIMUM PROTECTION FOR FLOOR AND ROOF SYSTEMS[a,1]—(Continued)

FLOOR OR ROOF CONSTRUCTION	ITEM NUMBER	CEILING CONSTRUCTION	THICKNESS OF FLOOR OR ROOF SLAB (inches)				MINIMUM THICKNESS OF CEILING (inches)			
			× 25.4 for mm				× 25.4 for mm			
			4 Hr.	3 Hr.	2 Hr.	1 Hr.	4 Hr.	3 Hr.	2 Hr.	1 Hr.
6. Steel joists constructed with a poured reinforced concrete slab on metal lath forms or steel form units.[4,5] (cont.)	6-4.1	Ceiling of $5/8''$ (15.9 mm) Type X wallboard[3] attached to $7/8''$ deep by $2 5/8''$ (22.2 mm deep by 66.7 mm) by 0.021 inch (0.53 mm) (No. 25 carbon sheet steel gage) hat-shaped furring channels $12''$ (305 mm) on center with $1''$ (25 mm) long No. 6 wallboard screws at $8''$ (203 mm) on center. Channels wire tied to bottom chord of joists with doubled 0.049 inch (1.24 mm) (No. 18 B.W. gage) wire or suspended below joists on wire hangers.[7]			$2 1/2$				$5/8$	
	6-5.1	Wood-fibered gypsum plaster mixed 1:1 by weight gypsum to sand aggregate applied over metal lath. Lath tied $6''$ (152 mm) on center to $3/4''$ (19 mm) channels spaced $13 1/2''$ (343 mm) on center. Channels secured to joists at each intersection with two strands of 0.049 inch (1.24 mm) (No. 18 B.W. gage) galvanized wire.			$2 1/2$				$3/4$	
7. Reinforced concrete slab and joists with hollow clay tile fillers laid end to end in rows $2 1/2''$ (63.5 mm) or more apart; reinforcement placed between rows and concrete cast around and over tile.	7-1.1	$5/8''$ (15.9 mm) gypsum plaster on bottom of floor or roof construction.			8^8				$5/8$	
	7-1.2	None.				$5 1/2^9$				
8. Steel joists constructed with a reinforced concrete slab on top poured on a $1/2''$-deep (12.7 mm) steel deck.[5]	8-1.1	Vermiculite gypsum plaster on metal lath attached to $3/4''$ (19 mm) cold-rolled channels with 0.049 inch (1.24 mm) (No. 18 B.W. gage) wire ties spaced $6''$ (152 mm) on center.	$2 1/2^{10}$				$3/4$			
9. $3''$ (76 mm) deep cellular steel deck with concrete slab on top. Slab thickness measured to top of cells.	9-1.1	Suspended ceiling of vermiculite gypsum plaster base coat and vermiculite acoustical plaster on metal lath attached at $6''$ (152 mm) intervals to $3/4''$ (19 mm) cold-rolled channels spaced $12''$ (305 mm) on center and secured to $1 1/2''$ (38 mm) cold-rolled channels spaced $36''$ (914 mm) on center with 0.065 inch (1.65 mm) (No. 16 B.W. gage) wire. $1 1/2''$ (38 mm) channels supported by No. 8 gage wire hangers at $36''$ (914 mm) on center. Beams within envelope and with a $2 1/2''$ (63.5 mm) air space between beam soffit and lath have a 4-hour rating.	$2 1/2$				$1 1/8^{11}$			
10. $1 1/2''$-deep (38 mm) steel roof deck on steel framing. Insulation board, 30 pcf density (480 kg/m^3), composed of wood fibers with cement binders of thickness shown bonded to deck with unified asphalt adhesive. Covered with a Class A or B roof covering.	10-1.1	Ceiling of gypsum plaster on metal lath. Lath attached to $3/4''$ (19 mm) furring channels with 0.049 inch (1.24 mm) (No. 18 B.W. gage) wire ties spaced $6''$ (152 mm) on center. $3/4''$ (19 mm) channel saddle-tied to $2''$ (51 mm) channels with doubled 0.065 inch (1.65 mm) (No. 16 B.W. gage) wire ties. $2''$ (51 mm) channels spaced $36''$ (914 mm) on center suspended $2''$ (51 mm) below steel framing and saddle-tied with 0.165 inch (4.19 mm) (No. 8 B.W. gage) wire. Plaster mixed 1:2 by weight, gypsum to sand aggregate.			$1 7/8$	1			$3/4^{12}$	$3/4^{12}$

(Continued)

TABLE 7-C—MINIMUM PROTECTION FOR FLOOR AND ROOF SYSTEMS[a,1]—(Continued)

FLOOR OR ROOF CONSTRUCTION	ITEM NUMBER	CEILING CONSTRUCTION	THICKNESS OF FLOOR OR ROOF SLAB (inches)				MINIMUM THICKNESS OF CEILING (inches)			
			× 25.4 for mm							
			4 Hr.	3 Hr.	2 Hr.	1 Hr.	4 Hr.	3 Hr.	2 Hr.	1 Hr.
11. $1^1/_2$″-deep (38 mm) steel roof deck on steel-framing wood fiber insulation board, 17.5 pcf density (280 kg/m^3) on top applied over a 15-lb. (6.8 kg) asphalt-saturated felt. Class A or B roof covering.	11-1.1	Ceiling of gypsum plaster on metal lath. Lath attached to $^3/_4$″ (19 mm) furring channels with 0.049 inch (1.24 mm) (No. 18 B.W. gage) wire ties spaced 6″ (152 mm) on center. $^3/_4$″ (19 mm) channels saddle tied to 2″ (51 mm) channels with doubled 0.065 inch (1.65 mm) (No. 16 B.W. gage) wire ties. 2″ (51 mm) channels spaced 36″ (914 mm) on center suspended 2″ (51 mm) below steel framing and saddle tied with 0.165 inch (4.19 mm) (No. 8 B.W. gage) wire. Plaster mixed 1:2 for scratch coat and 1:3 for brown coat, by weight, gypsum to sand aggregate for one-hour system. For two-hour system plaster mix is 1:2 by weight, gypsum to sand aggregate.			$1^1/_2$	1			$^7/_8$[7]	$^3/_4$[12]
12. $1^1/_2$″-deep (38 mm) steel roof deck on steel-framing insulation of rigid board consisting of expanded perlite and fibers impregnated with integral asphalt waterproofing; density 9 to 12 pcf (144 to 192 kg/m^3) secured to metal roof deck by $^1/_2$″ (12.7 mm) wide ribbons of waterproof, cold-process liquid adhesive spaced 6″ (152 mm) apart. Steel joist or light steel construction with metal roof deck, insulation, and Class A or B built-up roof covering.[5]	12-1.1	Gypsum-vermiculite plaster on metal lath wire tied at 6″ (152 mm) intervals to $^3/_4$″ (19 mm) furring channels spaced 12″ (305 mm) on center and wire tied to 2″ (51 mm) runner channels spaced 32″ (813 mm) on center. Runners wire tied to bottom chord of steel joists.			1				$^7/_8$	
13. Double wood floor over wood joists spaced 16″ (406 mm) on center.[13,14]	13-1.1	Gypsum plaster over $^3/_8$″ (9.5 mm) Type X gypsum lath. Lath initially applied with not less than four $1^1/_8$″ (28.6 mm) by No. 13 gage by $^{19}/_{64}$″ (7.5 mm) head plasterboard blued nails per bearing. Continuous stripping over lath along all joist lines. Stripping consists of 3″ (76 mm) wide strips of metal lath attached by $1^1/_2$″ (38 mm) by No. 11 gage by $^1/_2$″ (12.7 mm) head roofing nails spaced 6″ (152 mm) on center. Alternate stripping consists of 3″ wide 0.049″ (76 mm 1.24 mm) diameter wire stripping weighing 1 pound per square yard (0.38 kg/m^2) and attached by No. 16 gage by $1^1/_2$″ by $^3/_4$″ (38 mm by 19 mm) crown width staples, spaced 4″ (102 mm) on center. Where alternate stripping is used, the lath nailing may consist of two nails at each end and one nail at each intermediate bearing. Plaster mixed 1:2 by weight, gypsum to sand aggregate.								$^7/_8$
	13-1.2	Cement or gypsum plaster on metal lath. Lath fastened with $1^1/_2$″ (38 mm) by No. 11 gage by $^7/_{16}$″ (11.1 mm) head barbed shank roofing nails spaced 5″ (127 mm) on center. Plaster mixed 1:2 for scratch coat and 1:3 for brown coat, by weight, cement to sand aggregate.								$^5/_8$
	13-1.3	Perlite or vermiculite gypsum plaster on metal lath secured to joists with $1^1/_2$″ (38 mm) by No. 11 gage by $^7/_{16}$″ (11.1 mm) head barbed shank roofing nails spaced 5″ (127 mm) on center.								$^5/_8$

(Continued)

TABLE 7-C   1997 UNIFORM BUILDING CODE

TABLE 7-C—MINIMUM PROTECTION FOR FLOOR AND ROOF SYSTEMS[a,1]—(Continued)

FLOOR OR ROOF CONSTRUCTION	ITEM NUMBER	CEILING CONSTRUCTION	THICKNESS OF FLOOR OR ROOF SLAB (inches)				MINIMUM THICKNESS OF CEILING (inches)			
			× 25.4 for mm							
			4 Hr.	3 Hr.	2 Hr.	1 Hr.	4 Hr.	3 Hr.	2 Hr.	1 Hr.
13. Double wood floor over wood joists spaced 16″ (406 mm) on center.[13,14] (cont.)	13-1.4	$1/2$″ (12.7 mm) Type X gypsum wallboard[3] nailed to joists with 5d cooler[15] or wallboard[15] nails at 6″ (152 mm) on center. End joints of wallboard centered on joists.								$1/2$
14. Plywood stressed skin panels consisting of $5/8$″ (15.9 mm) thick interior C-D (exterior glue) top stressed skin on 2″ by 6″ (51 mm by 152 mm) nominal (minimum) stringers. Adjacent panel edges joined with 8d common wire nails spaced 6″ (152 mm) on center. Stringers spaced 12″ (305 mm) maximum on center.	14-1.1	$1/2$″-thick (12.7 mm) wood fiberboard weighing 15 to 18 pounds per cubic foot (240 to 288 kg/m^3) installed with long dimension parallel to stringers or $3/8$″ (9.5 mm) C-D (exterior glue) plywood glued and/or nailed to stringers. Nailing to be with 5d cooler[15] or wallboard[15] nails at 12″ (305 mm) on center. Second layer of $1/2$″ (12.7 mm) Type X gypsum wallboard[3] applied with long dimension perpendicular to joists and attached with 8d cooler[15] or wallboard[15] nails at 6″ (152 mm) on center at end joints and 8″ (203 mm) on center elsewhere. Wallboard joints staggered with respect to fiberboard joints.								1
15. Vermiculite concrete slab proportioned 1:4 (portland cement to vermiculite aggregate) on a $1^1/_2$″-deep (38 mm) steel deck supported on individually protected steel framing. Maximum span of deck 6′ 10″ (2083 mm) where deck is less than 0.019 inch (0.48 mm) (No. 26 carbon steel sheet gage) and 8′ 0″ (2438 mm) where deck is 0.019 inch (0.48 mm) (No. 26 carbon steel sheet gage) or greater. Slab reinforced with 4″ by 8″ (102 mm by 203 mm) 0.109/0.083 inch (0.277/0.211 mm) (No. 12/14 B.W. gage) welded wire mesh.	15-1.1	None.				3^{10}				
16. Perlite concrete slab proportioned 1:6 (portland cement to perlite aggregate) on a $1^1/_4$″-deep (32 mm) steel deck supported on individually protected steel framing. Slab reinforced with 4″ by 8″ (102 by 203 mm) 0.109/0.083 inch (0.277/0.211 mm) (No. 12/14 B.W. gage) welded wire mesh.	16-1.1	None.				$3^1/_2{}^{10}$				
17. Perlite concrete slab proportioned 1:6 (portland cement to perlite aggregate) on a $9/_{16}$″-deep (14 mm) steel deck supported by steel joists 4′ (1219 mm) on center. Class A or B roof covering on top.	17-1.1	Perlite gypsum plaster on metal lath wire tied to $3/_4$″ (19 mm) furring channels attached with 0.065 inch (1.65 mm) (No. 16 B.W. gage) wire ties to lower chord of joists.	2^{16}	2^{16}				$7/_8$	$3/_4$	

(Continued)

TABLE 7-C—MINIMUM PROTECTION FOR FLOOR AND ROOF SYSTEMS[a,1]—(Continued)

FLOOR OR ROOF CONSTRUCTION	ITEM NUMBER	CEILING CONSTRUCTION	THICKNESS OF FLOOR OR ROOF SLAB (inches)				MINIMUM THICKNESS OF CEILING (inches)			
			× 25.4 for mm							
			4 Hr.	3 Hr.	2 Hr.	1 Hr.	4 Hr.	3 Hr.	2 Hr.	1 Hr.
18. Perlite concrete slab proportioned 1:6 (portland cement to perlite aggregate) on $1^1/_4''$-deep (32 mm) steel deck supported on individually protected steel framing. Maximum span of deck 6' 10" (2083 mm) where deck is less than 0.019 inch (0.48 mm) (No. 26 carbon sheet steel gage) and 8' 0" (2438 mm) where deck is 0.019 inch (0.48 mm) (No. 26 carbon sheet steel gage) or greater. Slab reinforced with 0.042 inch (1.07 mm) (No. 19 B.W. gage) hexagonal wire mesh. Class A or B roof covering on top.	18-1.1	None.			$2^1/_4{}^{16}$					
19. Floor and beam construction consisting of 3"-deep (76 mm) cellular steel floor units mounted on steel members with 1:4 (proportion of portland cement to perlite aggregate) perlite-concrete floor slab on top.	19-1.1	Suspended envelope ceiling of perlite gypsum plaster on metal lath attached to $^3/_4''$ (19 mm) cold-rolled channels, secured to $1^1/_2''$ (38 mm) cold-rolled channels spaced 42" (1067 mm) on center supported by 0.203 inch (5.16 mm) (No. 6 B.W. gage) wire 36" (914 mm) on center. Beams in envelope with 3" (76 mm) minimum air space between beam soffit and lath have a 4-hour rating.	2^{16}				1^{12}			
20. Perlite concrete proportioned 1:6 (portland cement to perlite aggregate) poured to $^1/_8$-inch (3 mm) thickness above top of corrugations of $1^5/_{16}$-inch-deep (33 mm) galvanized steel deck maximum span 8' 0" (2438 mm) for 0.024 inch (0.61 mm) (No. 24 galvanized sheet gage) or 6' 0" (1829 mm) for 0.019 inch (0.48 mm) (No. 26 galvanized sheet gage) with deck supported by individually protected steel framing. Approved polystyrene foam plastic insulation board having a flame spread not exceeding 75 [1" to 4" (25 mm to 102 mm) thickness with vent holes that approximate 3 percent of the board surface area] placed on top of perlite slurry. A 2' by 4' (610 mm by 1219 mm) insulation board contains six $2^3/_4''$ (70 mm) diameter holes. Board covered with $2^1/_4''$ (57 mm) minimum perlite concrete slab. Slab reinforced with mesh consisting of 0.042 inch (1.07 mm) (No. 19 B.W. gage) galvanized steel wire twisted together to form 2" (51 mm) hexagons with straight 0.065 inch (1.65 mm) (No. 16 B.W. gage) galvanized steel wire woven into mesh and spaced 3" (76 mm). Alternate slab reinforcement may consist of 4" by 8" (102 mm by 203 mm), 0.109/0.238 inch (2.77/6.05 mm) (No. 12/4 B.W. gage), or 2" by 2" (51 mm by 51 mm), 0.083/0.083 inch (2.11/2.11 mm) (No. 14/14 B.W. gage) welded wire fabric. Class A or B roof covering on top.	20-1.1	None.			Varies					

(Continued)

TABLE 7-C
TABLE 7-D

1997 UNIFORM BUILDING CODE

TABLE 7-C—MINIMUM PROTECTION FOR FLOOR AND ROOF SYSTEMS[a,1]—(Continued)

FLOOR OR ROOF CONSTRUCTION	ITEM NUMBER	CEILING CONSTRUCTION	THICKNESS OF FLOOR OR ROOF SLAB (inches)				MINIMUM THICKNESS OF CEILING (inches)			
			$\times$ 25.4 for mm							
			4 Hr.	3 Hr.	2 Hr.	1 Hr.	4 Hr.	3 Hr.	2 Hr.	1 Hr.
21. Wood joist, floor trusses and roof trusses spaced 24″ (610 mm) o.c. with $1/2$″ (12.7 mm) wood structural panels with exterior glue applied at right angles to top of joist or truss with 8d nails. The wood structural panel thickness shall not be less than $1/2$″ (12.7 mm) nor less than required by Chapter 23.	21-1.1	Base layer $5/8$″ (15.9 mm) Type X gypsum wallboard applied at right angles to joist or truss 24″ (610 mm) o.c. with $1 1/4$″ (32 mm) Type S or Type W drywall screws 24″ (610 mm) o.c. Face layer $5/8$″ (15.9 mm) Type X gypsum wallboard or veneer base applied at right angles to joist or truss through base layer with $1 7/8$″ (48 mm) Type S or Type W drywall screws 12″ (305 mm) o.c. at joints and intermediate joist or truss. Face layer joints offset 24″ (610 mm) from base layer joints, $1 1/2$″ (38 mm) Type G drywall screws placed 2″ (51 mm) back on either side of face layer end joints, 12″ (305 mm) o.c.				Varies				$1 1/4$

[a]Generic fire-resistance ratings (those not designated as PROPRIETARY* in the listing) in the *Fire-Resistance Design Manual*, Fourteenth Edition, dated April 1994, as published by the Gypsum Association, may be accepted as if herein listed.

[1]Staples with equivalent holding power and penetration may be used as alternate fasteners to nails for attachment to wood framing.

[2]When the slab is in an unrestrained condition, minimum reinforcement cover shall not be less than $1 5/8$ inches (41 mm) for four-hour (siliceous aggregate only); $1 1/4$ inches (32 mm) for four- and three-hour; 1 inch (25 mm) for two-hour (siliceous aggregate only); and $3/4$ inch (19.1 mm) for all other restrained and unrestrained conditions.

[3]For all of the construction with gypsum wallboard described in this table, gypsum base for veneer plaster of the same size, thickness and core type may be substituted for gypsum wallboard, provided attachment is identical to that specified for the wallboard, and the joints on the face layer are reinforced and the entire surface is covered with a minimum of $1/16$-inch (1.6 mm) gypsum veneer plaster.

[4]Slab thickness over steel joists measured at the joists for metal lath form and at the top of the form for steel form units.

[5](a) The maximum allowable stress level for H-Series joists shall not exceed 22,000 psi (152 MPa).
  (b) The allowable stress for K-Series joists shall not exceed 26,000 psi (179 MPa), the nominal depth of such joist shall not be less than 10 inches (254 mm) and the nominal joist weight shall not be less than 5 pounds per lineal foot (7.4 kg/m).

[6]Cement plaster with 15 pounds (6.8 kg) of hydrated lime and 3 pounds (1.4 kg) of approved additives or admixtures per bag of cement.

[7]Gypsum wallboard ceilings attached to steel framing may be suspended with $1 1/2$-inch (38 mm) cold-formed carrying channels spaced 48 inches (1219 mm) on center, which are suspended with No. 8 SWG galvanized wire hangers spaced 48 inches (1219 mm) on center. Cross-furring channels are tied to the carrying channels with No. 18 SWG galvanized wire (double strand) and spaced as required for direct attachment to the framing. This alternative is also applicable to those steel framing assemblies recognized under Footnote a.

[8]Six-inch (152 mm) hollow clay tile with 2-inch (51 mm) concrete slab above.

[9]Four-inch (102 mm) hollow clay tile with $1 1/2$-inch (38 mm) concrete slab above.

[10]Thickness measured to bottom of steel form units.

[11]Five-eighths inch (15.9 mm) of vermiculite gypsum plaster plus $1/2$ inch (12.7 mm) of approved vermiculite acoustical plastic.

[12]Furring channels spaced 12 inches (305 mm) on center.

[13]Double wood floor may be either of the following:
  (a) Subfloor of 1-inch (25 mm) nominal boarding, a layer of asbestos paper weighing not less than 14 pounds per 100 square feet (0.7 kg/m^2) and a layer of 1-inch (25 mm) nominal tongue-and-groove finish flooring; or
  (b) Subfloor of 1-inch (25 mm) nominal tongue-and-groove boarding or $15/32$-inch (11.9 mm) wood structural panels with exterior glue and a layer of 1-inch (25 mm) nominal tongue-and-groove finish flooring or $19/32$-inch (15.1 mm) wood structural panel finish flooring or a layer of Type I Grade M-1 particleboard not less than $5/8$ inch (15.9 mm) thick.

[14]The ceiling may be omitted over unusable space, and flooring may be omitted where unusable space occurs above.

[15]For properties of cooler or wallboard nails, see approved nationally recognized standards.

[16]Thickness measured on top of steel deck unit.

TABLE 7-D—PRECONDITIONING CYCLES FOR FIRE-RESISTIVE JOINT SYSTEMS

TYPE OF JOINT SYSTEM	NUMBER OF CYCLES
Expansion/contraction	500
Seismic	100
Wind sway	500

Chapter 8
INTERIOR FINISHES

SECTION 801 — GENERAL

801.1 Scope. Interior wall and ceiling finish shall mean the exposed interior surfaces of buildings including, but not limited to, fixed or movable walls and partitions, interior wainscoting, paneling or other finish applied structurally or for decoration, acoustical correction, surface insulation, sanitation, structural fire resistance or similar purposes. Requirements for finishes in this chapter shall not apply to trim defined as picture molds, chair rails, baseboards and handrails; or to doors and windows or their frames; or to materials that are less than $1/28$ inch (0.9 mm) in thickness applied directly to the surface of walls or ceilings.

Foam plastics shall not be used as interior finish except as provided in Section 2602. For foam plastic trim, see Section 601.5.5.

See Section 1403 for veneer.

801.2 Standards of Quality. The standards listed below labeled a "UBC standard" are also listed in Chapter 35, Part II, and are part of this code.

1. UBC Standard 8-1, Test Method for Surface-burning Characteristics of Building Materials

2. UBC Standard 8-2, Standard Test Method for Evaluating Room Fire Growth Contribution of Textile Wall Covering

801.3 Veneer. Veneers shall comply with Section 1403.

SECTION 802 — TESTING AND CLASSIFICATION OF MATERIALS

802.1 Testing. Tests shall be made by an approved testing agency to establish surface-burning characteristics and to show that materials when cemented or otherwise fastened in place will not readily become detached when subjected to room temperatures of 300°F (149°C) for 25 minutes. Surface-burning characteristics shall be determined by one of the following methods:

1. The surface-burning characteristics as set forth in UBC Standard 8-1.

2. Any other recognized method of test procedure for determining the surface-burning characteristics of finish materials that will give comparable results to those specified in method Item 1.

3. The room fire growth contribution for textile wall coverings as set forth in UBC Standard 8-2.

802.2 Classification. The classes of materials based on their flame-spread index shall be as set forth in Table 8-A. The smoke density shall be no greater than 450 when tested in accordance with UBC Standard 8-1 in the way intended for use.

SECTION 803 — APPLICATION OF CONTROLLED INTERIOR FINISH

Interior finish materials applied to walls and ceilings shall be tested as specified in Section 802 and regulated for purposes of limiting surface-burning by the following provisions:

1. When walls and ceilings are required by any provision in this code to be of fire-resistive or noncombustible construction, the finish material shall be applied directly against such fire-resistive or noncombustible construction or to furring strips not exceeding

$1^3/4$ inches (44 mm) applied directly against such surfaces. The intervening spaces between such furring strips shall be filled with inorganic or Class I material or shall be fire blocked not to exceed 8 feet (2438 mm) in any direction. See Section 708 for fireblocking.

2. Where walls and ceilings are required to be of fire-resistive or noncombustible construction and walls are set out or ceilings are dropped distances greater than specified in paragraph 1 of this section, Class I finish materials shall be used except where the finish materials are protected on both sides by automatic sprinkler systems or are attached to a noncombustible backing or to furring strips installed as specified in Item 1. The hangers and assembly members of such dropped ceilings that are below the main ceiling line shall be of noncombustible materials except that in Types III and V construction, fire-retardant-treated wood may be used. The construction of each set-out wall shall be of fire-resistive construction as required elsewhere in this code. See Section 708 for fire blocks and draft stops.

3. Wall and ceiling finish materials of all classes as permitted in this chapter may be installed directly against the wood decking or planking of Type IV heavy-timber construction, or to wood furring strips applied directly to the wood decking or planking installed and fire blocked as specified in Item 1.

4. An interior wall or ceiling finish that is less than $1/4$ inch (6.4 mm) thick shall be applied directly against a noncombustible backing.

> **EXCEPTIONS:** 1. Class I materials.
>
> 2. Materials where the qualifying tests were made with the material suspended or furred out from the noncombustible backing.

SECTION 804 — MAXIMUM ALLOWABLE FLAME SPREAD

804.1 General. The maximum flame-spread class of finish materials used on interior walls and ceilings shall not exceed that set forth in Table 8-B.

> **EXCEPTIONS:** 1. Except in Group I Occupancies and in enclosed vertical exits, Class III may be used in other means of egress and rooms as wainscoting extending not more than 48 inches (1219 mm) above the floor and for tack and bulletin boards covering not more than 5 percent of the gross wall area of the room.
>
> 2. When a sprinkler system complying with UBC Standard 9-1 or 9-3 is provided, the flame-spread classification rating may be reduced one classification, but in no case shall materials having a classification greater than Class III be used.
>
> 3. The exposed faces of Type IV-H.T., structural members, and Type IV-H.T., decking and planking, where otherwise permissible under this code, are excluded from flame-spread requirements.

804.2 Carpeting on Ceilings. When used as interior ceiling finish, carpeting and similar materials having a napped, tufted, looped or similar surface shall have a Class I flame spread.

SECTION 805 — TEXTILE WALL COVERINGS

When used as interior wall finish, textile wall coverings, including materials such as those having a napped, tufted, looped, nonwoven, woven or similar surface shall comply with the following:

1. Textile wall coverings shall have a Class I flame spread and shall be protected by automatic sprinklers complying with UBC Standard 9-1 or 9-3, or

2. The textile wall covering shall meet the acceptance criteria of UBC Standard 8-2 when tested using a product mounting system, including adhesive, representative of actual use.

SECTION 806 — INSULATION

Thermal and acoustical insulation installed on walls or ceilings shall comply with Section 707.

SECTION 807 — SANITATION

807.1 Floors and Walls in Water Closet Compartment and Showers.

807.1.1 Floors. In other than dwelling units, toilet room floors shall have a smooth, hard nonabsorbent surface such as portland cement, concrete, ceramic tile or other approved material that extends upward onto the walls at least 5 inches (127 mm).

807.1.2 Walls. Walls within 2 feet (610 mm) of the front and sides of urinals and water closets shall have a smooth, hard nonabsorbent surface of portland cement, concrete, ceramic tile or other smooth, hard nonabsorbent surface to a height of 4 feet (1219 mm), and except for structural elements, the materials used in such walls shall be of a type that is not adversely affected by moisture. See Section 2512 for other limitations.

EXCEPTIONS: 1. Dwelling units and guest rooms.

2. Toilet rooms that are not accessible to the public and that have not more than one water closet.

In all occupancies, accessories such as grab bars, towel bars, paper dispensers and soap dishes, provided on or within walls, shall be installed and sealed to protect structural elements from moisture.

807.1.3 Showers. Showers in all occupancies shall be finished as specified in Sections 807.1.1 and 807.1.2 to a height of not less than 70 inches (1778 mm) above the drain inlet. Materials other than structural elements used in such walls shall be of a type that is not adversely affected by moisture. See Section 2512 for other limitations.

807.1.4 Shower doors. For shower doors, see Sections 2406.4 and 2407.

807.2 Water Closet Room Separation. See Section 302.6 for requirements to separate water closet rooms.

TABLE 8-A—FLAME-SPREAD CLASSIFICATION

MATERIAL QUALIFIED BY:	
Class	Flame-spread Index
I	0-25
II	26-75
III	76-200

TABLE 8-B—MAXIMUM FLAME-SPREAD CLASS[1]

OCCUPANCY GROUP	ENCLOSED VERTICAL EXITWAYS	OTHER EXITWAYS[2]	ROOMS OR AREAS
A	I	II	II[3]
B	I	II	III
E	I	II	III
F	II	III	III
H	I	II	III[4]
I-1.1, I-1.2, I-2	I	I[5]	II[6]
I-3	I	I[5]	I[6]
M	I	II	III
R-1	I	II	III
R-3	III	III	III[7]
S-1, S-2	II	II	III
S-3, S-4, S-5	I	II	III
U	NO RESTRICTIONS		

[1]Foam plastics shall comply with the requirements specified in Section 2602. Carpeting on ceilings and textile wall coverings shall comply with the requirements specified in Sections 804.2 and 805, respectively.

[2]Finish classification is not applicable to interior walls and ceilings of exterior exit balconies.

[3]In Group A, Divisions 3 and 4 Occupancies, Class III may be used.

[4]Over two stories shall be of Class II.

[5]In Group I, Divisions 2 and 3 Occupancies, Class II may be used.

[6]Class III may be used in administrative spaces.

[7]Flame-spread provisions are not applicable to kitchens and bathrooms of Group R, Division 3 Occupancies.

Chapter 9
FIRE-PROTECTION SYSTEMS

SECTION 901 — SCOPE

This chapter applies to the design and installation of fire-extinguishing systems, smoke-control systems and smoke and heat venting systems.

For requirements on fire alarm systems, see the following:

SECTION	SUBJECT
303.9	Group A, Divisions 1 and 2 Occupancies
305.2.3, 305.9	Group E Occupancies
307.9	Group H Occupancies
308.9	Group I Occupancies
310.10	Group R Occupancies
403.5	High-rise buildings
408.5	Amusement buildings
307.11.5.5	Group H, Division 6 Occupancies

For smoke detectors in Group R Occupancies, see Section 310.9.

SECTION 902 — STANDARDS OF QUALITY

Fire-extinguishing systems, including automatic sprinkler systems, Class I, Class II and Class III standpipe systems, special automatic extinguishing systems, basement pipe inlets, smoke-control systems, and smoke and heat vents shall be approved and shall be subject to such periodic tests as may be required.

The standards listed below labeled a "UBC standard" are also listed in Chapter 35, Part II, and are part of this code. The other standards listed below are recognized standards (see Sections 3503 and 3504).

1. **Fire-extinguishing system.**

    1.1 UBC Standard 9-1, Installation of Sprinkler Systems

    1.2 UBC Standard 9-3, Installation of Sprinkler Systems in Group R Occupancies Four Stories or Less

2. **Standpipe systems.**

    UBC Standard 9-2, Standpipe Systems

3. **Smoke control.**

    3.1 UBC Standard 7-2, Fire Tests of Door Assemblies

    3.2 UL 555, Fire Dampers

    3.3 UL 555C, Ceiling Dampers

    3.4 UL 555S, Leakage Rated Dampers for Use in Smoke Control Systems

    3.5 UL 33, Heat Response Links for Fire Protection Service

    3.6 UL 353, Limit Controls

4. **Smoke and heat vents.**

    UBC Standard 15-7, Automatic Smoke and Heat Vents

SECTION 903 — DEFINITIONS

For the purpose of this chapter, certain terms are defined as follows:

AUTOMATIC FIRE-EXTINGUISHING SYSTEM is an approved system of devices and equipment that automatically detects a fire and discharges an approved fire-extinguishing agent onto or in the area of a fire.

FIRE DEPARTMENT INLET CONNECTION is a connection through which the fire department can pump water into a standpipe system or sprinkler system.

PRESSURIZATION is the creation and maintenance of pressure levels in zones of a building, including elevator shafts and stairwells that are higher than the pressure level at the smoke source, such pressure levels being produced by positive pressures of a supply of uncontaminated air, by exhausting air and smoke at the smoke source, or by a combination of these methods.

PRESSURIZED STAIRWAY ENCLOSURE is a type of smoke-control system in which stairway enclosures are mechanically pressurized to minimize smoke contamination of them during a fire incident.

SMOKE is the airborne solid and liquid particulates and gases evolved when a material undergoes pyrolysis or combustion, including the quantity of air that is entrained or otherwise mixed into the mass.

SMOKE BARRIER is a continuous membrane, either vertical or horizontal, such as a wall, floor or ceiling assembly that is designed and constructed to restrict the movement of smoke.

SMOKE-CONTROL MODE is a predefined operational configuration of a system or device for the purpose of smoke control.

SMOKE-CONTROL SYSTEM, MECHANICAL, is an engineered system that uses mechanical fans to produce pressure differences across smoke barriers or establish airflows to limit and direct smoke movement.

SMOKE-CONTROL SYSTEM, PASSIVE, is a system of smoke barriers arranged to limit the migration of smoke.

SMOKE-CONTROL ZONE is a space within a building enclosed by smoke barriers.

SMOKE DAMPER is a device that meets the requirements of approved recognized standards, and is designed to resist the passage of air or smoke. A combination fire and smoke damper shall meet the requirements of approved recognized standards. See Chapter 35, Part IV.

SMOKE EXHAUST SYSTEM is a mechanical or gravity system intended to move smoke from the smoke zone to the exterior of the building, including smoke removal, purging and venting systems, as well as the function of exhaust fans utilized to reduce the pressure in a smoke zone.

STACK EFFECT is the vertical airflow within buildings caused by temperature differences.

STANDPIPE SYSTEM is a wet or dry system of piping, valves, outlets and related equipment designed to provide water at specified pressures and installed exclusively for the fighting of fires, including the following:

Class I is a standpipe system equipped with $2^1/_2$-inch (63.5 mm) outlets.

Class II is a standpipe system directly connected to a water supply and equipped with $1^1/_2$-inch (38.1 mm) outlets and hose.

Class III is a standpipe system directly connected to a water supply and equipped with $2^1/_2$-inch (63.5 mm) outlets or $2^1/_2$-inch (63.5 mm) and $1^1/_2$-inch (38.1 mm) outlets when a $1^1/_2$-inch (38.1 mm) hose is required. Hose connections for Class III systems may

F
F
F
be made through $2^1/_2$-inch (63.5 mm) hose valves with easily removable $2^1/_2$-inch by $1^1/_2$-inch (63.5 mm by 38.1 mm) reducers.

TENABLE ENVIRONMENT is an environment in which the quantity and location of smoke is limited or otherwise restricted to allow for ready evacuation through the space.

ZONED SMOKE CONTROL is a smoke-control system utilizing pressure differences between adjacent smoke-control zones.

SECTION 904 — FIRE-EXTINGUISHING SYSTEMS

904.1 Installation Requirements.

904.1.1 General. Fire-extinguishing systems required in this code shall be installed in accordance with the requirements of this section.

Fire hose threads used in connection with fire-extinguishing systems shall be national standard hose thread or as approved by the fire department.

The location of fire department hose connections shall be approved by the fire department.

In buildings used for high-piled combustible storage, fire protection shall be in accordance with the Fire Code.

904.1.2 Standards. Fire-extinguishing systems shall comply with UBC Standards 9-1 and 9-2.

> **EXCEPTIONS:** 1. Automatic fire-extinguishing systems not covered by UBC Standard 9-1 or 9-2 shall be approved and installed in accordance with approved standards.
>
> 2. Automatic sprinkler systems may be connected to the domestic water-supply main when approved by the building official, provided the domestic water supply is of adequate pressure, capacity and sizing for the combined domestic and sprinkler requirements. In such case, the sprinkler system connection shall be made between the public water main or meter and the building shutoff valve, and there shall not be intervening valves or connections. The fire department connection may be omitted when approved by the fire department.
>
> 3. Automatic sprinkler systems in Group R Occupancies four stories or less may be in accordance with UBC Standard 9-3.

904.1.3 Modifications. When residential sprinkler systems as set forth in UBC Standard 9-3 are provided, exceptions to, or reductions in, code requirements based on the installation of an automatic fire-extinguishing system are not allowed.

904.2 Automatic Fire-extinguishing Systems.

904.2.1 Where required. An automatic fire-extinguishing system shall be installed in the occupancies and locations as set forth in this section.

For provisions on special hazards and hazardous materials, see the Fire Code.

904.2.2 All occupancies except Group R, Division 3 and Group U Occupancies. Except for Group R, Division 3 and Group U Occupancies, an automatic sprinkler system shall be installed:

1. In every story or basement of all buildings when the floor area exceeds 1,500 square feet (139.4 m^2) and there is not provided at least 20 square feet (1.86 m^2) of opening entirely above the adjoining ground level in each 50 lineal feet (15 240 mm) or fraction thereof of exterior wall in the story or basement on at least one side of the building. Openings shall have a minimum dimension of not less than 30 inches (762 mm). Such openings shall be accessible to the fire department from the exterior and shall not be obstructed in a manner that firefighting or rescue cannot be accomplished from the exterior.

When openings in a story are provided on only one side and the opposite wall of such story is more than 75 feet (22 860 mm) from such openings, the story shall be provided with an approved automatic sprinkler system, or openings as specified above shall be provided on at least two sides of an exterior wall of the story.

If any portion of a basement is located more than 75 feet (22 860 mm) from openings required in this section, the basement shall be provided with an approved automatic sprinkler system.

2. At the top of rubbish and linen chutes and in their terminal rooms. Chutes extending through three or more floors shall have additional sprinkler heads installed within such chutes at alternate floors. Sprinkler heads shall be accessible for servicing.

3. In rooms where nitrate film is stored or handled.

4. In protected combustible fiber storage vaults as defined in the Fire Code.

5. Throughout all buildings with a floor level with an occupant load of 30 or more that is located 55 feet (16 764 mm) or more above the lowest level of fire department vehicle access.

> **EXCEPTIONS:** 1. Airport control towers.
> 2. Open parking structures.
> 3. Group F, Division 2 Occupancies.

904.2.3 Group A Occupancies.

904.2.3.1 Drinking establishments. An automatic sprinkler system shall be installed in rooms used by the occupants for the consumption of alcoholic beverages and unseparated accessory uses where the total area of such unseparated rooms and assembly uses exceeds 5,000 square feet (465 m^2). For uses to be considered as separated, the separation shall not be less than as required for a one-hour occupancy separation. The area of other uses shall be included unless separated by at least a one-hour occupancy separation.

904.2.3.2 Basements. An automatic sprinkler system shall be installed in basements classified as a Group A Occupancy when the basement is larger than 1,500 square feet (139.4 m^2) in floor area.

904.2.3.3 Exhibition and display rooms. An automatic sprinkler system shall be installed in Group A Occupancies that have more than 12,000 square feet (1115 m^2) of floor area that can be used for exhibition or display purposes.

904.2.3.4 Stairs. An automatic sprinkler system shall be installed in enclosed usable space below or over a stairway in Group A, Divisions 2, 2.1, 3 and 4 Occupancies. See Section 1005.3.3.6.

904.2.3.5 Multitheater complexes. An automatic sprinkler system shall be installed in every building containing a multitheater complex.

904.2.3.6 Amusement buildings. An automatic sprinkler system shall be installed in all amusement buildings. The main waterflow switch shall be electrically supervised. The sprinkler main cutoff valve shall be supervised. When the amusement building is temporary, the sprinkler water-supply system may be of an approved temporary type.

> **EXCEPTION:** An automatic sprinkler system need not be provided when the floor area of a temporary amusement building is less than 1,000 square feet (92.9 m^2) and the exit travel distance from any point is less than 50 feet (15 240 mm).

904.2.3.7 Stages. All stages shall be provided with an automatic sprinkler system. Such sprinklers shall be provided throughout the stage and in dressing rooms, workshops, storerooms and other accessory spaces contiguous to such stages.

EXCEPTIONS: 1. Sprinklers are not required for stages 1,000 square feet (92.9 m²) or less in area and 50 feet (15 240 mm) or less in height where curtains, scenery or other combustible hangings are not retractable vertically. Combustible hangings shall be limited to a single main curtain, borders, legs and a single backdrop.

2. Under stage areas less than 4 feet (1219 mm) in clear height used exclusively for chair or table storage and lined on the inside with ⁵/₈-inch (16 mm) Type X gypsum wallboard or an approved equal.

904.2.3.8 Smoke-protected assembly seating. All areas enclosed with walls and ceilings in buildings or structures containing smoke-protected assembly seating shall be protected with an approved automatic sprinkler system.

EXCEPTION: Press boxes and storage facilities less than 1,000 square feet (92.9 m²) in area and in conjunction with outdoor seating facilities where all means of egress in the seating area are essentially open to the outside.

904.2.4 Group E Occupancies.

904.2.4.1 General. An automatic fire sprinkler system shall be installed throughout all buildings containing a Group E, Division 1 Occupancy.

EXCEPTIONS: 1. When each room used for instruction has at least one exterior exit door at ground level and when rooms used for assembly purposes have at least one half of the required exits directly to the exterior ground level, a sprinkler system need not be provided.

2. When area separation walls, or occupancy separations having a fire-resistive rating of not less than two hours subdivide the building into separate compartments such that each compartment contains an aggregate floor area not greater than 20,000 square feet (1858 m²), an automatic sprinkler system need not be provided.

904.2.4.2 Basements. An automatic sprinkler system shall be installed in basements classified as Group E, Division 1 Occupancies.

904.2.4.3 Stairs. An automatic sprinkler system shall be installed in enclosed usable space below or over a stairway in Group E, Division 1 Occupancies. See Section 1005.3.3.6.

904.2.5 Group F Occupancies.

904.2.5.1 Woodworking occupancies. An automatic fire sprinkler system shall be installed in Group F woodworking occupancies over 2,500 square feet (232.3 m²) in area that use equipment, machinery or appliances that generate finely divided combustible waste or that use finely divided combustible materials.

904.2.6 Group H Occupancies.

904.2.6.1 General. An automatic fire-extinguishing system shall be installed in Group H, Divisions 1, 2, 3 and 7 Occupancies.

904.2.6.2 Group H, Division 4 Occupancies. An automatic fire-extinguishing system shall be installed in Group H, Division 4 Occupancies having a floor area of more than 3,000 square feet (279 m²).

904.2.6.3 Group H, Division 6 Occupancies. An automatic fire-extinguishing system shall be installed throughout buildings containing Group H, Division 6 Occupancies. The design of the sprinkler system shall not be less than that required under UBC Standard 9-1 for the occupancy hazard classifications as follows:

LOCATION	OCCUPANCY HAZARD CLASSIFICATION
Fabrication areas	Ordinary Hazard Group 2
Service corridors	Ordinary Hazard Group 2
Storage rooms without dispensing	Ordinary Hazard Group 2
Storage rooms with dispensing	Extra Hazard Group 2
Corridors	Ordinary Hazard Group 2[1]

[1]When the design area of the sprinkler system consists of a corridor protected by one row of sprinklers, the maximum number of sprinklers that needs to be calculated is 13.

904.2.7 Group I Occupancies. An automatic sprinkler system shall be installed in Group I Occupancies. In Group I, Division 1.1 and Group I, Division 2 Occupancies, approved quick-response or residential sprinklers shall be installed throughout patient sleeping areas.

EXCEPTION: In jails, prisons and reformatories, the piping system may be dry, provided a manually operated valve is installed at a continuously monitored location. Opening of the valve will cause the piping system to be charged. Sprinkler heads in such systems shall be equipped with fusible elements or the system shall be designed as required for deluge systems in UBC Standard 9-1.

904.2.8 Group M Occupancies. An automatic sprinkler system shall be installed in rooms classed as Group M Occupancies where the floor area exceeds 12,000 square feet (1115 m²) on any floor or 24,000 square feet (2230 m²) on all floors or in Group M Occupancies more than three stories in height. The area of mezzanines shall be included in determining the areas where sprinklers are required.

904.2.9 Group R, Division 1 Occupancies. An automatic sprinkler system shall be installed throughout every apartment house three or more stories in height or containing 16 or more dwelling units, every congregate residence three or more stories in height or having an occupant load of 20 or more, and every hotel three or more stories in height or containing 20 or more guest rooms. Residential or quick-response standard sprinklers shall be used in the dwelling units and guest room portions of the building.

904.3 Sprinkler System Monitoring and Alarms.

904.3.1 Where required. All valves controlling the water supply for automatic sprinkler systems and water-flow switches on all sprinkler systems shall be electrically monitored where the number of sprinklers are:

1. Twenty or more in Group I, Divisions 1.1 and 1.2 Occupancies.

2. One hundred or more in all other occupancies.

Valve monitoring and water-flow alarm and trouble signals shall be distinctly different and shall be automatically transmitted to an approved central station, remote station or proprietary monitoring station as defined by national standards, or, when approved by the building official with the concurrence of the chief of the fire department, sound an audible signal at a constantly attended location.

EXCEPTION: Underground key or hub valves in roadway boxes provided by the municipality or public utility need not be monitored.

904.3.2 Alarms. An approved audible sprinkler flow alarm shall be provided on the exterior of the building in an approved location. An approved audible sprinkler flow alarm to alert the occupants shall be provided in the interior of the building in a normally occupied location. Actuation of the alarm shall be as set forth in UBC Standard 9-1.

904.4 Permissible Sprinkler Omissions. Subject to the approval of the building official and with the concurrence of the chief of the fire department, sprinklers may be omitted in rooms or areas as follows:

1. When sprinklers are considered undesirable because of the nature of the contents or in rooms or areas that are of noncombustible construction with wholly noncombustible contents and that are not exposed by other areas. Sprinklers shall not be omitted from any room merely because it is damp, of fire-resistive construction or contains electrical equipment.

2. Sprinklers shall not be installed when the application of water or flame and water to the contents may constitute a serious life or fire hazard, as in the manufacture or storage of quantities of alu-

minum powder, calcium carbide, calcium phosphide, metallic sodium and potassium, quicklime, magnesium powder and sodium peroxide.

3. Safe deposit or other vaults of fire-resistive construction, when used for the storage of records, files and other documents, when stored in metal cabinets.

4. Communication equipment areas under the exclusive control of a public communication utility agency, provided:

4.1 The equipment areas are separated from the remainder of the building by one-hour fire-resistive occupancy separation;

4.2 Such areas are used exclusively for such equipment;

4.3 An approved automatic smoke-detection system is installed in such areas and is supervised by an approved central, proprietary or remote station service or a local alarm that will give an audible signal at a constantly attended location; and

4.4 Other approved fire-protection equipment such as portable fire extinguishers or Class II standpipes are installed in such areas.

5. Other approved automatic fire-extinguishing systems may be installed to protect special hazards or occupancies in lieu of automatic sprinklers.

904.5 Standpipes.

904.5.1 General. Standpipes shall comply with the requirements of this section and UBC Standard 9-2.

904.5.2 Where required. Standpipe systems shall be provided as set forth in Table 9-A.

904.5.3 Location of Class I standpipes. There shall be a Class I standpipe outlet connection at every floor-level landing of every required stairway above or below grade and on each side of the wall adjacent to the exit opening of a horizontal exit. Outlets at stairways shall be located within the exit enclosure or, in the case of pressurized enclosures, within the vestibule or exterior balcony, giving access to the stairway.

Risers and laterals of Class I standpipe systems not located within an enclosed stairway or pressurized enclosure shall be protected by a degree of fire resistance equal to that required for vertical enclosures in the building in which they are located.

> **EXCEPTION:** In buildings equipped with an approved automatic sprinkler system, risers and laterals that are not located within an enclosed stairway or pressurized enclosure need not be enclosed within fire-resistive construction.

There shall be at least one outlet above the roof line when the roof has a slope of less than 4 units vertical in 12 units horizontal (33.3% slope).

In buildings where more than one standpipe is provided, the standpipes shall be interconnected at the bottom.

904.5.4 Location of Class II standpipes. Class II standpipe outlets shall be accessible and shall be located so that all portions of the building are within 30 feet (9144 mm) of a nozzle attached to 100 feet (30 480 mm) of hose.

In Group A, Divisions 1 and 2.1 Occupancies, with occupant loads of more than 1,000, outlets shall be located on each side of any stage, on each side of the rear of the auditorium and on each side of the balcony.

Fire-resistant protection of risers and laterals of Class II standpipe systems is not required.

904.5.5 Location of Class III standpipes. Class III standpipe systems shall have outlets located as required for Class I standpipes in Section 904.5.3 and shall have Class II outlets as required in Section 904.5.4.

Risers and laterals of Class III standpipe systems shall be protected as required for Class I systems.

> **EXCEPTIONS:** 1. In buildings equipped with an approved automatic sprinkler system, risers and laterals that are not located within an enclosed stairway or pressurized enclosure need not be enclosed within fire-resistive construction.
>
> 2. Laterals for Class II outlets on Class III systems need not be protected.

In buildings where more than one Class III standpipe is provided, the standpipes shall be interconnected at the bottom.

904.6 Buildings under Construction.

904.6.1 General. During the construction of a building and until the permanent fire-extinguishing system has been installed and is in service, fire protection shall be provided in accordance with this section.

904.6.2 Where required. Every building four stories or more in height shall be provided with not less than one standpipe for use during construction. Such standpipes shall be installed when the progress of construction is not more than 35 feet (10 668 mm) in height above the lowest level of fire department access. Such standpipe shall be provided with fire department hose connections at accessible locations adjacent to usable stairs and the standpipe outlets shall be located adjacent to such usable stairs. Such standpipe systems shall be extended as construction progresses to within one floor of the highest point of construction having secured decking or flooring.

In each floor there shall be provided a $2^1/_2$-inch (63.5 mm) valve outlet for fire department use. Where construction height requires installation of a Class III standpipe, fire pumps and water main connections shall be provided to serve the standpipe.

904.6.3 Temporary standpipes. Temporary standpipes may be provided in place of permanent systems if they are designed to furnish a minimum of 500 gallons of water per minute (1893 L) at 50 pounds per square inch (345 kPa) pressure with a standpipe size of not less than 4 inches (102 mm). All outlets shall not be less than $2^1/_2$ inches (63.5 mm). Pumping equipment sufficient to provide this pressure and volume shall be available at all times when a Class III standpipe system is required.

904.6.4 Detailed requirements. Standpipe systems for buildings under construction shall be installed as required for permanent standpipe systems.

904.7 Basement Pipe Inlets. For basement pipe inlet requirements, see Appendix Section 907.

SECTION 905 — SMOKE CONTROL

905.1 Scope and Purpose. This section applies to mechanical or passive smoke-control systems when they are required by other provisions of this code. The purpose of this section is to establish minimum requirements for the design, installation and acceptance testing of smoke-control systems that are intended to provide a tenable environment for the evacuation or relocation of occupants. These provisions are not intended for the preservation of contents or for assistance in fire-suppression or overhaul activities. Smoke-control systems need not comply with the requirements of Section 609 in the Mechanical Code unless their normal use would otherwise require compliance. Nothing within these requirements is intended to apply when smoke control is not otherwise required by this code. Smoke-control systems are not a substitute for sprinkler protection.

905.2 Design Methods.

905.2.1 General. Buildings or portions thereof required by this code to have a smoke-control system shall have such systems designed in accordance with the requirements of this section.

> **EXCEPTION:** Smoke and heat venting required by Section 906.

905.2.2 Rationality.

905.2.2.1 General. Systems or methods of construction to be used in smoke control shall be based on a rational analysis in accordance with well-established principles of engineering. The analysis shall include, but not be limited by, Sections 905.2.2.2 through 905.2.2.6.

905.2.2.2 Stack effect. The system shall be designed such that the maximum probable normal or reverse stack effects will not adversely interfere with the system's capabilities. In determining the maximum probable stack effects, altitude, elevation, weather history and interior temperatures shall be used.

905.2.2.3 Temperature effect of fire. Buoyancy and expansion caused by the design fire (Section 905.6) shall be analyzed. The system shall be designed such that these effects do not adversely interfere with the system's capabilities.

905.2.2.4 Wind effect. The design shall consider the adverse effects of wind. Such consideration shall be consistent with the requirements of Chapter 16, Division III—Wind Design.

905.2.2.5 HVAC systems. The design shall consider the effects of the heating, ventilating and air-conditioning (HVAC) systems on both smoke and fire transport. The analysis shall include all permutations of systems status. The design shall consider the effects of the fire on the heating, ventilating and air-conditioning systems.

905.2.2.6 Climate. The design shall consider the effects of low temperatures on systems, property and occupants. Air inlets and exhausts shall be located so as to prevent snow or ice blockage.

905.2.3 Smoke barrier construction. A smoke barrier may or may not have a fire-resistive rating. Smoke barriers shall be constructed and sealed to limit leakage areas exclusive of protected openings. Maximum allowable leakage area shall be the aggregate area calculated using the following leakage area ratios:

1. Walls: $A/A_W = 0.00100$
2. Exit enclosures: $A/A_W = 0.00035$
3. All other shafts: $A/A_W = 0.00150$
4. Floors and roofs: $A/A_F = 0.00050$

WHERE:

A = total leakage area, square feet (m^2).

A_F = unit floor or roof area of barrier, square feet (m^2).

A_W = unit wall area of barrier, square feet (m^2).

Total leakage area of the barrier is the product of the smoke barrier gross area times the allowable leakage area ratio. Compliance shall be determined by achieving the minimum air pressure difference across the barrier with the system in the smoke-control mode for mechanical smoke-control systems. Passive smoke-control systems may be tested using other approved means such as door fan testing.

905.2.4 Opening protection. Openings in smoke barriers shall be protected by self-closing devices or automatic-closing devices

actuated by the required controls for the mechanical smoke-control system.

> **EXCEPTIONS:** 1. Passive smoke-control systems may have automatic-closing devices actuated by spot-type smoke detectors listed for releasing service.
>
> 2. The airflow method may be used to protect openings fixed in a permanently open position which are located between smoke zones.

Door openings shall be protected in accordance with Section 1004.3.4.3.2.

> **EXCEPTIONS:** 1. In Group I, Division 1 Occupancies when such doors are installed across corridors, a pair of opposite-swinging doors without a center mullion shall be installed having vision panels with approved fire-rated glazing materials in approved fire-rated frames, the area of which shall not exceed that tested. The doors shall be close fitting within operational tolerances, and shall not have undercuts, louvers or grilles. The doors shall have head and jamb stops, astragals or rabbets at meeting edges and automatic-closing devices. Positive latching devices may be omitted.
>
> 2. Group I, Division 3 Occupancies.

Duct and other heating, ventilating and air-conditioning openings shall be equipped with a minimum Class II, 250°F (121°C) smoke damper as defined and tested in accordance with approved recognized standards. See Chapter 35, Part IV.

905.2.5 Duration of operation. All portions of active or passive smoke-control systems shall be capable of continued operation after detection of the fire event for not less than 20 minutes.

905.3 Pressurization Method.

905.3.1 General. The primary means of controlling smoke shall be pressure differences across smoke barriers. Maintenance of a tenable environment is not required in the smoke-control zone of fire origin.

905.3.2 Minimum pressure difference. The minimum pressure difference across a smoke barrier shall be 0.05 inch water gage (12.4 Pa) in fully sprinklered buildings.

> **EXCEPTION:** Smoke-control systems serving other than fully sprinklered buildings may be approved by the building official, provided the system is designed to achieve pressure differences at least two times the maximum calculated pressure difference produced by the design fire.

905.3.3 Maximum pressure difference. The maximum air pressure difference across a smoke barrier shall be determined by required door-opening forces. The actual force required to open exit doors when the system is in the smoke-control mode shall be in accordance with Section 1003.3.1.5. The calculated force to set a side-hinged, swinging door in motion shall be determined by:

$$F = F_{dc} + K(WA\Delta P)/2(W-d) \qquad (5\text{-}1)$$

WHERE:

A = door area, square feet (m^2).

d = distance from door handle to latch edge of door, feet (m).

F = total door opening force, pounds (N).

F_{dc} = force required to overcome closing device, pounds (N).

K = 5.2 (9.6).

W = door width, feet (m).

ΔP = design pressure difference, inches water gage (Pa).

Opening forces for other doors shall be determined by standard engineering methods for the resolution of forces and reactions.

905.4 Airflow Method.

905.4.1 General. When approved by the building official, smoke may be prevented from migrating through openings fixed in a permanently open position, which are located between smoke-control zones by the use of the airflow method. The design airflows shall be in accordance with this section.

905.4.2 Velocity. The minimum average velocity through a fixed opening shall not be less than:

$$v = 217.2 \ [h \ (T_f - T_o)/(T_f + 460)]^{1/2} \qquad (5\text{-}2)$$

For **SI:** $\qquad v = 119.9 \ [h \ (T_f - T_o)/T_f]^{1/2}$

WHERE:

h = height of opening, feet (m).

T_f = temperature of smoke, °F (K).

T_o = temperature of ambient air, °F (K).

v = air velocity, feet per minute (m/s).

Airflow shall be directed to limit smoke migration from the fire zone. The geometry of openings shall be considered to prevent flow reversal from turbulent effects.

905.4.3 Prohibited conditions. This method shall not be employed where either the quantity of air or the velocity of the airflow will adversely affect other portions of the smoke-control system, unduly intensify the fire, disrupt plume dynamics or interfere with exiting. In no case shall airflows toward the fire exceed 200 feet per minute (60 960 mm per minute). Where Formula (5-2) requires airflows to exceed this limit, the airflow method shall not be used.

905.5 Exhaust Method.

905.5.1 General. When approved by the building official, for large enclosed volume, such as in atria or malls, the exhaust method may be used. The design exhaust volumes shall be in accordance with this section.

905.5.2 Exhaust rate.

905.5.2.1 General. The height of the lowest horizontal surface of the accumulating smoke layer shall be maintained at least 10 feet (3048 mm) above any walking surface within the smoke zone. The required exhaust rate for the zone shall be the largest of the calculated plume mass flow rates for the possible plume configurations. Provisions shall be made for natural or mechanical supply of outside air to make up an equal volume of the air exhausted at flow rates not to exceed 200 feet per minute (60 960 mm per minute) toward the fire.

905.5.2.2 Axisymmetric plumes. The plume mass flow rate [m_p, lbs./sec. (kg/s)] shall be determined by placing the design fire center on the axis of the space being analyzed. The limiting flame height shall be determined by:

$$z_l = 0.533Q_c^{2/5} \qquad (5\text{-}3)$$

For **SI:** $\qquad z_l = 0.166Q_c^{2/5}$

WHERE:

Q = total heat output.

Q_c = convective heat output, Btu/s (kW). (The value of Q_c shall not be taken as less than $0.70Q$.)

z = height from top of fuel surface to bottom of smoke layer, feet (m).

z_l = limiting flame height, feet (m). (z_l must be greater than the fuel equivalent diameter. See Section 905.6.)

for $z > z_l$

$$m_p = 0.022Q_c^{1/3}z^{5/3} + 0.0042Q_c \qquad (5\text{-}4)$$

For **SI:** $\qquad m_p = 0.071Q_c^{1/3}z^{5/3} + 0.0018Q_c$

for $z = z_l$

$$m_p = 0.011Q_c \qquad (5\text{-}5)$$

For **SI:** $\qquad m_p = 0.035Q_c$

for $z < z_l$

$$m_p = 0.0208Q_c^{3/5}z \qquad (5\text{-}6)$$

For **SI:** $\qquad m_p = 0.032Q_c^{3/5}z$

To convert m_p from pounds per second of mass flow to a volumetric rate, the following formula shall be used:

$$V = 60m_p/\rho \qquad (5\text{-}7)$$

WHERE:

V = volumetric flow rate, cubic feet per minute (m³/s).

ρ = density of air at the temperature of the smoke layer, lbs./ft.³ (T: in °F) [kg/m³ (T: in °C)].

905.5.2.3 Balcony spill plumes. The plume mass flow rate (m_p) for spill plumes shall be determined using the geometrically probable width based on architectural elements and projections in the following formula:

$$m_p = 0.124(QW^2)^{1/3}(z_b + 0.3H)$$
$$[1 + 0.063(z_b + 0.6H)/W]^{2/3} \qquad (5\text{-}8)$$

For **SI:** $\qquad m_p = 0.41(QW^2)^{1/3}(z_b + 0.3H)$
$$[1 + 0.063(z_b + 0.6H)/W]^{2/3}$$

WHERE:

H = height above fire to underside of balcony, feet (m).

W = plume width at point of spill, feet (m).

z_b = height from balcony, feet (m).

905.5.2.4 Window plumes. The plume mass flow rate (m_p) shall be determined from:

$$m_p = 0.077(A_wH_w^{1/2})^{1/3}(z_w + a)^{5/3} + 0.18A_wH_w^{1/2} \qquad (5\text{-}9)$$

For **SI:** $m_p = 0.68(A_wH_w^{1/2})^{1/3}(z_w + a)^{5/3} + 1.5A_wH_w^{1/2}$

WHERE:

A_w = area of the opening, square feet (m²).

H_w = height of the opening, feet (m).

z_w = height from the top of the window or opening to the bottom of the smoke layer, feet (m).

$$a = 2.4A_w^{2/5}H_w^{1/5} - 2.1H_w \qquad (5\text{-}10)$$

905.5.2.5 Plume contact with walls. When the axisymmetric plume contacts the surrounding walls, the mass flow rate may be considered to be constant from the point of contact and beyond provided that contact remains constant. Use of this provision requires calculation of the plume diameter, which shall be calculated by:

$$d = 0.48 \ [(T_c + 460)/(T_a + 460)]^{1/2}z \qquad (5\text{-}11)$$

For **SI:** $\qquad d = 0.48 \ (T_c/T_a)^{1/2}z$

WHERE:

d = plume diameter, feet (m).

T_a = ambient air temperature, °F (K).

T_c = plume center line temperature, °F (K).

$\quad = (318 \ Q_c^{2/3}H^{-5/3}) + T_a$

For **SI:** $(23.3 \ Q_c^{2/3}H^{-5/3} + 273.15) + T_a$

z = height at which T_c is determined, feet (m).

905.6 Design Fire.

905.6.1 General. The design fire shall be based on a Q of not less than 5,000 Btu per second (5275 kW) unless a rational analysis is performed by the designer and approved by the building official.

905.6.2 Rational analysis.

905.6.2.1 Factors considered. The engineering analysis shall include the characteristics of the fuel, fuel load, effects included by the fire, whether the fire is likely to be steady or unsteady.

905.6.2.2 Separation distance. Determination of the design fire shall include consideration of the type of fuel, fuel spacing and configuration. The design fire shall be increased if other combustibles are within the separation distance as determined by:

$$R = [Q/(12\pi q'')]^{1/2} \qquad (5\text{-}12)$$

WHERE:

Q = heat release from fire, Btu/s (kW).

q'' = incident radiant heat flux required for nonpiloted ignition, Btu/ft^2·s (W/m^2).

R = separation distance from target to center of fuel package, feet (m).

The ratio of the separation distance to the fuel equivalent radius shall not be less than 4. The fuel equivalent radius shall be the radius of a circle of equal area to floor area of the fuel package.

905.6.2.3 Heat-release assumptions. The analysis shall make use of best available data and shall not be based on excessively stringent limitations of combustible material. For offices, the heat release rate shall be 25 Btu/ft.^{2}·s (284 kW/m^2) or greater. For mercantile and residential occupancies, the heat release rate shall be 50 Btu/ft.^{2}·s (567 kW/m^2) or greater.

905.6.2.4 Sprinkler effectiveness assumptions. The effect of sprinklers may be assumed to have halted fire growth at time of activation only upon a documented engineering analysis.

905.7 Equipment.

905.7.1 General. Equipment such as, but not limited to, fans, ducts and balance dampers shall be suitable for their intended use, suitable for the probable temperatures to which they may be exposed and approved by the building official.

905.7.2 Exhaust fans. Components of exhaust fans shall be rated and certified by the manufacturer for the probable temperature rise to which the components may be exposed. This temperature rise shall be computed by:

$$T_s = (Q_c/mc) + (T_a) \qquad (5\text{-}13)$$

WHERE:

c = specific heat of smoke at smoke-layer temperature, Btu/lb.°F (kJ/kg-K).

m = exhaust rate, pounds per second (kg/s).

Q_c = convective heat output of fire, Btu/sec. (kW).

T_a = ambient temperature, °F (K).

T_s = smoke temperature, °F (K).

> **EXCEPTION:** T_s may be reduced if dilution air is ensured and the new T_s is calculated.

905.7.3 Ducts. Duct materials and joints shall be capable of withstanding the probable temperatures and pressures to which they are exposed as determined by Formula (5-13). Ducts shall be constructed and supported in accordance with the Mechanical Code. Ducts shall be leak tested to 1.5 times the maximum design pressure in accordance with nationally accepted practices. Measured leakage shall not exceed 5 percent of design flow. Results of such testing shall be a part of the documentation procedure. Ducts shall be supported by substantial, noncombustible supports.

> **EXCEPTION:** Flexible connections, for the purpose of vibration isolations complying with the Mechanical Code, may be used if constructed of approved fire-resistive materials.

905.7.4 Equipment, inlets and outlets. Equipment shall be located so as to not expose uninvolved portions of the building to an additional fire hazard. Outside air inlets shall be located so as to minimize the potential for introducing smoke or flame into the building. Exhaust outlets shall be located so as to minimize reintroduction of smoke into the building and to limit exposure of the building or adjacent buildings to an additional fire hazard.

905.7.5 Automatic dampers. Automatic dampers installed within the smoke-control system shall be listed and conform to the requirements of approved recognized standards. See Chapter 35, Part III.

905.7.6 Fans. In addition to other requirements, belt-driven fans shall have 1.5 times the number of belts required for the design duty with the minimum number of belts being two. Fans shall be selected for stable performance based on normal temperature and, where applicable, elevated temperature. Calculations and manufacturer's fan curves shall be part of the documentation procedures. Fans shall be supported and restrained by noncombustible devices in accordance with the requirements of Chapter 16. Motors driving fans shall not be operating beyond their name plate horsepower (kW) as determined from measurement of actual current draw. Motors driving fans shall have a minimum service factor of 1.15.

905.8 Power Systems.

905.8.1 General. The smoke-control system shall be supplied with two sources of power. Primary power shall be the normal building power systems. Secondary power shall be from an approved standby source complying with the Electrical Code. The standby power source and its transfer switches shall be in a separate room from the normal power transformers and switchgear and shall be enclosed in a room of not less than one-hour fire-resistive construction, ventilated directly to and from the exterior. Power distribution from the two sources shall be by independent routes.

Transfer to full standby power shall be automatic and within 60 seconds of failure of the primary power. The systems shall comply with the Electrical Code.

905.8.2 Power sources and power surges. Elements of the smoke-management system relying on volatile memories or the like shall be supplied with uninterruptable power sources of sufficient duration to span 15-minute primary power interruption. Elements of the smoke-management system susceptible to power surges shall be suitably protected by conditioners, suppressors or other approved means.

905.9 Detection and Control Systems.

905.9.1 General. Fire-detection and control systems for mechanical smoke-control systems shall be supervised in accordance with the Fire Code. Supervision shall further provide positive confirmation of actuation, testing of devices, manual override mechanisms, and the presence of power downstream of all disconnects. When supervision requires the sensing of damper position, it shall be accomplished by limit or proximity switches. When supervision requires sensing of air flow, it shall be by differential pressure transmitters. Required supervision shall be indicated at the Fire Fighter's Control Panel. The fire-detection and control system shall be listed.

905.9.2 Wiring. In addition to meeting requirements of the Electrical Code, all wiring, regardless of voltage, shall be fully enclosed within continuous raceways.

905.9.3 Activation. Smoke-control systems shall be activated as follows:

1. Mechanical smoke-control systems, using the pressurization method, serving buildings having no occupied floor more than 300 feet (91 440 mm) above or 75 feet (22 860 mm) below exit grade shall have automatic control of pressurized stairwell enclosure systems. All other portions of the smoke-control system may be manual in accordance with Section 905.13.

> **EXCEPTION:** When required in Group I Occupancies, they shall be entirely automatic.

2. Mechanical smoke-control systems, using the pressurization method, serving buildings having occupied floors more than 300 feet (91 440 mm) above or 75 feet (22 860 mm) below exit grade shall have completely automatic control.

3. Mechanical smoke-control systems using the airflow or exhaust method shall have completely automatic control.

4. Passive smoke-control systems may be actuated by approved spot-type detectors listed for releasing service.

905.9.4 Automatic control. Whenever completely automatic control is required or used, the automatic-control sequences shall be initiated from an appropriately zoned automatic sprinkler system meeting the requirements of UBC Standard 9-1 or from an appropriately zoned, total coverage smoke-detection system meeting the requirements of the Fire Code.

905.9.5 Smoke detection. Smoke detectors shall be listed and shall be installed in accordance with the Fire Code.

905.10 Control Air Tubing.

905.10.1 General. Control-air tubing shall be of sufficient size to meet the required response times. Tubing shall be flushed clean and dry prior to final connections. Tubing shall be adequately supported and protected from damage. Tubing passing through concrete or masonry shall be sleeved and protected from abrasion and electrolytic action.

905.10.2 Materials. Control-air tubing shall be hard drawn copper, Type L, ACR, see ASTM B 42-92, B 43-91, B 68-88, B 88-92, B 251-88 and B 280-92. Fittings shall be wrought copper or brass, solder type, see ANSI B16.22-89 or ANSI B16.18-84. Changes in direction may be made with appropriate tool bends. Brass, compression-type fittings may be used at final connection to devices; other joints shall be brazed using a BCuP$_5$ brazing alloy with solidus above 1,100°F (593°C) and liquidus below 1,500°F (816°C). Brazing flux shall be used on copper to brass joints only.

> **EXCEPTION:** Nonmetallic tubing may be used within control panels and at the final connection to devices, providing all of the following conditions are met:
>
> 1. Tubing shall be listed by an approved agency for flame and smoke characteristics.
>
> 2. Tubing and connected device shall be completely enclosed within galvanized or paint grade steel enclosure of not less than 0.030 inch (0.76 mm) (No. 22 galvanized sheet gage) thickness. Entry to the enclosure shall be by copper tubing with a protective grommet of neoprene or teflon or by suitable brass compression to male barbed adapter.
>
> 3. Tubing shall be identified by appropriately documented coding.
>
> 4. Tubing shall be neatly tied and supported within enclosure. Tubing bridging cabinet and door or movable device shall be of sufficient length to avoid tension and excessive stress. Tubing shall be protected against abrasion. Tubing serving devices on doors shall be fastened along hinges.

905.10.3 Isolation from other functions. All control tubing serving other than smoke-control functions shall be isolated by automatic isolation valves or shall be an independent system.

905.10.4 Testing. Test all control-air tubing at three times operating pressure for not less than 30 minutes without any noticeable loss in gage pressure prior to final connection to devices.

905.11 Marking and Identification. The detection and control systems shall be clearly marked at all junctions, accesses and terminations.

905.12 Control Diagrams. Identical control diagrams showing all devices in the system and identifying their location and function shall be maintained current and kept on file with the building official, the fire department and with the firefighter's control panel in an approved format and manner.

905.13 Firefighter's Control Panel.

905.13.1 General. A firefighter's control panel shall be provided for manual control or override of automatic control for mechanical smoke-control systems. Such panel shall be designed to graphically depict the building arrangement and smoke-control system zones served by the systems. The status of each smoke-control zone shall be indicated by lamps and appropriate legends.

Fans, major ducts and dampers within the building that are portions of the smoke-control systems shall be shown on the firefighter's control panel and shall be shown connected to their respective ducts with a clear indication of the direction of airflow.

Devices, switches, indicators and the like shall bear plain English identifying legends having a size and stroke equivalent to 12-point helvetica bold.

Status indicators shall be provided for all smoke-control equipment by pilot lamp-type indicators as follows:

1. Fans, dampers and other operating equipment in their normal status—GREEN.

2. Fans, dampers and other operating equipment in their off or closed status—RED.

3. Fans, dampers and other operating equipment in a fault status—YELLOW.

Provision for testing the pilot lamp on the firefighter's control panel by means of one or more "lamp test" momentary push buttons or other self-restoring means shall be included.

The fault status shall be further identified by pulsing the indicator lamp.

> **EXCEPTION:** Light-emitting diodes may be used in lieu of pilot lamps with prior approval.

The firefighter's control panel layout shall be submitted at full scale for approval prior to installation.

905.13.2 Smoke-control capability. The firefighter's control panel shall provide control capability over the complete smoke-control system equipment within the building as follows:

1. **ON-AUTO-OFF** control over each individual piece of operating smoke-control equipment that can also be controlled from other sources within the building. This includes stairway pressurization fans; smoke exhaust fans; supply, return and exhaust fans; elevator shaft fans; and other operating equipment used or intended for smoke-control purposes.

2. **OPEN-AUTO-CLOSE** control over all individual dampers relating to smoke control and that are also controlled from other sources within the building.

3. **ON-OFF** or **OPEN-CLOSE** control over all smoke-control and other critical equipment associated with a fire or smoke emer-

gency and that can only be controlled from the firefighter's control panel.

> **EXCEPTIONS:** 1. For complex systems, with prior approval, the controls and indicators may be combined to control and indicate all elements of a single smoke zone as a unit.
>
> 2. For complex systems, with prior approval, the control may be accomplished by computer interface using approved, plain English commands.

905.13.3 Control action and priorities. The firefighter's control panel actions shall be as follows:

1. **ON-OFF, OPEN-CLOSE** control actions shall have the highest priority of any control point within the building. Once issued from the firefighter's control panel, no automatic or manual control from any other control point within the building shall contradict the control action.

Where automatic means is provided to interrupt normal, nonemergency equipment operation or produce a specific result to safeguard the building or equipment (i.e., duct freezestats, duct smoke detectors, high-temperature cutouts, temperature-actuated linkage and similar devices), such means shall be capable of being overridden by the firefighter's control panel control action and the last control action as indicated by each firefighter's control panel switch position shall prevail.

> **EXCEPTION:** Power disconnects required by the Electrical Code.

2. Only the AUTO position of each three-position firefighter's control panel switch shall allow automatic or manual control action from other control points within the building. The AUTO position shall be the NORMAL, nonemergency, building control position. When a firefighter's control panel is in the AUTO position, the actual status of the device (on, off, open, closed) shall continue to be indicated by the status indicator described above.

905.14 Response Time. Smoke-control system activation shall be initiated immediately after receipt of an appropriate automatic or manual activation command. Smoke-control systems shall activate individual components (such as dampers and fans) in the sequence necessary to prevent physical damage to the fans, dampers, ducts and other equipment. The total response time for individual components to achieve their desired operating mode shall not exceed the following:

1.	Control air isolation valves	Immediately
2.	Smoke damper closing	15 seconds
3.	Smoke damper opening	15 seconds maximum
4.	Fan starting (energizing)	15 seconds maximum
5.	Fan stopping (de-energizing)	Immediately
6.	Fan volume modulation	30 seconds maximum
7.	Pressure control modulation	15 seconds maximum
8.	Temperature control safety override	Immediately
9.	Positive indication of status	15 seconds maximum

For purposes of smoke control, the firefighter's control panel response time shall be the same for automatic or manual smoke-control action initiated from any other building control point.

905.15 Acceptance Testing.

905.15.1 General. Devices, equipment, components and sequences shall be individually tested. These tests, in addition to those required above or by other provisions of this code, shall consist of determination of function, sequence and, where applicable, capacity of their installed condition.

905.15.2 Detection devices. Smoke or fire detectors that are a part of a smoke-control system shall be tested in accordance with the Fire Code in their installed condition. When applicable, this testing shall include verification of airflow in both minimum and maximum conditions.

905.15.3 Ducts. Ducts that are part of a smoke-control system shall be traversed using generally accepted practices to determine actual air quantities.

905.15.4 Dampers. Dampers shall be tested for function in their installed condition.

905.15.5 Inlets and outlets. Inlets and outlets shall be read using generally accepted practices to determine air quantities.

905.15.6 Fans. Fans shall be examined for correct rotation. Measurements of voltage, amperage, revolutions per minute and belt tension shall be made.

905.15.7 Smoke barriers. Measurements using inclined manometers shall be made of the pressure differences across smoke barriers. Such measurements shall be conducted for each possible smoke-control condition.

905.15.8 Controls. Each smoke zone, equipped with an automatic initiation device, shall be put into operation by the actuation of one such device. Each additional such device within the zone shall be verified to cause the same sequence but the operation of fan motors may be bypassed to prevent damage.

Control sequences shall be verified throughout the system, including verification of override from the firefighter's control panel and simulation of standby power conditions.

905.15.9 Reports. A complete report of testing shall be prepared by the required special inspector or special inspection agency. The report shall include identification of all devices by manufacturer, nameplate data, design values, measured values and identification tag or mark. The report shall be reviewed by the responsible designer, and when satisfied that the design intent has been achieved, the responsible designer shall affix the designer's signature and date to the report with a statement as follows:

> I have reviewed this report and by personal knowledge and on-site observation certify that the smoke-control system is in substantial compliance with the design intent, and to the best of my understanding complies with requirements of the code.

A copy of the final report shall be filed with the building official and an identical copy shall be maintained in an approved location at the building.

905.15.10 Identification and documentation. Charts, drawings and other documents identifying and locating each component of the smoke-control system, and describing their proper function and maintenance requirements shall be maintained on file at the building with the above-described report.

Devices shall have an approved identifying tag or mark on them consistent with the other required documentation and shall be dated indicating the last time they were successfully tested and by whom.

905.16 Acceptance. Buildings, or portions thereof, required by this code to comply with this section shall not be issued a certificate of occupancy until such time that the building official determines that the provisions of this section have been fully complied with and that the fire department has received satisfactory instruction on the operation, both automatic and manual, of the system.

> **EXCEPTION:** In buildings of phased construction, the building official may issue a temporary certificate of occupancy if those portions of the building to be occupied meet the requirements of this sec-

tion and that the remainder does not pose a significant hazard to the safety of the proposed occupants or adjacent buildings.

SECTION 906 — SMOKE AND HEAT VENTING

906.1 When Required. Smoke and heat vents complying with UBC Standard 15-7 or fixed openings shall be installed in accordance with the provisions of this section as follows:

1. In single-story Groups B, F, M and S, Divisions 1 and 2 Occupancies having over 50,000 square feet (4645 m²) in undivided area.

> **EXCEPTIONS:** 1. Office buildings and retail sales areas where storage does not exceed 12 feet (3658 mm) in height.
>
> 2. Group S, Division 2 Occupancies used for bulk frozen food storage when the building is protected by a complete automatic sprinkler system.

2. In Group H, Divisions 1, 2, 3, 4 or 5 Occupancies any of which are over 15,000 square feet (1394 m²) in single floor area.

For requirements on smoke and heat venting in buildings with high-piled combustible stock, see the Fire Code.

906.2 Mixed Occupancies. Venting facilities shall be installed in buildings of mixed occupancy on the basis of the individual occupancy involved.

906.3 Types of Vents. Vents shall be fixed in the open position or vents shall be activated by temperature and shall open automatically in the event of fire.

Fixed openings may consist of skylights or other openings that provide venting directly to exterior above the plane of the main roof in which they are located. Vents shall meet the design criteria of this section regarding elevation, and Section 906.5 regarding venting area, dimensions, spacing and venting ratios. The building official may require documentation of the design to ensure proper performance of required venting.

Temperature activation of vents shall be at or near the highest elevation of the ceiling and in no case lower than the upper one third of the smoke curtain. Where plain glass is used, provisions shall be made to protect the occupants from glass breakage. In no case shall vents be located closer than 20 feet (6096 mm) to an adjacent property line.

906.4 Releasing Devices. Release devices shall be in accordance with UBC Standard 15-7.

906.5 Size and Spacing of Vents.

906.5.1 Effective venting area. The effective venting area is the minimum cross-sectional area through which the hot gases must pass en route to atmosphere. The effective venting area shall not be less than 16 square feet (1.5 m²) with no dimension less than 4 feet (1219 mm), excluding ribs or gutters whose total width does not exceed 6 inches (152 mm).

906.5.2 Spacing. The maximum center-to-center spacing between vents within the building shall be:

1. In Groups B, F, M and S Occupancies: 120 feet (36 576 mm).

2. In Group H Occupancies: 100 feet (30 480 mm).

906.5.3 Venting ratios. The following ratios of effective area of vent openings to floor areas shall be:

1. In Groups B, F, M and S Occupancies: 1:100.

2. In Group H Occupancies: 1:50.

906.6 Curtain Boards.

906.6.1 General. Curtain boards shall be provided to subdivide a vented building in accordance with the provisions of this section.

906.6.2 Construction. Curtain boards shall be sheet metal, asbestos board, lath and plaster, gypsum wallboard or other approved materials that provide equivalent performance that will resist the passage of smoke. All joints and connections shall be smoke tight.

906.6.3 Location and depth. Curtain boards shall extend down from the ceiling for a minimum depth of 6 feet (1829 mm), but need not extend closer than 8 feet (2438 mm) to the floor. In Group H Occupancies, the minimum depth shall be 12 feet (3658 mm) except that it need not be closer than 8 feet (2436 mm) to the floor, provided the curtain is not less than 6 feet (1829 mm) in depth.

906.6.4 Spacing. The distance between curtain boards shall not exceed 250 feet (76 200 mm) and the curtained area shall be limited to 50,000 square feet (4645 m²). In Group H Occupancies, the distance between curtain boards shall not exceed 100 feet (30 480 mm) and the curtained area shall be limited to 15,000 square feet (1394 m²).

TABLE 9-A—STANDPIPE REQUIREMENTS

OCCUPANCY	NONSPRINKLERED BUILDING[1]		SPRINKLERED BUILDING[2,3]	
× 304.8 for mm × 0.0929 for m²	Standpipe Class	Hose Requirement	Standpipe Class	Hose Requirement
1. Occupancies exceeding 150 feet in height and more than one story	III	Yes	I	No
2. Occupancies four stories or more but less than 150 feet in height, except Group R, Division 3[6]	[I and II[4]] (or III)	[5] Yes	I	No
3. Group A Occupancies with occupant load exceeding 1,000[7]	II	Yes	No requirement	No
4. Group A, Division 2.1 Occupancies over 5,000 square feet in area used for exhibition	II	Yes	II	Yes
5. Groups I; H; B; S; M; F, Division 1 Occupancies less than four stories in height but greater than 20,000 square feet per floor[6]	II[4]	Yes	No requirement	No
6. Stages more than 1,000 square feet in area	II	No	III	No

[1]Except as otherwise specified in Item 4 of this table, Class II standpipes need not be provided in basements having an automatic fire-extinguishing system throughout.

[2]The standpipe system may be combined with the automatic sprinkler system.

[3]Portions of otherwise sprinklered buildings that are not protected by automatic sprinklers shall have Class II standpipes installed as required for the unsprinklered portions.

[4] In open structures where Class II standpipes may be damaged by freezing, the building official may authorize the use of Class I standpipes that are located as required for Class II standpipes.

[5]Hose is required for Class II standpipes only.

[6]For the purposes of this table, occupied roofs of parking structures shall be considered an additional story. In parking structures, a tier is a story.

[7]Class II standpipes need not be provided in assembly areas used solely for worship.

Chapter 10
MEANS OF EGRESS
NOTE: This chapter has been revised in its entirety.

SECTION 1001 — ADMINISTRATIVE

1001.1 Scope. Every building or portion thereof shall be provided with a means of egress as required by this chapter. A means of egress is an exit system that provides a continuous, unobstructed and undiminished path of exit travel from any occupied point in a building or structure to a public way. Such means of egress system consists of three separate and distinct elements:

1. The exit access,
2. The exit, and
3. The exit discharge.

1001.2 Standards of Quality. The standards listed below which are labeled a "UBC Standard" are also listed in Chapter 35, Part II, and are part of this code.

1. **Power doors.**

    1.1 UBC Standard 10-1, Power-operated Egress Doors

    1.2 UBC Standard 7-8, Horizontal Sliding Fire Doors Used in a Means of Egress

2. **Stairway numbering system.**

UBC Standard 10-2, Stairway Identification

3. **Hardware.**

UBC Standard 10-4, Panic Hardware

SECTION 1002 — DEFINITIONS

For the purpose of this chapter, certain terms are defined as follows:

AISLE ACCESSWAYS are that portion of an exit access that leads to an aisle.

EXIT. See Section 1005.1.

EXIT ACCESS. See Section 1004.1.

EXIT DISCHARGE. See Section 1006.1.

EXIT DOOR. See Section 1003.3.1.1.

MEANS OF EGRESS. See Section 1001.1.

MULTITHEATER COMPLEX is a building or portion thereof containing two or more motion picture auditoriums that are served by a common lobby.

PANIC HARDWARE is a door-latching assembly incorporating an unlatching device, the activating portion of which extends across at least one half the width of the door leaf on which it is installed.

PHOTOLUMINESCENT is the property of emitting light as the result of absorption of visible or invisible light, which continues for a length of time after excitation.

PRIVATE STAIRWAY is a stairway serving one tenant only.

PUBLIC WAY is any street, alley or similar parcel of land essentially unobstructed from the ground to the sky that is deeded, dedicated or otherwise permanently appropriated to the public for public use and having a clear width of not less than 10 feet (3048 mm).

SELF-LUMINOUS means powered continuously by a self-contained power source other than a battery or batteries, such as radioactive tritium gas. A self-luminous sign is independent of external power supplies or other energy for its operation.

SMOKE-PROTECTED ASSEMBLY SEATING is seating served by a means of egress system and is not subject to blockage by smoke accumulation within or under a structure.

SECTION 1003 — GENERAL

1003.1 Means of Egress. All portions of the means of egress shall comply with the applicable requirements of Section 1003.

1003.2 System Design Requirements. The general design requirements specified in this section shall apply to all three elements of the means of egress system, in addition to those specific design requirements for the exit access, the exit and the exit discharge detailed elsewhere in this chapter.

1003.2.1 Use.

1003.2.1.1 General. The building official shall assign a use category as set forth in Table 10-A to all portions of a building. When an intended use is not listed in Table 10-A, the building official shall establish a use based on a listed use that most nearly resembles the intended use.

1003.2.1.2 Change in use. No change in use or occupancy shall be made to any existing building or structure unless the means of egress system is made to comply with the requirements of this chapter for the new use or occupancy. See Section 3405.

1003.2.2 Occupant load.

1003.2.2.1 General. The basis for the design of the means of egress system is the occupant load served by the various components of such system.

1003.2.2.2 Determination of occupant load. Occupant loads shall be determined in accordance with the requirements of this section.

1003.2.2.2.1 Areas to be included. In determining the occupant load, all portions of a building shall be presumed to be occupied at the same time.

> **EXCEPTION:** Accessory use areas that ordinarily are used only by persons who occupy the main areas of an occupancy shall be provided with means of egress as though they are completely occupied, but their occupant load need not be included when computing the total occupant load of the building.

1003.2.2.2.2 Areas without fixed seats. For areas without fixed seats, the occupant load shall not be less than the number determined by dividing the floor area under consideration by the occupant load factor assigned to the use for such area as set forth in Table 10-A.

The occupant load for buildings or areas containing two or more uses or occupancies shall be determined by adding the occupant loads of the various use areas as computed in accordance with the applicable requirements of Section 1003.2.2.2.

Where an individual area has more than one proposed use, the occupant load for such area shall be determined based on that use that yields the largest occupant load.

1003.2.2.2.3 Areas with fixed seats. For areas having fixed seats, the occupant load for such areas shall be determined by the number of fixed seats installed therein.

For areas having fixed benches or pews, the occupant load shall not be less than the number of seats based on one person for each 18 inches (457 mm) of length of pew or bench. Where fixed booths are used in dining areas, the occupant load shall be based on one person for each 24 inches (610 mm) of booth length. Where fixed benches, pews or booths are curved, the larger radius shall determine the booth length.

1003.2.2.2.4 Outdoor areas. The occupant load of yards, patios, courts and similar outdoor areas shall be assigned by the building official in accordance with their anticipated use. Such outdoor areas accessible to and usable by the building occupants shall be provided with a means of egress as required by this chapter. Where an outdoor area exits only through a building, the occupant load of such outdoor area shall be considered in the design of the means of egress system of that building.

1003.2.2.2.5 Reviewing stands, grandstands and bleachers. The occupant load for reviewing stands, grandstands and bleachers shall be calculated in accordance with Section 1003.2.2.2 and the specific requirements contained in Section 1008.

1003.2.2.3 Maximum occupant load.

1003.2.2.3.1 Assembly occupancies. The maximum occupant load for an assembly occupancy shall not exceed the occupant load determined in accordance with Section 1003.2.2.2.

> **EXCEPTION:** When approved by the building official, the occupant load for an assembly occupancy may be increased, provided the maximum occupant load served does not exceed the capacity of the means of egress system for such increased number of occupants.

For temporary increases of occupant loads in places of assembly, see the Fire Code.

1003.2.2.3.2 Other occupancies. For other than assembly occupancies, an occupant load greater than that determined in accordance with Section 1003.2.2.2 is permitted; however, the means of egress system shall comply with the requirements of this chapter for such increased occupant load.

1003.2.2.4 Minimum occupant load. An occupant load less than that determined in accordance with Section 1003.2.2.2 shall not be used.

1003.2.2.5 Revised occupant load. No increase in occupant load shall be made to any existing building or structure unless the means of egress system is made to comply with the requirements of this chapter for such increased occupant load. See Section 3405.

1003.2.3 Width.

1003.2.3.1 General. The width of the means of egress system or any portion thereof shall be based on the occupant load served.

1003.2.3.2 Minimum width. The width, in inches (mm), of any component in the means of egress system shall not be less than the product determined by multiplying the total occupant load served by such component by the applicable factor set forth in Table 10-B. In no case shall the width of an individual means of egress component be less than the minimum required for such component as specified elsewhere in this chapter.

Where more than one exit or exit-access doorway serves a building or portion thereof, such calculated width shall be divided approximately equally among the means of egress components serving as exits or exit-access doorways for that area.

1003.2.3.3 Maintaining width. If the minimum required width of the means of egress system increases along the path of exit travel based on cumulative occupant loads served, such width shall not be reduced or otherwise diminished to less than the largest minimum width required to that point along the path of exit travel.

> **EXCEPTION:** In other than Group H, Divisions 1, 2, 3 and 7 Occupancies, the width of exterior exit doors from an exit enclosure may be based on the largest occupant load of all levels served by such exit enclosure multiplied by a factor of 0.2 (5.08).

1003.2.3.4 Exiting from adjacent levels. No cumulative or contributing occupant loads from adjacent building levels need be considered when determining the required width of means of egress components from a given level.

Where an exit enclosure from an upper floor and a lower floor converge at an intermediate floor, the width of the exit from the intermediate floor shall be based on the sum of the occupant loads of such upper and lower floors.

1003.2.3.5 Two-way exits. Where exit or exit-access doorways serve paths of exit travel from opposite directions, the width of such exit or exit-access doorways shall be based on the largest occupant load served. Where such exit or exit-access doorways are required to swing in the direction of exit travel by Section 1003.3.1.5, separate exit width for each path of exit travel shall be provided based on the occupant load of the area that is served.

1003.2.4 Height. Except as specified elsewhere in this chapter, the means of egress system shall have a clear height of not less than 7 feet (2134 mm) measured vertically from the walking surface to the lowest projection from the ceiling or overhead structure.

> **EXCEPTION:** Sloped ceilings permitted by Section 310.6.1.

1003.2.5 Exit continuity. The path of exit travel along a means of egress shall not be interrupted by any building element other than a means of egress component as specified in this chapter. Obstructions shall not be placed in the required width of a means of egress except projections permitted by this chapter. The required capacity of a means of egress system shall not be diminished along the path of exit travel.

1003.2.6 Changes in elevation. All exterior elevation changes and interior elevation changes of 12 inches (305 mm) or more along the path of exit travel shall be made by steps, stairs or stairways conforming with the requirements of Section 1003.3.3.3 or ramps conforming with the requirements of Section 1003.3.4.

Interior elevation changes of less than 12 inches (305 mm) along the path of exit travel serving an occupant load of 10 or more shall be by ramps conforming with the requirements of Section 1003.3.4.

> **EXCEPTIONS:** 1. In Group R, Division 3 Occupancies and within individual dwelling units of Group R, Division 1 Occupancies.
> 2. Along aisles adjoining seating areas.

1003.2.7 Elevators or escalators. Elevators or escalators shall not be used as a required means of egress component.

1003.2.8 Means of egress identification.

1003.2.8.1 General. For the purposes of Section 1003.2.8, the term "exit sign" shall mean those required signs that indicate the path of exit travel within the means of egress system.

1003.2.8.2 Where required. The path of exit travel to and within exits in a building shall be identified by exit signs conforming to the requirements of Section 1003.2.8. Exit signs shall be readily visible from any direction of approach. Exit signs shall be located as necessary to clearly indicate the direction of egress travel. No point shall be more than 100 feet (30 480 mm) from the nearest visible sign.

EXCEPTIONS: 1. Main exterior exit doors that obviously and clearly are identifiable as exit doors need not have exit signs when approved by the building official.

2. Rooms or areas that require only one exit or exit access.

3. In Group R, Division 3 Occupancies and within individual units of Group R, Division 1 Occupancies.

4. Exits or exit access from rooms or areas with an occupant load of less than 50 where located within a Group I, Division 1.1, 1.2 or 2 Occupancy or a Group E, Division 3 day-care occupancy.

1003.2.8.3 Graphics. The color and design of lettering, arrows and other symbols on exit signs shall be in high contrast with their background. Exit signs shall have the word "EXIT" on the sign in block capital letters not less than 6 inches (152 mm) in height with a stroke of not less than $^3/_4$ inch (19 mm). The word "EXIT" shall have letters having a width of not less than 2 inches (51 mm) except for the letter "I" and a minimum spacing between letters of not less than $^3/_8$ inch (9.5 mm). Signs with lettering larger than the minimum dimensions established herein shall have the letter width, stroke and spacing in proportion to their height.

1003.2.8.4 Illumination. Exit signs shall be internally or externally illuminated. When the face of an exit sign is illuminated from an external source, it shall have an intensity of not less than 5 footcandles (54 lx) from either of two electric lamps. Internally illuminated signs shall provide equivalent luminance and be listed for the purpose.

EXCEPTION: Approved self-luminous signs that provide evenly illuminated letters that have a minimum luminance of 0.06 foot lambert (0.21 cd/m^2).

1003.2.8.5 Power source. All exit signs shall be illuminated at all times. To ensure continued illumination for a duration of not less than $1^1/_2$ hours in case of primary power loss, the exit signs shall also be connected to an emergency electrical system provided from storage batteries, unit equipment or an on-site generator set, and the system shall be installed in accordance with the Electrical Code. For high-rise buildings, see Section 403.

EXCEPTION: Approved self-luminous signs that provide continuous illumination independent of an external power source.

1003.2.9 Means of egress illumination.

1003.2.9.1 General. Any time a building is occupied, the means of egress shall be illuminated at an intensity of not less than 1 footcandle (10.76 lx) at the floor level.

EXCEPTIONS: 1. In Group R, Division 3 Occupancies and within individual units of Group R, Division 1 Occupancies.

2. In auditoriums, theaters, concert or opera halls, and similar assembly uses, the illumination at the floor level may be reduced during performances to not less than 0.2 footcandle (2.15 lx), provided that the required illumination be automatically restored upon activation of a premise's fire alarm system when such system is provided.

1003.2.9.2 Power supply. The power supply for means of egress illumination shall normally be provided by the premises' electrical supply. In the event of its failure, illumination shall be automatically provided from an emergency system for Group I, Divisions 1.1 and 1.2 Occupancies and for all other occupancies where the means of egress system serves an occupant load of 100 or more. Such emergency systems shall be installed in accordance with the Electrical Code.

For high-rise buildings, see Section 403.

1003.2.10 Building accessibility. In addition to the requirements of this chapter, means of egress, which provide access to, or egress from, buildings for persons with disabilities, shall also comply with the requirements of Chapter 11.

1003.3 Means of egress components. Doors, gates, stairways and ramps that are incorporated into the design of any portion of the means of egress system shall comply with the requirements of this section. These means of egress components may be selectively included in the exit access, the exit or the exit discharge portions of the means of egress system.

1003.3.1 Doors.

1003.3.1.1 General. For the purposes of Section 1003.3.1, the term "exit door" shall mean all of those doors or doorways along the path of exit travel anywhere in a means of egress system.

Exit doors serving the means of egress system shall comply with the requirements of Section 1003.3.1. Where additional doors are installed for egress purposes, they shall conform to all requirements of this section. Buildings or structures used for human occupancy shall have at least one exterior exit door that meets the requirements of Section 1003.3.1.3.

Exit doors shall be readily distinguishable from the adjacent construction and shall be easily recognizable as exit doors. Mirrors or similar reflecting materials shall not be used on exit doors, and exit doors shall not be concealed by curtains, drapes, decorations and similar materials.

1003.3.1.2 Special doors. Revolving, sliding and overhead doors serving an occupant load of 10 or more shall not be used as required exit doors.

EXCEPTIONS: 1. Approved revolving doors having leaves that will collapse under opposing pressures may be used, provided

1.1 Such doors have a minimum width of 6 feet 6 inches (1981 mm).

1.2 At least one conforming exit door is located adjacent to each revolving door.

1.3 The revolving door shall not be considered to provide any required width when computing required means of egress width in accordance with Section 1003.2.3.

2. Horizontal sliding doors complying with UBC Standard 7-8 may be used

2.1 In elevator lobby separations.

2.2 In other than Groups A and H Occupancies, where smoke barriers are required.

2.3 In other than Group H Occupancies, where serving an occupant load of less than 50.

Power-operated doors complying with UBC Standard 10-1 may be used for egress purposes. Such doors, where swinging, shall have two guide rails installed on the swing side projecting out from the face of the door jambs for a distance not less than the widest door leaf. Guide rails shall not be less than 30 inches (762 mm) in height with solid or mesh panels to prevent penetration into door swing and shall be capable of resisting a horizontal load at top of rail of not less than 50 pounds per lineal foot (730 N/m).

EXCEPTIONS: 1. Walls or other types of separators may be used in lieu of the above guide rail, provided all the criteria are met.

2. Guide rails in industrial or commercial occupancies not accessible to the public may comply with the exception to Section 509.3.

3. Doors swinging toward flow of traffic shall not be permitted unless actuating devices start to function at least 8 feet 11 inches (2718 mm) beyond the door in an open position and guide rails extend 6 feet 5 inches (1956 mm) beyond the door in an open position.

Clearances for guide rails shall be as follows:

1. Six inches (152 mm) maximum between rails and leading edge of door at the closest point in its arc of travel.

2. Six inches (152 mm) maximum between rails and the door in an open position.

3. Two inches (51 mm) minimum between rail at hinge side and door in an open position.

4. Two inches (51 mm) maximum between freestanding rails and jamb or other adjacent surface.

1003.3.1.3 Width and height. Every required exit doorway serving an occupant load of 10 or more shall be of a size to permit the installation of a door not less than 3 feet (914 mm) in nominal width and not less than 6 feet 8 inches (2032 mm) in nominal height. Where installed, exit doors shall be capable of opening such that the clear width of the exit is not less than 32 inches (813 mm). In computing the exit width as required by Section 1003.2.3, the net dimension of the doorway shall be used.

1003.3.1.4 Door leaf width. A single leaf of an exit door serving an occupant load of 10 or more shall not exceed 4 feet (1219 mm) in width.

1003.3.1.5 Swing and opening force. Exit doors serving an occupant load of 10 or more shall be of the pivoted, balanced or side-hinged swinging type. Exit doors shall swing in the direction of the path of exit travel where the area served has an occupant load of 50 or more. The door shall swing to the fully open position when an opening force not to exceed 30 pounds (133.45 N) is applied to the latch side. For other door opening forces, see Section 905.3 and Chapter 11. See Section 3207 for doors swinging over public property.

> **EXCEPTIONS:** 1. Group I, Division 3 Occupancy used as a place of detention.
>
> 2. Doors within or serving an individual dwelling unit.
>
> 3. Special doors conforming to Section 1003.3.1.2.

Double-acting doors shall not be used as exits where any of the following conditions exist:

1. The occupant load served by the door is 100 or more.

2. The door is part of a fire assembly.

3. The door is part of a smoke- and draft-control assembly.

4. Panic hardware is required or provided on the door.

A double-acting door shall be provided with a view panel of not less than 200 square inches (0.129 m^2).

1003.3.1.6 Floor level at doors. Regardless of the occupant load served, there shall be a floor or a landing on each side of a door. Where access for persons with disabilities is required by Chapter 11, the floor or landing shall not be more than $^1/_2$ inch (12.7 mm) lower than the threshold of the doorway. Where such access is not required, the threshold shall not exceed 1 inch (25 mm). Landings shall be level except that exterior landings may have a slope not to exceed $^1/_4$ unit vertical in 12 units horizontal (2% slope).

> **EXCEPTIONS:** 1. In Group R, Division 3, and Group U Occupancies and within individual units of Group R, Division 1 Occupancies:
>
> 1.1 A door may open at the top step of an interior flight of stairs, provided the door does not swing over the top step.
>
> 1.2 A door may open at a landing that is not more than 8 inches (203 mm) lower than the floor level, provided the door does not swing over the landing.
>
> 1.3 Screen doors and storm doors may swing over stairs, steps or landings.
>
> 2. Doors serving building equipment rooms that are not normally occupied.

1003.3.1.7 Landings at doors. Regardless of the occupant load served, landings shall have a width not less than the width of the door or the width of the stairway served, whichever is greater. Doors in the fully open position shall not reduce a required dimension by more than 7 inches (178 mm). Where a landing serves an occupant load of 50 or more, doors in any position shall not reduce the landing dimension to less than one half its required width. Landings shall have a length measured in the direction of travel of not less than 44 inches (1118 mm).

> **EXCEPTION:** In Group R, Division 3, and Group U Occupancies and within individual units of Group R, Division 1 Occupancies, such length need not exceed 36 inches (914 mm).

A landing that has no adjoining door shall comply with the requirements of Section 1003.3.3.5.

1003.3.1.8 Type of lock or latch. Regardless of the occupant load served, exit doors shall be openable from the inside without the use of a key or any special knowledge or effort.

> **EXCEPTIONS:** 1. In Groups A, Division 3; B; F; M and S Occupancies and in all churches, key-locking hardware may be used on the main exit where the main exit consists of a single door or pair of doors where there is a readily visible, durable sign on or adjacent to the door stating, "THIS DOOR MUST REMAIN UNLOCKED DURING BUSINESS HOURS." The sign shall be in letters not less than 1 inch (25 mm) high on a contrasting background. When unlocked, the single door or both leaves of a pair of doors must be free to swing without operation of any latching device. The use of this exception may be revoked by the building official for due cause.
>
> 2. Exit doors from individual dwelling units; Group R, Division 3 congregate residences; and guest rooms of Group R Occupancies having an occupant load of 10 or less may be provided with a night latch, dead bolt or security chain, provided such devices are openable from the inside without the use of a key or tool and mounted at a height not to exceed 48 inches (1219 mm) above the finished floor.

Manually operated edge- or surface-mounted flush bolts and surface bolts or any other type of device that may be used to close or restrain the door other than by operation of the locking device shall not be used. Where exit doors are used in pairs and approved automatic flush bolts are used, the door leaf having the automatic flush bolts shall have no doorknob or surface-mounted hardware. The unlatching of any leaf shall not require more than one operation.

> **EXCEPTIONS:** 1. Group R, Division 3 Occupancies.
>
> 2. Where a pair of doors serving a room not normally occupied is needed for the movement of equipment, manually operated edge- or surface-mounted bolts may be used.

1003.3.1.9 Panic hardware. Panic hardware, where installed, shall comply with the requirements of UBC Standard 10-4. The activating member shall be mounted at a height of not less than 30 inches (762 mm) nor more than 44 inches (1118 mm) above the floor. The unlatching force shall not exceed 15 pounds (66.72 N) when applied in the direction of travel.

Where pivoted or balanced doors are used and panic hardware is required, panic hardware shall be of the push-pad type and the pad shall not extend across more than one half of the width of the door measured from the latch side.

1003.3.1.10 Special egress-control devices. When approved by the building official, exit doors in Group B; Group F; Group I, Division 2; Group M; Group R, Division 1 congregate residences serving as group-care facilities and Group S Occupancies may be equipped with approved listed special egress-control devices of the time-delay type, provided the building is protected throughout by an approved automatic sprinkler system and an approved automatic smoke-detection system. Such devices shall conform to all the following:

1. The egress-control device shall automatically deactivate upon activation of either the sprinkler system or the smoke-detection system.

2. The egress-control device shall automatically deactivate upon loss of electrical power to any one of the following:

2.1 The egress-control device itself.

2.2 The smoke-detection system.

2.3 Means of egress illumination as required by Section 1003.2.9.

3. The egress-control device shall be capable of being deactivated by a signal from a switch located in an approved location.

4. An irreversible process that will deactivate the egress-control device shall be initiated whenever a manual force of not more than 15 pounds (66.72 N) is applied for two seconds to the panic bar or other door-latching hardware. The egress-control device shall deactivate within an approved time period not to exceed a total of 15 seconds. The time delay established for each egress-control device shall not be field adjustable.

5. Actuation of the panic bar or other door-latching hardware shall activate an audible signal at the door.

6. The unlatching shall not require more than one operation.

A sign shall be provided on the door located above and within 12 inches (305 mm) of the panic bar or other door-latching hardware reading:

KEEP PUSHING. THIS DOOR WILL OPEN IN
_____ SECONDS. ALARM WILL SOUND.

Sign lettering shall be at least 1 inch (25 mm) in height and shall have a stroke of not less than $^1/_8$ inch (3.2 mm).

Regardless of the means of deactivation, relocking of the egress-control device shall be by manual means only at the door.

1003.3.1.11 Safety glazing identification. Regardless of the occupant load served, glass doors shall conform to the requirements specified in Section 2406.

1003.3.2 Gates.

1003.3.2.1 General. Gates serving a means of egress system shall comply with the requirements of Section 1003.3.2.

1003.3.2.2 Detailed requirements. Gates used as a component in a means of egress system shall conform to the applicable requirements of Section 1003.3.1.

> **EXCEPTION:** Gates surrounding stadiums may be of the horizontal sliding or swinging type and may exceed the 4-foot (1219 mm) maximum leaf width limitation.

1003.3.3 Stairways.

1003.3.3.1 General. Every stairway having two or more risers serving any building or portion thereof shall comply with the requirements of Section 1003.3.3. For the purposes of Section 1003.3.3, the term "stairway" shall include stairs, landings, handrails and guardrails as applicable. Where aisles in assembly rooms have steps, they shall comply with the requirements in Section 1004.3.2.

> **EXCEPTION:** Stairs or ladders used only to attend equipment or window wells are exempt from the requirements of this section.

For the purpose of this chapter, the term "step" shall mean those portions of the means of egress achieving a change in elevation by means of a single riser. Individual steps shall comply with the detailed requirements of this chapter that specify applicability to steps.

1003.3.3.2 Width. The width of stairways shall be determined as specified in Section 1003.2.3, but such width shall not be less than 44 inches (1118 mm), except as specified herein and in Chapter 11. Stairways serving an occupant load less than 50 shall not be less than 36 inches (914 mm) in width.

Handrails may project into the required width a distance of $3^1/_2$ inches (89 mm) from each side of a stairway. Stringers and other projections such as trim and similar decorative features may project into the required width $1^1/_2$ inches (38 mm) from each side.

1003.3.3.3 Rise and run. The rise of steps and stairs shall not be less than 4 inches (102 mm) nor more than 7 inches (178 mm). The greatest riser height within any flight of stairs shall not exceed the smallest by more than $^3/_8$ inch (9.5 mm). Except as permitted in Sections 1003.3.3.8.1, 1003.3.3.8.2 and 1003.3.3.8.3, the run shall not be less than 11 inches (279 mm) as measured horizontally between the vertical planes of the furthermost projection of adjacent treads or nosings. Stair treads shall be of uniform size and shape, except the largest tread run within any flight of stairs shall not exceed the smallest by more than $^3/_8$ inch (9.5 mm).

> **EXCEPTIONS:** 1. Private steps and stairways serving an occupant load of less than 10 and stairways to unoccupied roofs may be constructed with an 8-inch-maximum (203 mm) rise and a 9-inch-minimum (229 mm) run.
>
> 2. Where the bottom or top riser adjoins a sloping public way, walk or driveway having an established grade (other than natural earth) and serving as a landing, the bottom or top riser may be reduced along the slope to less than 4 inches (102 mm) in height with the variation in height of the bottom or top riser not to exceed 1 unit vertical in 12 units horizontal (8.3% slope) of stairway width.

1003.3.3.4 Headroom. Every stairway shall have a headroom clearance of not less than 6 feet 8 inches (2032 mm). Such clearances shall be measured vertically from a plane parallel and tangent to the stairway tread nosings to the soffit or other construction above at all points.

1003.3.3.5 Landings. There shall be a floor or a landing at the top and bottom of each stairway or stair run. Every landing shall have a dimension measured in the direction of travel not less than the width of the stairway. Such dimension need not exceed 44 inches (1118 mm) where the stair has a straight run. At least one intermediate landing shall be provided for each 12 feet (3658 mm) of vertical stairway rise measured between the horizontal planes of adjacent landings. Landings shall be level except that exterior landings may have a slope not to exceed $^1/_4$ unit vertical in 12 units horizontal (2% slope). For landings with adjoining doors, see Section 1003.3.1.7.

> **EXCEPTIONS:** 1. In Group R, Division 3, and Group U Occupancies and within individual units of Group R, Division 1 Occupancies, such length need not exceed 36 inches (914 mm) where the stair has a straight run.
>
> 2. Stairs serving an unoccupied roof are exempt from these requirements.

1003.3.3.6 Handrails. Stairways shall have handrails on each side, and every stairway required to be more than 88 inches (2235 mm) in width shall be provided with not less than one intermediate handrail for each 88 inches (2235 mm) of required width. Intermediate handrails shall be spaced approximately equally across with the entire width of the stairway.

> **EXCEPTIONS:** 1. Stairways less than 44 inches (1118 mm) in width or stairways serving one individual dwelling unit in Group R, Division 1 or 3 Occupancy or a Group R, Division 3 congregate residence may have one handrail.
>
> 2. Private stairways 30 inches (762 mm) or less in height may have a handrail on one side only.
>
> 3. Stairways having less than four risers and serving one individual dwelling unit in Group R, Division 1 or 3, or a Group R, Division 3 congregate residence or Group U Occupancies need not have handrails.

The top of handrails and handrail extensions shall not be placed less than 34 inches (864 mm) nor more than 38 inches (965 mm) above landings and the nosing of treads. Handrails shall be continuous the full length of the stairs and at least one handrail shall extend in the direction of the stair run not less than 12 inches (305 mm) beyond the top riser nor less than 12 inches (305 mm) beyond the bottom riser. Ends shall be returned or shall have rounded terminations or bends.

> **EXCEPTIONS:** 1. Private stairways do not require handrail extensions.

2. Handrails may have starting or volute newels within the first tread on stairways in Group R, Division 3 Occupancies and within individual dwelling units of Group R, Division 1 Occupancies.

The handgrip portion of handrails shall not be less than 1 1/4 inches (32 mm) nor more than 2 inches (51 mm) in cross-sectional dimension or the shape shall provide an equivalent gripping surface. The handgrip portion of handrails shall have a smooth surface with no sharp corners. Handrails projecting from a wall shall have a space of not less than 1 1/2 inches (38 mm) between the wall and the handrail.

1003.3.3.7 Guardrails. Stairways open on one or both sides shall have guardrails as required by Section 509.

1003.3.3.8 Alternative stairways.

1003.3.3.8.1 Circular stairways. Circular stairways conforming to the requirements of this section may be used as a means of egress component in any occupancy. The minimum width of run shall not be less than 10 inches (254 mm) and the smaller stairway radius shall not be less than twice the width of the stairway.

1003.3.3.8.2 Winding stairways. In Group R, Division 3 Occupancies and in private stairways in Group R, Division 1 Occupancies, winding stairways may be used if the required width of run is provided at a point not more than 12 inches (305 mm) from the side of the stairway where the treads are narrower, but in no case shall the width of run be less than 6 inches (152 mm) at any point.

1003.3.3.8.3 Spiral stairways. In Group R, Division 3 Occupancies and in private stairways within individual units of Group R, Division 1 Occupancies, spiral stairways may be installed. A spiral stairway is a stairway having a closed circular form in its plan view with uniform section shaped treads attached to and radiating about a minimum diameter supporting column. Such stairways may be used as a required means of egress component where the area served is limited to 400 square feet (37.16 m^2).

The tread shall provide a clear walking area measuring at least 26 inches (660 mm) from the outer edge of the supporting column to the inner edge of the handrail. The effective tread is delineated by the nosing radius line, the exterior arc (inner edge of railing) and the overlap radius line (nosing radius line of tread above). Effective tread dimensions are taken along a line perpendicular to the center line of the tread. A run of at least 7 1/2 inches (191 mm) shall be provided at a point 12 inches (305 mm) from where the tread is the narrowest. The rise shall be sufficient to provide a headroom clearance of not less than 6 feet 6 inches (1981 mm); however, such rise shall not exceed 9 1/2 inches (241 mm).

1003.3.3.9 Interior stairway construction. Interior stairways shall be constructed based on type of construction requirements as specified in Sections 602.4, 603.4, 604.4, 605.4 and 606.4.

Except where enclosed usable space under stairs is prohibited by Section 1005.3.3.6, the walls and soffits of such enclosed space shall be protected on the enclosed side as required for one-hour fire-resistive construction.

Stairways exiting directly to the exterior of a building four or more stories in height shall be provided with a means for emergency entry for fire department access. (See the Fire Code.)

1003.3.3.10 Protection of exterior wall openings. All openings in the exterior wall below and within 10 feet (3048 mm), measured horizontally, of openings in an interior exit stairway serving a building over two stories in height or a floor level having such openings in two or more floors below, shall be protected by fixed or self-closing fire assemblies having a three-fourths-hour fire-protection rating. See Section 1006.3.3.1.

EXCEPTIONS: 1. Group R, Division 3 Occupancies.

2. Protection of exterior wall openings is not required where the exterior openings in the interior stairway are protected by fixed or self-closing fire assemblies having a three-fourths-hour fire-protection rating.

3. Protection of openings is not required for open parking garages conforming to Section 405.

1003.3.3.11 Stairway to roof. In buildings four or more stories in height, one stairway shall extend to the roof surface, unless the roof has a slope steeper than 4 units vertical in 12 units horizontal (33% slope).

1003.3.3.12 Roof hatches. All required interior stairways that extend to the top floor in any building four or more stories in height shall have, at the highest point of the stair shaft, an approved hatch openable to the exterior not less than 16 square feet (1.5 m^2) in area and having a minimum dimension of 2 feet (610 mm).

EXCEPTION: A roof hatch need not be provided on pressurized enclosures or on stairways that extend to the roof with an opening onto that roof.

1003.3.3.13 Stairway identification. Stairway identification signs shall be located at each floor level in all enclosed stairways in buildings four or more stories in height. Such signs shall identify the stairway, indicate whether or not there is roof access, the floor level, and the upper and lower terminus of the stairway. The sign shall be located approximately 5 feet (1524 mm) above the landing floor in a position that is readily visible when the door is in either the open or closed position. Signs shall comply with requirements of UBC Standard 10-2.

1003.3.4 Ramps.

1003.3.4.1 General. Ramps used as a component in a means of egress system shall conform to the requirements of Section 1003.3.4.

EXCEPTION: Ramped aisles within assembly rooms shall conform to the requirements in Section 1004.3.2.

1003.3.4.2 Width. The width of ramps shall be determined as specified in Section 1003.2.3, but shall not be less than 44 inches (1118 mm), except as specified herein and in Chapter 11. Ramps serving an occupant load of less than 50 shall not be less than 36 inches (914 mm) in width.

Handrails may project into the required width a distance of 3 1/2 inches (89 mm) from each side of a ramp. Other projections, such as trim and similar decorative features, may project into the required width 1 1/2 inches (38 mm) from each side.

1003.3.4.3 Slope. The slope of ramps required by Chapter 11 that are located within an accessible route of travel shall not be steeper than 1 unit vertical in 12 units horizontal (8.3% slope). The slope of other ramps shall not be steeper than 1 unit vertical in 8 units horizontal (12.5% slope).

1003.3.4.4 Landings. Ramps having slopes steeper than 1 unit vertical in 20 units horizontal (5% slope) shall have landings at the top and bottom, and at least one intermediate landing shall be provided for each 5 feet (1524 mm) of vertical rise measured between the horizontal planes of adjacent landings. Top landings and intermediate landings shall have a dimension measured in the direction of ramp run of not less than 5 feet (1524 mm). Landings at the bottom of ramps shall have a dimension in the direction of ramp run of not less than 6 feet (1829 mm).

Doors in any position shall not reduce the minimum dimension of the landing to less than 42 inches (1067 mm) and shall not reduce the required width by more than 7 inches (178 mm) when fully open.

Where ramp access is provided to comply with the requirements of Chapter 11 and a door swings over a landing, the landing shall extend at least 24 inches (610 mm) beyond the latch edge of the door, measured parallel to the door in the closed position, and shall have a length measured in the direction of travel through the doorway of not less than 5 feet (1524 mm).

1003.3.4.5 Handrails. Ramps having slopes steeper than 1 unit vertical in 20 units horizontal (5% slope) shall have handrails as required for stairways, except that intermediate handrails shall not be required. Ramped aisles serving fixed seating shall have handrails as required in Section 1004.3.2.

1003.3.4.6 Guardrails. Ramps open on one or both sides shall have guardrails as required by Section 509.

1003.3.4.7 Construction. Ramps shall be constructed as required for stairways.

1003.3.4.8 Surface. The surface of ramps shall be roughened or shall be of slip-resistant materials.

SECTION 1004 — THE EXIT ACCESS

1004.1 General. The exit access is that portion of a means of egress system between any occupied point in a building or structure and a door of the exit. Components that may be selectively included in the exit access include aisles, hallways and corridors, in addition to those means of egress components described in Section 1003.3.

1004.2 Exit-access Design Requirements.

1004.2.1 General. The exit access portion of the means of egress system shall comply with the applicable design requirements of Section 1004.2. For the purposes of Section 1004.2, the term "exit-access doorway" shall mean the point of entry to one portion of the building or structure from another along the path of exit travel. An exit-access doorway occurs where access to all exits is not direct (see Section 1004.2.3). An exit-access doorway does not necessarily include a door. When a detailed requirement specifies an "exit-access door," however, then a door shall be included as a portion of the doorway.

1004.2.2 Travel through intervening rooms. The required access to exits from any portion of a building shall be directly from the space under consideration to an exit or to a corridor that provides direct access to an exit. Exit access shall not be interrupted by intervening rooms.

> **EXCEPTIONS:** 1. Access to exits may occur through foyers, lobbies and reception rooms.
>
> 2. Where access to only one exit is required from a space under consideration, exit access may occur through an adjoining or intervening room, which in turn provides direct access to an exit or to a corridor that provides direct access to an exit.
>
> 3. Rooms with a cumulative occupant load of less than 10 may access exits through more than one intervening room.
>
> 4. Where access to more than one exit is required from a space under consideration, such spaces may access one required exit through an adjoining or intervening room, which in turn provides direct access to an exit or to a corridor that provides direct access to an exit. All other required access to exits shall be directly from the space under consideration to an exit or to a corridor that provides direct access to an exit.
>
> 5. In a one- or two-story building classified as a Group F, Group S or Group H, Division 5 Occupancy, offices and similar administrative areas may have access to two required exits through an adjoining or intervening room, which in turn provides direct access to an exit or to a corridor that provides direct access to an exit, if the building is equipped with an automatic sprinkler system throughout and is pro-

vided with smoke and heat ventilation as specified in Section 906. Such areas shall not exceed 25 percent of the floor area of the major use.

> 6. Rooms within dwelling units may access exits through more than one intervening room.

Hallways shall be considered as intervening rooms.

Interior courts enclosed on all sides shall be considered as interior intervening rooms.

> **EXCEPTION:** Such courts not less than 10 feet (3048 mm) in width and not less than the width determined as specified in Section 1003.2.3 and providing direct access to the exit need not be considered intervening rooms.

In other than dwelling units, a means of egress shall not pass through kitchens, storerooms, restrooms, closets or spaces used for similar purposes.

A means of egress serving other than Group H Occupancies shall not pass through rooms that contain Group H Occupancies.

1004.2.3 Access to exits.

1004.2.3.1 General. Exits shall be provided from each building level. Additionally, access to such exits shall be provided from all occupied areas within building levels. The maximum number of exits required from any story, basement or individual space shall be maintained until arrival at grade or the public way.

1004.2.3.2 From individual floors. For the purposes of Section 1004.2, floors, stories, occupied roofs and similar designations of building levels other than basements and mezzanines shall be considered synonymous.

Every occupant on the first story shall have access to not less than one exit and not less than two exits when required by Table 10-A. Every occupant in basements and on stories other than the first story shall have access to not less than two exits.

> **EXCEPTIONS:** 1. Second stories having an occupant load less than 10 may be provided with access to only one exit.
>
> 2. Two or more dwelling units on the second story or in a basement may have access to only one exit where the total occupant load served by that exit does not exceed 10.
>
> 3. Except as provided in Table 10-A, access to only one exit need be provided from the second floor or a basement within an individual dwelling unit or a Group R, Division 3 congregate residence.
>
> 4. Where the third floor within an individual dwelling unit or a Group R, Division 3 congregate residence does not exceed 500 square feet (46.45 m²), access to only one exit need be provided from that floor.
>
> 5. Occupied roofs on Group R, Division 3 Occupancies may have access to only one exit where such occupied areas are less than 500 square feet (46.45 m²) and are located no higher than immediately above the second story.
>
> 6. Floors and basements used exclusively for the service of the building may have access to only one exit. For the purposes of this exception, storage rooms, laundry rooms, maintenance offices and similar uses shall not be considered as providing service to the building.

No cumulative or contributing occupant loads from adjacent levels need be considered when determining the number of required exits from a given level.

1004.2.3.3 From individual spaces. All occupied portions of the building shall have access to not less than one exit or exit-access doorway. Access to not less than two exits, exit-access doorways or combination thereof shall be provided when the individual or cumulative occupant load served by a portion of the exit access is equal to, or greater than, that listed in Table 10-A.

> **EXCEPTIONS:** 1. Elevator lobbies may have access to only one exit or exit-access doorway provided the use of such exit or exit-access doorway does not require keys, tools, special knowledge or effort.
>
> 2. Storage rooms, laundry rooms and maintenance offices not exceeding 300 square feet (27.87 m²) in floor area may be provided with access to only one exit or exit-access doorway.

1004.2.3.4 Additional access to exits. Access to not less than three exits, exit-access doorways or combination thereof shall be provided when the individual or cumulative occupant load served by the exit access is 501 to 1,000.

Access to not less than four exits, exit-access doorways or combination thereof shall be provided when the individual or cumulative occupant load served by the exit access exceeds 1,000.

1004.2.4 Separation of exits or exit-access doorways. Where two or more exits or exit-access doorways are required from any level or portion of the building, at least two of the exits or exit-access doorways shall be placed a distance apart equal to not less than one half of the length of the maximum overall diagonal dimension of the area served measured in a straight line between the center of such exits or exit-access doorways. Additional exits or exit-access doorways shall be arranged a reasonable distance apart so that if one becomes blocked, the others will be available.

> **EXCEPTION:** The separation distance determined in accordance with this section may be measured along a direct path of exit travel within a fire-resistive corridor complying with Section 1004.3.4.3.1 serving exit enclosures. The walls of any such exit enclosure shall not be less than 30 feet (9144 mm), measured in a straight line, from the walls of another exit enclosure.

1004.2.5 Travel distance.

1004.2.5.1 General. Travel distance is that distance an occupant must travel from any point within occupied portions of the exit access to the door of the nearest exit. Travel distance shall be measured in a straight line along the path of exit travel from the most remote point through the center of exit-access doorways to the center of the exit door. Travel distance shall include that portion of the path of exit travel through or around permanent construction features and building elements. Travel around tables, chairs, furnishings, cabinets and similar temporary or movable fixtures or equipment need not be considered as the normal presence of such items is factored into the permitted travel distance.

Unless prohibited elsewhere in this chapter, travel within the exit access may occur on multiple levels by way of unenclosed stairways or ramps. Where the path of exit travel includes unenclosed stairways or ramps within the exit access, the distance of travel on such means of egress components shall also be included in the travel distance measurement. The measurement along stairways shall be made on a plane parallel and tangent to the stair tread nosings in the center of the stairway.

1004.2.5.2 Maximum travel distance. The travel distance to at least one exit shall not exceed that specified in this section.

Special travel distance requirements are contained in other sections of this code as follows:

1. For atria, see Section 402.5.
2. For Group E Occupancies, see Section 1007.3.
3. For Group H Occupancies, see Section 1007.4.
4. For malls, see Sections 404.4.3 and 404.4.5.

1004.2.5.2.1 Nonsprinklered buildings. In buildings not equipped with an automatic sprinkler system throughout, the travel distance shall not exceed 200 feet (60 960 mm).

1004.2.5.2.2 Sprinklered buildings. In buildings equipped with an automatic sprinkler system throughout, the travel distance shall not exceed 250 feet (76 200 mm).

1004.2.5.2.3 Corridor increases. The travel distances specified in Sections 1004.2.5.2.1, 1004.2.5.2.2, 1004.2.5.2.4 and 1004.2.5.2.5 may be increased up to an additional 100 feet (30 480 mm) provided that the last portion of exit access leading to the exit occurs within a fire-resistive corridor complying with

Section 1004.3.4.3.1. The length of such corridor shall not be less than the amount of the increase taken, in feet (mm).

1004.2.5.2.4 Open parking garages. In a Group S, Division 4 open parking garage as defined in Section 311.9, the travel distance shall not exceed 300 feet (91 440 mm) in a building not equipped with an automatic sprinkler system throughout and 400 feet (121 920 mm) in a building equipped with an automatic sprinkler system throughout. The travel distance may be measured to open stairways, which are permitted in accordance with Section 1005.3.3.1.

1004.2.5.2.5 Factory, hazardous and storage occupancies. In a one-story building classified as a Group H, Division 5 aircraft repair hangar, or as a Group F or Group S Occupancy, the travel distance shall not exceed 300 feet (91 440 mm) and may be increased to 400 feet (121 920 mm) if the building is equipped with an automatic sprinkler system throughout and is also provided with smoke and heat ventilation as specified in Section 906.

1004.2.6 Dead ends. Where more than one exit or exit-access doorway is required, the exit access shall be arranged such that there are no dead ends in hallways and corridors more than 20 feet (6096 mm) in length.

1004.3 Exit-access Components.

1004.3.1 General. Exit-access components incorporated into the design of the exit-access portion of the means of egress system shall comply with the requirements of Section 1004.3.

1004.3.2 Aisles.

1004.3.2.1 General. Aisles serving as a portion of an exit access in the means of egress system shall comply with the requirements of Section 1004.3.2. Aisles shall be provided from all occupied portions of the exit access that contain seats, tables, furnishings, displays, and similar fixtures or equipment.

1004.3.2.2 Width in occupancies without fixed seats. The width of aisles in occupancies without fixed seats shall be determined in accordance with the following:

1. In areas serving employees only, the minimum aisle width shall be 24 inches (610 mm), but not less than the width determined as specified in Section 1003.2.3.

2. In public areas of Groups B and M Occupancies, and in assembly occupancies without fixed seats, the minimum clear aisle width shall be 36 inches (914 mm) where seats, tables, furnishings, displays and similar fixtures or equipment are placed on only one side of the aisle and 44 inches (1118 mm) where such fixtures or equipment are placed on both sides of the aisle.

The required width of aisles shall be unobstructed.

> **EXCEPTION:** Handrails and doors, when fully opened, shall not reduce the required width by more than 7 inches (178 mm). Doors in any position shall not reduce the required width by more than one half. Other nonstructural projections such as trim and similar decorative features may project into the required width 1 1/2 inches (38 mm) from each side.

1004.3.2.3 Occupancies with fixed seats. Aisles in occupancies with fixed seats shall comply with the requirements of this section.

1004.3.2.3.1 Width. The clear width of aisles shall be based on the number of fixed seats served by such aisles. The required width of aisles serving fixed seats shall not be used for any other purpose.

The minimum clear width of aisles in buildings without smoke-protected assembly seating shall be in accordance with Table 10-C.

The minimum clear width of aisles in buildings where smoke-protected assembly seating has been provided, and for which an approved life-safety evaluation has also been conducted, shall be in accordance with Table 10-D. For Table 10-D, the number of seats specified must be within a single assembly place, and interpolation shall be permitted between the specified values shown.

For both tables, the minimum clear widths shown shall be modified in accordance with the following:

1. Where risers exceed 7 inches (178 mm) in height, multiply the stairway width in the tables by factor A, where:

$$A = 1 + \frac{(\text{riser height} - 7.0 \text{ inches})}{5} \quad (4\text{-}1)$$

For **SI:** $\quad A = 1 + \frac{(\text{riser height} - 178 \text{ mm})}{127}$

Where risers do not exceed 7 inches (178 mm) in height, $A = 1$.

2. Stairways not having a handrail within a 30-inch (762 mm) horizontal distance shall be 25 percent wider than otherwise calculated, i.e., multiply by $B = 1.25$. For all other stairs, $B = 1$.

3. Ramps steeper than 1 unit vertical in 10 units horizontal (10% slope) where used in ascent shall have their width increased by 10 percent, i.e., multiply by $C = 1.10$. For ramps not steeper than 1 unit vertical in 10 units horizontal (10% slope), $C = 1$. Where fixed seats are arranged in rows, the clear width of aisles shall not be less than set forth above or less than the following minimum widths:

3.1 Forty-eight inches (1219 mm) for stairways having seating on both sides.

3.2 Thirty-six inches (914 mm) for stairways having seating on one side.

3.3 Twenty-three inches (584 mm) between a stairway handrail and seating where the aisles are subdivided by the handrail.

3.4 Forty-two inches (1067 mm) for level or ramped aisles having seating on both sides.

3.5 Thirty-six inches (914 mm) for level or ramped aisles having seating on one side.

3.6 Twenty-three inches (584 mm) between a stairway handrail and seating where an aisle does not serve more than five rows on one side.

Where exit access is possible in two directions, the width of such aisles shall be uniform throughout their length. Where aisles converge to form a single path of exit travel, the aisle width shall not be less than the combined required width of the converging aisles.

1004.3.2.3.2 Seat spacing. Where seating rows have 14 or less seats, the minimum clear width of aisle accessways shall not be less than 12 inches (305 mm) measured as the clear horizontal distance from the back of the row or guardrail ahead and the nearest projection of the row behind. Where seats are automatic or self-rising, measurement may be made with the seats in the raised position. Where seats are not automatic or self-rising, the minimum clear width shall be measured with the seat in the down position.

The clear width shall be increased as follows:

1. For rows of seating served by aisles or doorways at both ends, there shall be no more than 100 seats per row. The minimum clear width of 12 inches (305 mm) for aisle accessways shall be increased by 0.3 inch (7.6 mm) for every additional seat beyond 14, but the minimum clear width need not exceed 22 inches (559 mm). If the aisles are dead-ended, see Section 1004.3.2.4 for further limitations.

EXCEPTION: For smoke-protected assembly seating, the row length limits, beyond which the minimum clear width of 12 inches (305 mm) must be increased, may be in accordance with Table 10-E.

2. For rows of seating served by an aisle or doorway at one end only, the minimum clear width of 12 inches (305 mm) for aisle accessways shall be increased by 0.6 inch (15 mm) for every additional seat beyond seven, but the minimum clear width need not exceed 22 inches (559 mm).

EXCEPTION: For smoke-protected assembly seating, the row length limits, beyond which the minimum clear width of 12 inches (305 mm) must be increased, may be in accordance with Table 10-E.

In addition, the distance to the point where the occupant has a choice of two directions of travel to an exit shall not exceed 30 feet (9144 mm) from the point where the occupant is seated.

EXCEPTION: For smoke-protected assembly seating, the distance to the point where the occupant has a choice of two directions of travel to an exit may be increased to 50 feet (15 240 mm) from the point where the occupant is seated.

1004.3.2.4 Aisle termination. Aisles shall terminate at a cross aisle, vomitory, foyer or doorway. Aisles shall not have a dead end more than 20 feet (6096 mm) in length.

EXCEPTIONS: 1. A longer dead-end aisle is permitted where seats served by the dead-end aisle are not more than 24 seats from another aisle measured along a row of seats having a minimum clear width of 12 inches (305 mm) plus 0.6 inch (15 mm) for each additional seat above seven in a row.

2. When seats are without backrests, dead ends in vertical aisles shall not exceed a distance of 16 rows.

3. For smoke-protected assembly seating, the dead ends in vertical aisles shall not exceed a distance of 21 rows.

4. For smoke-protected assembly seating, a longer dead-end aisle is permitted where seats served by the dead-end aisle are no more than 40 seats from another aisle, measured along a row of seats having an aisle accessway with a minimum clear width of 12 inches (305 mm) plus 0.3 inch (7.6 mm) for each additional seat above seven in the row.

Each end of a cross aisle shall terminate at an aisle, vomitory, foyer or doorway.

1004.3.2.5 Aisle steps.

1004.3.2.5.1 Where prohibited. Steps shall not be used in aisles having a slope of 1 unit vertical in 8 units horizontal (12.5% slope) or less.

1004.3.2.5.2 Where required. Aisles with a slope steeper than 1 unit vertical in 8 units horizontal (12.5% slope) shall consist of a series of risers and treads extending across the entire width of the aisle.

The height of risers shall be not more than 8 inches (203 mm) nor less than 4 inches (102 mm) and the tread run shall not be less than 11 inches (279 mm). The riser height shall be uniform within each flight and the tread run shall be uniform throughout the aisle. Variations in run or height between adjacent treads or risers shall not exceed $^3/_{16}$ inch (4.8 mm).

EXCEPTION: Where the slope of aisle steps and the adjoining seating area is the same, the riser heights may be increased to a maximum of 9 inches (229 mm) and may be nonuniform, but only to the extent necessitated by changes in the slope of the adjoining seating area to maintain adequate sight lines. Variations may exceed $^3/_{16}$ inch (4.8 mm) between adjacent risers, provided the exact location of such variations is identified with a marking stripe on each tread at the nosing or leading edge adjacent to the nonuniform riser. The marking stripe shall be distinctively different from the contrasting marking stripe.

A contrasting marking stripe or other approved marking shall be provided on each tread at the nosing or leading edge such that the location of each tread is readily apparent when viewed in descent. Such stripe shall be a minimum of 1 inch (25 mm) wide and a maximum of 2 inches (51 mm) wide.

EXCEPTION: The marking stripe may be omitted where tread surfaces are such that the location of each tread is readily apparent when viewed in descent.

1004.3.2.6 Ramp slope. The slope of ramped aisles shall not be more than 1 unit vertical in 8 units horizontal (12.5% slope). Ramped aisles shall have a slip-resistant surface.

1004.3.2.7 Handrails. Handrails shall comply with the height, size and shape dimensions set forth in Section 1003.3.3.6, and ends shall be returned or shall have rounded terminations or bends. Ramped aisles having a slope steeper than 1 unit vertical in 15 units horizontal (6.7% slope) and aisle stairs (two or more adjacent steps) shall have handrails located either at the side or within the aisle width. Handrails may project into the required aisle width a distance of $3^1/_2$ inches (89 mm).

EXCEPTIONS: 1. Handrails may be omitted on ramped aisles having a slope not steeper than 1 unit vertical in 8 units horizontal (12.5% slope) and having fixed seats on both sides of the aisle.

2. Handrails may be omitted where a guardrail is at the side of an aisle that conforms to the size and shape requirements for handrails.

Handrails located within the aisle width shall be discontinuous with gaps or breaks at intervals not to exceed five rows. These gaps or breaks shall have a clear width of not less than 22 inches (559 mm) nor more than 36 inches (914 mm) measured horizontally. Such handrails shall have an additional intermediate handrail located 12 inches (305 mm) below the main handrail.

1004.3.3 Hallways.

1004.3.3.1 General. Hallways serving as a portion of the exit access in the means of egress system shall comply with the requirements of Section 1004.3.3. Hallways may be used as an exit-access component unless specifically prohibited based on requirements specified elsewhere in this chapter. For exit-access design purposes, hallways shall be considered as intervening rooms.

1004.3.3.2 Width. The width of hallways shall be determined as specified in Section 1003.2.3, but such width shall not be less than 44 inches (1118 mm), except as specified herein. Hallways serving an occupant load of less than 50 shall not be less than 36 inches (914 mm) in width.

The required width of hallways shall be unobstructed.

EXCEPTION: Doors, when fully opened, and handrails shall not reduce the required width by more than 7 inches (178 mm). Doors in any position shall not reduce the required width by more than one half. Other nonstructural projections such as trim and similar decorative features may project into the required width $1^1/_2$ inches (38 mm) from each side.

1004.3.3.3 Construction. Hallways are not required to be of fire-resistive construction unless a building element of the hallway is required to be of fire-resistive construction by some other provision of this code.

Hallways in buildings of Types I or II construction shall be of noncombustible construction, except where combustible materials are permitted in applicable building elements by other provisions of this code. Hallways in buildings of Types III, IV or V construction may be of combustible or noncombustible construction.

Hallways may have walls of any height. Partitions, rails, counters and similar space dividers not over 6 feet (1829 mm) in height above the floor shall not be construed to form a hallway.

1004.3.3.4 Openings. There is no restriction as to the amount and type of openings permitted in hallways, unless protection of openings is required by some other provision of this code.

1004.3.3.5 Elevator lobbies. Elevators opening into hallways need not be provided with elevator lobbies unless smoke- and draft-control assemblies are required for the protection of elevator door openings by some other provision of this code.

1004.3.4 Corridors.

1004.3.4.1 General. Corridors serving as a portion of an exit access in the means of egress system shall comply with the requirements of Section 1004.3.4.

For restrictions on the use of corridors to convey air, see Chapter 6 of the Mechanical Code.

1004.3.4.2 Width. The width of corridors shall be determined as specified in Section 1003.2.3, but such width shall not be less than 44 inches (1118 mm), except as specified herein. Corridors serving an occupant load of less than 50 shall not be less than 36 inches (914 mm) in width.

The required width of corridors shall be unobstructed.

EXCEPTION: Doors, when fully opened, and handrails shall not reduce the required width by more than 7 inches (178 mm). Doors in any position shall not reduce the required width by more than one half. Other nonstructural projections such as trim and similar decorative features may project into the required width $1^1/_2$ inches (38 mm) from each side.

1004.3.4.3 Construction. Corridors shall be fully enclosed by walls, a floor, a ceiling and permitted protected openings. The walls and ceilings of corridors shall be constructed of fire-resistive materials as specified in Section 1004.3.4.3.1.

EXCEPTIONS: 1. One-story buildings housing Group F, Division 2 and Group S, Division 2 Occupancies.

2. Corridors more than 30 feet (9144 mm) in width where occupancies served by such corridors have at least one exit independent from the corridor. (See Chapter 4 for covered malls.)

3. In Group I, Division 3 Occupancies such as jails, prisons, reformatories and similar buildings with open-barred cells forming corridor walls, the corridors and cell doors need not be fire-resistive.

4. Corridor walls and ceilings need not be of fire-resistive construction within office spaces having an occupant load of 100 or less when the entire story in which the space is located is equipped with an automatic sprinkler system throughout and an automatic smoke-detection system installed within the corridor. The actuation of any detector shall activate alarms audible in all areas served by the corridor.

5. Corridor walls and ceilings need not be of fire-resistive construction within office spaces having an occupant load of 100 or less when the building in which the space is located is equipped with an automatic sprinkler system throughout.

6. In Group B office buildings of Type I, Type II-FR and Type II-one-hour construction, corridor walls and ceilings need not be of fire-resistive construction within office spaces of a single tenant when the entire story in which the space is located is equipped with an approved automatic sprinkler system and an automatic smoke-detection system is installed within the corridor. The actuation of any detector shall activate alarms audible in all areas served by the corridor.

Corridor floors are not required to be of fire-resistive construction unless specified by some other provision of this code.

Corridors in buildings of Type I or II construction shall be of noncombustible construction, except where combustible materials are permitted in applicable building elements by other provisions of this code. Corridors in buildings of Type III, IV or V construction may be of combustible or noncombustible construction.

1004.3.4.3.1 Fire-resistive materials. Corridor walls shall be constructed of materials approved for one-hour fire-resistive construction on each side. Corridor walls shall extend vertically to a floor-ceiling or roof-ceiling constructed in accordance with one of the following:

1. The corridor-side fire-resistive membrane of the corridor wall shall terminate at the corridor ceiling membrane constructed

of materials approved for a one-hour fire-resistive floor-ceiling or roof-ceiling assembly to include suspended ceilings, dropped ceilings and lay-in roof-ceiling panels, which are a portion of a fire-resistive assembly.

The room-side fire-resistive membrane of the corridor wall shall terminate at the underside of a floor or roof constructed of materials approved for a one-hour fire-resistive floor-ceiling or roof-ceiling assembly.

> **EXCEPTION:** Where the corridor ceiling is an element of not less than a one-hour fire-resistive floor-ceiling or roof-ceiling assembly at the entire story, both sides of corridor walls may terminate at the ceiling membrane.

2. The corridor ceiling may be constructed of materials approved for a fire-resistive wall assembly. When this method is utilized, the corridor-side fire-resistive membrane of the corridor wall shall terminate at the lower ceiling membrane and the room-side fire-resistive membrane of the corridor wall shall terminate at the upper ceiling membrane.

Corridor ceilings of noncombustible construction may be suspended below the fire-resistive ceiling membrane.

For wall and ceiling finish requirements, see Table 8-B.

1004.3.4.3.2 Openings. Openings in corridors shall be protected in accordance with the requirements of this section.

> **EXCEPTIONS:** 1. Corridors that are excepted from fire-resistive requirements by Section 1004.3.4.3.
>
> 2. Corridors on the exterior walls of buildings may have unprotected openings to the exterior when permitted by Table 5-A.
>
> 3. Corridors in multitheater complexes may have unprotected openings where each motion picture auditorium has at least one half of its required exit or exit-access doorways opening directly to the exterior or into an exit passageway.

1004.3.4.3.2.1 Doors. All exit-access doorways and doorways from unoccupied areas to a corridor shall be protected by tightfitting smoke- and draft-control assemblies having a fire-protection rating of not less than 20 minutes when tested in accordance with UBC Standard 7-2, Part II. Such doors shall not have louvers, mail slots or similar openings. The door and frame shall bear an approved label or other identification showing the rating thereof, followed by the letter "S," the name of the manufacturer and the identification of the service conducting the inspection of materials and workmanship at the factory during fabrication and assembly. Doors shall be maintained self-closing or shall be automatic closing by actuation of a smoke detector in accordance with Section 713.2. Smoke- and draft-control door assemblies shall be provided with a gasket installed so as to provide a seal where the door meets the stop on both sides and across the top.

> **EXCEPTION:** View ports may be installed if they require a hole not larger than 1 inch (25 mm) in diameter through the door, have at least a $^1/_4$-inch-thick (6.4 mm) glass disc and the holder is of metal that will not melt out when subject to temperatures of 1,700°F (927°C).

Exit doors from a corridor shall comply with the requirements for the individual exit component being accessed as specified elsewhere in this chapter.

1004.3.4.3.2.2 Windows. Windows in corridor walls shall be protected by fixed glazing listed and labeled or marked for a fire-protection rating of at least three-fourths hour and complying with Sections 713.8 and 713.9. The total area of windows in a corridor shall not exceed 25 percent of the area of a common wall with any room.

1004.3.4.3.2.3 Duct openings. For duct openings in corridors, see Sections 713.10 and 713.11. Where both smoke dampers and fire dampers are required by Sections 713.10 and 713.11, combination fire/smoke dampers shall be used.

1004.3.4.4 Intervening rooms. Corridors shall not be interrupted by intervening rooms.

> **EXCEPTIONS:** 1. Foyers, lobbies or reception rooms constructed as required for corridors shall not be construed as intervening rooms.
>
> 2. In fully sprinklered office buildings, corridors may lead through enclosed elevator lobbies if all areas of the building have access to at least one required exit without passing through the elevator lobby.

1004.3.4.5 Elevators. Elevators opening into a corridor shall be provided with an elevator lobby at each floor containing such a corridor. The lobby shall completely separate the elevators from the corridor by construction conforming to Section 1004.3.4.3.1 and all openings into the lobby wall contiguous with the corridor shall be protected as required by Section 1004.3.4.3.2.

> **EXCEPTIONS:** 1. In office buildings, separations need not be provided from a street floor elevator lobby, provided the entire street floor is protected with an automatic sprinkler system.
>
> 2. Elevators not required to meet the shaft enclosure requirements of Section 711.
>
> 3. Where additional doors are provided in accordance with Section 3007.

Elevator lobbies shall comply with the requirements of Section 3002.

SECTION 1005 — THE EXIT

1005.1 General. The exit is that portion of the means of egress system between the exit access and the exit discharge or the public way. Components that may be selectively included in the exit include exterior exit doors, exit enclosures, exit passageways and horizontal exits, in addition to those common means of egress components described in Section 1003.3.

1005.2 Exit Design Requirements. The exit portion of the means of egress system shall comply with the applicable design requirements of this section.

1005.2.1 Separation of exits. Exits shall be separated in accordance with the requirements of Section 1004.2.4.

1005.2.2 Travel distance. Travel distance shall not be limited within an exit enclosure or exit passageway, which complies with the applicable requirements of Section 1005.3.

1005.2.3 Travel through intervening rooms. Exits shall not be interrupted by intervening rooms.

> **EXCEPTIONS:** 1. Horizontal exits may lead to an exit-access element complying with the requirements of Section 1004.
>
> 2. In office buildings, and Group I, Division 1.1 hospitals and nursing homes, a maximum of 50 percent of the exits may pass through a street-floor lobby, provided the entire street floor is protected with an automatic sprinkler system.

1005.3 Exit Components.

1005.3.1 General. Exit components incorporated into the design of the exit portion of the means of egress system shall comply with the requirements of Section 1005.3.

Once a given level of fire-resistive protection is achieved in an exit component, the fire-resistive time-period of such component shall not be reduced until arrival at the exit discharge or the public way.

> **EXCEPTION:** Horizontal exits may lead to an exit-access element complying with the requirements of Section 1004.

Doors of exit components that open directly to the exterior of a building shall not be located in areas where openings are not permitted due to location on property by Table 5-A.

1005.3.2 Exterior exit doors.

1005.3.2.1 General. Exterior exit doors serving as an exit in a means of egress system shall comply with the requirements of

Section 1005.3.2. Buildings or structures used for human occupancy shall have at least one exterior exit door that meets the requirements of Section 1003.3.1.3.

1005.3.2.2 Detailed requirements. Exterior exit doors shall comply with the applicable requirements of Section 1003.3.1.

1005.3.2.3 Arrangement. Exterior exit doors shall lead directly to the exit discharge or the public way.

1005.3.3 Exit enclosures.

1005.3.3.1 General. Exit enclosures serving as an exit in a means of egress system shall comply with the requirements of Section 1005.3.3. Exit enclosures shall not be used for any purpose other than as a means of egress.

Interior stairways, ramps or escalators shall be enclosed as specified in this section.

> EXCEPTIONS: 1. In other than Groups H and I Occupancies, an exit enclosure need not be provided for a stairway, ramp or escalator serving only one adjacent floor. Any two such atmospherically interconnected floors shall not communicate with other floors. For enclosure of escalators serving Groups B, F, M and S Occupancies, see Sections 304.6, 306.6, 309.6 and 311.6.
>
> 2. Stairways in Group R, Division 3 Occupancies and stairways within individual dwelling units in Group R, Division 1 Occupancies need not be enclosed.
>
> 3. Stairs in open parking garages, as defined in Section 311.9, need not be enclosed.

1005.3.3.2 Construction. Exit enclosures shall be of fire-resistive construction as follows:

1. In buildings of other than Type I- or Type II-F.R. construction and less than four stories in height, exit enclosures shall not be of less than one-hour fire-resistive construction.

2. In buildings of Type I- or Type II-F.R. construction of any height, exit enclosures shall not be of less than two-hour fire-resistive construction.

3. In buildings of any type of construction and four or more stories in height, exit enclosures shall not be of less than two-hour fire-resistive construction.

> EXCEPTION: In sprinkler-protected parking garages restricted to the storage of private or pleasure-type motor vehicles, exit enclosures may be enclosed with glazing meeting the requirements of Sections 713.7, 713.8 and 713.9.

Exit enclosures in buildings of Type I or II construction shall be of noncombustible construction except where combustible materials are permitted in applicable building elements by other provisions of this code. Exit enclosures in buildings of Type III, IV or V construction may be of combustible or noncombustible construction.

1005.3.3.3 Extent of enclosure. Exit enclosures shall be continuous and fully enclose all portions of the stairway or ramp to include parts of floors connecting stairway flights. Exit enclosures shall exit directly to the exterior of the building or shall include an exit passageway on the ground floor leading from the exit enclosure directly to the exterior of the building. Openings into the exit passageway shall comply with the requirements of Section 1005.3.3.5.

> EXCEPTIONS: 1. Exit passageways are not required from unenclosed stairways or ramps.
>
> 2. In office buildings, and Group I, Division 1.1 hospitals and nursing homes, a maximum of 50 percent of the exits may pass through a street-floor lobby, provided the entire street floor is protected with an automatic sprinkler system.

1005.3.3.4 Barrier. A stairway in an exit enclosure shall not continue below the grade level exit unless an approved barrier is provided at the ground-floor level to prevent persons from accidentally continuing into the basement. Directional exit signs shall be provided as specified in Section 1003.2.8.

1005.3.3.5 Openings and penetrations. Openings in exit enclosures shall be limited to those necessary for egress from normally occupied spaces into the enclosure and those necessary for egress from the enclosure.

> EXCEPTION: Exit enclosures on the exterior walls of buildings may have unprotected openings to the exterior when permitted by Table 5-A.

All interior exit doors in an exit enclosure shall be protected by a fire assembly having a fire-protection rating of not less than one hour where one-hour enclosure construction is permitted in Section 1005.3.3.2 and one and one-half hours where two-hour enclosure construction is required by Section 1005.3.3.2. Such doors shall be maintained self-closing or shall be automatic closing by actuation of a smoke detector as specified in Section 713.2. All hold-open devices shall be listed for the intended purpose and shall close or release the fire assembly to the closed position in the event of a power failure. The maximum transmitted temperature end point for such doors shall not exceed 450°F (232°C) above ambient at the end of 30 minutes of the fire exposure specified in UBC Standard 7-2.

Penetrations into or through an exit enclosure are prohibited except for those serving the exit enclosure such as ductwork and equipment necessary for independent stairway pressurization, sprinkler piping, standpipes and electrical conduit terminating in a listed box not exceeding 16 square inches (10 323 mm²) in area. Penetrations and communicating openings between exit enclosures in the same building are not permitted regardless of their protection.

1005.3.3.6 Use of space under stairway or ramp. There shall not be enclosed usable space under stairways or ramps in an exit enclosure. The open space under such stairways shall not be used for any purpose.

1005.3.3.7 Pressurized enclosure. In a building having a floor level used for human occupancy located more than 75 feet (22 860 mm) above the lowest level of fire department vehicle access, all required exit enclosures shall be pressurized in accordance with Section 905 and this section. Pressurization shall occur automatically upon activation of an approved fire alarm system.

> EXCEPTION: If the building is not equipped with a fire alarm system, pressurization shall be upon activation of a spot-type smoke detector listed for releasing service located within 5 feet (1524 mm) of each vestibule entry.

A controlled relief vent capable of discharging a minimum of 2,500 cubic feet per minute (1180 L/s) of air at the design pressure difference shall be located in the upper portion of such pressurized exit enclosures.

1005.3.3.7.1 Vestibules. Pressurized exit enclosures shall be provided with a pressurized entrance vestibule that complies with the requirements of this section.

1005.3.3.7.1.1 Vestibule size. Vestibules shall not be less than 44 inches (1118 mm) in width and not less than 72 inches (1829 mm) in the direction of travel.

1005.3.3.7.1.2 Vestibule construction. Vestibules shall have walls, floors and ceilings of not less than two-hour fire-resistive construction.

1005.3.3.7.1.3 Vestibule doors. The door assembly from the building into the vestibule shall not have less than a one and one-half hour fire-protection rating, and the door assembly from the vestibule to the exit enclosure shall be a smoke- and draft-control assembly having not less than a 20-minute fire-protection rating.

Doors shall be maintained self-closing or shall be automatic closing by activation of a smoke detector installed in accordance with Section 713. All hold-open devices shall be listed for the intended purpose and shall close or release the fire assembly to the closed position in the event of a power failure. The maximum transmitted temperature end point for the vestibule entry doors shall not exceed 450°F (232°C) above ambient at the end of 30 minutes of the fire exposure specified in UBC Standard 7-2.

1005.3.3.7.1.4 Pressure differences. The minimum pressure differences within the vestibule with the doors closed shall be 0.05-inch water gage (12.44 Pa) positive pressure relative to the fire floor and 0.05-inch water gage (12.44 Pa) negative pressure relative to the exit enclosure. No pressure difference is required relative to a nonfire floor.

1005.3.3.7.1.5 Standpipes. Fire department standpipe connections and valves serving the floor shall be within the vestibule and located in such a manner so as not to obstruct egress where hose lines arc connected and charged.

1005.3.4 Exit passageways.

1005.3.4.1 General. Exit passageways serving as an exit in a means of egress system shall comply with the requirements of Section 1005.3.4. Exit passageways shall not be used for any purpose other than as a means of egress.

1005.3.4.2 Width. The width of exit passageways shall be determined as specified in Section 1003.2.3, but such width shall not be less than 44 inches (1118 mm), except as specified herein. Exit passageways serving an occupant load of less than 50 shall not be less than 36 inches (914 mm) in width.

The required width of exit passageways shall be unobstructed.

> **EXCEPTION:** Doors, when fully opened, and handrails shall not reduce the required width by more than 7 inches (178 mm). Doors in any position shall not reduce the required width by more than one half. Other nonstructural projections such as trim and similar decorative features may project into the required width 1 1/2 inches (38 mm) on each side.

1005.3.4.3 Construction. Exit passageways less than 400 feet (121 920 mm) in length shall have walls, floors and ceilings of not less than one-hour fire-resistive construction. Exit passageways 400 feet (121 920 mm) or more in length shall have walls, floors and ceilings of not less than two-hour fire-resistive construction.

Exit passageways in buildings of Type I or II construction shall be of noncombustible construction except where combustible materials are permitted in applicable building elements by other provisions of this code. Exit passageways in buildings of Type III, IV or V construction may be of combustible or noncombustible construction.

1005.3.4.4 Openings and penetrations. Openings into exit passageways shall be limited to those necessary for egress from normally occupied spaces into the exit passageway and those necessary for egress from the exit passageway. Elevators shall not open into an exit passageway.

All interior exit doors in an exit passageway shall be protected by a fire assembly having a fire-protection rating of not less than one hour where one-hour exit passageway construction is permitted in Section 1005.3.4.3 and not less than one and one-half hours where two-hour exit passageway construction is required by Section 1005.3.4.3. Such doors shall be maintained self-closing or shall be automatic closing by actuation of a smoke detector as specified in Section 713.2. All hold-open devices shall be listed for the intended purpose and shall close or release the fire assembly to the closed position in the event of a power failure. The maximum

transmitted temperature end point for such doors shall not exceed 450°F (232°C) above ambient at the end of 30 minutes of the fire exposure specified in UBC Standard 7-2.

Penetrations into or through an exit passageway are prohibited except for those serving the exit passageway such as sprinkler piping, standpipes and electrical conduit terminating in a listed box not exceeding 16 square inches (10 323 mm^2) in area.

1005.3.4.5 Intervening rooms. Exit passageways shall not be interrupted by intervening rooms.

> **EXCEPTION:** In office buildings, a maximum of 50 percent of the exits may discharge through a street-floor lobby provided the entire street floor is protected with an automatic sprinkler system.

1005.3.4.6 Dead ends. Where an exit passageway is used and more than one exit is required, exit doors shall be arranged so that it is possible to go in either direction from any point in the exit passageway to a separate exit door, except for dead ends not exceeding 20 feet (6096 mm) in length.

1005.3.5 Horizontal exits.

1005.3.5.1 General. Horizontal exits serving as an exit in a means of egress system shall comply with the requirements of Section 1005.3.5. A horizontal exit is a wall that completely divides a floor of a building into two or more separate exit-access areas to afford safety from fire and smoke in the exit-access area of incident origin.

It is permissible for a horizontal exit to serve as an exit for each adjacent exit-access area (e.g., a two-way exit), providing that the exit-access design requirements for each exit-access area are independently satisfied.

A horizontal exit shall not serve as the only exit from the exit access. Where two or more exits are required from the exit access, not more than one half of the total number of exits or total exit width may be provided by horizontal exits.

1005.3.5.2 Construction. The wall containing a horizontal exit shall be constructed as required for an occupancy separation having a fire-resistive rating of not less than two hours. The horizontal exit wall shall be continuous from exterior wall to exterior wall and shall extend from the floor to the underside of the floor or roof directly above so as to completely divide the floor that is served by the horizontal exit. Structural members supporting a horizontal exit shall be protected by equivalent fire-resistive construction.

Horizontal exits in buildings of Type I, II or III construction shall be of noncombustible construction. Horizontal exits in buildings of Type IV or V construction may be of combustible or noncombustible construction.

1005.3.5.3 Openings and penetrations. Openings in a horizontal exit shall be protected by a fire assembly having a fire-protection rating of not less than one and one-half hours. Such fire assemblies shall be maintained self-closing or shall be automatic closing by actuation of a smoke detector as specified in Section 713.2. All hold-open devices shall be listed for the intended purpose and shall close or release the fire assembly to the closed position in the event of a power failure. The maximum transmitted temperature end point for such doors shall not exceed 450°F (232°C) above ambient at the end of 30 minutes of the fire exposure specified in UBC Standard 7-2.

1005.3.5.4 Refuge area. The floor area of the exit access to which a horizontal exit leads shall be of sufficient size to accommodate 100 percent of the occupant load of the exit access from which refuge is sought, plus 100 percent of the normal occupant load of the exit access serving as the refuge area. The capacity of such refuge floor area shall be determined by allowing 3 square feet (0.28 m^2) of net clear floor area of aisles, hallways and corridors per occupant. The area of stairs, elevators and other shafts shall not be used. In Group I, Division 1.1 Occupancies, the capacity of the refuge area shall be determined by allowing 15 square feet (1.4 m^2) of net clear floor area per ambulatory occupant and 30 square feet (2.8 m^2) of net clear floor area per nonambulatory occupant.

The design of the exit access serving as the refuge area shall comply with the requirements of Section 1004.2 based on the normal occupant load served and need not consider the increased occupant load imposed by persons entering such refuge area through horizontal exits.

SECTION 1006 — THE EXIT DISCHARGE

1006.1 General. The exit discharge is that portion of the means of egress system between the exit and the public way. Components that may be selectively included in the exit discharge include exterior exit balconies, exterior exit stairways, exterior exit ramps and exit courts, in addition to those common means of egress components described in Section 1003.3.

> **EXCEPTION:** When approved by the building official, the exit discharge may lead to a safe dispersal area on the same property as the building being exited. The proximity and size of such safe dispersal area shall be based on such factors as the occupant load served, the mobility of occupants, the type of construction of the building, the fire-protection features of the building, the height of the building and the degree of hazard of the occupancy. In any case, such safe dispersal areas shall not be located less than 50 feet (15 240 mm) from the building served. (See Section 1007 for means of egress from safe dispersal areas.)

Grade level areas designated as an exit discharge component for a building shall be permanently maintained. Such areas shall not be developed or otherwise altered in their capacity to provide for a continuous, unobstructed and undiminished means of egress for building occupants. If such areas are sold independent of the building they serve, an exit discharge complying with the requirements of Section 1006 shall be provided for such building.

1006.2 Exit Discharge Design Requirements. The exit discharge portion of the means of egress system shall comply with the applicable design requirements of this section.

1006.2.1 Location. The exit discharge shall be at grade or shall provide direct access to grade. The exit discharge shall not reenter the exit access. Exterior exit balconies, exterior exit stairways and exterior exit ramps shall not be located in areas where building openings are prohibited or openings are required to be protected by Table 5-A.

1006.2.2 Access to grade. Where the exit from a building discharges at other than grade level, there shall not be less than two separate paths of exit travel to grade level. Such paths of exit travel shall be arranged so that there are no dead ends more than 20 feet (6096 mm) in length.

> **EXCEPTIONS:** 1. Where the occupant load served by such exit is less than 10, only one path of exit travel to grade level need be provided.
>
> 2. Where exits discharge to an exterior exit stairway, such stairway may serve as a single path of exit travel directly to grade.

1006.2.3 Travel distance. Travel distance in the exit discharge at grade level shall not be limited.

Travel distance in the exit discharge at other than grade level shall not exceed the following:

1. In buildings not equipped with an automatic sprinkler system throughout, the travel distance to grade shall not exceed 200 feet (60 960 mm).

2. In buildings equipped with an automatic sprinkler system throughout, the travel distance to grade shall not exceed 250 feet (76 200 mm).

Where the path of exit travel includes unenclosed stairways or ramps within the exit discharge, the distance of travel on such means of egress components shall also be included in the travel distance measurement. The measurement along stairways shall be made on a plane parallel and tangent to the stair tread nosings in the center of the stairway.

1006.3 Exit Discharge Components.

1006.3.1 General. Exit discharge components incorporated into the design of the exit discharge portion of the means of egress system shall comply with the requirements of Section 1006.3. In all cases, components of the exit discharge shall be sufficiently open to the exterior to prevent the accumulation of smoke and toxic gases.

1006.3.2 Exterior exit balconies.

1006.3.2.1 General. Exterior exit balconies serving as a portion of the exit discharge in the means of egress system shall comply with the requirements of Section 1006.3.2. An exterior exit balcony is a balcony, landing or porch projecting from the wall of a building and serves as an exit discharge component in a means of egress system.

1006.3.2.2 Width. The width of exterior exit balconies shall be determined as specified in Section 1003.2.3, but such width shall not be less than 44 inches (1118 mm), except as specified herein. Exterior exit balconies serving an occupant load of less than 50 shall not be less than 36 inches (914 mm) in width.

The required width of exterior exit balconies shall be unobstructed.

> **EXCEPTION:** Doors, when fully opened, and handrails shall not reduce the required width by more than 7 inches (178 mm). Doors in any position shall not reduce the required width by more than one half. Other nonstructural projections such as trim and similar decorative features may project into the required width 1$^1/_2$ inches (38 mm) from each side.

1006.3.2.3 Construction. Exterior exit balconies projecting from the walls of buildings of Type I or II construction shall be of noncombustible construction. Exterior exit balconies projecting from the walls of buildings of Type III, IV or V construction may be of combustible or noncombustible construction.

Walls of exterior exit balconies serving a Group R, Division 1 or Group I Occupancy having an occupant load of 10 or more shall not be less than one-hour fire-resistive construction and ceilings shall not be less than that required for a one-hour fire-resistive floor or roof system.

> **EXCEPTIONS:** 1. Exterior sides of exterior exit balconies.
>
> 2. In other than Type I or II construction, exterior exit balcony roof assemblies may be of heavy-timber construction without concealed spaces.

1006.3.2.4 Openness. The long side of an exterior exit balcony shall be at least 50 percent open, and the open area above the guardrail shall be distributed to prevent the accumulation of smoke or toxic gases.

1006.3.3 Exterior exit stairways.

1006.3.3.1 General. Exterior exit stairways serving as a portion of the exit discharge in the means of egress system shall comply with the requirements of Section 1006.3.3. An exterior exit stairway serves as an exit discharge component in a means of egress system and is open on not less than two adjacent sides, except for required structural columns and open-type handrails and guardrails. The adjoining open areas shall be either yards, exit courts or public ways; the remaining sides may be enclosed by the exterior walls of the building. Any stairway not meeting the definition of an exterior stairway shall comply with the requirements for interior stairways.

1006.3.3.2 Construction. Exterior exit stairways shall be constructed based on type of construction requirements as specified in Sections 602.4, 603.4, 604.4, 605.4 and 606.4.

There shall be no enclosed usable space under exterior exit stairways. The open space under such stairways shall not be used for any purpose.

1006.3.3.3 Protection of exterior wall openings. All openings in the exterior wall below and within 10 feet (3048 mm), measured horizontally, of an exterior exit stairway serving a building over two stories in height or a floor level having such openings in two or more floors below shall be protected by fixed or self-closing fire assemblies having a three-fourths-hour fire-protection rating.

> **EXCEPTIONS:** 1. Group R, Division 3 Occupancies.
>
> 2. Openings may be unprotected where two separated exterior stairways are served by a common exterior exit balcony.
>
> 3. Protection of openings is not required for open parking garages conforming to Section 311.9.

1006.3.3.4 Detailed requirements. Except for construction and opening protection as specified in Sections 1006.3.3.2 and 1006.3.3.3, exterior exit stairways shall comply with the applicable requirements for stairways as specified in Section 1003.3.3.

1006.3.4 Exterior exit ramps.

1006.3.4.1 General. Exterior exit ramps serving as a portion of the exit discharge in the means of egress system shall comply with the requirements of Section 1006.3.4. An exterior exit ramp serves as an exit discharge component in a means of egress system and is open on not less than two adjacent sides, except for required structural columns and open-type handrails and guardrails. The adjoining open areas shall be either yards, exit courts or public way; the remaining sides may be enclosed by the exterior walls of the building. Any ramp not meeting the definition of an exterior ramp shall comply with the requirements for interior ramps.

1006.3.4.2 Construction. Exterior exit ramps shall be constructed based on type of construction requirements as specified in Sections 602.4, 603.4, 604.4, 605.4 and 606.4.

There shall be no enclosed usable space under exterior exit ramps. The open space under such ramps shall not be used for any purpose.

1006.3.4.3 Protection of exterior wall openings. All openings in the exterior wall below and within 10 feet (3048 mm), measured horizontally, of an exterior exit ramp serving a building over two stories in height or a floor level having such openings in two or more floors below shall be protected by fixed or self-closing fire assemblies having a three-fourths-hour fire-protection rating.

> **EXCEPTIONS:** 1. Group R, Division 3 Occupancies.
>
> 2. Openings may be unprotected where two separated exterior ramps are served by a common exterior exit balcony.
>
> 3. Protection of openings is not required for open parking garages conforming to Section 405.

1006.3.4.4 Detailed requirements. Except for construction and opening protection as specified in Sections 1006.3.4.2 and 1006.3.4.3, exterior exit ramps shall comply with the applicable requirements for ramps as specified in Section 1003.3.4.

1006.3.5 Exit courts.

1006.3.5.1 General. Exit courts serving as a portion of the exit discharge in the means of egress system shall comply with the requirements of Section 1006.3.5. An exit court is a court or yard that provides access to a public way for one or more required exits.

1006.3.5.2 Width. The width of exit courts shall be determined as specified in Section 1003.2.3, but such width shall not be less than 44 inches (1118 mm), except as specified herein. Exit courts serving Group R, Division 3 and Group U Occupancies shall not be less than 36 inches (914 mm) in width.

The required width of exit courts shall be unobstructed to a height of 7 feet (2134 mm).

> **EXCEPTION:** Doors, when fully opened, and handrails shall not reduce the required width by more than 7 inches (178 mm). Doors in any position shall not reduce the required width by more than one half. Other nonstructural projections such as trim and similar decorative features may project into the required width $1^1/_2$ inches (38 mm) from each side.

Where an exit court exceeds the minimum required width and the width of such exit court is then reduced along the path of exit travel, the reduction in width shall be gradual. The transition in width shall be affected by a guardrail not less than 36 inches (914 mm) in height and shall not create an angle of more than 30 degrees with respect to the axis of the exit court along the path of exit travel. In no case shall the width of the exit court be less than the required minimum.

1006.3.5.3 Construction and openings. Where an exit court serving a building or portion thereof having an occupant load of 10 or more is less than 10 feet (3048 mm) in width, the exit court walls shall not be less than one-hour fire-resistive construction for a distance of 10 feet (3048 mm) above the floor of the court, and all openings therein shall be protected by fixed or self-closing fire assemblies having a three-fourths-hour fire-protection rating.

SECTION 1007 — MEANS OF EGRESS REQUIREMENTS BASED ON OCCUPANCY

1007.1 General. In addition to the general means of egress requirements specified elsewhere in this chapter, the detailed requirements of this section shall apply to those occupancies described herein.

1007.2 Group A Occupancies.

1007.2.1 Main exit. Group A, Division 1, 2 and 2.1 Occupancies shall be provided with a main exit. The main exit shall be of sufficient width to accommodate not less than one half of the total occupant load, but such width shall not be less than the total required width of all means of egress components leading thereto.

1007.2.2 Side exits. Auditoriums, theaters and similar assembly rooms of Group A, Division 1, 2 or 2.1 Occupancies shall be provided with exits on each side. The exits on each side of such assembly rooms shall be of sufficient width to accommodate not less than one third of the total occupant load served. Side exits shall open directly to a public way or into an exit or exit discharge leading to a public way. Side exits shall be accessible from a cross aisle.

1007.2.3 Balcony exits. Balconies, mezzanines and similar areas having an occupant load of 10 or more shall be provided with access to a minimum of two exits. Balconies shall directly access an exterior stairway or other approved stairway or ramp. Where there is more than one level of balconies, balconies shall directly access an exit enclosure or an exterior stairway or ramp. Balcony exits or exit access shall be accessible from a cross aisle. The number and distribution of exits and exit access shall be as specified elsewhere in this chapter.

1007.2.4 Multitheater complex. The main exit from a multitheater complex shall be of sufficient width to accommodate one half of the total occupant load of such complex.

1007.2.5 Panic hardware. Exit and exit-access doors serving Group A Occupancies shall not be provided with a latch or lock unless it is panic hardware.

> **EXCEPTIONS:** 1. In Group A, Division 3 Occupancies and in all churches, panic hardware may be omitted from the main exit where the main exit consists of a single door or pair of doors. A key-locking device may be used in place of the panic hardware, provided there is a readily visible durable sign adjacent to the doorway stating, "THIS DOOR MUST REMAIN UNLOCKED DURING BUSINESS HOURS." The sign shall be in letters not less than 1 inch (25 mm) high on a contrasting background. When unlocked, the single door or both leaves of a pair of doors must be free to swing without operation of any latching device. Manually operated edge- or surface-mounted flush bolts and surface bolts or any other type of device that may be used to close or restrain the door other than by operation of the locking device shall not be used. The use of this exception may be revoked by the building official for due cause.
>
> 2. Panic hardware may be waived on gates surrounding stadiums where such gates are under constant immediate supervision while the public is present, and further provided that safe dispersal areas based on 3 square feet (0.28 m^2) per occupant are located between the stadium and the fence. Such required safe dispersal areas shall not be located less than 50 feet (15 240 mm) from the stadium. Gates may be of the horizontal sliding or swinging type and may exceed the 4-foot (1219 mm) maximum leaf width limitation.

1007.2.6 Posting of room capacity. Any room that is used for an assembly purpose where fixed seats are not installed shall have the capacity of the room posted in a conspicuous place on an approved sign near the main exit or exit-access doorway from the room. Such signs shall indicate the number of occupants permitted for each room use.

1007.2.7 Amusement building exit marking. Approved exit signs and directional exit marking that complies with the provisions of Section 1003.2.8 shall be provided in amusement buildings.

Additional approved low-level exit signs that are internally or externally illuminated, photoluminescent or self-luminous shall be provided. The bottom of such sign shall not be less than 6 inches (152 mm) nor more than 8 inches (203 mm) above the walking surface and shall indicate the path of exit travel. For exit and exit-access doors, the sign shall be on the door or adjacent to the door with the closest edge of the sign within 4 inches (102 mm) of the door frame.

1007.3 Group E Occupancies.

1007.3.1 Definitions. For the purpose of Section 1007.3, certain terms are defined as follows:

INTERIOR ROOM is a room whose only exit access is through an adjoining or intervening room and not a corridor.

ROOM is a space or area enclosed on more than 80 percent of the perimeter of such space or area. When determining the enclosed area, openings less than 3 feet (914 mm) in clear width and less than 6 feet 8 inches (2032 mm) high need not be considered.

SEPARATE MEANS OF EGRESS SYSTEM is not less than two paths of exit travel, which are separated in such a manner to provide an atmospheric separation that precludes contamination of both paths of exit travel by the same fire.

1007.3.2 Separate means of egress systems required. Every room with an occupant load of 300 or more shall have one of its exits or exit-access doorways lead directly into a separate means of egress system. Not more than two required exits or exit-access doorways shall enter into the same means of egress system.

1007.3.3 Travel distance.

1007.3.3.1 In rooms. The travel distance from any point in a room shall not exceed 75 feet (22 860 mm) to a corridor or an exit.

> **EXCEPTIONS:** 1. In buildings not more than two stories in height and protected throughout by smoke detectors, the travel distance may be increased to 90 feet (27 432 mm).
>
> 2. In buildings equipped with an automatic sprinkler system throughout, the travel distance may be increased to 110 feet (33 528 mm).

1007.3.3.2 From any location. In buildings not equipped with an automatic sprinkler system throughout, the travel distance shall not exceed 150 feet (45 720 mm).

> **EXCEPTIONS:** 1. In buildings not more than two stories in height and protected throughout by smoke detectors, the travel distance may be increased to 175 feet (53 340 mm).
>
> 2. In buildings equipped with an automatic sprinkler system throughout, the travel distance may be increased to 225 feet (68 580 mm).

The travel distances specified above may be increased up to an additional 100 feet (30 480 mm), provided that the last portion of travel leading to the exit occurs within a corridor. The length of such corridor shall not be less than the amount of the increase taken.

1007.3.4 Travel through intervening rooms. The path of exit travel shall not pass through laboratories using hazardous materials, industrial shops or other similar places.

Where only one exit access is required from an interior room and the path of exit travel is through an adjoining or intervening room, smoke detectors shall be installed throughout the common atmosphere of the exit access through which the path of exit travel passes. Such smoke detectors shall actuate alarms audible in the interior room and shall be connected to the school fire alarm system.

> **EXCEPTIONS:** 1. Where the aggregate occupant load of the interior room or rooms is 10 or less.
>
> 2. Where the enclosures forming interior rooms are less than two thirds of the floor-to-ceiling height and do not exceed 8 feet (2438 mm).
>
> 3. Rooms used exclusively for mechanical or public utility service to the buildings.

1007.3.5 Hallways, corridors and exterior exit balconies. The width of hallways and corridors in a Group E, Division 1 Occupancy shall be determined as specified in Section 1003.2.3, plus 2 feet (610 mm), but shall not be less than 6 feet (1829 mm).

> **EXCEPTION:** Where the total number of occupants served is less than 100, such hallway or corridor may be 44 inches (1118 mm) wide.

Any change in elevation of less than 2 feet (610 mm) in a hallway, corridor or exterior exit balcony shall be by means of a ramp.

1007.3.6 Stairways. The width of stairways shall be determined as specified in Section 1003.2.3, but stairways serving an occupant load of 100 or more shall not be less than 5 feet (1524 mm) in width.

1007.3.7 Exits serving auditoriums in Group E, Division 1 Occupancies. In determining the means of egress design requirements, an auditorium may be considered an accessory use area in accordance with the provisions of Section 1003.2.2.2.1 if the auditorium is not to be used simultaneously with other rooms.

1007.3.8 Laboratories. Laboratories having a floor area of 200 square feet (18.6 m^2) or more shall have access to not less than two separate exits or exit-access doorways. All portions of such laboratories shall be within 75 feet (22 860 mm) of an exit or exit-access door.

1007.3.9 Basement rooms. Exit stairways from a basement shall open directly to the exterior of the building without entering the first floor.

1007.3.10 Panic hardware. Exit and exit-access doors from rooms having an occupant load of 50 or more and from corridors shall not be provided with a latch or lock unless it is panic hardware.

1007.3.11 Fences and gates. School grounds may be fenced and gates therein may be equipped with locks, provided that safe dispersal areas based on 3 square feet (0.28 m^2) per occupant are located between the school and the fence. Such required safe dispersal areas shall not be located less than 50 feet (15 240 mm) from school buildings.

1007.4 Group H Occupancies.

1007.4.1 Access to exits. Every portion of a Group H Occupancy having a floor area of 200 square feet (18.6 m^2) or more shall have access to not less than two separate exits or exit-access doors.

> **EXCEPTION:** Group H, Division 4 Occupancies having a floor area of less than 1,000 square feet (92.9 m^2) may have one exit or exit-access door.

1007.4.2 Travel distance. In Group H, Divisions 1, 2 and 3 Occupancies, the travel distance specified in Section 1004.2.5 shall not exceed 75 feet (22 860 mm) to an exit or an exit-access door.

The travel distances specified above may be increased up to an additional 100 feet (30 480 mm), provided that the last portion of exit access leading to the exit occurs within a corridor. The length of such corridor shall not be less than the amount of the increase taken.

In Group H, Division 7, and within fabrication areas of Group H, Division 6 Occupancies, the travel distance specified in Section 1004.2.5 shall not exceed 100 feet (30 480 mm) to an exit or a corridor.

1007.4.3 Corridor doors. Corridor doors shall be protected by a fire assembly having a fire-protection rating of not less than three-fourths-hour, shall not have more than 100 square inches (64 516 mm^2) of wired glass set in steel frames and shall be maintained self-closing or shall be automatic closing as specified in Section 713.2.

1007.4.4 Door swing. All exit and exit-access doors serving hazardous occupancies shall swing in the exit travel, regardless of the occupant load served.

1007.4.5 Panic hardware. Exit and exit-access doors from rooms in Group H, Divisions 1, 2, 3, 6 and 7 Occupancies shall not be provided with a latch or lock unless it is panic hardware.

1007.4.6 Incinerator rooms. Interior openings between a Group H Occupancy and an incinerator room are prohibited.

1007.5 Group I Occupancies.

1007.5.1 Minimum size of means of egress. The clear width of means of egress components in areas serving bed or litter patients shall be such to allow ready passage of beds, gurneys and similar equipment, but shall not be less than 44 inches (1118 mm). Other aisles shall have a clear width of not less than 32 inches (813 mm).

1007.5.2 Travel distance. All portions of Group I, Division 1.1 or 3 Occupancies shall be within 200 feet (60 960 mm) of an exit.

1007.5.3 Hallways. Hallways in Group I Occupancies that serve an occupant load of 10 or more shall comply with the requirements of Sections 1004.3.4 and 1007.5.4 for corridors.

1007.5.4 Corridors. Corridors serving any area caring for one or more nonambulatory persons shall not be less than 8 feet (2438 mm) in width.

> **EXCEPTION:** Corridors serving surgical areas of Group I, Division 1.2 Occupancies shall not be less than 6 feet (1829 mm) in width.

Any change in elevation in a corridor serving nonambulatory persons shall be by means of a ramp.

Corridors shall comply with the requirements of Section 1004.3.4, except that in hospitals and nursing homes classified as Group I, Division 1.1 Occupancies, the following exceptions apply:

1. Nurses' stations, including space for doctors' and nurses' charting and communications, constructed as required for corridors need not be separated from corridors.

2. Waiting areas and similar spaces constructed as required for corridors need not be separated from corridors, provided:

    2.1   Where the aggregate of waiting areas in each smoke compartment does not exceed 600 square feet (55.7 m^2).

        2.1.1 Each area is located to permit direct visual supervision by the facility staff;

        2.1.2 Each area is equipped with an electrically supervised automatic smoke-detection system; and

        2.1.3 Each area is arranged not to obstruct access to required exits.

    2.2   Where such spaces may be unlimited in size and open to the corridor.

        2.2.1 The spaces are not used for patient sleeping rooms, treatment rooms, hazardous areas or special use areas listed in Table 3-C;

        2.2.2 Each space is located to permit direct visual supervision by the facility staff;

        2.2.3 The space and corridors that the space open onto in the same smoke compartment are protected by an electrically supervised automatic smoke-detection system; and

        2.2.4 The space is arranged not to obstruct access to required exits.

3. In fully sprinklered buildings, door closers need not be installed on doors to sleeping or treatment rooms.

4. Fixed fully tempered or laminated glass in wood or metal frames may be used in corridor walls, provided the glazed area does not exceed 25 percent of the area of the corridor wall of the room.

5. The total area of glass in corridor walls is not limited when the glazing is fixed $^1/_4$-inch-thick (6.4 mm) wired glass in steel frames and the size of individual glazed panel does not exceed 1,296 square inches (0.836 m^2).

6. Corridor doors other than those required to be rated by Section 308.8 or for the enclosing of a vertical opening or an exit are not required to be fire-rated, provided the doors are tightfitting, smoke- and draft-control assemblies and are provided with positive latches. Roller latches are prohibited.

1007.5.5 Exterior exit doors. All required exterior exit doors shall open in the direction of exit travel regardless of the occupant load served.

1007.5.6 Basement exits. All rooms below grade shall have not less than one exit that leads directly to the exterior at grade level.

1007.5.7 Ramps. Where the first story of Group I, Divisions 1.1 and 1.2 Occupancies is at other than grade level, such occupancies housing nonambulatory patients shall have a ramp leading from the first story to the exterior of the building at grade level.

1007.5.8 Hardware. Exit and exit-access doors serving an area having an occupant load of 50 or more shall not be provided with a latch or lock unless it is panic hardware. Patient room doors shall be readily openable from either side without the use of keys.

> **EXCEPTIONS:** 1. In Group I, Division 1.1 hospitals and nursing homes, locking devices, when approved, may be installed on patient sleeping rooms, provided such devices are readily openable from the patient room side and are readily operable by the facility staff on the other side. Where key locks are used on patient room doors, keys shall be located on the floor involved at a prominent location accessible to the staff.
>
> 2. In Group I, Division 3 Occupancies, approved locks or safety devices may be used where it is necessary to forcibly restrain the personal liberties of inmates or patients.

1007.5.9 Suites.

1007.5.9.1 General. A group of rooms in a Group I, Division 1.1, Division 1.2 or Division 2 Occupancy may be considered a suite when it complies with the following:

1. **Size.** Suites or rooms, other than suites containing patient sleeping rooms, shall not exceed 10,000 square feet (929 m²) in area. Suites containing patient sleeping rooms shall not exceed 5,000 square feet (464.5 m²) in area.

2. **Occupancy separation.** Each suite of rooms shall be separated from the remainder of the building by not less than a one-hour fire-resistive occupancy separation.

3. **Visual supervision.** Each patient sleeping room in the suite shall be located to permit direct and constant visual supervision by the facility staff.

4. **Other exits.** Exiting for portions of the building outside a suite shall not require passage through the suite.

1007.5.9.2 Corridors. One-hour fire-resistive corridor construction is not required within a suite.

1007.5.9.3 Travel through adjoining rooms. Rooms within suites may have access to exits through one adjoining room if there is not more than 100 feet (30 480 mm) of travel distance within the suite to an exit or to a corridor that provides direct access to an exit. Rooms other than patient sleeping rooms may access exits through two adjoining rooms where there is not more than 50 feet (15 240 mm) of travel distance within the suite to an exit or to a corridor that provides direct access to an exit.

Other portions of the exit access shall not pass through suites.

1007.6 Group R Occupancies.

1007.6.1 Hallways. Hallways in Group R, Division 1 Occupancies which serve an occupant load of 10 or more shall comply with the requirements of Section 1004.3.4 for corridors.

1007.6.2 Floor-level exit signs. Where exit signs are required by Section 1003.2.8.2, additional approved low-level exit signs that are internally or externally illuminated, photoluminescent or self-luminous, shall be provided in all corridors serving guest rooms of hotels in Group R, Division 1 Occupancies.

The bottom of such sign shall be not less than 6 inches (152 mm) nor more than 8 inches (203 mm) above the floor level and shall indicate the path of exit travel. For exit and exit-access doors, the sign shall be on the door or adjacent to the door with the closest edge of the sign within 4 inches (102 mm) of the door frame.

1007.7 Special Hazards.

1007.7.1 Rooms containing fuel-fired equipment. All rooms containing a boiler, furnace, incinerator or other fuel-fired equipment shall be provided with access to two exits or exit-access doors when both of the following conditions exist:

1. The area of the room exceeds 500 square feet (46.45 m²), and

2. The largest piece of fuel-fired equipment exceeds 400,000 Btu per hour (117 228 W) input capacity.

> **EXCEPTIONS:** 1. In Group R, Division 3 Occupancies.
>
> 2. If access to two exits or exit-access doors are required, one such access may be by a fixed ladder.

1007.7.2 Refrigeration machinery rooms.

1007.7.2.1 Access to exits. Machinery rooms larger than 1,000 square feet (92.9 m²) shall have access to not less than two exits as required in Section 1007.7.1.

1007.7.2.2 Travel distance. Travel distance shall be determined as specified in Section 1004.2.5, but all portions of machinery rooms shall be within 150 feet (45 720 mm) of an exit or exit-access doorway. Travel distance may be increased in accordance with Section 1004.2.5.

1007.7.2.3 Doors. Doors shall swing in the direction of exit travel, regardless of the occupant load served. Doors shall be tight-fitting and self-closing.

1007.7.3 Refrigerated rooms or spaces.

1007.7.3.1 Access to exits. Rooms or spaces having a floor area of 1,000 square feet (92.9 m²) or more, containing a refrigerant evaporator and maintained at a temperature below 68°F (20°C), shall have access to not less than two exits or exit-access doors.

1007.7.3.2 Travel distance. Travel distance shall be determined as specified in Section 1004.2.5, but all portions of the refrigerated room or space shall be within 150 feet (45 720 mm) of an exit or exit-access door where such rooms are not protected by an approved automatic sprinkler system. Travel distance may be increased in accordance with Section 1004.2.5. Egress is allowed through adjoining refrigerated rooms or spaces.

> **EXCEPTION:** Where using refrigerants in quantities limited to the amounts based on the volume set forth in the Mechanical Code.

1007.7.4 Cellulose nitrate film handling. Where cellulose nitrate film is handled in film laboratories, projection rooms and film processing rooms, access to not less than two exits or exit-access doors shall be provided. Doors to such rooms shall be protected by a fire assembly having a fire-protection rating of not less than one hour and shall be maintained self-closing.

SECTION 1008 — REVIEWING STANDS, GRANDSTANDS, BLEACHERS, AND FOLDING AND TELESCOPING SEATING

1008.1 Scope. The requirements of Section 1008 shall apply to reviewing stands, grandstands, bleachers, and folding and telescoping seating.

1008.2 Definitions. For the purpose of Section 1008, certain terms are defined as follows:

BLEACHERS are tiered or stepped seating facilities without backrests.

FOLDING AND TELESCOPING SEATING is a structure that is used for tiered seating of persons, and having an overall shape and size that may be reduced without being dismantled, for purposes of moving or storing.

FOOTBOARDS are that part of a raised seating facility other than an aisle or cross aisle upon which the occupant walks to reach a seat.

GRANDSTANDS are tiered or stepped seating facilities.

PERMANENT STANDS are those seating facilities that remain at a location for more than 90 days.

REVIEWING STANDS are elevated platforms accommodating not more than 50 persons. Seating facilities, if provided, are normally in the nature of loose chairs. Reviewing stands accommodating more than 50 persons shall be regulated as grandstands.

SAFE DISPERSAL AREA is an area that will accommodate a number of persons equal to the total capacity of the stand and building that it serves, such that a person within the area will not be closer than 50 feet (15 240 mm) from the stand or building. Safe dispersal area capacity shall be determined by allowing 3 square feet (0.28 m^2) of net clear area per person.

TEMPORARY SEATING FACILITIES are those that are intended for use at a location for not more than 90 days.

1008.3 Height of Reviewing Stands, Grandstands, Bleachers, and Folding and Telescoping Seating. See Section 303.2.

1008.4 Design Requirements. See Chapter 16 and Section 1806.10.

1008.5 General Requirements.

1008.5.1 Row spacing. There shall be a clear space of not less than 12 inches (305 mm) measured horizontally between the back or backrest of each seat and the front of the seat immediately behind it. The minimum spacing of rows of seats measured from back to back shall be:

1. Twenty-two inches (559 mm) for seats without backrests.

2. Thirty inches (762 mm) for seats with backrests.

3. Thirty-three inches (838 mm) for chair seating.

1008.5.2 Rise between rows. The maximum rise from one row of seats to the next shall not exceed 16 inches (406 mm) unless the seat spacing from back to back measured horizontally is 40 inches (1016 mm) or more.

> **EXCEPTION:** Where automatic- or self-rising seats are installed, the rise between rows may be increased to 24 inches (610 mm) with the horizontal spacing back to back of 33 inches (838 mm).

1008.5.3 Seating capacity determination. Where bench-type seating is used, the number of seats shall be based on one person for each 18 inches (457 mm) of length of the bench.

1008.5.4 Aisles.

1008.5.4.1 Aisles required. Aisles shall be provided in all seating facilities, except that aisles may be omitted when all the following conditions exist:

1. Seats are without backrests.

2. The rise from row to row does not exceed 6 inches (152 mm) per row.

3. The row spacing does not exceed 28 inches (711 mm) unless the seat boards and footboards are at the same elevation.

4. The number of rows does not exceed 16 in height.

5. The first seating board is not more than 12 inches (305 mm) above grade or floor below or a cross aisle.

6. Seat boards are continuous flat surfaces.

7. Seat boards provide a walking surface with a minimum width of 11 inches (279 mm).

1008.5.4.2 Obstructions. No obstruction shall be placed in the required width of any aisle or other means of egress component.

1008.5.4.3 Width. Aisles serving seats on both sides shall have a minimum width of 44 inches (1118 mm). Where serving seats on only one side, the aisle shall have a minimum width of 36 inches (914 mm). Except for temporary seating facilities, the required width for aisles shall equal the greater of the minimum required widths determined in accordance with Section 1004.3.2.3 and this section.

1008.5.5 Cross aisles and vomitories. Cross aisles and vomitories shall not be less than 54 inches (1372 mm) in clear width and shall extend to an exit or an exterior perimeter ramp. Except for temporary seating facilities, the required width for cross aisles shall equal the greatest of the minimum required widths determined as specified in Section 1004.3.2 and this section.

1008.5.6 Stairways and ramps. Except as otherwise provided in this section, grandstands, bleachers, and folding and telescoping seating shall comply with other applicable sections of this chapter. Stairways and ramps shall have a maximum rise and run as provided in Sections 1003.3.3.3 and 1003.3.4, except those within the seating facility that serve as aisles at right angles to the rows of seats where the rise shall not exceed 8 inches (203 mm). Where an aisle terminates at an elevation more than 8 inches (203 mm) above grade or floor below, the aisle shall be provided with a stairway or ramp with a width not less than the width of the aisle.

Stairways and ramps shall have handrails as provided in Sections 1003.3.3.6 and 1003.3.4.5, except stairways within the seating facility that serve as aisles at right angles where handrails shall be provided at one side or along the center line. A minimum clear width of 48 inches (1219 mm) between seats shall be provided for aisle stairways having center-aisle handrails. Where there is seating on both sides of the aisle, handrails shall be discontinuous with openings at intervals not exceeding five rows for access to seating. The opening shall have a clear width of at least 22 inches (559 mm) and not more than 36 inches (914 mm) measured horizontally, and the handrail shall have rounded terminations. Where handrails are provided in the middle of the aisle stairs, there shall be an additional intermediate rail located approximately 12 inches (305 mm) below the top of the handrail.

> **EXCEPTION:** Temporary seating facility stairways within the seating area that serve as aisles at right angles need not be provided with handrails.

1008.5.7 Guardrails. Perimeter guardrails, enclosing walls or fencing shall be provided for all portions of elevated seating facilities that are more than 30 inches (762 mm) above grade or the floor. Construction of guardrails shall comply with the requirements of Section 509 and Table 16-B. Guardrails shall be 42 inches (1067 mm) in height measured vertically above the leading edge of the tread adjacent walking surface, adjacent walking surface or adjacent seatboards.

> **EXCEPTION:** Guardrails at the front of the front row of seats, which are not located at the end of an aisle and where there is no cross aisle, may have a height of 26 inches (660 mm) and need not meet the 4-inch-maximum (102 mm) spacing specified in Section 509; however, a midrail shall be installed.

The open vertical space between footboards and seats shall not exceed 9 inches (229 mm) when footboards are more than 30 inches (762 mm) above grade.

1008.5.8 Toeboards. A 4-inch-high (102 mm) vertical barrier shall be installed along the edge of walking platforms whenever guardrails are required.

> **EXCEPTION:** Toeboards shall not be required at the ends of footboards.

1008.5.9 Footboards. Footboards shall be provided for all rows of seats above the third row or beginning at such a point where the seat is more than 2 feet (610 mm) above the grade or floor below. Where the same platform is used for both seating and footrests, footboards are not required, provided each level or platform is not less than 24 inches (610 mm) wide. When projected on a horizontal plane, there shall not be horizontal gaps exceeding $1/4$ inch (6.4 mm) between footboards and seatboards. At aisles, there shall not be horizontal gaps exceeding $1/4$ inch (6.4 mm) between footboards.

1008.6 Grandstands, Bleachers, and Folding and Telescoping Seating within Buildings. Except as otherwise provided in this section and Section 1008.7, grandstands, bleachers, and folding and telescoping seating within a building shall comply with the other applicable sections of this chapter.

> **EXCEPTIONS:** 1. Where seats are without backrests, there may be nine seats between any seat and an aisle.
>
> 2. Where seats are without backrests, dead ends in vertical aisles shall not exceed a depth of 16 rows.

1008.7 Smoke-protected Assembly Seating.

1008.7.1 General. To be considered smoke protected, an assembly seating facility shall comply with the following requirements.

1008.7.2 Roof height. A smoke-protected assembly seating area with a roof shall have the lowest portion of the roof not less than 15 feet (4572 mm) above the level of the highest aisle or aisle accessway.

1008.7.3 Smoke control. All means of egress serving a smoke-protected assembly seating area shall be provided with completely automatic smoke control complying with Section 905.

> **EXCEPTION:** Automatic smoke control is not required when a natural venting system design can be demonstrated to accomplish equivalent results.

1008.7.4 Travel distance. In a smoke-protected assembly seating area, the travel distance from each seat to the nearest entrance to an egress concourse shall not exceed 200 feet (60 960 mm). The travel distance from the entrance to vomitory portal or egress concourse to an approved egress stair, ramp or walk at the building exterior shall not exceed 200 feet (60 960 mm).

In outdoor assembly seating facilities where all portions of the means of egress are open to the outside, the distance of travel to an approved egress stair, ramp or walk at the building exterior shall not exceed 400 feet (121 920 mm). In outdoor assembly seating facilities of Type I or II construction where all portions of the means of egress are essentially open to the outside, the distance shall not be limited.

SECTION 1009 — BUILDING SECURITY

See Appendix Chapter 10 for requirements covering building security.

TABLE 10-A—MINIMUM EGRESS REQUIREMENTS[1]

USE[2]	MINIMUM OF TWO MEANS OF EGRESS ARE REQUIRED WHERE NUMBER OF OCCUPANTS IS AT LEAST	OCCUPANT LOAD FACTOR[3] (square feet) × 0.0929 for m²
1. Aircraft hangars (no repair)	10	500
2. Auction rooms	30	7
3. Assembly areas, concentrated use (without fixed seats) 　Auditoriums 　Churches and chapels 　Dance floors 　Lobby accessory to assembly occupancy 　Lodge rooms 　Reviewing stands 　Stadiums 　Waiting area	50 50	7 3
4. Assembly areas, less-concentrated use 　Conference rooms 　Dining rooms 　Drinking establishments 　Exhibit rooms 　Gymnasiums 　Lounges 　Stages 　Gaming: keno, slot machine and live games area	50 50	15 11
5. Bowling alley (assume no occupant load for bowling lanes)	50	4
6. Children's homes and homes for the aged	6	80
7. Classrooms	50	20
8. Congregate residences	10	200
9. Courtrooms	50	40
10. Dormitories	10	50
11. Dwellings	10	300
12. Exercising rooms	50	50
13. Garage, parking	30	200
14. Health care facilities— 　Sleeping rooms 　Treatment rooms	 8 10	 120 240
15. Hotels and apartments	10	200
16. Kitchen—commercial	30	200
17. Library— 　Reading rooms 　Stack areas	 50 30	 50 100
18. Locker rooms	30	50
19. Malls (see Chapter 4)	—	—
20. Manufacturing areas	30	200
21. Mechanical equipment room	30	300
22. Nurseries for children (day care)	7	35
23. Offices	30	100
24. School shops and vocational rooms	50	50
25. Skating rinks	50	50 on the skating area; 15 on the deck
26. Storage and stock rooms	30	300
27. Stores—retail sales rooms 　Basements and ground floor 　Upper floors	 50 50	 30 60
28. Swimming pools	50	50 for the pool area; 15 on the deck
29. Warehouses[5]	30	500
30. All others	50	100

[1]Access to, and egress from, buildings for persons with disabilities shall be provided as specified in Chapter 11.

[2]For additional provisions on number of exits from Groups H and I Occupancies and from rooms containing fuel-fired equipment or cellulose nitrate, see Sections 1007.4, 1007.5 and 1007.7, respectively.

[3]This table shall not be used to determine working space requirements per person.

[4]Occupant load based on five persons for each alley, including 15 feet (4572 mm) of runway.

[5]Occupant load for warehouses containing approved high rack storage systems designed for mechanical handling may be based on the floor area exclusive of the rack area rather than the gross floor area.

TABLE 10-B
TABLE 10-E

1997 UNIFORM BUILDING CODE

TABLE 10-B—EGRESS WIDTH PER PERSON SERVED

USE	STAIRWAYS (inches per person)	OTHER EGRESS COMPONENTS (inches per person) ($\times$ 25.4 for mm/person)
Hazardous: H-1, H-2, H-3 and H-7	0.7	0.4
Institutional: I-1	0.3	0.2
Institutional: I-2	0.4	0.2
All other uses	0.3	0.2

TABLE 10-C—CALCULATION FOR MINIMUM WIDTH IN BUILDINGS WITHOUT SMOKE-PROTECTED ASSEMBLY SEATING[1]

NUMBER OF SEATS	CLEAR WIDTH PER SEAT SERVED FOR STAIRS (inches)	CLEAR WIDTH PER SEAT SERVED FOR PASSAGEWAY, RAMPS AND DOORWAYS (inches)
		$\times$ 25.4 for mm
Unlimited	0.300 AB	0.220 C

[1]See Section 1004.3.2.3.1 for determination of values A, B and C.

TABLE 10-D—CALCULATION FOR MINIMUM WIDTH IN BUILDINGS WITH SMOKE-PROTECTED ASSEMBLY SEATING[1]

NUMBER OF SEATS	CLEAR WIDTH PER SEAT SERVED FOR STAIRS (inches)	CLEAR WIDTH PER SEAT SERVED FOR PASSAGEWAYS, RAMPS AND DOORWAYS (inches)
		$\times$ 25.4 for mm
2,000	0.300 AB	0.220 C
5,000	0.200 AB	0.150 C
10,000	0.130 AB	0.100 C
15,000	0.096 AB	0.070 C
20,000	0.076 AB	0.056 C
25,000 or more	0.060 AB	0.044 C

[1]See Section 1004.3.2.3.1 for determination of values A, B and C.

TABLE 10-E—MAXIMUM NUMBER OF SEATS ALLOWED TO HAVE THE MINIMUM 12 INCH (305 mm) CLEAR WIDTH

TOTAL NUMBER OF SEATS IN THE SPACE	NUMBER OF SEATS PER ROW PERMITTED TO HAVE A MINIMUM 12-INCH (305 mm) CLEAR WIDTH AISLE ACCESSWAY	
	Aisle or Doorway at Both Ends of Row	Aisle or Doorway at One End of Row
< 4,000	14	7
4,000-6,999	15	7
7,000-9,999	16	8
10,000-12,999	17	8
13,000-15,999	18	9
16,000-18,999	19	9
19,000-21,999	20	10
$\geq$ 22,000	21	11

Chapter 11
ACCESSIBILITY

SECTION 1101 — SCOPE

1101.1 General. Buildings or portions of buildings shall be accessible to persons with disabilities as required by this chapter.

See also Appendix Chapter 11 for requirements governing the provision of accessible site facilities not regulated by this chapter. See Section 101.3 for applicability of appendix.

1101.2 Standards of Quality. The standard listed below labeled an "Adopted Standard" is also listed in Chapter 35, Part III, and is part of this code.

1. Accessible Design

Adopted Standard—CABO/ANSI A117.1-1992

1101.3 Design. The design and construction of accessible buildings and building elements shall be in accordance with this chapter and CABO/ANSI A117.1-1992. For a building to be considered accessible, it shall be designed and constructed to the minimum provisions of this chapter and CABO/ANSI A117.1.

> **EXCEPTION:** Type B dwelling units shall comply with Section 1106.

SECTION 1102 — DEFINITIONS

For the purpose of this chapter, certain terms are defined as follows:

ACCESSIBLE describes a site, building, facility or portion thereof that complies with this chapter and that can be approached, entered and used by persons with physical disabilities.

ACCESSIBLE MEANS OF EGRESS is a path of travel, usable by a mobility-impaired person, that leads to a public way.

ACCESSIBLE ROUTE is a continuous path connecting accessible elements and spaces in a building or facility that is usable by persons with disabilities.

ADAPTABILITY is the capability of altering or adding to certain building spaces and elements, such as kitchen counters, sinks and grab bars, to accommodate the needs of persons with and without disabilities, or to accommodate the needs of persons with different types or degrees of disability.

AREA OF REFUGE is an area with direct access to an exit or an elevator where persons unable to use stairs can remain temporarily in safety to await instructions or assistance during emergency evacuation.

CABO/ANSI A117.1 is American National Standard A117.1-1992 published by the Council of American Building Officials.

COMMON-USE AREAS are rooms, spaces or elements that are made available for use by a specific group of people.

DWELLING UNIT—TYPE A is a dwelling unit that is designed and constructed for accessibility in accordance with CABO/ANSI A117.1.

DWELLING UNIT—TYPE B is a dwelling unit that is designed and constructed for accessibility in accordance with Section 1106.

ELEMENT is an architectural or mechanical component of a building, facility, space or site that is used in making spaces accessible.

FACILITY is all or any portion of a building, structure or area, including the site on which such building, structure or area is located, wherein specific services are provided or activities are performed.

GROUND FLOOR DWELLING UNIT is a dwelling unit with a primary entrance and habitable space at grade.

MULTISTORY DWELLING UNIT is a dwelling unit with habitable or bathroom space located on more than one story.

PERSON WITH DISABILITY is an individual who has an impairment, including a mobility, sensory or cognitive impairment, that results in a functional limitation in access to and use of a building or facility.

PUBLIC-USE AREAS are rooms or spaces that are made available to the general public.

SITE is a parcel of land bounded by a property line or a designated portion of a public right-of-way.

SECTION 1103 — BUILDING ACCESSIBILITY

1103.1 Where Required.

1103.1.1 General. Accessibility to temporary or permanent buildings, or portions thereof, shall be provided for all occupancy classifications except as modified by this chapter. See also Appendix Chapter 11.

> **EXCEPTIONS:** 1. Floors or portions of floors not customarily occupied, including, but not limited to, elevator pits; observation galleries used primarily for security purposes; elevator penthouses; nonoccupiable spaces accessed only by ladders, catwalks, crawl spaces or freight elevators; piping and equipment catwalks; and machinery, mechanical and electrical equipment rooms.
>
> 2. Subject to the approval of the building official, areas where work cannot reasonably be performed by persons having a severe impairment (mobility, sight or hearing) need not have specific features which provide accessibility to such persons.
>
> 3. Temporary structures, sites and equipment directly associated with the construction process such as construction site trailers, scaffolding, bridging or material hoists are not required to be accessible. This exception does not include walkways or pedestrian protection required by Chapter 33.

1103.1.2 Group A Occupancies. Group A Occupancies shall be accessible as provided in this chapter.

> **EXCEPTION:** In the assembly area of dining and drinking establishments that are located within nonelevator buildings, when the area of mezzanine seating is not more than 25 percent of the total seating, an accessible means of vertical access to the mezzanine is not required, provided the same services are provided in an accessible space.

1103.1.3 Group B Occupancies. Group B Occupancies shall be accessible as provided in this chapter.

1103.1.4 Group E Occupancies. Group E Occupancies shall be accessible as provided in this chapter.

1103.1.5 Group F Occupancies. Group F Occupancies shall be accessible as provided in this chapter.

1103.1.6 Group H Occupancies. Group H Occupancies shall be accessible as provided in this chapter.

1103.1.7 Group I Occupancies. Group I Occupancies shall be accessible in public-use, common-use and employee-use areas, and shall have accessible patient rooms, cells, and treatment or examination rooms as follows:

1. In Group I, Division 1.1 patient-care units within hospitals that specialize in treating conditions that affect mobility, all patient rooms, including associated toilet rooms and bathrooms.

2. In Group I, Division 1.1 patient-care units within hospitals that do not specialize in treating conditions that affect mobility, at least one in every 10 patient rooms, or fraction thereof, including associated toilet rooms and bathrooms.

3. In Group I, Divisions 1.1 and 2 nursing homes and long-term care facilities, at least one in every two patient rooms, or fraction thereof, including associated toilet rooms and bathrooms.

4. In Group I, Division 3 mental health occupancies, at least one in every 10 patient rooms, or fraction thereof, including associated toilet rooms and bathrooms.

5. In Group I, Division 3 jail, prison and similar occupancies, at least one in every 20 rooms or cells, or fraction thereof, including associated toilet rooms and bathrooms.

6. In Group I Occupancies, all treatment and examination rooms shall be accessible.

1103.1.8 Group M Occupancies. Group M Occupancies shall be accessible as provided in this chapter.

1103.1.9 Group R Occupancies.

1103.1.9.1 General. Group R Occupancies shall be accessible as provided in this chapter. Rooms and spaces available to the general public and spaces available for the use of residents that serve Group R, Division 1 Occupancy accessible dwelling units shall be accessible.

Where recreational facilities are provided serving accessible dwelling units, 25 percent, but not less than one of each type in each group of such facilities, shall be accessible. All recreational facilities of each type on a site shall be considered to determine the total number of each type that are required to be accessible.

1103.1.9.2 Hotels, lodging houses and congregate residences. In hotels, lodging houses and congregate residence occupancies containing six or more guest rooms, multibed rooms or spaces for more than six occupants, one for the first 30 guest rooms or spaces and one additional for each additional 100 guest rooms or spaces, or fraction thereof, shall be accessible. In hotels with more than 50 sleeping rooms or suites, roll-in-type showers shall be provided in one half, but not less than one, of the required accessible sleeping rooms or suites.

In addition to the accessible guest rooms required above, guest rooms for persons with hearing impairments shall be provided in accordance with Table 11-B. Guest rooms for persons with hearing impairments shall be provided with visible and audible alarm-indicating appliances, activated by both the in-room smoke detector and the building fire protective signaling system.

1103.1.9.3 Multi-unit dwellings. In Group R, Division 1 Occupancy apartments containing four or more dwelling units and Group R, Division 3 Occupancies where there are four or more dwelling units in a single structure, all dwelling units shall be Type B dwelling units. In Group R, Division 1 apartment occupancies containing more than 20 dwelling units, at least 2 percent, but not less than one, of the dwelling units shall be Type A dwelling units. All dwelling units on a site shall be considered to determine the total number of accessible dwelling units.

EXCEPTIONS: 1. Where no elevator service is provided in a building, Type B dwelling units need not be provided on floors other than the ground floor.

2. Where no elevator service is provided in a building and the ground floor does not contain dwelling units, only those dwelling units located on the first floor of either Group R, Division 1 apartment occupancies or Group R, Division 3 Occupancies need comply with the requirements of this section.

3. A multistory dwelling unit not provided with elevator service is not required to comply with requirements for Type B dwelling units. Where a multistory dwelling unit is provided with elevator service to only one floor, the floor provided with elevator service shall comply with the requirements for a Type B dwelling unit, and a toilet facility shall be provided on that floor.

4. The number of Type B dwelling units provided in multiple non-elevator buildings on a single site may be reduced to a percentage of the ground floor units that is equal to the percentage of the entire site having grades, prior to development, that are 10 percent or less; but in no case shall the number of Type B units be less than 20 percent of the ground floor dwelling units on the entire site.

5. The required number of Type A and Type B dwelling units shall not apply to a site where the lowest floor or the lowest structural building members is required to be at or above the base flood elevation resulting in:

   5.1 A difference in elevation between the minimum required floor elevation at the primary entrances and all vehicular and pedestrian arrival points within 50 feet (15 240 mm) exceeding 30 inches (762 mm).

   5.2 A slope exceeding 10 percent between the minimum required floor elevation at the primary entrances and all vehicular and pedestrian arrival points within 50 feet (15 240 mm).

Where no such arrival points are within 50 feet (15 240 mm) of the primary entrances, the closest arrival point shall be used.

1103.1.10 Group S Occupancies. Group S Occupancies shall be accessible as provided in this chapter.

1103.1.11 Group U Occupancies. Group U, Division 1 Occupancies shall be accessible as follows:

1. Private garages and carports that contain accessible parking.

2. In Group U, Division 1 agricultural buildings, access need be provided only to paved work areas and areas open to the general public.

1103.2 Design and Construction.

1103.2.1 General. When accessibility is required by this chapter, it shall be designed and constructed in accordance with this chapter and CABO/ANSI A117.1.

EXCEPTION: Type B dwelling units shall comply with Section 1106.

1103.2.2 Accessible route. When a building, or portion of a building, is required to be accessible, an accessible route shall be provided to all portions of the building, to accessible building entrances, connecting accessible pedestrian walkways and the public way.

EXCEPTION: In other than the offices of health-care providers, transportation facilities, airports and Group M Occupancies with five or more tenants, floors above and below accessible levels that have an aggregate area of not more than 3,000 square feet (278.7 m²) need not be served by an accessible route from an accessible level.

When floor levels are required to be connected by an accessible route, and an interior path of travel is provided between the levels, an interior accessible route between the levels shall be provided. When only one accessible route is provided, it shall not pass through kitchens, storage rooms, toilet rooms, bathrooms, closets or other similar spaces.

EXCEPTION: A single accessible route may pass through a kitchen or storage room in a Type A dwelling unit.

When more than one building or facility is located on a site, accessible routes shall be provided connecting accessible buildings and accessible site facilities.

EXCEPTION: For Group R, Division 1 apartment occupancies, when the slope of the finished grade between accessible buildings and facilities exceeds 1 unit vertical in 12 units horizontal (8.33% slope), or when physical barriers of the site prevent the installation of an accessible route, a vehicular route with parking at each accessible building or facility may be provided in place of an accessible route.

1103.2.3 Accessible entrances. Each building and structure, and each separate tenancy within a building or structure, shall be provided with at least one entrance that complies with the accessible route provisions of CABO/ANSI A117.1. At least 50 percent of all entrances shall be accessible.

EXCEPTIONS: 1. Entrances used exclusively for loading and service.

2. Entrances to spaces not required to be accessible as provided for in Section 1103.

When a building or facility has entrances that normally serve accessible parking facilities, transportation facilities, passenger loading zones, taxi stands, public streets and sidewalks, or accessible interior vertical access, at least one of the entrances serving each such function shall comply with the accessible route provisions of CABO/ANSI A117.1.

The primary entrance to either a Type A or Type B dwelling unit shall be located on an accessible route from public or common areas. The primary entrance to the dwelling unit shall not be to a bedroom.

1103.2.4 Signs.

1103.2.4.1 International symbol of accessibility. The following elements and spaces of accessible facilities shall be identified by the international symbol of accessibility:

1. Accessible parking spaces, except where the total parking spaces provided are five or less.

2. Accessible areas of refuge.

3. Accessible passenger loading zones.

4. Accessible toilet and bathing facilities.

1103.2.4.2 Other signs. Inaccessible building entrances, inaccessible public toilets and bathing facilities, and elevators not on an accessible route shall be provided with directional signage indicating the route to the nearest similar accessible element.

In assembly areas, a sign notifying the general public of the availability of assistive listening systems shall be provided at ticket offices or similar locations.

Each door to an exit stairway shall have a tactile sign, including raised letters and Braille, stating EXIT and shall comply with CABO/ANSI A117.1.

At exits and elevators serving a required accessible space, but not providing an approved accessible means of egress, signs shall be installed indicating the location of accessible means of egress.

In addition to the international symbol of accessibility, each unisex toilet or bathing room shall be identified by a tactile sign including raised letters and Braille. Directional signage shall be provided at all separate-sex toilet or bathing facilities indicating the location of the nearest unisex room.

SECTION 1104 — EGRESS AND AREAS OF REFUGE

1104.1 Means of Egress.

1104.1.1 General. All required accessible spaces shall be provided with not less than one accessible means of egress. When more than one exit or exit-access door is required from any accessible space, each accessible portion of the space shall be served by not less than two accessible means of egress. The maximum travel distance from any accessible space to an area of refuge shall not exceed the travel distance set forth in Chapter 10.

Each accessible means of egress shall be continuous from each required accessible occupied area to a public way and shall include accessible routes, ramps, exit stairs, elevators, horizontal exits or smoke barriers.

1104.1.2 Stairways. When an exit stairway is part of an accessible means of egress, the stairway shall have a clear width of not less than 48 inches (1219 mm) between handrails. The stairway shall either incorporate an area of refuge within an enlarged story-level landing or shall be accessed from an area of refuge complying with Section 1104.2 or a horizontal exit.

EXCEPTIONS: 1. Exit stairways serving a single dwelling unit or guest room.

2. Exit stairways serving buildings protected throughout by an approved automatic sprinkler system.

3. The clear width of 48 inches (1219 mm) between handrails is not required for exit stairways accessed from a horizontal exit.

4. Areas of refuge are not required in open parking garages.

1104.1.3 Elevators. When an accessible floor is four or more stories above or below the level of exit discharge serving that floor, at least one elevator shall serve as one required accessible means of egress.

EXCEPTION: In fully sprinklered buildings, the elevator need not be provided to floors provided with a horizontal exit and located at or above the level of exit discharge.

When an elevator is part of an accessible means of egress, standby power shall be provided. The elevator shall be accessed from either an area of refuge complying with Section 1104.2 or a horizontal exit.

EXCEPTIONS: 1. Elevators are not required to be accessed by an area of refuge or a horizontal exit in buildings protected throughout by an approved automatic sprinkler system.

2. Areas of refuge are not required in open parking garages.

1104.1.4 Platform lifts. Platform (wheelchair) lifts shall not serve as part of an accessible means of egress.

EXCEPTION: Within a dwelling unit.

1104.2 Areas of Refuge.

1104.2.1 Access. Required areas of refuge shall be accessible from the space served by an accessible means of egress. Required areas of refuge shall have direct access to a stairway or an elevator complying with Section 1104.1.

1104.2.2 Pressurization. When an elevator lobby is used as an area of refuge, the elevator shaft and lobby shall be pressurized in accordance with the requirements of Section 905.

EXCEPTION: When elevators are in an area of refuge formed by a horizontal exit or smoke barrier.

1104.2.3 Size. Each area of refuge shall be sized to accommodate one wheelchair space not less than 30 inches by 48 inches (762 mm by 1219 mm) for each 200 occupants, or portion thereof, based on the occupant load of the area of refuge and areas served by the area of refuge.

Wheelchair spaces shall not reduce the required exit width or interfere with access to or use of fire department hose connections and valves. Access to required wheelchair spaces in an area of refuge shall not be obstructed by more than one adjoining wheelchair space.

1104.2.4 Construction. Each area of refuge shall be separated from the remainder of the story by a smoke barrier having at least a one-hour fire-resistance rating. Smoke barriers shall extend to the roof or floor deck above. Doors in the smoke barrier shall be tightfitting smoke- and draft-control assemblies having a fire-protection rating of not less than 20 minutes. Doors shall be self-closing or automatic closing by smoke detection. An approved damper designed to resist the passage of smoke shall be provided at each point a duct penetrates the smoke barrier.

EXCEPTION: Areas of refuge located within a stairway enclosure.

1104.2.5 Two-way communication. Areas of refuge shall be provided with a two-way communication system between the area of refuge and a central control point. If the central control point is not constantly attended, the area of refuge shall also have controlled access to a public telephone system. Location of the central control point shall be approved by the fire department.

EXCEPTION: Buildings four stories or less in height.

1104.2.6 Instructions. In areas of refuge that have a two-way emergency communication system, instructions on the use of the area under emergency conditions shall be posted adjoining the communications system. The instructions shall include:

1. Directions to find other exits,

2. Advice that persons able to use the exit stairway do so as soon as possible, unless they are assisting others,

3. Information on planned availability of assistance in the use of stairs or supervised operation of elevators and how to summon such assistance, and

4. Directions for use of the emergency communications system.

1104.2.7 Identification. Each area of refuge shall be identified by a sign stating AREA OF REFUGE and the international symbol of accessibility. The sign shall be located at each door providing access to the area of refuge. The sign shall be illuminated as required for exit signs when exit sign illumination is required. Tactile signage shall be located at each door to an area of refuge.

SECTION 1105 — FACILITY ACCESSIBILITY

1105.1 General. When buildings or portions of buildings are required to be accessible, building facilities shall be accessible as provided in this section.

Building facilities or elements required by this section to be accessible shall be designed and constructed in accordance with CABO/ANSI A117.1.

EXCEPTION: Type B dwelling units shall comply with Section 1106.

1105.2 Bathing and Toilet Facilities.

1105.2.1 Bathing facilities. When bathing facilities are provided, at least one of each type of fixture or element shall be accessible.

EXCEPTION: A bathing facility for a single occupant and not for common or public use may be adaptable.

In recreational facilities, where separate-sex bathing facilities are provided, an accessible unisex bathing room shall be provided.

EXCEPTION: Where each separate-sex bathing facility has only one shower fixture, unisex bathing facilities need not be provided.

1105.2.2 Toilet facilities. Toilet facilities located within accessible dwelling units, guest rooms and congregate residences shall comply with CABO/ANSI A117.1.

In other occupancies, each toilet room shall be accessible. At least one of each type of fixture or element in each accessible toilet room shall be accessible. When toilet stalls are provided in a

toilet room, at least one toilet stall shall be wheelchair accessible. When six or more toilet stalls are provided in a toilet room, at least one ambulatory accessible toilet stall shall be provided in addition to the wheelchair accessible toilet stall.

EXCEPTION: A toilet facility for a single occupant and not for common or public use may be adaptable.

In Groups A and M Occupancies, an accessible unisex toilet room shall be provided where an aggregate of six or more male and female water closets are required. In buildings of mixed occupancy, only those water closets required for the Group A or M Occupancy shall be used to determine the unisex toilet room requirement.

1105.2.3 Lavatories, mirrors and towel fixtures. At least one accessible lavatory shall be provided within toilet facilities. When mirrors, towel fixtures, and other toilet and bathroom accessories are provided, at least one of each shall be accessible.

1105.2.4 Unisex bathing and toilet rooms.

1105.2.4.1 General. Unisex bathing and toilet rooms shall comply with this section and CABO/ANSI A117.1.

1105.2.4.2 Location. Unisex toilet and bathing rooms shall be located on an accessible route. Unisex toilet rooms shall be located not more than one story above or below separate-sex toilet facilities. The accessible route from any separate-sex toilet room to a unisex toilet room shall not exceed 500 feet (152 400 mm).

Additionally, in passenger transportation facilities and airports, the accessible route from separate-sex toilet facilities to a unisex toilet room shall not pass through security checkpoints.

1105.2.4.3 Clear floor space. Where doors swing into a unisex toilet or bathing room, a clear floor space not less than 30 inches by 48 inches (762 mm by 1219 mm) shall be provided, within the room, beyond the area of the door swing.

1105.2.4.4 Privacy. Doors to unisex toilet and bathing rooms shall be securable from within the room.

1105.2.4.5 Required fixtures.

1105.2.4.5.1 Unisex toilet rooms. Unisex toilet rooms shall include only one water closet and only one lavatory. Where a bathing facility is provided within a unisex toilet room, only one shower shall be provided.

EXCEPTION: A separate-sex toilet room containing not more than two water closets without urinals, or containing only one water closet and one urinal may be considered a unisex toilet room.

1105.2.4.5.2 Unisex bathing rooms. Unisex bathing rooms shall include only one shower fixture. Unisex bathing rooms shall also include one water closet and one lavatory. Where storage facilities are provided for separate-sex bathing facilities, accessible storage facilities shall be provided for unisex bathing rooms.

1105.3 Elevators and Stairway and Platform Lifts. Elevators on an accessible route shall be accessible.

EXCEPTION: Private elevators serving only one dwelling unit.

Elevators required to be accessible shall be designed and constructed to comply with CABO/ANSI A117.1.

Stairways in buildings, or portions of buildings, required to be accessible shall be designed and constructed to comply with CABO/ANSI A117.1.

Platform lifts may be used in lieu of an elevator under one of the following conditions subject to approval by the building official:

1. To provide an accessible route of travel to a performing area in a Group A Occupancy.

2. To provide unobstructed sight lines and distribution for wheelchair viewing positions in Group A Occupancies.

3. To provide access to spaces with an occupant load of less than five.

4. To provide access where existing site constraints or other constraints make use of a ramp or elevator infeasible.

All platform lifts used in lieu of an elevator shall be capable of independent operation.

1105.4 Other Building Facilities.

1105.4.1 Drinking fountains. On any floor where drinking fountains are provided, at least 50 percent, but not less than one fountain, shall be accessible.

1105.4.2 Fixed or built-in seating or tables. When fixed or built-in seating or tables are provided, at least 5 percent, but not less than one, shall be accessible. In dining and drinking establishments, such seating or tables shall be distributed throughout the facility.

1105.4.3 Storage. When storage facilities such as cabinets, shelves, closets, lockers and drawers are provided in required accessible or adaptable spaces, at least one of each type provided shall contain storage space complying with CABO/ANSI A117.1.

1105.4.4 Customer service facilities.

1105.4.4.1 Dressing and fitting rooms. When dressing or fitting rooms are provided, at least 5 percent, but not less than one, in each group of rooms serving distinct and different functions shall be accessible.

1105.4.4.2 Counters and windows. Where customer sales and service counters or windows are provided, a portion of the counter or at least one window shall be accessible.

1105.4.4.3 Checkout aisles. Accessible checkout aisles shall be installed in accordance with Table 11-C. Traffic control devices, security devices and turnstiles located in accessible checkout aisles or lanes shall be accessible.

1105.4.5 Controls, operating mechanisms and hardware. Controls, operating mechanisms and hardware intended for operation by the occupant, including switches that control lighting and ventilation and electrical convenience outlets, in accessible spaces, along accessible routes or as parts of accessible elements shall be accessible.

1105.4.6 Alarms. Alarm systems, when provided, shall include both audible and visible alarms. The alarm devices shall be located in hotel guest rooms as required by Section 1103.1.9.2; accessible public- and common-use areas, including toilet rooms and bathing facilities; hallways; and lobbies.

1105.4.7 Rail transit platforms. Rail transit platform edges bordering a drop-off and not protected by platform screens or guardrails shall be provided with detectable warnings in accordance with CABO/ANSI A117.1.

1105.4.8 Assembly areas.

1105.4.8.1 Wheelchair spaces. Stadiums, theaters, auditoriums and similar occupancies shall be provided with wheelchair spaces in accordance with Table 11-A. Removable seats shall be permitted in the wheelchair positions.

When the seating capacity of an individual assembly area exceeds 300, wheelchair spaces shall be provided in more than one location and shall be on an accessible route of travel. Disper-

sion of wheelchair spaces shall be based on the availability of accessible routes to various seating areas, including seating at various levels in multilevel facilities. Services provided in inaccessible areas shall also be provided on an accessible level and shall be accessible.

1105.4.8.2 Assistive listening systems. Assistive listening systems complying with CABO/ANSI A117.1 shall be installed in stadiums, theaters, auditoriums, lecture halls and similar areas when these areas have fixed seats and where audible communications are integral to the use of the space as follows:

1. Areas with an occupant load of 50 or more.

2. Areas where an audio-amplification system is installed.

Receivers for assistive-listening systems shall be provided at a rate of 4 percent of the total number of seats, but in no case less than two receivers.

Stadiums, theaters, auditoriums, lecture halls and similar areas not equipped with an audio-amplification system or having an occupant load of less than 50 shall have a permanently installed assistive-listening system, or shall have electrical outlets or other supplementary wiring necessary to support a portable assistive-listening system.

Signage shall be installed to notify patrons of the availability of the listening system.

SECTION 1106 — TYPE B DWELLING UNITS

1106.1 General. Type B dwelling units, when required, shall comply with this section.

> **EXCEPTION:** Type B dwelling units designed and constructed as Type A dwelling units.

1106.2 Type B Accessible Route.

1106.2.1 General. At least one accessible route complying with this section shall connect all spaces and elements that are a part of the dwelling unit. Where only one accessible route is provided, it shall not pass through bathrooms, closets or similar spaces.

> **EXCEPTION:** One of the following is not required to be on an accessible route:
>
> 1. A raised floor area in a portion of a living, dining or sleeping room;
>
> 2. A sunken floor area in a portion of a living, dining or sleeping room; or
>
> 3. A mezzanine that does not have plumbing fixtures or an enclosed habitable space.

1106.2.2 Clear width. Clear width of the accessible route shall be 36 inches (914 mm) minimum, except at doors.

1106.2.3 Changes in level. Changes in level of not more than $1/2$ inch (12.7 mm) in height shall comply with CABO/ANSI A117.1. Changes in level greater than $1/2$ inch (12.7 mm) in height shall be accomplished by a ramp, elevator or wheelchair lift complying with CABO/ANSI A117.1.

> **EXCEPTION:** Where exterior deck, patio or balcony surface materials are impervious, the finished exterior impervious surface shall be 4 inches (102 mm) maximum below the finished floor level of the adjacent interior spaces of the dwelling unit.

1106.3 Operating Controls.

1106.3.1 General. Lighting controls, electrical receptacles, environmental controls, and user controls for security or intercom systems shall comply with this section.

> **EXCEPTIONS:** 1. Electrical receptacles serving a dedicated use.
>
> 2. Appliance-mounted controls or switches.
>
> 3. A single receptacle located above a portion of countertop uninterrupted by a sink or appliance need not be accessible, provided:

3.1 At least one receptacle complying with this section is provided for the portion of countertop and

3.2 All other receptacles provided for the portion of countertop comply with this section.

4. Floor electrical receptacles.

5. Plumbing fixture controls.

1106.3.2 Clear floor space. A 30-inch-by-48-inch (762 mm by 1219 mm) minimum clear floor space positioned for forward or parallel approach shall be provided at each accessible operating control. Where a parallel approach is provided to an operating control located above an obstruction, the offset between the center lines of the clear floor space and the operating control shall be 12 inches (305 mm) maximum.

1106.3.3 Height. Operable parts of operating controls shall be 48 inches (1219 mm) maximum and 15 inches (381 mm) minimum above the floor. Operable parts located above an obstruction shall comply with Section 4.2.5.2 or 4.2.6.2 of CABO/ANSI A117.1.

1106.4 Doors.

1106.4.1 Primary entrance door. The primary entrance door to the dwelling unit shall comply with Section 4.13 of CABO/ANSI A117.1.

> **EXCEPTION:** Maneuvering clearances required by Section 4.13.6 of CABO/ANSI A117.1 are not required on the dwelling unit side of the door.

1106.4.2 Other doorways. Doorways intended for user passage shall comply with this section.

1106.4.2.1 Clear width. Doorways shall have a clear opening of 32 inches (813 mm) minimum. The clear opening of swinging doors shall be measured between the face of the door and the stop, with the door open 90 degrees.

> **EXCEPTION:** A tolerance of minus $^1/_4$ inch (6.4 mm) is permitted.

1106.4.2.2 Double leaf doorways. Where an inactive leaf with operable parts of hardware located more than 48 inches (1219 mm) above the floor is provided, the active leaf shall provide the required clear width.

1106.4.2.3 Thresholds. Thresholds, if provided, shall be $^1/_2$ inch (12.7 mm) high maximum and shall comply with CABO/ANSI A117.1.

> **EXCEPTION:** Thresholds at exterior sliding doors may be $^3/_4$ inch (19 mm) high maximum, provided they are beveled with a slope of not greater than 1 unit vertical in 2 units horizontal (50% slope).

1106.5 Kitchens.

1106.5.1 Clearances. Clearances between all opposing base cabinets, counter tops, appliances or walls within kitchen work areas shall be 40 inches (1016 mm) minimum.

In kitchens with counters, appliances or cabinets located on three contiguous sides, clearance between all opposing base cabinets, counter tops, appliances or walls within kitchen work areas shall be 60 inches (1524 mm) minimum.

1106.5.2 Clear floor space. A 30-inch-by-48-inch (762 mm by 1219 mm) minimum clear floor space shall be provided at the sink and at each appliance.

1. The clear floor space at the sink shall be positioned for parallel approach. The clear floor space shall extend 15 inches (381 mm) minimum from each side of the sink center line.

> **EXCEPTION:** Sinks complying with Section 4.33.4.5 of CABO/ANSI A117.1.

2. Where provided, the dishwasher, range, cooktop, oven, refrigerator/freezer and trash compactor shall have a clear floor space positioned for either parallel or forward approach.

1106.6 Toilet and Bathing Facilities.

1106.6.1 General. Toilet and bathing facilities in Type B dwelling units shall comply with Sections 1106.2 through 1106.4 and this section.

> **EXCEPTION:** Facilities on levels not required to be accessible.

1106.6.2 Clear floor space. Doors shall not swing into the clear floor space or clearance required for any fixture.

> **EXCEPTION:** Where a 30-inch-by-48-inch (762 mm by 1219 mm) minimum clear floor space is provided within the room, beyond the arc of the door swing.

Clear floor space shall be permitted to include knee and toe clearances in accordance with Section 4.2.4.3 of CABO/ANSI A117.1.

Clear floor spaces and clearances may overlap.

1106.6.3 Grab bar and seat reinforcement. Where walls are located to permit installation of grab bars and seats complying with Section 4.17.4, 4.21.4, 4.22.3 or 4.22.4 of CABO/ANSI A117.1, reinforcement shall be provided for the installation of grab bars and seats meeting those requirements.

> **EXCEPTION:** Reinforcement is not required in a room containing only a lavatory and a water closet, provided that the room does not contain the only lavatory or water closet on the accessible level of the dwelling unit.

1106.6.4 Toilet and bathing fixtures. Toilet and bathing fixtures shall comply with either Section 1106.6.4.1 or 1106.6.4.2.

1106.6.4.1 Option A. Each fixture provided shall comply with this section.

> **EXCEPTION:** A lavatory and a water closet located in a room containing only a lavatory and water closet, provided that the room does not contain the only lavatory or water closet on the accessible level of the dwelling unit.

1106.6.4.1.1 Lavatory. A 30-inch-by-48-inch (762 mm by 1219 mm) minimum clear floor space positioned for parallel approach shall be provided.

> **EXCEPTION:** A lavatory complying with Section 4.20 of CABO/ANSI A117.1.

Clear floor space shall extend 15 inches (381 mm) minimum from each side of the lavatory center line.

1106.6.4.1.2 Water closet. The lateral distance from the center line of the water closet to a bathtub, lavatory or wall shall be 18 inches (457 mm) minimum on one side and 15 inches (381 mm) minimum on the other side. Where the water closet is located adjacent to a wall, the lateral distance from the center line of the water closet to the wall shall be 18 inches (457 mm) and 15 inches (381 mm) minimum to a lavatory or bathtub. Where the water closet is not located adjacent to a wall, the water closet shall be located to allow for the installation of a grab bar on the side with 18-inch (457 mm) clearance. Clearance areas around the water closet shall comply with one of the following:

1. **Parallel approach.**

   1.1 Fifty-six inches (1422 mm) minimum, measured from the wall behind the water closet.

   1.2 Forty-eight inches (1219 mm) minimum, measured from a point 18 inches (457 mm) from the center line of the water closet on the side designated for the installation of grab bars.

1.3 Vanities or lavatories located on the wall behind the water closet are permitted to overlap the clear floor space.

2. **Forward approach.**

2.1 Sixty-six inches (1676 mm) minimum, measured from the wall behind the water closet.

2.2 Forty-eight inches (1219 mm) minimum, measured from a point 18 inches (457 mm) from the center line of the water closet on the side designated for the installation of grab bars.

2.3 Vanities or lavatories located on the wall behind the water closet are permitted to overlap the clear floor space.

3. **Parallel or forward approach.**

3.1 Fifty-six inches (1422 mm) minimum, measured from the wall behind the water closet.

3.2 Sixty inches (1524 mm) minimum, measured from a point 18 inches (457 mm) from the center line of the water closet on the side designated for the installation of grab bars.

1106.6.4.1.3 Bathing fixtures. Where bathing fixtures are provided, at least one bathing fixture in each toilet/bathing area shall comply with the following:

1. **Parallel approach bathtubs.** Bathtubs with a parallel approach shall have a clearance 30 inches (762 mm) wide by 60 inches (1524 mm) long minimum adjacent to the bathtub. A lavatory may extend into the clearance at the control end of the tub if the 30-inch-by-48-inch (762 mm by 1219 mm) clearance remains.

EXCEPTION: Lavatories complying with Section 4.20 of CABO/ANSI A117.1 may be placed in the clearance.

2. **Forward approach bathtubs.** Bathtubs with a forward approach shall have a clearance 48 inches (1219 mm) wide by 60 inches (1524 mm) long minimum adjacent to the bathtub. A water closet may be placed in the clearance at the control end of the tub.

1106.6.4.1.4 Showers. If a stall shower is the only bathing fixture, the stall shower shall have minimum dimensions of 36 inches by 36 inches (914 mm by 914 mm). A clear floor space of not less than 30 inches (762 mm) measured perpendicular from the face of the shower stall by 48 inches (1219 mm) measured parallel from the shower head wall shall be provided.

1106.6.4.2 Option B. One of each type of fixture provided shall comply with Section 1106.6.4.2. The accessible fixtures shall be located in a single toilet/bathing area.

1106.6.4.2.1 Lavatory. A 30-inch-by-48-inch (762 mm by 1219 mm) minimum clear floor space positioned for parallel approach shall be provided.

EXCEPTION: A lavatory complying with Section 4.20 of CABO/ANSI A117.1.

The clear floor space shall extend 15 inches (381 mm) minimum from each side of the lavatory center line.

The fixture rim shall be 34 inches (864 mm) maximum above the finished floor.

1106.6.4.2.2 Water closet. The water closet shall comply with Section 1106.6.4.1.2.

1106.6.4.2.3 Bathing fixtures. Where bathing fixtures are provided, at least one bathing fixture shall comply with the following:

1. **Bathtub.** A 30-inch-by-48-inch (762 mm by 1219 mm) minimum clear floor space positioned for parallel approach shall be provided adjacent to the bathtub. The front edge of the clear floor space shall align with the control end of the bathtub.

2. **Stall showers.** If a stall shower is the only bathing fixture, the stall shower shall have minimum dimensions of 36 inches by 36 inches (914 mm by 914 mm). A clear floor space of not less than 30 inches (762 mm) measured perpendicular from the face of the shower stall by 48 inches (1219 mm) measured parallel from the shower head wall shall be provided.

TABLE 11-A
TABLE 11-C

1997 UNIFORM BUILDING CODE

TABLE 11-A—WHEELCHAIR SPACES REQUIRED IN ASSEMBLY AREAS

CAPACITY OF SEATING	NUMBER OF REQUIRED WHEELCHAIR SPACES
4 to 25	1
26 to 50	2
51 to 300	4
301 to 500	6
over 500	6 plus 1 for each 200 over 500

TABLE 11-B—NUMBER OF ROOMS FOR PERSONS WITH HEARING IMPAIRMENTS

TOTAL NUMBER OF ROOMS	MINIMUM REQUIRED NUMBER
6-25	1
26-50	2
51-75	3
76-100	4
101-150	5
151-200	6
201-300	7
301-400	8
401-500	9
501-1,000	2% of total rooms
Over 1,000	20 plus 1 for every 100 rooms, or fraction thereof, over 1,000

TABLE 11-C—REQUIRED CHECKOUT AISLES

TOTAL CHECKOUT AISLES	MINIMUM NUMBER OF ACCESSIBLE CHECKOUT AISLES
1-4	1
5-8	2
9-15	3
Over 15	3 plus 20% of additional aisles over 15

Chapter 12
INTERIOR ENVIRONMENT

SECTION 1201 — GENERAL

Buildings and portions thereof shall provide occupants with light and ventilation as set forth in this chapter. For ventilation of hazardous vapors or fumes, see Section 306.5 and the Mechanical Code.

SECTION 1202 — LIGHT AND VENTILATION IN GROUPS A, B, E, F, H, I, M AND S OCCUPANCIES

1202.1 Light. All enclosed portions of Groups A, B, E, F, H, I, M and S Occupancies customarily occupied by human beings shall be provided with natural light by means of exterior glazed openings with an area not less than one tenth of the total floor area, or shall be provided with artificial light. Such exterior openings shall open directly onto a public way or a yard or court as set forth in Section 1203.4. See Section 1003.2.9 for required means of egress illumination.

1202.2 Ventilation.

1202.2.1 General. All enclosed portions of Groups A, B, E, F, H, I, M and S Occupancies customarily occupied by human beings shall be provided with natural ventilation by means of openable exterior openings with an area not less than $^1/_{20}$ of the total floor area or shall be provided with a mechanically operated ventilation system. Such exterior openings shall open directly onto a public way or a yard or court as set forth in Section 1203.4. Such mechanically operated ventilation system shall be capable of supplying a minimum of 15 cubic feet per minute (7 L/s) of outside air per occupant in all portions of the building during such time as the building is occupied. If the velocity of the air at a register exceeds 10 feet per second (3 m/s), the register shall be placed more than 8 feet (2438 mm) above the floor directly beneath.

Toilet rooms shall be provided with a fully openable exterior window with an area not less than 3 square feet (0.279 m²), or a vertical duct not less than 100 square inches (64 516 mm²) in area for the first water closet plus 50 square inches (32 258 mm²) additional of area for each additional water closet, or a mechanically operated exhaust system capable of providing a complete change of air every 15 minutes. Such mechanically operated exhaust systems shall be connected directly to the outside, and the point of discharge shall be at least 3 feet (914 mm) from any opening that allows air entry into occupied portions of the building.

For ventilation of hazardous vapors or fumes in Group H Occupancies, see Sections 307.5.2 and 1202.2.3. For Group S, Division 3 Occupancies, see Section 1202.2.7.

1202.2.2 Groups B, F, M and S Occupancies. In all buildings classified as Groups B, F, M and S Occupancies or portions thereof where Class I, II or III-A liquids are used, a mechanically operated exhaust ventilation shall be provided sufficient to produce six air changes per hour. Such exhaust ventilation shall be taken from a point at or near the floor level.

1202.2.3 Group H Occupancies. Rooms, areas or spaces of Group H Occupancies in which explosive, corrosive, combustible, flammable or highly toxic dusts, mists, fumes, vapors or gases are or may be emitted due to the processing, use, handling or storage of materials shall be mechanically ventilated as required by the Fire Code and the Mechanical Code.

Ducts conveying explosives or flammable vapors, fumes or dusts shall extend directly to the exterior of the building without entering other spaces. Exhaust ducts shall not extend into or through ducts and plenums.

> **EXCEPTION:** Ducts conveying vapor or fumes having flammable constituents less than 25 percent of their lower flammability limit may pass through other spaces.

Emissions generated at work stations shall be confined to the area in which they are generated as specified in the Fire Code and the Mechanical Code.

The location of supply and exhaust openings shall be in accordance with the Mechanical Code. Exhaust air contaminated by highly toxic material shall be treated in accordance with the Fire Code.

A manual shutoff control shall be provided outside of the room in a position adjacent to the access door to the room or in a location approved by the chief. The switch shall be of the break-glass type and shall be labeled VENTILATION SYSTEM EMERGENCY SHUTOFF.

1202.2.4 Group H, Division 4 Occupancies. In all buildings classified as Group H, Division 4 Occupancies used for the repair or handling of motor vehicles operating under their own power, mechanical ventilation shall be provided capable of exhausting a minimum of 1 cubic foot per minute per square foot (0.044 L/s/m²) of floor area. Each engine repair stall shall be equipped with an exhaust pipe extension duct, extending to the outside of the building, which, if over 10 feet (3048 mm) in length, shall mechanically exhaust 300 cubic feet per minute (141.6 L/s). Connecting offices and waiting rooms shall be supplied with conditioned air under positive pressure.

> **EXCEPTION:** When approved, ventilating equipment may be omitted in repair garages, enclosed heliports and aircraft hangars when well-distributed unobstructed openings to the outer air of sufficient size to supply necessary ventilation are furnished.

1202.2.5 Group H, Division 6 Occupancies. In Group H, Division 6 Occupancies, mechanical exhaust ventilation shall be provided throughout the fabrication area at the rate of not less than 1 cubic foot per minute per square foot (0.044 L/s/m²) of floor area. The exhaust air duct system of one fabrication area shall not connect to another duct system outside that fabrication area within the building.

Ventilation systems shall comply with the Mechanical Code except that the automatic shutoffs need not be installed on air-moving equipment. However, smoke detectors shall be installed in the circulating airstream and shall initiate a signal at the emergency control station.

Except for exhaust systems, at least one manually operated remote control switch that will shut down the fabrication area ventilation system shall be installed at an approved location outside the fabrication area.

A ventilation system shall be provided to capture and exhaust fumes and vapors at work stations. Two or more operations shall not be connected to the same exhaust system when either one or the combination of the substances removed could constitute a fire, explosion or hazardous chemical reaction within the exhaust duct system.

Exhaust ducts penetrating occupancy separations shall be contained in a shaft of equivalent fire-resistive construction. Exhaust ducts shall not penetrate area separation walls. Fire dampers shall not be installed in exhaust ducts.

1202.2.6 Group S repair and storage garages and aircraft hangars. In Group S, Division 3 repair garages and storage garages and in Division 5 aircraft hangars, the mechanical ventilating system required by Section 1202.2.1 may be omitted when, in the opinion of the building official, the building is supplied with unobstructed openings to the outer air that are sufficient to provide the necessary ventilation.

1202.2.7 Group S parking garages. In Group S, Division 3 parking garages, other than open parking garages, used for storing or handling automobiles operating under their own power and on loading platforms in bus terminals, ventilation shall be provided capable of exhausting a minimum of 1.5 cubic feet per minute (cfm) per square foot (0.761 L/s/m^2) of gross floor area. The building official may approve an alternate ventilation system designed to exhaust a minimum of 14,000 cfm (6608 L/s) for each operating vehicle. Such system shall be based on the anticipated instantaneous movement rate of vehicles, but not less than 2.5 percent (or one vehicle) of the garage capacity. Automatic carbon monoxide-sensing devices may be employed to modulate the ventilation system to maintain a maximum average concentration of carbon monoxide of 50 parts per million during any eight-hour period, with a maximum concentration not greater than 200 parts per million for a period not exceeding one hour. Connecting offices, waiting rooms, ticket booths and similar uses shall be supplied with conditioned air under positive pressure.

> **EXCEPTION:** Mechanical ventilation need not be provided within a Group S, Division 3 parking garage when openings complying with Section 311.9.2.2 are provided.

SECTION 1203 — LIGHT AND VENTILATION IN GROUP R OCCUPANCIES

1203.1 General. For the purpose of determining the light or ventilation for Group R Occupancies required by this section, any room may be considered as a portion of an adjoining room when one half of the area of the common wall is open and unobstructed and provides an opening of not less than one tenth of the floor area of the interior room or 25 square feet (2.3 m^2), whichever is greater.

Exterior openings for natural light or ventilation required by this section shall open directly onto a public way or a yard or court as set forth in Section 1203.4.

> **EXCEPTIONS:** 1. Required exterior openings may open into a roofed porch where the porch:
>
> 1.1 Abuts a public way, yard or court;
>
> 1.2 Has a ceiling height of not less than 7 feet (2134 mm); and
>
> 1.3 Has a longer side at least 65 percent open and unobstructed.
>
> 2. Skylights.

1203.2 Light. Guest rooms and habitable rooms within a dwelling unit or congregate residence shall be provided with natural light by means of exterior glazed openings with an area not less than one tenth of the floor area of such rooms with a minimum of 10 square feet (0.93 m^2).

> **EXCEPTION:** Kitchens in Group R Occupancies may be provided with artificial light.

1203.3 Ventilation. Guest rooms and habitable rooms within a dwelling unit or congregate residence shall be provided with natural ventilation by means of openable exterior openings with an area of not less than $^1/_{20}$ of the floor area of such rooms with a minimum of 5 square feet (0.46 m^2).

In lieu of required exterior openings for natural ventilation, a mechanical ventilating system may be provided. Such system shall be capable of providing two air changes per hour in guest rooms, dormitories, habitable rooms and in public corridors with a minimum of 15 cubic feet per minute (7 L/s) of outside air per occupant during such time as the building is occupied.

Bathrooms, water closet compartments, laundry rooms and similar rooms shall be provided with natural ventilation by means of openable exterior openings with an area not less than $^1/_{20}$ of the floor area of such rooms with a minimum of $1^1/_2$ square feet (0.14 m^2).

> **EXCEPTION:** Laundry rooms in Group R, Division 3 Occupancies.

In lieu of required exterior openings for natural ventilation in bathrooms containing a bathtub, shower or combination thereof; laundry rooms; and similar rooms, a mechanical ventilation system connected directly to the outside capable of providing five air changes per hour shall be provided. Such systems shall be connected directly to the outside, and the point of discharge shall be at least 3 feet (914 mm) from any opening that allows air entry into occupied portions of the building. Bathrooms that contain only a water closet, lavatory or combination thereof and similar rooms may be ventilated with an approved mechanical recirculating fan or similar device designed to remove odors from the air.

1203.4 Yards or Courts.

1203.4.1 General. This section shall apply to yards and courts adjacent to exterior openings that provide required natural light or ventilation. Such yards and courts shall be on the same property as the building.

1203.4.2 Yards. Yards shall not be less than 3 feet (914 mm) in width for one- and two-story buildings. For buildings more than two stories in height, the minimum width of the yard shall be increased at the rate of 1 foot (305 mm) for each additional story. For buildings exceeding 14 stories in height, the required width of the yard shall be computed on the basis of 14 stories.

1203.4.3 Courts. Courts shall not be less than 3 feet (914 mm) in width. Courts having windows opening on opposite sides shall not be less than 6 feet (1829 mm) in width. Courts bounded on three or more sides by the walls of the building shall not be less than 10 feet (3048 mm) in length unless bounded on one end by a public way or yard. For buildings more than two stories in height, the court shall be increased 1 foot (305 mm) in width and 2 feet (610 mm) in length for each additional story. For buildings exceeding 14 stories in height, the required dimensions shall be computed on the basis of 14 stories.

Adequate access shall be provided to the bottom of all courts for cleaning purposes. Every court more than two stories in height shall be provided with a horizontal air intake at the bottom not less than 10 square feet (0.93 m^2) in area and leading to the exterior of the building unless abutting a yard or public way. The construction of the air intake shall be as required for the court walls of the building, but in no case shall be less than one-hour fire resistive.

SECTION 1204 — EAVES

Where eaves extend over required windows, they shall project no closer than 30 inches (762 mm) to any side or rear property line. See also Sections 503.2 and 705.

SECTION 1205 — ALTERNATE VENTILATION WHEN APPLICABLE

1205.1 General. Requirements for ventilation are included in Appendix Chapter 12 of this code. When adopted (see Section

101.3) the appendix criteria shall take precedence over the ventilation requirements set forth in Sections 1202 and 1203 of this code.

1205.2 Standards. The standard listed below is a recognized standard (see Sections 3503 and 3504).

ANSI/ASHRAE 62-1989 including ANSI/ASHRAE Addendum 62a-1990, Ventilation for Acceptable Indoor Air Quality

Chapter 13
ENERGY CONSERVATION

SECTION 1301 — SOLAR ENERGY COLLECTORS

Collectors that function as building components shall comply with the applicable provisions of the code.

Collectors located above or upon a roof and not functioning as building components shall not reduce the required fire-resistance or fire-retardancy classification of the roof-covering materials.

EXCEPTIONS: 1. Collectors installed on one- and two-family dwellings.

2. Noncombustible collectors located on buildings not over three stories in height or 9,000 square feet (836 m^2) in total floor area.

3. Collectors that comply with the provisions of Section 2603.14.

A complete code for energy conservation in new buildings is contained in Appendix Chapter 13. When adopted, as set forth in Section 101.3, Appendix Chapter 13 applies.

Chapter 14
EXTERIOR WALL COVERINGS

SECTION 1401 — GENERAL

1401.1 Applicability. Exterior wall coverings for the building shall provide weather protection for the building at its exterior boundaries.

Exterior wall covering shall be in accordance with this chapter and as specified by the applicable provisions elsewhere in this code. For additional provisions, see Chapter 19 for concrete, Chapter 20 for lightweight metals, Chapter 21 for masonry, Chapter 22 for steel, Chapter 23 for wood, Chapter 25 for gypsum wallboard and plaster, and Chapter 26 for plastics. Also, see the following:

SECTION	SUBJECT
601.5.4	Walls fronting on streets
602.1	Materials in Type I construction
603.1	Materials in Type II construction
604.3.1	Exterior walls in Type III construction
605.3.1	Exterior walls in Type IV construction
606.1	Materials in Type V construction

1401.2 Standards. The standards listed below labeled a "UBC standard" are also listed in Chapter 35, Part II, and are part of this code.

1. UBC Standard 14-1, Kraft Waterproof Building Paper

2. UBC Standard 14-2, Vinyl Siding

SECTION 1402 — WEATHER PROTECTION

1402.1 Weather-resistive Barriers. All weather-exposed surfaces shall have a weather-resistive barrier to protect the interior wall covering. Such barrier shall be equal to that provided for in UBC Standard 14-1 for kraft waterproof building paper or asphalt-saturated rag felt. Building paper and felt shall be free from holes and breaks other than those created by fasteners and construction system due to attaching of the building paper, and shall be applied over studs or sheathing of all exterior walls. Such felt or paper shall be applied horizontally, with the upper layer lapped over the lower layer not less than 2 inches (51 mm). Where vertical joints occur, felt or paper shall be lapped not less than 6 inches (152 mm).

A weather-resistive barrier may be omitted in the following cases:

1. When exterior covering is of approved weatherproof panels.

2. In back-plastered construction.

3. When there is no human occupancy.

4. Over water-repellent panel sheathing.

5. Under approved paperbacked metal or wire fabric lath.

6. Behind lath and portland cement plaster applied to the underside of roof and eave projections.

1402.2 Flashing and Counterflashing. Exterior openings exposed to the weather shall be flashed in such a manner as to make them weatherproof.

All parapets shall be provided with coping of approved materials. All flashing, counterflashing and coping, when of metal, shall have a minimum thickness of 0.019-inch (0.48 mm) (No. 26 galvanized sheet metal gage) corrosion-resistant metal.

1402.3 Waterproofing Weather-exposed Areas. Balconies, landings, exterior stairways, occupied roofs and similar surfaces exposed to the weather and sealed underneath shall be waterproofed and sloped a minimum of $1/4$ unit vertical in 12 units horizontal (2% slope) for drainage.

1402.4 Dampproofing Foundation Walls. Unless otherwise approved by the building official, foundation walls enclosing a basement below finished grade shall be dampproofed outside by approved methods and materials.

1402.5 Window Wells. All window wells shall extend below the window sill height.

SECTION 1403 — VENEER

1403.1 Scope.

1403.1.1 General. All veneer and its application shall conform to the requirements of this code. Wainscots not exceeding 4 feet (1219 mm) in height measured above the adjacent ground elevation for exterior veneer or the finish floor elevation for interior veneer may be exempted from the provisions of this chapter if approved by the building official.

1403.1.2 Limitations. Exterior veneer shall not be attached to wood-frame construction at a point more than 30 feet (9144 mm) in height above the noncombustible foundation, except the 30-foot (9144 mm) limit may be increased when special construction is designed to provide for differential movement and when approved by the building official.

1403.2 Definitions. For the purpose of this chapter, certain terms are defined as follows:

BACKING as used in this chapter is the surface or assembly to which veneer is attached.

VENEER is nonstructural facing of brick, concrete, stone, tile, metal, plastic or other similar approved material attached to a backing for the purpose of ornamentation, protection or insulation.

Adhered Veneer is veneer secured and supported through adhesion to an approved bonding material applied over an approved backing.

Anchored Veneer is veneer secured to and supported by approved connectors attached to an approved backing.

Exterior Veneer is veneer applied to weather-exposed surfaces as defined in Section 224.

Interior Veneer is veneer applied to surfaces other than weather-exposed surfaces as defined in Section 224.

1403.3 Materials. Materials used in the application of veneer shall conform to the applicable requirements for such materials as set forth elsewhere in this code.

For masonry units and mortar, see Chapter 21.

For precast concrete units, see Chapter 19.

For portland cement plaster, see Chapter 25.

Anchors, supports and ties shall be noncombustible and corrosion resistant.

When the terms "corrosion resistant" or "noncorrosive" are used in this chapter, they shall mean having a corrosion resistance equal to or greater than a hot-dipped galvanized coating of 1.5 ounces of zinc per square foot (458 g/m^2) of surface area. When an element is required to be corrosion resistant or noncorrosive, all of its parts, such as screws, nails, wire, dowels, bolts, nuts, washers, shims, anchors, ties and attachments, shall be corrosion resistant.

1403.4 Design.

1403.4.1 General. The design of all veneer shall comply with the requirements of Chapter 16 and this section.

Veneer shall support no load other than its own weight and the vertical dead load of veneer above.

Surfaces to which veneer is attached shall be designed to support the additional vertical and lateral loads imposed by the veneer.

Consideration shall be given for differential movement of supports, including that caused by temperature changes, shrinkage, creep and deflection.

1403.4.2 Adhered veneer. With the exception of ceramic tile, adhered veneer and its backing shall be designed to have a bond to the supporting element sufficient to withstand a shearing stress of 50 psi (345 kPa).

1403.4.3 Anchored veneer. Anchored veneer and its attachments shall be designed to resist a horizontal force equal to at least twice the weight of the veneer.

1403.5 Adhered Veneer.

1403.5.1 Permitted backing. Backing shall be continuous and may be of any material permitted by this code. It shall have surfaces prepared to secure and support the imposed loads of veneer.

Exterior veneer, including its backing, shall provide a weatherproof covering.

For additional backing requirements, see Section 1402.

1403.5.2 Area limitations. The height and length of veneered areas shall be unlimited except as required to control expansion and contraction and as limited by Section 1403.1.2.

1403.5.3 Unit size limitations. Veneer units shall not exceed 36 inches (914 mm) in the greatest dimension or more than 720 square inches (0.46 m^2) in total area and shall not weigh more than 15 pounds per square foot (psf) (73.2 kg/m^2) unless approved by the building official.

> **EXCEPTION:** Veneer units weighing less than 3 psf (14.6 kg/m^2) shall not be limited in dimension or area.

1403.5.4 Application. In lieu of the design required by Sections 1403.4.1 and 1403.4.2, adhered veneer may be applied by one of the following application methods:

1. A paste of neat portland cement shall be brushed on the backing and the back of the veneer unit. Type S mortar then shall be applied to the backing and the veneer unit. Sufficient mortar shall be used to create a slight excess to be forced out the edges of the units. The units shall be tapped into place so as to completely fill the space between the units and the backing. The resulting thickness of mortar in back of the units shall not be less than $^1/_2$ inch (12.7 mm) or more than $1^1/_4$ inches (32 mm).

2. Units of tile, masonry, stone or terra cotta, not over 1 inch (25 mm) in thickness, shall be restricted to 81 square inches (52 258 mm^2) in area unless the back side of each unit is ground or box screeded to true up any deviations from plane. These units and glass mosaic units of tile not over 2 inches by 2 inches by $^3/_8$ inch (51 mm by 51 mm by 9.5 mm) in size may be adhered by means of

portland cement. Backing may be of masonry, concrete or portland cement plaster on metal lath. Metal lath shall be fastened to the supports in accordance with the requirements of Chapter 25. Mortar as described in Table 14-A shall be applied to the backing as a setting bed. The setting bed shall be a minimum of $^3/_8$ inch (10 mm) thick and a maximum of $^3/_4$ inch (19 mm) thick. A paste of neat portland cement or one half portland cement and one half graded sand shall be applied to the back of the exterior veneer units and to the setting bed and the veneer pressed and tapped into place to provide complete coverage between the mortar bed and veneer unit. A cement mortar shall be used to point the veneer.

1403.5.5 Ceramic tile. Portland cement mortars for installing ceramic tile on walls, floors and ceilings shall be as set forth in Table 14-A.

1403.6 Anchored Veneer.

1403.6.1 Permitted backing. Backing may be of any material permitted by this code. Exterior veneer including its backing shall provide a weatherproof covering.

1403.6.2 Height and support limitations. Anchored veneers shall be supported on footings, foundations or other noncombustible support except as provided under Section 2307.

In Seismic Zones 2, 3 and 4, the weight of all anchored veneers installed on structures more than 30 feet (9144 mm) in height above the noncombustible foundation or support shall be supported by noncombustible, corrosion-resistant structural framing. The structural framing shall have horizontal supports spaced not more than 12 feet (3658 mm) vertically above the initial 30-foot (9144 mm) height. The vertical spacing between horizontal supports may be increased when special design techniques, approved by the building official, are used in the construction.

Noncombustible, noncorrosive lintels and noncombustible supports shall be provided over all openings where the veneer unit is not self-spanning. The deflections of all structural lintels and horizontal supports required by this subsection shall not exceed $^1/_{600}$ of the span under full load of the veneer.

1403.6.3 Area limitations. The area and length of anchored veneer walls shall be unlimited, except as required to control expansion and contraction and by Section 1403.1.2.

1403.6.4 Application.

1403.6.4.1 General. In lieu of the design required by Sections 1403.4.1 and 1403.4.3, anchored veneer may be applied in accordance with this section.

1403.6.4.2 Masonry and stone units [5 inches (127 mm) maximum in thickness]. Masonry and stone veneer not exceeding 5 inches (127 mm) in thickness may be anchored directly to structural masonry, concrete or studs in one of the following manners:

1. Wall ties shall be corrosion resistant, and if made of sheet metal, shall have a minimum thickness of 0.030 inch (0.76 mm) (No. 22 galvanized sheet gage) by $^3/_4$ inch (19.1 mm) or, if of wire, shall have a minimum diameter of 0.148 inch (3.76 mm) (No. 9 B.W. gage). Wall ties shall be spaced so as to support not more than 2 square feet (0.19 m^2) of wall area but shall be not more than 24 inches (610 mm) on center horizontally. In Seismic Zones 3 and 4, wall ties shall have a lip or hook on the extended leg that will engage or enclose a horizontal joint reinforcement wire having a diameter of 0.148 inch (3.76 mm) (No. 9 B.W. gage) or equivalent. The joint reinforcement shall be continuous with butt splices between ties permitted.

When applied over stud construction, 2-inch-by-4-inch (51 mm by 102 mm) stud spacing shall not exceed 16 inches (406 mm) on center and 2-inch-by-6-inch (51 mm by 152 mm) stud spacing shall not exceed 24 inches (610 mm) on center. Approved paper shall first be applied over the sheathing or wires between studs except as otherwise provided in Section 1402, and mortar shall be slushed into the 1-inch (25 mm) space between facing and paper.

As an alternate to approved paper with slush fill, an air space of at least 1 inch (25 mm) may be maintained between the backing and the veneer in which case spot bedding at all ties shall be of cement mortar.

2. Veneer may be applied with 1-inch-minimum (25 mm) grouted backing space reinforced by not less than 2-inch-by-2-inch (51 mm by 51 mm) 0.065-inch (1.65 mm) (No. 16 B.W. gage) galvanized wire mesh placed over waterproof paper backing and anchored directly to stud construction.

Two-inch-by-4-inch (51 mm by 102 mm) stud spacing shall not exceed 16 inches (406 mm) on center and 2-inch-by-6-inch (51 mm by 152 mm) stud spacing shall not exceed 24 inches (610 mm) on center. The galvanized wire mesh shall be anchored to wood studs by galvanized steel wire furring nails at 4 inches (102 mm) on center or by barbed galvanized nails at 6 inches (152 mm) on center with a $1^1/_8$-inch-minimum (29 mm) penetration. The galvanized wire mesh may be attached to steel studs by equivalent wire ties. If this method is applied over solid sheathing, the mesh must be furred for embedment in grout. The wire mesh must be attached at the top and bottom with not less than 8-penny (64 mm) common wire nails. The grout fill shall be placed to fill the space intimately around the mesh and veneer facing.

1403.6.4.3 Stone units [10 inches (254 mm) maximum in thickness]. Stone veneer units not exceeding 10 inches (254 mm) in thickness may be anchored directly to structural masonry, concrete or to studs:

1. **With concrete or masonry backing.** Anchor ties shall not be less than 0.109-inch (2.77 mm) (No. 12 B.W. gage) galvanized wire, or approved equal, formed as an exposed eye and extending not less than $^1/_2$ inch (12.7 mm) beyond the face of the backing. The legs of the loops shall not be less than 6 inches (152 mm) in length bent at right angles and laid in the masonry mortar joint and spaced so that the eyes or loops are 12 inches (254 mm) maximum on center in both directions. There shall be provided not less than a 0.109-inch (2.77 mm) (No. 12 B.W. gage) galvanized wire tie, or approved equal, threaded through the exposed loops for every 2 square feet (0.19 m²) of stone veneer. This tie shall be a loop having legs not less than 15 inches (381 mm) in length bent so that it will lie in the stone veneer mortar joint. The last 2 inches (51 mm) of each wire leg shall have a right angle bend. One inch (25 mm) of cement grout shall be placed between the backing and the stone veneer.

2. **With stud backing.** A 2-inch-by-2-inch (51 mm by 51 mm) 0.065-inch (1.65 mm) (No. 16 B.W. gage) galvanized wire mesh with two layers of waterproof paper backing shall be applied directly to 2-inch-by-4-inch (51 mm by 102 mm) wood studs spaced a maximum of 16 inches (406 mm) on center or 2-inch-by-6-inch (51 mm by 152 mm) wood studs spaced a maximum of 24 inches (610 mm) on center. On studs, the mesh shall be attached with 2-inch-long (51 mm) galvanized steel wire furring nails at 4 inches (102 mm) on center providing a minimum $1^1/_8$-inch (29 mm) penetration into each stud and with 8-penny (64 mm) common nails at 8 inches (203 mm) on center into top and bottom plates. The galvanized wire mesh may be attached to steel studs with equivalent wire ties. There shall not be less than 0.109-inch (2.77 mm) (No. 12 B.W. gage) galvanized wire, or approved equal, looped through the mesh for every 2 square feet (0.19 m²) of stone veneer. This tie shall be a loop having legs not less than 15 inches (381 mm) in length, bent so that it will lie in the stone veneer mortar joint.

The last 2 inches (51 mm) of each wire leg shall have a right angle bend. One-inch-minimum (25 mm) thickness of cement grout shall be placed between the backing and the stone veneer.

1403.6.4.4 Slab-type units [2 inches (51 mm) maximum in thickness]. For veneer units of marble, travertine, granite or other stone units of slab form, ties of corrosion-resistant dowels shall engage drilled holes located in the middle of the edge of the units spaced a maximum of 24 inches (610 mm) apart around the periphery of each unit with not less than four ties per veneer unit. Units shall not exceed 20 square feet (1.9 m²) in area.

If the dowels are not tightfitting, the holes may be drilled not more than $^1/_{16}$ inch (1.6 mm) larger in diameter than the dowel with the hole countersunk to a diameter and depth equal to twice the diameter of the dowel in order to provide a tightfitting key of cement mortar at the dowel locations when the mortar in the joint has set.

All veneer ties shall be corrosion-resistant metal capable of resisting in tension or compression a force equal to two times the weight of the attached veneer.

If made of sheet metal, veneer ties shall not be smaller in area than 0.030 inch (0.76 mm) (No. 22 galvanized sheet gage) by 1 inch (25 mm) or, if made of wire, not smaller in diameter than 0.148-inch (3.76 mm) (No. 9 B.W. gage) wire.

1403.6.4.5 Terra cotta or ceramic units. Tied terra cotta or ceramic veneer units shall not be less than $1^1/_4$ inches (32 mm) in thickness with projecting dovetail webs on the back surface spaced approximately 8 inches (203 mm) on centers. The facing shall be tied to the backing wall with noncorrosive metal anchors of not less than 0.165-inch (4.19 mm) (No. 8 B.W. gage) wire installed at the top of each piece in horizontal bed joints not less than 12 inches (305 mm) or more than 18 inches (457 mm) on centers; these anchors shall be secured to $^1/_4$-inch (6.4 mm) galvanized pencil rods that pass through the vertical aligned loop anchors in the backing wall. The veneer ties shall have sufficient strength to support the full weight of the veneer in tension. The facing shall be set with not less than a 2-inch (51 mm) space from the backing wall and the space shall be filled solidly with portland cement grout and pea gravel. Immediately prior to setting, the backing wall and the facing shall be drenched with clean water and shall be distinctly damp when the grout is poured.

SECTION 1404 — VINYL SIDING

1404.1 General. Vinyl siding conforming to the requirements of this section and complying with UBC Standard 14-2 may be installed on exterior walls of buildings of Type V construction located in areas where the wind speed specified in Figure 16-1 does not exceed 80 miles per hour (129 km/h) and the building height is less than 40 feet (12 192 mm) in Exposure C. If construction is located in areas where wind speed exceeds 80 miles per hour (129 km/h) or building heights are in excess of 40 feet (12 192 mm), data indicating compliance with Chapter 16 must be submitted. Vinyl siding shall be secured to the building to provide weather protection for the exterior walls of the building.

1404.2 Application. The siding shall be applied over sheathing or materials listed in Section 2310. Siding shall be applied to conform with the weather-resistive barrier requirements in Section 1402.1. Siding and accessories shall be installed in accordance with approved manufacturer's instructions.

Nails used to fasten the siding and accessories shall have a minimum $^3/_8$-inch (9.5 mm) head diameter and 0.120-inch (3.05 mm) shank diameter. The nails shall be corrosion resistant and shall be long enough to penetrate the studs or nailing strip at least $^3/_4$ inch (19 mm). Where the siding is installed horizontally, the fastener spacing shall not exceed 16 inches (406 mm) horizontally and 12 inches (305 mm) vertically. Where the siding is installed vertically, the fastener spacing shall not exceed 12 inches (305 mm) horizontally and 12 inches (305 mm) vertically.

TABLE 14-A—CERAMIC TILE SETTING MORTARS

COAT		VOLUME TYPE 1 PORTLAND CEMENT	VOLUME TYPE S HYDRATED LIME	VOLUME SAND		MAXIMUM THICKNESS OF COAT (inches)	MINIMUM INTERVAL BETWEEN COATS (hours)
				Dry	Damp	× 25.4 for mm	
1. Walls and ceilings over 10 square feet (0.93 m^2)	Scratch	1	$1/2$	4	5	$3/8$	24
		1	0	3	4	$3/8$	24
		1	$1/2$	4	5	$3/4$	24
	Float or leveling	1	1	6	7	$3/4$	24
2. Walls and ceilings 10 square feet (0.93 m^2) or less	Scratch and float	1	$1/2$	$2^1/2$	3	$3/8$ $3/4$	24
3. Floors	Setting bed	1	0	5	6	$1^1/4$	—
		1	$1/10$	5	6	$1^1/4$	—

Chapter 15
ROOFING AND ROOF STRUCTURES

SECTION 1501 — SCOPE

1501.1 General. Roofing assemblies, roof coverings and roof structures shall be as specified in this code and as otherwise required by this chapter.

Subject to the requirements of this chapter, combustible roofing and roof insulation may be used in any type of construction.

Skylights shall be constructed as required in Chapter 24.

For use of plastics in roofs, see Chapter 26.

For solar energy collectors located above or upon a roof, see Chapter 13.

1501.2 Standards of Quality. The standards listed below labeled a "UBC Standard" are also listed in Chapter 35, Part II, and are part of this code. The other standards listed below are recognized standards (see Sections 3503 and 3504).

1. **Roof coverings.**

    1.1 UL 55-A, Materials for Use in Construction of Built-up Roof Coverings

    1.2 UL 55-B, Class C Sheet Roofing and Shingles Made from Organic Felt

    1.3 ASTM A 570 and A 611, Sheet Metals

    1.4 UBC Standard 15-3, Wood Shakes

    1.5 ASTM C 222, Asbestos-Cement Shingles

    1.6 ASTM C 406, Slate Shingles

    1.7 UBC Standard 15-4, Wood Shingles

    1.8 UBC Standard 15-5, Roof Tile

    1.9 UBC Standard 15-6, Modified Bitumen, Thermoplastic and Thermoset Membranes Used for Roof Coverings

2. **Roofing materials.**

    2.1 ASTM D 312 and D 450, Roofing Asphalt and Coal Tar Bitumen

    2.2 UBC Standard 15-1, Roofing Aggregates

    2.3 ASTM A 219 and A 239, Corrosion-resistant Metals

    2.4 ASTM B 134, B 211 and B 250, Wire

    2.5 ASTM D 1970, Self-adhering Polymer Modified Bituminous Sheet Materials Used as Steep Roofing Underlayment for Ice Dam Protection

3. **Roofing test.**

    UBC Standard 15-2, Test Standard for Determining the Fire Retardancy of Roof Assemblies

SECTION 1502 — DEFINITIONS

For purposes of this chapter, certain terms are designated as follows:

BASE PLY is one layer of felt secured to the deck over which a built-up roof is applied.

BASE SHEET is a product used as the base ply in a built-up roofing membrane.

BUILT-UP ROOFING is two or more layers of felt cemented together and surfaced with a cap sheet, mineral aggregate, smooth coating or similar surfacing material.

BUILT-UP ROOFING PLY is a layer of felt in built-up roofing.

CAP SHEET is roof covering made of organic or inorganic fibers, saturated and coated on both sides with a bituminous compound, surfaced with mineral granules, mica, talc, ilmenite, inorganic fibers or similar materials.

CEMENTING is solidly mopped application of asphalt, cold liquid asphalt compound, coal tar pitch or other approved cementing material.

COMBINATION SHEET is a glass fiber felt integrally attached to kraft paper.

CORROSION-RESISTANT is any nonferrous metal or any metal having an unbroken surfacing of nonferrous metal, or steel with not less than 10 percent chromium or with not less than 0.20 percent copper.

EQUIVISCOUS TEMPERATURE (EVT) is the temperature determined by the manufacturer at which a bitumen attains the proper viscosity for built-up membrane applications.

FELT is matted organic or inorganic fibers, saturated or coated with bituminous compound.

FELT, NONBITUMINOUS SATURATED, is a felt for special-purpose roofing weighing no less than 12 pounds per 100 square feet (0.6 kg/m^2), not less than 0.022 inch (0.56 mm) in thickness, containing a fire- and water-retardant binder and reinforced with glass fibers running lengthwise of the sheet not more than $^1/_4$ inch (6.4 mm) apart.

INTERLAYMENT is a layer of felt or nonbituminous saturated felt not less than 18 inches (457 mm) wide, shingled between each course of roofing material.

INTERLOCKING ROOFING TILES are individual units, typically of clay or concrete, possessing matching riffed or interlocking vertical side joints that have been designed to restrict lateral movement and water penetration.

METAL ROOF COVERING is metal shingles or sheets for application on solid roof surfaces, and corrugated or otherwise shaped metal sheets or sections for application on roof frameworks or on solid roof surfaces.

MODIFIED BITUMEN MEMBRANE ROOF COVERING is one or more layers of polymer modified asphalt sheet membranes complying with UBC Standard 15-6. The sheet materials may be fully adhered or mechanically attached to the substrate or held in place with an appropriate ballast layer.

ROOF COVERING is a durable exterior surface material that provides weather protection for the building at the roof.

ROOFING ASSEMBLY includes the roof deck, substrate or thermal barrier, insulation, vapor retarder, underlayment, interlayment, base plies, roofing plies, and roof covering that is assigned a roofing classification.

ROOFING ASSEMBLY, FIRE RETARDANT, is a roofing assembly complying with UBC Standard 15-2 and listed as a Class A, Class B or Class C roofing assembly.

ROOFING CLASSIFICATION is the classification by Section 1504 assigned to a roof covering or roofing assembly.

ROOFING SQUARE is 100 square feet (9.3 m²) of roof surface.

SPOT CEMENTING is discontinuous application of asphalt, cold liquid asphalt compound, coal tar pitch or other approved cementing material.

THERMOPLASTIC MEMBRANE ROOF COVERING is a sheet membrane composed of polymers and other proprietary ingredients, in compliance with UBC Standard 15-6, whose chemical composition allows the sheet to be welded together by either heat or solvent throughout its service life.

THERMOSET MEMBRANE ROOF COVERING is a sheet membrane composed of polymers and other proprietary ingredients, in compliance with UBC Standard 15-6, whose chemical composition vulcanizes or cross-links during manufacture or during its service life.

TILES are roof covering units, typically clay, concrete or cement-based material, that comply with UBC Standard 15-5.

UNDERLAYMENT is one or more layers of felt, sheathing paper, nonbituminous saturated felt or other approved material over which a roofing system is applied.

VAPOR RETARDER is a layer of material or a laminate used to appreciably reduce the flow of water vapor into the roofing system.

WOOD SHAKES are split or sawn tapered or nontapered pieces of approved durable wood or taper-sawn pieces of approved preservative treated wood complying with UBC Standard 15-3.

WOOD SHAKES AND SHINGLES, FIRE-RETARDANT (treated), are wood shakes and shingles complying with UBC Standard 15-3 or 15-4 impregnated by the full-cell vacuum-pressure process with fire-retardant chemicals, complying with UBC Standard 15-2 for use on Class A, B or C roofs.

WOOD SHINGLES are tapered pieces of approved durable wood sawn both sides complying with UBC Standard 15-4.

SECTION 1503 — ROOFING REQUIREMENTS

The roof covering or roofing assembly on any structure regulated by this code shall be as specified in Table 15-A and as classified in Section 1504. Noncombustible roof covering as defined in Section 1504.2 may be applied in accordance with the manufacturer's requirements in lieu of a fire-retardant roofing assembly.

Roofing shall be secured or fastened to the supporting roof construction and shall provide weather protection for the building at the roof.

SECTION 1504 — ROOFING CLASSIFICATION

1504.1 Fire-retardant Roofing. Fire-retardant roofs are roofing assemblies complying with UBC Standard 15-2 and listed as Class A, B or C roofs.

1504.2 Noncombustible Roof Covering. Noncombustible roof covering shall be one of the following:

1. Cement shingles or sheets.
2. Exposed concrete slab roof.
3. Ferrous or copper shingles or sheets.
4. Slate shingles.
5. Clay or concrete roofing tile.

6. Approved roof covering of noncombustible material.

1504.3 Nonrated Roofing. Nonrated roofing is approved material that is not listed as a Class A, B or C roofing assembly.

SECTION 1505 — ATTICS: ACCESS, DRAFT STOPS AND VENTILATION

1505.1 Access. An attic access opening shall be provided to attics of buildings with combustible ceiling or roof construction.

> **EXCEPTION:** Attics with a maximum vertical height of less than 30 inches (762 mm).

The opening shall not be less than 22 inches (559 mm) by 30 inches (762 mm) and shall be located in a corridor, hallway or other readily accessible location. Thirty-inch-minimum (762 mm) unobstructed headroom in the attic space shall be provided at or above the access opening.

1505.2 Draft Stops. Attics, mansards, overhangs and other concealed roof spaces formed of combustible construction shall be draft stopped as specified in Section 708.3.

1505.3 Ventilation. Where determined necessary by the building official due to atmospheric or climatic conditions, enclosed attics and enclosed rafter spaces formed where ceilings are applied directly to the underside of roof rafters shall have cross ventilation for each separate space by ventilating openings protected against the entrance of rain and snow. The net free ventilating area shall not be less than $1/150$ of the area of the space ventilated.

> **EXCEPTIONS:** 1. The opening area may be $1/300$ of the area of the space ventilated provided 50 percent of the required opening area is provided by ventilators located in the upper portion of the space to be ventilated at least 3 feet (914 mm) above eave or cornice vents with the balance of the required ventilation provided by eave or cornice vents.
>
> 2. The opening area may be $1/300$ of the area of the space ventilated provided a vapor barrier not exceeding 1 perm [5.7 × 10⁻¹¹ kg/(Pa · s · m²)] is installed on the warm side of the attic insulation.

Where eave or cornice vents are installed, insulation shall not block the free flow of air. A minimum of 1 inch (25 mm) of air space shall be provided between the insulation and roof sheathing.

Openings for ventilation shall be covered with corrosion-resistant metal mesh with mesh openings of $1/4$ inch (6.4 mm) in dimension.

Smoke and heat venting shall be in accordance with Section 906.

SECTION 1506 — ROOF DRAINAGE

1506.1 General. Roofs shall be sloped a minimum of 1 unit vertical in 48 units horizontal (2% slope) for drainage unless designed for water accumulation in accordance with Section 1611 and approved by the building official.

1506.2 Roof Drains. Unless roofs are sloped to drain over roof edges, roof drains shall be installed at each low point of the roof.

Roof drains shall be sized and discharged in accordance with the Plumbing Code.

1506.3 Overflow Drains and Scuppers. Where roof drains are required, overflow drains having the same size as the roof drains shall be installed with the inlet flow line located 2 inches (51 mm) above the low point of the roof, or overflow scuppers having three times the size of the roof drains and having a minimum opening height of 4 inches (102 mm) may be installed in the adjacent parapet walls with the inlet flow line located 2 inches (51 mm) above the low point of the adjacent roof.

Overflow drains shall discharge to an approved location and shall not be connected to roof drain lines.

1506.4 Concealed Piping. Roof drains and overflow drains, where concealed within the construction of the building, shall be installed in accordance with the Plumbing Code.

1506.5 Over Public Property. Roof drainage water from a building shall not be permitted to flow over public property.

> **EXCEPTION:** Group R, Division 3 and Group U Occupancies.

SECTION 1507 — ROOF-COVERING MATERIALS AND APPLICATION

1507.1 Materials. The quality and design of roofing materials and their fastenings shall conform to the applicable standards listed in Chapter 35, Part II.

1507.2 Identification. All material shall be delivered in packages bearing the manufacturer's label or identifying mark.

Each package of asphalt shingles, mineral surfaced roll roofing, fire-retardant-treated wood shingles and shakes, modified bitumen, thermoplastic and thermoset membranes, and built-up roofing ply materials shall bear the label of an approved agency having a service for the inspection of material and finished products during manufacture.

Each bundle of wood shakes or shingles shall comply with UBC Standard 15-3 or 15-4, respectively. Each bundle of wood shakes or shingles and slate shingles shall bear the label or identification mark of an approved inspection bureau or agency showing the grade.

Asphalt shall be delivered in cartons indicating the name of the manufacturer, the flash point and the type of product. Bulk shipments shall be accompanied with the same information issued in the form of a certification or on the bill of lading by the manufacturer. Coal tar pitch shall bear the manufacturer's name and type. Additional information such as equiviscous temperature (EVT) may be furnished.

1507.3 Asbestos-cement Roofing. Corrugated asbestos-cement roofing shall be applied in an approved manner.

1507.4 Asbestos-cement Shingles. Asbestos-cement shingles shall be installed in an approved manner.

1507.5 Asphalt Shingles. Asphalt shingles shall be fastened according to the manufacturer's instructions and Table 15-B-1.

1507.6 Built-up Roofs. Built-up roofing shall be applied in accordance with the manufacturer's instructions and Tables 15-E through 15-G.

1507.7 Clay or Concrete Tile. Tile of clay or concrete shall comply with UBC Standard 15-5 and shall be installed in accordance with the manufacturer's instructions and Tables 15-D-1 and 15-D-2.

1507.8 Metal Roof Covering. Metal roof covering exposed to the weather shall be corrosion resistant.

Corrugated or ribbed steel shall not be less than 0.013 inch (0.33 mm) (No. 30 galvanized sheet gage).

Flat steel sheets shall not be less than 0.013 inch (0.33 mm) (No. 30 galvanized sheet gage). Other ferrous sections or shapes shall not be less than No. 26 galvanized sheet gage.

Flat nonferrous sheets shall not be less than 0.0159 inch (0.40 mm) (No. 28 B.&S. gage). Other nonferrous sections or shapes shall not be less than 0.0179 inch (0.45 mm) (No. 25 B.&S. gage).

Corrugated or otherwise shaped sheets or sections shall be designed to support the loads required by Chapter 16.

Ferrous sheets or sections shall comply with Chapter 22, Division V.

1507.9 Metal Shingles. Metal shingles shall be applied in an approved manner. Nonferrous shingles shall not be less than 0.0159 inch (0.40 mm) (No. 28 B.&S. gage).

1507.10 Sheet Roof Covering. Sheet roof covering shall be installed in an approved manner.

1507.11 Slate Shingles. Slate shingles shall be installed in an approved manner.

1507.12 Wood Shakes. Shakes shall comply with UBC Standard 15-3 and shall be installed in accordance with Table 15-B-2.

1507.13 Wood Shingles. Shingles shall comply with UBC Standard 15-4 and shall be installed in accordance with Table 15-B-2.

1507.14 Modified Bitumen, Thermoplastic and Thermoset Membranes. Modified bitumen, thermoplastic and thermoset roof membranes shall be applied in accordance with the manufacturer's instructions.

SECTION 1508 — VALLEY FLASHING

1508.1 Valleys. Roof valley flashings shall be as noted in this section. Shingle application shall be consistent with applicable Table 15-B-1, 15-B-2, 15-D-1 or 15-D-2.

1508.2 Asphalt Shingles. The roof valley flashing shall not be provided of less than 0.016-inch (0.41 mm) (No. 28 galvanized sheet gage) corrosion-resistant metal, and shall extend at least 8 inches (203 mm) from the center line each way. Sections of flashing shall have an end lap of not less than 4 inches (102 mm). Alternatively, the valley shall consist of woven asphalt shingles applied in accordance with the manufacturer's printed instructions.

In each case, the roof valley flashing shall have a 36-inch-wide (914 mm) underlayment directly under it consisting of one layer of Type 15 felt running the full length of the valley, in addition to the underlayment specified in Table 15-B-1. In severe climates, the metal valley flashing underlayment shall be solid cemented to the roof underlayment for slopes under 7 units vertical in 12 units horizontal (58.3% slope).

1508.3 Metal Shingles. The roof valley flashing shall not be provided of less than 0.016-inch (0.41 mm) (No. 28 galvanized sheet gage) corrosion-resistant metal, which shall extend at least 8 inches (203 mm) from the center line each way and shall have a splash diverter rib not less than $^3/_4$ inch (19 mm) high at the flow line formed as part of the flashing. Sections of flashing shall have an end lap of not less than 4 inches (102 mm). The metal valley flashing shall have a 36-inch-wide (914 mm) underlayment directly under it consisting of one layer of Type 15 felt running the full length of the valley, in addition to underlayment required for metal shingles. In severe climates, the metal valley flashing underlayment shall be solid cemented to the roofing underlayment for roof slopes under 7 units vertical in 12 units horizontal (58.3% slope).

1508.4 Asbestos-cement Shingles, Slate Shingles, and Clay and Concrete Tile. The roof valley flashing shall not be provided of less than 0.016-inch (0.41 mm) (No. 28 galvanized sheet gage) corrosion-resistant metal, which shall extend at least 11 inches (279 mm) from the center line each way and shall have a splash diverter rib not less than 1 inch (25 mm) high at the flow line formed as part of the flashing. Sections of flashing shall have an end lap of not less than 4 inches (102 mm). For roof slopes of 3 units vertical in 12 units horizontal (25% slope) and over, the metal valley flashing shall have a 36-inch-wide (914 mm) underlayment directly under it consisting of one layer of Type 15 felt running the full length of the valley, in addition to the underlayment specified in Tables 15-D-1 and 15-D-2. In severe climates, the metal valley flashing underlayment shall be solid cemented to the roofing underlayment for slopes under 7 units vertical in 12 units horizontal (58.3% slope).

1508.5 Wood Shingles and Wood Shakes. The roof valley flashing shall not be provided of less than 0.016-inch (0.41 mm) (No. 28 galvanized sheet gage) corrosion-resistant metal, which shall extend at least 8 inches (203 mm) from the center line each way for wood shingles and 11 inches (279 mm) from the center line each way for wood shakes. Sections of flashing shall have an overlap of not less than 4 inches (102 mm). The metal valley flashing shall have a 36-inch-wide (914 mm) underlayment directly under it consisting of one layer of Type 15 felt running the full length of the valley, in addition to underlayment specified in Table 15-B-2. In severe climates, the metal valley flashing underlayment shall be solid cemented to the roofing underlayment for roof slopes under 7 units vertical in 12 units horizontal (58.3% slope).

> **EXCEPTION:** Where local practice indicates satisfactory performance, the building official may permit valley flashing without underlayment.

SECTION 1509 — OTHER FLASHING

At the juncture of the roof and vertical surfaces, flashing and counterflashing shall be provided per the roofing manufacturer's instructions and, when of metal, shall not be less than 0.019-inch (0.48 mm) (No. 26 galvanized sheet gage) corrosion-resistant metal.

SECTION 1510 — ROOF INSULATION

Roof insulation shall be of a rigid type suitable as a base for application of a roof covering. Foam plastic roof insulation shall conform to the requirements of Section 2602. The use of insulation in fire-resistive construction shall comply with Section 710.1.

The roof insulation, deck material and roof covering shall meet the fire-retardancy requirements of Section 1504 and Table 15-A.

Insulation for built-up roofs shall be applied in accordance with Table 15-E. Insulation for modified bitumen, thermoplastic and thermoset membrane roofs shall be applied in accordance with the roofing manufacturer's recommendations. For other roofing materials such as shingles or tile, the insulation shall be covered with a suitable nailing base secured to the structure.

SECTION 1511 — PENTHOUSES AND ROOF STRUCTURES

1511.1 Height. In buildings other than Type I construction, penthouses or other roof structures shall not exceed 28 feet (8534 mm) in height above the roof surface.

1511.2 Area. The aggregate area of all penthouses and other roof structures shall not exceed $33^1/_3$ percent of the area of the supporting roof.

1511.3 Prohibited Uses. No penthouse, bulkhead or any other similar projection above the roof shall be used for purposes other than shelter of mechanical equipment or shelter of vertical shaft openings in the roof. Penthouses or bulkheads used for purposes other than permitted by this section shall conform to the requirements of this code for an additional story.

1511.4 Construction. Roof structures shall be constructed with walls, floors and roof as required for the main portion of the building.

> **EXCEPTIONS:** 1. On Types I and II-F.R. buildings, the exterior walls and roofs of penthouses that are 5 feet (1524 mm) or more from an adjacent property line may be of one-hour fire-resistive noncombustible construction.
>
> 2. On Types III and IV buildings, walls not less than 5 feet (1524 mm) from an adjacent property line may be of one-hour fire-resistive noncombustible construction.
>
> 3. Enclosures housing only mechanical equipment and located at least 20 feet (6096 mm) from adjacent property lines may be of unprotected noncombustible construction.
>
> 4. On one-story buildings, unroofed mechanical equipment screens, fences or similar enclosures may be of combustible construction when located at least 20 feet (6096 mm) from adjacent property lines and when not exceeding 4 feet (1219 mm) in height above the roof surface.

The restrictions of this section shall not prohibit the placing of wood flagpoles or similar structures on the roof of any building.

SECTION 1512 — TOWERS AND SPIRES

Towers or spires when enclosed shall have exterior walls as required for the building to which they are attached. Towers not enclosed and which extend more than 75 feet (22 860 mm) above grade shall have their framework constructed of iron, steel or reinforced concrete. No tower or spire shall occupy more than one fourth of the street frontage of any building to which it is attached and in no case shall the base area exceed 1,600 square feet (149 m²) unless it conforms entirely to the type of construction requirements of the building to which it is attached and is limited in height as a main part of the building. If the area of the tower or spire exceeds 100 square feet (9.29 m²) at any horizontal cross section, its supporting frame shall extend directly to the ground. The roof covering of spires shall be as required for the main roof of the rest of the structure.

Skeleton towers used as radio masts and placed on the roof of any building shall be constructed entirely of noncombustible materials when more than 25 feet (7620 mm) in height and shall be directly supported on a noncombustible framework to the ground. They shall be designed to withstand a wind load from any direction as specified in Chapter 16, Division III, in addition to any other loads.

SECTION 1513 — ACCESS TO ROOFTOP EQUIPMENT

Access shall be provided to all mechanical equipment located on the roof as required by the Mechanical Code.

TABLE 15-A—MINIMUM ROOF CLASSES

OCCUPANCY	TYPES OF CONSTRUCTION								
	I	II			III		IV	V	
	F.R.	F.R.	One-hour	N	One-hour	N	H.T.	One-hour	N
A-1	B	B	—	—	—	—	—	—	—
A) 2-2.1	B	B	B	—	B	—	B	B	—
A-3	B	B	B	B	B[1]	C	B[1]	B[1]	C
A-4	B	B	B	B	B	B	B	B	B[1]
B	B	B	B	B	B[1]	C	B[1]	B[1]	C
E	B	B	B	B	B	B	B	B	B[1]
F	B	B	B	B	B[1]	C	B[1]	B[1]	C
H-1	A	A	A	A	—	—	—	—	—
H) 2-3-4-5-6-7	A	B	B	B	B	B	B	B	B
I) 1.1-1.2-2	A	B	B	—	B	—	B	B	—
I-3	A	B	B[1]	—	B[2]	—	—	B[3]	—
M	B	B	B	B	B[1]	C	B[1]	B[1]	C
R-1	B	B	B	B	B[1,3]	C[3]	B[1,3]	B[1,3]	C[3]
R-3	B	B	B	B	NR	NR	NR	NR	NR
S-1, S-3	B	B	B	B	B[1]	C	B[1]	B[1]	C
S-2, S-5	B	B	B	B	B	B	B	B	B[1]
S-4	B	B	B	B	—	—	—	—	—
U	B	B	B	B	NR[4]	NR[4]	NR[4]	NR[4]	NR[4]

A—Class A roofing.
B—Class B roofing.
C—Class C roof covering.
F.R.—Fire resistive.
H.T.—Heavy timber.
N—No requirements for fire resistance.
NR—Nonrated roof coverings.

[1]Buildings that are not more than two stories in height and have not more than 6,000 square feet (557 m^2) of projected roof area and where there is a minimum of 10 feet (3048 mm) from the extremity of the roof to the property line or assumed property line on all sides except for street fronts may have Class C roof coverings that comply with UBC Standard 15-2.

[2]See Section 308.2.2.

[3]Nonrated roof coverings may be used on buildings that are not more than two stories in height and have not more than 3,000 square feet (279 m^2) of projected roof area and where there is a minimum of 10 feet (3048 mm) from the extremity of the roof to the property line on all sides except for street fronts.

[4]Unless otherwise required because of location, Group U, Division 1 roof coverings shall consist of not less than one layer of cap sheet, or built-up roofing consisting of two layers of felt and a surfacing material of 300 pounds per roofing square (14.6 kg/m^2) of gravel or other approved surfacing material, or 250 pounds (12.2 kg/m^2) of crushed slag.

TABLE 15-B-1   1997 UNIFORM BUILDING CODE

TABLE 15-B-1—ASPHALT SHINGLE APPLICATION

	ASPHALT SHINGLES	
	Not Permitted below 2 Units Vertical in 12 Units Horizontal (16.7% Slope)	
Roof Slope	**2 Units Vertical in 12 Units Horizontal (16.7% Slope) to Less than 4 Units Vertical in 12 Units Horizontal (33.3% Slope)**	**4 Units Vertical in 12 Units Horizontal (33.3% Slope) and Over**
1. Deck requirement	Asphalt shingles shall be fastened to solidly sheathed roofs. Sheathing shall conform to Sections 2312.2 and 2320.12.9.	
2. Underlayment Temperate climate	Asphalt strip shingles may be installed on slopes as low as 2 units vertical in 12 units horizontal (16.7% slope), provided the shingles are approved self-sealing or are hand sealed and are installed with an underlayment consisting of two layers of nonperforated Type 15 felt applied shingle fashion. Starting with an 18-inch-wide (457 mm) sheet and a 36-inch-wide (914 mm) sheet over it at the eaves, each subsequent sheet shall be lapped 19 inches (483 mm) horizontally.	One layer nonperforated Type 15 felt lapped 2 inches (51 mm) horizontally and 4 inches (102 mm) vertically to shed water.
Severe climate: In areas subject to wind-driven snow or roof ice buildup	Same as for temperate climate, and the two layers shall be solid cemented together with approved cementing material between the plies extending from the eave up the roof to a line 24 inches (610 mm) inside the exterior wall line of the building. As an alternative to the two layers of cemented Type 15 felt, an approved self-adhering, polymer modified, bituminous sheet may be used.	Same as for temperate climate, except that one layer No. 40 coated roofing or coated glass base shall be applied from the eaves to a line 12 inches (305 mm) inside the exterior wall line with all laps cemented together. As an alternative to the layer of No. 40 felt, a self-adhering, polymer modified, bituminous sheet may be used.
3. Attachment Combined systems, type of fasteners	Corrosion-resistant nails, minimum 12-gage $^3/_8$-inch (9.5 mm) head, or approved corrosion-resistant staples, minimum 16-gage $^{15}/_{16}$-inch (23.8 mm) crown width. Fasteners shall comply with the requirements of Chapter 23, Division III, Part III. Fasteners shall be long enough to penetrate into the sheathing $^3/_4$ inch (19 mm) or through the thickness of the sheathing, whichever is less.	
No. of fasteners[1]	4 per 36-inch to 40-inch (914 mm to 1016 mm) strip 2 per 9-inch to 18-inch (229 mm to 457 mm) shingle	
Exposure Field of roof Hips and ridges	Per manufacturer's instructions included with packages of shingles. Hip and ridge weather exposures shall not exceed those permitted for the field of the roof.	
Method	Per manufacturer's instructions included with packages of shingles.	
4. Flashing Valleys Other flashing	Per Section 1508.2 Per Section 1509	

[1]Figures shown are for normal application. For special conditions, such as mansard application and where roofs are in special wind regions, shingles shall be attached per the manufacturer's instructions.

TABLE 15-B-2—WOOD SHINGLE OR SHAKE APPLICATION

ROOF SLOPE	WOOD SHINGLES	WOOD SHAKES
	Not Permitted below 3 Units Vertical in 12 Units Horizontal (25% Slope)	Not Permitted below 4 Units Vertical in 12 Units Horizontal (33.3% Slope)[1]
	See Table 15-C	See Table 15-C
1. Deck requirement	Shingles and shakes shall be applied to roofs with solid or spaced sheathing. When spaced sheathing is used, sheathing boards shall not be less than 1 inch by 4 inches (25 mm by 102 mm) nominal dimensions and shall be spaced on centers equal to the weather exposure to coincide with the placement of fasteners. When 1-inch-by-4-inch (25 mm by 102 mm) spaced sheathing is installed at 10 inches (254 mm) on center, additional 1-inch-by-4-inch (25 mm by 102 mm) boards must be installed between the sheathing boards. Sheathing shall conform to Sections 2312.2 and 2320.12.9.	
2. Interlayment	No requirements.	One 18-inch-wide (457 mm) interlayment of Type 30 felt shingled between each course in such a manner that no felt is exposed to the weather below the shake butts and in the keyways (between the shakes).
3. Underlayment Temperate climate	No requirements.	No requirements.

Severe climate:
      In areas subject to
      wind-driven snow or
      roof ice buildup
	Two layers of nonperforated Type 15 felt applied shingle fashion shall be installed and solid cemented together with approved cementing material between the plies extending from the eave up the roof to a line 36 inches (914 mm) inside the exterior wall line of the building.	Sheathing shall be solid and, in addition to the interlayment of felt shingled between each course in such a manner that no felt is exposed to the weather below the shake butts, the shakes shall be applied over a layer of nonperforated Type 15 felt applied shingle fashion. Two layers of nonperforated Type 15 felt applied shingle fashion shall be installed and solid cemented together with approved cementing material between the plies extending from the eave up the roof to a line 36 inches (914 mm) inside the exterior wall line of the building.

4. Attachment Type of fasteners	Corrosion-resistant nails, minimum No. $14^1/_2$-gage, $^7/_{32}$-inch (5.6 mm) head, or corrosion-resistant staples, when approved by the building official.	Corrosion-resistant nails, minimum No. 13-gage, $^7/_{32}$-inch (5.6 mm) head, or corrosion-resistant staples, when approved by the building official.
	Fasteners shall comply with the requirements of Chapter 23, Division III, Part III. Fasteners shall be long enough to penetrate into the sheathing $^3/_4$ inch (19 mm) or through the thickness of the sheathing, whichever is less.	
No. of fasteners	2 per shingle	2 per shake

Exposure
      Field of roof
      Hips and ridges
	Weather exposures shall not exceed those set forth in Table 15-C.
	Hip and ridge weather exposure shall not exceed those permitted for the field of the roof.	

Method	Shingles shall be laid with a side lap of not less than $1^1/_2$ inches (38 mm) between joints in adjacent courses, and not in direct alignment in alternate courses. Spacing between shingles shall be approximately $^1/_4$ inch (6.4 mm). Each shingle shall be fastened with two nails only, positioned approximately $^3/_4$ inch (19 mm) from each edge and approximately 1 inch (25 mm) above the exposure line. Starter course at the eaves shall be doubled.	Shakes shall be laid with a side lap of not less than $1^1/_2$ inches (38 mm) between joints in adjacent courses. Spacing between shakes shall not be less than $^3/_8$ inch (9.5 mm) or more than $^5/_8$ inch (15.9 mm) except for preservative-treated wood shakes, which shall have a spacing not less than $^1/_4$ inch (6.4 mm) or more than $^3/_8$ inch (9.5 mm). Shakes shall be fastened to the sheathing with two nails only, positioned approximately 1 inch (25 mm) from each edge and approximately 2 inches (51 mm) above the exposure line. The starter course at the eaves shall be doubled. The bottom or first layer may be either shakes or shingles. Fifteen-inch or 18-inch (381 mm or 457 mm) shakes may be used for the starter course at the eaves and final course at the ridge.

5. Flashing
      Valleys
      Other flashing
	Per Section 1508.5
	Per Section 1509	

[1]When approved by the building official, wood shakes may be installed on a slope of not less than 3 units vertical in 12 units horizontal (25% slope) when an underlayment of not less than nonperforated Type 15 felt is installed.

TABLE 15-C
TABLE 15-D-1

1997 UNIFORM BUILDING CODE

TABLE 15-C—MAXIMUM WEATHER EXPOSURE

GRADE LENGTH	3 UNITS VERTICAL TO LESS THAN 4 UNITS VERTICAL IN 12 UNITS HORIZONTAL (25% ≤ 33.3% SLOPE)	4 UNITS VERTICAL IN 12 UNITS HORIZONTAL (33.3% SLOPE)
	× 25.4 for mm	
	Wood Shingles	
1. No. 1 16-inch	$3^3/_4$	5
2. No. 2[1] 16-inch	$3^1/_2$	4
3. No. 3[1] 16-inch	3	$3^1/_2$
4. No. 1 18-inch	$4^1/_4$	$5^1/_2$
5. No. 2[1] 18-inch	4	$4^1/_2$
6. No. 3[1] 18-inch	$3^1/_2$	4
7. No. 1 24-inch	$5^3/_4$	$7^1/_2$
8. No. 2[1] 24-inch	$5^1/_2$	$6^1/_2$
9. No. 3[1] 24-inch	5	$5^1/_2$
	Wood Shakes[2]	
10. No. 1 18-inch	$7^1/_2$	$7^1/_2$
11. No. 1 24-inch	10	10
12. No. 2 18-inch tapersawn shakes	—	$5^1/_2$
13. No. 2 24-inch tapersawn shakes	—	$7^1/_2$

[1]To be used only when specifically permitted by the building official.
[2]Exposure of 24-inch by $3/_8$-inch (610 mm by 9.5 mm) resawn handsplit shakes shall not exceed 5 inches (127 mm) regardless of the roof slope.

TABLE 15-D-1—ROOFING TILE APPLICATION[1] FOR ALL TILES

		ROOF SLOPE $2^1/_2$ UNITS VERTICAL IN 12 UNITS HORIZONTAL (21% Slope) TO LESS THAN 3 UNITS VERTICAL IN 12 UNITS HORIZONTAL (25% Slope)	ROOF SLOPE 3 UNITS VERTICAL IN 12 UNITS HORIZONTAL (25% Slope) AND OVER
1.	Deck requirements	Solid sheathing per Sections 2312.2 and 2320.12.9	
2.	Underlayment In climate areas subject to wind-driven snow, roof ice damming or special wind regions as shown in Chapter 16, Figure 16-1.	Built-up roofing membrane, three plies minimum, applied per Section 1507.6. Surfacing not required.	Same as for other climate areas, except that extending from the eaves up the roof to a line 24 inches (610 mm) inside the exterior wall line of the building, two layers of underlayment shall be applied shingle fashion and solidly cemented together with an approved cementing material.
	Other climate areas		One layer heavy-duty felt or Type 30 felt side lapped 2 inches (51 mm) and end lapped 6 inches (153 mm).
3.	Attachment[2] Type of fasteners	Corrosion-resistant nails not less than No. 11 gage, $5/_{16}$-inch (7.9 mm) head. Fasteners shall comply with the requirements of Chapter 23, Division III, Part III. Fasteners shall be long enough to penetrate into the sheathing $3/_4$ inch (19 mm) or through the thickness of the sheathing, whichever is less. Attaching wire for clay or concrete tile shall not be smaller than 0.083 inch (2.11 mm) (No. 14 B.W. gage).	
	Number of fasteners[2,3]	One fastener per tile. Flat tile without vertical laps, two fasteners per tile.	Two fasteners per tile. Only one fastener on slopes of 7 units vertical in 12 units horizontal (58.3% slope) and less for tiles with installed weight exceeding 7.5 pounds per square foot (36.6 kg/m^2) having a width no greater than 16 inches (406 mm).[4]
4.	Tile headlap	3 inches (76 mm) minimum.	
5.	Flashing	Per Sections 1508.4 and 1509.	

[1]In snow areas, a minimum of two fasteners per tile are required.
[2]In areas designated by the building official as being subject to repeated wind velocities in excess of 80 miles per hour (129 km/h) or where the roof height exceeds 40 feet (12 192 mm) above grade, all tiles shall be attached as follows:
  2.1 The heads of all tiles shall be nailed.
  2.2 The noses of all eave course tiles shall be fastened with approved clips.
  2.3 All rake tiles shall be nailed with two nails.
  2.4 The noses of all ridge, hip and rake tiles shall be set in a bead of approved roofer's mastic.
[3]In snow areas, a minimum of two fasteners per tile are required, or battens and one fastener.
[4]On slopes over 24 units vertical in 12 units horizontal (200% slope), the nose end of all tiles shall be securely fastened.

TABLE 15-D-2—CLAY OR CONCRETE ROOFING TILE APPLICATION INTERLOCKING TILE WITH PROJECTING ANCHOR LUGS—MINIMUM ROOF SLOPE 4 UNITS VERTICAL IN 12 UNITS HORIZONTAL (33.3% Slope)

ROOF SLOPE	4 UNITS VERTICAL IN 12 UNITS HORIZONTAL (33.3% Slope) AND OVER
1. Deck requirements	Spaced structural sheathing boards or solid roof sheathing.
2. Underlayment In climate areas subject to wind-driven snow, roof ice or special wind regions as shown in Chapter 16, Figure 16-1.	Solid sheathing one layer of Type 30 felt lapped 2 inches (51 mm) horizontally and 6 inches (152 mm) vertically, except that extending from the eaves up the roof to line 24 inches (610 mm) inside the exterior wall line of the building, two layers of the underlayment shall be applied shingle fashion and solid cemented together with approved cementing material.
Other climates	For spaced sheathing, approved reinforced membrane. For solid sheathing, one layer heavy-duty felt or Type 30 felt lapped 2 inches (51 mm) horizontally and 6 inches (152 mm) vertically.
3. Attachment[1] Type of fasteners	Corrosion-resistant nails not less than No. 11 gage, $5/16$-inch (7.9 mm) head. Fasteners shall comply with the requirements of Chapter 23, Division III, Part III. Fasteners shall be long enough to penetrate into the battens[2] or sheathing $3/4$ inch (19 mm) or through the thickness of the sheathing, whichever is less. Attaching wire for clay or concrete tile shall not be smaller than 0.083 inch (2.11 mm) (No. 14 B.W. gage). Horizontal battens are required on solid sheathing for slopes 7 units vertical in 12 units horizontal (58.3% slope) and over.[2] Horizontal battens are required for slopes over 7 units vertical in 12 units horizontal (58.3% slope).[2]
No. of fasteners with: Spaced/solid sheathing with battens, or spaced sheathing[3]	Below 5 units vertical in 12 units horizontal (41.7% slope), fasteners not required. Five units vertical in 12 units horizontal (41.7% slope) to less than 12 units vertical in 12 units horizontal (100% slope), one fastener per tile every other row. Twelve units vertical in 12 units horizontal (100% slope) to 24 units vertical in 12 units horizontal (200% slope), one fastener every tile.[4] All perimeter tiles require one fastener.[5] Tiles with installed weight less than 9 pounds per square foot (4.4 kg/m²) require a minimum of one fastener per tile regardless of roof slope.
Solid sheathing without battens[3]	One fastener per tile.
4. Tile headlap	3-inch (76 mm) minimum.
5. Flashing	Per Sections 1508.4 and 1509.

[1]In areas designated by the building official as being subject to repeated wind velocities in excess of 80 miles per hour (129 km/h), or where the roof height exceeds 40 feet (12 192 mm) above grade, all tiles shall be attached as set forth below:
  1.1 The heads of all tiles shall be nailed.
  1.2 The noses of all eave course tiles shall be fastened with a special clip.
  1.3 All rake tiles shall be nailed with two nails.
  1.4 The noses of all ridge, hip and rake tiles shall be set in a bead of approved roofer's mastic.
[2]Battens shall not be less than 1-inch-by-2-inch (25 mm by 51 mm) nominal. Provisions shall be made for drainage beneath battens by a minimum of $1/8$-inch (3.2 mm) risers at each nail or by 4-foot-long (1219 mm) battens with at least $1/2$-inch (12.7 mm) separation between battens. Battens shall be fastened with approved fasteners spaced at not more than 24 inches (610 mm) on center.
[3]In snow areas, a minimum of two fasteners per tile are required, or battens and one fastener.
[4]Slopes over 24 units vertical in 12 units horizontal (200% slope), nose ends of all tiles must be securely fastened.
[5]Perimeter fastening areas include three tile courses but not less than 36 inches (914 mm) from either side of hips or ridges and edges of eaves and gable rakes.

TABLE 15-E

1997 UNIFORM BUILDING CODE

TABLE 15-E—BUILT-UP ROOF-COVERING APPLICATION

	MECHANICALLY FASTENED SYSTEMS	ADHESIVELY FASTENED SYSTEMS
1. Deck conditions	Decks shall be firm, broom-clean, smooth and dry. Insulated decks shall have wood insulation stops at all edges of the deck, unless an alternative suitable curbing is provided. Insulated decks with slopes greater than 2 units vertical in 12 units horizontal (16.7% slope) shall have wood insulation stops at not more than 8 feet (2438 mm) face to face. Wood nailers shall be provided where nailing is required for roofing plies.	
	Solid wood sheathing shall conform to Sections 2312.2 and 2320.12.9.	Provide wood nailers where nailing is required for roofing plies (see below).
2. Underlayment	One layer of sheathing paper, Type 15 felt or other approved underlayment nailed sufficiently to hold in place, is required over board decks where openings between boards would allow bitumen to drip through. No underlayment requirements for plywood decks. Underlayment on other decks shall be in accordance with deck manufacturer's recommendations.	Not required.
3. Base ply requirements Over noninsulated decks	Over approved decks, the base ply shall be nailed using not less than one fastener for each $1^1/_3$ square feet (0.124 m^2).	Decks shall be primed in accordance with the roofing manufacturer's instructions. The base ply shall be solidly cemented or spot mopped as required by the type of deck material using adhesive application rates shown in Table 15-F.
4. Mechanical fasteners	Fasteners shall be long enough to penetrate $^3/_4$ inch (19 mm) into the sheathing or through the thickness of the sheathing, whichever is less. Built-up roofing nails for wood board decks shall be minimum No. 12 gage, $^7/_{16}$-inch (11.1 mm) head driven through tin caps or approved nails with integral caps. For plywood, No. 11 gage ring-shank nails driven through tin caps or approved nails with integral caps shall be used. For gypsum, insulating concrete, cementitious wood fiber and other decks, fasteners recommended by the manufacturer shall be used.	When mechanical fasteners are required for attachment of roofing plies to wood nailers or insulation stops (see below), they shall be as required for wood board decks.
5. Vapor retarder Over insulated decks	A vapor retarder shall be installed where the average January temperature is below 45°F (7°C), or where excessive moisture conditions are anticipated within the building. It shall be applied as for a base ply.	
6. Insulation	When no vapor retarder is required, roof insulation shall be fastened in an approved manner. When a vapor retarder is required, roof insulation is to be solidly mopped to the vapor retarder using the adhesive application rate specified in Table 15-F. See manufacturer's instructions for the attachment of insulation over steel decks.	When no vapor retarder is required, roof insulation shall be solid mopped to the deck using the adhesive application rate specified in Table 15-F. When a vapor retarder is required, roof insulation is to be solidly mopped to the vapor retarder, using the adhesive application rate specified in Table 15-F. See manufacturer's installation instructions for attachment of insulation over steel decks.
7. Roofing plies	Successive layers shall be solidly cemented together and to the base ply or the insulation using the adhesive rates shown in Table 15-F. On slopes greater than 1 unit vertical in 12 units horizontal (8.3% slope) for aggregate-surfaced, or 2 units vertical in 12 units horizontal (16.7% slope) for smooth-surfaced or cap sheet surfaced roofs, mechanical fasteners are required. Roofing plies shall be blind-nailed to the deck, wood nailers or wood insulation stops in accordance with the roofing manufacturer's recommendations. On slopes exceeding 3 units vertical in 12 units horizontal (25% slope), plies shall be laid parallel to the slope of the deck (strapping method).	
8. Cementing materials	See Table 15-G.	
9. Curbs and walls	Suitable cant strips shall be used at all vertical intersections. Adequate attachment shall be provided for both base flashing and counterflashing on all vertical surfaces. Reglets shall be provided in wall or parapets receiving metal counterflashing.	
10. Surfacing	Mineral aggregate surfaced roofs shall comply with the requirements of UBC Standard 15-1 and Table 15-F. Cap sheets shall be cemented to the roofing plies as set forth in Table 15-F.	

TABLE 15-F—BUILT-UP ROOFING CEMENTING ADHESIVE AND SURFACING APPLICATION RATES

| | MINIMUM APPLICATION RATE, MATERIAL/100 FT.[2] (9.3 m[2]) ROOF AREA | | |
| | Hot Asphalt (pounds) | Hot Coal-tar (pounds) | Cold-process Cement (gallons) |
MATERIAL TO BE ADHERED	× 0.45 for kg	× 0.45 for kg	× 3.785 for liters
Base ply or vapor retarder 1. Spot mopping 2. Solid cementing	 15 20	 15 20	 1 $1^1/_2$
Insulation 1. Solid cementing	 20	 20	 $1^1/_2$
Roofing plies (and between layers of vapor retarder) 1. Felts 2. Coated felts	 20 20	 20 20	 Not permitted $1^1/_2$
Cap sheets 1. Solid cementing	 20	 Not permitted	 $1^1/_2$

Mineral aggregate[1,2]
1. Fire-retardant roof coverings
  1.1 Gravel, 400 lb./sq. (20.1 kg/m[2])
  1.2 Slag, 300 lb./sq. (15.1 kg/m[2])
  1.3 Granules, 60 lb./sq. (3 kg/m[2])
2. Nonrated roof coverings
  2.1 Gravel, 300 lb./sq. (15.1 kg/m[2])
  2.2 Slag, 250 lb./sq. (12.6 kg/m[2])
  2.3 Granules, 60 lb./sq. (3 kg/m[2])
	
	50
	50
	—
	
	40
	40
	—
		
		60
		60
		—
		
		50
		50
		—
			
			4
			4
			3
			
			4
			4
			3

[1]Mineral aggregate shall not be used for built-up roofing membranes at roof slopes greater than 3 units vertical in 12 units horizontal (25% slope).
[2]A minimum of 50 percent of the required aggregate shall be embedded in the pour coat.

TABLE 15-G—APPLICATION OF CEMENTING MATERIALS

| | MAXIMUM SLOPE, VERTICAL UNITS PER 12 UNITS HORIZONTAL | | | | |
| | Asphalt Type | | | | |
APPLICATION	Type I	Type II	Type III	Type IV	Coal-tar Pitch
1. Insulation to deck	—	—	All	All	—
2. Felt or vapor retarder to deck	—	$^1/_2$ (4% slope) or less	3 (25% slope) or less	All	$^1/_2$ (4% slope) or less
3. Felt to felt	—	$^1/_2$ (4% slope) or less	$^1/_2$ -3 (4%-25% slope)	All	$^1/_2$ (4% slope) or less
4. Cap sheet to felt	—	—	3 (25% slope) or less	All	—
5. Gravel to felts	$^1/_2$ (4% slope) or less	$^1/_2$ (4% slope) or less	$^1/_2$ -3 (4%-25% slope)	N.P.	$^1/_2$ (4% slope) or less
6. Heating of cementing material,[1] °F Temperature at kettle[2] (maximum)	475 (246°C)	525 (274°C)	525 (274°C)	525 (274°C)	425 (218°C)
Application temperature,[3] °F	375-425 (190-218°C)	375-425 (190-218°C)	375-425 (190-218°C)	400-450 (204-232°C)	350-400 (177-204°C)

N.P.—Not permitted.

[1]Bulk tanker temperatures shall be reduced to 320°F to 350°F (160°C to 177°C) at night or during periods when no roofing will occur.
[2]Cementing material shall not be heated above a temperature that is 25°F (14°C) below its flash point.
[3]Bitumen identified with the equiviscous temperature (EVT) shall be applied at the EVT ± 25°F (14°C).

Chapter 16 is printed in its entirety in Volume 2 of the *Uniform Building Code.*
Excerpts from Chapter 16 are reprinted herein.

Excerpts from Chapter 16
STRUCTURAL DESIGN REQUIREMENTS

SECTION 1601 — SCOPE

This chapter prescribes general design requirements applicable to all structures regulated by this code.

SECTION 1602 — DEFINITIONS

The following terms are defined for use in this code:

ALLOWABLE STRESS DESIGN is a method of proportioning structural elements such that computed stresses produced in the elements by the allowable stress load combinations do not exceed specified allowable stress (also called working stress design).

BALCONY, EXTERIOR, is an exterior floor system projecting from a structure and supported by that structure, with no additional independent supports.

DEAD LOADS consist of the weight of all materials and fixed equipment incorporated into the building or other structure.

DECK is an exterior floor system supported on at least two opposing sides by an adjoining structure and/or posts, piers, or other independent supports.

FACTORED LOAD is the product of a load specified in Sections 1606 through 1611 and a load factor. See Section 1612.2 for combinations of factored loads.

LIMIT STATE is a condition in which a structure or component is judged either to be no longer useful for its intended function (serviceability limit state) or to be unsafe (strength limit state).

LIVE LOADS are those loads produced by the use and occupancy of the building or other structure and do not include dead load, construction load, or environmental loads such as wind load, snow load, rain load, earthquake load or flood load.

LOAD AND RESISTANCE FACTOR DESIGN (LRFD) is a method of proportioning structural elements using load and resistance factors such that no applicable limit state is reached when the structure is subjected to all appropriate load combinations. The term "LRFD" is used in the design of steel and wood structures.

STRENGTH DESIGN is a method of proportioning structural elements such that the computed forces produced in the elements by the factored load combinations do not exceed the factored element strength. The term "strength design" is used in the design of concrete and masonry structures.

SECTION 1604 — STANDARDS

The standards listed below are recognized standards (see Section 3504).

1. Wind Design.

    1.1   ASCE 7, Chapter 6, Minimum Design Loads for Buildings and Other Structures

    1.2   ANSI EIA/TIA 222-E, Structural Standards for Steel Antenna Towers and Antenna Supporting Structures

    1.3   ANSI/NAAMM FP1001, Guide Specifications for the Design Loads of Metal Flagpoles

SECTION 1605 — DESIGN

1605.1 General. Buildings and other structures and all portions thereof shall be designed and constructed to sustain, within the limitations specified in this code, all loads set forth in Chapter 16 and elsewhere in this code, combined in accordance with Section 1612. Design shall be in accordance with Strength Design, Load and Resistance Factor Design or Allowable Stress Design methods, as permitted by the applicable materials chapters.

> **EXCEPTION:** Unless otherwise required by the building official, buildings or portions thereof that are constructed in accordance with the conventional light-framing requirements specified in Chapter 23 of this code shall be deemed to meet the requirements of this section.

1605.2 Rationality. Any system or method of construction to be used shall be based on a rational analysis in accordance with well-established principles of mechanics. Such analysis shall result in a system that provides a complete load path capable of transferring all loads and forces from their point of origin to the load-resisting elements. The analysis shall include, but not be limited to, the provisions of Sections 1605.2.1 through 1605.2.3.

1605.2.1 Distribution of horizontal shear. The total lateral force shall be distributed to the various vertical elements of the lateral-force-resisting system in proportion to their rigidities considering the rigidity of the horizontal bracing system or diaphragm. Rigid elements that are assumed not to be part of the lateral-force-resisting system may be incorporated into buildings, provided that their effect on the action of the system is considered and provided for in the design.

Provision shall be made for the increased forces induced on resisting elements of the structural system resulting from torsion due to eccentricity between the center of application of the lateral forces and the center of rigidity of the lateral-force-resisting system. For accidental torsion requirements for seismic design, see Section 1630.6.

1605.2.2 Stability against overturning. Every structure shall be designed to resist the overturning effects caused by the lateral forces specified in this chapter. See Section 1611.6 for retaining walls, Section 1615 for wind and Section 1626 for seismic.

1605.2.3 Anchorage. Anchorage of the roof to walls and columns, and of walls and columns to foundations, shall be provided to resist the uplift and sliding forces that result from the application of the prescribed forces.

Concrete and masonry walls shall be anchored to all floors, roofs and other structural elements that provide lateral support for the wall. Such anchorage shall provide a positive direct connection capable of resisting the horizontal forces specified in this chapter but not less than the minimum forces in Section 1611.4. In addition, in Seismic Zones 3 and 4, diaphragm to wall anchorage using embedded straps shall have the straps attached to or hooked around the reinforcing steel or otherwise terminated so as to effectively transfer forces to the reinforcing steel. Walls shall be designed to resist bending between anchors where the anchor spacing exceeds 4 feet (1219 mm). Required anchors in masonry walls of hollow units or cavity walls shall be embedded in a reinforced grouted structural element of the wall. See Sections 1632, 1633.2.8 and 1633.2.9 for earthquake design requirements.

1605.3 Erection of Structural Framing. Walls and structural framing shall be erected true and plumb in accordance with the design.

SECTION 1606 — DEAD LOADS

1606.1 General. Dead loads shall be as defined in Section 1602 and this section.

1606.2 Partition Loads. Floors in office buildings and other buildings where partition locations are subject to change shall be designed to support, in addition to all other loads, a uniformly distributed dead load equal to 20 pounds per square foot (psf) (0.96 kN/m^2) of floor area.

> **EXCEPTION:** Access floor systems shall be designed to support, in addition to all other loads, a uniformly distributed dead load not less than 10 psf (0.48 kN/m^2) of floor area.

SECTION 1607 — LIVE LOADS

1607.1 General. Live loads shall be the maximum loads expected by the intended use or occupancy but in no case shall be less than the loads required by this section.

1607.2 Critical Distribution of Live Loads. Where structural members are arranged to create continuity, members shall be designed using the loading conditions, which would cause maximum shear and bending moments. This requirement may be satisfied in accordance with the provisions of Section 1607.3.2 or 1607.4.2, where applicable.

1607.3 Floor Live Loads.

1607.3.1 General. Floors shall be designed for the unit live loads as set forth in Table 16-A. These loads shall be taken as the minimum live loads in pounds per square foot of horizontal projection to be used in the design of buildings for the occupancies listed, and loads at least equal shall be assumed for uses not listed in this section but that create or accommodate similar loadings.

Where it can be determined in designing floors that the actual live load will be greater than the value shown in Table 16-A, the actual live load shall be used in the design of such buildings or portions thereof. Special provisions shall be made for machine and apparatus loads.

1607.3.2 Distribution of uniform floor loads. Where uniform floor loads are involved, consideration may be limited to full dead load on all spans in combination with full live load on adjacent spans and alternate spans.

1607.3.3 Concentrated loads. Provision shall be made in designing floors for a concentrated load, L, as set forth in Table 16-A placed upon any space 2^1/$_2$ feet (762 mm) square, wherever this load upon an otherwise unloaded floor would produce stresses greater than those caused by the uniform load required therefor.

Provision shall be made in areas where vehicles are used or stored for concentrated loads, L, consisting of two or more loads spaced 5 feet (1524 mm) nominally on center without uniform live loads. Each load shall be 40 percent of the gross weight of the maximum-size vehicle to be accommodated. Parking garages for the storage of private or pleasure-type motor vehicles with no repair or refueling shall have a floor system designed for a concentrated load of not less than 2,000 pounds (8.9 kN) acting on an area of 20 square inches (12 903 mm^2) without uniform live load. The condition of concentrated or uniform live load, combined in accordance with Section 1612.2 or 1612.3 as appropriate, producing the greatest stresses shall govern.

1607.3.4 Special loads. Provision shall be made for the special vertical and lateral loads as set forth in Table 16-B.

1607.3.5 Live loads posted. The live loads for which each floor or portion thereof of a commercial or industrial building is or has been designed shall have such design live loads conspicuously posted by the owner in that part of each story in which they apply, using durable metal signs, and it shall be unlawful to remove or deface such notices. The occupant of the building shall be responsible for keeping the actual load below the allowable limits.

1607.4 Roof Live Loads.

1607.4.1 General. Roofs shall be designed for the unit live loads, L_r, set forth in Table 16-C. The live loads shall be assumed to act vertically upon the area projected on a horizontal plane.

1607.4.2 Distribution of loads. Where uniform roof loads are involved in the design of structural members arranged to create continuity, consideration may be limited to full dead loads on all spans in combination with full roof live loads on adjacent spans and on alternate spans.

> **EXCEPTION:** Alternate span loading need not be considered where the uniform roof live load is 20 psf (0.96 kN/m^2) or more or where load combinations, including snow load, result in larger members or connections.

For those conditions where light-gage metal preformed structural sheets serve as the support and finish of roofs, roof structural members arranged to create continuity shall be considered adequate if designed for full dead loads on all spans in combination with the most critical one of the following superimposed loads:

1. Snow load in accordance with Section 1614.

2. The uniform roof live load, L_r, set forth in Table 16-C on all spans.

3. A concentrated gravity load, L_r, of 2,000 pounds (8.9 kN) placed on any span supporting a tributary area greater than 200 square feet (18.58 m^2) to create maximum stresses in the member, whenever this loading creates greater stresses than those caused by the uniform live load. The concentrated load shall be placed on the member over a length of 2^1/$_2$ feet (762 mm) along the span. The concentrated load need not be applied to more than one span simultaneously.

4. Water accumulation as prescribed in Section 1611.7.

1607.4.3 Unbalanced loading. Unbalanced loads shall be used where such loading will result in larger members or connections. Trusses and arches shall be designed to resist the stresses caused by unit live loads on one half of the span if such loading results in reverse stresses, or stresses greater in any portion than the stresses produced by the required unit live load on the entire span. For roofs whose structures are composed of a stressed shell, framed or solid, wherein stresses caused by any point loading are distributed throughout the area of the shell, the requirements for unbalanced unit live load design may be reduced 50 percent.

1607.4.4 Special roof loads. Roofs to be used for special purposes shall be designed for appropriate loads as approved by the building official.

Greenhouse roof bars, purlins and rafters shall be designed to carry a 100-pound-minimum (444.8 N) concentrated load, L_r, in addition to the uniform live load.

1607.5 Reduction of Live Loads. The design live load determined using the unit live loads as set forth in Table 16-A for floors and Table 16-C, Method 2, for roofs may be reduced on any member supporting more than 150 square feet (13.94 m^2), including flat slabs, except for floors in places of public assembly and for live loads greater than 100 psf (4.79 kN/m^2), in accordance with the following formula:

$$R = r(A - 150) \qquad (7\text{-}1)$$

For **SI:**

$$R = r(A - 13.94)$$

The reduction shall not exceed 40 percent for members receiving load from one level only, 60 percent for other members or R, as determined by the following formula:

$$R = 23.1(1 + D/L) \qquad (7\text{-}2)$$

WHERE:

A = area of floor or roof supported by the member, square feet (m^2).

D = dead load per square foot (m^2) of area supported by the member.

L = unit live load per square foot (m^2) of area supported by the member.

R = reduction in percentage.

r = rate of reduction equal to 0.08 percent for floors. See Table 16-C for roofs.

For storage loads exceeding 100 psf (4.79 kN/m^2), no reduction shall be made, except that design live loads on columns may be reduced 20 percent.

The live load reduction shall not exceed 40 percent in garages for the storage of private pleasure cars having a capacity of not more than nine passengers per vehicle.

1607.6 Alternate Floor Live Load Reduction. As an alternate to Formula (7-1), the unit live loads set forth in Table 16-A may be reduced in accordance with Formula (7-3) on any member, including flat slabs, having an influence area of 400 square feet (37.2 m^2) or more.

$$L = L_o\left(0.25 + \frac{15}{\sqrt{A_I}}\right) \qquad (7\text{-}3)$$

For **SI:**

$$L = L_o\left[0.25 + 4.57\left(\frac{1}{\sqrt{A_I}}\right)\right]$$

WHERE:

A_I = influence area, in square feet (m^2). The influence area A_I is four times the tributary area for a column, two times the tributary area for a beam, equal to the panel area for a two-way slab, and equal to the product of the span and the full flange width for a precast T-beam.

L = reduced design live load per square foot (m^2) of area supported by the member.

L_o = unreduced design live load per square foot (m^2) of area supported by the member (Table 16-A).

The reduced live load shall not be less than 50 percent of the unit live load L_o for members receiving load from one level only, nor less than 40 percent of the unit live load L_o for other members.

SECTION 1608 — SNOW LOADS

Snow loads shall be determined in accordance with Chapter 16, Division II.

SECTION 1609 — WIND LOADS

Wind loads shall be determined in accordance with Chapter 16, Division III.

SECTION 1610 — EARTHQUAKE LOADS

Earthquake loads shall be determined in accordance with Chapter 16, Division IV.

SECTION 1611 — OTHER MINIMUM LOADS

1611.1 General. In addition to the other design loads specified in this chapter, structures shall be designed to resist the loads specified in this section and the special loads set forth in Table 16-B.

1611.2 Other Loads. Buildings and other structures and portions thereof shall be designed to resist all loads due to applicable fluid pressures, F, lateral soil pressures, H, ponding loads, P, and self-straining forces, T. See Section 1611.7 for ponding loads for roofs.

1611.3 Impact Loads. Impact loads shall be included in the design of any structure where impact loads occur.

1611.4 Anchorage of Concrete and Masonry Walls. Concrete and masonry walls shall be anchored as required by Section 1605.2.3. Such anchorage shall be capable of resisting the load combinations of Section 1612.2 or 1612.3 using the greater of the wind or earthquake loads required by this chapter or a minimum horizontal force of 280 pounds per linear foot (4.09 kN/m) of wall, substituted for E.

1611.5 Interior Wall Loads. Interior walls, permanent partitions and temporary partitions that exceed 6 feet (1829 mm) in height shall be designed to resist all loads to which they are subjected but not less than a load, L, of 5 psf (0.24 kN/m^2) applied perpendicular to the walls. The 5 psf (0.24 kN/m^2) load need not be applied simultaneously with wind or seismic loads. The deflection of such walls under a load of 5 psf (0.24 kN/m^2) shall not exceed $1/240$ of the span for walls with brittle finishes and $1/120$ of the span for walls with flexible finishes. See Table 16-O for earthquake design requirements where such requirements are more restrictive.

> **EXCEPTION:** Flexible, folding or portable partitions are not required to meet the load and deflection criteria but must be anchored to the supporting structure to meet the provisions of this code.

1611.6 Retaining Walls. Retaining walls shall be designed to resist loads due to the lateral pressure of retained material in accordance with accepted engineering practice. Walls retaining drained soil, where the surface of the retained soil is level, shall be designed for a load, H, equivalent to that exerted by a fluid weighing not less than 30 psf per foot of depth (4.71 kN/m^2/m) and having a depth equal to that of the retained soil. Any surcharge shall be in addition to the equivalent fluid pressure.

Retaining walls shall be designed to resist sliding by at least 1.5 times the lateral force and overturning by at least 1.5 times the overturning moment, using allowable stress design loads.

1611.7 Water Accumulation. All roofs shall be designed with sufficient slope or camber to ensure adequate drainage after the long-term deflection from dead load or shall be designed to resist ponding load, P, combined in accordance with Section 1612.2 or 1612.3. Ponding load shall include water accumulation from any source, including snow, due to deflection. See Section 1506 and Table 16-C, Footnote 3, for drainage slope. See Section 1615 for deflection criteria.

1611.8 Hydrostatic Uplift. All foundations, slabs and other footings subjected to water pressure shall be designed to resist a uniformly distributed uplift load, F, equal to the full hydrostatic pressure.

1611.9 Flood-resistant Construction. For flood-resistant construction requirements, where specifically adopted, see Appendix Chapter 31, Division I.

1611.10 Heliport and Helistop Landing Areas. In addition to other design requirements of this chapter, heliport and helistop landing or touchdown areas shall be designed for the following loads, combined in accordance with Section 1612.2 or 1612.3:

1. Dead load plus actual weight of the helicopter.

2. Dead load plus a single concentrated impact load, L, covering 1 square foot (0.093 m^2) of 0.75 times the fully loaded weight of the helicopter if it is equipped with hydraulic-type shock absorbers, or 1.5 times the fully loaded weight of the helicopter if it is equipped with a rigid or skid-type landing gear.

3. The dead load plus a uniform live load, L, of 100 psf (4.8 kN/m^2). The required live load may be reduced in accordance with Section 1607.5 or 1607.6.

1611.11 Prefabricated Construction.

1611.11.1 Connections. Every device used to connect prefabricated assemblies shall be designed as required by this code and shall be capable of developing the strength of the members connected, except in the case of members forming part of a structural frame designed as specified in this chapter. Connections shall be capable of withstanding uplift forces as specified in this chapter.

1611.11.2 Pipes and conduit. In structural design, due allowance shall be made for any material to be removed for the installation of pipes, conduits or other equipment.

1611.11.3 Tests and inspections. See Section 1704 for requirements for tests and inspections of prefabricated construction.

TABLE 16-A—UNIFORM AND CONCENTRATED LOADS

USE OR OCCUPANCY		UNIFORM LOAD[1] (psf)	CONCENTRATED LOAD (pounds)
Category	Description	× 0.0479 for kN/m²	× 0.004 48 for kN
1. Access floor systems	Office use	50	2,000[2]
	Computer use	100	2,000[2]
2. Armories		150	0
3. Assembly areas[3] and auditoriums and balconies therewith	Fixed seating areas	50	0
	Movable seating and other areas	100	0
	Stage areas and enclosed platforms	125	0
4. Cornices and marquees		60[4]	0
5. Exit facilities[5]		100	0[6]
6. Garages	General storage and/or repair	100	7
	Private or pleasure-type motor vehicle storage	50	7
7. Hospitals	Wards and rooms	40	1,000[2]
8. Libraries	Reading rooms	60	1,000[2]
	Stack rooms	125	1,500[2]
9. Manufacturing	Light	75	2,000[2]
	Heavy	125	3,000[2]
10. Offices		50	2,000[2]
11. Printing plants	Press rooms	150	2,500[2]
	Composing and linotype rooms	100	2,000[2]
12. Residential[8]	Basic floor area	40	0[6]
	Exterior balconies	60[4]	0
	Decks	40[4]	0
	Storage	40	0
13. Restrooms[9]			
14. Reviewing stands, grandstands, bleachers, and folding and telescoping seating		100	0
15. Roof decks	Same as area served or for the type of occupancy accommodated		
16. Schools	Classrooms	40	1,000[2]
17. Sidewalks and driveways	Public access	250	7
18. Storage	Light	125	
	Heavy	250	
19. Stores		100	3,000[2]
20. Pedestrian bridges and walkways		100	

[1]See Section 1607 for live load reductions.

[2]See Section 1607.3.3, first paragraph, for area of load application.

[3]Assembly areas include such occupancies as dance halls, drill rooms, gymnasiums, playgrounds, plazas, terraces and similar occupancies that are generally accessible to the public.

[4]When snow loads occur that are in excess of the design conditions, the structure shall be designed to support the loads due to the increased loads caused by drift buildup or a greater snow design as determined by the building official. See Section 1614. For special-purpose roofs, see Section 1607.4.4.

[5]Exit facilities shall include such uses as corridors serving an occupant load of 10 or more persons, exterior exit balconies, stairways, fire escapes and similar uses.

[6]Individual stair treads shall be designed to support a 300-pound (1.33 kN) concentrated load placed in a position that would cause maximum stress. Stair stringers may be designed for the uniform load set forth in the table.

[7]See Section 1607.3.3, second paragraph, for concentrated loads. See Table 16-B for vehicle barriers.

[8]Residential occupancies include private dwellings, apartments and hotel guest rooms.

[9]Restroom loads shall not be less than the load for the occupancy with which they are associated, but need not exceed 50 pounds per square foot (2.4 kN/m²).

TABLE 16-B—SPECIAL LOADS[1]

USE		VERTICAL LOAD	LATERAL LOAD
Category	**Description**	(pounds per square foot unless otherwise noted)	
		$\times$ 0.0479 for kN/m^2	
1. Construction, public access at site (live load)	Walkway, see Section 3303.6	150	
	Canopy, see Section 3303.7	150	
2. Grandstands, reviewing stands, bleachers, and folding and telescoping seating (live load)	Seats and footboards	120[2]	See Footnote 3
3. Stage accessories (live load)	Catwalks	40	
	Followspot, projection and control rooms	50	
4. Ceiling framing (live load)	Over stages	20	
	All uses except over stages	10[4]	
5. Partitions and interior walls, see Sec. 1611.5 (live load)			5
6. Elevators and dumbwaiters (dead and live loads)		$2 \times$ total loads[5]	
7. Mechanical and electrical equipment (dead load)		Total loads	
8. Cranes (dead and live loads)	Total load including impact increase	$1.25 \times$ total load[6]	$0.10 \times$ total load[7]
9. Balcony railings and guardrails	Exit facilities serving an occupant load greater than 50		50[8]
	Other than exit facilities		20[8]
	Components		25[9]
10. Vehicle barriers	See Section 311.2.3.5		6,000[10]
11. Handrails		See Footnote 11	See Footnote 11
12. Storage racks	Over 8 feet (2438 mm) high	Total loads[12]	See Table 16-O
13. Fire sprinkler structural support		250 pounds (1112 N) plus weight of water-filled pipe[13]	See Table 16-O
14. Explosion exposure	Hazardous occupancies, see Section 307.10		

[1]The tabulated loads are minimum loads. Where other vertical loads required by this code or required by the design would cause greater stresses, they shall be used.

[2]Pounds per lineal foot ($\times$ 14.6 for N/m).

[3]Lateral sway bracing loads of 24 pounds per foot (350 N/m) parallel and 10 pounds per foot (145.9 N/m) perpendicular to seat and footboards.

[4]Does not apply to ceilings that have sufficient total access from below, such that access is not required within the space above the ceiling. Does not apply to ceilings if the attic areas above the ceiling are not provided with access. This live load need not be considered as acting simultaneously with other live loads imposed upon the ceiling framing or its supporting structure.

[5]Where Appendix Chapter 30 has been adopted, see reference standard cited therein for additional design requirements.

[6]The impact factors included are for cranes with steel wheels riding on steel rails. They may be modified if substantiating technical data acceptable to the building official is submitted. Live loads on crane support girders and their connections shall be taken as the maximum crane wheel loads. For pendant-operated traveling crane support girders and their connections, the impact factors shall be 1.10.

[7]This applies in the direction parallel to the runway rails (longitudinal). The factor for forces perpendicular to the rail is 0.20 $\times$ the transverse traveling loads (trolley, cab, hooks and lifted loads). Forces shall be applied at top of rail and may be distributed among rails of multiple rail cranes and shall be distributed with due regard for lateral stiffness of the structures supporting these rails.

[8]A load per lineal foot ($\times$ 14.6 for N/m) to be applied horizontally at right angles to the top rail.

[9]Intermediate rails, panel fillers and their connections shall be capable of withstanding a load of 25 pounds per square foot (1.2 kN/m^2) applied horizontally at right angles over the entire tributary area, including openings and spaces between rails. Reactions due to this loading need not be combined with those of Footnote 8.

[10]A horizontal load in pounds (N) applied at right angles to the vehicle barrier at a height of 18 inches (457 mm) above the parking surface. The force may be distributed over a 1-foot-square (304.8-millimeter-square) area.

[11]The mounting of handrails shall be such that the completed handrail and supporting structure are capable of withstanding a load of at least 200 pounds (890 N) applied in any direction at any point on the rail. These loads shall not be assumed to act cumulatively with Item 9.

[12]Vertical members of storage racks shall be protected from impact forces of operating equipment, or racks shall be designed so that failure of one vertical member will not cause collapse of more than the bay or bays directly supported by that member.

[13]The 250-pound (1.11 kN) load is to be applied to any single fire sprinkler support point but not simultaneously to all support joints.

TABLE 16-C—MINIMUM ROOF LIVE LOADS[1]

	METHOD 1			METHOD 2		
	Tributary Loaded Area in Square Feet for Any Structural Member					
	× 0.0929 for m²					
	0 to 200	201 to 600	Over 600			
	Uniform Load (psf)			Uniform Load[2] (psf)	Rate of Reduction *r* (percentage)	Maximum Reduction *R* (percentage)
ROOF SLOPE	× 0.0479 for kN/m²					
1. Flat[3] or rise less than 4 units vertical in 12 units horizontal (33.3% slope). Arch or dome with rise less than one eighth of span	20	16	12	20	.08	40
2. Rise 4 units vertical to less than 12 units vertical in 12 units horizontal (33% to less than 100% slope). Arch or dome with rise one eighth of span to less than three eighths of span	16	14	12	16	.06	25
3. Rise 12 units vertical in 12 units horizontal (100% slope) and greater. Arch or dome with rise three eighths of span or greater	12	12	12	12	No reductions permitted	
4. Awnings except cloth covered[4]	5	5	5	5		
5. Greenhouses, lath houses and agricultural buildings[5]	10	10	10	10		

[1]Where snow loads occur, the roof structure shall be designed for such loads as determined by the building official. See Section 1614. For special-purpose roofs, see Section 1607.4.4.

[2]See Sections 1607.5 and 1607.6 for live load reductions. The rate of reduction *r* in Section 1607.5 Formula (7-1) shall be as indicated in the table. The maximum reduction *R* shall not exceed the value indicated in the table.

[3]A flat roof is any roof with a slope of less than $1/4$ unit vertical in 12 units horizontal (2% slope). The live load for flat roofs is in addition to the ponding load required by Section 1611.7.

[4]As defined in Section 3206.

[5]See Section 1607.4.4 for concentrated load requirements for greenhouse roof members.

Chapter 17 is printed in its entirety in Volume 2 of the *Uniform Building Code*.
Excerpts from Chapter 17 are reprinted herein.

Excerpts from Chapter 17
STRUCTURAL TESTS AND INSPECTIONS

SECTION 1701 — SPECIAL INSPECTIONS

1701.1 General. In addition to the inspections required by Section 108, the owner or the engineer or architect of record acting as the owner's agent shall employ one or more special inspectors who shall provide inspections during construction on the types of work listed under Section 1701.5.

> **EXCEPTION:** The building official may waive the requirement for the employment of a special inspector if the construction is of a minor nature.

1701.2 Special Inspector. The special inspector shall be a qualified person who shall demonstrate competence, to the satisfaction of the building official, for inspection of the particular type of construction or operation requiring special inspection.

1701.3 Duties and Responsibilities of the Special Inspector. The special inspector shall observe the work assigned for conformance to the approved design drawings and specifications.

The special inspector shall furnish inspection reports to the building official, the engineer or architect of record, and other designated persons. All discrepancies shall be brought to the immediate attention of the contractor for correction, then, if uncorrected, to the proper design authority and to the building official.

The special inspector shall submit a final signed report stating whether the work requiring special inspection was, to the best of the inspector's knowledge, in conformance to the approved plans and specifications and the applicable workmanship provisions of this code.

1701.4 Standards of Quality. The standards listed below labeled a "UBC Standard" are also listed in Chapter 35, Part II, and are part of this code. The other standards listed below are recognized standards. (See Sections 3503 and 3504.)

1. Concrete.

ASTM C 94, Ready-mixed Concrete

2. Connections.

Specification for Structural Joints Using ASTM A 325 or A 490 Bolts-Load and Resistance Factor Design, Research Council of Structural Connections, Section 1701.5, Item 6.

Specification for Structural Joints Using ASTM A 325 or A 490 Bolts-Allowable Stress Design, Research Council of Structural Connections, Section 1701.5, Item 6.

3. Spray-applied Fire-resistive Materials.

UBC Standard 7-6, Thickness and Density Determination for Spray-applied Fire-resistive Materials

1701.5 Types of Work. Except as provided in Section 1701.1, the types of work listed below shall be inspected by a special inspector.

1. Concrete. During the taking of test specimens and placing of reinforced concrete. See Item 12 for shotcrete.

> **EXCEPTIONS:** 1. Concrete for foundations conforming to minimum requirements of Table 18-I-C or for Group R, Division 3 or Group U, Division 1 Occupancies, provided the building official finds that a special hazard does not exist.

> 2. For foundation concrete, other than cast-in-place drilled piles or caissons, where the structural design is based on an f'_c no greater than 2,500 pounds per square inch (psi) (17.2 MPa).

> 3. Nonstructural slabs on grade, including prestressed slabs on grade when effective prestress in concrete is less than 150 psi (1.03 MPa).

> 4. Site work concrete fully supported on earth and concrete where no special hazard exists.

2. Bolts installed in concrete. Prior to and during the placement of concrete around bolts when stress increases permitted by Footnote 5 of Table 19-D or Section 1923 are utilized.

3. Special moment-resisting concrete frame. For moment frames resisting design seismic load in structures within Seismic Zones 3 and 4, the special inspector shall provide reports to the person responsible for the structural design and shall provide continuous inspection of the placement of the reinforcement and concrete.

4. Reinforcing steel and prestressing steel tendons.

    4.1 During all stressing and grouting of tendons in prestressed concrete.

    4.2 During placing of reinforcing steel and prestressing tendons for all concrete required to have special inspection by Item 1.

> **EXCEPTION:** The special inspector need not be present continuously during placing of reinforcing steel and prestressing tendons, provided the special inspector has inspected for conformance to the approved plans prior to the closing of forms or the delivery of concrete to the jobsite.

5. Structural welding.

    5.1 **General.** During the welding of any member or connection that is designed to resist loads and forces required by this code.

> **EXCEPTIONS:** 1. Welding done in an approved fabricator's shop in accordance with Section 1701.7.

> 2. The special inspector need not be continuously present during welding of the following items, provided the materials, qualifications of welding procedures and welders are verified prior to the start of work; periodic inspections are made of work in progress; and a visual inspection of all welds is made prior to completion or prior to shipment of shop welding:

>   2.1 Single-pass fillet welds not exceeding $5/16$ inch (7.9 mm) in size.

>   2.2 Floor and roof deck welding.

>   2.3 Welded studs when used for structural diaphragm or composite systems.

>   2.4 Welded sheet steel for cold-formed steel framing members such as studs and joists.

>   2.5 Welding of stairs and railing systems.

    5.2 **Special moment-resisting steel frames.** During the welding of special moment-resisting steel frames. In addition to Item 5.1 requirements, nondestructive testing as required by Section 1703 of this code.

    5.3 **Welding of reinforcing steel.** During the welding of reinforcing steel.

> **EXCEPTION:** The special inspector need not be continuously present during the welding of ASTM A 706 reinforcing steel not larger

than No. 5 bars used for embedments, provided the materials, qualifications of welding procedures and welders are verified prior to the start of work; periodic inspections are made of work in progress; and a visual inspection of all welds is made prior to completion or prior to shipment of shop welding.

6. **High-strength bolting.** The inspection of high-strength A 325 and A 490 bolts shall be in accordance with approved nationally recognized standards and the requirements of this section.

While the work is in progress, the special inspector shall determine that the requirements for bolts, nuts, washers and paint; bolted parts; and installation and tightening in such standards are met. Such inspections may be performed on a periodic basis in accordance with the requirements of Section 1701.6. The special inspector shall observe the calibration procedures when such procedures are required by the plans or specifications and shall monitor the installation of bolts to determine that all plies of connected materials have been drawn together and that the selected procedure is properly used to tighten all bolts.

7. **Structural masonry.**

7.1 For masonry, other than fully grouted open-end hollow-unit masonry, during preparation and taking of any required prisms or test specimens, placing of all masonry units, placement of reinforcement, inspection of grout space, immediately prior to closing of cleanouts, and during all grouting operations.

EXCEPTION: For hollow-unit masonry where the f'_m is no more than 1,500 psi (10.34 MPa) for concrete units or 2,600 psi (17.93 MPa) for clay units, special inspection may be performed as required for fully grouted open-end hollow-unit masonry specified in Item 7.2.

7.2 For fully grouted open-end hollow-unit masonry during preparation and taking of any required prisms or test specimens, at the start of laying units, after the placement of reinforcing steel, grout space prior to each grouting operation, and during all grouting operations.

EXCEPTION: Special inspection as required in Items 7.1 and 7.2 need not be provided when design stresses have been adjusted as specified in Chapter 21 to permit noncontinuous inspection.

8. **Reinforced gypsum concrete.** When cast-in-place Class B gypsum concrete is being mixed and placed.

9. **Insulating concrete fill.** During the application of insulating concrete fill when used as part of a structural system.

EXCEPTION: The special inspections may be limited to an initial inspection to check the deck surface and placement of reinforcing. The special inspector shall supervise the preparation of compression test specimens during this initial inspection.

10. **Spray-applied fire-resistive materials.** As required by UBC Standard 7-6.

11. **Piling, drilled piers and caissons.** During driving and testing of piles and construction of cast-in-place drilled piles or caissons. See Items 1 and 4 for concrete and reinforcing steel inspection.

12. **Shotcrete.** During the taking of test specimens and placing of all shotcrete and as required by Sections 1924.10 and 1924.11.

EXCEPTION: Shotcrete work fully supported on earth, minor repairs and when, in the opinion of the building official, no special hazard exists.

13. **Special grading, excavation and filling.** During earth-work excavations, grading and filling operations inspection to satisfy requirements of Chapter 18 and Appendix Chapter 33.

14. **Smoke-control system.**

14.1 During erection of ductwork and prior to concealment for the purposes of leakage testing and recording of device location.

14.2 Prior to occupancy and after sufficient completion for the purposes of pressure difference testing, flow measurements, and detection and control verification.

15. **Special cases.** Work that, in the opinion of the building official, involves unusual hazards or conditions.

1701.6 Continuous and Periodic Special Inspection.

1701.6.1 Continuous special inspection. Continuous special inspection means that the special inspector is on the site at all times observing the work requiring special inspection.

1701.6.2 Periodic special inspection. Some inspections may be made on a periodic basis and satisfy the requirements of continuous inspection, provided this periodic scheduled inspection is performed as outlined in the project plans and specifications and approved by the building official.

1701.7 Approved Fabricators. Special inspections required by this section and elsewhere in this code are not required where the work is done on the premises of a fabricator registered and approved by the building official to perform such work without special inspection. The certificate of registration shall be subject to revocation by the building official if it is found that any work done pursuant to the approval is in violation of this code. The approved fabricator shall submit a certificate of compliance that the work was performed in accordance with the approved plans and specifications to the building official and to the engineer or architect of record. The approved fabricator's qualifications shall be contingent on compliance with the following:

1. The fabricator has developed and submitted a detailed fabrication procedural manual reflecting key quality control procedures that will provide a basis for inspection control of workmanship and the fabricator plant.

2. Verification of the fabricator's quality control capabilities, plant and personnel as outlined in the fabrication procedural manual shall be by an approved inspection or quality control agency.

3. Periodic plant inspections shall be conducted by an approved inspection or quality control agency to monitor the effectiveness of the quality control program.

4. It shall be the responsibility of the inspection or quality control agency to notify the approving authority in writing of any change to the procedural manual. Any fabricator approval may be revoked for just cause. Reapproval of the fabricator shall be contingent on compliance with quality control procedures during the past year.

SECTION 1702 — STRUCTURAL OBSERVATION

Structural observation shall be provided in Seismic Zone 3 or 4 when one of the following conditions exists:

1. The structure is defined in Table 16-K as Occupancy Category 1, 2 or 3,

2. The structure is required to comply with Section 403,

3. The structure is in Seismic Zone 4, N_a as set forth in Table 16-S is greater than one, and a lateral design is required for the entire structure,

EXCEPTION: One- and two-story Group R, Division 3 and Group U Occupancies and one- and two-story Groups B, F, M and S Occupancies.

4. When so designated by the architect or engineer of record, or

5. When such observation is specifically required by the building official.

The owner shall employ the engineer or architect responsible for the structural design, or another engineer or architect designated by the engineer or architect responsible for the structural design, to perform structural observation as defined in Section 220. Observed deficiencies shall be reported in writing to the owner's representative, special inspector, contractor and the building official. The structural observer shall submit to the building official a written statement that the site visits have been made and identifying any reported deficiencies that, to the best of the structural observer's knowledge, have not been resolved.

SECTION 1703 — NONDESTRUCTIVE TESTING

In Seismic Zones 3 and 4, welded, fully restrained connections between the primary members of ordinary moment frames and special moment-resisting frames shall be tested by nondestructive methods for compliance with approved standards and job specifications. This testing shall be a part of the special inspection requirements of Section 1701.5. A program for this testing shall be established by the person responsible for structural design and as shown on plans and specifications.

As a minimum, this program shall include the following:

1. All complete penetration groove welds contained in joints and splices shall be tested 100 percent either by ultrasonic testing or by radiography.

> **EXCEPTIONS:** 1. When approved, the nondestructive testing rate for an individual welder or welding operator may be reduced to 25 percent, provided the reject rate is demonstrated to be 5 percent or less of the welds tested for the welder or welding operator. A sampling of at least 40 completed welds for a job shall be made for such reduction evaluation. Reject rate is defined as the number of welds containing rejectable defects divided by the number of welds completed. For evaluating the reject rate of continuous welds over 3 feet (914 mm) in length where the effective throat thickness is 1 inch (25 mm) or less, each 12-inch increment (305 mm) or fraction thereof shall be considered as one weld. For evaluating the reject rate on continuous welds over 3 feet (914 mm) in length where the effective throat thickness is greater than 1 inch (25 mm), each 6 inches (152 mm) of length or fraction thereof shall be considered one weld.
>
> 2. For complete penetration groove welds on materials less than $5/16$ inch (7.9 mm) thick, nondestructive testing is not required; for this welding, continuous inspection is required.
>
> 3. When approved by the building official and outlined in the project plans and specifications, this nondestructive ultrasonic testing may be performed in the shop of an approved fabricator utilizing qualified test techniques in the employment of the fabricator.

2. Partial penetration groove welds when used in column splices shall be tested either by ultrasonic testing or radiography when required by the plans and specifications. For partial penetration groove welds when used in column splices, with an effective throat less than $3/4$ inch (19.1 mm) thick, nondestructive testing is not required; for this welding, continuous special inspection is required.

3. Base metal thicker than $1^1/2$ inches (38 mm), when subjected to through-thickness weld shrinkage strains, shall be ultrasonically inspected for discontinuities directly behind such welds after joint completion.

Any material discontinuities shall be accepted or rejected on the basis of the defect rating in accordance with the (larger reflector) criteria of approved national standards.

SECTION 1704 — PREFABRICATED CONSTRUCTION

1704.1 General.

1704.1.1 Purpose. The purpose of this section is to regulate materials and establish methods of safe construction where any structure or portion thereof is wholly or partially prefabricated.

1704.1.2 Scope. Unless otherwise specifically stated in this section, all prefabricated construction and all materials used therein shall conform to all the requirements of this code. (See Section 104.2.8.)

1704.1.3 Definition.

PREFABRICATED ASSEMBLY is a structural unit, the integral parts of which have been built up or assembled prior to incorporation in the building.

1704.2 Tests of Materials. Every approval of a material not specifically mentioned in this code shall incorporate as a proviso the kind and number of tests to be made during prefabrication.

1704.3 Tests of Assemblies. The building official may require special tests to be made on assemblies to determine their durability and weather resistance.

1704.4 Connections. See Section 1611.11.1 for design requirements of connections for prefabricated assemblies.

1704.5 Pipes and Conduits. See Section 1611.11.2 for design requirements for removal of material for pipes, conduit and other equipment.

1704.6 Certificate and Inspection.

1704.6.1 Materials. Materials and the assembly thereof shall be inspected to determine compliance with this code. Every material shall be graded, marked or labeled where required elsewhere in this code.

1704.6.2 Certificate. A certificate of approval shall be furnished with every prefabricated assembly, except where the assembly is readily accessible to inspection at the site. The certificate of approval shall certify that the assembly in question has been inspected and meets all the requirements of this code. When mechanical equipment is installed so that it cannot be inspected at the site, the certificate of approval shall certify that such equipment complies with the laws applying thereto.

1704.6.3 Certifying agency. To be acceptable under this code, every certificate of approval shall be made by an approved agency.

1704.6.4 Field erection. Placement of prefabricated assemblies at the building site shall be inspected by the building official to determine compliance with this code.

1704.6.5 Continuous inspection. If continuous inspection is required for certain materials where construction takes place on the site, it shall also be required where the same materials are used in prefabricated construction.

> **EXCEPTION:** Continuous inspection will not be required during prefabrication if the approved agency certifies to the construction and furnishes evidence of compliance.

Chapter 18 is printed in its entirety in Volume 2 of the *Uniform Building Code*.
Excerpts from Chapter 18 are reprinted herein.

Excerpts from Chapter 18
FOUNDATIONS AND RETAINING WALLS
Division I—GENERAL

SECTION 1801 — SCOPE

1801.1 General. This chapter sets forth requirements for excavation and fills for any building or structure and for foundations and retaining structures.

Reference is made to Appendix Chapter 33 for requirements governing excavation, grading and earthwork construction, including fills and embankments.

1801.2 Standards of Quality. The standards listed below labeled a "UBC Standard" are also listed in Chapter 35, Part II, and are part of this code.

1. **Testing.**

    1.1 UBC Standard 18-1, Soils Classification

    1.2 UBC Standard 18-2, Expansion Index Test

SECTION 1802 — QUALITY AND DESIGN

The quality and design of materials used structurally in excavations, footings and foundations shall conform to the requirements specified in Chapters 16, 19, 21, 22 and 23.

Excavations and fills shall comply with Chapter 33.

Allowable bearing pressures, allowable stresses and design formulas provided in this chapter shall be used with the allowable stress design load combinations specified in Section 1612.3.

SECTION 1803 — SOIL CLASSIFICATION— EXPANSIVE SOIL

1803.1 General. For the purposes of this chapter, the definition and classification of soil materials for use in Table 18-I-A shall be according to UBC Standard 18-1.

1803.2 Expansive Soil. When the expansive characteristics of a soil are to be determined, the procedures shall be in accordance with UBC Standard 18-2 and the soil shall be classified according to Table 18-I-B. Foundations for structures resting on soils with an expansion index greater than 20, as determined by UBC Standard 18-2, shall require special design consideration. If the soil expansion index varies with depth, the variation is to be included in the engineering analysis of the expansive soil effect upon the structure.

SECTION 1804 — FOUNDATION INVESTIGATION

1804.1 General. The classification of the soil at each building site shall be determined when required by the building official. The building official may require that this determination be made by an engineer or architect licensed by the state to practice as such.

1804.2 Investigation. The classification shall be based on observation and any necessary tests of the materials disclosed by borings or excavations made in appropriate locations. Additional studies may be necessary to evaluate soil strength, the effect of moisture variation on soil-bearing capacity, compressibility, liquefaction and expansiveness.

In Seismic Zones 3 and 4, when required by the building official, the potential for seismically induced soil liquefaction and soil instability shall be evaluated as described in Section 1804.5.

> **EXCEPTIONS:** 1. The building official may waive this evaluation upon receipt of written opinion of a qualified geotechnical engineer or geologist that liquefaction is not probable.
>
> 2. A detached, single-story dwelling of Group R, Division 3 Occupancy with or without attached garages.
>
> 3. Group U, Division 1 Occupancies.
>
> 4. Fences.

1804.3 Reports. The soil classification and design-bearing capacity shall be shown on the plans, unless the foundation conforms to Table 18-I-C. The building official may require submission of a written report of the investigation, which shall include, but need not be limited to, the following information:

1. A plot showing the location of all test borings and/or excavations.

2. Descriptions and classifications of the materials encountered.

3. Elevation of the water table, if encountered.

4. Recommendations for foundation type and design criteria, including bearing capacity, provisions to mitigate the effects of expansive soils, provisions to mitigate the effects of liquefaction and soil strength, and the effects of adjacent loads.

5. Expected total and differential settlement.

1804.4 Expansive Soils. When expansive soils are present, the building official may require that special provisions be made in the foundation design and construction to safeguard against damage due to this expansiveness. The building official may require a special investigation and report to provide these design and construction criteria.

1804.5 Liquefaction Potential and Soil Strength Loss. When required by Section 1804.2, the potential for soil liquefaction and soil strength loss during earthquakes shall be evaluated during the geotechnical investigation. The geotechnical report shall assess potential consequences of any liquefaction and soil strength loss, including estimation of differential settlement, lateral movement or reduction in foundation soil-bearing capacity, and discuss mitigating measures. Such measures shall be given consideration in the design of the building and may include, but are not limited to, ground stabilization, selection of appropriate foundation type and depths, selection of appropriate structural systems to accommodate anticipated displacements, or any combination of these measures.

The potential for liquefaction and soil strength loss shall be evaluated for a site peak ground acceleration that, as a minimum, conforms to the probability of exceedance specified in Section 1631.2. Peak ground acceleration may be determined based on a site-specific study taking into account soil amplification effects. In the absence of such a study, peak ground acceleration may be assumed equal to the seismic zone factor in Table 16-I.

1804.6 Adjacent Loads. Where footings are placed at varying elevations, the effect of adjacent loads shall be included in the foundation design.

1804.7 Drainage. Provisions shall be made for the control and drainage of surface water around buildings. (See also Section 1806.5.5.)

SECTION 1805 — ALLOWABLE FOUNDATION AND LATERAL PRESSURES

The allowable foundation and lateral pressures shall not exceed the values set forth in Table 18-I-A unless data to substantiate the use of higher values are submitted. Table 18-I-A may be used for design of foundations on rock or nonexpansive soil for Type II One-hour, Type II-N and Type V buildings that do not exceed three stories in height or for structures that have continuous footings having a load of less than 2,000 pounds per lineal foot (29.2 kN/m) and isolated footings with loads of less than 50,000 pounds (222.4 kN).

Allowable bearing pressures provided in Table 18-I-A shall be used with the allowable stress design load combinations specified in Section 1612.3.

SECTION 1806 — FOOTINGS

1806.1 General. Footings and foundations shall be constructed of masonry, concrete or treated wood in conformance with Division II and shall extend below the frost line. Footings of concrete and masonry shall be of solid material. Foundations supporting wood shall extend at least 6 inches (152 mm) above the adjacent finish grade. Footings shall have a minimum depth as indicated in Table 18-I-C, unless another depth is recommended by a foundation investigation.

The provisions of this section do not apply to building and foundation systems in those areas subject to scour and water pressure by wind and wave action. Buildings and foundations subject to such loads shall be designed in accordance with approved national standards. See Section 3302 for subsoil preparation and wood form removal.

1806.2 Footing Design. Except for special provisions of Section 1808 covering the design of piles, all portions of footings shall be designed in accordance with the structural provisions of this code and shall be designed to minimize differential settlement when necessary and the effects of expansive soils when present.

Slab-on-grade and mat-type footings for buildings located on expansive soils may be designed in accordance with the provisions of Division III or such other engineering design based on geotechnical recommendation as approved by the building official.

1806.3 Bearing Walls. Bearing walls shall be supported on masonry or concrete foundations or piles or other approved foundation system that shall be of sufficient size to support all loads. Where a design is not provided, the minimum foundation requirements for stud bearing walls shall be as set forth in Table 18-I-C, unless expansive soils of a severity to cause differential movement are known to exist.

> **EXCEPTIONS:** 1. A one-story wood- or metal-frame building not used for human occupancy and not over 400 square feet (37.2 m²) in floor area may be constructed with walls supported on a wood foundation plate when approved by the building official.
>
> 2. The support of buildings by posts embedded in earth shall be designed as specified in Section 1806.8. Wood posts or poles embedded

in earth shall be pressure treated with an approved preservative. Steel posts or poles shall be protected as specified in Section 1807.9.

1806.4 Stepped Foundations. Foundations for all buildings where the surface of the ground slopes more than 1 unit vertical in 10 units horizontal (10% slope) shall be level or shall be stepped so that both top and bottom of such foundation are level.

1806.5 Footings on or Adjacent to Slopes.

1806.5.1 Scope. The placement of buildings and structures on or adjacent to slopes steeper than 1 unit vertical in 3 units horizontal (33.3% slope) shall be in accordance with this section.

1806.5.2 Building clearance from ascending slopes. In general, buildings below slopes shall be set a sufficient distance from the slope to provide protection from slope drainage, erosion and shallow failures. Except as provided for in Section 1806.5.6 and Figure 18-I-1, the following criteria will be assumed to provide this protection. Where the existing slope is steeper than 1 unit vertical in 1 unit horizontal (100% slope), the toe of the slope shall be assumed to be at the intersection of a horizontal plane drawn from the top of the foundation and a plane drawn tangent to the slope at an angle of 45 degrees to the horizontal. Where a retaining wall is constructed at the toe of the slope, the height of the slope shall be measured from the top of the wall to the top of the slope.

1806.5.3 Footing setback from descending slope surface. Footing on or adjacent to slope surfaces shall be founded in firm material with an embedment and setback from the slope surface sufficient to provide vertical and lateral support for the footing without detrimental settlement. Except as provided for in Section 1806.5.6 and Figure 18-I-1, the following setback is deemed adequate to meet the criteria. Where the slope is steeper than 1 unit vertical in 1 unit horizontal (100% slope), the required setback shall be measured from an imaginary plane 45 degrees to the horizontal, projected upward from the toe of the slope.

1806.5.4 Pools. The setback between pools regulated by this code and slopes shall be equal to one half the building footing setback distance required by this section. That portion of the pool wall within a horizontal distance of 7 feet (2134 mm) from the top of the slope shall be capable of supporting the water in the pool without soil support.

1806.5.5 Foundation elevation. On graded sites, the top of any exterior foundation shall extend above the elevation of the street gutter at point of discharge or the inlet of an approved drainage device a minimum of 12 inches (305 mm) plus 2 percent. The building official may approve alternate elevations, provided it can be demonstrated that required drainage to the point of discharge and away from the structure is provided at all locations on the site.

1806.5.6 Alternate setback and clearance. The building official may approve alternate setbacks and clearances. The building official may require an investigation and recommendation of a qualified engineer to demonstrate that the intent of this section has been satisfied. Such an investigation shall include consideration of material, height of slope, slope gradient, load intensity and erosion characteristics of slope material.

1806.6 Foundation Plates or Sills. Wood plates or sills shall be bolted to the foundation or foundation wall. Steel bolts with a minimum nominal diameter of $^1/_2$ inch (12.7 mm) shall be used in Seismic Zones 0 through 3. Steel bolts with a minimum nominal diameter of $^5/_8$ inch (16 mm) shall be used in Seismic Zone 4. Bolts shall be embedded at least 7 inches (178 mm) into the concrete or masonry and shall be spaced not more than 6 feet (1829 mm) apart. There shall be a minimum of two bolts per piece with one bolt located not more than 12 inches (305 mm) or less than seven bolt diameters from each end of the piece. A properly sized nut and washer shall be tightened on each bolt to the plate. Foundation plates and sills shall be the kind of wood specified in Section 2306.4.

1806.6.1 Additional requirements in Seismic Zones 3 and 4. The following additional requirements shall apply in Seismic Zones 3 and 4.

1. Sill bolt diameter and spacing for three-story raised wood floor buildings shall be specifically designed.

2. Plate washers a minimum of 2 inch by 2 inch by $^3/_{16}$ inch (51 mm by 51 mm by 4.8 mm) thick shall be used on each bolt.

1806.7 Seismic Zones 3 and 4. In Seismic Zones 3 and 4, horizontal reinforcement in accordance with Sections 1806.7.1 and 1806.7.2 shall be placed in continuous foundations to minimize differential settlement. Foundation reinforcement shall be provided with cover in accordance with Section 1907.7.1.

1806.7.1 Foundations with stemwalls. Foundations with stemwalls shall be provided with a minimum of one No. 4 bar at the top of the wall and one No. 4 bar at the bottom of the footing.

1806.7.2 Slabs–on–ground with turned–down footings. Slabs–on–ground with turned-down footings shall have a minimum of one No. 4 bar at the top and bottom.

> **EXCEPTION:** For slabs-on-ground cast monolithically with a footing, one No. 5 bar may be located at either the top or bottom.

1806.8 Designs Employing Lateral Bearing.

1806.8.1 General. Construction employing posts or poles as columns embedded in earth or embedded in concrete footings in the earth may be used to resist both axial and lateral loads. The depth to resist lateral loads shall be determined by means of the design criteria established herein or other methods approved by the building official.

1806.8.2 Design criteria.

1806.8.2.1 Nonconstrained. The following formula may be used in determining the depth of embedment required to resist lateral loads where no constraint is provided at the ground surface, such as rigid floor or rigid ground surface pavement.

$$d = \frac{A}{2}\left(1 + \sqrt{1 + \frac{4.36h}{A}}\right) \qquad (6\text{-}1)$$

WHERE:

$A = \dfrac{2.34P}{S_1 b}$

b = diameter of round post or footing or diagonal dimension of square post or footing, feet (m).

d = depth of embedment in earth in feet (m) but not over 12 feet (3658 mm) for purpose of computing lateral pressure.

h = distance in feet (m) from ground surface to point of application of "P."

P = applied lateral force in pounds (kN).

S_1 = allowable lateral soil-bearing pressure as set forth in Table 18-I-A based on a depth of one third the depth of embedment (kPa).

S_3 = allowable lateral soil-bearing pressure as set forth in Table 18-I-A based on a depth equal to the depth of embedment (kPa).

1806.8.2.2 Constrained. The following formula may be used to determine the depth of embedment required to resist lateral loads

where constraint is provided at the ground surface, such as a rigid floor or pavement.

$$d^2 = 4.25\frac{Ph}{S_3 b} \qquad (6\text{-}2)$$

1806.8.2.3 Vertical load. The resistance to vertical loads is determined by the allowable soil-bearing pressure set forth in Table 18-I-A.

1806.8.3 Backfill. The backfill in the annular space around columns not embedded in poured footings shall be by one of the following methods:

1. Backfill shall be of concrete with an ultimate strength of 2,000 pounds per square inch (13.79 MPa) at 28 days. The hole shall not be less than 4 inches (102 mm) larger than the diameter of the column at its bottom or 4 inches (102 mm) larger than the diagonal dimension of a square or rectangular column.

2. Backfill shall be of clean sand. The sand shall be thoroughly compacted by tamping in layers not more than 8 inches (203 mm) in depth.

1806.8.4 Limitations. The design procedure outlined in this section shall be subject to the following limitations:

The frictional resistance for retaining walls and slabs on silts and clays shall be limited to one half of the normal force imposed on the soil by the weight of the footing or slab.

Posts embedded in earth shall not be used to provide lateral support for structural or nonstructural materials such as plaster, masonry or concrete unless bracing is provided that develops the limited deflection required.

1806.9 Grillage Footings. When grillage footings of structural steel shapes are used on soils, they shall be completely embedded in concrete with at least 6 inches (152 mm) on the bottom and at least 4 inches (102 mm) at all other points.

1806.10 Bleacher Footings. Footings for open-air seating facilities shall comply with Chapter 18.

> **EXCEPTIONS:** Temporary open-air portable bleachers as defined in Section 1008.2 may be supported upon wood sills or steel plates placed directly upon the ground surface, provided soil pressure does not exceed 1,200 pounds per square foot (57.5 kPa).

SECTION 1807 — PILES — GENERAL REQUIREMENTS

1807.1 General. Pile foundations shall be designed and installed on the basis of a foundation investigation as defined in Section 1804 where required by the building official.

The investigation and report provisions of Section 1804 shall be expanded to include, but not be limited to, the following:

1. Recommended pile types and installed capacities.

2. Driving criteria.

3. Installation procedures.

4. Field inspection and reporting procedures (to include procedures for verification of the installed bearing capacity where required).

5. Pile load test requirements.

The use of piles not specifically mentioned in this chapter shall be permitted, subject to the approval of the building official upon submission of acceptable test data, calculations or other information relating to the properties and load-carrying capacities of such piles.

1807.2 Interconnection. Individual pile caps and caissons of every structure subjected to seismic forces shall be interconnected by ties. Such ties shall be capable of resisting, in tension or compression, a minimum horizontal force equal to 10 percent of the larger column vertical load.

> **EXCEPTION:** Other approved methods may be used where it can be demonstrated that equivalent restraint can be provided.

1807.3 Determination of Allowable Loads. The allowable axial and lateral loads on piles shall be determined by an approved formula, by load tests or by a foundation investigation.

1807.4 Static Load Tests. When the allowable axial load of a single pile is determined by a load test, one of the following methods shall be used:

Method 1. It shall not exceed 50 percent of the yield point under test load. The yield point shall be defined as that point at which an increase in load produces a disproportionate increase in settlement.

Method 2. It shall not exceed one half of the load which causes a net settlement, after deducting rebound, of 0.01 inch per ton (0.000565 mm/N) of test load which has been applied for a period of at least 24 hours.

Method 3. It shall not exceed one half of that load under which, during a 40-hour period of continuous load application, no additional settlement takes place.

1807.5 Column Action. All piles standing unbraced in air, water or material not capable of lateral support, shall conform with the applicable column formula as specified in this code. Such piles driven into firm ground may be considered fixed and laterally supported at 5 feet (1524 mm) below the ground surface and in soft material at 10 feet (3048 mm) below the ground surface unless otherwise prescribed by the building official after a foundation investigation by an approved agency.

1807.6 Group Action. Consideration shall be given to the reduction of allowable pile load when piles are placed in groups. Where soil conditions make such load reductions advisable or necessary, the allowable axial load determined for a single pile shall be reduced by any rational method or formula approved by the building official.

1807.7 Piles in Subsiding Areas. Where piles are driven through subsiding fills or other subsiding strata and derive support from underlying firmer materials, consideration shall be given to the downward frictional forces which may be imposed on the piles by the subsiding upper strata.

Where the influence of subsiding fills is considered as imposing loads on the pile, the allowable stresses specified in this chapter may be increased if satisfactory substantiating data are submitted.

1807.8 Jetting. Jetting shall not be used except where and as specifically permitted by the building official. When used, jetting shall be carried out in such a manner that the carrying capacity of existing piles and structures shall not be impaired. After withdrawal of the jet, piles shall be driven down until the required resistance is obtained.

1807.9 Protection of Pile Materials. Where the boring records of site conditions indicate possible deleterious action on pile materials because of soil constituents, changing water levels or other factors, such materials shall be adequately protected by methods or processes approved by the building official. The effectiveness of such methods or processes for the particular purpose shall have been thoroughly established by satisfactory service records or other evidence which demonstrates the effectiveness of such protective measures.

1807.10 Allowable Loads. The allowable loads based on soil conditions shall be established in accordance with Section 1807.

> **EXCEPTION:** Any uncased cast-in-place pile may be assumed to develop a frictional resistance equal to one sixth of the bearing value of the soil material at minimum depth as set forth in Table 18-I-A but not to exceed 500 pounds per square foot (24 kPa) unless a greater value is allowed by the building official after a soil investigation as specified in Section 1804 is submitted. Frictional resistance and bearing resistance shall not be assumed to act simultaneously unless recommended after a foundation investigation as specified in Section 1804.

1807.11 Use of Higher Allowable Pile Stresses. Allowable compressive stresses greater than those specified in Section 1808 shall be permitted when substantiating data justifying such higher stresses are submitted to and approved by the building official. Such substantiating data shall include a foundation investigation including a report in accordance with Section 1807.1 by a soils engineer defined as a civil engineer experienced and knowledgeable in the practice of soils engineering.

SECTION 1808 — SPECIFIC PILE REQUIREMENTS

1808.1 Round Wood Piles.

1808.1.1 Material. Except where untreated piles are permitted, wood piles shall be pressure treated. Untreated piles may be used only when it has been established that the cutoff will be below lowest groundwater level assumed to exist during the life of the structure.

1808.1.2 Allowable stresses. The allowable unit stresses for round wood piles shall not exceed those set forth in Chapter 23, Division III, Part I.

The allowable values listed in Chapter 23, Division III, Part I, for compression parallel to the grain at extreme fiber in bending are based on load sharing as occurs in a pile cluster. For piles which support their own specific load, a safety factor of 1.25 shall be applied to compression parallel to the grain values and 1.30 to extreme fiber in bending values.

1808.2 Uncased Cast-in-place Concrete Piles.

1808.2.1 Material. Concrete piles cast in place against earth in drilled or bored holes shall be made in such a manner as to ensure the exclusion of any foreign matter and to secure a full-sized shaft. The length of such pile shall be limited to not more than 30 times the average diameter. Concrete shall have a specified compressive strength f'_c of not less than 2,500 psi (17.24 MPa).

> **EXCEPTION:** The length of pile may exceed 30 times the diameter provided the design and installation of the pile foundation is in accordance with an approved investigation report.

1808.2.2 Allowable stresses. The allowable compressive stress in the concrete shall not exceed $0.33f'_c$. The allowable compressive stress of reinforcement shall not exceed 34 percent of the yield strength of the steel or 25,500 psi (175.7 MPa).

1808.3 Metal-cased Concrete Piles.

1808.3.1 Material. Concrete used in metal-cased concrete piles shall have a specified compressive strength f'_c of not less than 2,500 psi (17.24 MPa).

1808.3.2 Installation. Every metal casing for a concrete pile shall have a sealed tip with a diameter of not less than 8 inches (203 mm).

Concrete piles cast in place in metal shells shall have shells driven for their full length in contact with the surrounding soil and left permanently in place. The shells shall be sufficiently strong to resist collapse and sufficiently watertight to exclude water and foreign material during the placing of concrete.

Piles shall be driven in such order and with such spacing as to ensure against distortion of or injury to piles already in place. No pile shall be driven within four and one-half average pile diameters of a pile filled with concrete less than 24 hours old unless approved by the building official.

1808.3.3 Allowable stresses. Allowable stresses shall not exceed the values specified in Section 1808.2.2, except that the allowable concrete stress may be increased to a maximum value of $0.40f'_c$ for that portion of the pile meeting the following conditions:

1. The thickness of the metal casing is not less than 0.068 inch (1.73 mm) (No. 14 carbon sheet steel gage).

2. The casing is seamless or is provided with seams of equal strength and is of a configuration that will provide confinement to the cast-in-place concrete.

3. The specified compressive strength f'_c shall not exceed 5,000 psi (34.47 MPa) and the ratio of steel minimum specified yield strength f_y to concrete specified compressive strength f'_c shall not be less than 6.

4. The pile diameter is not greater than 16 inches (406 mm).

1808.4 Precast Concrete Piles.

1808.4.1 Materials. Precast concrete piles shall have a specified compressive strength f'_c of not less than 3,000 psi (20.68 MPa), and shall develop a compressive strength of not less than 3,000 psi (20.68 MPa) before driving.

1808.4.2 Reinforcement ties. The longitudinal reinforcement in driven precast concrete piles shall be laterally tied with steel ties or wire spirals. Ties and spirals shall not be spaced more than 3 inches (76 mm) apart, center to center, for a distance of 2 feet (610 mm) from the ends and not more than 8 inches (203 mm) elsewhere. The gage of ties and spirals shall be as follows:

For piles having a diameter of 16 inches (406 mm) or less, wire shall not be smaller than 0.22 inch (5.6 mm) (No. 5 B.W. gage).

For piles having a diameter of more than 16 inches (406 mm) and less than 20 inches (508 mm), wire shall not be smaller than 0.238 inch (6.0 mm) (No. 4 B.W. gage).

For piles having a diameter of 20 inches (508 mm) and larger, wire shall not be smaller than $1/4$ inch (6.4 mm) round or 0.259 inch (6.6 mm) (No. 3 B.W. gage).

1808.4.3 Allowable stresses. Precast concrete piling shall be designed to resist stresses induced by handling and driving as well as by loads. The allowable stresses shall not exceed the values specified in Section 1808.2.2.

1808.5 Precast Prestressed Concrete Piles (Pretensioned).

1808.5.1 Materials. Precast prestressed concrete piles shall have a specified compressive strength f'_c of not less than 5,000 psi (34.48 MPa) and shall develop a compressive strength of not less than 4,000 psi (27.58 MPa) before driving.

1808.5.2 Reinforcement. The longitudinal reinforcement shall be high-tensile seven-wire strand. Longitudinal reinforcement shall be laterally tied with steel ties or wire spirals.

Ties or spiral reinforcement shall not be spaced more than 3 inches (76 mm) apart, center to center, for a distance of 2 feet (610 mm) from the ends and not more than 8 inches (203 mm) elsewhere.

At each end of the pile, the first five ties or spirals shall be spaced 1 inch (25 mm) center to center.

For piles having a diameter of 24 inches (610 mm) or less, wire shall not be smaller than 0.22 inch (5.6 mm) (No. 5 B.W. gage). For piles having a diameter greater than 24 inches (610 mm) but less than 36 inches (914 mm), wire shall not be smaller than 0.238 inch (6.0 mm) (No. 4 B.W. gage). For piles having a diameter greater than 36 inches (914 mm), wire shall not be smaller than $1/4$ inch (6.4 mm) round or 0.259 inch (6.6 mm) (No. 3 B.W. gage).

1808.5.3 Allowable stresses. Precast prestressed piling shall be designed to resist stresses induced by handling and driving as well as by loads. The effective prestress in the pile shall not be less than 400 psi (2.76 MPa) for piles up to 30 feet (9144 mm) in length, 550 psi (3.79 MPa) for piles up to 50 feet (15 240 mm) in length, and 700 psi (4.83 MPa) for piles greater than 50 feet (15 240 mm) in length.

The compressive stress in the concrete due to externally applied load shall not exceed:

$$f_c = 0.33f'_c - 0.27fp_c$$

WHERE:

fp_c = effective prestress stress on the gross section.

Effective prestress shall be based on an assumed loss of 30,000 psi (206.85 MPa) in the prestressing steel. The allowable stress in the prestressing steel shall not exceed the values specified in Section 1918.

1808.6 Structural Steel Piles.

1808.6.1 Material. Structural steel piles, steel pipe piles and fully welded steel piles fabricated from plates shall conform to UBC Standard 22-1 and be identified in accordance with Section 2202.2.

1808.6.2 Allowable stresses. The allowable axial stresses shall not exceed 0.35 of the minimum specified yield strength F_y or 12,600 psi (86.88 MPa), whichever is less.

> **EXCEPTION:** When justified in accordance with Section 1807.11, the allowable axial stress may be increased above 12,600 psi (86.88 MPa) and $0.35F_y$, but shall not exceed $0.5F_y$.

1808.6.3 Minimum dimensions. Sections of driven H-piles shall comply with the following:

1. The flange projection shall not exceed 14 times the minimum thickness of metal in either the flange or the web, and the flange widths shall not be less than 80 percent of the depth of the section.

2. The nominal depth in the direction of the web shall not be less than 8 inches (203 mm).

3. Flanges and webs shall have a minimum nominal thickness of $3/8$ inch (9.5 mm).

Sections of driven pipe piles shall have an outside diameter of not less than 10 inches (254 mm) and a minimum thickness of not less than $1/4$ inch (6.4 mm).

1808.7 Concrete-filled Steel Pipe Piles.

1808.7.1 Material. The concrete-filled steel pipe piles shall conform to UBC Standard 22-1 and shall be identified in accordance with Section 2202.2. The concrete-filled steel pipe piles shall have a specified compressive strength f'_c of not less than 2,500 psi (17.24 MPa).

1808.7.2 Allowable stresses. The allowable axial stresses shall not exceed 0.35 of the minimum specified yield strength F_y of the steel plus 0.33 of the specified compressive strength f'_c of concrete, provided F_y shall not be assumed greater than 36,000 psi (248.22 MPa) for computational purposes.

EXCEPTION: When justified in accordance with Section 2807.11, the allowable stresses may be increased to $0.50 F_y$.

1808.7.3 Minimum dimensions. Driven piles of uniform section shall have a nominal outside diameter of not less than 8 inches (203 mm).

SECTION 1809 — FOUNDATION CONSTRUCTION—SEISMIC ZONES 3 AND 4

1809.1 General. In Seismic Zones 3 and 4 the further requirements of this section shall apply to the design and construction of foundations, foundation components and the connection of superstructure elements thereto.

1809.2 Soil Capacity. The foundation shall be capable of transmitting the design base shear and overturning forces prescribed in Section 1630 from the structure into the supporting soil. The short-term dynamic nature of the loads may be taken into account in establishing the soil properties.

1809.3 Superstructure-to-Foundation Connection. The connection of superstructure elements to the foundation shall be adequate to transmit to the foundation the forces for which the elements were required to be designed.

1809.4 Foundation-Soil Interface. For regular buildings, the force F_t as provided in Section 1630.5 may be omitted when determining the overturning moment to be resisted at the foundation-soil interface.

1809.5 Special Requirements for Piles and Caissons.

1809.5.1 General. Piles, caissons and caps shall be designed according to the provisions of Section 1603, including the effects of lateral displacements. Special detailing requirements as described in Section 1809.5.2 shall apply for a length of piles equal to 120 percent of the flexural length. Flexural length shall be considered as a length of pile from the first point of zero lateral deflection to the underside of the pile cap or grade beam.

1809.5.2 Steel piles, nonprestressed concrete piles and prestressed concrete piles.

1809.5.2.1 Steel piles. Piles shall conform to width-thickness ratios of stiffened, unstiffened and tubular compression elements as shown in Chapter 22, Division VIII.

1809.5.2.2 Nonprestressed concrete piles. Piles shall have transverse reinforcement meeting the requirements of Section 1921.4.

EXCEPTION: Transverse reinforcement need not exceed the amount determined by Formula (21-2) in Section 1921.4.4.1 for spiral or circular hoop reinforcement or by Formula (21-4) in Section 1921.4.4.1 for rectangular hoop reinforcement.

1809.5.2.3 Prestressed concrete piles. Piles shall have a minimum volumetric ratio of spiral reinforcement no less than 0.021 for 14-inch (356 mm) square and smaller piles, and 0.012 for 24-inch (610 mm) square and larger piles unless a smaller value can be justified by rational analysis. Interpolation may be used between the specified ratios for intermediate sizes.

TABLE 18-I-A—ALLOWABLE FOUNDATION AND LATERAL PRESSURE

CLASS OF MATERIALS[1]	ALLOWABLE FOUNDATION PRESSURE (psf)[2] $\times$ 0.0479 for kPa	LATERAL BEARING LBS./SQ./FT./FT. OF DEPTH BELOW NATURAL GRADE[3] $\times$ 0.157 for kPa per meter	LATERAL SLIDING[4] Coefficient[5]	LATERAL SLIDING[4] Resistance (psf)[6] $\times$ 0.0479 for kPa
1. Massive crystalline bedrock	4,000	1,200	0.70	
2. Sedimentary and foliated rock	2,000	400	0.35	
3. Sandy gravel and/or gravel (GW and GP)	2,000	200	0.35	
4. Sand, silty sand, clayey sand, silty gravel and clayey gravel (SW, SP, SM, SC, GM and GC)	1,500	150	0.25	
5. Clay, sandy clay, silty clay and clayey silt (CL, ML, MH and CH)	1,000[7]	100		130

[1]For soil classifications OL, OH and PT (i.e., organic clays and peat), a foundation investigation shall be required.

[2]All values of allowable foundation pressure are for footings having a minimum width of 12 inches (305 mm) and a minimum depth of 12 inches (305 mm) into natural grade. Except as in Footnote 7, an increase of 20 percent shall be allowed for each additional foot (305 mm) of width or depth to a maximum value of three times the designated value. Additionally, an increase of one third shall be permitted when considering load combinations, including wind or earthquake loads, as permitted by Section 1612.3.2.

[3]May be increased the amount of the designated value for each additional foot (305 mm) of depth to a maximum of 15 times the designated value. Isolated poles for uses such as flagpoles or signs and poles used to support buildings that are not adversely affected by a $1/2$-inch (12.7 mm) motion at ground surface due to short-term lateral loads may be designed using lateral bearing values equal to two times the tabulated values.

[4]Lateral bearing and lateral sliding resistance may be combined.

[5]Coefficient to be multiplied by the dead load.

[6]Lateral sliding resistance value to be multiplied by the contact area. In no case shall the lateral sliding resistance exceed one half the dead load.

[7]No increase for width is allowed.

TABLE 18-I-B—CLASSIFICATION OF EXPANSIVE SOIL

EXPANSION INDEX	POTENTIAL EXPANSION
0-20	Very low
21-50	Low
51-90	Medium
91-130	High
Above 130	Very high

TABLE 18-I-C—FOUNDATIONS FOR STUD BEARING WALLS—MINIMUM REQUIREMENTS[1,2,3,4]

NUMBER OF FLOORS SUPPORTED BY THE FOUNDATION[5]	THICKNESS OF FOUNDATION WALL (inches) $\times$ 25.4 for mm Concrete	THICKNESS OF FOUNDATION WALL (inches) $\times$ 25.4 for mm Unit Masonry	WIDTH OF FOOTING (inches) $\times$ 25.4 for mm	THICKNESS OF FOOTING (inches) $\times$ 25.4 for mm	DEPTH BELOW UNDISTURBED GROUND SURFACE (inches) $\times$ 25.4 for mm
1	6	6	12	6	12
2	8	8	15	7	18
3	10	10	18	8	24

[1]Where unusual conditions or frost conditions are found, footings and foundations shall be as required in Section 1806.1.

[2]The ground under the floor may be excavated to the elevation of the top of the footing.

[3]Interior stud bearing walls may be supported by isolated footings. The footing width and length shall be twice the width shown in this table and the footings shall be spaced not more than 6 feet (1829 mm) on center.

[4]In Seismic Zone 4, continuous footings shall be provided with a minimum of one No. 4 bar top and bottom.

[5]Foundations may support a roof in addition to the stipulated number of floors. Foundations supporting roofs only shall be as required for supporting one floor.

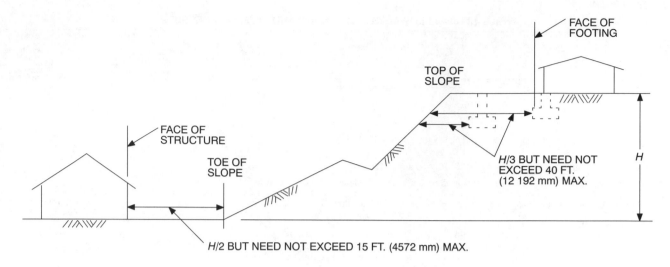

FIGURE 18-I-1—SETBACK DIMENSIONS

**Chapter 19 is printed in its entirety in Volume 2 of the *Uniform Building Code*.
Excerpts from Chapter 19 are reprinted herein.**

Excerpts from Chapter 19
CONCRETE

NOTE: This is a new division.

Division I — GENERAL

SECTION 1900 — GENERAL

1900.1 Scope. The design of concrete structures of cast-in-place or precast construction, plain, reinforced or prestressed shall conform to the rules and principles specified in this chapter.

1900.2 General Requirements. All concrete structures shall be designed and constructed in accordance with the requirements of Division II and the additional requirements contained in Section 1900.4 of this division.

1900.3 Design Methods. The design of concrete structures shall be in accordance with one of the following methods.

1900.3.1 Strength design (load and resistance factor design). The design of concrete structures using the strength design method shall be in accordance with the requirements of Division II.

1900.3.2 Allowable stress design. The design of concrete structures using the Allowable Stress Design Method shall be in accordance with the requirements of Division VI, Section 1926.

1900.4 Additional Design and Construction Requirements.

1900.4.1 Anchorage. Anchorage of bolts and headed stud anchors to concrete shall be in accordance with Division III.

1900.4.2 Shotcrete. In addition to the requirements of Division II, design and construction of shotcrete structures shall meet the requirements of Division IV.

1900.4.3 Reinforced gypsum concrete. Reinforced gypsum concrete shall be in accordance with Division V.

1900.4.4 Minimum slab thickness. The minimum thickness of concrete floor slabs supported directly on the ground shall not be less than $3^1/_2$ inches (89 mm).

1900.4.5 Unified design provisions for reinforced and prestressed concrete flexural and compression members. It shall be permitted to use the alternate flexural and axial load design provisions in accordance with Division VII, Section 1927.

1900.4.6 Alternative load-factor combination and strength-reduction factors. It shall be permitted to use the alternative load-factor and strength-reduction factors in accordance with Division VIII, Section 1928.

Division II

Copyright © by the American Concrete Institute and reproduced
with their consent. All rights reserved.

The contents of this division are patterned after, and in general conformity with, the provisions of Building Code Requirements for Reinforced Concrete (ACI 318-95) and commentary—ACI 318 R-95. For additional background information and research data, see the referenced American Concrete Institute (ACI) publication.

To make reference to the ACI commentary easier for users of the code, the section designations of this division have been made similar to those found in ACI 318. The first two digits of a section number indicates this chapter number and the balance matches the ACI chapter and section designation wherever possible. Italics are used in this chapter to indicate where the *Uniform Building Code* differs substantively from the ACI standard.

SECTION 1901 — SCOPE

The design of structures in concrete of cast-in-place or precast construction, plain, reinforced or prestressed, shall conform to the rules and principles specified in this chapter.

SECTION 1902 — DEFINITIONS

The following terms are defined for general use in this code. Specialized definitions appear in individual *sections*.

ADMIXTURE is material other than water, aggregate, or hydraulic cement used as an ingredient of concrete and added to concrete before or during its mixing to modify its properties.

AGGREGATE is granular material, such as sand, gravel, crushed stone and iron blast-furnace slag, and when used with a cementing medium forms a hydraulic cement concrete or mortar.

AGGREGATE, LIGHTWEIGHT, is aggregate with a dry, loose weight of 70 pounds per cubic foot (pcf) (1120 kg/m^3) or less.

AIR-DRY WEIGHT is the unit weight of a lightweight concrete specimen cured for seven days with neither loss nor gain of moisture at 60°F to 80°F (15.6°C to 26.7°C) and dried for 21 days in 50 ± 7 percent relative humidity at 73.4°F ± 2°F (23.0°C ± 1.1°C).

ANCHORAGE in posttensioning is a device used to anchor tendons to concrete member; in pretensioning, a device used to anchor tendons during hardening of concrete.

BONDED TENDON is a prestressing tendon that is bonded to concrete either directly or through grouting.

CEMENTITIOUS MATERIALS are materials as specified in Section 1903 which have cementing value when used in concrete either by themselves, such as portland cement, blended hydraulic cements and expansive cement, or such materials in combination with fly ash, raw or other calcined natural pozzolans, silica fume, or ground granulated blast-furnace slag.

COLUMN is a member with a ratio of height-to-least-lateral dimension of 3 or greater used primarily to support axial compressive load.

COMPOSITE CONCRETE FLEXURAL MEMBERS are concrete flexural members of precast and cast-in-place concrete elements or both constructed in separate placements but so interconnected that all elements respond to loads as a unit.

COMPRESSION-CONTROLLED SECTION is a cross section in which the net tensile strain in the extreme tension steel at nominal strength is less than or equal to the compression-controlled strain limit.

COMPRESSION-CONTROLLED STRAIN LIMIT is the net tensile strain at balanced strain conditions. (See *Section B1910.3.2.*)

CONCRETE is a mixture of portland cement or any other hydraulic cement, fine aggregate, coarse aggregate and water, with or without admixtures.

CONCRETE, SPECIFIED COMPRESSIVE STRENGTH OF (f'_c), is the compressive strength of concrete used in design and evaluated in accordance with provisions of Section 1905, expressed in pounds per square inch (psi) (MPa). Whenever the quantity f'_c is under a radical sign, square root of numerical value only is intended, and result has units of psi (MPa).

CONCRETE, STRUCTURAL LIGHTWEIGHT, is concrete containing lightweight aggregate having an air-dry unit weight as determined by definition above, not exceeding 115 pcf (1840 kg/m^3). In this code, a lightweight concrete without natural sand is termed "all-lightweight concrete" and lightweight concrete in which all fine aggregate consists of normal-weight sand is termed "sand-lightweight concrete."

CONTRACTION JOINT is a formed, sawed, or tooled groove in a concrete structure to create a weakened plane and regulate the location of cracking resulting from the dimensional change of different parts of the structure.

CURVATURE FRICTION is friction resulting from bends or curves in the specified prestressing tendon profile.

DEFORMED REINFORCEMENT is deformed reinforcing bars, bar and rod mats, deformed wire, welded smooth wire fabric and welded deformed wire fabric.

DEVELOPMENT LENGTH is the length of embedded reinforcement required to develop the design strength of reinforcement at a critical section. See Section 1909.3.3.

EFFECTIVE DEPTH OF SECTION (*d*) is the distance measured from extreme compression fiber to centroid of tension reinforcement.

EFFECTIVE PRESTRESS is the stress remaining in prestressing tendons after all losses have occurred, excluding effects of dead load and superimposed load.

EMBEDMENT LENGTH is the length of embedded reinforcement provided beyond a critical section.

EXTREME TENSION STEEL is the reinforcement (prestressed or nonprestressed) that is the farthest from the extreme compression fiber.

ISOLATION JOINT is a separation between adjoining parts of a concrete structure, usually a vertical plane, at a designed location such as to interfere least with performance of the structure, yet such as to allow relative movement in three directions and avoid formation of cracks elsewhere in the concrete and through which all or part of the bonded reinforcement is interrupted.

JACKING FORCE is the temporary force exerted by device that introduces tension into prestressing tendons in prestressed concrete.

LOAD, DEAD, is the dead weight supported by a member, as defined by *Section 1602* (without load factors).

LOAD, FACTORED, is the load, multiplied by appropriate load factors, used to proportion members by the strength design method of this code. See Sections 1908.1.1 and 1909.2.

LOAD, LIVE, *is the live load specified by Section 1602* (without load factors).

LOAD, SERVICE, *is the live and dead loads* (without load factors).

MODULUS OF ELASTICITY is the ratio of normal stress to corresponding strain for tensile or compressive stresses below proportional limit of material. See Section 1908.5.

NET TENSILE STRAIN is the tensile strain at nominal strength exclusive of strains due to effective prestress, creep, shrinkage and temperature.

PEDESTAL is an upright compression member with a ratio of unsupported height to average least lateral dimension of 3 or less.

PLAIN CONCRETE is structural concrete with no reinforcement or with less reinforcement than the minimum amount specified for reinforced concrete.

PLAIN REINFORCEMENT is reinforcement that does not conform to definition of deformed reinforcement.

POSTTENSIONING is a method of prestressing in which tendons are tensioned after concrete has hardened.

PRECAST CONCRETE is a structural concrete element cast in other than its final position in the structure.

PRESTRESSED CONCRETE is structural concrete in which internal stresses have been introduced to reduce potential tensile stresses in concrete resulting from loads.

PRETENSIONING is a method of prestressing in which tendons are tensioned before concrete is placed.

REINFORCED CONCRETE is structural concrete reinforced with no less than the minimum amounts of prestressing tendons or nonprestressed reinforcement specified in this code.

REINFORCEMENT is material that conforms to Section 1903.5.1, excluding prestressing tendons unless specifically included.

RESHORES are shores placed snugly under a concrete slab or other structural member after the original forms and shores have been removed from a larger area, thus requiring the new slab or structural member to deflect and support its own weight and existing construction loads applied prior to the installation of the reshores.

SHORES are vertical or inclined support members designed to carry the weight of the formwork, concrete and construction loads above.

SPAN LENGTH. See Section 1908.7.

SPIRAL REINFORCEMENT is continuously wound reinforcement in the form of a cylindrical helix.

SPLITTING TENSILE STRENGTH (f_{ct}) is the tensile strength of concrete. See Section 1905.1.4.

STIRRUP is reinforcement used to resist shear and torsion stresses in a structural member; typically bars, wires, or welded wire fabric (smooth or deformed) bent into L, U or rectangular shapes and located perpendicular to or at an angle to longitudinal reinforcement. (The term "stirrups" is usually applied to lateral reinforcement in flexural members and the term "ties" to those in compression members.) See "tie."

STRENGTH, DESIGN, is the nominal strength multiplied by a strength-reduction factor ϕ. See Section 1909.3.

STRENGTH, NOMINAL, is the strength of a member or cross section calculated in accordance with provisions and assumptions of the strength design method of this code before application of any strength-reduction factors. See Section 1909.3.1.

STRENGTH, REQUIRED, is the strength of a member or cross section required to resist factored loads or related internal moments and forces in such combinations as are stipulated in this code. See Section 1909.1.1.

STRESS is the intensity of force per unit area.

STRUCTURAL CONCRETE is all concrete used for structural purposes, including plain and reinforced concrete.

TENDON is a steel element such as wire, cable, bar, rod or strand, or a bundle of such elements, used to impart prestress to concrete.

TENSION-CONTROLLED SECTION is a cross section in which the net tensile strain in the extreme tension steel at nominal strength is greater than or equal to 0.005.

TIE is a loop of reinforcing bar or wire enclosing longitudinal reinforcement. A continuously wound bar or wire in the form of a circle, rectangle or other polygon shape without re-entrant corners is acceptable. See "stirrup."

TRANSFER is the act of transferring stress in prestressing tendons from jacks or pretensioning bed to concrete member.

WALL is a member, usually vertical, used to enclose or separate spaces.

WOBBLE FRICTION in prestressed concrete, is friction caused by unintended deviation of prestressing sheath or duct from its specified profile.

YIELD STRENGTH is the specified minimum yield strength or yield point of reinforcement in psi.

SECTION 1903 — SPECIFICATIONS FOR TESTS AND MATERIALS

1903.0 Notation.

f_y = specified yield strength of nonprestressed reinforcement, psi (MPa).

1903.1 Tests of Materials.

1903.1.1 The building official may require the testing of any materials used in concrete construction to determine if materials are of quality specified.

1903.1.2 Tests of materials and of concrete shall be made *by an approved agency and at no expense to the jurisdiction. Such tests shall be made in accordance with the standards listed in Section 1903.*

1903.1.3 A complete record of tests of materials and of concrete shall be available for inspection during progress of work and for two years after completion of the project, and shall be preserved by the inspecting engineer or architect for that purpose.

1903.1.4 Material and test standards. *The standards listed in this chapter labeled a "UBC Standard" are also listed in Chapter 35, Part II, and are part of this code. The other standards listed in this chapter are recognized standards. (See Sections 3503 and 3504.)*

1903.2 Cement.

1. ASTM C 845, Expansive Hydraulic Cement

2. ASTM C 150, Portland Cement

3. ASTM C 595 or ASTM C 1157, Blended Hydraulic Cements

1903.3 Aggregates.

1903.3.1 Recognized standards.

1. ASTM C 33, Concrete Aggregates

2. ASTM C 330, Lightweight Aggregates for Structural Concrete

3. ASTM C 332, Lightweight Aggregates for Insulating Concrete

4. ASTM C 144, Aggregate for Masonry Mortar

5. Aggregates failing to meet the above specifications but which have been shown by special test or actual service to produce concrete of adequate strength and durability may be used where authorized by the building official.

1903.3.2 The nominal maximum size of coarse aggregate shall not be larger than:

1. One fifth the narrowest dimension between sides of forms, or

2. One third the depth of slabs, or

3. Three fourths the minimum clear spacing between individual reinforcing bars or wires, bundles of bars, or prestressing tendons or ducts.

These limitations may be waived if, in the judgment of the *building official*, workability and methods of consolidation are such that concrete can be placed without honeycomb or voids.

1903.4 Water.

1903.4.1 Water used in mixing concrete shall be clean and free from injurious amounts of oils, acids, alkalis, salts, organic materials or other substances deleterious to concrete or reinforcement.

1903.4.2 Mixing water for prestressed concrete or for concrete that will contain aluminum embedments, including that portion of mixing water contributed in the form of free moisture on aggregates, shall not contain deleterious amounts of chloride ions. See Section 1904.4.1.

1903.4.3 Nonpotable water shall not be used in concrete unless the following are satisfied:

1903.4.3.1 Selection of concrete proportions shall be based on concrete mixes using water from the same source.

1903.4.3.2 Mortar test cubes made with nonpotable mixing water shall have seven-day and 28-day strengths equal to at least 90 percent of strengths of similar specimens made with potable water. Strength test comparison shall be made on mortars, identical except for the mixing water, prepared and tested in accordance with ASTM C 109 (Compressive Strength of Hydraulic Cement Mortars).

1903.5 Steel Reinforcement.

1903.5.1 Reinforcement shall be deformed reinforcement, except that plain reinforcement may be used for spirals or tendons, and reinforcement consisting of structural steel, steel pipe or steel tubing may be used as specified in this chapter.

1903.5.2 Welding of reinforcing bars shall conform to *approved nationally recognized standards*. Type and location of welded splices and other required welding of reinforcing bars shall be indicated on the design drawings or in the project specifications. ASTM reinforcing bar specifications, except for A 706, shall be supplemented to require a report of material properties necessary to conform to requirements in UBC Standard 19-1.

1903.5.3 Deformed reinforcements.

1903.5.3.1 ASTM A 615, A 616, A 617, A 706, A 767 and A 775, Reinforcing Bars for Concrete.

1903.5.3.2 Deformed reinforcing bars with a specified yield strength f_y exceeding 60,000 psi (413.7 MPa) may be used, provided f_y shall be the stress corresponding to a strain of 0.35 percent and the bars otherwise conform to approved national standards, see *ASTM A 615, A 616, A 617, A 706, A 767 and A 775*. See Section 1909.4.

1903.5.3.3 ASTM A 184, Fabricated Deformed Steel Bar Mats. For reinforced bars used in bar mats, see ASTM A 615, A 616, A 617, A 706, A 767 or A 775.

1903.5.3.4 ASTM A 496, Steel Wire, Deformed, for Concrete Reinforcement.

For deformed wire for concrete reinforcement, see *ASTM A 496,* except that wire shall not be smaller than size D4, and for wire with a specified yield strength f_y exceeding 60,000 psi (413.7 MPa), f_y shall be the stress corresponding to a strain of 0.35 percent, if the yield strength specified in design exceeds 60,000 psi (413.7 MPa).

1903.5.3.5 ASTM A 185, Steel Welded Wire, Fabric, Plain for Concrete Reinforcement.

For welded plain wire fabric for concrete reinforcement, see *ASTM 185,* except that for wire with a specified yield strength f_y exceeding 60,000 psi (413.7 MPa), f_y shall be the stress corresponding to a strain of 0.35 percent, if the yield strength specified in design exceeds 60,000 psi (413.7 MPa). Welded intersections shall not be spaced farther apart than 12 inches (305 mm) in direction of calculated stress, except for wire fabric used as stirrups in accordance with Section 1912.14.

1903.5.3.6 ASTM A 497, Welded Deformed Steel Wire Fabric for Concrete Reinforcement.

For welded deformed wire fabric for concrete reinforcement, see *ASTM A 497,* except that for wire with a specified yield strength f_y exceeding 60,000 psi (413.7 MPa), f_y shall be the stress corresponding to a strain of 0.35 percent, if the yield strength specified in design exceeds 60,000 psi (413.7 MPa). Welded intersections shall not be spaced farther apart than 16 inches (406 mm) in direction of calculated stress, except for wire fabric used as stirrups in accordance with Section 1912.13.2.

1903.5.3.7 Deformed reinforcing bars may be galvanized or epoxy coated. For zinc or epoxy-coated reinforcement, see ASTM A 615, A 616, A 617, A 706, A 767 and A 775 and ASTM A 934 *(Epoxy-Coated Steel Reinforcing Bars).*

1903.5.3.8 Epoxy-coated wires and welded wire fabric shall comply with ASTM A 884 *(Standard Specification for Epoxy-Coated Steel Wire and Welded Wire Fabric for Reinforcement).* Epoxy-coated wires shall conform to Section 1903.5.3.4 and epoxy-coated welded wire fabric shall conform to Section 1903.5.3.5 or 1903.5.3.6.

1903.5.4 Plain reinforcement.

1903.5.4.1 Plain bars for spiral reinforcement shall conform to approved national standards, see ASTM *A 615, A 616 and A 617.*

1903.5.4.2 For plain wire for spiral reinforcement, see *ASTM A 82* except that for wire with a specified yield strength f_y exceeding 60,000 psi (413.7 MPa), f_y shall be the stress corresponding to a strain of 0.35 percent, if the yield strength specified in design exceeds 60,000 psi (413.7 MPa).

1903.5.5 Prestressing tendons.

1903.5.5.1 1. ASTM A 416, Uncoated Seven-wire Stress-relieved Steel Strand for Prestressed Concrete

2. ASTM A 421, Uncoated Stress-relieved Wire for Prestressed Concrete

3. ASTM A 722, Uncoated High-strength Steel Bar for Prestressing Concrete

1903.5.5.2 Wire, strands and bars not specifically listed in *ASTM A 416, A 421 and A 722* may be used, provided they conform to minimum requirements of these specifications and do not have properties that make them less satisfactory than those listed.

1903.5.6 Structural steel, steel pipe or tubing.

1903.5.6.1 For structural steel used with reinforcing bars in composite compression members meeting requirements of Section 1910.16.7 or 1910.16.8, see *ASTM A 36, A 242, A 572 and A 588.*

1903.5.6.2 For steel pipe or tubing for composite compression members composed of a steel-encased concrete core meeting requirements of Section 1910.16.4, see *ASTM A 53, A 500 and A 501.*

1903.5.7 *UBC Standard 19-1, Welding Reinforcing Steel, Metal Inserts and Connections in Reinforced Concrete Construction*

1903.6 Admixtures.

1903.6.1 Admixtures to be used in concrete shall be subject to prior approval by the *building official.*

1903.6.2 An admixture shall be shown capable of maintaining essentially the same composition and performance throughout the work as the product used in establishing concrete proportions in accordance with Section 1905.2.

1903.6.3 Calcium chloride or admixtures containing chloride from other than impurities from admixture ingredients shall not be used in prestressed concrete, in concrete containing embedded aluminum, or in concrete cast against stay-in-place galvanized steel forms. See Sections 1904.3.2 and 1904.4.1.

1903.6.4 ASTM C 260, Air-entraining Admixtures for Concrete

1903.6.5 ASTM C 494 and C 1017, Chemical Admixtures for Concrete

1903.6.6 ASTM C 618, Fly Ash and Raw or Calcined Natural Pozzolans for Use as Admixtures in Portland Cement Concrete

1903.6.7 *ASTM C 989,* Ground-iron Blast-furnace Slag for Use in Concrete and Mortars

1903.6.8 Admixtures used in concrete containing ASTM C 845 expansive cements shall be compatible with the cement and produce no deleterious effects.

1903.6.9 Silica fume used as an admixture shall conform to ASTM C 1240 (*Silica Fume for Use in Hydraulic Cement Concrete and Mortar*).

1903.7 Storage of Materials.

1903.7.1 Cementitious materials and aggregate shall be stored in such manner as to prevent deterioration or intrusion of foreign matter.

1903.7.2 Any material that has deteriorated or has been contaminated shall not be used for concrete.

1903.8 Concrete Testing.

1. ASTM C 192, Making and Curing Concrete Test Specimens in the Laboratory

2. ASTM C 31, Making and Curing Concrete Test Specimens in the Field

3. ASTM C 42, Obtaining and Testing Drilled Cores and Sawed Beams of Concrete

4. ASTM C 39, Compressive Strength of Cylindrical Concrete Specimens

5. ASTM C 172, Sampling Freshly Mixed Concrete

6. ASTM C 496, Splitting Tensile Strength of Cylindrical Concrete Specimens

7. ASTM C 1218, Water-Soluble Chloride in Mortar and Concrete

1903.9 Concrete Mix.

1. *ASTM C 94, Ready-mixed Concrete*

2. ASTM C 685, Concrete Made by Volumetric Batching and Continuous Mixing

3. *UBC Standard 19-2, Mill-mixed Gypsum Concrete and Poured Gypsum Roof Diaphragms*

4. ASTM C 109, Compressive Strength of Hydraulic Cement Mortars

5. ASTM C 567, Unit Weight of Structural Lightweight Concrete

1903.10 Welding. *The welding of reinforcing steel, metal inserts and connections in reinforced concrete construction shall conform to UBC Standard 19-1.*

1903.11 Glass Fiber Reinforced Concrete. Recommended Practice for Glass Fiber Reinforced Concrete Panels, Manual 128.

SECTION 1904 — DURABILITY REQUIREMENTS

1904.0 Notation.

f'_c = specified compressive strength of concrete, psi (MPa).

1904.1 Water-Cementitious Materials Ratio.

1904.1.1 The water-cementitious materials ratios specified in Tables 19-A-2 and 19-A-4 shall be calculated using the weight of cement meeting ASTM C 150, C 595 or C 845 plus the weight of fly ash and other pozzolans meeting ASTM C 618, slag meeting ASTM C 989, and silica fume meeting ASTM C 1240, if any, except that when concrete is exposed to deicing chemicals, Section 1904.2.3 further limits the amount of fly ash, pozzolans, silica fume, slag or the combination of these materials.

1904.2 Freezing and Thawing Exposures.

1904.2.1 Normal-weight and lightweight concrete exposed to freezing and thawing or deicing chemicals shall be air entrained with air content indicated in Table 19-A-1. Tolerance on air content as delivered shall be ± 1.5 percent. For specified compressive strength f'_c greater than 5,000 psi (34.47 MPa), reduction of air content indicated in Table 19-A-1 by 1.0 percent shall be permitted.

1904.2.2 Concrete that will be subjected to the exposures given in Table 19-A-2 shall conform to the corresponding maximum water-cementitious materials ratios and minimum specified concrete compressive strength requirements of that table. In addition, concrete that will be exposed to deicing chemicals shall conform to the limitations of Section 1904.2.3.

1904.2.3 For concrete exposed to deicing chemicals, the maximum weight of fly ash, other pozzolans, silica fume or slag that is included in the concrete shall not exceed the percentages of the total weight of cementitious materials given in Table 19-A-3.

1904.3 Sulfate Exposure.

1904.3.1 Concrete to be exposed to sulfate-containing solutions or soils shall conform to the requirements of Table 19-A-4 or shall be concrete made with a cement that provides sulfate resistance and that has a maximum water-cementitious materials ratio and minimum compressive strength set forth in Table 19-A-4.

1904.3.2 Calcium chloride as an admixture shall not be used in concrete to be exposed to severe or very severe sulfate-containing solutions, as defined in Table 19-A-4.

1904.4 Corrosion Protection of Reinforcement.

1904.4.1 For corrosion protection of reinforcement in concrete, maximum water soluble chloride ion concentrations in hardened concrete at ages from 28 to 42 days contributed from the ingredients, including water, aggregates, cementitious materials and admixtures shall not exceed the limits of Table 19-A-5. When testing is performed to determine water soluble chloride ion content, test procedures shall conform to ASTM C 1218.

1904.4.2 If concrete with reinforcement will be exposed to chlorides from deicing chemicals, salt, salt water, brackish water, sea water or spray from these sources, requirements of Table 19-A-2 for water-cementitious materials ratio and concrete strength and the minimum concrete cover requirements of Section 1907.7 shall be satisfied. In addition, see Section 1918.14 for unbonded prestressed tendons.

SECTION 1905 — CONCRETE QUALITY, MIXING AND PLACING

1905.0 Notations.

f'_c = specified compressive strength of concrete, psi (MPa).

f'_{cr} = required average compressive strength of concrete used as the basis for selection of concrete proportions, psi (MPa).

f_{ct} = average splitting tensile strength of lightweight aggregate concrete, psi (MPa).

s = standard deviation, psi (MPa).

1905.1 General.

1905.1.1 Concrete shall be proportioned to provide an average compressive strength as prescribed in Section 1905.3.2, as well as satisfy the durability criteria of Section 1904. Concrete shall be produced to minimize frequency of strengths below f'_c as prescribed in Section 1905.6.2.3.

1905.1.2 Requirements for f'_c shall be based on tests of cylinders made and tested as prescribed in Section 1905.6.2.

1905.1.3 Unless otherwise specified, f'_c shall be based on 28-day tests. If other than 28 days, test age for f'_c shall be as indicated in design drawings or specifications.

Design drawings shall show specified compressive strength of concrete f'_c for which each part of structure is designed.

1905.1.4 Where design criteria in Sections 1909.5.2.3, 1911.2; and 1912.2.4, provide for use of a splitting tensile strength value of concrete, laboratory tests shall be made to establish value of f_{ct} corresponding to specified values of f'_c.

1905.1.5 Splitting tensile strength tests shall not be used as a basis for field acceptance of concrete.

1905.2 Selection of Concrete Proportions.

1905.2.1 Proportions of materials for concrete shall be established to provide:

1. Workability and consistency to permit concrete to be worked readily into forms and around reinforcement under conditions of placement to be employed without segregation or excessive bleeding.

2. Resistance to special exposures as required by Section 1904.

3. Conformance with strength test requirements of Section 1905.6.

1905.2.2 Where different materials are to be used for different portions of proposed work, each combination shall be evaluated.

1905.2.3 Concrete proportions, including water-cementitious materials ratio, shall be established on the basis of field experience and/or trial mixtures with materials to be employed (see Section 1905.3), except as permitted in Section 1905.4 or required by Section 1904.

1905.3 Proportioning on the Basis of Field Experience and Trial Mixtures.

1905.3.1 Standard deviation.

1905.3.1.1 Where a concrete production facility has test records, a standard deviation shall be established. Test records from which a standard deviation is calculated:

1. Must represent materials, quality control procedures and conditions similar to those expected, and changes in materials and proportions within the test records shall not have been more restricted than those for proposed work.

2. Must represent concrete produced to meet a specified strength or strengths f'_c within 1,000 psi (6.89 MPa) of that specified for proposed work.

3. Must consist of at least 30 consecutive tests or two groups of consecutive tests totaling at least 30 tests as defined in Section 1905.6.1.4, except as provided in Section 1905.3.1.2.

1905.3.1.2 Where a concrete production facility does not have test records meeting requirements of Section 1905.3.1.1, but does have a record based on 15 to 29 consecutive tests, a standard deviation may be established as the product of the calculated standard deviation and the modification factor of Table 19-A-6. To be acceptable, the test record must meet the requirements of Section 1905.3.1.1, Items 1 and 2, and represent only a single record of consecutive tests that span a period of not less than 45 calendar days.

1905.3.2 Required average strength.

1905.3.2.1 Required average compressive strength f'_{cr} used as the basis for selection of concrete proportions shall be the larger of Formula (5-1) or (5-2) using a standard deviation calculated in accordance with Section 1905.3.1.1 or 1905.3.1.2.

$$f'_{cr} = f'_c + 1.34s \qquad (5\text{-}1)$$

or

$$f'_{cr} = f'_c + 2.33s - 500 \qquad (5\text{-}2)$$

For **SI:** $\qquad f'_{cr} = f'_c + 2.33s - 3.45$

1905.3.2.2 When a concrete production facility does not have field strength test records for calculation of standard deviation meeting requirements of Section 1905.3.1.1 or 1905.3.1.2, required average strength f'_{cr} shall be determined from Table 19-B and documentation of average strength shall be in accordance with requirements of Section 1905.3.3.

1905.3.3 Documentation of average strength. Documentation that proposed concrete proportions will produce an average compressive strength equal to or greater than required average compressive strength (see Section 1905.3.2) shall consist of a field strength test record, several strength test records, or trial mixtures.

1905.3.3.1 When test records are used to demonstrate that proposed concrete proportions will produce the required average strength f'_{cr} (see Section 1905.3.2), such records shall represent materials and conditions similar to those expected. Changes in materials, conditions and proportions within the test records shall not have been more restricted than those for proposed work. For the purpose of documenting average strength potential, test records consisting of less than 30 but not less than 10 consecutive tests may be used, provided test records encompass a period of time not less than 45 days. Required concrete proportions may be established by interpolation between the strengths and proportions of two or more test records each of which meets other requirements of this section.

1905.3.3.2 When an acceptable record of field test results is not available, concrete proportions established from trial mixtures meeting the following restrictions shall be permitted:

1. Combination of materials shall be those for proposed work.

2. Trial mixtures having proportions and consistencies required for proposed work shall be made using at least three different water-cementitious materials ratios or cementitious materials contents that will produce a range of strengths encompassing the required average strength f'_{cr}.

3. Trial mixture shall be designed to produce a slump within ± 0.75 inch (± 19 mm) of maximum permitted, and for air-entrained concrete, within ± 0.5 percent of maximum allowable air content.

4. For each water-cementitious materials ratio or cementitious materials content, at least three test cylinders for each test age shall be made and cured. Cylinders shall be tested at 28 days or at test age designated for determination of f'_c.

5. From results of cylinder tests, a curve shall be plotted showing relationship between water-cementitious materials ratio or cementitious materials content and compressive strength at designated test age.

6. Maximum water-cementitious materials ratio or minimum cementitious materials content for concrete to be used in proposed work shall be that shown by the curve to produce the average strength required by Section 1905.3.2, unless a lower water-cementitious materials ratio or higher strength is required by Section 1904.

1905.4 Proportioning without Field Experience or Trial Mixtures.

1905.4.1 If data required by Section 1905.3 are not available, concrete proportions shall be based upon other experience or information, if approved by the building official. The required average compressive strength f'_{cr} of concrete produced with materials similar to those proposed for use shall be at least 1,200 psi (8.3 MPa) greater than the specified compressive strength, f'_c. This alternative shall not be used for specified compressive strength greater than 4,000 psi (27.58 MPa).

1905.4.2 Concrete proportioned by Section 1905.4 shall conform to the durability requirements of Section 1904 and to compressive strength test criteria of Section 1905.6.

1905.5 Average Strength Reduction. As data become available during construction, it shall be permitted to reduce the amount by which f'_{cr} must exceed the specified value of f'_c, provided:

1. Thirty or more test results are available and average of test results exceeds that required by Section 1905.3.2.1, using a standard deviation calculated in accordance with Section 1905.3.1.1, or

2. Fifteen to 29 test results are available and average of test results exceeds that required by Section 1905.3.2.1, using a standard deviation calculated in accordance with Section 1905.3.1.2, and

3. Special exposure requirements of Section 1904 are met.

1905.6 Evaluation and Acceptance of Concrete.

1905.6.1 Frequency of testing.

1905.6.1.1 Samples for strength tests of each class of concrete placed each day shall be taken not less than once a day, or not less than once for each 150 cubic yards (115 m³) of concrete, or not less than once for each 5,000 square feet (465 m²) of surface area for slabs or walls.

1905.6.1.2 On a given project, if the total volume of concrete is such that the frequency of testing required by Section 1905.6.1.1 would provide less than five strength tests for a given class of concrete, tests shall be made from at least five randomly selected batches or from each batch if fewer than five batches are used.

1905.6.1.3 When total quantity of a given class of concrete is less than 50 cubic yards (38 m³), strength tests are not required when evidence of satisfactory strength is submitted to and approved by the building official.

1905.6.1.4 A strength test shall be the average of the strengths of two cylinders made from the same sample of concrete and tested at 28 days or at test age designated for determination of f'_c.

1905.6.2 Laboratory-cured specimens.

1905.6.2.1 Samples for strength tests shall be taken.

1905.6.2.2 Cylinders for strength tests shall be molded and laboratory cured and tested.

1905.6.2.3 Strength level of an individual class of concrete shall be considered satisfactory if both the following requirements are met:

1. Every arithmetic average of any three consecutive strength tests equals or exceeds f'_c.

2. No individual strength test (average of two cylinders) falls below f'_c by more than 500 psi (3.45 MPa).

1905.6.2.4 If either of the requirements of Section 1905.6.2.3 are not met, steps shall be taken to increase the average of subsequent strength test results. Requirements of Section 1905.6.4 shall be observed if the requirement of Item 2 of Section 1905.6.2.3 is not met.

1905.6.3 Field-cured specimens.

1905.6.3.1 If required by the building official, results of strength tests of cylinders cured under field conditions shall be provided.

1905.6.3.2 Field-cured cylinders shall be cured under field conditions, in accordance with Section 1903.8.

1905.6.3.3 Field-cured test cylinders shall be molded at the same time and from the same samples as laboratory-cured test cylinders.

1905.6.3.4 Procedures for protecting and curing concrete shall be improved when strength of field-cured cylinders at test age designated for determination of f'_c is less than 85 percent of that of companion laboratory-cured cylinders. The 85 percent limitation shall not apply if field-cured strength exceeds f'_c by more than 500 psi (3.45 MPa).

1905.6.4 Investigation of low-strength test results.

1905.6.4.1 If any strength test (see Section 1905.6.1.4) of laboratory-cured cylinders falls below specified values of f'_c by more than 500 psi (3.45 MPa) (see Section 1905.6.2.3, Item 2) or if tests of field-cured cylinders indicate deficiencies in protection and curing (see Section 1905.6.3.4), steps shall be taken to ensure that load-carrying capacity of the structure is not jeopardized.

1905.6.4.2 If the likelihood of low-strength concrete is confirmed and calculations indicate that load-carrying capacity is significantly reduced, tests of cores drilled from the area in question shall be permitted. In such case, three cores shall be taken for each strength test more than 500 psi (3.45 MPa) below specified value of f'_c.

1905.6.4.3 If concrete in the structure will be dry under service conditions, cores shall be air dried [temperatures 60°F to 80°F (15.6°C to 26.7°C), relative humidity less than 60 percent] for seven days before test and shall be tested dry. If concrete in the structure will be more than superficially wet under service conditions, cores shall be immersed in water for at least 40 hours and be tested wet.

1905.6.4.4 Concrete in an area represented by core tests shall be considered structurally adequate if the average of three cores is equal to at least 85 percent of f'_c and if no single core is less than 75 percent of f'_c. Additional testing of cores extracted from locations represented by erratic core strength results shall be permitted.

1905.6.4.5 If criteria of Section 1905.6.4.4 are not met, and if structural adequacy remains in doubt, the responsible authority shall be permitted to order a strength evaluation in accordance with Section 1920 for the questionable portion of the structure, or take other appropriate action.

1905.7 Preparation of Equipment and Place of Deposit.

1905.7.1 Preparation before concrete placement shall include the following:

1. All equipment for mixing and transporting concrete shall be clean.

2. All debris and ice shall be removed from spaces to be occupied by concrete.

3. Forms shall be properly coated.

4. Masonry filler units that will be in contact with concrete shall be well drenched.

5. Reinforcement shall be thoroughly clean of ice or other deleterious coatings.

6. Water shall be removed from place of deposit before concrete is placed unless a tremie is to be used or unless otherwise permitted by the building official.

7. All laitance and other unsound material shall be removed before additional concrete is placed against hardened concrete.

1905.8 Mixing.

1905.8.1 All concrete shall be mixed until there is a uniform distribution of materials and shall be discharged completely before mixer is recharged.

1905.8.2 Ready-mixed concrete shall be mixed and delivered in accordance with requirements of *ASTM C 94 (Ready-Mixed Concrete)* or *ASTM C 685 (Concrete Made by Volumetric Batching and Continuous Mixing)*.

1905.8.3 Job-mixed concrete shall be mixed in accordance with the following:

1. Mixing shall be done in a batch mixer of an approved type.

2. Mixer shall be rotated at a speed recommended by the manufacturer.

3. Mixing shall be continued for at least $1^1/_2$ minutes after all materials are in the drum, unless a shorter time is shown to be satisfactory by the mixing uniformity tests of *ASTM C 94 (Ready-Mixed Concrete)*.

4. Materials handling, batching and mixing shall conform to applicable provisions of *ASTM C 94 (Ready-Mixed Concrete)*.

5. A detailed record shall be kept to identify:

　5.1　Number of batches produced;

　5.2　Proportions of materials used;

　5.3　Approximate location of final deposit in structure;

　5.4　Time and date of mixing and placing.

1905.9 Conveying.

1905.9.1 Concrete shall be conveyed from mixer to place of final deposit by methods that will prevent separation or loss of materials.

1905.9.2 Conveying equipment shall be capable of providing a supply of concrete at site of placement without separation of ingredients and without interruptions sufficient to permit loss of plasticity between successive increments.

1905.10 Depositing.

1905.10.1 Concrete shall be deposited as nearly as practicable in its final position to avoid segregation due to rehandling or flowing.

1905.10.2 Concreting shall be carried on at such a rate that concrete is at all times plastic and flows readily into spaces between reinforcement.

1905.10.3 Concrete that has partially hardened or been contaminated by foreign materials shall not be deposited in the structure.

1905.10.4 Retempered concrete or concrete that has been remixed after initial set shall not be used unless approved by the *building official*.

1905.10.5 After concreting is started, it shall be carried on as a continuous operation until placing of a panel or section, as defined by its boundaries or predetermined joints, is completed, except as permitted or prohibited by Section 1906.4.

1905.10.6 Top surfaces of vertically formed lifts shall be generally level.

1905.10.7 When construction joints are required, joints shall be made in accordance with Section 1906.4.

1905.10.8 All concrete shall be thoroughly consolidated by suitable means during placement and shall be thoroughly worked around reinforcement and embedded fixtures and into corners of forms.

1905.11 Curing.

1905.11.1 Concrete (other than high-early-strength) shall be maintained above 50°F (10.0°C) and in a moist condition for at least the first seven days after placement, except when cured in accordance with Section 1905.11.3.

1905.11.2 High-early-strength concrete shall be maintained above 50°F (10.0°C) and in a moist condition for at least the first three days, except when cured in accordance with Section 1905.11.3.

1905.11.3 Accelerated curing.

1905.11.3.1 Curing by high-pressure steam, steam at atmospheric pressure, heat and moisture or other accepted processes, may be employed to accelerate strength gain and reduce time of curing.

1905.11.3.2 Accelerated curing shall provide a compressive strength of the concrete at the load stage considered at least equal to required design strength at that load stage.

1905.11.3.3 Curing process shall be such as to produce concrete with a durability at least equivalent to the curing method of Section 1905.11.1 or 1905.11.2.

1905.11.3.4 When required by the building official, supplementary strength tests in accordance with Section 1905.6.3 shall be performed to assure that curing is satisfactory.

1905.12 Cold Weather Requirements.

1905.12.1 Adequate equipment shall be provided for heating concrete materials and protecting concrete during freezing or near-freezing weather.

1905.12.2 All concrete materials and all reinforcement, forms, fillers and ground with which concrete is to come in contact shall be free from frost.

1905.12.3 Frozen materials or materials containing ice shall not be used.

1905.13 Hot Weather Requirements. During hot weather, proper attention shall be given to ingredients, production methods, handling, placing, protection and curing to prevent excessive concrete temperatures or water evaporation that may impair required strength or serviceability of the member or structure.

SECTION 1906 — FORMWORK, EMBEDDED PIPES AND CONSTRUCTION JOINTS

1906.1 Design of Formwork.

1906.1.1 Forms shall result in a final structure that conforms to shapes, lines and dimensions of the members as required by the design drawings and specifications.

1906.1.2 Forms shall be substantial and sufficiently tight to prevent leakage of mortar.

1906.1.3 Forms shall be properly braced or tied together to maintain position and shape.

1906.1.4 Forms and their supports shall be designed so as not to damage previously placed structure.

1906.1.5 Design of formwork shall include consideration of the following factors:

1. Rate and method of placing concrete.

2. Construction loads, including vertical, horizontal and impact loads.

3. Special form requirements for construction of shells, folded plates, domes, architectural concrete or similar types of elements.

1906.1.6 Forms for prestressed concrete members shall be designed and constructed to permit movement of the member without damage during application of prestressing force.

1906.2 Removal of Forms, Shores and Reshoring.

1906.2.1 Removal of forms. Forms shall be removed in such a manner as not to impair safety and serviceability of the structure. Concrete to be exposed by form removal shall have sufficient strength not to be damaged by removal operation.

1906.2.2 Removal of shores and reshoring. The provisions of Section 1906.2.2.1 through 1906.2.2.3 shall apply to slabs and beams except where cast on the ground.

1906.2.2.1 Before starting construction, the contractor shall develop a procedure and schedule for removal of shores and installation of reshores and for calculating the loads transferred to the structure during the process.

1. The structural analysis and concrete strength data used in planning and implementing form removal and shoring shall be furnished by the contractor to the building official when so requested.

2. Construction loads shall *not* be supported on, or any shoring removed from, any part of the structure under construction except when that portion of the structure in combination with remaining forming and shoring system has sufficient strength to support safely its weight and loads placed thereon.

3. Sufficient strength shall be demonstrated by structural analysis considering proposed loads, strength of forming and shoring system and concrete strength data. Concrete strength data may be based on tests of field-cured cylinders or, when approved by the building official, on other procedures to evaluate concrete strength.

1906.2.2.2 Construction loads exceeding the combination of superimposed dead load plus specified live load shall *not* be supported on any unshored portion of the structure under construction, unless analysis indicates adequate strength to support such additional loads.

1906.2.2.3 Form supports for prestressed concrete members shall not be removed until sufficient prestressing has been applied to enable prestressed members to carry their dead load and anticipated construction loads.

1906.3 Conduits and Pipes Embedded in Concrete.

1906.3.1 Conduits, pipes and sleeves of any material not harmful to concrete and within limitations of this subsection may be embedded in concrete with approval of the *building official*, provided they are not considered to replace structurally the displaced concrete.

1906.3.2 Conduits and pipes of aluminum shall not be embedded in structural concrete unless effectively coated or covered to prevent aluminum-concrete reaction or electrolytic action between aluminum and steel.

1906.3.3 Conduits, pipes and sleeves passing through a slab, wall or beam shall not impair significantly the strength of the construction.

1906.3.4 Conduits and pipes, with their fittings, embedded within a column shall not displace more than 4 percent of the area of cross section on which strength is calculated or which is required for fire protection.

1906.3.5 Except when plans for conduits and pipes are approved by the *building official*, conduits and pipes embedded within a slab, wall or beam (other than those merely passing through) shall satisfy the following:

1906.3.5.1 They shall not be larger in outside dimension than one third the overall thickness of slab, wall or beam in which they are embedded.

1906.3.5.2 They shall be spaced not closer than three diameters or widths on center.

1906.3.5.3 They shall not impair significantly the strength of the construction.

1906.3.6 Conduits, pipes and sleeves may be considered as replacing structurally in compression the displaced concrete, provided:

1906.3.6.1 They are not exposed to rusting or other deterioration.

1906.3.6.2 They are of uncoated or galvanized iron or steel not thinner than standard Schedule 40 steel pipe.

1906.3.6.3 They have a nominal inside diameter not over 2 inches (51 mm) and are spaced not less than three diameters on centers.

1906.3.7 Pipes and fittings shall be designed to resist effects of the material, pressure and temperature to which they will be subjected.

1906.3.8 No liquid, gas or vapor, except water not exceeding $90°F$ ($32.2°C$) or 50 psi (0.34 MPa) pressure, shall be placed in the pipes until the concrete has attained its design strength.

1906.3.9 In solid slabs, piping, unless it is used for radiant heating or snow melting, shall be placed between top and bottom reinforcement.

1906.3.10 Concrete cover for pipes, conduit and fittings shall not be less than $1^1/_2$ inches (38 mm) for concrete exposed to earth or weather, or less than $^3/_4$ inch (19 mm) for concrete not exposed to weather or in contact with ground.

1906.3.11 Reinforcement with an area not less than 0.002 times the area of concrete section shall be provided normal to the piping.

1906.3.12 Piping and conduit shall be so fabricated and installed that cutting, bending or displacement of reinforcement from its proper location will not be required.

1906.4 Construction Joints.

1906.4.1 Surface of concrete construction joints shall be cleaned and laitance removed.

1906.4.2 Immediately before new concrete is placed, all construction joints shall be wetted and standing water removed.

1906.4.3 Construction joints shall be so made and located as not to impair the strength of the structure. Provision shall be made for transfer of shear and other forces through construction joints. See Section 1911.7.9.

1906.4.4 Construction joints in floors shall be located within the middle third of spans of slabs, beams and girders. Joints in girders shall be offset a minimum distance of two times the width of intersecting beams.

1906.4.5 Beams, girders or slabs supported by columns or walls shall not be cast or erected until concrete in the vertical support members is no longer plastic.

1906.4.6 Beams, girders, haunches, drop panels and capitals shall be placed monolithically as part of a slab system, unless otherwise shown in design drawings or specifications.

SECTION 1907 — DETAILS OF REINFORCEMENT

1907.0 Notations.

d = distance from extreme compression fiber to centroid of tension reinforcement, inches (mm).

d_b = nominal diameter of bar, wire or prestressing strand, inches (mm).

f_y = specified yield strength of nonprestressed reinforcement, psi (MPa).

l_d = development length, inches (mm). See Section 1912.

1907.1 Standard Hooks. "Standard hook" as used in this code is one of the following:

1907.1.1 One-hundred-eighty-degree bend plus $4d_b$ extension, but not less than $2^1/_2$ inches (64 mm) at free end of bar.

1907.1.2 Ninety-degree bend plus $12d_b$ extension at free end of bar.

1907.1.3 For stirrup and tie hooks:

1. No. 5 bar and smaller, 90-degree bend plus $6d_b$ extension at free end of bar, or

2. No. 6, No. 7 and No. 8 bar, 90-degree bend, plus $12d_b$ extension at free end of bar, or

3. No. 8 bar and smaller, 135-degree bend plus $6d_b$ extension at free end of bar.

4. *For stirrups and tie hooks in Seismic Zones 3 and 4, refer to the hoop and crosstie provisions of Section 1921.1.*

1907.2 Minimum Bend Diameters.

1907.2.1 Diameter of bend measured on the inside of the bar, other than for stirrups and ties in sizes No. 3 through No. 5, shall not be less than the values in Table 19-B.

1907.2.2 Inside diameter of bends for stirrups and ties shall not be less than $4d_b$ for No. 5 bar and smaller. For bars larger than No. 5, diameter of bend shall be in accordance with Table 19-B.

1907.2.3 Inside diameter of bends in welded wire fabric (plain or deformed) for stirrups and ties shall not be less than $4d_b$ for deformed wire larger than D6 and $2d_b$ for all other wires. Bends with inside diameter of less than $8d_b$ shall not be less than $4d_b$ from nearest welded intersection.

1907.3 Bending.

1907.3.1 All reinforcement shall be bent cold, unless otherwise permitted by the *building official*.

1907.3.2 Reinforcement partially embedded in concrete shall not be field bent, except as shown on the design drawings or permitted by the *building official*.

1907.4 Surface Conditions of Reinforcement.

1907.4.1 At the time concrete is placed, reinforcement shall be free from mud, oil or other nonmetallic coatings that decrease bond. Epoxy coatings of bars in accordance with Section 1903.5.3.7 shall be permitted.

1907.4.2 Reinforcement, except prestressing tendons, with rust, mill scale or a combination of both, shall be considered satisfactory, provided the minimum dimensions (including height of deformations) and weight of a hand-wire-brushed test specimen are not less than applicable specification requirements.

1907.4.3 Prestressing tendons shall be clean and free of oil, dirt, scale, pitting and excessive rust. A light oxide shall be permitted.

1907.5 Placing Reinforcement.

1907.5.1 Reinforcement, prestressing tendons and ducts shall be accurately placed and adequately supported before concrete is placed, and shall be secured against displacement within tolerances of this section.

1907.5.2 Unless otherwise *approved by the building official,* reinforcement, prestressing tendons and prestressing ducts shall be placed within the following tolerances:

1907.5.2.1 Tolerance for depth d, and minimum concrete cover in flexural members, walls and compression members shall be as follows:

	TOLERANCE ON d	TOLERANCE ON MINIMUM CONCRETE COVER
$d \leq 8$ in. (203 mm)	$\pm\,^3/_8$ in. (9.5 mm)	$-\,^3/_8$ in. (9.5 mm)
$d > 8$ in. (203 mm)	$\pm\,^1/_2$ in. (12.7 mm)	$-\,^1/_2$ in. (12.7 mm)

except that tolerance for the clear distance to formed soffits shall be minus $^1/_4$ inch (6.4 mm) and tolerance for cover shall not exceed minus one third the minimum concrete cover required by the approved plans or specifications.

1907.5.2.2 Tolerance for longitudinal location of bends and ends of reinforcement shall be $\pm\,2$ inches ($\pm\,51$ mm) except at discontinuous ends of members where tolerance shall be $\pm\,^1/_2$ inch ($\pm\,12.7$ mm).

1907.5.3 Welded wire fabric (with wire size not greater than W5 or D5) used in slabs not exceeding 10 feet (3048 mm) in span shall be permitted to be curved from a point near the top of slab over the support to a point near the bottom of slab at midspan, provided such reinforcement is either continuous over, or securely anchored at, support.

1907.5.4 Welding of crossing bars shall not be permitted for assembly of reinforcement.

> **EXCEPTIONS:** *1. Reinforcing steel not required by design.*
>
> *2. When specifically approved by the building official, welding of crossing bars for assembly purposes in Seismic Zones 0, 1 and 2 may be permitted, provided that data are submitted to the building official to show that there is no detrimental effect on the action of the structural member as a result of welding of the crossing bars.*

1907.6 Spacing Limits for Reinforcement.

1907.6.1 The minimum clear spacing between parallel bars in a layer shall be d_b but not less than 1 inch (25 mm). See also Section 1903.3.2.

1907.6.2 Where parallel reinforcement is placed in two or more layers, bars in the upper layers shall be placed directly above bars in the bottom layer with clear distance between layers not less than 1 inch (25 mm).

1907.6.3 In spirally reinforced or tied reinforced compression members, clear distance between longitudinal bars shall not be less than $1.5d_b$ or less than $1^1/_2$ inches (38 mm). See also Section 1903.3.2.

1907.6.4 Clear distance limitation between bars shall apply also to the clear distance between a contact lap splice and adjacent splices or bars.

1907.6.5 In walls and slabs other than concrete joist construction, primary flexural reinforcement shall not be spaced farther apart than three times the wall or slab thickness, or 18 inches (457 mm).

1907.6.6 Bundled bars.

1907.6.6.1 Groups of parallel reinforcing bars bundled in contact to act as a unit shall be limited to four bars in one bundle.

1907.6.6.2 Bundled bars shall be enclosed within stirrups or ties.

1907.6.6.3 Bars larger than No. 11 shall not be bundled in beams.

1907.6.6.4 Individual bars within a bundle terminated within the span of flexural members shall terminate at different points with at least $40d_b$ stagger.

1907.6.6.5 Where spacing limitations and minimum concrete cover are based on bar diameter d_b, a unit of bundled bars shall be treated as a single bar of a diameter derived from the equivalent total area.

1907.6.7 Prestressing tendons and ducts.

1907.6.7.1 Clear distance between pretensioning tendons at each end of a member shall be not less than $4d_b$ for wire, or $3d_b$ for strands. See also Section 1903.3.2. Closer vertical spacing and bundling of tendons shall be permitted in the middle portion of a span.

1907.6.7.2 Bundling of posttensioning ducts shall be permitted if it is shown that concrete can be satisfactorily placed and if provision is made to prevent the tendons, when tensioned, from breaking through the duct.

1907.7 Concrete Protection for Reinforcement.

1907.7.1 Cast-in-place concrete (nonprestressed). The following minimum concrete cover shall be provided for reinforcement:

		MINIMUM COVER, inches (mm)
1.	Concrete cast against and permanently exposed to earth	3 (76)
2.	Concrete exposed to earth or weather:	
	No. 6 through No. 18 bar	2 (51)
	No. 5 bar, W31 or D31 wire, and smaller	$1^1/_2$ (38)
3.	Concrete not exposed to weather or in contact with ground:	
	Slabs, walls, joists:	
	No. 14 and No. 18 bar	$1^1/_2$ (38)
	No. 11 bar and smaller	$^3/_4$ (19)
	Beams, columns:	
	Primary reinforcement, ties, stirrups, spirals	$1^1/_2$ (38)
	Shells, folded plate members:	
	No. 6 bar and larger	$^3/_4$ (19)
	No. 5 bar, W31 or D31 wire, and smaller	$^1/_2$ (12.7)
4.	*Concrete tilt-up panels cast against a rigid horizontal surface, such as a concrete slab, exposed to the weather:*	
	No. 8 and smaller	*1 (25)*
	No. 9 through No. 18	*2 (51)*

1907.7.2 Precast concrete (manufactured under plant control conditions). The following minimum concrete cover shall be provided for reinforcement:

	MINIMUM COVER, inches (mm)
1. Concrete exposed to earth or weather:	
Wall panels:	
No. 14 and No. 18 bar	$1^1/_2$ (38)
No. 11 bar and smaller	$^3/_4$ (19)
Other members:	
No. 14 and No. 18 bar	2 (51)
No. 6 through No. 11 bar	$1^1/_2$ (38)
No. 5 bar W31 or D31 wire, and smaller	$1^1/_4$ (32)
2. Concrete not exposed to weather or in contact with ground:	
Slabs, walls, joists:	
No. 14 and No. 18 bar	$1^1/_4$ (32)
No. 11 bar and smaller	$^5/_8$ (16)
Beams, columns:	
Primary reinforcement	d_b but not less than $^5/_8$ (16) and need not exceed $1^1/_2$ (38)
Ties, stirrups, spirals	$^3/_8$ (9.5)
Shells, folded plate members:	
No. 6 bar and larger	$^5/_8$ (16)
No. 5 bar, W31 or D31 wire, and smaller	$^3/_8$ (9.5)

1907.7.3 Prestressed concrete.

1907.7.3.1 The following minimum concrete cover shall be provided for prestressed and nonprestressed reinforcement, ducts and end fittings, except as provided in Sections 1907.7.3.2 and 1907.7.3.3.

	MINIMUM COVER, inches (mm)
1. Concrete cast against and permanently exposed to earth	3 (76)
2. Concrete exposed to earth or weather:	
Wall panels, slabs, joists	1 (25)
Other members	$1^1/_2$ (32)
3. Concrete not exposed to weather or in contact with ground:	
Slabs, walls, joists	$^3/_4$ (19)
Beams, columns:	
Primary reinforcement	$1^1/_2$ (38)
Ties, stirrups, spirals	1 (25)
Shells, folded plate members:	
No. 5 bars, W31 or D31 wire, and smaller	$^3/_8$ (9.5)
Other reinforcement	d_b but not less than $^3/_4$ (19)

1907.7.3.2 For prestressed concrete members exposed to earth, weather or corrosive environments, and in which permissible tensile stress of Section 1918.4.2, Item 3, is exceeded, minimum cover shall be increased 50 percent.

1907.7.3.3 For prestressed concrete members manufactured under plant control conditions, minimum concrete cover for nonprestressed reinforcement shall be as required in Section 1907.7.2.

1907.7.4 Bundled bars. For bundled bars, minimum concrete cover shall be equal to the equivalent diameter of the bundle, but need not be greater than 2 inches (51 mm); except for concrete cast against and permanently exposed to earth, minimum cover shall be 3 inches (76 mm).

1907.7.5 Corrosive environments. In corrosive environments or other severe exposure conditions, amount of concrete protec-

tion shall be suitably increased, and denseness and nonporosity of protecting concrete shall be considered, or other protection shall be provided.

1907.7.6 Future extensions. Exposed reinforcement, inserts and plates intended for bonding with future extensions shall be protected from corrosion.

1907.7.7 Fire protection. When a thickness of cover for fire protection greater than the minimum concrete cover specified in Section 1907.7 is required, such greater thickness shall be used.

1907.8 Special Reinforcement Details for Columns.

1907.8.1 Offset bars. Offset bent longitudinal bars shall conform to the following:

1907.8.1.1 Slope of inclined portion of an offset bar with axis of column shall not exceed 1 in 6.

1907.8.1.2 Portions of bar above and below an offset shall be parallel to axis of column.

1907.8.1.3 Horizontal support at offset bends shall be provided by lateral ties, spirals or parts of the floor construction. Horizontal support provided shall be designed to resist one and one-half times the horizontal component of the computed force in the inclined portion of an offset bar. Lateral ties or spirals, if used, shall be placed not more than 6 inches (152 mm) from points of bend.

1907.8.1.4 Offset bars shall be bent before placement in the forms. See Section 1907.3.

1907.8.1.5 Where a column face is offset 3 inches (76 mm) or greater, longitudinal bars shall not be offset bent. Separate dowels, lap spliced with the longitudinal bars adjacent to the offset column faces, shall be provided. Lap splices shall conform to Section 1912.17.

1907.8.2 Steel cores. Load transfer in structural steel cores of composite compression members shall be provided by the following:

1907.8.2.1 Ends of structural steel cores shall be accurately finished to bear at end-bearing splices, with positive provision for alignment of one core above the other in concentric contact.

1907.8.2.2 At end-bearing splices, bearing shall be considered effective to transfer not more than 50 percent of the total compressive stress in the steel core.

1907.8.2.3 Transfer of stress between column base and footing shall be designed in accordance with Section 1915.8.

1907.8.2.4 Base of structural steel section shall be designed to transfer the total load from the entire composite member to the footing; or, the base may be designed to transfer the load from the steel core only, provided ample concrete section is available for transfer of the portion of the total load carried by the reinforced concrete section to the footing by compression in the concrete and by reinforcement.

1907.9 Connections.

1907.9.1 At connections of principal framing elements (such as beams and columns), enclosure shall be provided for splices of continuing reinforcement and for anchorage of reinforcement terminating in such connections.

1907.9.2 Enclosure at connections may consist of external concrete or internal closed ties, spirals or stirrups.

1907.10 Lateral Reinforcement for Compression Members.

1907.10.1 Lateral reinforcement for compression members shall conform to the provisions of Sections 1907.10.4 and 1907.10.5

and, where shear or torsion reinforcement is required, shall also conform to provisions of Section 1911.

1907.10.2 Lateral reinforcement requirements for composite compression members shall conform to Section 1910.16. Lateral reinforcement requirements for prestressing tendons shall conform to Section 1918.11.

1907.10.3 It shall be permitted to waive the lateral reinforcement requirements of Sections 1907.10, 1910.16 and 1918.11 where tests and structural analyses show adequate strength and feasibility of construction.

1907.10.4 Spirals. Spiral reinforcement for compression members shall conform to Section 1910.9.3 and to the following:

1907.10.4.1 Spirals shall consist of evenly spaced continuous bar or wire of such size and so assembled as to permit handling and placing without distortion from designed dimensions.

1907.10.4.2 For cast-in-place construction, size of spirals shall not be less than $^3/_8$-inch (9.5 mm) diameter.

1907.10.4.3 Clear spacing between spirals shall not exceed 3 inches (76 mm) or be less than 1 inch (25 mm). See also Section 1903.3.2.

1907.10.4.4 Anchorage of spiral reinforcement shall be provided by one and one-half extra turns of spiral bar or wire at each end of a spiral unit.

1907.10.4.5 Splices in spiral reinforcement shall be lap splices of $48d_b$, but not less than 12 inches (305 mm) or welded.

1907.10.4.6 Spirals shall extend from top of footing or slab in any story to level of lowest horizontal reinforcement in members supported above.

1907.10.4.7 Where beams or brackets do not frame into all sides of a column, ties shall extend above termination of spiral to bottom of slab or drop panel.

1907.10.4.8 In columns with capitals, spirals shall extend to a level at which the diameter or width of capital is two times that of the column.

1907.10.4.9 Spirals shall be held firmly in place and true to line.

1907.10.5 Ties. Tie reinforcement for compression members shall conform to the following:

1907.10.5.1 All nonprestressed bars shall be enclosed by lateral ties, at least No. 3 in size for longitudinal bars No. 10 or smaller, and at least No. 4 in size for Nos. 11, 14 and 18 and bundled longitudinal bars. Deformed wire or welded wire fabric of equivalent area shall be permitted.

1907.10.5.2 Vertical spacing of ties shall not exceed 16 longitudinal bar diameters, 48 tie bar or wire diameters, or least dimension of the compression member.

1907.10.5.3 Ties shall be arranged such that every corner and alternate longitudinal bar shall have lateral support provided by the corner of a tie with an included angle of not more than 135 degrees and a bar shall be not farther than 6 inches (152 mm) clear on each side along the tie from such a laterally supported bar. Where longitudinal bars are located around the perimeter of a circle, a complete circular tie shall be permitted.

1907.10.5.4 Ties shall be located vertically not more than one half a tie spacing above the top of footing or slab in any story and shall be spaced as provided herein to not more than one half a tie spacing below the lowest horizontal reinforcement *in members supported above.*

1907.10.5.5 Where beams or brackets frame from four directions into a column, termination of ties not more than 3 inches (76 mm) below reinforcement in shallowest of such beams or brackets shall be permitted.

1907.10.5.6 *Column ties shall have hooks as specified in Section 1907.1.3.*

1907.11 Lateral Reinforcement for Flexural Members.

1907.11.1 Compression reinforcement in beams shall be enclosed by ties or stirrups satisfying the size and spacing limitations in Section 1907.10.5 or by welded wire fabric of equivalent area. Such ties or stirrups shall be provided throughout the distance where compression reinforcement is required.

1907.11.2 Lateral reinforcement for flexural framing members subject to stress reversals or to torsion at supports shall consist of closed ties, closed stirrups, or spirals extending around the flexural reinforcement.

1907.11.3 Closed ties or stirrups may be formed in one piece by overlapping standard stirrup or tie end hooks around a longitudinal bar, or formed in one or two pieces lap spliced with a Class B splice (lap of 1.3 l_d), or anchored in accordance with Section 1912.13.

1907.12 Shrinkage and Temperature Reinforcement.

1907.12.1 Reinforcement for shrinkage and temperature stresses normal to flexural reinforcement shall be provided in structural slabs where the flexural reinforcement extends in one direction only.

1907.12.1.1 Shrinkage and temperature reinforcement shall be provided in accordance with either Section 1907.12.2 or 1907.12.3 below.

1907.12.1.2 Where shrinkage and temperature movements are significantly restrained, the requirements of Sections 1908.2.4 and 1909.2.7 shall be considered.

1907.12.2 Deformed reinforcement conforming to Section 1903.5.3 used for shrinkage and temperature reinforcement shall be provided in accordance with the following:

1907.12.2.1 Area of shrinkage and temperature reinforcement shall provide at least the following ratios of reinforcement area to gross concrete area, but not less than 0.0014:

1. Slabs where Grade 40 or 50 deformed bars are used   0.0020

2. Slabs where Grade 60 deformed bars or welded wire fabric (smooth or deformed) are used   0.0018

3. Slabs where reinforcement with yield stress exceeding 60,000 psi (413.7 MPa) measured at a yield strain of 0.35 percent is used

$$\frac{0.0018 \times 60,000}{f_y}$$

For SI:

$$\frac{0.0018 \times 413.7}{f_y}$$

1907.12.2.2 Shrinkage and temperature reinforcement shall be spaced not farther apart than five times the slab thickness, or 18 inches (457 mm).

1907.12.2.3 At all sections where required, reinforcement for shrinkage and temperature stresses shall develop the specified yield strength f_y in tension in accordance with Section 1912.

1907.12.3 Prestressing tendons conforming to Section 1903.5.5 used for shrinkage and temperature reinforcement shall be provided in accordance with the following:

1907.12.3.1 Tendons shall be proportioned to provide a minimum average compressive stress of 100 psi (0.69 MPa) on gross concrete area using effective prestress, after losses, in accordance with Section 1918.6.

1907.12.3.2 Spacing of prestressed tendons shall not exceed 6 feet (1829 mm).

1907.12.3.3 When the spacing of prestressed tendons exceeds 54 inches (1372 mm), additional bonded shrinkage and temperature reinforcement conforming with Section 1907.12.2 shall be provided between the tendons at slab edges extending from the slab edge for a distance equal to the tendon spacing.

1907.13 Requirements for Structural Integrity.

1907.13.1 In the detailing of reinforcement and connections, members of a structure shall be effectively tied together to improve integrity of the overall structure.

1907.13.2 For cast-in-place construction, the following shall constitute minimum requirements:

1907.13.2.1 In joist construction, at least one bottom bar shall be continuous or shall be spliced over the support with a Class A tension splice and at noncontinuous supports be terminated with a standard hook.

1907.13.2.2 Beams at the perimeter of the structure shall have at least one sixth of the tension reinforcement required for negative moment at the support and one-quarter of the positive moment reinforcement required at midspan made continuous around the perimeter and tied with closed stirrups or stirrups anchored around the negative moment reinforcement with a hook having a bend of at least 135 degrees. Stirrups need not be extended through any joints. When splices are needed, the required continuity shall be provided with top reinforcement spliced at midspan and bottom reinforcement spliced at or near the support with Class A tension splices.

1907.13.2.3 In other than perimeter beams, when closed stirrups are not provided, at least one-quarter of the positive moment reinforcement required at midspan shall be continuous or shall be spliced over the support with a Class A tension splice and at noncontinuous supports be terminated with a standard hook.

1907.13.2.4 For two-way slab construction, see Section 1913.3.8.5.

1907.13.3 For precast concrete construction, tension ties shall be provided in the transverse, longitudinal, and vertical directions and around the perimeter of the structure to effectively tie elements together. The provisions of Section 1916.5 shall apply.

1907.13.4 For lift-slab construction, see Sections 1913.3.8.6 and 1918.12.6.

SECTION 1915 — FOOTINGS

1915.7 Minimum Footing Depth. Depth of footing above bottom reinforcement shall not be less than 6 inches (152 mm) for footings on soil, or not less than 12 inches (305 mm) for footings on piles.

SECTION 1922 — STRUCTURAL PLAIN CONCRETE

1922.1 Scope.

1922.1.1 This section provides minimum requirements for design and construction of structural plain concrete members (cast-in-place or precast) except as specified in Sections 1922.1.1.1 and 1922.1.1.2.

> **EXCEPTION:** *The design is not required when the minimum foundation for stud walls is in accordance with Table 18-I-C.*

1922.1.1.1 Structural plain concrete basement walls shall be exempted from the requirements for special exposure conditions of Section 1904.2.2.

1922.1.1.2 Design and construction of soil-supported slabs, such as sidewalks and slabs on grade shall not be regulated by this code unless they transmit vertical loads from other parts of the structure to the soil.

1922.1.2 For special structures, such as arches, underground utility structures, gravity walls, and shielding walls, provisions of this section shall govern where applicable.

1922.2 Limitations.

1922.2.1 Provisions of this section shall apply for design of structural plain concrete members defined as either unreinforced or containing less reinforcement than the minimum amount specified in this code for reinforced concrete.

1922.2.2 Use of structural plain concrete shall be limited to (1) members that are continuously supported by soil or supported by other structural members capable of providing continuous vertical support, (2) members for which arch action provides compression under all conditions of loading, or (3) walls and pedestals. See Sections 1922.6 and 1922.8. The use of structural plain concrete columns is not permitted.

1922.2.3 This section does not govern design and installation of cast-in-place concrete piles and piers embedded in ground.

1922.2.4 Minimum strength. Specified compressive strength of concrete, f'_c, used in structural plain concrete elements shall not be less than 2,500 psi (17.2 MPa).

1922.2.5 Seismic Zones 2, 3 and 4. Plain concrete shall not be used in Seismic Zone 2, 3 or 4 except where specifically permitted by Section 1922.10.3.

1922.3 Joints.

1922.3.1 Contraction or isolation joints shall be provided to divide structural plain concrete members into flexurally discontinuous elements. Size of each element shall be limited to control buildup of excessive internal stresses within each element caused by restraint to movements from creep, shrinkage and temperature effects.

1922.3.2 In determining the number and location of contraction or isolation joints, consideration shall be given to: influence of climatic conditions; selection and proportioning of materials; mixing, placing and curing of concrete; degree of restraint to movement; stresses due to loads to which an element is subject; and construction techniques.

TABLE 19-B—MINIMUM DIAMETERS OF BEND

BAR SIZE	MINIMUM DIAMETER
Nos. 3 through 8	$6d_b$
Nos. 9, 10 and 11	$8d_b$
Nos. 14 and 18	$10d_b$

Chapter 20

Chapter 20 is printed in Volume 2 of the *Uniform Building Code*.

Chapter 21 is printed in its entirety in Volume 2 of the *Uniform Building Code.*
Excerpts from Chapter 21 are reprinted herein.

Excerpts from Chapter 21
MASONRY

SECTION 2101 — GENERAL

2101.1 Scope. The materials, design, construction and quality assurance of masonry shall be in accordance with this chapter.

2101.2 Design Methods. Masonry shall comply with the provisions of one of the following design methods in this chapter as well as the requirements of Sections 2101 through 2105.

2101.2.1 Working stress design. Masonry designed by the working stress design method shall comply with the provisions of Sections 2106 and 2107.

2101.2.2 Strength design. Masonry designed by the strength design method shall comply with the provisions of Sections 2106 and 2108.

2101.2.3 Empirical design. Masonry designed by the empirical design method shall comply with the provisions of Sections 2106.1 and 2109.

2101.2.4 Glass masonry. Glass masonry shall comply with the provisions of Section 2110.

2101.3 Definitions. For the purpose of this chapter, certain terms are defined as follows:

AREAS:

Bedded Area is the area of the surface of a masonry unit which is in contact with mortar in the plane of the joint.

Effective Area of Reinforcement is the cross-sectional area of reinforcement multiplied by the cosine of the angle between the reinforcement and the direction for which effective area is to be determined.

Gross Area is the total cross-sectional area of a specified section.

Net Area is the gross cross-sectional area minus the area of ungrouted cores, notches, cells and unbedded areas. Net area is the actual surface area of a cross section of masonry.

Transformed Area is the equivalent area of one material to a second based on the ratio of moduli of elasticity of the first material to the second.

BOND:

Adhesion Bond is the adhesion between masonry units and mortar or grout.

Reinforcing Bond is the adhesion between steel reinforcement and mortar or grout.

BOND BEAM is a horizontal grouted element within masonry in which reinforcement is embedded.

CELL is a void space having a gross cross-sectional area greater than $1^1/_2$ square inches (967 mm^2).

CLEANOUT is an opening to the bottom of a grout space of sufficient size and spacing to allow the removal of debris.

COLLAR JOINT is the mortared or grouted space between wythes of masonry.

COLUMN, REINFORCED, is a vertical structural member in which both the reinforcement and masonry resist compression.

COLUMN, UNREINFORCED, is a vertical structural member whose horizontal dimension measured at right angles to the thickness does not exceed three times the thickness.

DIMENSIONS:

Actual Dimensions are the measured dimensions of a designated item. The actual dimension shall not vary from the specified dimension by more than the amount allowed in the appropriate standard of quality in Section 2102.

Nominal Dimensions of masonry units are equal to its specified dimensions plus the thickness of the joint with which the unit is laid.

Specified Dimensions are the dimensions specified for the manufacture or construction of masonry, masonry units, joints or any other component of a structure.

GROUT LIFT is an increment of grout height within the total grout pour.

GROUT POUR is the total height of masonry wall to be grouted prior to the erection of additional masonry. A grout pour will consist of one or more grout lifts.

GROUTED MASONRY:

Grouted Hollow-unit Masonry is that form of grouted masonry construction in which certain designated cells of hollow units are continuously filled with grout.

Grouted Multiwythe Masonry is that form of grouted masonry construction in which the space between the wythes is solidly or periodically filled with grout.

JOINTS:

Bed Joint is the mortar joint that is horizontal at the time the masonry units are placed.

Head Joint is the mortar joint having a vertical transverse plane.

MASONRY UNIT is brick, tile, stone, glass block or concrete block conforming to the requirements specified in Section 2102.

Hollow-masonry Unit is a masonry unit whose net cross-sectional areas (solid area) in any plane parallel to the surface containing cores, cells or deep frogs is less than 75 percent of its gross cross-sectional area measured in the same plane.

Solid-masonry Unit is a masonry unit whose net cross-sectional area in any plane parallel to the surface containing the cores or cells is at least 75 percent of the gross cross-sectional area measured in the same plane.

PRISM is an assemblage of masonry units and mortar with or without grout used as a test specimen for determining properties of the masonry.

REINFORCED MASONRY is that form of masonry construction in which reinforcement acting in conjunction with the masonry is used to resist forces.

SHELL is the outer portion of a hollow masonry unit as placed in masonry.

WALLS

Bonded Wall is a masonry wall in which two or more wythes are bonded to act as a structural unit.

Cavity Wall is a wall containing continuous air space with a minimum width of 2 inches (51 mm) and a maximum width of $4^1/_2$ inches (114 mm) between wythes which are tied with metal ties.

WALL TIE is a mechanical metal fastener which connects wythes of masonry to each other or to other materials.

WEB is an interior solid portion of a hollow-masonry unit as placed in masonry.

WYTHE is the portion of a wall which is one masonry unit in thickness. A collar joint is not considered a wythe.

SECTION 2102 — MATERIAL STANDARDS

2102.1 Quality. Materials used in masonry shall conform to the requirements stated herein. If no requirements are specified in this section for a material, quality shall be based on generally accepted good practice, subject to the approval of the building official.

Reclaimed or previously used masonry units shall meet the applicable requirements as for new masonry units of the same material for their intended use.

2102.2 Standards of Quality. The standards listed below labeled a "UBC Standard" are also listed in Chapter 35, Part II, and are part of this code. The other standards listed below are recognized standards. See Sections 3503 and 3504.

1. **Aggregates.**

    1.1 ASTM C 144, Aggregates for Masonry Mortar

    1.2 ASTM C 404, Aggregates for Grout

2. **Cement.**

    2.1 UBC Standard 21-11, Cement, Masonry. (Plastic cement conforming to the requirements of UBC Standard 25-1 may be used in lieu of masonry cement when it also conforms to UBC Standard 21-11.)

    2.2 ASTM C 150, Portland Cement

    2.3 UBC Standard 21-14, Mortar Cement

3. **Lime.**

    3.1 UBC Standard 21-12, Quicklime for Structural Purposes

    3.2 UBC Standard 21-13, Hydrated Lime for Masonry Purposes. When Types N and NA hydrated lime are used in masonry mortar, they shall comply with the provisions of UBC Standard 21-15, Section 21.1506.7, excluding the plasticity requirement.

4. **Masonry units of clay or shale.**

    4.1 ASTM C 34, Structural Clay Load-bearing Wall Tile

    4.2 ASTM C 56, Structural Clay Nonload-bearing Tile

    4.3 UBC Standard 21-1, Section 21.101, Building Brick (solid units)

    4.4 ASTM C 126, Ceramic Glazed Structural Clay Facing Tile, Facing Brick and Solid Masonry Units. Load-bearing glazed brick shall conform to the weathering and structural requirements of UBC Standard 21-1, Section 21.106, Facing Brick

    4.5 UBC Standard 21-1, Section 21.106, Facing Brick (solid units)

    4.6 UBC Standard 21-1, Section 21.107, Hollow Brick

    4.7 ASTM C 67, Sampling and Testing Brick and Structural Clay Tile

    4.8 ASTM C 212, Structural Clay Facing Tile

    4.9 ASTM C 530, Structural Clay Non-Loadbearing Screen Tile

5. **Masonry units of concrete.**

    5.1 UBC Standard 21-3, Concrete Building Brick

    5.2 UBC Standard 21-4, Hollow and Solid Load-bearing Concrete Masonry Units

    5.3 UBC Standard 21-5, Nonload-bearing Concrete Masonry Units

    5.4 ASTM C 140, Sampling and Testing Concrete Masonry Units

    5.5 ASTM C 426, Standard Test Method for Drying Shrinkage of Concrete Block

6. **Masonry units of other materials.**

    6.1 **Calcium silicate.**

        UBC Standard 21-2, Calcium Silicate Face Brick (Sand-lime Brick)

    6.2 UBC Standard 21-9, Unburned Clay Masonry Units and Standard Methods of Sampling and Testing Unburned Clay Masonry Units

    6.3 ACI-704, Cast Stone

    6.4 UBC Standard 21-17, Test Method for Compressive Strength of Masonry Prisms

7. **Connectors.**

    7.1 Wall ties and anchors made from steel wire shall conform to UBC Standard 21-10, Part II, and other steel wall ties and anchors shall conform to A 36 in accordance with UBC Standard 22-1. Wall ties and anchors made from copper, brass or other nonferrous metal shall have a minimum tensile yield strength of 30,000 psi (207 MPa).

    7.2 All such items not fully embedded in mortar or grout shall either be corrosion resistant or shall be coated after fabrication with copper, zinc or a metal having at least equivalent corrosion-resistant properties.

8. **Mortar.**

    8.1 UBC Standard 21-15, Mortar for Unit Masonry and Reinforced Masonry other than Gypsum

    8.2 UBC Standard 21-16, Field Tests Specimens for Mortar

    8.3 UBC Standard 21-20, Standard Test Method for Flexural Bond Strength of Mortar Cement

9. **Grout.**

    9.1 UBC Standard 21-18, Method of Sampling and Testing Grout

    9.2 UBC Standard 21-19, Grout for Masonry

10. **Reinforcement.**

    10.1 UBC Standard 21-10, Part I, Joint Reinforcement for Masonry

    10.2 ASTM A 615, A 616, A 617, A 706, A 767, and A 775, Deformed and Plain Billet-steel Bars, Rail-steel Deformed and Plain Bars, Axle-steel Deformed and Plain Bars, and Deformed Low-alloy Bars for Concrete Reinforcement

10.3 UBC Standard 21-10, Part II, Cold-drawn Steel Wire for Concrete Reinforcement

SECTION 2103 — MORTAR AND GROUT

2103.1 General. Mortar and grout shall comply with the provisions of this section. Special mortars, grouts or bonding systems may be used, subject to satisfactory evidence of their capabilities when approved by the building official.

2103.2 Materials. Materials used as ingredients in mortar and grout shall conform to the applicable requirements in Section 2102. Cementitious materials for grout shall be one or both of the following: lime and portland cement. Cementitious materials for mortar shall be one or more of the following: lime, masonry cement, portland cement and mortar cement. Cementitious materials or additives shall not contain epoxy resins and derivatives, phenols, asbestos fibers or fireclays.

Water used in mortar or grout shall be clean and free of deleterious amounts of acid, alkalies or organic material or other harmful substances.

2103.3 Mortar.

2103.3.1 General. Mortar shall consist of a mixture of cementitious materials and aggregate to which sufficient water and approved additives, if any, have been added to achieve a workable, plastic consistency.

2103.3.2 Selecting proportions. Mortar with specified proportions of ingredients that differ from the mortar proportions of Table 21-A may be approved for use when it is demonstrated by laboratory or field experience that this mortar with the specified proportions of ingredients, when combined with the masonry units to be used in the structure, will achieve the specified compressive strength f'_m. Water content shall be adjusted to provide proper workability under existing field conditions. When the proportion of ingredients is not specified, the proportions by mortar type shall be used as given in Table 21-A.

2103.4 Grout.

2103.4.1 General. Grout shall consist of a mixture of cementitious materials and aggregate to which water has been added such that the mixture will flow without segregation of the constituents. The specified compressive strength of grout, f'_g, shall not be less than 2,000 psi (13.8 MPa).

2103.4.2 Selecting proportions. Water content shall be adjusted to provide proper workability and to enable proper placement under existing field conditions, without segregation. Grout shall be specified by one of the following methods:

1. Proportions of ingredients and any additives shall be based on laboratory or field experience with the grout ingredients and the masonry units to be used. The grout shall be specified by the proportion of its constituents in terms of parts by volume, or

2. Minimum compressive strength which will produce the required prism strength, or

3. Proportions by grout type shall be used as given in Table 21-B.

2103.5 Additives and Admixtures.

2103.5.1 General. Additives and admixtures to mortar or grout shall not be used unless approved by the building official.

2103.5.2 Antifreeze compounds. Antifreeze liquids, chloride salts or other such substances shall not be used in mortar or grout.

2103.5.3 Air entrainment. Air-entraining substances shall not be used in mortar or grout unless tests are conducted to determine compliance with the requirements of this code.

2103.5.4 Colors. Only pure mineral oxide, carbon black or synthetic colors may be used. Carbon black shall be limited to a maximum of 3 percent of the weight of the cement.

SECTION 2104 — CONSTRUCTION

2104.1 General. Masonry shall be constructed according to the provisions of this section.

2104.2 Materials: Handling, Storage and Preparation. All materials shall comply with applicable requirements of Section 2102. Storage, handling and preparation at the site shall conform also to the following:

1. Masonry materials shall be stored so that at the time of use the materials are clean and structurally suitable for the intended use.

2. All metal reinforcement shall be free from loose rust and other coatings that would inhibit reinforcing bond.

3. At the time of laying, burned clay units and sand lime units shall have an initial rate of absorption not exceeding 0.035 ounce per square inch (1.6 L/m^2) during a period of one minute. In the absorption test, the surface of the unit shall be held $1/8$ inch (3 mm) below the surface of the water.

4. Concrete masonry units shall not be wetted unless otherwise approved.

5. Materials shall be stored in a manner such that deterioration or intrusion of foreign materials is prevented and that the material will be capable of meeting applicable requirements at the time of mixing or placement.

6. The method of measuring materials for mortar and grout shall be such that proportions of the materials can be controlled.

7. Mortar or grout mixed at the jobsite shall be mixed for a period of time not less than three minutes or more than 10 minutes in a mechanical mixer with the amount of water required to provide the desired workability. Hand mixing of small amounts of mortar is permitted. Mortar may be retempered. Mortar or grout which has hardened or stiffened due to hydration of the cement shall not be used. In no case shall mortar be used two and one-half hours, nor grout used one and one-half hours, after the initial mixing water has been added to the dry ingredients at the jobsite.

> **EXCEPTION:** Dry mixes for mortar and grout which are blended in the factory and mixed at the jobsite shall be mixed in mechanical mixers until workable, but not to exceed 10 minute

2104.3 Cold-weather Construction.

2104.3.1 General. All materials shall be delivered in a usable condition and stored to prevent wetting by capillary action, rain and snow.

The tops of all walls not enclosed or sheltered shall be covered with a strong weather-resistive material at the end of each day or shutdown.

Partially completed walls shall be covered at all times when work is not in progress. Covers shall be draped over the wall and extend a minimum of 2 feet (600 mm) down both sides and shall be securely held in place, except when additional protection is required in Section 2104.3.4.

2104.3.2 Preparation. If ice or snow has inadvertently formed on a masonry bed, it shall be thawed by application of heat carefully applied until top surface of the masonry is dry to the touch.

A section of masonry deemed frozen and damaged shall be removed before continuing construction of that section.

2104.3.3 Construction. Masonry units shall be dry at time of placement. Wet or frozen masonry units shall not be laid.

Special requirements for various temperature ranges are as follows:

1. Air temperature 40°F to 32°F (4.5°C to 0°C): Sand or mixing water shall be heated to produce mortar temperatures between 40°F and 120°F (4.5°C and 49°C).

2. Air temperature 32°F to 25°F (0°C to –4°C): Sand and mixing water shall be heated to produce mortar temperatures between 40°F and 120°F (4.5°C and 49°C). Maintain temperatures of mortar on boards above freezing.

3. Air temperature 25°F to 20°F (–4°C to –7°C): Sand and mixing water shall be heated to produce mortar temperatures between 40°F and 120°F (4.5°C and 49°C). Maintain mortar temperatures on boards above freezing. Salamanders or other sources of heat shall be used on both sides of walls under construction. Windbreaks shall be employed when wind is in excess of 15 miles per hour (24 km/h).

4. Air temperature 20°F (–7°C) and below: Sand and mixing water shall be heated to produce mortar temperatures between 40°F and 120°F (4.5°C and 49°C). Enclosure and auxiliary heat shall be provided to maintain air temperature above freezing. Temperature of units when laid shall not be less than 20°F (–7°C).

2104.3.4 Protection. When the mean daily air temperature is 40°F to 32°F (4.5°C to 0°C), masonry shall be protected from rain or snow for 24 hours by covering with a weather-resistive membrane.

When the mean daily air temperature is 32°F to 25°F (0°C to –4°C), masonry shall be completely covered with a weather-resistive membrane for 24 hours.

When the mean daily air temperature is 25°F to 20°F (–4°C to –7°C), masonry shall be completely covered with insulating blankets or equally protected for 24 hours.

When the mean daily air temperature is 20°F (–7°C) or below, masonry temperature shall be maintained above freezing for 24 hours by enclosure and supplementary heat, by electric heating blankets, infrared heat lamps or other approved methods.

2104.3.5 Placing grout and protection of grouted masonry. When air temperatures fall below 40°F (4.5°C), grout mixing water and aggregate shall be heated to produce grout temperatures between 40°F and 120°F (4.5°C and 49°C).

Masonry to be grouted shall be maintained above freezing during grout placement and for at least 24 hours after placement.

When atmospheric temperatures fall below 20°F (–7°C), enclosures shall be provided around the masonry during grout placement and for at least 24 hours after placement.

2104.4 Placing Masonry Units.

2104.4.1 Mortar. The mortar shall be sufficiently plastic and units shall be placed with sufficient pressure to extrude mortar from the joint and produce a tight joint. Deep furrowing which produces voids shall not be used.

The initial bed joint thickness shall not be less than 1/4 inch (6 mm) or more than 1 inch (25 mm); subsequent bed joints shall not be less than 1/4 inch (6 mm) or more than 5/8 inch (16 mm) in thickness.

2104.4.2 Surfaces. Surfaces to be in contact with mortar or grout shall be clean and free of deleterious materials.

2104.4.3 Solid masonry units. Solid masonry units shall have full head and bed joints.

2104.4.4 Hollow-masonry units. All head and bed joints shall be filled solidly with mortar for a distance in from the face of the unit not less than the thickness of the shell.

Head joints of open-end units with beveled ends that are to be fully grouted need not be mortared. The beveled ends shall form a grout key which permits grout within 5/8 inch (16 mm) of the face of the unit. The units shall be tightly butted to prevent leakage of grout.

2104.5 Reinforcement Placing. Reinforcement details shall conform to the requirements of this chapter. Metal reinforcement shall be located in accordance with the plans and specifications. Reinforcement shall be secured against displacement prior to grouting by wire positioners or other suitable devices at intervals not exceeding 200 bar diameters.

Tolerances for the placement of reinforcement in walls and flexural elements shall be plus or minus 1/2 inch (12.7 mm) for d equal to 8 inches (200 mm) or less, ± 1 inch (± 25 mm) for d equal to 24 inches (600 mm) or less but greater than 8 inches (200 mm), and ± 1 1/4 inches (32 mm) for d greater than 24 inches (600 mm).

Tolerance for longitudinal location of reinforcement shall be ± 2 inches (51 mm).

2104.6 Grouted Masonry.

2104.6.1 General conditions. Grouted masonry shall be constructed in such a manner that all elements of the masonry act together as a structural element.

Prior to grouting, the grout space shall be clean so that all spaces to be filled with grout do not contain mortar projections greater than 1/2 inch (12.7 mm), mortar droppings or other foreign material. Grout shall be placed so that all spaces designated to be grouted shall be filled with grout and the grout shall be confined to those specific spaces.

Grout materials and water content shall be controlled to provide adequate fluidity for placement without segregation of the constituents, and shall be mixed thoroughly.

The grouting of any section of wall shall be completed in one day with no interruptions greater than one hour.

Between grout pours, a horizontal construction joint shall be formed by stopping all wythes at the same elevation and with the grout stopping a minimum of 1 1/2 inches (38 mm) below a mortar joint, except at the top of the wall. Where bond beams occur, the grout pour shall be stopped a minimum of 1/2 inch (12.7 mm) below the top of the masonry.

Size and height limitations of the grout space or cell shall not be less than shown in Table 21-C. Higher grout pours or smaller cavity widths or cell size than shown in Table 21-C may be used when approved, if it is demonstrated that grout spaces will be properly filled.

Cleanouts shall be provided for all grout pours over 5 feet (1524 mm) in height.

Where required, cleanouts shall be provided in the bottom course at every vertical bar but shall not be spaced more than 32 inches (813 mm) on center for solidly grouted masonry. When cleanouts are required, they shall be sealed after inspection and before grouting.

Where cleanouts are not provided, special provisions must be made to keep the bottom and sides of the grout spaces, as well as the minimum total clear area as required by Table 21-C, clean and clear prior to grouting.

Units may be laid to the full height of the grout pour and grout shall be placed in a continuous pour in grout lifts not exceeding 6 feet (1830 mm). When approved, grout lifts may be greater than 6 feet (1830 mm) if it can be demonstrated the grout spaces can be properly filled.

All cells and spaces containing reinforcement shall be filled with grout.

2104.6.2 Construction requirements. Reinforcement shall be placed prior to grouting. Bolts shall be accurately set with templates or by approved equivalent means and held in place to prevent dislocation during grouting.

Segregation of the grout materials and damage to the masonry shall be avoided during the grouting process.

Grout shall be consolidated by mechanical vibration during placement before loss of plasticity in a manner to fill the grout space. Grout pours greater than 12 inches (300 mm) in height shall be reconsolidated by mechanical vibration to minimize voids due to water loss. Grout pours 12 inches (300 mm) or less in height shall be mechanically vibrated or puddled.

In one-story buildings having wood-frame exterior walls, foundations not over 24 inches (600 mm) high measured from the top of the footing may be constructed of hollow-masonry units laid in running bond without mortared head joints. Any standard shape unit may be used, provided the masonry units permit horizontal flow of grout to adjacent units. Grout shall be solidly poured to the full height in one lift and shall be puddled or mechanically vibrated.

In nonstructural elements which do not exceed 8 feet (2440 mm) in height above the highest point of lateral support, including fireplaces and residential chimneys, mortar of pouring consistency may be substituted for grout when the masonry is constructed and grouted in pours of 12 inches (300 mm) or less in height.

In multiwythe grouted masonry, vertical barriers of masonry shall be built across the grout space the entire height of the grout pour and spaced not more than 30 feet (9144 mm) horizontally. The grouting of any section of wall between barriers shall be completed in one day with no interruption longer than one hour.

2104.7 Aluminum Equipment. Grout shall not be handled nor pumped utilizing aluminum equipment unless it can be demonstrated with the materials and equipment to be used that there will be no deleterious effect on the strength of the grout.

2104.8 Joint Reinforcement. Wire joint reinforcement used in the design as principal reinforcement in hollow-unit construction shall be continuous between supports unless splices are made by lapping:

1. Fifty-four wire diameters in a grouted cell, or

2. Seventy-five wire diameters in the mortared bed joint, or

3. In alternate bed joints of running bond masonry a distance not less than 54 diameters plus twice the spacing of the bed joints, or

4. As required by calculation and specific location in areas of minimum stress, such as points of inflection.

Side wires shall be deformed and shall conform to UBC Standard 21-10, Part I, Joint Reinforcement for Masonry.

SECTION 2105 — QUALITY ASSURANCE

2105.1 General. Quality assurance shall be provided to ensure that materials, construction and workmanship are in compliance with the plans and specifications, and the applicable requirements of this chapter. When required, inspection records shall be maintained and made available to the building official.

2105.2 Scope. Quality assurance shall include, but is not limited to, assurance that:

1. Masonry units, reinforcement, cement, lime, aggregate and all other materials meet the requirements of the applicable standards of quality and that they are properly stored and prepared for use.

2. Mortar and grout are properly mixed using specified proportions of ingredients. The method of measuring materials for mortar and grout shall be such that proportions of materials are controlled.

3. Construction details, procedures and workmanship are in accordance with the plans and specifications.

4. Placement, splices and reinforcement sizes are in accordance with the provisions of this chapter and the plans and specifications.

2105.3 Compliance with f'_m.

2105.3.1 General. Compliance with the requirements for the specified compressive strength of masonry f'_m shall be in accordance with one of the sections in this subsection.

2105.3.2 Masonry prism testing. The compressive strength of masonry determined in accordance with UBC Standard 21-17 for each set of prisms shall equal or exceed f'_m. Compressive strength of prisms shall be based on tests at 28 days. Compressive strength at seven days or three days may be used provided a relationship between seven-day and three-day and 28-day strength has been established for the project prior to the start of construction. Verification by masonry prism testing shall meet the following:

1. A set of five masonry prisms shall be built and tested in accordance with UBC Standard 21-17 prior to the start of construction. Materials used for the construction of the prisms shall be taken from those specified to be used in the project. Prisms shall be constructed under the observation of the engineer or special inspector or an approved agency and tested by an approved agency.

2. When full allowable stresses are used in design, a set of three prisms shall be built and tested during construction in accordance with UBC Standard 21-17 for each 5,000 square feet (465 m^2) of wall area, but not less than one set of three masonry prisms for the project.

3. When one half the allowable masonry stresses are used in design, testing during construction is not required. A letter of certification from the supplier of the materials used to verify the f'_m in accordance with Section 2105.3.2, Item 1, shall be provided at the time of, or prior to, delivery of the materials to the jobsite to ensure the materials used in construction are representative of the materials used to construct the prisms prior to construction.

2105.3.3 Masonry prism test record. Compressive strength verification by masonry prism test records shall meet the following:

1. A masonry prism test record approved by the building official of at least 30 masonry prisms which were built and tested in accordance with UBC Standard 21-17. Prisms shall have been constructed under the observation of an engineer or special inspector or an approved agency and shall have been tested by an approved agency.

2. Masonry prisms shall be representative of the corresponding construction.

3. The average compressive strength of the test record shall equal or exceed $1.33 f'_m$.

4. When full allowable stresses are used in design, a set of three masonry prisms shall be built during construction in accordance with UBC Standard 21-17 for each 5,000 square feet (465 m^2) of wall area, but not less than one set of three prisms for the project.

5. When one half the allowable masonry stresses are used in design, field testing during construction is not required. A letter of certification from the supplier of the materials to the jobsite shall be provided at the time of, or prior to, delivery of the materials to assure the materials used in construction are representative of the materials used to develop the prism test record in accordance with Section 2105.3.3, Item 1.

2105.3.4 Unit strength method. Verification by the unit strength method shall meet the following:

1. When full allowable stresses are used in design, units shall be tested prior to construction and test units during construction for each 5,000 square feet (465 m^2) of wall area for compressive strength to show compliance with the compressive strength required in Table 21-D; and

> **EXCEPTION:** Prior to the start of construction, prism testing may be used in lieu of testing the unit strength. During construction, prism testing may also be used in lieu of testing the unit strength and the grout as required by Section 2105.3.4, Item 4.

2. When one half the allowable masonry stresses are used in design, testing is not required for the units. A letter of certification from the manufacturer of the units shall be provided at the time of, or prior to, delivery of the units to the jobsite to assure the units comply with the compressive strength required in Table 21-D; and

3. Mortar shall comply with the mortar type required in Table 21-D; and

4. When full stresses are used in design for concrete masonry, grout shall be tested for each 5,000 square feet (465 m^2) of wall area, but not less than one test per project, to show compliance with the compressive strength required in Table 21-D, Footnote 4.

5. When one half the allowable masonry stresses are used in design for concrete masonry, testing is not required for the grout. A letter of certification from the supplier of the grout shall be provided at the time of, or prior to, delivery of the grout to the jobsite to assure the grout complies with the compressive strength required in Table 21-D, Footnote 4; or

6. When full allowable stresses are used in design for clay masonry, grout proportions shall be verified by the engineer or special inspector or an approved agency to conform with Table 21-B.

7. When one half the allowable masonry stresses are used in design for clay masonry, a letter of certification from the supplier of the grout shall be provided at the time of, or prior to, delivery of the grout to the jobsite to assure the grout conforms to the proportions of Table 21-B.

2105.3.5 Testing prisms from constructed masonry. When approved by the building official, acceptance of masonry which does not meet the requirements of Section 2105.3.2, 2105.3.3 or 2105.3.4 shall be permitted to be based on tests of prisms cut from the masonry construction in accordance with the following:

1. A set of three masonry prisms that are at least 28 days old shall be saw cut from the masonry for each 5,000 square feet (465 m^2) of the wall area that is in question but not less than one set of three masonry prisms for the project. The length, width and height dimensions of the prisms shall comply with the requirements of UBC Standard 21-17. Transporting, preparation and testing of prisms shall be in accordance with UBC Standard 21-17.

2. The compressive strength of prisms shall be the value calculated in accordance with UBC Standard 21-17, Section 21.1707.2, except that the net cross-sectional area of the prism shall be based on the net mortar bedded area.

3. Compliance with the requirement for the specified compressive strength of masonry, f'_m, shall be considered satisfied provided the modified compressive strength equals or exceeds the specified f'_m. Additional testing of specimens cut from locations in question shall be permitted.

2105.4 Mortar Testing. When required, mortar shall be tested in accordance with UBC Standard 21-16.

2105.5 Grout Testing. When required, grout shall be tested in accordance with UBC Standard 21-18.

SECTION 2107 — WORKING STRESS DESIGN OF MASONRY

2107.2.2.1 Maximum reinforcement size. The maximum size of reinforcement shall be No. 11 bars. Maximum reinforcement area in cells shall be 6 percent of the cell area without splices and 12 percent of the cell area with splices.

2107.2.2.2 Cover. All reinforcing bars, except joint reinforcement, shall be completely embedded in mortar or grout and have a minimum cover, including the masonry unit, of at least $^3/_4$ inch (19 mm), $1^1/_2$ inches (38 mm) of cover when the masonry is exposed to weather and 2 inches (51 mm) of cover when the masonry is exposed to soil.

SECTION 2109 — EMPIRICAL DESIGN OF MASONRY

2109.1 General. The design of masonry structures using empirical design located in those portions of Seismic Zones 0 and 1 as defined in Part III of Chapter 16 where the basic wind speed is less than 80 miles per hour as defined in Part II of Chapter 16 shall comply with the provisions of Section 2106 and this section, subject to approval of the building official.

2109.2 Height. Buildings relying on masonry walls for lateral load resistance shall not exceed 35 feet (10 668 mm) in height.

2109.3 Lateral Stability. Where the structure depends on masonry walls for lateral stability, shear walls shall be provided parallel to the direction of the lateral forces resisted.

Minimum nominal thickness of masonry shear walls shall be 8 inches (203 mm).

In each direction in which shear walls are required for lateral stability, the minimum cumulative length of shear walls provided shall be 0.4 times the long dimension of the building. The cumulative length of shear walls shall not include openings.

The maximum spacing of shear walls shall not exceed the ratio listed in Table 21-L.

2109.4 Compressive Stresses.

2109.4.1 General. Compressive stresses in masonry due to vertical dead loads plus live loads, excluding wind or seismic loads, shall be determined in accordance with Section 2109.4.3. Dead and live loads shall be in accordance with this code with permitted live load reductions.

2109.4.2 Allowable stresses. The compressive stresses in masonry shall not exceed the values set forth in Table 21-M. The allowable stresses given in Table 21-M for the weakest combination of the units and mortar used in any load wythe shall be used for all loaded wythes of multiwythe walls.

2109.4.3 Stress calculations. Stresses shall be calculated based on specified rather than nominal dimensions. Calculated compressive stresses shall be determined by dividing the design load by the gross cross-sectional area of the member. The area of openings, chases or recesses in walls shall not be included in the gross cross-sectional area of the wall.

2109.4.4 Anchor bolts. Bolt values shall not exceed those set forth in Table 21-N.

2109.5 Lateral Support. Masonry walls shall be laterally supported in either the horizontal or vertical direction not exceeding the intervals set forth in Table 21-O.

Lateral support shall be provided by cross walls, pilasters, buttresses or structural framing members horizontally or by floors, roof or structural framing members vertically.

Except for parapet walls, the ratio of height to nominal thickness for cantilever walls shall not exceed 6 for solid masonry or 4 for hollow masonry.

In computing the ratio for cavity walls, the value of thickness shall be the sums of the nominal thickness of the inner and outer wythes of the masonry. In walls composed of different classes of units and mortars, the ratio of height or length to thickness shall not exceed that allowed for the weakest of the combinations of units and mortar of which the member is composed.

2109.6 Minimum Thickness.

2109.6.1 General. The nominal thickness of masonry bearing walls in buildings more than one story in height shall not be less than 8 inches (203 mm). Solid masonry walls in one-story buildings may be of 6-inch nominal thickness when not over 9 feet (2743 mm) in height, provided that when gable construction is used, an additional 6 feet (1829 mm) is permitted to the peak of the gable.

> **EXCEPTION:** The thickness of unreinforced grouted brick masonry walls may be 2 inches (51 mm) less than required by this section, but in no case less than 6 inches (152 mm).

2109.6.2 Variation in thickness. Where a change in thickness due to minimum thickness occurs between floor levels, the greater thickness shall be carried up to the higher floor level.

2109.6.3 Decrease in thickness. Where walls of masonry of hollow units or masonry-bonded hollow walls are decreased in thickness, a course or courses of solid masonry shall be constructed between the walls below and the thinner wall above, or special units or construction shall be used to transmit the loads from face shells or wythes to the walls below.

2109.6.4 Parapets. Parapet walls shall be at least 8 inches (203 mm) in thickness and their height shall not exceed three times their thickness. The parapet wall shall not be thinner than the wall below.

2109.6.5 Foundation walls. Mortar used in masonry foundation walls shall be either Type M or S.

Where the height of unbalanced fill (height of finished grade above basement floor or inside grade) and the height of the wall between lateral support does not exceed 8 feet (2438 mm), and when the equivalent fluid weight of unbalanced fill does not exceed 30 pounds per cubic foot (480 kg/m^2), the minimum thickness of foundation walls shall be as set forth in Table 21-P. Maximum depths of unbalanced fill permitted in Table 21-P may be increased with the approval of the building official when local soil conditions warrant such an increase.

Where the height of unbalanced fill, height between lateral supports or equivalent fluid weight of unbalanced fill exceeds that set forth above, foundation walls shall be designed in accordance with Chapter 18.

2109.7 Bond.

2109.7.1 General. The facing and backing of multiwythe masonry walls shall be bonded in accordance with this section.

2109.7.2 Masonry headers. Where the facing and backing of solid masonry construction are bonded by masonry headers, not less than 4 percent of the wall surface of each face shall be composed of headers extending not less than 3 inches (76 mm) into the backing. The distance between adjacent full-length headers shall not exceed 24 inches (610 mm) either vertically or horizontally. In walls in which a single header does not extend through the wall, headers from opposite sides shall overlap at least 3 inches (76 mm), or headers from opposite sides shall be covered with another header course overlapping the header below at least 3 inches (76 mm).

Where two or more hollow units are used to make up the thickness of the wall, the stretcher courses shall be bonded at vertical intervals not exceeding 34 inches (864 mm) by lapping at least 3 inches (76 mm) over the unit below, or by lapping at vertical intervals not exceeding 17 inches (432 mm) with units which are at least 50 percent greater in thickness than the units below.

2109.7.3 Wall ties. Where the facing and backing of masonry walls are bonded with $^3/_{16}$-inch-diameter (4.8 mm) wall ties or metal ties of equivalent stiffness embedded in the horizontal mortar joints, there shall be at least one metal tie for each $4^1/_2$ square feet (0.42 m^2) of wall area. Ties in alternate courses shall be staggered, the maximum vertical distance between ties shall not exceed 24 inches (610 mm), and the maximum horizontal distance shall not exceed 36 inches (914 mm). Rods bent to rectangular shape shall be used with hollow-masonry units laid with the cells vertical. In other walls, the ends of ties shall be bent to 90-degree angles to provide hooks not less than 2 inches (51 mm) long. Additional ties shall be provided at all openings, spaced not more than 3 feet (914 mm) apart around the perimeter and within 12 inches (305 mm) of the opening.

The facing and backing of masonry walls may be bonded with prefabricated joint reinforcement. There shall be at least one cross wire serving as a tie for each $2^2/_3$ square feet (0.25 m^2) of wall area. The vertical spacing of the joint reinforcement shall not exceed 16 inches (406 mm). Cross wires of prefabricated joint reinforcement shall be at least No. 9 gage wire. The longitudinal wire shall be embedded in mortar.

2109.7.4 Longitudinal bond. In each wythe of masonry, head joints in successive courses shall be offset at least one fourth of the unit length or the walls shall be reinforced longitudinally as required in Section 2106.1.12.3, Item 4.

2109.8 Anchorage.

2109.8.1 Intersecting walls. Masonry walls depending on one another for lateral support shall be anchored or bonded at locations where they meet or intersect by one of the following methods:

1. Fifty percent of the units at the intersection shall be laid in an overlapping pattern, with alternating units having a bearing of not less than 3 inches (76 mm) on the unit below.

2. Walls shall be anchored by steel connectors having a minimum section of $^1/_4$ inch by $1^1/_2$ inches (6.4 mm by 38 mm) with ends bent up at least 2 inches (51 mm), or with cross pins to form anchorage. Such anchors shall be at least 24 inches (610 mm) long and the maximum spacing shall be 4 feet (1219 mm) vertically.

3. Walls shall be anchored by joint reinforcement spaced at a maximum distance of 8 inches (203 mm) vertically. Longitudinal

rods of such reinforcement shall be at least No. 9 gage and shall extend at least 30 inches (762 mm) in each direction at the intersection.

4. Interior nonbearing walls may be anchored at their intersection, at vertical spacing of not more than 16 inches (406 mm) with joint reinforcement or $1/4$-inch (6.4 mm) mesh galvanized hardware cloth.

5. Other metal ties, joint reinforcement or anchors may be used, provided they are spaced to provide equivalent area of anchorage to that required by this section.

2109.8.2 Floor and roof anchorage. Floor and roof diaphragms providing lateral support to masonry walls shall be connected to the masonry walls by one of the following methods:

1. Wood floor joists bearing on masonry walls shall be anchored to the wall by approved metal strap anchors at intervals not exceeding 6 feet (1829 mm). Joists parallel to the wall shall be anchored with metal straps spaced not more than 6 feet (1829 mm) on center extending over and under and secured to at least three joists. Blocking shall be provided between joists at each strap anchor.

2. Steel floor joists shall be anchored to masonry walls with No. 3 bars, or their equivalent, spaced not more than 6 feet (1829 mm) on center. Where joists are parallel to the wall, anchors shall be located at joist cross bridging.

3. Roof structures shall be anchored to masonry walls with $1/2$-inch-diameter (12.7 mm) bolts at 6 feet (1829 mm) on center or their equivalent. Bolts shall extend and be embedded at least 15 inches (381 mm) into the masonry, or be hooked or welded to not less than 0.2 square inch (129 mm^2) of bond beam reinforcement placed not less than 6 inches (152 mm) from the top of the wall.

2109.8.3 Walls adjoining structural framing. Where walls are dependent on the structural frame for lateral support, they shall be anchored to the structural members with metal anchors or keyed to the structural members. Metal anchors shall consist of $1/2$-inch-diameter (12.7 mm) bolts spaced at a maximum of 4 feet (1219 mm) on center and embedded at least 4 inches (102 mm) into the masonry, or their equivalent area.

2109.9 Unburned Clay Masonry.

2109.9.1 General. Masonry of stabilized unburned clay units shall not be used in any building more than one story in height. The unsupported height of every wall of unburned clay units shall not be more than 10 times the thickness of such walls. Bearing walls shall in no case be less than 16 inches (406 mm) in thickness. All footing walls which support masonry of unburned clay units shall extend to an elevation not less than 6 inches (152 mm) above the adjacent ground at all points.

2109.9.2 Bolts. Bolt values shall not exceed those set forth in Table 21-Q.

2109.10 Stone Masonry.

2109.10.1 General. Stone masonry is that form of construction made with natural or cast stone in which the units are laid and set in mortar with all joints filled.

2109.10.2 Construction. In ashlar masonry, bond stones uniformly distributed shall be provided to the extent of not less than 10 percent of the area of exposed facets. Rubble stone masonry 24 inches (610 mm) or less in thickness shall have bond stones with a maximum spacing of 3 feet (914 mm) vertically and 3 feet

(914 mm) horizontally and, if the masonry is of greater thickness than 24 inches (610 mm), shall have one bond stone for each 6 square feet (0.56 m^2) of wall surface on both sides.

2109.10.3 Minimum thickness. The thickness of stone masonry bearing walls shall not be less than 16 inches (406 mm).

SECTION 2110 — GLASS MASONRY

2110.1 General. Masonry of glass blocks may be used in non-load-bearing exterior or interior walls and in openings which might otherwise be filled with windows, either isolated or in continuous bands, provided the glass block panels have a minimum thickness of 3 inches (76 mm) at the mortar joint and the mortared surfaces of the blocks are treated for mortar bonding. Glass block may be solid or hollow and may contain inserts.

2110.2 Mortar Joints. Glass block shall be laid in Type S or N mortar. Both vertical and horizontal mortar joints shall be at least $1/4$ inch (6 mm) and not more than $3/8$ inch (9.5 mm) thick and shall be completely filled. All mortar contact surfaces shall be treated to ensure adhesion between mortar and glass.

2110.3 Lateral Support. Glass panels shall be laterally supported along each end of the panel.

Lateral support shall be provided by panel anchors spaced not more than 16 inches (406 mm) on center or by channels. The lateral support shall be capable of resisting the horizontal design forces determined in Chapter 16 or a minimum of 200 pounds per lineal foot (2920 N per linear meter) of wall, whichever is greater. The connection shall accommodate movement requirements of Section 2110.6.

2110.4 Reinforcement. Glass block panels shall have joint reinforcement spaced not more than 16 inches (406 mm) on center and located in the mortar bed joint extending the entire length of the panel. A lapping of longitudinal wires for a minimum of 6 inches (152 mm) is required for joint reinforcement splices. Joint reinforcement shall also be placed in the bed joint immediately below and above openings in the panel. Joint reinforcement shall conform to UBC Standard 21-10, Part I. Joint reinforcement in exterior panels shall be hot-dip galvanized in accordance with UBC Standard 21-10, Part I.

2110.5 Size of Panels. Glass block panels for exterior walls shall not exceed 144 square feet (13.4 m^2) of unsupported wall surface or 15 feet (4572 mm) in any dimension. For interior walls, glass block panels shall not exceed 250 square feet (23.2 m^2) of unsupported area or 25 feet (7620 mm) in any dimension.

2110.6 Expansion Joints. Glass block shall be provided with expansion joints along the sides and top, and these joints shall have sufficient thickness to accommodate displacements of the supporting structure, but not less than $3/8$ inch (9.5 mm). Expansion joints shall be entirely free of mortar and shall be filled with resilient material.

2110.7 Reuse of Units. Glass block units shall not be reused after being removed from an existing panel.

SECTION 2111 — CHIMNEYS, FIREPLACES AND BARBECUES

Chimneys, flues, fireplaces and barbecues and their connections carrying products of combustion shall be designed, anchored, supported and reinforced as set forth in Chapter 31 and any applicable provisions of this chapter.

TABLE 21-A—MORTAR PROPORTIONS FOR UNIT MASONRY

MORTAR	TYPE	Portland Cement or Blended Cement	Masonry Cement[1] M	S	N	Mortar Cement[2] M	S	N	Hydrated Lime or Lime Putty	AGGREGATE MEASURED IN A DAMP, LOOSE CONDITION
Cement-lime	M	1	—	—	—	—	—	—	$1/4$	
	S	1	—	—	—	—	—	—	over $1/4$ to $1/2$	
	N	1	—	—	—	—	—	—	over $1/2$ to $1\frac{1}{4}$	
	O	1	—	—	—	—	—	—	over $1\frac{1}{4}$ to $2\frac{1}{2}$	Not less than $2\frac{1}{4}$ and not more than 3 times the sum of the separate volumes of cementitious materials.
Mortar cement	M	1	—	—	—	—	—	1	—	
	M	—	—	—	—	1	—	—	—	
	S	$1/2$	—	—	—	—	—	1	—	
	S	—	—	—	—	—	1	—	—	
	N	—	—	—	—	—	—	1	—	
Masonry cement	M	1	—	—	1	—	—	—	—	
	M	—	1	—	—	—	—	—	—	
	S	$1/2$	—	—	1	—	—	—	—	
	S	—	—	1	—	—	—	—	—	
	N	—	—	—	1	—	—	—	—	
	O	—	—	—	1	—	—	—	—	

[1]Masonry cement conforming to the requirements of UBC Standard 21-11.

[2]Mortar cement conforming to the requirements of UBC Standard 21-14.

TABLE 21-B—GROUT PROPORTIONS BY VOLUME[1]

TYPE	PARTS BY VOLUME OF PORTLAND CEMENT OR BLENDED CEMENT	PARTS BY VOLUME OF HYDRATED LIME OR LIME PUTTY	AGGREGATE MEASURED IN A DAMP, LOOSE CONDITION Fine	Coarse
Fine grout	1	0 to $1/10$	$2\frac{1}{4}$ to 3 times the sum of the volumes of the cementitious materials	
Coarse grout	1	0 to $1/10$	$2\frac{1}{4}$ to 3 times the sum of the volumes of the cementitious materials	1 to 2 times the sum of the volumes of the cementitious materials

[1]Grout shall attain a minimum compressive strength at 28 days of 2,000 psi (13.8 MPa). The building official may require a compressive field strength test of grout made in accordance with UBC Standard 21-18.

TABLE 21-C—GROUTING LIMITATIONS

GROUT TYPE	GROUT POUR MAXIMUM HEIGHT (feet)[1] × 304.8 for mm	MINIMUM DIMENSIONS OF THE TOTAL CLEAR AREAS WITHIN GROUT SPACES AND CELLS[2,3] × 25.4 for mm Multiwythe Masonry	Hollow-unit Masonry
Fine	1	$3/4$	$1\frac{1}{2} \times 2$
Fine	5	$1\frac{1}{2}$	$1\frac{1}{2} \times 2$
Fine	8	$1\frac{1}{2}$	$1\frac{1}{2} \times 3$
Fine	12	$1\frac{1}{2}$	$1\frac{3}{4} \times 3$
Fine	24	2	3×3
Coarse	1	$1\frac{1}{2}$	$1\frac{1}{2} \times 3$
Coarse	5	2	$2\frac{1}{2} \times 3$
Coarse	8	2	3×3
Coarse	12	$2\frac{1}{2}$	3×3
Coarse	24	3	3×4

[1]See also Section 2104.6.

[2]The actual grout space or grout cell dimensions must be larger than the sum of the following items: (1) The required minimum dimensions of total clear areas in Table 21-C; (2) The width of any mortar projections within the space; and (3) The horizontal projections of the diameters of the horizontal reinforcing bars within a cross section of the grout space or cell.

[3]The minimum dimensions of the total clear areas shall be made up of one or more open areas, with at least one area being $3/4$ inch (19 mm) or greater in width.

TABLE 21-G—MINIMUM DIAMETERS OF BEND

BAR SIZE	MINIMUM DIAMETER
No. 3 through No. 8 No. 9 through No. 11	6 bar diameters 8 bar diameters

TABLE 21-H-1—RADIUS OF GYRATION[1] FOR CONCRETE MASONRY UNITS[2]

GROUT SPACING (inches) × 25.4 for mm	NOMINAL WIDTH OF WALL (inches) × 25.4 for mm				
	4	6	8	10	12
Solid grouted	1.04	1.62	2.19	2.77	3.34
16	1.16	1.79	2.43	3.04	3.67
24	1.21	1.87	2.53	3.17	3.82
32	1.24	1.91	2.59	3.25	3.91
40	1.26	1.94	2.63	3.30	3.97
48	1.27	1.96	2.66	3.33	4.02
56	1.28	1.98	2.68	3.36	4.05
64	1.29	1.99	2.70	3.38	4.08
72	1.30	2.00	2.71	3.40	4.10
No grout	1.35	2.08	2.84	3.55	4.29

[1]For single-wythe masonry or for an individual wythe of a cavity wall.

$$r = \sqrt{I/A_e}$$

[2]The radius of gyration shall be based on the specified dimensions of the masonry units or shall be in accordance with the values shown which are based on the minimum dimensions of hollow concrete masonry unit face shells and webs in accordance with UBC Standard 21-4 for two cell units.

TABLE 21-L—SHEAR WALL SPACING REQUIREMENTS FOR EMPIRICAL DESIGN OF MASONRY

FLOOR OR ROOF CONSTRUCTION	MAXIMUM RATIO Shear Wall Spacing to Shear Wall Length
Cast-in-place concrete	5:1
Precast concrete	4:1
Metal deck with concrete fill	3:1
Metal deck with no fill	2:1
Wood diaphragm	2:1

Chapter 22 is printed in its entirety in Volume 2 of the *Uniform Building Code.*
Excerpts from Chapter 22 are reprinted herein.

Excerpts from Chapter 22
STEEL

Division I—GENERAL

SECTION 2201 — SCOPE

The quality, testing and design of steel used structurally in buildings or structures shall conform to the requirements specified in this chapter.

SECTION 2202 — STANDARDS OF QUALITY

The standards listed below labeled a "UBC Standard" are also listed in Chapter 35, Part II, and are part of this code. The other standards listed below are recognized standards. (See Sections 3503 and 3504.)

2202.1 Material Standards.

UBC Standard 22-1, Material Specifications for Structural Steel

2202.2 Design Standards.

ANSI/ASCE 8, Specification for the Design of Cold-formed Stainless Steel Structural Members, American Society of Civil Engineers

2202.3 Connectors.

ASTM A 502, Structural Rivet Steel

SECTION 2203 — MATERIAL IDENTIFICATION

2203.1 General. Steel furnished for structural load-carrying purposes shall be properly identified for conformity to the ordered grade in accordance with approved national standards, the provisions of this chapter and the appropriate UBC standards. Steel which is not readily identifiable as to grade from marking and test records shall be tested to determine conformity to such standards.

2203.2 Structural Steel. structural steel shall be identified by the mill in accordance with approved national standards. When such steel is furnished to a specified minimum yield point greater than 36,000 pounds per square inch (psi) (248 MPa), the American Society for Testing and Materials (ASTM) or other specification designation shall be so indicated.

The fabricator shall maintain identity of the material and shall maintain suitable procedures and records attesting that the specified grade has been furnished in conformity with the applicable standard. The fabricator's identification mark system shall be established and on record prior to fabrication.

When structural steel is furnished to a specified minimum yield point greater than 36,000 psi (248 MPa), the ASTM or other specification designation shall be included near the erection mark on each shipping assembly or important construction component over any shop coat of paint prior to shipment from the fabricator's plant. Pieces of such steel which are to be cut to smaller sizes shall, before cutting, be legibly marked with the fabricator's identification mark on each of the smaller-sized pieces to provide continuity of identification. When subject to fabrication operations, prior to assembling into members, which might obliterate paint marking, such as blast cleaning, galvanizing or heating for forming, such pieces of steel shall be marked by steel die stamping or by a substantial tag firmly attached.

Individual pieces of steel having a minimum specified yield point in excess of 36,000 psi (248 MPa), which are received by the fabricator in a tagged bundle or lift or which have only the top shape or plate in the bundle or lift marked by the mill shall be marked by the fabricator prior to use in accordance with the fabricator's established identification marking system.

2203.3 Cold-formed Carbon and Low-alloy Steel. Cold-formed carbon and low-alloy steel used for structural purposes shall be identified by the mill in accordance with approved national standards. When such steel is furnished to a specified minimum yield point greater than 33,000 psi (228 MPa), the fabricator shall indicate the ASTM or other specification designation, by painting, decal, tagging or other suitable means, on each lift or bundle of fabricated elements.

When cold-formed carbon and low-alloy steel used for structural purposes has a specified yield point equal to or greater than 33,000 psi (228 MPa), which was obtained through additional treatment, the resulting minimum yield point shall be identified in addition to the specification designation.

2203.4 Cold-formed Stainless Steel. Cold-formed stainless steel structural members designed in accordance with recognized standards shall be identified as to grade through mill test reports. (See reference to ANSI/ASCE 8 in Chapter 35.) A certification shall be furnished that the chemical and mechanical properties of the material supplied equals or exceeds that considered in the design. Each lift or bundle of fabricated elements shall be identified by painting, decal, tagging or other suitable means.

2203.5 Open-web Steel Joists. Open-web steel joists and similar fabricated light steel load-carrying members shall be identified in accordance with Division II as to type, size and manufacturer by tagging or other suitable means at the time of manufacture or fabrication. Such identification shall be maintained continuously to the point of their installation in a structure.

SECTION 2205 — DESIGN AND CONSTRUCTION PROVISIONS

2205.11 Bolts. The use of high-strength A 325 and A 490 bolts shall be in accordance with the requirements of Divisions II and III.

Anchor bolts shall be set accurately to the pattern and dimensions called for on the plans. The protrusion of the threaded ends through the connected material shall be sufficient to fully engage the threads of the nuts, but shall not be greater than the length of threads on the bolts. Base plate holes for anchor bolts may be oversized as follows:

Bolt Size, inches (mm)	Hole Size, inches (mm)
$^3/_4$ (19.1)	$^5/_{16}$ (7.9) oversized
$^7/_8$ (22.2)	$^5/_{16}$ (7.9) oversized
$1 < 2$ (25.4 < 50.8)	$^1/_2$ (12.7) oversized
> 2 (> 50.8)	1 (25.4) $>$ bolt diameter

Chapter 23 is printed in its entirety in Volume 2 of the *Uniform Building Code*. Excerpts from Chapter 23 are reprinted herein.

Excerpts from Chapter 23
WOOD
NOTE: This chapter has been revised in its entirety.

Division I—GENERAL DESIGN REQUIREMENTS

SECTION 2301 — GENERAL

2301.1 Scope. The quality and design of wood members and their fastenings shall conform to the provisions of this chapter.

2301.2 Design Methods. Design shall be based on one of the following methods.

2301.2.1 Allowable stress design. Design using allowable stress design methods shall resist the load combinations of Section 1612.3, in accordance with the applicable requirements of Section 2305.

2301.2.2 Conventional light-frame construction. The design and construction of conventional light-frame wood structures shall be in accordance with the applicable requirements of Section 2305.

SECTION 2302 — DEFINITIONS

2302.1 Definitions. The following terms used in this chapter shall have the meanings indicated in this section:

AFPA is the American Forest and Paper Association, 1111 19th Street, N.W., Suite 800, Washington, D.C. 20036 (formerly NFoPA, National Forest Products Association).

AHA is the American Hardboard Association, Inc., 1210 W. Northwest Highway, Palatine, Illinois 60067.

AITC is the American Institute of Timber Construction, 7012 S. Revere Parkway, Suite 140, Englewood, Colorado 80112.

ALSC is the American Lumber Standard Committee, Post Office Box 210, Germantown, Maryland 20875-0210.

APA is the American Plywood Association, 7011 South 19th Street, Tacoma, Washington 98411.

AWPA is the American Wood Preservers Association, Post Office Box 286, Woodstock, Maryland 21163-0286.

BLOCKED DIAPHRAGM is a diaphragm in which all sheathing edges not occurring on framing members are supported on and connected to blocking.

BRACED WALL LINE is a series of braced wall panels in a single story that meets the requirements of Section 2320.11.3.

BRACED WALL PANEL is a section of wall braced in accordance with Section 2320.11.3.

CONVENTIONAL LIGHT-FRAME CONSTRUCTION is a type of construction whose primary structural elements are formed by a system of repetitive wood-framing members. Refer to Section 2320 for conventional light-frame construction provisions.

DIAPHRAGM is a horizontal or nearly horizontal system acting to transmit lateral forces to the vertical-resisting elements. When the term "diaphragm" is used, it includes horizontal bracing systems.

FIBERBOARD is a fibrous-felted, homogeneous panel made from lignocellulosic fibers (usually wood or cane) and having a density of less than 31 pounds per cubic foot (497 kg/m^3) but more than 10 pounds per cubic foot (160 kg/m^3).

GLUED BUILT-UP MEMBERS are structural elements, the sections of which are composed of built-up lumber, wood structural panels or wood structural panels in combination with lumber, all parts bonded together with adhesives.

GRADE (Lumber) is the classification of lumber in regard to strength and utility in accordance with UBC Standard 23-1 and the grading rules of an approved lumber grading agency.

HARDBOARD is a fibrous-felted, homogeneous panel made from lignocellulosic fibers consolidated under heat and pressure in a hot press to a density not less than 31 pounds per cubic foot (497 kg/m^3).

NELMA is the Northeastern Lumber Manufacturers Association, 272 Tuttle Road, Post Office Box 87 A, Cumberland Center, Maine 04021.

NLGA is the National Lumber Grades Authority, 103-4000 Dominion Street, Burnaby B.C., Canada V5G 4G3.

NSLB is the Northern Softwood Lumber Bureau (serviced by NELMA), 272 Tuttle Road, Post Office Box 87 A, Cumberland Center, Maine 04021.

NOMINAL LOADING is a design load that stresses a member of fastening to the full allowable stress tabulated in this chapter. This loading may be applied for approximately 10 years, either continuously or cumulatively, and 90 percent of this load may be applied for the remainder of the life of the member or fastening.

NOMINAL SIZE (Lumber) is the commercial size designation of width and depth, in standard sawn lumber and glued-laminated lumber grades; somewhat larger than the standard net size of dressed lumber, in accordance with UBC Standard 23-1 for sawn lumber.

PARTICLEBOARD is a manufactured panel product consisting of particles of wood or combinations of wood particles and wood fibers bonded together with synthetic resins or other suitable bonding system by a bonding process in accordance with approved nationally recognized standards.

PLYWOOD is a panel of laminated veneers conforming to UBC Standard 23-2 or 23-3.

RIS is the Redwood Inspection Service, 405 Enfrente Drive, Suite 200, Novato, California 94949.

ROTATION is the torsional movement of a diaphragm about a vertical axis.

SPIB is the Southern Pine Inspection Bureau, 4709 Scenic Highway, Pensacola, Florida 32504.

STRUCTURAL GLUED-LAMINATED TIMBER is any member comprising an assembly of laminations of lumber in which the grain of all laminations is approximately parallel longitudinally, in which the laminations are bonded with adhesives.

SUBDIAPHRAGM is a portion of a larger wood diaphragm designed to anchor and transfer local forces to primary diaphragm struts and the main diaphragm.

TREATED WOOD is wood treated with an approved preservative under treating and quality control procedures.

WCLIB is the West Coast Lumber Inspection Bureau, 6980 S.W. Varnes Road, Post Office Box 23145, Portland, Oregon 97223.

WOOD OF NATURAL RESISTANCE TO DECAY OR TERMITES is the heartwood of the species set forth below. Corner sapwood is permitted on 5 percent of the pieces provided 90 percent or more of the width of each side on which it occurs is heartwood. Recognized species are:

Decay resistant: Redwood, Cedars, Black Locust

Termite resistant: Redwood, Eastern Red Cedar

WOOD STRUCTURAL PANEL is a structural panel product composed primarily of wood and meeting the requirements of UBC Standard 23-2 or 23-3. Wood structural panels include all-veneer plywood, composite panels containing a combination of veneer and wood-based material, and matformed panels such as oriented strand board and waferboard.

WWPA is the Western Wood Products Association, Yeon Building, 522 S. W. Fifth Avenue, Portland, Oregon 97204-2122.

SECTION 2303 — STANDARDS OF QUALITY

The standards listed below labeled a "UBC Standard" are also listed in Chapter 35, Part II, and are part of this code. The other standards listed below are recognized standards. (See Sections 3503 and 3504.)

1. **Grading rules.**

    1.1 UBC Standard 23-1, Classification, Definition; Methods of Grading and Development of Design values for All Species of Lumber

    1.2 Standard Grading Rules for Canadian Lumber, United States Edition, NLGA

    1.3 Standard Grading Rules No. 17, WCLIB

    1.4 Standard Grading Rules, WWPA

    1.5 Grading Rules, NHPMA

    1.6 Grading Rules, SPIB

    1.7 Standard Specifications for Grades of California Redwood Lumber, RIS

    1.8 Standard Grading Rules, NELMA

2. **Structural glued-laminated timber.**

    2.1 ANSI/AITC Standard A190.1 and ASTM D 3737, Design and Manufacture of Structural Glued-laminated Timber

    2.2 Standard Specifications for Structural Glued-laminated Timber of Softwood Species, AITC 117; Manufacturing, AITC 117; Design and Standard Specifications for Hardwood Glued-laminated Timber, AITC 119.

    2.3 Inspection Manual AITC 200 of the American Institute of Timber Construction, Tests for Structural Glued-laminated Timber.

    2.4 AITC 500, Determination of Design Values for Structural Glued-laminated Timber in accordance with ASTM D 3737, American Institute of Timber Construction.

3. **Preservative treatment by pressure process and quality control.**

    3.1 Standard Specifications C1, C2, C3, C4, C9, C14, C15, C16, C22, C23, C24, C28 and M4, AWPA

4. **Product standards.**

    4.1 UBC Standard 23-2, Construction and Industrial Plywood

    4.2 UBC Standard 23-3, Performance Standard for Wood-Based Structural-Use Panels

    4.3 ANSI A208.1, Particleboard

    4.4 ASTM D 1037, Evaluating the Properties of Wood-based Fiber and Particle Panel Materials

    4.5 ASTM D 1333, Determining Formaldehyde Levels from Wood-based Products Under Defined Test Conditions Using a Large Chamber

    4.6 ANSI 05.1, Wood Poles—Specifications and Dimensions

    4.7 ASTM D 25, Round Timber Piles

    4.8 ANSI/AHA A194.1, Cellulosic Fiber Insulating Board (Fiberboard)

    4.9 ANSI/AHA 135.6, Hardboard Siding

5. **Design standards.**

    5.1 ASTM D 5055, Structural Capacities of Prefabricated Wood I-Joists

    5.2 ANSI/TPI 1 National Design Standard for Metal Plate Connected Wood Truss Construction

    5.3 ANSI/TPI 2 Standard for Testing Performance for Metal Plate Connected Wood Trusses

    5.4 ASCE 16, Load and Resistance Factor Design Standard for Engineered Wood Construction

6. **Fire retardancy.**

    6.1 UBC Standard 23-4, Fire-retardant-treated Wood Tests on Durability and Hygroscopic Properties

    6.2 UBC Standard 23-5, Fire-retardant-treated Wood

7. **Adhesives and glues.**

    7.1 ASTM D 3024, Dry Use Adhesive with Protein Base, Casein Type

    7.2 ASTM D 2559, Wet Use Adhesives

    7.3 APA Specification AFG-01, Adhesives for Field Gluing Plywood to Wood Framing

    7.4 ASTM D 1101 and AITC 200 in Testing of Glue Joints in Laminated Wood Product

8. **Design values.**

    8.1 ASTM D 1990, Establishing Allowable Properties for Visually-Graded Dimension Lumber from In-Grade Tests of Full-Size Specimens

    8.2 ASTM D 245, Establishing Structural Grades and Related Allowable Properties for Visually Graded Lumber

    8.3 ASTM D 2555, Standard Test Methods for Establishing Clear Wood Strength Values

SECTION 2304 — MINIMUM QUALITY

2304.1 Quality and Identification. All lumber, wood structural panels, particleboard, structural glued-laminated timber, end-jointed lumber, fiberboard sheathing (when used structurally), hardboard siding (when used structurally), piles and poles regulated by this chapter shall conform to the applicable standards and grading rules specified in this code and shall be so identified by the grade mark or certificate of inspection issued by an approved agency.

All preservatively treated wood required to be treated under Section 2306 shall be identified by the quality mark of an inspection agency which has been accredited by an accreditation body which complies with the requirements of the American Lumber Standard Committee Treated Wood Program, or equivalent.

2304.2 Minimum Capacity or Grade. Minimum capacity of structural framing members may be established by performance tests. When tests are not made, capacity shall be based on allowable stresses and design criteria specified in this code.

Studs, joists, rafters, foundation plates or sills, planking 2 inches (51 mm) or more in depth, beams, stringers, posts, structural sheathing and similar load-bearing members shall be of at least the minimum grades set forth in the tables in this chapter.

Approved end-jointed lumber may be used interchangeably with solid-sawn members of the same species and grade. Such use shall include, but not be limited to, light-framing joists, planks and decking.

Wood structural panels shall be of the grades specified in UBC Standard 23-2 or 23-3.

2304.3 Timber Connectors and Fasteners. Safe loads and design practices for types of connectors and fasteners not mentioned or fully covered in Division III, Part III, may be determined in a manner approved by the building official.

The number and size of nails connecting wood members shall not be less than that set forth in Tables 23-II-B-1 and 23-II-B-2. Other connections shall be fastened to provide equivalent strength. End and edge distances and nail penetrations shall be in accordance with the applicable provisions of Division III, Part III.

Fasteners for pressure-preservative treated and fire-retardant treated wood shall be of hot-dipped zinc coated galvanized, stainless steel, silicon bronze or copper. Fasteners for wood foundations shall be as required in Chapter 18, Division II. Fasteners required to be corrosion resistant shall be either zinc-coated fasteners, aluminum alloy wire fasteners or stainless steel fasteners.

Connections depending on joist hangers or framing anchors, ties, and other mechanical fastenings not otherwise covered may be used where approved.

2304.4 Fabrication, Installation and Manufacture.

2304.4.1 General. Preparation, fabrication and installation of wood members and their fastenings shall conform to accepted engineering practices and to the requirements of this code. All members shall be framed, anchored, tied and braced to develop the strength and rigidity necessary for the purposes for which they are used.

2304.4.2 Timber connectors and fasteners. The installation of timber connectors and fasteners shall be in accordance with the provisions set forth in Division III, Part III.

2304.4.3 Structural glued-laminated timber. The manufacture and fabrication of structural glued-laminated timber shall be under the supervision of qualified personnel.

2304.4.4 Metal-plate-connected wood trusses. Metal-plate-connected wood trusses shall conform to the provisions of Division V. Each manufacturer of trusses using metal plate connectors shall retain an approved agency having no financial interest in the plant being inspected to make nonscheduled inspections of truss fabrication, delivery, and operations. The inspection shall cover all phases of truss operation, including lumber storage, handling, cutting, fixtures, presses or rollers, fabrication, bundling and banding, handling and delivery.

2304.5 Dried Fire-retardant-treated Wood. Approved fire-retardant-treated wood shall be dried, following treatment, to a maximum moisture content as follows: solid-sawn lumber 2 inches (51 mm) in thickness or less to 19 percent, and plywood to 15 percent.

2304.6 Size of Structural Members. Sizes of lumber and structural glued-laminated timber referred to in this code are nominal sizes. Computations to determine the required sizes of members shall be based on the net dimensions (actual sizes) and not the nominal sizes.

2304.7 Shrinkage. Consideration shall be given in design to the possible effect of cross-grain dimensional changes considered vertically which may occur in lumber fabricated in a green condition.

2304.8 Rejection. The building official may deny permission for the use of a wood member where permissible grade characteristics or defects are present in such a combination that they affect the serviceability of the member.

SECTION 2305 — DESIGN AND CONSTRUCTION REQUIREMENTS

2305.1 General. The following design requirements apply.

2305.2 All wood structures shall be designed and constructed in accordance with the requirements of Division I and Division II, Part I.

2305.3 Wind and earthquake load-resisting systems for all engineered wood structures shall be designed and constructed in accordance with the requirements of Division II, Part II.

2305.4 The design and construction of wood structures using allowable stress design methods shall be in accordance with Division III.

2305.5 The design and construction of conventional light-frame wood structures shall be in accordance with Division IV.

2305.6 The design and installation of timber connectors and fasteners shall be in accordance with Division III, Part III.

2305.7 Metal-plate-connected wood trusses shall conform to the provisions of Division V.

2305.8 Design of structural glued built-up members with plywood components shall be in accordance with Division VI.

2305.9 Design of joists and rafters shall be permitted to be in accordance with Division VII.

2305.10 Design of plank and beam flooring shall be permitted to be in accordance with Division VIII.

Division II—GENERAL REQUIREMENTS

Part I—REQUIREMENTS APPLICABLE TO ALL DESIGN METHODS

SECTION 2306 — DECAY AND TERMITE PROTECTION

2306.1 Preparation of Building Site. Site preparation shall be in accordance with Section 3302.

2306.2 Wood Support Embedded in Ground. Wood embedded in the ground or in direct contact with the earth and used for the support of permanent structures shall be treated wood unless continuously below the groundwater line or continuously submerged in fresh water. Round or rectangular posts, poles and sawn timber columns supporting permanent structures that are embedded in concrete or masonry in direct contact with earth or embedded in concrete or masonry exposed to the weather shall be treated wood. The wood shall be treated for ground contact.

2306.3 Under-floor Clearance. When wood joists or the bottom of wood structural floors without joists are located closer than 18 inches (457 mm) or wood girders are located closer than 12 inches (305 mm) to exposed ground in crawl spaces or unexcavated areas located within the periphery of the building foundation, the floor assembly, including posts, girders, joists and subfloor, shall be approved wood of natural resistance to decay as listed in Section 2306.4 or treated wood.

When the above under-floor clearances are required, the under-floor area shall be accessible. Accessible under-floor areas shall be provided with a minimum 18-inch-by-24-inch (457 mm by 610 mm) opening unobstructed by pipes, ducts and similar construction. All under-floor access openings shall be effectively screened or covered. Pipes, ducts and other construction shall not interfere with the accessibility to or within under-floor areas.

2306.4 Plates, Sills and Sleepers. All foundation plates or sills and sleepers on a concrete or masonry slab, which is in direct contact with earth, and sills that rest on concrete or masonry foundations, shall be treated wood or Foundation redwood, all marked or branded by an approved agency. Foundation cedar or No. 2 Foundation redwood marked or branded by an approved agency may be used for sills in territories subject to moderate hazard, where termite damage is not frequent and when specifically approved by the building official. In territories where hazard of termite damage is slight, any species of wood permitted by this code may be used for sills when specifically approved by the building official.

2306.5 Columns and Posts. Columns and posts located on concrete or masonry floors or decks exposed to the weather or to water splash or in basements and that support permanent structures shall be supported by concrete piers or metal pedestals projecting above floors unless approved wood of natural resistance to decay or treated wood is used. The pedestals shall project at least 6 inches (152 mm) above exposed earth and at least 1 inch (25 mm) above such floors.

Individual concrete or masonry piers shall project at least 8 inches (203 mm) above exposed ground unless the columns or posts that they support are of approved wood of natural resistance to decay or treated wood is used.

2306.6 Girders Entering Masonry or Concrete Walls. Ends of wood girders entering masonry or concrete walls shall be provided with a $^1/_2$-inch (12.7 mm) air space on tops, sides and ends unless approved wood of natural resistance to decay or treated wood is used.

2306.7 Under-floor Ventilation. Under-floor areas shall be ventilated by an approved mechanical means or by openings into the under-floor area walls. Such openings shall have a net area of not less than 1 square foot for each 150 square feet (0.067 m^2 for each 10 m^2) of under-floor area. Openings shall be located as close to corners as practical and shall provide cross ventilation. The required area of such openings shall be approximately equally distributed along the length of at least two opposite sides. They shall be covered with corrosion-resistant wire mesh with mesh openings of $^1/_4$ inch (6.4 mm) in dimension. Where moisture due to climate and groundwater conditions is not considered excessive, the building official may allow operable louvers and may allow the required net area of vent openings to be reduced to 10 percent of the above, provided the under-floor ground surface area is covered with an approved vapor retarder.

2306.8 Wood and Earth Separation. Protection of wood against deterioration as set forth in the previous sections for specified applications is required. In addition, wood used in construction of permanent structures and located nearer than 6 inches (152 mm) to earth shall be treated wood or wood of natural resistance to decay, as defined in Section 2302.1. Where located on concrete slabs placed on earth, wood shall be treated wood or wood of natural resistance to decay. Where not subject to water splash or to exterior moisture and located on concrete having a minimum thickness of 3 inches (76 mm) with an impervious membrane installed between concrete and earth, the wood may be untreated and of any species.

Where planter boxes are installed adjacent to wood frame walls, a 2-inch-wide (51 mm) air space shall be provided between the planter and the wall. Flashings shall be installed when the air space is less than 6 inches (152 mm) in width. Where flashing is used, provisions shall be made to permit circulation of air in the air space. The wood-frame wall shall be provided with an exterior wall covering conforming to the provisions of Section 2310.

2306.9 Wood Supporting Roofs and Floors. Wood structural members supporting moisture-permeable floors or roofs that are exposed to the weather, such as concrete or masonry slabs, shall be approved wood of natural resistance to decay or treated wood unless separated from such floors or roofs by an impervious moisture barrier.

2306.10 Moisture Content of Treated Wood. When wood pressure treated with a water-borne preservative is used in enclosed locations where drying in service cannot readily occur, such wood shall be at a moisture content of 19 percent or less before being covered with insulation, interior wall finish, floor covering or other material.

2306.11 Retaining Walls. Wood used in retaining or crib walls shall be treated wood.

2306.12 Weather Exposure. Those portions of glued-laminated timbers that form the structural supports of a building or other structure and are exposed to weather and not properly protected by a roof, eave overhangs of similar covering shall be pressure treated with an approved preservative or be manufactured from wood of natural resistance to decay.

All wood structural panels, when designed to be exposed in outdoor applications, shall be of exterior type, except as provided in Section 2306.2. In geographical areas where experience has demonstrated a specific need, approved wood of natural resistance to decay or treated wood shall be used for those portions of wood members which form the structural supports of buildings, balconies, porches or similar permanent building appurtenances when such members are exposed to the weather without adequate protection from a roof, eave, overhang or other covering to prevent moisture or water accumulation on the surface or at joints between members. Depending on local experience, such members may include horizontal members such as girders, joists and decking; or vertical members such as posts, poles and columns; or both horizontal and vertical members.

2306.13 Water Splash. Where wood-frame walls and partitions are covered on the interior with plaster, tile or similar materials and are subject to water splash, the framing shall be protected with approved waterproof paper conforming to Section 1402.1.

SECTION 2307 — WOOD SUPPORTING MASONRY OR CONCRETE

Wood members shall not be used to permanently support the dead load of any masonry or concrete.

> **EXCEPTIONS:** 1. Masonry or concrete nonstructural floor or roof surfacing not more than 4 inches (102 mm) thick may be supported by wood members.
>
> 2. Any structure may rest upon wood piles constructed in accordance with the requirements of Chapter 18.
>
> 3. Veneer of brick, concrete or stone applied as specified in Section 1403.6.2 may be supported by approved treated wood foundations when the maximum height of veneer does not exceed 30 feet (9144 mm) above the foundations. Such veneer used as an interior wall finish may also be supported on wood floors that are designed to support the additional load and designed to limit the deflection and shrinkage to $1/600$ of the span of the supporting members.
>
> 4. Glass block masonry having an installed weight of 20 pounds per square foot (97.6 kg/m^2) or less and installed with the provisions of Section 2109.5. When glass block is supported on wood floors, the floors shall be designed to limit deflection and shrinkage to $1/600$ of the span of the supporting members and the allowable stresses for the framing members shall be reduced in accordance with Division III, Part I.

See Division II, Part II for wood members resisting horizontal forces contributed by masonry or concrete.

SECTION 2308 — WALL FRAMING

The framing of exterior and interior walls shall be in accordance with provisions specified in Division IV unless a specific design is furnished.

Wood stud walls and bearing partitions shall not support more than two floors and a roof unless an analysis satisfactory to the building official shows that shrinkage of the wood framing will not have adverse effects on the structure or any plumbing, electrical or mechanical systems, or other equipment installed therein due to excessive shrinkage or differential movements caused by shrinkage. The analysis shall also show that the roof drainage system and the foregoing systems or equipment will not be adversely affected or, as an alternate, such systems shall be designed to accommodate the differential shrinkage or movements.

SECTION 2309 — FLOOR FRAMING

Wood-joisted floors shall be framed and constructed and anchored to supporting wood stud or masonry walls as specified in Chapter 16.

Fire block and draft stops shall be in accordance with Section 708.

SECTION 2310 — EXTERIOR WALL COVERINGS

2310.1 General. Exterior wood stud walls shall be covered on the outside with the materials and in the manner specified in this section or elsewhere in this code. Studs or sheathing shall be covered on the outside face with a weather-resistive barrier when required by Section 1402.1. Exterior wall coverings of the minimum thickness specified in this section are based on a maximum stud spacing of 16 inches (406 mm) unless otherwise specified.

2310.2 Siding. Solid wood siding shall have an average thickness of $3/8$ inch (9.5 mm) unless placed over sheathing permitted by this code.

Siding patterns known as rustic, drop siding or shiplap shall have an average thickness in place of not less than $19/32$ inch (15 mm) and shall have a minimum thickness of not less than $3/8$ inch (9.5 mm). Bevel siding shall have a minimum thickness measured at the butt section of not less than $7/16$ inch (11 mm) and a tip thickness of not less than $3/16$ inch (4.8 mm). Siding of lesser dimensions may be used, provided such wall covering is placed over sheathing which conforms to the provisions specified elsewhere in this code.

All weatherboarding or siding shall be securely nailed to each stud with not less than one nail, or to solid 1-inch (25 mm) nominal wood sheathing or $15/32$-inch (12 mm) wood structural panel sheathing or $1/2$-inch (13 mm) particleboard sheathing with not less than one line of nails spaced not more than 24 inches (610 mm) on center in each piece of the weatherboarding or siding.

Wood board sidings applied horizontally, diagonally or vertically shall be fastened to studs, nailing strips or blocking set at a maximum 24 inches (610 mm) on center. Fasteners shall be nails or screws with a penetration of not less than $1^{1}/_{2}$ inches (38 mm) into studs, studs and wood sheathing combined, or blocking. Distance between such fastenings shall not exceed 24 inches (610 mm) for horizontally or vertically applied sidings and 32 inches (813 mm) for diagonally applied sidings.

2310.3 Plywood. When plywood is used for covering the exterior of outside walls, it shall be of the exterior type not less than $3/8$ inch (9.5 mm) thick. Plywood panel siding shall be installed in accordance with Table 23-II-A-1. Unless applied over 1-inch (25 mm) wood sheathing or $15/32$-inch (12 mm) wood structural panel sheathing or $1/2$-inch (13 mm) particleboard sheathing, joints shall occur over framing members and shall be protected with a continuous wood batten, approved caulking, flashing, vertical or horizontal shiplaps; or joints shall be lapped horizontally or otherwise made waterproof.

2310.4 Shingles or Shakes. Wood shingles or shakes and asbestos cement shingles may be used for exterior wall covering, provided the frame of the structure is covered with building paper as specified in Section 1402.1. All shingles or shakes attached to sheathing other than wood sheathing shall be secured with approved corrosion-resistant fasteners or on furring strips attached to the studs. Wood shingles or shakes may be applied over fiberboard shingle backer and sheathing with annular grooved nails. The thickness of wood shingles or shakes between wood nailing boards shall not be less than $3/8$ inch (9.5 mm). Wood shingles or shakes and asbestos shingles or siding may be nailed directly to approved fiberboard nailbase sheathing not less than $1/2$-inch (13 mm) nominal thickness with annular grooved nails.

The weather exposure of wood shingle or shake siding used on exterior walls shall not exceed maximums set forth in Table 23-II-K.

2310.5 Particleboard. When particleboard is used for covering the exterior of outside walls, it shall be of the M-1, M-S and M-2 Exterior Glue grades. Particleboard panel siding shall be installed in accordance with Tables 23-II-A-2 and 23-II-B-1. Panels shall be gapped $1/8$ inch (3.2 mm) and nails shall be spaced not less than $3/8$ inch (9.5 mm) from edges and ends of sheathing. Unless applied over $5/8$-inch (16 mm) net wood sheathing or $1/2$-inch (13 mm) plywood sheathing or $1/2$-inch (13 mm) particleboard sheathing, joints shall occur over framing members and shall be covered with a continuous wood batt; or joints shall be lapped horizontally or otherwise made waterproof to the satisfaction of the building official. Particleboard shall be sealed and protected with exterior quality finishes.

2310.6 Hardboard. When hardboard siding is used for covering the outside of exterior walls, it shall conform to Table 23-II-C. Lap siding shall be installed horizontally and applied to sheathed or unsheathed walls. Corner bracing shall be installed in conformance with Division IV. A weather-resistive barrier shall be installed under the lap siding as required by Section 1402.1.

Square-edged nongrooved panels and shiplap grooved or nongrooved siding shall be applied vertically to sheathed or unsheathed walls. Siding that is grooved shall not be less than $1/4$ inch (6.4 mm) thick in the groove.

Nail size and spacing shall follow Table 23-II-C and shall penetrate framing $1^1/2$ inches (38 mm). Lap siding shall overlap 1 inch (25 mm) minimum and be nailed through both courses and into framing members with nails located $1/2$ inch (13 mm) from bottom of the overlapped course. Square-edged nongrooved panels shall be nailed $3/8$ inch (9.5 mm) from the perimeter of the panel and intermediately into studs. Shiplap edge panel siding with $3/8$-inch (9.5 mm) shiplap shall be nailed $3/8$ inch (9.5 mm) from the edges on both sides of the shiplap. The $3/4$-inch (19 mm) shiplap shall be nailed $3/8$ inch (9.5 mm) from the edge and penetrate through both the overlap and underlap. Top and bottom edges of the panel shall be nailed $3/8$ inch (9.5 mm) from the edge. Shiplap and lap siding shall not be force fit. Square-edged panels shall maintain a $1/16$-inch (1.6 mm) gap at joints. All joints and edges of siding shall be over framing members, and shall be made resistant to weather penetration with battens, horizontal overlaps or shiplaps to the satisfaction of the building official. A $1/8$-inch (3.2 mm) gap shall be provided around all openings.

2310.7 Nailing. All fasteners used for the attachment of siding shall be of a corrosion-resistant type.

SECTION 2311 — INTERIOR PANELING

All softwood wood structural panels shall conform with the provisions of Chapter 8 and shall be installed in accordance with Table 23-II-B-1. Panels shall comply with UBC Standard 23-3.

SECTION 2312 — SHEATHING

2312.1 Structural Floor Sheathing. Structural floor sheathing shall be designed in accordance with the general provisions of this code and the special provisions in this section.

Sheathing used as subflooring shall be designed to support all loads specified in this code and shall be capable of supporting concentrated loads of not less than 300 pounds (1334 N) without failure. The concentrated load shall be applied by a loaded disc, 3 inches (76 mm) or smaller in diameter.

Flooring, including the finish floor, underlayment and subfloor, where used, shall meet the following requirements:

1. Deflection under uniform design load limited to $1/360$ of the span between supporting joists or beams.

2. Deflection of flooring relative to joists under a 1-inch-diameter (25 mm) concentrated load of 200 pounds (890 N) limited to 0.125 inch (3.2 mm) or less when loaded midway between

supporting joists or beams not over 24 inches (610 mm) on center and $1/360$ of the span for spans over 24 inches (610 mm).

Floor sheathing conforming to the provisions of Table 23-II-D-1, 23-II-D-2, 23-II-E-1, 23-II-F-1 or 23-II-F-2 shall be deemed to meet the requirements of this section.

2312.2 Structural Roof Sheathing. Structural roof sheathing shall be designed in accordance with the general provisions of this code and the special provisions in this section. Structural roof sheathing shall be designed to support all loads specified in this code and shall be capable of supporting concentrated loads of not less than 300 pounds (1334 N) without failure. The concentrated load shall be applied by a loaded disc, 3 inches (76 mm) or smaller in diameter. Structural roof sheathing shall meet the following requirement:

1. Deflection under uniform design live and dead load limited to $1/180$ of the span between supporting rafters or beams and $1/240$ under live load only.

Roof sheathing conforming to the provisions of Tables 23-II-D-1 and 23-II-D-2 or 23-II-E-1 and 23-II-E-2 shall be deemed to meet the requirements of this section.

Wood structural panel roof sheathing shall be bonded by intermediate or exterior glue. Wood structural panel roof sheathing exposed on the underside shall be bonded with exterior glue.

SECTION 2313 — MECHANICALLY LAMINATED FLOORS AND DECKS

A laminated lumber floor or deck built up of wood members set on edge, when meeting the following requirements, may be designed as a solid floor or roof deck of the same thickness, and continuous spans may be designed on the basis of the full cross section using the simple span moment coefficient.

Nail length shall not be less than two and one-half times the net thickness of each lamination. When deck supports are 4 feet (1219 mm) on center or less, side nails shall be spaced not more than 30 inches (762 mm) on center and staggered one third of the spacing in adjacent laminations. When supports are spaced more than 4 feet (1219 mm) on center, side nails shall be spaced not more than 18 inches (457 mm) on center alternately near top and bottom edges, and also staggered one third of the spacing in adjacent laminations. Two side nails shall be used at each end of butt-jointed pieces.

Laminations shall be toenailed to supports with 20d or larger common nails. When the supports are 4 feet (1219 mm) on center or less, alternate laminations shall be toenailed to alternate supports; when supports are spaced more than 4 feet (1219 mm) on center, alternate laminations shall be toenailed to every support.

A single-span deck shall have all laminations full length.

A continuous deck of two spans shall not have more than every fourth lamination spliced within quarter points adjoining supports.

Joints shall be closely butted over supports or staggered across the deck but within the adjoining quarter spans.

No lamination shall be spliced more than twice in any span.

SECTION 2314 — POST-BEAM CONNECTIONS

Where post and beam or girder construction is used, the design shall be in accordance with the provisions of this code. Positive connection shall be provided to ensure against uplift and lateral displacement.

TABLE 23-II-A-1—EXPOSED PLYWOOD PANEL SIDING

MINIMUM THICKNESS[1] (inch)	MINIMUM NUMBER OF PLIES	STUD SPACING (inches) PLYWOOD SIDING APPLIED DIRECTLY TO STUDS OR OVER SHEATHING
× 25.4 for mm		× 25.4 for mm
$3/8$	3	16^2
$1/2$	4	24

[1]Thickness of grooved panels is measured at bottom of grooves.
[2]May be 24 inches (610 mm) if plywood siding applied with face grain perpendicular to studs or over one of the following: (1) 1-inch (25 mm) board sheathing, (2) $7/16$-inch (11 mm) wood structural panel sheathing or (3) $3/8$-inch (9.5 mm) wood structural panel sheathing with strength axis (which is the long direction of the panel unless otherwise marked) of sheathing perpendicular to studs.

TABLE 23-II-A-2—ALLOWABLE SPANS FOR EXPOSED PARTICLEBOARD PANEL SIDING

GRADE	STUD SPACING (inches)	MINIMUM THICKNESS (inches)		Exterior Ceilings and Soffits
		Siding		
	× 25.4 for mm	Direct to Studs	Continuous Support	Direct to Supports
M-1 M-S M-2 "Exterior Glue"	16	$5/8$	$3/8$	$3/8$
	24	$5/8$	$3/8$	$3/8$

TABLE 23-II-B-1—NAILING SCHEDULE

CONNECTION	NAILING[1]
1. Joist to sill or girder, toenail	3-8d
2. Bridging to joist, toenail each end	2-8d
3. 1″ × 6″ (25 mm × 152 mm) subfloor or less to each joist, face nail	2-8d
4. Wider than 1″ × 6″ (25 mm × 152 mm) subfloor to each joist, face nail	3-8d
5. 2″ (51 mm) subfloor to joist or girder, blind and face nail	2-16d
6. Sole plate to joist or blocking, typical face nail Sole plate to joist or blocking, at braced wall panels	16d at 16″ (406 mm) o.c. 3-16d per 16″ (406 mm)
7. Top plate to stud, end nail	2-16d
8. Stud to sole plate	4-8d, toenail or 2-16d, end nail
9. Double studs, face nail	16d at 24″ (610 mm) o.c.
10. Doubled top plates, typical face nail Double top plates, lap splice	16d at 16″ (406 mm) o.c. 8-16d
11. Blocking between joists or rafters to top plate, toenail	3-8d
12. Rim joist to top plate, toenail	8d at 6″ (152 mm) o.c.
13. Top plates, laps and intersections, face nail	2-16d
14. Continuous header, two pieces	16d at 16″ (406 mm) o.c. along each edge
15. Ceiling joists to plate, toenail	3-8d
16. Continuous header to stud, toenail	4-8d
17. Ceiling joists, laps over partitions, face nail	3-16d
18. Ceiling joists to parallel rafters, face nail	3-16d
19. Rafter to plate, toenail	3-8d
20. 1″ (25 mm) brace to each stud and plate, face nail	2-8d
21. 1″ × 8″ (25 mm × 203 mm) sheathing or less to each bearing, face nail	2-8d
22. Wider than 1″ × 8″ (25 mm × 203 mm) sheathing to each bearing, face nail	3-8d
23. Built-up corner studs	16d at 24″ (610 mm) o.c.
24. Built-up girder and beams	20d at 32″ (813 mm) o.c. at top and bottom and staggered 2-20d at ends and at each splice
25. 2″ (51 mm) planks	2-16d at each bearing

26. Wood structural panels and particleboard[2]:
    Subfloor and wall sheathing (to framing):
        $1/2$″ (12.7 mm) and less
        $19/32$″-$3/4$″ (15 mm-19 mm)
        $7/8$″-1″ (22 mm-25 mm)
        $1 1/8$″-$1 1/4$″ (29 mm-32 mm)
    Combination subfloor-underlayment (to framing):
        $3/4$″ (19 mm) and less
        $7/8$″-1″ (22 mm-25 mm)
        $1 1/8$″-$1 1/4$″ (29 mm-32 mm)
	
	6d[3]
	8d[4] or 6d[5]
	8d[3]
	10d[4] or 8d[5]
	
	6d[5]
	8d[5]
	10d[4] or 8d[5]

27. Panel siding (to framing):[2]
    $1/2$″ (12.7 mm) or less
    $5/8$″ (16 mm)
	
	6d[6]
	8d[6]

28. Fiberboard sheathing:[7]
    $1/2$″ (12.7 mm)
 	
    $25/32$″ (20 mm)
	
	No. 11 ga.[8]
	6d[4]
	No. 16 ga.[9]
	No. 11 ga.[8]
	8d[4]
	No. 16 ga.[9]

29. Interior paneling
    $1/4$″ (6.4 mm)
    $3/8$″ (9.5 mm)
	
	4d[10]
	6d[11]

[1]Common or box nails may be used except where otherwise stated.

[2]Nails spaced at 6 inches (152 mm) on center at edges, 12 inches (305 mm) at intermediate supports except 6 inches (152 mm) at all supports where spans are 48 inches (1219 mm) or more. For nailing of wood structural panel and particleboard diaphragms and shear walls, refer to Sections 2315.3.3 and 2315.4. Nails for wall sheathing may be common, box or casing.

[3]Common or deformed shank.

[4]Common.

[5]Deformed shank.

[6]Corrosion-resistant siding or casing nails conforming to the requirements of Section 2304.3.

[7]Fasteners spaced 3 inches (76 mm) on center at exterior edges and 6 inches (152 mm) on center at intermediate supports.

[8]Corrosion-resistant roofing nails with $7/16$-inch-diameter (11 mm) head and $1 1/2$-inch (38 mm) length for $1/2$-inch (12.7 mm) sheathing and $1 3/4$-inch (44 mm) length for $25/32$-inch (20 mm) sheathing conforming to the requirements of Section 2304.3.

[9]Corrosion-resistant staples with nominal $7/16$-inch (11 mm) crown and $1 1/8$-inch (29 mm) length for $1/2$-inch (12.7 mm) sheathing and $1 1/2$-inch (38 mm) length for $25/32$-inch (20 mm) sheathing conforming to the requirements of Section 2304.3.

[10]Panel supports at 16 inches (406 mm) [20 inches (508 mm) if strength axis in the long direction of the panel, unless otherwise marked]. Casing or finish nails spaced 6 inches (152 mm) on panel edges, 12 inches (305 mm) at intermediate supports.

[11]Panel supports at 24 inches (610 mm). Casing or finish nails spaced 6 inches (152 mm) on panel edges, 12 inches (305 mm) at intermediate supports.

TABLE 23-II-B-2—WOOD STRUCTURAL PANEL ROOF SHEATHING NAILING SCHEDULE[1]

WIND REGION	NAILS	PANEL LOCATION	ROOF FASTENING ZONE[2]		
			1	2	3
			Fastening Schedule (inches on center)		
			× 25.4 for mm		
Greater than 90 mph (145 km/h)	8d common	Panel edges[3]	6	6	4[4]
		Panel field	6	6	6[4]
Greater than 80 mph (129 km/h) to 90 mph (145 km/h)	8d common	Panel edges[3]	6	6	4
		Panel field	12	6	6
80 mph (129 km/h) or less	8d common	Panel edges[3]	6	6	6
		Panel field	12	12	12

[1]Applies only to mean roof heights up to 35 feet (10 700 mm). For mean roof heights over 35 feet (10 700 mm), the nailing shall be designed.

[2]The roof fastening zones are shown below:

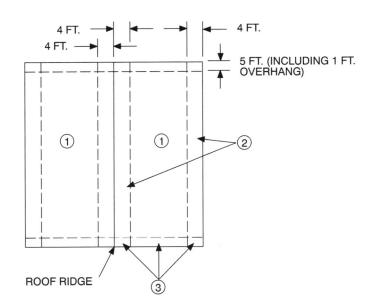

ROOF FASTENING ZONES

For **SI:** 1 foot = 304.8 mm.

[3]Edge spacing also applies over roof framing at gable-end walls.

[4]Use 8d ring-shank nails in this zone if mean roof height is greater than 25 feet (7600 mm).

TABLE 23-II-C—HARDBOARD SIDING

SIDING	MINIMAL NOMINAL THICKNESS (inch)	FRAMING (2″ x 4″) MAXIMUM SPACING	NAIL SIZE[1,2]	NAIL SPACING	
				General	Bracing Panels[3]
				× 25.4 for mm	
1. LAP SIDING					
Direct to studs	$3/8$	16″ o.c.	8d	16″ o.c.	Not applicable
Over sheathing	$3/8$	16″ o.c.	10d	16″ o.c.	Not applicable
2. SQUARE EDGE PANEL SIDING					
Direct to studs	$3/8$	24″ o.c.	6d	6″ o.c. edges; 12″ o.c. at intermed. supports	4″ o.c. edges; 8″ o.c. intermed. supports
Over sheathing	$3/8$	24″ o.c.	8d	6″ o.c. edges; 12″ o.c. at intermed. supports	4″ o.c. edges; 8″ o.c. intermed. supports
3. SHIPLAP EDGE PANEL SIDING					
Direct to studs	$3/8$	16″ o.c.	6d	6″ o.c. edges; 12″ o.c. at intermed. supports	4″ o.c. edges; 8″ o.c. intermed. supports
Over sheathing	$3/8$	16″ o.c.	8d	6″ o.c. edges; 12″ o.c. at intermed. supports	4″ o.c. edges; 8″ o.c. intermed. supports

[1]Nails shall be corrosion resistant in accordance with Division III, Part III.
[2]Minimum acceptable nail dimensions (inches).

	Panel Siding (inch)	Lap Siding (inch)
	× 25.4 for mm	
Shank diameter	0.092	0.099
Head diameter	0.225	0.240

[3]When used to comply with Division IV, Section 2320.11.3.

TABLE 23-II-D-1—ALLOWABLE SPANS FOR LUMBER FLOOR AND ROOF SHEATHING[1,2]

SPAN (inches)	MINIMUM NET THICKNESS (inches) OF LUMBER PLACED			
	Perpendicular to Supports		Diagonally to Supports	
	× 25.4 for mm			
× 25.4 for mm	Surfaced Dry[3]	Surfaced Unseasoned	Surfaced Dry[3]	Surfaced Unseasoned
Floors				
1. 24	$3/4$	$25/32$	$3/4$	$25/32$
2. 16	$5/8$	$11/16$	$5/8$	$11/16$
Roofs				
3. 24	$5/8$	$11/16$	$3/4$	$25/32$

[1]Installation details shall conform to Sections 2320.9.1 and 2320.12.8 for floor and roof sheathing, respectively.
[2]Floor or roof sheathing conforming with this table shall be deemed to meet the design criteria of Section 2312.
[3]Maximum 19 percent moisture content.

TABLE 23-II-D-2—SHEATHING LUMBER SHALL MEET THE FOLLOWING MINIMUM GRADE REQUIREMENTS: BOARD GRADE

SOLID FLOOR OR ROOF SHEATHING	SPACED ROOF SHEATHING	GRADING RULES
1. Utility	Standard	NLGA, WCLIB, WWPA
2. 4 common or utility	3 common or standard	NLGA, WCLIB, WWPA, NHPMA or NELMA
3. No. 3	No. 2	SPIB
4. Merchantable	Construction common	RIS

TABLE 23-II-E-1—ALLOWABLE SPANS AND LOADS FOR WOOD STRUCTURAL PANEL SHEATHING AND SINGLE-FLOOR GRADES CONTINUOUS OVER TWO OR MORE SPANS WITH STRENGTH AXIS PERPENDICULAR TO SUPPORTS[1,2]

SHEATHING GRADES		ROOF[3]				FLOOR[4]
		Maximum Span (inches)		Load[5] (pounds per square foot)		
		× 25.4 for mm		× 0.0479 for kN/m^2		Maximum Span (inches)
Panel Span Rating	Panel Thickness (inches)	With Edge Support[6]	Without Edge Support	Total Load	Live Load	
Roof/Floor Span	× 25.4 for mm					× 25.4 for mm
12/0	$^5/_{16}$	12	12	40	30	0
16/0	$^5/_{16}, ^3/_8$	16	16	40	30	0
20/0	$^5/_{16}, ^3/_8$	20	20	40	30	0
24/0	$^3/_8, ^7/_{16}, ^1/_2$	24	20[7]	40	30	0
24/16	$^7/_{16}, ^1/_2$	24	24	50	40	16
32/16	$^{15}/_{32}, ^1/_2, ^5/_8$	32	28	40	30	16[8]
40/20	$^{19}/_{32}, ^5/_8, ^3/_4, ^7/_8$	40	32	40	30	20[8,9]
48/24	$^{23}/_{32}, ^3/_4, ^7/_8$	48	36	45	35	24
54/32	$^7/_8, 1$	54	40	45	35	32
60/48	$^7/_8, 1, 1^1/_8$	60	48	45	35	48
SINGLE-FLOOR GRADES		ROOF[3]				FLOOR[4]
		Maximum Span (inches)		Load[5] (pounds per square foot)		
		× 25.4 for mm		× 0.0479 for kN/m^2		Maximum Span (inches)
Panel Span Rating (inches)	Panel Thickness (inches)	With Edge Support[6]	Without Edge Support	Total Load	Live Load	
× 25.4 for mm						× 25.4 for mm
16 oc	$^1/_2, ^{19}/_{32}, ^5/_8$	24	24	50	40	16[8]
20 oc	$^{19}/_{32}, ^5/_8, ^3/_4$	32	32	40	30	20[8,9]
24 oc	$^{23}/_{32}, ^3/_4$	48	36	35	25	24
32 oc	$^7/_8, 1$	48	40	50	40	32
48 oc	$1^3/_{32}, 1^1/_8$	60	48	50	50	48

[1]Applies to panels 24 inches (610 mm) or wider.
[2]Floor and roof sheathing conforming with this table shall be deemed to meet the design criteria of Section 2312.
[3]Uniform load deflection limitations $^1/_{180}$ of span under live load plus dead load, $^1/_{240}$ under live load only.
[4]Panel edges shall have approved tongue-and-groove joints or shall be supported with blocking unless $^1/_4$-inch (6.4 mm) minimum thickness underlayment or $1^1/_2$ inches (38 mm) of approved cellular or lightweight concrete is placed over the subfloor, or finish floor is $^3/_4$-inch (19 mm) wood strip. Allowable uniform load based on deflection of $^1/_{360}$ of span is 100 pounds per square foot (psf) (4.79 kN/m^2) except the span rating of 48 inches on center is based on a total load of 65 psf (3.11 kN/m).
[5]Allowable load at maximum span.
[6]Tongue-and-groove edges, panel edge clips [one midway between each support, except two equally spaced between supports 48 inches (1219 mm) on center], lumber blocking, or other. Only lumber blocking shall satisfy blocked diaphragms requirements.
[7]For $^1/_2$-inch (12.7 mm) panel, maximum span shall be 24 inches (610 mm).
[8]May be 24 inches (610 mm) on center where $^3/_4$-inch (19 mm) wood strip flooring is installed at right angles to joist.
[9]May be 24 inches (610 mm) on center for floors where $1^1/_2$ inches (38 mm) of cellular or lightweight concrete is applied over the panels.

TABLE 23-II-E-2—ALLOWABLE LOAD (PSF) FOR WOOD STRUCTURAL PANEL ROOF SHEATHING CONTINUOUS OVER TWO OR MORE SPANS AND STRENGTH AXIS PARALLEL TO SUPPORTS
(Plywood structural panels are five-ply, five-layer unless otherwise noted.)[1,2]

PANEL GRADE	THICKNESS (inch)	MAXIMUM SPAN (inches)	LOAD AT MAXIMUM SPAN (psf)	
			× 0.0479 for kN/m^2	
	× 25.4 for mm		Live	Total
Structural I	$^7/_{16}$	24	20	30
	$^{15}/_{32}$	24	35[3]	45[3]
	$^1/_2$	24	40[3]	50[3]
	$^{19}/_{32}, ^5/_8$	24	70	80
	$^{23}/_{32}, ^3/_4$	24	90	100
Other grades covered in UBC Standard 23-2 or 23-3	$^7/_{16}$	16	40	50
	$^{15}/_{32}$	24	20	25
	$^1/_2$	24	25	30
	$^{19}/_{32}$	24	40[3]	50[3]
	$^5/_8$	24	45[3]	55[3]
	$^{23}/_{32}, ^3/_4$	24	60[3]	65[3]

[1]Roof sheathing conforming with this table shall be deemed to meet the design criteria of Section 2312.
[2]Uniform load deflection limitations: $^1/_{180}$ of span under live load plus dead load, $^1/_{240}$ under live load only. Edges shall be blocked with lumber or other approved type of edge supports.
[3]For composite and four-ply plywood structural panel, load shall be reduced by 15 pounds per square foot (0.72 kN/m^2).

TABLE 23-II-F-1—ALLOWABLE SPAN FOR WOOD STRUCTURAL PANEL COMBINATION SUBFLOOR-UNDERLAYMENT (SINGLE FLOOR)[1,2] Panels Continuous over Two or More Spans and Strength Axis Perpendicular to Supports

IDENTIFICATION	MAXIMUM SPACING OF JOISTS (inches)				
	× 25.4 for mm				
	16	20	24	32	48
Species Group[3]	Thickness (inches)				
	× 25.4 for mm				
1	$1/2$	$5/8$	$3/4$	—	—
2, 3	$5/8$	$3/4$	$7/8$	—	—
4	$3/4$	$7/8$	1	—	—
Span rating[4]	16 o.c.	20 o.c.	24 o.c.	32 o.c.	48 o.c.

[1]Spans limited to value shown because of possible effects of concentrated loads. Allowable uniform loads based on deflection of $1/360$ of span is 100 pounds per square foot (psf) (4.79 kN/m^2), except allowable total uniform load for $11/8$-inch (29 mm) wood structural panels over joists spaced 48 inches (1219 mm) on center is 65 psf (3.11 kN/m^2). Panel edges shall have approved tongue-and-groove joints or shall be supported with blocking, unless $1/4$-inch (6.4 mm) minimum thickness underlayment or $11/2$ inches (38 mm) of approved cellular or lightweight concrete is placed over the subfloor, or finish floor is $3/4$-inch (19 mm) wood strip.

[2]Floor panels conforming with this table shall be deemed to meet the design criteria of Section 2312.

[3]Applicable to all grades of sanded exterior-type plywood. See UBC Standard 23-2 for plywood species groups.

[4]Applicable to underlayment grade and C-C (plugged) plywood, and single floor grade wood structural panels.

TABLE 23-II-F-2—ALLOWABLE SPANS FOR PARTICLEBOARD SUBFLOOR AND COMBINED SUBFLOOR-UNDERLAYMENT[1,2]

GRADE	THICKNESS (inches)	MAXIMUM SPACING OF SUPPORTS (inches)[3]	
	× 25.4 for mm	× 25.4 for mm	
		Subfloor	Combined Subfloor-Underlayment[4,5]
2-M-W	$1/2$	16	—
	$5/8$	20	16
	$3/4$	24	24
2-M-3	$3/4$	20	20

[1]All panels are continuous over two or more spans.

[2]Floor sheathing conforming with this table shall be deemed to meet the design criteria of Section 2312.

[3]Uniform deflection limitation: $1/360$ of the span under 100 pounds per square foot (4.79 kN/m^2) minimum load.

[4]Edges shall have tongue-and-groove joints or shall be supported with blocking. The tongue-and-groove panels are installed with the long dimension perpendicular to supports.

[5]A finish wearing surface is to be applied to the top of the panel.

TABLE 23-II-G—MAXIMUM DIAPHRAGM DIMENSION RATIOS

MATERIAL	HORIZONTAL DIAPHRAGMS	SHEAR WALLS
	Maximum Span-Width Ratios	Maximum Height-Width Ratios
1. Diagonal sheathing, conventional	3:1	1:1[1]
2. Diagonal sheathing, special	4:1	2:1[2]
3. Wood structural panels and particleboard, nailed all edges	4:1	2:1[2,3]
4. Wood structural panels and particleboard, blocking omitted at intermediate joints.	4:1	[4]

[1]In Seismic Zones 0, 1, 2 and 3, the maximum ratio may be 2:1.

[2]In Seismic Zones 0, 1, 2 and 3, the maximum ratio may be $31/2$:1.

[3]In Seismic Zone 4, the maximum ratio may be $31/2$:1 for walls not exceeding 10 feet (3048 mm) in height on one side of the door to a one-story Group U Occupancy.

[4]Not permitted.

TABLE 23-II-K—WOOD SHINGLE AND SHAKE SIDE WALL EXPOSURES

SHINGLE OR SHAKE	MAXIMUM WEATHER EXPOSURES (inches)			
	× 25.4 for mm			
	Single-Coursing		Double-Coursing	
Length and Type	No. 1	No. 2	No. 1	No. 2
16-inch (405 mm) shingles	$71/2$	$71/2$	12	10
18-inch (455 mm) shingles	$81/2$	$81/2$	14	11
24-inch (610 mm) shingles	$111/2$	$111/2$	16	14
18-inch (455 mm) resawn shakes	$81/2$	—	14	—
18-inch (455 mm) straight-split shakes	$81/2$	—	16	—
24-inch (610 mm) resawn shakes	$111/2$	—	20	—

Division IV—CONVENTIONAL LIGHT-FRAME CONSTRUCTION

SECTION 2320 — CONVENTIONAL LIGHT-FRAME CONSTRUCTION DESIGN PROVISIONS

2320.1 General. The requirements in this section are intended for conventional light-frame construction. Other methods may be used provided a satisfactory design is submitted showing compliance with other provisions of this code.

Only the following occupancies may be constructed in accordance with this division:

1. One-, two- or three-story buildings housing Group R Occupancies.

2. One-story Occupancy Category 4 buildings, as defined in Table 16-K, when constructed on a slab-on-grade floor.

3. Group U Occupancies.

4. Top-story walls and roofs of Occupancy Category 4 buildings not exceeding two stories of wood framing.

5. Interior nonload-bearing partitions, ceilings and curtain walls in all occupancies.

When total loads exceed those specified in Tables 23-IV-J-1, 23-IV-J-3, and 23-IV-R-1, 23-IV-R-2, 23-IV-R-3, 23-IV-R-4, 23-IV-R-7, 23-IV-R-8, 23-IV-R-9, 23-IV-R-10, 23-IV-R-11 and 23-IV-R-12; 23-VII-R-1, 23-VII-R-3, 23-VII-R-7, 23-VII-R-9, 23-VIII-A, 23-VIII-B, 23-VIII-C, 23-VIII-D, an engineering design shall be provided for the gravity load system.

Other approved repetitive wood members may be used in lieu of solid-sawn lumber in conventional construction provided these members comply with the provisions of this code.

2320.2 Design of Portions. When a building of otherwise conventional construction contains nonconventional structural elements, those elements shall be designed in accordance with Section 1605.2.

2320.3 Additional Requirements for Conventional Construction in High-wind Areas. Appendix Chapter 23 provisions for conventional construction in high-wind areas shall apply when specifically adopted.

2320.4 Additional Requirements for Conventional Construction in Seismic Zones 0, 1, 2 and 3.

2320.4.1 Braced wall lines. Where the basic wind speed is not greater than 80 miles per hour (mph) (129 km/h), buildings shall be provided with exterior and interior braced wall lines. Spacing shall not exceed 34 feet (10 363 mm) on center in both the longitudinal and transverse directions in each story.

2320.4.2 Braced wall lines for high wind. Where the basic wind speed exceeds 80 mph (129 km/h), buildings shall be provided with exterior and interior braced wall lines. Spacing shall not exceed 25 feet (7620 mm) on center in both the longitudinal and transverse directions in each story.

> **EXCEPTION:** In one- and two-story Group R, Division 3 buildings, interior braced wall line spacing may be increased to not more than 34 feet (10 363 mm) on center in order to accommodate one single room per dwelling unit not exceeding 900 square feet (83.6 m²). The building official may require additional walls to contain braced panels when this exception is used.

2320.4.3 Veneer. Anchored masonry and stone wall veneer shall not exceed 5 inches (127 mm) in thickness and shall conform to the requirements of Chapter 14.

2320.4.4 Lateral force-resisting system. Buildings in Seismic Zone 3 that are not provided with braced wall lines in accordance with Section 2320.4 or that are of unusual shape as described in Section 2320.5.4 shall have a lateral-force-resisting system designed to resist the forces specified in Chapter 16.

2320.5 Additional Requirements for Conventional Construction in Seismic Zone 4.

2320.5.1 Braced wall lines. Buildings shall be provided with exterior and interior braced wall lines. Spacing shall not exceed 25 feet (7620 mm) on center in both the longitudinal and transverse directions in each story.

> **EXCEPTION:** In one- and two-story Group R, Division 3 buildings, interior braced wall line spacing may be increased to not more than 34 feet (10 363 mm) on center in order to accommodate one single room per dwelling unit not exceeding 900 square feet (83.61 m²). The building official may require additional walls to contain braced panels when this exception is used.

2320.5.2 Lateral-force-resisting system. When total loads supported on wood framing exceed those specified in Tables 23-IV-J-1, 23-IV-J-3, 23-IV-R-1, 23-IV-R-2, 23-IV-R-3, 23-IV-R-4, 23-IV-R-7, 23-IV-R-8, 23-IV-R-9 and 23-IV-R-10, 23-VII-R-1. 23-VII-R-3, 23-VII-R-7, 23-VII-R-9, 23-VIII-A, 23-VIII-B, 23-VIII-C and 23-VIII-D, an engineering design shall be provided for the lateral-force-resisting system.

2320.5.3 Veneer. Anchored masonry and stone wall veneer shall not exceed 5 inches (127 mm) in thickness, shall conform to the requirements of Chapter 14 and shall not extend above the first story.

2320.5.4 Unusually shaped buildings. When of unusual shape, buildings of light-frame construction shall have a lateral-force-resisting system designed to resist the forces specified in Chapter 16. Buildings shall be considered to be of unusual shape when the building official determines that the structure has framing irregularities, offsets, split levels or any configuration that creates discontinuities in the seismic load path and may include one or more of the following:

2320.5.4.1 When exterior braced wall panels, as required by Section 2320.11.3, are not in one plane vertically from the foundation to the uppermost story in which they are required.

> **EXCEPTION:** Floors with cantilevers or setbacks not exceeding four times the nominal depth of the floor joists may support braced wall panels provided:
> 1. Floor joists are 2 inches by 10 inches (51 mm by 254 mm) or larger and spaced at not more than 16 inches (406 mm) on center.
> 2. The ratio of the back span to the cantilever is at least 2 to 1.
> 3. Floor joists at ends of braced wall panels are doubled.
> 4. A continuous rim joist is connected to ends of all cantilevered joists. The rim joist may be spliced using a metal tie not less than 0.058 inch (1.47 mm) (16 galvanized gage) and 1¹/₂ inches (38 mm) wide fastened with six 16d nails.
> 5. Gravity loads carried at the end of cantilevered joists are limited to uniform wall and roof load and the reactions from headers having a span of 8 feet (2438 mm) or less.

2320.5.4.2 When a section of floor or roof is not laterally supported by braced wall lines on all edges.

> **EXCEPTION:** Portions of roofs or floors which do not support braced wall panels above may extend up to 6 feet (1829 mm) beyond a braced wall line.

2320.5.4.3 When the end of a required braced wall panel extends more than 1 foot (305 mm) over an opening in the wall below. This provision is applicable to braced wall panels offset in plane and to braced wall panels offset out of plane as permitted by Section 2320.5.4.1, exception.

EXCEPTION: Braced wall panels may extend over an opening not more than 8 feet (2438 mm) in width when the header is a 4-inch by 12-inch (102 mm by 305 mm) or larger member.

2320.5.4.4 When an opening in a floor or roof exceeds the lesser of 12 feet (3657 mm) or 50 percent of the least floor or roof dimension.

2320.5.4.5 Construction where portions of a floor level are vertically offset such that the framing members on either side of the offset cannot be lapped or tied together in an approved manner as required by Section 2320.8.3.

EXCEPTION: Framing supported directly by foundations.

2320.5.4.6 When braced wall lines do not occur in two perpendicular directions.

2320.5.5 Lumber roof decks. Lumber roof decks shall have solid sheathing.

2320.5.6 Interior braced wall support. In one-story buildings, interior braced wall lines shall be supported on continuous foundations at intervals not exceeding 50 feet (15 240 mm). In buildings more than one story in height, all interior braced wall panels shall be supported on continuous foundations.

EXCEPTION: Two-story buildings may have interior braced wall lines supported on continuous foundations at intervals not exceeding 50 feet (15 240 mm) provided:

1. Cripple wall height does not exceed 4 feet (1219 mm).
2. First-floor braced wall panels are supported on doubled floor joists, continuous blocking or floor beams.
3. Distance between bracing lines does not exceed twice the building width parallel to the braced wall line.

2320.6 Foundation Plates or Sills. Foundations and footings shall be as specified in Chapter 18. Foundation plates or sills resting on concrete or masonry foundations shall be bolted as required by Section 1806.6.

2320.7 Girders. Girders for single-story construction or girders supporting loads from a single floor shall not be less than 4 inches by 6 inches (102 mm by 153 mm) for spans 6 feet (1829 mm) or less, provided that girders are spaced not more than 8 feet (2438 mm) on center. Other girders shall be designed to support the loads specified in this code. Girder end joints shall occur over supports. When a girder is spliced over a support, an adequate tie shall be provided. The end of beams or girders supported on masonry or concrete shall not have less than 3 inches (76 mm) of bearing.

2320.8 Floor Joists.

2320.8.1 General. Spans for joists shall be in accordance with Tables 23-IV-J-1 and 23-IV-J-2.

2320.8.2 Bearing. Except where supported on a 1-inch by 4-inch (25 mm by 102 mm) ribbon strip and nailed to the adjoining stud, the ends of each joist shall not have less than $1^1/_2$ inches (38 mm) of bearing on wood or metal, or less than 3 inches (76 mm) on masonry.

2320.8.3 Framing details. Joists shall be supported laterally at the ends and at each support by solid blocking except where the ends of joists are nailed to a header, band or rim joist or to an adjoining stud or by other approved means. Solid blocking shall not be less than 2 inches (51 mm) in thickness and the full depth of joist.

Notches on the ends of joists shall not exceed one fourth the joist depth. Holes bored in joists shall not be within 2 inches (51 mm) of the top or bottom of the joist, and the diameter of any such hole shall not exceed one third the depth of the joist. Notches in the top or bottom of joists shall not exceed one sixth the depth and shall not be located in the middle third of the span.

Joist framing from opposite sides of a beam, girder or partition shall be lapped at least 3 inches (76 mm) or the opposing joists shall be tied together in an approved manner.

Joists framing into the side of a wood girder shall be supported by framing anchors or on ledger strips not less than 2 inches by 2 inches (51 mm by 51 mm).

2320.8.4 Framing around openings. Trimmer and header joists shall be doubled, or of lumber of equivalent cross section, when the span of the header exceeds 4 feet (1219 mm). The ends of header joists more than 6 feet (1829 mm) long shall be supported by framing anchors or joist hangers unless bearing on a beam, partition or wall. Tail joists over 12 feet (3658 mm) long shall be supported at header by framing anchors or on ledger strips not less than 2 inches by 2 inches (51 mm by 51 mm).

2320.8.5 Supporting bearing partitions. Bearing partitions perpendicular to joists shall not be offset from supporting girders, walls or partitions more than the joist depth.

Joists under and parallel to bearing partitions shall be doubled.

2320.8.6 Blocking. Floor joists shall be blocked when required by the provisions of Division III, Part I or Section 2320.8.3.

2320.9 Subflooring.

2320.9.1 Lumber subfloor. Sheathing used as a structural subfloor shall conform to the limitations set forth in Tables 23-II-D-1 and 23-II-D-2.

Joints in subflooring shall occur over supports unless end-matched lumber is used, in which case each piece shall bear on at least two joists.

Subflooring may be omitted when joist spacing does not exceed 16 inches (406 mm) and 1-inch (25 mm) nominal tongue-and-groove wood strip flooring is applied perpendicular to the joists.

2320.9.2 Wood structural panels. Where used as structural subflooring, wood structural panels shall be as set forth in Tables 23-II-E-1 and 23-II-E-2. Wood structural panel combination subfloor underlayment shall have maximum spans as set forth in Table 23-II-F-1.

When wood structural panel floors are glued to joists with an adhesive in accordance with the adhesive manufacturer's directions, fasteners may be spaced a maximum of 12 inches (305 mm) on center at all supports.

2320.9.3 Plank flooring. Plank flooring shall be designed in accordance with the general provisions of this code.

In lieu of such design, 2-inch (51 mm) tongue-and-groove planking may be used in accordance with Table 23-IV-A. Joints in such planking may be randomly spaced, provided the system is applied to not less than three continuous spans, planks are center-matched and end-matched or splined, each plank bears on at least one support and joints are separated by at least 24 inches (610 mm) in adjacent pieces. One-inch (25 mm) nominal strip square-edged flooring, $1/_2$-inch (12.7 mm) tongue-and-groove flooring or $3/_8$-inch (9.5 mm) wood structural panel shall be applied over random-length decking used as a floor. The strip and tongue-and-groove flooring shall be applied at right angles to the span of the planks. The $3/_8$-inch (9.5 mm) plywood shall be applied with the face grain at right angles to the span of the planks.

2320.9.4 Particleboard. Where used as structural subflooring or as combined subfloor underlayment, particleboard shall be as set forth in Table 23-II-F-2.

2320.10 Particleboard Underlayment. In accordance with approved recognized standards, particleboard floor underlayment shall conform to Type PBU. Underlayment shall not be less than $1/4$ inch (6.4 mm) in thickness and shall be identified by the grade mark of an approved inspection agency. Underlayment shall be installed in accordance with this code and as recommended by the manufacturer.

2320.11 Wall Framing.

2320.11.1 Size, height and spacing. The size, height and spacing of studs shall be in accordance with Table 23-IV-B except that Utility grade studs shall not be spaced more than 16 inches (406 mm) on center, or support more than a roof and ceiling, or exceed 8 feet (2438 mm) in height for exterior walls and load-bearing walls or 10 feet (3048 mm) for interior nonload-bearing walls.

2320.11.2 Framing details. Studs shall be placed with their wide dimension perpendicular to the wall. Not less than three studs shall be installed at each corner of an exterior wall.

> **EXCEPTION:** At corners, a third stud may be omitted through the use of wood spacers or backup cleats of $3/8$-inch-thick (9.5 mm) wood structural panel, $3/8$-inch (9.5 mm) Type M "Exterior Glue" particleboard, 1-inch-thick (25 mm) lumber or other approved devices that will serve as an adequate backing for the attachment of facing materials. Where fire-resistance ratings or shear values are involved, wood spacers, backup cleats or other devices shall not be used unless specifically approved for such use.

Bearing and exterior wall studs shall be capped with double top plates installed to provide overlapping at corners and at intersections with other partitions. End joints in double top plates shall be offset at least 48 inches (2438 mm).

> **EXCEPTION:** A single top plate may be used, provided the plate is adequately tied at joints, corners and intersecting walls by at least the equivalent of 3-inch by 6-inch (76 mm by 152 mm) by 0.036-inch-thick (0.9 mm) galvanized steel that is nailed to each wall or segment of wall by six 8d nails or equivalent, provided the rafters, joists or trusses are centered over the studs with a tolerance of no more than 1 inch (25 mm).

When bearing studs are spaced at 24-inch (610 mm) intervals and top plates are less than two 2-inch by 6-inch (51 mm by 152 mm) or two 3-inch by 4-inch (76 mm by 102 mm) members and when the floor joists, floor trusses or roof trusses which they support are spaced at more than 16-inch (406 mm) intervals, such joists or trusses shall bear within 5 inches (127 mm) of the studs beneath or a third plate shall be installed.

Interior nonbearing partitions may be capped with a single top plate installed to provide overlapping at corners and at intersections with other walls and partitions. The plate shall be continuously tied at joints by solid blocking at least 16 inches (406 mm) in length and equal in size to the plate or by $1/8$-inch by $1^{1}/2$-inch (3.2 mm by 38 mm) metal ties with spliced sections fastened with two 16d nails on each side of the joint.

Studs shall have full bearing on a plate or sill not less than 2 inches (51 mm) in thickness having a width not less than that of the wall studs.

2320.11.3 Bracing. Braced wall lines shall consist of braced wall panels which meet the requirements for location, type and amount of bracing specified in Table 23-IV-C-1 and are in line or offset from each other by not more than 4 feet (1219 mm). Braced wall panels shall start at not more than 8 feet (2438 mm) from each end of a braced wall line. All braced wall panels shall be clearly indicated on the plans. Construction of braced wall panels shall be by one of the following methods:

1. Nominal 1-inch by 4-inch (25 mm by 102 mm) continuous diagonal braces let into top and bottom plates and intervening studs, placed at an angle not more than 60 degrees or less than 45 degrees from the horizontal, and attached to the framing in conformance with Table 23-II-B-1.

2. Wood boards of $5/8$-inch (16 mm) net minimum thickness applied diagonally on studs spaced not over 24 inches (610 mm) on center.

3. Wood structural panel sheathing with a thickness not less than $5/16$ inch (7.9 mm) for 16-inch (406 mm) stud spacing and not less than $3/8$ inch (9.5 mm) for 24-inch (610 mm) stud spacing in accordance with Tables 23-II-A-1 and 23-IV-D-1.

4. Fiberboard sheathing 4-foot by 8-foot (1219 mm by 2438 mm) panels not less than $1/2$ inch (13 mm) thick applied vertically on studs spaced not over 16 inches (406 mm) on center when installed in accordance with Section 2315.6 and Table 23-II-J.

5. Gypsum board [sheathing $1/2$ inch (13 mm) thick by 4 feet (1219 mm) wide, wallboard or veneer base] on studs spaced not over 24 inches (610 mm) on center and nailed at 7 inches (178 mm) on center with nails as required by Table 25-I.

6. Particleboard wall sheathing panels where installed in accordance with Table 23-IV-D-2.

7. Portland cement plaster on studs spaced 16 inches (406 mm) on center installed in accordance with Table 25-I.

8. Hardboard panel siding when installed in accordance with Section 2310.6 and Table 23-II-C.

Method 1 is not permitted in Seismic Zones 2B, 3 and 4. For cripple wall bracing, see Section 2320.11.5. For Methods 2, 3, 4, 6, 7 and 8, each braced panel must be at least 48 inches (1219 mm) in length, covering three stud spaces where studs are spaced 16 inches (406 mm) apart and covering two stud spaces where studs are spaced 24 inches (610 mm) apart.

For Method 5, each braced wall panel must be at least 96 inches (2438 mm) in length when applied to one face of a braced wall panel and 48 inches (1219 mm) when applied to both faces.

All vertical joints of panel sheathing shall occur over studs. Horizontal joints shall occur over blocking equal in size to the studding except where waived by the installation requirements for the specific sheathing materials.

Braced wall panel sole plates shall be nailed to the floor framing and top plates shall be connected to the framing above in accordance with Table 23-II-B-1. Sills shall be bolted to the foundation or slab in accordance with Section 1806.6. Where joists are perpendicular to braced wall lines above, blocking shall be provided under and in line with the braced wall panels.

2320.11.4 Alternate braced wall panels. Any braced wall panel required by Section 2320.11.3 may be replaced by an alternate braced wall panel constructed in accordance with the following:

1. In one-story buildings, each panel shall have a length of not less than 2 feet 8 inches (813 mm) and a height of not more than 10 feet (3048 mm). Each panel shall be sheathed on one face with $3/8$-inch-minimum-thickness (9.5 mm) plywood sheathing nailed with 8d common or galvanized box nails in accordance with Table 23-II-B-1 and blocked at all plywood edges. Two anchor bolts installed in accordance with Section 1806.6, shall be provided in each panel. Anchor bolts shall be placed at panel quarter points. Each panel end stud shall have a tie-down device fastened to the foundation, capable of providing an approved uplift capacity of not less than 1,800 pounds (816.5 kg). The tie-down device shall

be installed in accordance with the manufacturer's recommendations. The panels shall be supported directly on a foundation or on floor framing supported directly on a foundation which is continuous across the entire length of the braced wall line. This foundation shall be reinforced with not less than one No. 4 bar top and bottom.

2. In the first story of two-story buildings, each braced wall panel shall be in accordance with Section 2320.11.4, Item 1, except that the plywood sheathing shall be provided on both faces, three anchor bolts shall be placed at one-fifth points, and tie-down device uplift capacity shall not be less than 3,000 pounds (1360.8 kg).

2320.11.5 Cripple walls. Foundation cripple walls shall be framed of studs not less in size than the studding above with a minimum length of 14 inches (356 mm), or shall be framed of solid blocking. When exceeding 4 feet (1219 mm) in height, such walls shall be framed of studs having the size required for an additional story.

Cripple walls having a stud height exceeding 14 inches (356 mm) shall be braced in accordance with Table 23-IV-C-2. Solid blocking or wood structural panel sheathing may be used to brace cripple walls having a stud height of 14 inches (356 mm) or less. In Seismic Zone 4, Method 7 is not permitted for bracing any cripple wall studs.

Spacing of boundary nailing for required wall bracing shall not exceed 6 inches (152 mm) on center along the foundation plate and the top plate of the cripple wall. Nail size, nail spacing for field nailing and more restrictive boundary nailing requirements shall be as required elsewhere in the code for the specific bracing material used.

2320.11.6 Headers. Headers and lintels shall conform to the requirements set forth in this paragraph and together with their supporting systems shall be designed to support the loads specified in this code. All openings 4 feet (1219 mm) wide or less in bearing walls shall be provided with headers consisting of either two pieces of 2-inch (51 mm) framing lumber placed on edge and securely fastened together or 4-inch (102 mm) lumber of equivalent cross section. All openings more than 4 feet (1219 mm) wide shall be provided with headers or lintels. Each end of a lintel or header shall have a length of bearing of not less than $1^1/_2$ inches (38 mm) for the full width of the lintel.

2320.11.7 Pipes in walls. Stud partitions containing plumbing, heating, or other pipes shall be so framed and the joists underneath so spaced as to give proper clearance for the piping. Where a partition containing such piping runs parallel to the floor joists, the joists underneath such partitions shall be doubled and spaced to permit the passage of such pipes and shall be bridged. Where plumbing, heating or other pipes are placed in or partly in a partition, necessitating the cutting of the soles or plates, a metal tie not less than 0.058 inch (1.47 mm) (16 galvanized gage) and $1^1/_2$ inches (38 mm) wide shall be fastened to each plate across and to each side of the opening with not less than six 16d nails.

2320.11.8 Bridging. Unless covered by interior or exterior wall coverings or sheathing meeting the minimum requirements of this code, all stud partitions or walls with studs having a height-to-least-thickness ratio exceeding 50 shall have bridging not less than 2 inches (51 mm) in thickness and of the same width as the studs fitted snugly and nailed thereto to provide adequate lateral support.

2320.11.9 Cutting and notching. In exterior walls and bearing partitions, any wood stud may be cut or notched to a depth not exceeding 25 percent of its width. Cutting or notching of studs to a depth not greater than 40 percent of the width of the stud is permitted in nonbearing partitions supporting no loads other than the weight of the partition.

2320.11.10 Bored holes. A hole not greater in diameter than 40 percent of the stud width may be bored in any wood stud. Bored holes not greater than 60 percent of the width of the stud are permitted in nonbearing partitions or in any wall where each bored stud is doubled, provided not more than two such successive doubled studs are so bored.

In no case shall the edge of the bored hole be nearer than $5/_8$ inch (16 mm) to the edge of the stud. Bored holes shall not be located at the same section of stud as a cut or notch.

2320.12 Roof and Ceiling Framing.

2320.12.1 General. The framing details required in this section apply to roofs having a minimum slope of 3 units vertical in 12 units horizontal (25% slope) or greater. When the roof slope is less than 3 units vertical in 12 units horizontal (25% slope), members supporting rafters and ceiling joists such as ridge board, hips and valleys shall be designed as beams.

2320.12.2 Spans. Allowable spans for ceiling joists shall be in accordance with Tables 23-IV-J-3 and 23-IV-J-4. Allowable spans for rafters shall be in accordance with Tables 23-IV-R-1 through 23-IV-R-12, where applicable.

2320.12.3 Framing. Rafters shall be framed directly opposite each other at the ridge. There shall be a ridge board at least 1-inch (25 mm) nominal thickness at all ridges and not less in depth than the cut end of the rafter. At all valleys and hips there shall be a single valley or hip rafter not less than 2-inch (51 mm) nominal thickness and not less in depth than the cut end of the rafter.

2320.12.4 Notches and holes. Notching at the ends of rafters or ceiling joists shall not exceed one fourth the depth. Notches in the top or bottom of the rafter or ceiling joist shall not exceed one sixth the depth and shall not be located in the middle one third of the span, except that a notch not exceeding one third of the depth is permitted in the top of the rafter or ceiling joist not further from the face of the support than the depth of the member.

Holes bored in rafters or ceiling joists shall not be within 2 inches (51 mm) of the top and bottom and their diameter shall not exceed one third the depth of the member.

2320.12.5 Framing around openings. Trimmer and header rafters shall be doubled, or of lumber of equivalent cross section, when the span of the header exceeds 4 feet (1219 mm). The ends of header rafters more than 6 feet (1829 mm) long shall be supported by framing anchors or rafter hangers unless bearing on a beam, partition or wall.

2320.12.6 Rafter ties. Rafters shall be nailed to adjacent ceiling joists to form a continuous tie between exterior walls when such joists are parallel to the rafters. Where not parallel, rafters shall be tied to 1-inch by 4-inch (25 mm by 102 mm) (nominal) minimum-size crossties. Rafter ties shall be spaced not more than 4 feet (1219 mm) on center.

2320.12.7 Purlins. Purlins to support roof loads may be installed to reduce the span of rafters within allowable limits and shall be supported by struts to bearing walls. The maximum span of 2-inch by 4-inch (51 mm by 102 mm) purlins shall be 4 feet (1219 mm). The maximum span of the 2-inch by 6-inch (51 mm by 152 mm) purlin shall be 6 feet (1829 mm) but in no case shall the purlin be smaller than the supported rafter. Struts shall not be smaller than 2-inch by 4-inch (51 mm by 102 mm) members. The unbraced length of struts shall not exceed 8 feet (2438 mm) and the minimum slope of the struts shall not be less than 45 degrees from the horizontal.

2320.12.8 Blocking. Roof rafters and ceiling joists shall be supported laterally to prevent rotation and lateral displacement when required by Division III, Part I, Section 4.4.1.2. Roof trusses shall be supported laterally at points of bearing by solid blocking to prevent rotation and lateral displacement.

2320.12.9 Roof sheathing. Roof sheathing shall be in accordance with Tables 23-II-E-1 and 23-II-E-2 for wood structural panels, and Tables 23-II-D-1 and 23-II-D-2 for lumber.

Joints in lumber sheathing shall occur over supports unless approved end-matched lumber is used, in which case each piece shall bear on at least two supports.

Wood structural panels used for roof sheathing shall be bonded by intermediate or exterior glue. Wood structural panel roof sheathing exposed on the underside shall be bonded with exterior glue.

2320.12.10 Roof planking. Planking shall be designed in accordance with the general provisions of this code.

In lieu of such design, 2-inch (51 mm) tongue-and-groove planking may be used in accordance with Table 23-IV-A. Joints in such planking may be randomly spaced, provided the system is applied to not less than three continuous spans, planks are center-matched and end-matched or splined, each plank bears on at least one support, and joints are separated by at least 24 inches (610 mm) in adjacent pieces.

2320.13 Exit Facilities. In Seismic Zones 3 and 4, exterior exit balconies, stairs and similar exit facilities shall be positively anchored to the primary structure at not over 8 feet (2438 mm) on center or shall be designed for lateral forces. Such attachment shall not be accomplished by use of toenails or nails subject to withdrawal.

TABLE 23-IV-A—ALLOWABLE SPANS FOR 2-INCH (51 mm) TONGUE-AND-GROOVE DECKING

SPAN[1] (feet) × 304.8 for mm	LIVE LOAD × 0.0479 for kN/m²	DEFLECTION LIMIT	f (psi) × 0.00689 for N/mm²	E (psi)
		Roofs		
4	20	1/240 1/360	160	170,000 256,000
	30	1/240 1/360	210	256,000 384,000
	40	1/240 1/360	270	340,000 512,000
4.5	20	1/240 1/360	200	242,000 305,000
	30	1/240 1/360	270	363,000 405,000
	40	1/240 1/360	350	484,000 725,000
5.0	20	1/240 1/360	250	332,000 500,000
	30	1/240 1/360	330	495,000 742,000
	40	1/240 1/360	420	660,000 1,000,000
5.5	20	1/240 1/360	300	442,000 660,000
	30	1/240 1/360	400	662,000 998,000
	40	1/240 1/360	500	884,000 1,330,000
6.0	20	1/240 1/360	360	575,000 862,000
	30	1/240 1/360	480	862,000 1,295,000
	40	1/240 1/360	600	1,150,000 1,730,000
6.5	20	1/240 1/360	420	595,000 892,000
	30	1/240 1/360	560	892,000 1,340,000
	40	1/240 1/360	700	1,190,000 1,730,000
7.0	20	1/240 1/360	490	910,000 1,360,000
	30	1/240 1/360	650	1,370,000 2,000,000
	40	1/240 1/360	810	1,820,000 2,725,000
7.5	20	1/240 1/360	560	1,125,000 1,685,000
	30	1/240 1/360	750	1,685,000 2,530,000
	40	1/240 1/360	930	2,250,000 3,380,000
8.0	20	1/240 1/360	640	1,360,000 2,040,000
	30	1/240 1/360	850	2,040,000 3,060,000
		Floors		
4 4.5 5.0	40	1/360	840 950 1060	1,000,000 1,300,000 1,600,000

[1]Spans are based on simple beam action with 10 pounds per square foot (0.48 kN/m²) dead load and provisions for a 300-pound (1334 N) concentrated load on a 12-inch (305 mm) width of floor decking. Random lay-up permitted in accordance with the provisions of Section 2320.9.3 or 2320.12.9. Lumber thickness assumed at 1¹/₂ inches (38 mm), net.

TABLE 23-IV-B—SIZE, HEIGHT AND SPACING OF WOOD STUDS

STUD SIZE (inches)	BEARING WALLS				NONBEARING WALLS	
	Laterally Unsupported Stud Height[1] (feet)	Supporting Roof and Ceiling Only	Supporting One Floor, Roof and Ceiling	Supporting Two Floors, Roof and Ceiling	Laterally Unsupported Stud Height[1] (feet)	Spacing (inches)
		Spacing (inches)				
× 25.4 for mm	× 304.8 for mm	× 25.4 for mm			× 304.8 for mm	× 25.4 for mm
1. 2 × 3[2]	—	—	—	—	10	16
2. 2 × 4	10	24	16	—	14	24
3. 3 × 4	10	24	24	16	14	24
4. 2 × 5	10	24	24	—	16	24
5. 2 × 6	10	24	24	16	20	24

[1]Listed heights are distances between points of lateral support placed perpendicular to the plane of the wall. Increases in unsupported height are permitted where justified by an analysis.
[2]Shall not be used in exterior walls.

TABLE 23-IV-C-1—BRACED WALL PANELS[1]

SEISMIC ZONE	CONDITION	CONSTRUCTION METHOD[2,3]								BRACED PANEL LOCATION AND LENGTH[4]
		1	2	3	4	5	6	7	8	
0, 1 and 2A	One story, top of two or three story	X	X	X	X	X	X	X	X	Each end and not more than 25 feet (7620 mm) on center
	First story of two story or second story of three story	X	X	X	X	X	X	X	X	
	First story of three story		X	X	X	X[5]	X	X	X	
2B, 3 and 4	One story, top of two story or three story		X	X	X	X	X	X[6]	X	Each end and not more than 25 feet (7620 mm) on center
	First story of two story or second of three story		X	X	X	X[5]	X	X[6]	X	Each end and not more than 25 feet (7620 mm) on center but not less than 25% of building length[7]
	First story of three story		X	X	X	X[5]	X	X[6]	X	Each end and not more than 25 feet (7620 mm) on center but not less than 40% of building length[7]

[1]This table specifies minimum requirements for braced panels which form interior or exterior braced wall lines.
[2]See Section 2320.11.3 for full description.
[3]See Section 2320.11.4 for alternate braced panel requirement.
[4]Building length is the dimension parallel to the braced wall length.
[5]Gypsum wallboard applied to supports at 16 inches (406 mm) on center.
[6]Not permitted for bracing cripple walls in Seismic Zone 4. See Section 2320.11.5.
[7]The required lengths shall be doubled for gypsum board applied to only one face of a braced wall panel.

TABLE 23-IV-C-2—CRIPPLE WALL BRACING

SEISMIC ZONE	CONDITION	AMOUNT OF CRIPPLE WALL BRACING[1,2]
		× 25.4 for mm
4	One story above cripple wall	$3/8''$ wood structural panel with 8d at $6''/12''$ nailing on 60 percent of wall length minimum
	Two story above cripple wall	$3/8''$ wood structural panel with 8d at $4''/12''$ nailing on 50 percent of wall length minimum or $3/8''$ wood structural panel with 8d at $6''/12''$ nailing on 75 percent of wall length minimum
3	One story above cripple wall	$3/8''$ wood structural panel with 8d at $6''/12''$ nailing on 40 percent of wall length minimum
0, 1 and 2	One story above cripple wall	$3/8''$ wood structural panel with 8d at $6''/12''$ nailing on 30 percent of wall length minimum
0, 1, 2 and 3	Two story above cripple wall	$3/8''$ wood structural panel with 8d at $4''/12''$ nailing on 40 percent of wall length minimum or $3/8''$ wood structural panel with 8d at $6''/12''$ nailing on 60 percent of wall length minimum

[1]Braced panel length shall be at least two times the height of the cripple wall, but not less than 48 inches (1219 mm).
[2]All panels along a wall shall be nearly equal in length and shall be nearly equally spaced along the length of the wall.

TABLE 23-IV-D-1—WOOD STRUCTURAL PANEL WALL SHEATHING[1]
(Not exposed to the weather, strength axis parallel or perpendicular to studs)

MINIMUM THICKNESS (inch) × 25.4 for mm	PANEL SPAN RATING	STUD SPACING (inches) × 25.4 for mm		
		Siding Nailed to Studs	Sheathing under Coverings Specified in Section 2310.4	
			Sheathing Parallel to Studs	Sheathing Perpendicular to Studs
5/16	12/0, 16/0, 20/0 Wall—16 o.c.	16	—	16
3/8, 15/32, 1/2	16/0, 20/0, 24/0, 32/16 Wall—24 o.c.	24	16	24
7/16, 15/32, 1/2	24/0, 24/16, 32/16 Wall—24 o.c.	24	24[2]	24

[1]In reference to Section 2320.11.3, blocking of horizontal joints is not required.
[2]Plywood shall consist of four or more plies.

TABLE 23-IV-D-2—ALLOWABLE SPANS FOR PARTICLEBOARD WALL SHEATHING[1]
(Not exposed to the weather, long dimension of the panel parallel or perpendicular to studs)

GRADE	THICKNESS (Inch)	STUD SPACING (inches) × 25.4 for mm	
		Siding Nailed to Studs	Sheathing under Coverings Specified in Section 2310.4 Parallel or Perpendicular to Studs
		× 25.4 for mm	
M-1 M-S M-2 "Exterior Glue"	3/8	16	16
	1/2	16	16

[1]In reference to Section 2320.11.3, blocking of horizontal joints is not required.

TABLE 23-IV-J-1—FLOOR JOISTS WITH *L*/360 DEFLECTION LIMITS
The allowable bending stress (*F_b*) and modulus of elasticity (*E*) used in this table shall be from Tables 23-IV-V-1 and 23-IV-V-2 only.

DESIGN CRITERIA:
Deflection — For 40 psf (1.92 kN/m^2) live load.
Limited to span in inches (mm) divided by 360.
Strength — Live load of 40 psf (1.92 kN/m^2) plus dead load of 10 psf (0.48 kN/m^2) determines the required bending design value.

| Joist Size (in) | Spacing (in) | Modulus of Elasticity, *E*, in 1,000,000 psi | | | | | | | | | | | | | | | | |
| | | × 0.00689 for N/mm^2 | | | | | | | | | | | | | | | | |
× 25.4 for mm		0.8	0.9	1.0	1.1	1.2	1.3	1.4	1.5	1.6	1.7	1.8	1.9	2.0	2.1	2.2	2.3	2.4
2 × 6	12.0	8-6	8-10	9-2	9-6	9-9	10-0	10-3	10-6	10-9	10-11	11-2	11-4	11-7	11-9	11-11	12-1	12-3
	16.0	7-9	8-0	8-4	8-7	8-10	9-1	9-4	9-6	9-9	9-11	10-2	10-4	10-6	10-8	10-10	11-0	11-2
	19.2	7-3	7-7	7-10	8-1	8-4	8-7	8-9	9-0	9-2	9-4	9-6	9-8	9-10	10-0	10-2	10-4	10-6
	24.0	6-9	7-0	7-3	7-6	7-9	7-11	8-2	8-4	8-6	8-8	8-10	9-0	9-2	9-4	9-6	9-7	9-9
2 × 8	12.0	11-3	11-8	12-1	12-6	12-10	13-2	13-6	13-10	14-2	14-5	14-8	15-0	15-3	15-6	15-9	15-11	16-2
	16.0	10-2	10-7	11-0	11-4	11-8	12-0	12-3	12-7	12-10	13-1	13-4	13-7	13-10	14-1	14-3	14-6	14-8
	19.2	9-7	10-0	10-4	10-8	11-0	11-3	11-7	11-10	12-1	12-4	12-7	12-10	13-0	13-3	13-5	13-8	13-10
	24.0	8-11	9-3	9-7	9-11	10-2	10-6	10-9	11-0	11-3	11-5	11-8	11-11	12-1	12-3	12-6	12-8	12-10
2 × 10	12.0	14-4	14-11	15-5	15-11	16-5	16-10	17-3	17-8	18-0	18-5	18-9	19-1	19-5	19-9	20-1	20-4	20-8
	16.0	13-0	13-6	14-0	14-6	14-11	15-3	15-8	16-0	16-5	16-9	17-0	17-4	17-8	17-11	18-3	18-6	18-9
	19.2	12-3	12-9	13-2	13-7	14-0	14-5	14-9	15-1	15-5	15-9	16-0	16-4	16-7	16-11	17-2	17-5	17-8
	24.0	11-4	11-10	12-3	12-8	13-0	13-4	13-8	14-0	14-4	14-7	14-11	15-2	15-5	15-8	15-11	16-2	16-5
2 × 12	12.0	17-5	18-1	18-9	19-4	19-11	20-6	21-0	21-6	21-11	22-5	22-10	23-3	23-7	24-0	24-5	24-9	25-1
	16.0	15-10	16-5	17-0	17-7	18-1	18-7	19-1	19-6	19-11	20-4	20-9	21-1	21-6	21-10	22-2	22-6	22-10
	19.2	14-11	15-6	16-0	16-7	17-0	17-6	17-11	18-4	18-9	19-2	19-6	19-10	20-2	20-6	20-10	21-2	21-6
	24.0	13-10	14-4	14-11	15-4	15-10	16-3	16-8	17-0	17-5	17-9	18-1	18-5	18-9	19-1	19-4	19-8	19-11
F_b	12.0	718	777	833	888	941	993	1,043	1,092	1,140	1,187	1,233	1,278	1,323	1,367	1,410	1,452	1,494
	16.0	790	855	917	977	1,036	1,093	1,148	1,202	1,255	1,306	1,357	1,407	1,456	1,504	1,551	1,598	1,644
	19.2	840	909	975	1,039	1,101	1,161	1,220	1,277	1,333	1,388	1,442	1,495	1,547	1,598	1,649	1,698	1,747
	24.0	905	979	1,050	1,119	1,186	1,251	1,314	1,376	1,436	1,496	1,554	1,611	1,667	1,722	1,776	1,829	1,882

NOTE: The required bending design value, *F_b*, in pounds per square inch (× 0.00689 for N/mm^2) is shown at the bottom of this table and is applicable to all lumber sizes shown. Spans are shown in feet-inches (1 foot = 304.8 mm, 1 inch = 25.4 mm) and are limited to 26 feet (7925 mm) and less.

TABLE 23-IV-J-2—FLOOR JOISTS WITH *L*/360 DEFLECTION LIMITS
The allowable bending stress (F_b) and modulus of elasticity *(E)* used in this table shall be from Tables 23-IV-V-1 and 23-IV-V-2 only.

DESIGN CRITERIA:
Deflection — For 40 psf (1.92 kN/m^2) live load.
Limited to span in inches (mm) divided by 360.
Strength — Live load of 40 psf (1.92 kN/m^2) plus dead load of 20 psf (0.96 kN/m^2) determines the required bending design value.

Joist Size (in)	Spacing (in)	Modulus of Elasticity, *E*, in 1,000,000 psi × 0.00689 for N/mm^2																
× 25.4 for mm		0.8	0.9	1.0	1.1	1.2	1.3	1.4	1.5	1.6	1.7	1.8	1.9	2.0	2.1	2.2	2.3	2.4
2 × 6	12.0	8-6	8-10	9-2	9-6	9-9	10-0	10-3	10-6	10-9	10-11	11-2	11-4	11-7	11-9	11-11	12-1	12-3
	16.0	7-9	8-0	8-4	8-7	8-10	9-1	9-4	9-6	9-9	9-11	10-2	10-4	10-6	10-8	10-10	11-0	11-2
	19.2	7-3	7-7	7-10	8-1	8-4	8-7	8-9	9-0	9-2	9-4	9-6	9-8	9-10	10-0	10-2	10-4	10-6
	24.0	6-9	7-0	7-3	7-6	7-9	7-11	8-2	8-4	8-6	8-8	8-10	9-0	9-2	9-4	9-6	9-7	9-9
2 × 8	12.0	11-3	11-8	12-1	12-6	12-10	13-2	13-6	13-10	14-2	14-5	14-8	15-0	15-3	15-6	15-9	15-11	16-2
	16.0	10-2	10-7	11-0	11-4	11-8	12-0	12-3	12-7	12-10	13-1	13-4	13-7	13-10	14-1	14-3	14-6	14-8
	19.2	9-7	10-0	10-4	10-8	11-0	11-3	11-7	11-10	12-1	12-4	12-7	12-10	13-0	13-3	13-5	13-8	13-10
	24.0	8-11	9-3	9-7	9-11	10-2	10-6	10-9	11-0	11-3	11-5	11-8	11-11	12-1	12-3	12-6	12-8	12-10
2 × 10	12.0	14-4	14-11	15-5	15-11	16-5	16-10	17-3	17-8	18-0	18-5	18-9	19-1	19-5	19-9	20-1	20-4	20-8
	16.0	13-0	13-6	14-0	14-6	14-11	15-3	15-8	16-0	16-5	16-9	17-0	17-4	17-8	17-11	18-3	18-6	18-9
	19.2	12-3	12-9	13-2	13-7	14-0	14-5	14-9	15-1	15-5	15-9	16-0	16-4	16-7	16-11	17-2	17-5	17-8
	24.0	11-4	11-10	12-3	12-8	13-0	13-4	13-8	14-0	14-4	14-7	14-11	15-2	15-5	15-8	15-11	16-2	16-5
2 × 12	12.0	17-5	18-1	18-9	19-4	19-11	20-6	21-0	21-6	21-11	22-5	22-10	23-3	23-7	24-0	24-5	24-9	25-1
	16.0	15-10	16-5	17-0	17-7	18-1	18-7	19-1	19-6	19-11	20-4	20-9	21-1	21-6	21-10	22-2	22-6	22-10
	19.2	14-11	15-6	16-0	16-7	17-0	17-6	17-11	18-4	18-9	19-2	19-6	19-10	20-2	20-6	20-10	21-2	21-6
	24.0	13-10	14-4	14-11	15-4	15-10	16-3	16-8	17-0	17-5	17-9	18-1	18-5	18-9	19-1	19-4	19-8	19-11
F_b	12.0	862	932	1,000	1,066	1,129	1,191	1,251	1,310	1,368	1,424	1,480	1,534	1,587	1,640	1,692	1,742	1,793
	16.0	949	1,026	1,101	1,173	1,243	1,311	1,377	1,442	1,506	1,568	1,629	1,688	1,747	1,805	1,862	1,918	1,973
	19.2	1,008	1,090	1,170	1,246	1,321	1,393	1,464	1,533	1,600	1,666	1,731	1,794	1,857	1,918	1,978	2,038	2,097
	24.0	1,086	1,174	1,260	1,343	1,423	1,501	1,577	1,651	1,724	1,795	1,864	1,933	2,000	2,066	2,131	2,195	2,258

NOTE: The required bending design value, F_b, in pounds per square inch (× 0.00689 for N/mm^2) is shown at the bottom of this table and is applicable to all lumber sizes shown. Spans are shown in feet-inches (1 foot = 304.8 mm, 1 inch = 25.4 mm) and are limited to 26 feet (7925 mm) and less.

TABLE 23-IV-J-3—CEILING JOISTS WITH *L*/240 DEFLECTION LIMITS
The allowable bending stress (F_b) and modulus of elasticity *(E)* used in this table shall be from Tables 23-IV-V-1 and 23-IV-V-2 only.

DESIGN CRITERIA:
Deflection — For 10 psf (0.48 kN/m^2) live load.
Limited to span in inches (mm) divided by 240.
Strength — Live load of 10 psf (0.48 kN/mm^2) plus dead load of 5 psf (0.24 kN/m^2) determines the required fiber stress value.

Joist Size (in)	Spacing (in)	Modulus of Elasticity, *E*, in 1,000,000 psi × 0.00689 for N/mm^2																
× 25.4 for mm		0.8	0.9	1.0	1.1	1.2	1.3	1.4	1.5	1.6	1.7	1.8	1.9	2.0	2.1	2.2	2.3	2.4
2 × 4	12.0	9-10	10-3	10-7	10-11	11-3	11-7	11-10	12-2	12-5	12-8	12-11	13-2	13-4	13-7	13-9	14-0	14-2
	16.0	8-11	9-4	9-8	9-11	10-3	10-6	10-9	11-0	11-3	11-6	11-9	11-11	12-2	12-4	12-6	12-9	12-11
	19.2	8-5	8-9	9-1	9-4	9-8	9-11	10-2	10-4	10-7	10-10	11-0	11-3	11-5	11-7	11-9	12-0	12-2
	24.0	7-10	8-1	8-5	8-8	8-11	9-2	9-5	9-8	9-10	10-0	10-3	10-5	10-7	10-9	10-11	11-1	11-3
2 × 6	12.0	15-6	16-1	16-8	17-2	17-8	18-2	18-8	19-1	19-6	19-11	20-3	20-8	21-0	21-4	21-8	22-0	22-4
	16.0	14-1	14-7	15-2	15-7	16-1	16-6	16-11	17-4	17-8	18-1	18-5	18-9	19-1	19-5	19-8	20-0	20-3
	19.2	13-3	13-9	14-3	14-8	15-2	15-7	15-11	16-4	16-8	17-0	17-4	17-8	17-11	18-3	18-6	18-10	19-1
	24.0	12-3	12-9	13-3	13-8	14-1	14-5	14-9	15-2	15-6	15-9	16-1	16-4	16-8	16-11	17-2	17-5	17-8
2 × 8	12.0	20-5	21-2	21-11	22-8	23-4	24-0	24-7	25-2	25-8								
	16.0	18-6	19-3	19-11	20-7	21-2	21-9	22-4	22-10	23-4	23-10	24-3	24-8	25-2	25-7	25-11		
	19.2	17-5	18-1	18-9	19-5	19-11	20-6	21-0	21-6	21-11	22-5	22-10	23-3	23-8	24-0	24-5	24-9	25-2
	24.0	16-2	16-10	17-5	18-0	18-6	19-0	19-6	19-11	20-5	20-10	21-2	21-7	21-11	22-4	22-8	23-0	23-4
2 × 10	12.0	26-0																
	16.0	23-8	24-7	25-5														
	19.2	22-3	23-1	23-11	24-9	25-5												
	24.0	20-8	21-6	22-3	22-11	23-8	24-3	24-10	25-5	26-0								
F_b	12.0	711	769	825	880	932	983	1,033	1,082	1,129	1,176	1,221	1,266	1,310	1,354	1,396	1,438	1,480
	16.0	783	847	909	968	1,026	1,082	1,137	1,191	1,243	1,294	1,344	1,394	1,442	1,490	1,537	1,583	1,629
	19.2	832	900	965	1,029	1,090	1,150	1,208	1,265	1,321	1,375	1,429	1,481	1,533	1,583	1,633	1,682	1,731
	24.0	896	969	1,040	1,108	1,174	1,239	1,302	1,363	1,423	1,481	1,539	1,595	1,651	1,706	1,759	1,812	1,864

NOTE: The required bending design value, F_b, in pounds per square inch (× 0.00689 for N/mm^2) is shown at the bottom of this table and is applicable to all lumber sizes shown. Spans are shown in feet-inches (1 foot = 304.8 mm, 1 inch = 25.4 mm) and are limited to 26 feet (7925 mm) and less.

TABLE 23-IV-J-4—CEILING JOISTS WITH L/240 DEFLECTION LIMITS
The allowable bending stress (F_b) and modulus of elasticity (E) used in this table shall be from Tables 23-IV-V-1 and 23-IV-V-2 only.

DESIGN CRITERIA:
Deflection — For 20 psf (0.96 kN/m²) live load.
Limited to span in inches (mm) divided by 240.
Strength — Live load of 20 psf (0.96 kN/m²) plus dead load of 10 psf (0.48 kN/m²) determines the required bending design value.

Joist Size (in) × 25.4 for mm	Spacing (in)	Modulus of Elasticity, E, in 1,000,000 psi × 0.00689 for N/mm²																
		0.8	0.9	1.0	1.1	1.2	1.3	1.4	1.5	1.6	1.7	1.8	1.9	2.0	2.1	2.2	2.3	2.4
2 × 4	12.0	7-10	8-1	8-5	8-8	8-11	9-2	9-5	9-8	9-10	10-0	10-3	10-5	10-7	10-9	10-11	11-1	11-3
	16.0	7-1	7-5	7-8	7-11	8-1	8-4	8-7	8-9	8-11	9-1	9-4	9-6	9-8	9-9	9-11	10-1	10-3
	19.2	6-8	6-11	7-2	7-5	7-8	7-10	8-1	8-3	8-5	8-7	8-9	8-11	9-1	9-3	9-4	9-6	9-8
	24.0	6-2	6-5	6-8	6-11	7-1	7-3	7-6	7-8	7-10	8-0	8-1	8-3	8-5	8-7	8-8	8-10	8-11
2 × 6	12.0	12-3	12-9	13-3	13-8	14-1	14-5	14-9	15-2	15-6	15-9	16-1	16-4	16-8	16-11	17-2	17-5	17-8
	16.0	11-2	11-7	12-0	12-5	12-9	13-1	13-5	13-9	14-1	14-4	14-7	14-11	15-2	15-5	15-7	15-10	16-1
	19.2	10-6	10-11	11-4	11-8	12-0	12-4	12-8	12-11	13-3	13-6	13-9	14-0	14-3	14-6	14-8	14-11	15-2
	24.0	9-9	10-2	10-6	10-10	11-2	11-5	11-9	12-0	12-3	12-6	12-9	13-0	13-3	13-5	13-8	13-10	14-1
2 × 8	12.0	16-2	16-10	17-5	18-0	18-6	19-0	19-6	19-11	20-5	20-10	21-2	21-7	21-11	22-4	22-8	23-0	23-4
	16.0	14-8	15-3	15-10	16-4	16-10	17-3	17-9	18-1	18-6	18-11	19-3	19-7	19-11	20-3	20-7	20-11	21-2
	19.2	13-10	14-5	14-11	15-5	15-10	16-3	16-8	17-1	17-5	17-9	18-1	18-5	18-9	19-1	19-5	19-8	19-11
	24.0	12-10	13-4	13-10	14-3	14-8	15-1	15-6	15-10	16-2	16-6	16-10	17-2	17-5	17-9	18-0	18-3	18-6
2 × 10	12.0	20-8	21-6	22-3	22-11	23-8	24-3	24-10	25-5	26-0								
	16.0	18-9	19-6	20-2	20-10	21-6	22-1	22-7	23-1	23-8	24-1	24-7	25-0	25-5	25-10			
	19.2	17-8	18-4	19-0	19-7	20-2	20-9	21-3	21-9	22-3	22-8	23-1	23-7	23-11	24-4	24-9	25-1	25-5
	24.0	16-5	17-0	17-8	18-3	18-9	19-3	19-9	20-2	20-8	21-1	21-6	21-10	22-3	22-7	22-11	23-4	23-8
F_b	12.0	896	969	1,040	1,108	1,174	1,239	1,302	1,363	1,423	1,481	1,539	1,595	1,651	1,706	1,759	1,812	1,864
	16.0	986	1,067	1,145	1,220	1,293	1,364	1,433	1,500	1,566	1,631	1,694	1,756	1,817	1,877	1,936	1,995	2,052
	19.2	1,048	1,134	1,216	1,296	1,374	1,449	1,522	1,594	1,664	1,733	1,800	1,866	1,931	1,995	2,058	2,120	2,181
	24.0	1,129	1,221	1,310	1,396	1,480	1,561	1,640	1,717	1,793	1,866	1,939	2,010	2,080	2,149	2,217	2,283	2,349

NOTE: The required bending design value, F_b, in pounds per square inch (× 0.00689 for N/mm²) is shown at the bottom of this table and is applicable to all lumber sizes shown. Spans are shown in feet-inches (1 foot = 304.8 mm, 1 inch = 25.4 mm) and are limited to 26 feet (7925 mm) and less.

TABLE 23-IV-R-1—RAFTERS WITH L/240 DEFLECTION LIMITATION
The allowable bending stress (F_b) and modulus of elasticity (E) used in this table shall be from Tables 23-IV-V-1 and 23-IV-V-2 only.

DESIGN CRITERIA:
Strength — Live load of 20 psf (0.96 kN/m²) plus dead load of 10 psf (0.48 kN/m²) determines the required bending design value.
Deflection — For 20 psf (0.96 kN/m²) live load.
Limited to span in inches (mm) divided by 240.

Rafter Size (in) × 25.4 for mm	Spacing (in)	Bending Design Value, F_b (psi) × 0.00689 for N/mm²										
		300	400	500	600	700	800	900	1000	1100	1200	1300
2 × 6	12.0	7-1	8-2	9-2	10-0	10-10	11-7	12-4	13-0	13-7	14-2	14-9
	16.0	6-2	7-1	7-11	8-8	9-5	10-0	10-8	11-3	11-9	12-4	12-10
	19.2	5-7	6-6	7-3	7-11	8-7	9-2	9-9	10-3	10-9	11-3	11-8
	24.0	5-0	5-10	6-6	7-1	7-8	8-2	8-8	9-2	9-7	10-0	10-5
2 × 8	12.0	9-4	10-10	12-1	13-3	14-4	15-3	16-3	17-1	17-11	18-9	19-6
	16.0	8-1	9-4	10-6	11-6	12-5	13-3	14-0	14-10	15-6	16-3	16-10
	19.2	7-5	8-7	9-7	10-6	11-4	12-1	12-10	13-6	14-2	14-10	15-5
	24.0	6-7	7-8	8-7	9-4	10-1	10-10	11-6	12-1	12-8	13-3	13-9
2 × 10	12.0	11-11	13-9	15-5	16-11	18-3	19-6	20-8	21-10	22-10	23-11	24-10
	16.0	10-4	11-11	13-4	14-8	15-10	16-11	17-11	18-11	19-10	20-8	21-6
	19.2	9-5	10-11	12-2	13-4	14-5	15-5	16-4	17-3	18-1	18-11	19-8
	24.0	8-5	9-9	10-11	11-11	12-11	13-9	14-8	15-5	16-2	16-11	17-7
2 × 12	12.0	14-6	16-9	18-9	20-6	22-2	23-9	25-2				
	16.0	12-7	14-6	16-3	17-9	19-3	20-6	21-9	23-0	24-1	25-2	
	19.2	11-6	13-3	14-10	16-3	17-6	18-9	19-11	21-0	22-0	23-0	23-11
	24.0	10-3	11-10	13-3	14-6	15-8	16-9	17-9	18-9	19-8	20-6	21-5
E	12.0	0.15	0.24	0.33	0.44	0.55	0.67	0.80	0.94	1.09	1.24	1.40
	16.0	0.13	0.21	0.29	0.38	0.48	0.58	0.70	0.82	0.94	1.07	1.21
	19.2	0.12	0.19	0.26	0.35	0.44	0.53	0.64	0.75	0.86	0.98	1.10
	24.0	0.11	0.17	0.24	0.31	0.39	0.48	0.57	0.67	0.77	0.88	0.99

Rafter Size (in) × 25.4 for mm	Spacing (in)	Bending Design Value, F_b (psi) × 0.00689 for N/mm²										
		1400	1500	1600	1700	1800	1900	2000	2100	2200	2300	2400
2 × 6	12.0	15-4	15-11	16-5	16-11	17-5	17-10					
	16.0	13-3	13-9	14-2	14-8	15-1	15-6	15-11	16-3			
	19.2	12-2	12-7	13-0	13-4	13-9	14-2	14-6	14-10	15-2	15-7	
	24.0	10-10	11-3	11-7	11-11	12-4	12-8	13-0	13-3	13-7	13-11	14-2
2 × 8	12.0	20-3	20-11	21-7	22-3	22-11	23-7					
	16.0	17-6	18-1	18-9	19-4	19-10	20-5	20-11	21-5			
	19.2	16-0	16-7	17-1	17-7	18-1	18-7	19-1	19-7	20-0	20-6	
	24.0	14-4	14-10	15-3	15-9	16-3	16-8	17-1	17-6	17-11	18-4	18-9
2 × 10	12.0	25-10										
	16.0	22-4	23-1	23-11	24-7	25-4	26-0					
	19.2	20-5	21-1	21-10	22-6	23-1	23-9	24-5	25-0	25-7		
	24.0	18-3	18-11	19-6	20-1	20-8	21-3	21-10	22-4	22-10	23-5	23-11
2 × 12	12.0											
	16.0											
	19.2	24-10	25-8									
	24.0	22-2	23-0	23-9	24-5	25-2	25-10					
E	12.0	1.56	1.73	1.91	2.09	2.28	2.47					
	16.0	1.35	1.50	1.65	1.81	1.97	2.14	2.31	2.48			
	19.2	1.23	1.37	1.51	1.65	1.80	1.95	2.11	2.27	2.43	2.60	
	24.0	1.10	1.22	1.35	1.48	1.61	1.75	1.89	2.03	2.18	2.33	2.48

NOTE: The required modulus of elasticity, E, in 1,000,000 pounds per square inch (psi) (× 0.00689 for N/mm²) is shown at the bottom of this table, is limited to 2.6 million psi (17 914 N/mm²) and less, and is applicable to all lumber sizes shown. Spans are shown in feet-inches (1 foot = 304.8 mm, 1 inch = 25.4 mm) and are limited to 26 feet (7925 mm) and less.

TABLE 23-IV-R-2—RAFTERS WITH L/240 DEFLECTION LIMITATION
The allowable bending stress (F_b) and modulus of elasticity (E) used in this table shall be from Tables 23-IV-V-1 and 23-IV-V-2 only.

DESIGN CRITERIA:
Strength — Live load of 30 psf (1.44 kN/m²) plus dead load of 10 psf (0.48 kN/m²) determines the required bending design value.
Deflection — For 30 psf (1.44 kN/m²) live load.
Limited to span in inches (mm) divided by 240.

Rafter Size (in)	Spacing (in)	Bending Design Value, F_b (psi) × 0.00689 for N/mm²										
× 25.4 for mm		300	400	500	600	700	800	900	1000	1100	1200	1300
2 × 6	12.0	6-2	7-1	7-11	8-8	9-5	10-0	10-8	11-3	11-9	12-4	12-10
	16.0	5-4	6-2	6-10	7-6	8-2	8-8	9-3	9-9	10-2	10-8	11-1
	19.2	4-10	5-7	6-3	6-10	7-5	7-11	8-5	8-11	9-4	9-9	10-1
	24.0	4-4	5-0	5-7	6-2	6-8	7-1	7-6	7-11	8-4	8-8	9-1
2 × 8	12.0	8-1	9-4	10-6	11-6	12-5	13-3	14-0	14-10	15-6	16-3	16-10
	16.0	7-0	8-1	9-1	9-11	10-9	11-6	12-2	12-10	13-5	14-0	14-7
	19.2	6-5	7-5	8-3	9-1	9-9	10-6	11-1	11-8	12-3	12-10	13-4
	24.0	5-9	6-7	7-5	8-1	8-9	9-4	9-11	10-6	11-0	11-6	11-11
2 × 10	12.0	10-4	11-11	13-4	14-8	15-10	16-11	17-11	18-11	19-10	20-8	21-6
	16.0	8-11	10-4	11-7	12-8	13-8	14-8	15-6	16-4	17-2	17-11	18-8
	19.2	8-2	9-5	10-7	11-7	12-6	13-4	14-2	14-11	15-8	16-4	17-0
	24.0	7-4	8-5	9-5	10-4	11-2	11-11	12-8	13-4	14-0	14-8	15-3
2 × 12	12.0	12-7	14-6	16-3	17-9	19-3	20-6	21-9	23-0	24-1	25-2	
	16.0	10-11	12-7	14-1	15-5	16-8	17-9	18-10	19-11	20-10	21-9	22-8
	19.2	9-11	11-6	12-10	14-1	15-2	16-3	17-3	18-2	19-0	19-11	20-8
	24.0	8-11	10-3	11-6	12-7	13-7	14-6	15-5	16-3	17-0	17-9	18-6
E	12.0	0.15	0.23	0.32	0.43	0.54	0.66	0.78	0.92	1.06	1.21	1.36
	16.0	0.13	0.20	0.28	0.37	0.47	0.57	0.68	0.80	0.92	1.05	1.18
	19.2	0.12	0.18	0.26	0.34	0.43	0.52	0.62	0.73	0.84	0.95	1.08
	24.0	0.11	0.16	0.23	0.30	0.38	0.46	0.55	0.65	0.75	0.85	0.96

Rafter Size (in)	Spacing (in)	Bending Design Value, F_b (psi) × 0.00689 for N/mm²										
× 25.4 for mm		1400	1500	1600	1700	1800	1900	2000	2100	2200	2300	2400
2 × 6	12.0	13-3	13-9	14-2	14-8	15-1	15-6	15-11				
	16.0	11-6	11-11	12-4	12-8	13-1	13-5	13-9	14-1	14-5		
	19.2	10-6	10-10	11-3	11-7	11-11	12-3	12-7	12-10	13-2	13-6	
	24.0	9-5	9-9	10-0	10-4	10-8	10-11	11-3	11-6	11-9	12-0	12-4
2 × 8	12.0	17-6	18-1	18-9	19-4	19-10	20-5	20-11				
	16.0	15-2	15-8	16-3	16-9	17-2	17-8	18-1	18-7	19-0		
	19.2	13-10	14-4	14-10	15-3	15-8	16-2	16-7	16-11	17-4	17-9	
	24.0	12-5	12-10	13-3	13-8	14-0	14-5	14-10	15-2	15-6	15-10	16-3
2 × 10	12.0	22-4	23-1	23-11	24-7	25-4	26-0					
	16.0	19-4	20-0	20-8	21-4	21-11	22-6	23-1	23-8	24-3		
	19.2	17-8	18-3	18-11	19-6	20-0	20-7	21-1	21-8	22-2	22-8	
	24.0	15-10	16-4	16-11	17-5	17-11	18-5	18-11	19-4	19-10	20-3	20-8
2 × 12	12.0											
	16.0	23-6	24-4	25-2	25-11							
	19.2	21-6	22-3	23-0	23-8	24-4	25-0	25-8				
	24.0	19-3	19-11	20-6	21-2	21-9	22-5	23-0	23-6	24-1	24-8	25-2
E	12.0	1.52	1.69	1.86	2.04	2.22	2.41	2.60				
	16.0	1.32	1.46	1.61	1.76	1.92	2.08	2.25	2.42	2.60		
	19.2	1.20	1.33	1.47	1.61	1.75	1.90	2.05	2.21	2.37	2.53	
	24.0	1.08	1.19	1.31	1.44	1.57	1.70	1.84	1.98	2.12	2.27	2.41

NOTE: The required modulus of elasticity, E, in 1,000,000 pounds per square inch (psi) (× 0.00689 for N/mm²) is shown at the bottom of this table, is limited to 2.6 million psi (17 914 N/mm²) and less, and is applicable to all lumber sizes shown. Spans are shown in feet-inches (1 foot = 304.8 mm, 1 inch = 25.4 mm) and are limited to 26 feet (7925 mm) and less.

TABLE 23-IV-R-3—RAFTERS WITH L/240 DEFLECTION LIMITATION
The allowable bending stress (Fb) and modulus of elasticity (E) used in this table shall be from Tables 23-IV-V-1 and 23-IV-V-2 only.

DESIGN CRITERIA:
Strength — Live load of 20 psf (0.96 kN/m²) plus dead load of 15 psf (0.72 kN/m²) determines the required bending design value.
Deflection — For 20 psf (0.96 kN/m²) live load.
Limited to span in inches (mm) divided by 240.

Rafter Size (in)	Spacing (in)	Bending Design Value, F_b (psi) $\times$ 0.00689 for N/mm²												
× 25.4 for mm		300	400	500	600	700	800	900	1000	1100	1200	1300	1400	1500
2 × 6	12.0	6-7	7-7	8-6	9-4	10-0	10-9	11-5	12-0	12-7	13-2	13-8	14-2	14-8
	16.0	5-8	6-7	7-4	8-1	8-8	9-4	9-10	10-5	10-11	11-5	11-10	12-4	12-9
	19.2	5-2	6-0	6-9	7-4	7-11	8-6	9-0	9-6	9-11	10-5	10-10	11-3	11-7
	24.0	4-8	5-4	6-0	6-7	7-1	7-7	8-1	8-6	8-11	9-4	9-8	10-0	10-5
2 × 8	12.0	8-8	10-0	11-2	12-3	13-3	14-2	15-0	15-10	16-7	17-4	18-0	18-9	19-5
	16.0	7-6	8-8	9-8	10-7	11-6	12-3	13-0	13-8	14-4	15-0	15-7	16-3	16-9
	19.2	6-10	7-11	8-10	9-8	10-6	11-2	11-10	12-6	13-1	13-8	14-3	14-10	15-4
	24.0	6-2	7-1	7-11	8-8	9-4	10-0	10-7	11-2	11-9	12-3	12-9	13-3	13-8
2 × 10	12.0	11-1	12-9	14-3	15-8	16-11	18-1	19-2	20-2	21-2	22-1	23-0	23-11	24-9
	16.0	9-7	11-1	12-4	13-6	14-8	15-8	16-7	17-6	18-4	19-2	19-11	20-8	21-5
	19.2	8-9	10-1	11-3	12-4	13-4	14-3	15-2	15-11	16-9	17-6	18-2	18-11	19-7
	24.0	7-10	9-0	10-1	11-1	11-11	12-9	13-6	14-3	15-0	15-8	16-3	16-11	17-6
2 × 12	12.0	13-5	15-6	17-4	19-0	20-6	21-11	23-3	24-7	25-9				
	16.0	11-8	13-5	15-0	16-6	17-9	19-0	20-2	21-3	22-4	23-3	24-3	25-2	26-0
	19.2	10-8	12-3	13-9	15-0	16-3	17-4	18-5	19-5	20-4	21-3	22-2	23-0	23-9
	24.0	9-6	11-0	12-3	13-5	14-6	15-6	16-6	17-4	18-2	19-0	19-10	20-6	21-3
E	12.0	0.12	0.19	0.26	0.35	0.44	0.54	0.64	0.75	0.86	0.98	1.11	1.24	1.37
	16.0	0.11	0.16	0.23	0.30	0.38	0.46	0.55	0.65	0.75	0.85	0.96	1.07	1.19
	19.2	0.10	0.15	0.21	0.27	0.35	0.42	0.51	0.59	0.68	0.78	0.88	0.98	1.09
	24.0	0.09	0.13	0.19	0.25	0.31	0.38	0.45	0.53	0.61	0.70	0.78	0.88	0.97

Rafter Size (in)	Spacing (in)	Bending Design Value, F_b (psi) $\times$ 0.00689 for N/mm²											
× 25.4 for mm		1600	1700	1800	1900	2000	2100	2200	2300	2400	2500	2600	2700
2 × 6	12.0	15-2	15-8	16-1	16-7	17-0	17-5	17-10					
	16.0	13-2	13-7	13-11	14-4	14-8	15-1	15-5	15-9	16-1	16-5		
	19.2	12-0	12-4	12-9	13-1	13-5	13-9	14-1	14-5	14-8	15-0	15-4	
	24.0	10-9	11-1	11-5	11-8	12-0	12-4	12-7	12-10	13-2	13-5	13-8	13-11
2 × 8	12.0	20-0	20-8	21-3	21-10	22-4	22-11	23-6					
	16.0	17-4	17-10	18-5	18-11	19-5	19-10	20-4	20-9	21-3	21-8		
	19.2	15-10	16-4	16-9	17-3	17-8	18-1	18-7	19-0	19-5	19-9	20-2	
	24.0	14-2	14-7	15-0	15-5	15-10	16-3	16-7	17-0	17-4	17-8	18-0	18-5
2 × 10	12.0	25-6											
	16.0	22-1	22-10	23-5	24-1	24-9	25-4	25-11					
	19.2	20-2	20-10	21-5	22-0	22-7	23-1	23-8	24-2	24-9	25-3	25-9	
	24.0	18-1	18-7	19-2	19-8	20-2	20-8	21-2	21-8	22-1	22-7	23-0	23-5
2 × 12	12.0												
	16.0												
	19.2	24-7	25-4	26-0									
	24.0	21-11	22-8	23-3	23-11	24-7	25-2	25-9					
E	12.0	1.51	1.66	1.81	1.96	2.12	2.28	2.44					
	16.0	1.31	1.44	1.56	1.70	1.83	1.97	2.11	2.26	2.41	2.56		
	19.2	1.20	1.31	1.43	1.55	1.67	1.80	1.93	2.06	2.20	2.34	2.48	
	24.0	1.07	1.17	1.28	1.39	1.50	1.61	1.73	1.85	1.97	2.09	2.22	2.35

NOTE: The required modulus of elasticity, E, in 1,000,000 pounds per square inch (psi) (× 0.00689 for N/mm²) is shown at the bottom of this table, is limited to 2.6 million psi (17 914 N/mm²) and less, and is applicable to all lumber sizes shown. Spans are shown in feet-inches (1 foot = 304.8 mm, 1 inch = 25.4 mm) and are limited to 26 feet (7925 mm) and less.

TABLE 23-IV-R-4—RAFTERS WITH *L*/240 DEFLECTION LIMITATION
The allowable bending stress (*F_b*) and modulus of elasticity *(E)* used in this table shall be from Tables 23-IV-V-1 and 23-IV-V-2 only.

DESIGN CRITERIA:
Strength — Live load of 30 psf (1.44 kN/m^2) plus dead load of 15 psf (0.72 kN/m^2) determines the required bending design value.
Deflection — For 30 psf (1.44 kN/m^2) live load.
Limited to span in inches (mm) divided by 240.

Rafter Size (in) × 25.4 for mm	Spacing (in)	Bending Design Value, F_b (psi) × 0.00689 for N/mm^2												
		300	400	500	600	700	800	900	1000	1100	1200	1300	1400	1500
2 × 6	12.0	5-10	6-8	7-6	8-2	8-10	9-6	10-0	10-7	11-1	11-7	12-1	12-6	13-0
	16.0	5-0	5-10	6-6	7-1	7-8	8-2	8-8	9-2	9-7	10-0	10-5	10-10	11-3
	19.2	4-7	5-4	5-11	6-6	7-0	7-6	7-11	8-4	8-9	9-2	9-6	9-11	10-3
	24.0	4-1	4-9	5-4	5-10	6-3	6-8	7-1	7-6	7-10	8-2	8-6	8-10	9-2
2 × 8	12.0	7-8	8-10	9-10	10-10	11-8	12-6	13-3	13-11	14-8	15-3	15-11	16-6	17-1
	16.0	6-7	7-8	8-7	9-4	10-1	10-10	11-6	12-1	12-8	13-3	13-9	14-4	14-10
	19.2	6-0	7-0	7-10	8-7	9-3	9-10	10-6	11-0	11-7	12-1	12-7	13-1	13-6
	24.0	5-5	6-3	7-0	7-8	8-3	8-10	9-4	9-10	10-4	10-10	11-3	11-8	12-1
2 × 10	12.0	9-9	11-3	12-7	13-9	14-11	15-11	16-11	17-10	18-8	19-6	20-4	21-1	21-10
	16.0	8-5	9-9	10-11	11-11	12-11	13-9	14-8	15-5	16-2	16-11	17-7	18-3	18-11
	19.2	7-8	8-11	9-11	10-11	11-9	12-7	13-4	14-1	14-9	15-5	16-1	16-8	17-3
	24.0	6-11	8-0	8-11	9-9	10-6	11-3	11-11	12-7	13-2	13-9	14-4	14-11	15-5
2 × 12	12.0	11-10	13-8	15-4	16-9	18-1	19-4	20-6	21-8	22-8	23-9	24-8	25-7	
	16.0	10-3	11-10	13-3	14-6	15-8	16-9	17-9	18-9	19-8	20-6	21-5	22-2	23-0
	19.2	9-4	10-10	12-1	13-3	14-4	15-4	16-3	17-1	17-11	18-9	19-6	20-3	21-0
	24.0	8-5	9-8	10-10	11-10	12-10	13-8	14-6	15-4	16-1	16-9	17-5	18-1	18-9
E	12.0	0.13	0.19	0.27	0.36	0.45	0.55	0.66	0.77	0.89	1.01	1.14	1.28	1.41
	16.0	0.11	0.17	0.24	0.31	0.39	0.48	0.57	0.67	0.77	0.88	0.99	1.10	1.22
	19.2	0.10	0.15	0.22	0.28	0.36	0.44	0.52	0.61	0.70	0.80	0.90	1.01	1.12
	24.0	0.09	0.14	0.19	0.25	0.32	0.39	0.46	0.54	0.63	0.72	0.81	0.90	1.00

Rafter Size (in) × 25.4 for mm	Spacing (in)	Bending Design Value, F_b (psi) × 0.00689 for N/mm^2												
		1600	1700	1800	1900	2000	2100	2200	2300	2400	2500	2600	2700
2 × 6	12.0	13-5	13-10	14-2	14-7	15-0	15-4	15-8					
	16.0	11-7	11-11	12-4	12-8	13-0	13-3	13-7	13-11	14-2			
	19.2	10-7	10-11	11-3	11-6	11-10	12-2	12-5	12-8	13-0	13-3	13-6	
	24.0	9-6	9-9	10-0	10-4	10-7	10-10	11-1	11-4	11-7	11-10	12-1	12-4
2 × 8	12.0	17-8	18-2	18-9	19-3	19-9	20-3	20-8					
	16.0	15-3	15-9	16-3	16-8	17-1	17-6	17-11	18-4	18-9			
	19.2	13-11	14-5	14-10	15-2	15-7	16-0	16-4	16-9	17-1	17-5	17-9	
	24.0	12-6	12-10	13-3	13-7	13-11	14-4	14-8	15-0	15-3	15-7	15-11	16-3
2 × 10	12.0	22-6	23-3	23-11	24-6	25-2	25-10						
	16.0	19-6	20-1	20-8	21-3	21-10	22-4	22-10	23-5	23-11			
	19.2	17-10	18-4	18-11	19-5	19-11	20-5	20-10	21-4	21-10	22-3	22-8	
	24.0	15-11	16-5	16-11	17-4	17-10	18-3	18-8	19-1	19-6	19-11	20-4	20-8
2 × 12	12.0												
	16.0	23-9	24-5	25-2	25-10								
	19.2	21-8	22-4	23-0	23-7	24-2	24-10	25-5	25-11				
	24.0	19-4	20-0	20-6	21-1	21-8	22-2	22-8	23-3	23-9	24-2	24-8	25-2
E	12.0	1.56	1.71	1.86	2.02	2.18	2.34	2.51					
	16.0	1.35	1.48	1.61	1.75	1.89	2.03	2.18	2.33	2.48			
	19.2	1.23	1.35	1.47	1.59	1.72	1.85	1.99	2.12	2.26	2.41	2.55	
	24.0	1.10	1.21	1.31	1.43	1.54	1.66	1.78	1.90	2.02	2.15	2.28	2.41

NOTE: The required modulus of elasticity, *E*, in 1,000,000 pounds per square inch (psi) (× 0.00689 for N/mm^2) is shown at the bottom of this table, is limited to 2.6 million psi (17 914 N/mm^2) and less, and is applicable to all lumber sizes shown. Spans are shown in feet-inches (1 foot = 304.8 mm, 1 inch = 25.4 mm) and are limited to 26 feet (7925 mm) and less.

TABLE 23-IV-R-5—RAFTERS WITH *L*/240 DEFLECTION LIMITATION
The allowable bending stress (*F$_b$*) and modulus of elasticity (*E*) used in this table shall be from Tables 23-IV-V-1 and 23-IV-V-2 only.

DESIGN CRITERIA:
Strength — Live load of 20 psf (0.96 kN/m²) plus dead load of 20 psf (0.96 kN/m²) determines the required bending design value.
Deflection — For 20 psf (0.96 kN/m²) live load.
Limited to span in inches (mm) divided by 240.

Rafter Size (in)	Spacing (in)	Bending Design Value, F_b (psi) × 0.00689 for N/mm²												
× 25.4 for mm		300	400	500	600	700	800	900	1000	1100	1200	1300	1400	1500
2×6	12.0	6-2	7-1	7-11	8-8	9-5	10-0	10-8	11-3	11-9	12-4	12-10	13-3	13-9
	16.0	5-4	6-2	6-10	7-6	8-2	8-8	9-3	9-9	10-2	10-8	11-1	11-6	11-11
	19.2	4-10	5-7	6-3	6-10	7-5	7-11	8-5	8-11	9-4	9-9	10-1	10-6	10-10
	24.0	4-4	5-0	5-7	6-2	6-8	7-1	7-6	7-11	8-4	8-8	9-1	9-5	9-9
2×8	12.0	8-1	9-4	10-6	11-6	12-5	13-3	14-0	14-10	15-6	16-3	16-10	17-6	18-1
	16.0	7-0	8-1	9-1	9-11	10-9	11-6	12-2	12-10	13-5	14-0	14-7	15-2	15-8
	19.2	6-5	7-5	8-3	9-1	9-9	10-6	11-1	11-8	12-3	12-10	13-4	13-10	14-4
	24.0	5-9	6-7	7-5	8-1	8-9	9-4	9-11	10-6	11-0	11-6	11-11	12-5	12-10
2×10	12.0	10-4	11-11	13-4	14-8	15-10	16-11	17-11	18-11	19-10	20-8	21-6	22-4	23-1
	16.0	8-11	10-4	11-7	12-8	13-8	14-8	15-6	16-4	17-2	17-11	18-8	19-4	20-0
	19.2	8-2	9-5	10-7	11-7	12-6	13-4	14-2	14-11	15-8	16-4	17-0	17-8	18-3
	24.0	7-4	8-5	9-5	10-4	11-2	11-11	12-8	13-4	14-0	14-8	15-3	15-10	16-4
2×12	12.0	12-7	14-6	16-3	17-9	19-3	20-6	21-9	23-0	24-1	25-2			
	16.0	10-11	12-7	14-1	15-5	16-8	17-9	18-10	19-11	20-10	21-9	22-8	23-6	24-4
	19.2	9-11	11-6	12-10	14-1	15-2	16-3	17-3	18-2	19-0	19-11	20-8	21-6	22-3
	24.0	8-11	10-3	11-6	12-7	13-7	14-6	15-5	16-3	17-0	17-9	18-6	19-3	19-11
E	12.0	0.10	0.15	0.22	0.28	0.36	0.44	0.52	0.61	0.71	0.80	0.91	1.01	1.13
	16.0	0.09	0.13	0.19	0.25	0.31	0.38	0.45	0.53	0.61	0.70	0.79	0.88	0.97
	19.2	0.08	0.12	0.17	0.23	0.28	0.35	0.41	0.48	0.56	0.64	0.72	0.80	0.89
	24.0	0.07	0.11	0.15	0.20	0.25	0.31	0.37	0.43	0.50	0.57	0.64	0.72	0.80

Rafter Size (in)	Spacing (in)	Bending Design Value, F_b (psi) × 0.00689 for N/mm²											
× 25.4 for mm		1600	1700	1800	1900	2000	2100	2200	2300	2400	2500	2600	2700
2×6	12.0	14-2	14-8	15-1	15-6	15-11	16-3	16-8	17-0	17-5	17-9	18-1	
	16.0	12-4	12-8	13-1	13-5	13-9	14-1	14-5	14-9	15-1	15-4	15-8	16-0
	19.2	11-3	11-7	11-11	12-3	12-7	12-10	13-2	13-6	13-9	14-0	14-4	14-7
	24.0	10-0	10-4	10-8	10-11	11-3	11-6	11-9	12-0	12-4	12-7	12-10	13-1
2×8	12.0	18-9	19-4	19-10	20-5	20-11	21-5	21-11	22-5	22-11	23-5	23-10	
	16.0	16-3	16-9	17-2	17-8	18-1	18-7	19-0	19-5	19-10	20-3	20-8	21-1
	19.2	14-10	15-3	15-8	16-2	16-7	16-11	17-4	17-9	18-1	18-6	18-10	19-3
	24.0	13-3	13-8	14-0	14-5	14-10	15-2	15-6	15-10	16-3	16-7	16-10	17-2
2×10	12.0	23-11	24-7	25-4	26-0								
	16.0	20-8	21-4	21-11	22-6	23-1	23-8	24-3	24-10	25-4	25-10		
	19.2	18-11	19-6	20-0	20-7	21-1	21-8	22-2	22-8	23-1	23-7	24-1	24-6
	24.0	16-11	17-5	17-11	18-5	18-11	19-4	19-10	20-3	20-8	21-1	21-6	21-11
2×12	12.0												
	16.0	25-2	25-11										
	19.2	23-0	23-8	24-4	25-0	25-8							
	24.0	20-6	21-2	21-9	22-5	23-0	23-6	24-1	24-8	25-2	25-8		
E	12.0	1.24	1.36	1.48	1.60	1.73	1.86	2.00	2.14	2.28	2.42	2.57	
	16.0	1.07	1.18	1.28	1.39	1.50	1.61	1.73	1.85	1.97	2.10	2.22	2.35
	19.2	0.98	1.07	1.17	1.27	1.37	1.47	1.58	1.69	1.80	1.91	2.03	2.15
	24.0	0.88	0.96	1.05	1.13	1.22	1.32	1.41	1.51	1.61	1.71	1.82	1.92

NOTE: The required modulus of elasticity, *E*, in 1,000,000 pounds per square inch (psi) (× 0.00689 for N/mm²) is shown at the bottom of this table, is limited to 2.6 million psi (17 914 N/mm²) and less, and is applicable to all lumber sizes shown. Spans are shown in feet-inches (1 foot = 304.8 mm, 1 inch = 25.4 mm) and are limited to 26 feet (7925 mm) and less.

TABLE 23-IV-R-6—RAFTERS WITH L/240 DEFLECTION LIMITATION
The allowable bending stress (F_b) and modulus of elasticity (E) used in this table shall be from Tables 23-IV-V-1 and 23-IV-V-2 only.

DESIGN CRITERIA:
Strength — Live load of 30 psf (1.44 kN/m²) plus dead load of 20 psf (0.96 kN/m²) determines the required bending design value.
Deflection — For 30 psf (1.44 kN/m²) live load.
Limited to span in inches (mm) divided by 240.

Rafter Size (in) × 25.4 for mm	Spacing (in)	Bending Design Value, F_b (psi) × 0.00689 for N/mm²												
		300	400	500	600	700	800	900	1000	1100	1200	1300	1400	1500
2×6	12.0	5-6	6-4	7-1	7-9	8-5	9-0	9-6	10-0	10-6	11-0	11-5	11-11	12-4
	16.0	4-9	5-6	6-2	6-9	7-3	7-9	8-3	8-8	9-1	9-6	9-11	10-3	10-8
	19.2	4-4	5-0	5-7	6-2	6-8	7-1	7-6	7-11	8-4	8-8	9-1	9-5	9-9
	24.0	3-11	4-6	5-0	5-6	5-11	6-4	6-9	7-1	7-5	7-9	8-1	8-5	8-8
2×8	12.0	7-3	8-4	9-4	10-3	11-1	11-10	12-7	13-3	13-11	14-6	15-1	15-8	16-3
	16.0	6-3	7-3	8-1	8-11	9-7	10-3	10-10	11-6	12-0	12-7	13-1	13-7	14-0
	19.2	5-9	6-7	7-5	8-1	8-9	9-4	9-11	10-6	11-0	11-6	11-11	12-5	12-10
	24.0	5-2	5-11	6-7	7-3	7-10	8-4	8-11	9-4	9-10	10-3	10-8	11-1	11-6
2×10	12.0	9-3	10-8	11-11	13-1	14-2	15-1	16-0	16-11	17-9	18-6	19-3	20-0	20-8
	16.0	8-0	9-3	10-4	11-4	12-3	13-1	13-10	14-8	15-4	16-0	16-8	17-4	17-11
	19.2	7-4	8-5	9-5	10-4	11-2	11-11	12-8	13-4	14-0	14-8	15-3	15-10	16-4
	24.0	6-6	7-7	8-5	9-3	10-0	10-8	11-4	11-11	12-6	13-1	13-7	14-2	14-8
2×12	12.0	11-3	13-0	14-6	15-11	17-2	18-4	19-6	20-6	21-7	22-6	23-5	24-4	25-2
	16.0	9-9	11-3	12-7	13-9	14-11	15-11	16-10	17-9	18-8	19-6	20-3	21-1	21-9
	19.2	8-11	10-3	11-6	12-7	13-7	14-6	15-5	16-3	17-0	17-9	18-6	19-3	19-11
	24.0	7-11	9-2	10-3	11-3	12-2	13-0	13-9	14-6	15-3	15-11	16-7	17-2	17-9
E	12.0	0.11	0.17	0.23	0.31	0.38	0.47	0.56	0.66	0.76	0.86	0.97	1.09	1.21
	16.0	0.09	0.14	0.20	0.26	0.33	0.41	0.49	0.57	0.66	0.75	0.84	0.94	1.05
	19.2	0.09	0.13	0.18	0.24	0.30	0.37	0.44	0.52	0.60	0.68	0.77	0.86	0.95
	24.0	0.08	0.12	0.16	0.22	0.27	0.33	0.40	0.46	0.54	0.61	0.69	0.77	0.85

Rafter Size (in) × 25.4 for mm	Spacing (in)	Bending Design Value, F_b (psi) × 0.00689 for N/mm²											
		1600	1700	1800	1900	2000	2100	2200	2300	2400	2500	2600	2700
2×6	12.0	12-8	13-1	13-6	13-10	14-2	14-7	14-11	15-3	15-7	15-11		
	16.0	11-0	11-4	11-8	12-0	12-4	12-7	12-11	13-2	13-6	13-9	14-0	14-3
	19.2	10-0	10-4	10-8	10-11	11-3	11-6	11-9	12-0	12-4	12-7	12-10	13-1
	24.0	9-0	9-3	9-6	9-9	10-0	10-3	10-6	10-9	11-0	11-3	11-5	11-8
2×8	12.0	16-9	17-3	17-9	18-3	18-9	19-2	19-8	20-1	20-6	20-11		
	16.0	14-6	14-11	15-5	15-10	16-3	16-7	17-0	17-5	17-9	18-1	18-6	18-10
	19.2	13-3	13-8	14-0	14-5	14-10	15-2	15-6	15-10	16-3	16-7	16-10	17-2
	24.0	11-10	12-2	12-7	12-11	13-3	13-7	13-11	14-2	14-6	14-10	15-1	15-5
2×10	12.0	21-4	22-0	22-8	23-3	23-11	24-6	25-1	25-7				
	16.0	18-6	19-1	19-7	20-2	20-8	21-2	21-8	22-2	22-8	23-1	23-7	24-0
	19.2	16-11	17-5	17-11	18-5	18-11	19-4	19-10	20-3	20-8	21-1	21-6	21-11
	24.0	15-1	15-7	16-0	16-6	16-11	17-4	17-9	18-1	18-6	18-11	19-3	19-7
2×12	12.0	26-0											
	16.0	22-6	23-2	23-10	24-6	25-2	25-9						
	19.2	20-6	21-2	21-9	22-5	23-0	23-6	24-1	24-8	25-2	25-8		
	24.0	18-4	18-11	19-6	20-0	20-6	21-1	21-7	22-0	22-6	23-0	23-5	23-10
E	12.0	1.33	1.46	1.59	1.72	1.86	2.00	2.14	2.29	2.44	2.60		
	16.0	1.15	1.26	1.37	1.49	1.61	1.73	1.86	1.99	2.12	2.25	2.39	2.53
	19.2	1.05	1.15	1.25	1.36	1.47	1.58	1.70	1.81	1.93	2.05	2.18	2.31
	24.0	0.94	1.03	1.12	1.22	1.31	1.41	1.52	1.62	1.73	1.84	1.95	2.06

NOTE: The required modulus of elasticity, E, in 1,000,000 pounds per square inch (psi) (× 0.00689 for N/mm²) is shown at the bottom of this table, is limited to 2.6 million psi (17 914 N/mm²) and less, and is applicable to all lumber sizes shown. Spans are shown in feet-inches (1 foot = 304.8 mm, 1 inch = 25.4 mm) and are limited to 26 feet (7925 mm) and less.

TABLE 23-IV-R-7—RAFTERS WITH L/180 DEFLECTION LIMITATION
The allowable bending stress (F_b) and modulus of elasticity (E) used in this table shall be from Tables 23-IV-V-1 and 23-IV-V-2 only.

DESIGN CRITERIA:
Strength — Live load of 20 psf (0.96 kN/m²) plus dead load of 10 psf (0.48 kN/m²) determines the required bending design value.
Deflection — For 20 psf (0.96 kN/m²) live load.
Limited to span in inches (mm) divided by 180.

Rafter Size (in) × 25.4 for mm	Spacing (in)	Bending Design Value, F_b (psi) × 0.00689 for N/mm²														
		200	300	400	500	600	700	800	900	1000	1100	1200	1300	1400	1500	1600
2×4	12.0	3-8	4-6	5-3	5-10	6-5	6-11	7-5	7-10	8-3	8-8	9-0	9-5	9-9	10-1	10-5
	16.0	3-2	3-11	4-6	5-1	5-6	6-0	6-5	6-9	7-2	7-6	7-10	8-2	8-5	8-9	9-0
	19.2	2-11	3-7	4-1	4-7	5-1	5-5	5-10	6-2	6-6	6-10	7-2	7-5	7-9	8-0	8-3
	24.0	2-7	3-2	3-8	4-1	4-6	4-11	5-3	5-6	5-10	6-1	6-5	6-8	6-11	7-2	7-5
2×6	12.0	5-10	7-1	8-2	9-2	10-0	10-10	11-7	12-4	13-0	13-7	14-2	14-9	15-4	15-11	16-5
	16.0	5-0	6-2	7-1	7-11	8-8	9-5	10-0	10-8	11-3	11-9	12-4	12-10	13-3	13-9	14-2
	19.2	4-7	5-7	6-6	7-3	7-11	8-7	9-2	9-9	10-3	10-9	11-3	11-8	12-2	12-7	13-0
	24.0	4-1	5-0	5-10	6-6	7-1	7-8	8-2	8-8	9-2	9-7	10-0	10-5	10-10	11-3	11-7
2×8	12.0	7-8	9-4	10-10	12-1	13-3	14-4	15-3	16-3	17-1	17-11	18-9	19-6	20-3	20-11	21-7
	16.0	6-7	8-1	9-4	10-6	11-6	12-5	13-3	14-0	14-10	15-6	16-3	16-10	17-6	18-1	18-9
	19.2	6-0	7-5	8-7	9-7	10-6	11-4	12-1	12-10	13-6	14-2	14-10	15-5	16-0	16-7	17-1
	24.0	5-5	6-7	7-8	8-7	9-4	10-1	10-10	11-6	12-1	12-8	13-3	13-9	14-4	14-10	15-3
2×10	12.0	9-9	11-11	13-9	15-5	16-11	18-3	19-6	20-8	21-10	22-10	23-11	24-10	25-10		
	16.0	8-5	10-4	11-11	13-4	14-8	15-10	16-11	17-11	18-11	19-10	20-8	21-6	22-4	23-1	23-11
	19.2	7-8	9-5	10-11	12-2	13-4	14-5	15-5	16-4	17-3	18-1	18-11	19-8	20-5	21-1	21-10
	24.0	6-11	8-5	9-9	10-11	11-11	12-11	13-9	14-8	15-5	16-2	16-11	17-7	18-3	18-11	19-6
E	12.0	0.06	0.12	0.18	0.25	0.33	0.41	0.51	0.60	0.71	0.82	0.93	1.05	1.17	1.30	1.43
	16.0	0.05	0.10	0.15	0.22	0.28	0.36	0.44	0.52	0.61	0.71	0.80	0.91	1.01	1.13	1.24
	19.2	0.05	0.09	0.14	0.20	0.26	0.33	0.40	0.48	0.56	0.64	0.73	0.83	0.93	1.03	1.13
	24.0	0.04	0.08	0.13	0.18	0.23	0.29	0.36	0.43	0.50	0.58	0.66	0.74	0.83	0.92	1.01

Rafter Size (in) × 25.4 for mm	Spacing (in)	Bending Design Value, F_b (psi) × 0.00689 for N/mm²													
		1700	1800	1900	2000	2100	2200	2300	2400	2500	2600	2700	2800	2900	3000
2×4	12.0	10-9	11-1	11-4	11-8	11-11	12-3	12-6							
	16.0	9-4	9-7	9-10	10-1	10-4	10-7	10-10	11-1	11-4	11-6				
	19.2	8-6	8-9	9-0	9-3	9-5	9-8	9-11	10-1	10-4	10-6	10-9			
	24.0	7-7	7-10	8-0	8-3	8-5	8-8	8-10	9-0	9-3	9-5	9-7	9-9	9-11	10-1
2×6	12.0	16-11	17-5	17-10	18-4	18-9	19-3	19-8							
	16.0	14-8	15-1	15-6	15-11	16-3	16-8	17-0	17-5	17-9	18-1				
	19.2	13-4	13-9	14-2	14-6	14-10	15-2	15-7	15-11	16-2	16-6	16-10			
	24.0	11-11	12-4	12-8	13-0	13-3	13-7	13-11	14-2	14-6	14-9	15-1	15-4	15-7	15-11
2×8	12.0	22-3	22-11	23-7	24-2	24-9	25-4	25-11							
	16.0	19-4	19-10	20-5	20-11	21-5	21-11	22-5	22-11	23-5	23-10				
	19.2	17-7	18-1	18-7	19-1	19-7	20-0	20-6	20-11	21-4	21-9	22-2			
	24.0	15-9	16-3	16-8	17-1	17-6	17-11	18-4	18-9	19-1	19-6	19-10	20-3	20-7	20-11
2×10	12.0														
	16.0	24-7	25-4	26-0											
	19.2	22-6	23-1	23-9	24-5	25-0	25-7								
	24.0	20-1	20-8	21-3	21-10	22-4	22-10	23-5	23-11	24-5	24-10	25-4	25-10		
E	12.0	1.57	1.71	1.85	2.00	2.15	2.31	2.47							
	16.0	1.36	1.48	1.60	1.73	1.86	2.00	2.14	2.28	2.42	2.57				
	19.2	1.24	1.35	1.46	1.58	1.70	1.82	1.95	2.08	2.21	2.34	2.48			
	24.0	1.11	1.21	1.31	1.41	1.52	1.63	1.74	1.86	1.98	2.10	2.22	2.34	2.47	2.60

NOTE: The required modulus of elasticity, E, in 1,000,000 pounds per square inch (psi) (× 0.00689 for N/mm²) is shown at the bottom of this table, is limited to 2.6 million psi (17 914 N/mm²) and less, and is applicable to all lumber sizes shown. Spans are shown in feet-inches (1 foot = 304.8 mm, 1 inch = 25.4 mm) and are limited to 26 feet (7925 mm) and less.

TABLE 23-IV-R-8—RAFTERS WITH L/180 DEFLECTION LIMITATION
The allowable bending stress (F_b) and modulus of elasticity (E) used in this table shall be from Tables 23-IV-V-1 and 23-IV-V-2 only.

DESIGN CRITERIA:
Strength — Live load of 30 psf (1.44 kN/m^2) plus dead load of 10 psf (0.48 kN/m^2) determines the required bending design value.
Deflection — For 30 psf (1.44 kN/m^2) live load.
Limited to span in inches (mm) divided by 180.

Rafter Size (in) × 25.4 for mm	Spacing (in)	Bending Design Value, F_b (psi) × 0.00689 for N/mm^2														
		200	300	400	500	600	700	800	900	1000	1100	1200	1300	1400	1500	1600
2×4	12.0	3-2	3-11	4-6	5-1	5-6	6-0	6-5	6-9	7-2	7-6	7-10	8-2	8-5	8-9	9-0
	16.0	2-9	3-5	3-11	4-4	4-10	5-2	5-6	5-10	6-2	6-6	6-9	7-1	7-4	7-7	7-10
	19.2	2-6	3-1	3-7	4-0	4-4	4-9	5-1	5-4	5-8	5-11	6-2	6-5	6-8	6-11	7-2
	24.0	2-3	2-9	3-2	3-7	3-11	4-3	4-6	4-10	5-1	5-4	5-6	5-9	6-0	6-2	6-5
2×6	12.0	5-0	6-2	7-1	7-11	8-8	9-5	10-0	10-8	11-3	11-9	12-4	12-10	13-3	13-9	14-2
	16.0	4-4	5-4	6-2	6-10	7-6	8-2	8-8	9-3	9-9	10-2	10-8	11-1	11-6	11-11	12-4
	19.2	4-0	4-10	5-7	6-3	6-10	7-5	7-11	8-5	8-11	9-4	9-9	10-1	10-6	10-10	11-3
	24.0	3-7	4-4	5-0	5-7	6-2	6-8	7-1	7-6	7-11	8-4	8-8	9-1	9-5	9-9	10-0
2×8	12.0	6-7	8-1	9-4	10-6	11-6	12-5	13-3	14-0	14-10	15-6	16-3	16-10	17-6	18-1	18-9
	16.0	5-9	7-0	8-1	9-1	9-11	10-9	11-6	12-2	12-10	13-5	14-0	14-7	15-2	15-8	16-3
	19.2	5-3	6-5	7-5	8-3	9-1	9-9	10-6	11-1	11-8	12-3	12-10	13-4	13-10	14-4	14-10
	24.0	4-8	5-9	6-7	7-5	8-1	8-9	9-4	9-11	10-6	11-0	11-6	11-11	12-5	12-10	13-3
2×10	12.0	8-5	10-4	11-11	13-4	14-8	15-10	16-11	17-11	18-11	19-10	20-8	21-6	22-4	23-1	23-11
	16.0	7-4	8-11	10-4	11-7	12-8	13-8	14-8	15-6	16-4	17-2	17-11	18-8	19-4	20-0	20-8
	19.2	6-8	8-2	9-5	10-7	11-7	12-6	13-4	14-2	14-11	15-8	16-4	17-0	17-8	18-3	18-11
	24.0	6-0	7-4	8-5	9-5	10-4	11-2	11-11	12-8	13-4	14-0	14-8	15-3	15-10	16-4	16-11
E	12.0	0.06	0.11	0.17	0.24	0.32	0.40	0.49	0.59	0.69	0.79	0.91	1.02	1.14	1.27	1.39
	16.0	0.05	0.10	0.15	0.21	0.28	0.35	0.43	0.51	0.60	0.69	0.78	0.88	0.99	1.10	1.21
	19.2	0.05	0.09	0.14	0.19	0.25	0.32	0.39	0.47	0.54	0.63	0.72	0.81	0.90	1.00	1.10
	24.0	0.04	0.08	0.12	0.17	0.23	0.29	0.35	0.42	0.49	0.56	0.64	0.72	0.81	0.89	0.99

Rafter Size (in) × 25.4 for mm	Spacing (in)	Bending Design Value, F_b (psi) × 0.00689 for N/mm^2													
		1700	1800	1900	2000	2100	2200	2300	2400	2500	2600	2700	2800	2900	3000
2×4	12.0	9-4	9-7	9-10	10-1	10-4	10-7	10-10	11-1						
	16.0	8-1	8-4	8-6	8-9	9-0	9-2	9-5	9-7	9-9	10-0				
	19.2	7-4	7-7	7-9	8-0	8-2	8-5	8-7	8-9	8-11	9-1	9-3	9-5		
	24.0	6-7	6-9	7-0	7-2	7-4	7-6	7-8	7-10	8-0	8-2	8-4	8-5	8-7	8-9
2×6	12.0	14-8	15-1	15-6	15-11	16-3	16-8	17-0	17-5						
	16.0	12-8	13-1	13-5	13-9	14-1	14-5	14-9	15-1	15-4	15-8				
	19.2	11-7	11-11	12-3	12-7	12-10	13-2	13-6	13-9	14-0	14-4	14-7	14-10		
	24.0	10-4	10-8	10-11	11-3	11-6	11-9	12-0	12-4	12-7	12-10	13-1	13-3	13-6	13-9
2×8	12.0	19-4	19-10	20-5	20-11	21-5	21-11	22-5	22-11						
	16.0	16-9	17-2	17-8	18-1	18-7	19-0	19-5	19-10	20-3	20-8				
	19.2	15-3	15-8	16-2	16-7	16-11	17-4	17-9	18-1	18-6	18-10	19-3	19-7		
	24.0	13-8	14-0	14-5	14-10	15-2	15-6	15-10	16-3	16-7	16-10	17-2	17-6	17-10	18-1
2×10	12.0	24-7	25-4	26-0											
	16.0	21-4	21-11	22-6	23-1	23-8	24-3	24-10	25-4	25-10					
	19.2	19-6	20-0	20-7	21-1	21-8	22-2	22-8	23-1	23-7	24-1	24-6	25-0		
	24.0	17-5	17-11	18-5	18-11	19-4	19-10	20-3	20-8	21-1	21-6	21-11	22-4	22-9	23-1
E	12.0	1.53	1.66	1.80	1.95	2.10	2.25	2.40	2.56						
	16.0	1.32	1.44	1.56	1.69	1.82	1.95	2.08	2.22	2.36	2.50				
	19.2	1.21	1.32	1.43	1.54	1.66	1.78	1.90	2.03	2.15	2.28	2.42	2.55		
	24.0	1.08	1.18	1.28	1.38	1.48	1.59	1.70	1.81	1.93	2.04	2.16	2.28	2.41	2.53

NOTE: The required modulus of elasticity, E, in 1,000,000 pounds per square inch (psi) (× 0.00689 for N/mm^2) is shown at the bottom of this table, is limited to 2.6 million psi (17 914 N/mm^2) and less, and is applicable to all lumber sizes shown. Spans are shown in feet-inches (1 foot = 304.8 mm, 1 inch = 25.4 mm) and are limited to 26 feet (7925 mm) and less.

TABLE 23-IV-R-9—RAFTERS WITH *L*/180 DEFLECTION LIMITATION
The allowable bending stress *(F_b)* and modulus of elasticity *(E)* used in this table shall be from Tables 23-IV-V-1 and 23-IV-V-2 only.

DESIGN CRITERIA:
Strength — Live load of 20 psf (0.96 kN/m²) plus dead load of 15 psf (0.72 kN/m²) determines the required bending design value.
Deflection — For 20 psf (0.96 kN/m²) live load.
Limited to span in inches (mm) divided by 180.

Rafter Size (in) × 25.4 for mm	Spacing (in)	Bending Design Value, F_b (psi) × 0.00689 for N/mm²														
		200	300	400	500	600	700	800	900	1000	1100	1200	1300	1400	1500	1600
2×4	12.0	3-5	4-2	4-10	5-5	5-11	6-5	6-10	7-3	7-8	8-0	8-4	8-8	9-0	9-4	9-8
	16.0	2-11	3-7	4-2	4-8	5-1	5-6	5-11	6-3	6-7	6-11	7-3	7-6	7-10	8-1	8-4
	19.2	2-8	3-4	3-10	4-3	4-8	5-1	5-5	5-9	6-0	6-4	6-7	6-11	7-2	7-5	7-8
	24.0	2-5	2-11	3-5	3-10	4-2	4-6	4-10	5-1	5-5	5-8	5-11	6-2	6-5	6-7	6-10
2×6	12.0	5-4	6-7	7-7	8-6	9-4	10-0	10-9	11-5	12-0	12-7	13-2	13-8	14-2	14-8	15-2
	16.0	4-8	5-8	6-7	7-4	8-1	8-8	9-4	9-10	10-5	10-11	11-5	11-10	12-4	12-9	13-2
	19.2	4-3	5-2	6-0	6-9	7-4	7-11	8-6	9-0	9-6	9-11	10-5	10-10	11-3	11-7	12-0
	24.0	3-10	4-8	5-4	6-0	6-7	7-1	7-7	8-1	8-6	8-11	9-4	9-8	10-0	10-5	10-9
2×8	12.0	7-1	8-8	10-0	11-2	12-3	13-3	14-2	15-0	15-10	16-7	17-4	18-0	18-9	19-5	20-0
	16.0	6-2	7-6	8-8	9-8	10-7	11-6	12-3	13-0	13-8	14-4	15-0	15-7	16-3	16-9	17-4
	19.2	5-7	6-10	7-11	8-10	9-8	10-6	11-2	11-10	12-6	13-1	13-8	14-3	14-10	15-4	15-10
	24.0	5-0	6-2	7-1	7-11	8-8	9-4	10-0	10-7	11-2	11-9	12-3	12-9	13-3	13-8	14-2
2×10	12.0	9-0	11-1	12-9	14-3	15-8	16-11	18-1	19-2	20-2	21-2	22-1	23-0	23-11	24-9	25-6
	16.0	7-10	9-7	11-1	12-4	13-6	14-8	15-8	16-7	17-6	18-4	19-2	19-11	20-8	21-5	22-1
	19.2	7-2	8-9	10-1	11-3	12-4	13-4	14-3	15-2	15-11	16-9	17-6	18-2	18-11	19-7	20-2
	24.0	6-5	7-10	9-0	10-1	11-1	11-11	12-9	13-6	14-3	15-0	15-8	16-3	16-11	17-6	18-1
E	12.0	0.05	0.09	0.14	0.20	0.26	0.33	0.40	0.48	0.56	0.65	0.74	0.83	0.93	1.03	1.14
	16.0	0.04	0.08	0.12	0.17	0.23	0.28	0.35	0.41	0.49	0.56	0.64	0.72	0.80	0.89	0.98
	19.2	0.04	0.07	0.11	0.16	0.21	0.26	0.32	0.38	0.44	0.51	0.58	0.66	0.73	0.81	0.90
	24.0	0.04	0.07	0.10	0.14	0.18	0.23	0.28	0.34	0.40	0.46	0.52	0.59	0.66	0.73	0.80

Rafter Size (in) × 25.4 for mm	Spacing (in)	Bending Design Value, F_b (psi) × 0.00689 for N/mm²														
		1700	1800	1900	2000	2100	2200	2300	2400	2500	2600	2700	2800	2900	3000
2×4	12.0	9-11	10-3	10-6	10-10	11-1	11-4	11-7	11-10	12-1	12-4	12-7			
	16.0	8-7	8-10	9-1	9-4	9-7	9-10	10-0	10-3	10-5	10-8	10-10	11-1	11-3	11-5
	19.2	7-10	8-1	8-4	8-6	8-9	8-11	9-2	9-4	9-7	9-9	9-11	10-1	10-3	10-5
	24.0	7-0	7-3	7-5	7-8	7-10	8-0	8-2	8-4	8-6	8-8	8-10	9-0	9-2	9-4
2×6	12.0	15-8	16-1	16-7	17-0	17-5	17-10	18-2	18-7	19-0	19-4	19-9			
	16.0	13-7	13-11	14-4	14-8	15-1	15-5	15-9	16-1	16-5	16-9	17-1	17-5	17-8	18-0
	19.2	12-4	12-9	13-1	13-5	13-9	14-1	14-5	14-8	15-0	15-4	15-7	15-11	16-2	16-5
	24.0	11-1	11-5	11-8	12-0	12-4	12-7	12-10	13-2	13-5	13-8	13-11	14-2	14-5	14-8
2×8	12.0	20-8	21-3	21-10	22-4	22-11	23-6	24-0	24-6	25-0	25-6	26-0			
	16.0	17-10	18-5	18-11	19-5	19-10	20-4	20-9	21-3	21-8	22-1	22-6	22-11	23-4	23-9
	19.2	16-4	16-9	17-3	17-8	18-1	18-7	19-0	19-5	19-9	20-2	20-7	20-11	21-4	21-8
	24.0	14-7	15-0	15-5	15-10	16-3	16-7	17-0	17-4	17-8	18-0	18-5	18-9	19-1	19-5
2×10	12.0														
	16.0	22-10	23-5	24-1	24-9	25-4	25-11								
	19.2	20-10	21-5	22-0	22-7	23-1	23-8	24-2	24-9	25-3	25-9				
	24.0	18-7	19-2	19-8	20-2	20-8	21-2	21-8	22-1	22-7	23-0	23-5	23-11	24-4	24-9
E	12.0	1.24	1.36	1.47	1.59	1.71	1.83	1.96	2.09	2.22	2.35	2.49			
	16.0	1.08	1.17	1.27	1.37	1.48	1.59	1.70	1.81	1.92	2.04	2.16	2.28	2.40	2.53
	19.2	0.98	1.07	1.16	1.25	1.35	1.45	1.55	1.65	1.75	1.86	1.97	2.08	2.19	2.31
	24.0	0.88	0.96	1.04	1.12	1.21	1.29	1.38	1.48	1.57	1.66	1.76	1.86	1.96	2.06

NOTE: The required modulus of elasticity, *E*, in 1,000,000 pounds per square inch (psi) (× 0.00689 for N/mm²) is shown at the bottom of this table, is limited to 2.6 million psi (17 914 N/mm²) and less, and is applicable to all lumber sizes shown. Spans are shown in feet-inches (1 foot = 304.8 mm, 1 inch = 25.4 mm) and are limited to 26 feet (7925 mm) and less.

TABLE 23-IV-R-10—RAFTERS WITH $L/180$ DEFLECTION LIMITATION

The allowable bending stress (F_b) and modulus of elasticity (E) used in this table shall be from Tables 23-IV-V-1 and 23-IV-V-2 only.

DESIGN CRITERIA:
Strength — Live load of 30 psf (1.44 kN/m²) plus dead load of 15 psf (0.72 kN/m²) determines the required bending design value.
Deflection — For 30 psf (1.44 kN/m²) live load.
Limited to span in inches (mm) divided by 180.

Rafter Size (in) × 25.4 for mm	Spacing (in)	Bending Design Value, F_b (psi) × 0.00689 for N/mm²														
		200	300	400	500	600	700	800	900	1000	1100	1200	1300	1400	1500	1600
2×4	12.0	3-0	3-8	4-3	4-9	5-3	5-8	6-0	6-5	6-9	7-1	7-5	7-8	8-0	8-3	8-6
	16.0	2-7	3-2	3-8	4-1	4-6	4-11	5-3	5-6	5-10	6-1	6-5	6-8	6-11	7-2	7-5
	19.2	2-5	2-11	3-4	3-9	4-1	4-5	4-9	5-1	5-4	5-7	5-10	6-1	6-4	6-6	6-9
	24.0	2-2	2-7	3-0	3-4	3-8	4-0	4-3	4-6	4-9	5-0	5-3	5-5	5-8	5-10	6-0
2×6	12.0	4-9	5-10	6-8	7-6	8-2	8-10	9-6	10-0	10-7	11-1	11-7	12-1	12-6	13-0	13-5
	16.0	4-1	5-0	5-10	6-6	7-1	7-8	8-2	8-8	9-2	9-7	10-0	10-5	10-10	11-3	11-7
	19.2	3-9	4-7	5-4	5-11	6-6	7-0	7-6	7-11	8-4	8-9	9-2	9-6	9-11	10-3	10-7
	24.0	3-4	4-1	4-9	5-4	5-10	6-3	6-8	7-1	7-6	7-10	8-2	8-6	8-10	9-2	9-6
2×8	12.0	6-3	7-8	8-10	9-10	10-10	11-8	12-6	13-3	13-11	14-8	15-3	15-11	16-6	17-1	17-8
	16.0	5-5	6-7	7-8	8-7	9-4	10-1	10-10	11-6	12-1	12-8	13-3	13-9	14-4	14-10	15-3
	19.2	4-11	6-0	7-0	7-10	8-7	9-3	9-10	10-6	11-0	11-7	12-1	12-7	13-1	13-6	13-11
	24.0	4-5	5-5	6-3	7-0	7-8	8-3	8-10	9-4	9-10	10-4	10-10	11-3	11-8	12-1	12-6
2×10	12.0	8-0	9-9	11-3	12-7	13-9	14-11	15-11	16-11	17-10	18-8	19-6	20-4	21-1	21-10	22-6
	16.0	6-11	8-5	9-9	10-11	11-11	12-11	13-9	14-8	15-5	16-2	16-11	17-7	18-3	18-11	19-6
	19.2	6-4	7-8	8-11	9-11	10-11	11-9	12-7	13-4	14-1	14-9	15-5	16-1	16-8	17-3	17-10
	24.0	5-8	6-11	8-0	8-11	9-9	10-6	11-3	11-11	12-7	13-2	13-9	14-4	14-11	15-5	15-11
E	12.0	0.05	0.09	0.15	0.20	0.27	0.34	0.41	0.49	0.58	0.67	0.76	0.86	0.96	1.06	1.17
	16.0	0.04	0.08	0.13	0.18	0.23	0.29	0.36	0.43	0.50	0.58	0.66	0.74	0.83	0.92	1.01
	19.2	0.04	0.08	0.12	0.16	0.21	0.27	0.33	0.39	0.46	0.53	0.60	0.68	0.76	0.84	0.92
	24.0	0.04	0.07	0.10	0.14	0.19	0.24	0.29	0.35	0.41	0.47	0.54	0.61	0.68	0.75	0.83

Rafter Size (in) × 25.4 for mm	Spacing (in)	Bending Design Value, F_b (psi) × 0.00689 for N/mm²													
		1700	1800	1900	2000	2100	2200	2300	2400	2500	2600	2700	2800	2900	3000
2×4	12.0	8-9	9-0	9-3	9-6	9-9	10-0	10-3	10-5	10-8	10-10	11-1			
	16.0	7-7	7-10	8-0	8-3	8-5	8-8	8-10	9-0	9-3	9-5	9-7	9-9	9-11	10-1
	19.2	6-11	7-2	7-4	7-6	7-9	7-11	8-1	8-3	8-5	8-7	8-9	8-11	9-1	9-3
	24.0	6-3	6-5	6-7	6-9	6-11	7-1	7-3	7-5	7-6	7-8	7-10	8-0	8-1	8-3
2×6	12.0	13-10	14-2	14-7	15-0	15-4	15-8	16-1	16-5	16-9	17-1	17-5			
	16.0	11-11	12-4	12-8	13-0	13-3	13-7	13-11	14-2	14-6	14-9	15-1	15-4	15-7	15-11
	19.2	10-11	11-3	11-6	11-10	12-2	12-5	12-8	13-0	13-3	13-6	13-9	14-0	14-3	14-6
	24.0	9-9	10-0	10-4	10-7	10-10	11-1	11-4	11-7	11-10	12-1	12-4	12-6	12-9	13-0
2×8	12.0	18-2	18-9	19-3	19-9	20-3	20-8	21-2	21-7	22-1	22-6	22-11			
	16.0	15-9	16-3	16-8	17-1	17-6	17-11	18-4	18-9	19-1	19-6	19-10	20-3	20-7	20-11
	19.2	14-5	14-10	15-2	15-7	16-0	16-4	16-9	17-1	17-5	17-9	18-1	18-5	18-9	19-1
	24.0	12-10	13-3	13-7	13-11	14-4	14-8	15-0	15-3	15-7	15-11	16-3	16-6	16-10	17-1
2×10	12.0	23-3	23-11	24-6	25-2	25-10									
	16.0	20-1	20-8	21-3	21-10	22-4	22-10	23-5	23-11	24-5	24-10	25-4	25-10		
	19.2	18-4	18-11	19-5	19-11	20-5	20-10	21-4	21-10	22-3	22-8	23-1	23-7	24-0	24-5
	24.0	16-5	16-11	17-4	17-10	18-3	18-8	19-1	19-6	19-11	20-4	20-8	21-1	21-5	21-10
E	12.0	1.28	1.39	1.51	1.63	1.76	1.88	2.01	2.15	2.28	2.42	2.56			
	16.0	1.11	1.21	1.31	1.41	1.52	1.63	1.74	1.86	1.98	2.10	2.22	2.34	2.47	2.60
	19.2	1.01	1.10	1.20	1.29	1.39	1.49	1.59	1.70	1.80	1.91	2.03	2.14	2.25	2.37
	24.0	0.90	0.99	1.07	1.15	1.24	1.33	1.42	1.52	1.61	1.71	1.81	1.91	2.02	2.12

NOTE: The required modulus of elasticity, E, in 1,000,000 pounds per square inch (psi) (× 0.00689 for N/mm²) is shown at the bottom of this table, is limited to 2.6 million psi (17 914 N/mm²) and less, and is applicable to all lumber sizes shown. Spans are shown in feet-inches (1 foot = 304.8 mm, 1 inch = 25.4 mm) and are limited to 26 feet (7925 mm) and less.

TABLE 23-IV-R-11—RAFTERS WITH L/180 DEFLECTION LIMITATION
The allowable bending stress (F_b) and modulus of elasticity (E) used in this table shall be from Tables 23-IV-V-1 and 23-IV-V-2 only.

DESIGN CRITERIA:
Strength — Live load of 20 psf (0.96 kN/m²) plus dead load of 20 psf (0.96 kN/m²) determines the required bending design value.
Deflection — For 20 psf (0.96 kN/m²) live load.
Limited to span in inches (mm) divided by 180.

Rafter Size (in) ×25.4 for mm	Spacing (in)	Bending Design Value, F_b (psi) × 0.00689 for N/mm²														
		200	300	400	500	600	700	800	900	1000	1100	1200	1300	1400	1500	1600
2×4	12.0	3-2	3-11	4-6	5-1	5-6	6-0	6-5	6-9	7-2	7-6	7-10	8-2	8-5	8-9	9-0
	16.0	2-9	3-5	3-11	4-4	4-10	5-2	5-6	5-10	6-2	6-6	6-9	7-1	7-4	7-7	7-10
	19.2	2-6	3-1	3-7	4-0	4-4	4-9	5-1	5-4	5-8	5-11	6-2	6-5	6-8	6-11	7-2
	24.0	2-3	2-9	3-2	3-7	3-11	4-3	4-6	4-10	5-1	5-4	5-6	5-9	6-0	6-2	6-5
2×6	12.0	5-0	6-2	7-1	7-11	8-8	9-5	10-0	10-8	11-3	11-9	12-4	12-10	13-3	13-9	14-2
	16.0	4-4	5-4	6-2	6-10	7-6	8-2	8-8	9-3	9-9	10-2	10-8	11-1	11-6	11-11	12-4
	19.2	4-0	4-10	5-7	6-3	6-10	7-5	7-11	8-5	8-11	9-4	9-9	10-1	10-6	10-10	11-3
	24.0	3-7	4-4	5-0	5-7	6-2	6-8	7-1	7-6	7-11	8-4	8-8	9-1	9-5	9-9	10-0
2×8	12.0	6-7	8-1	9-4	10-6	11-6	12-5	13-3	14-0	14-10	15-6	16-3	16-10	17-6	18-1	18-9
	16.0	5-9	7-0	8-1	9-1	9-11	10-9	11-6	12-2	12-10	13-5	14-0	14-7	15-2	15-8	16-3
	19.2	5-3	6-5	7-5	8-3	9-1	9-9	10-6	11-1	11-8	12-3	12-10	13-4	13-10	14-4	14-10
	24.0	4-8	5-9	6-7	7-5	8-1	8-9	9-4	9-11	10-6	11-0	11-6	11-11	12-5	12-10	13-3
2×10	12.0	8-5	10-4	11-11	13-4	14-8	15-10	16-11	17-11	18-11	19-10	20-8	21-6	22-4	23-1	23-11
	16.0	7-4	8-11	10-4	11-7	12-8	13-8	14-8	15-6	16-4	17-2	17-11	18-8	19-4	20-0	20-8
	19.2	6-8	8-2	9-5	10-7	11-7	12-6	13-4	14-2	14-11	15-8	16-4	17-0	17-8	18-3	18-11
	24.0	6-0	7-4	8-5	9-5	10-4	11-2	11-11	12-8	13-4	14-0	14-8	15-3	15-10	16-4	16-11
E	12.0	0.04	0.08	0.12	0.16	0.21	0.27	0.33	0.39	0.46	0.53	0.60	0.68	0.76	0.84	0.93
	16.0	0.04	0.07	0.10	0.14	0.18	0.23	0.28	0.34	0.40	0.46	0.52	0.59	0.66	0.73	0.80
	19.2	0.03	0.06	0.09	0.13	0.17	0.21	0.26	0.31	0.36	0.42	0.48	0.54	0.60	0.67	0.73
	24.0	0.03	0.05	0.08	0.11	0.15	0.19	0.23	0.28	0.32	0.37	0.43	0.48	0.54	0.60	0.66

Rafter Size (in) ×25.4 for mm	Spacing (in)	Bending Design Value, F_b (psi) × 0.00689 for N/mm²													
		1700	1800	1900	2000	2100	2200	2300	2400	2500	2600	2700	2800	2900	3000
2×4	12.0	9-4	9-7	9-10	10-1	10-4	10-7	10-10	11-1	11-4	11-6	11-9	11-11	12-2	12-4
	16.0	8-1	8-4	8-6	8-9	9-0	9-2	9-5	9-7	9-9	10-0	10-2	10-4	10-6	10-9
	19.2	7-4	7-7	7-9	8-0	8-2	8-5	8-7	8-9	8-11	9-1	9-3	9-5	9-7	9-9
	24.0	6-7	6-9	7-0	7-2	7-4	7-6	7-8	7-10	8-0	8-2	8-4	8-5	8-7	8-9
2×6	12.0	14-8	15-1	15-6	15-11	16-3	16-8	17-0	17-5	17-9	18-1	18-5	18-9	19-1	19-5
	16.0	12-8	13-1	13-5	13-9	14-1	14-5	14-9	15-1	15-4	15-8	16-0	16-3	16-7	16-10
	19.2	11-7	11-11	12-3	12-7	12-10	13-2	13-6	13-9	14-0	14-4	14-7	14-10	15-1	15-4
	24.0	10-4	10-8	10-11	11-3	11-6	11-9	12-0	12-4	12-7	12-10	13-1	13-3	13-6	13-9
2×8	12.0	19-4	19-10	20-5	20-11	21-5	21-11	22-5	22-11	23-5	23-10	24-4	24-9	25-2	25-8
	16.0	16-9	17-2	17-8	18-1	18-7	19-0	19-5	19-10	20-3	20-8	21-1	21-5	21-10	22-2
	19.2	15-3	15-8	16-2	16-7	16-11	17-4	17-9	18-1	18-6	18-10	19-3	19-7	19-11	20-3
	24.0	13-8	14-0	14-5	14-10	15-2	15-6	15-10	16-3	16-7	16-10	17-2	17-6	17-10	18-1
2×10	12.0	24-7	25-4	26-0											
	16.0	21-4	21-11	22-6	23-1	23-8	24-3	24-10	25-4	25-10					
	19.2	19-6	20-0	20-7	21-1	21-8	22-2	22-8	23-1	23-7	24-1	24-6	25-0	25-5	25-10
	24.0	17-5	17-11	18-5	18-11	19-4	19-10	20-3	20-8	21-1	21-6	21-11	22-4	22-9	23-1
E	12.0	1.02	1.11	1.20	1.30	1.40	1.50	1.60	1.71	1.82	1.93	2.04	2.15	2.27	2.39
	16.0	0.88	0.96	1.04	1.13	1.21	1.30	1.39	1.48	1.57	1.67	1.76	1.86	1.96	2.07
	19.2	0.80	0.88	0.95	1.03	1.10	1.18	1.27	1.35	1.44	1.52	1.61	1.70	1.79	1.89
	24.0	0.72	0.78	0.85	0.92	0.99	1.06	1.13	1.21	1.28	1.36	1.44	1.52	1.60	1.69

NOTE: The required modulus of elasticity, E, in 1,000,000 pounds per square inch (psi) (× 0.00689 for N/mm²) is shown at the bottom of this table, is limited to 2.6 million psi (17 914 N/mm²) and less, and is applicable to all lumber sizes shown. Spans are shown in feet-inches (1 foot = 304.8 mm, 1 inch = 25.4 mm) and are limited to 26 feet (7925 mm) and less.

TABLE 23-IV-R-12—RAFTERS WITH $L/180$ DEFLECTION LIMITATION
The allowable bending stress (F_b) and modulus of elasticity (E) used in this table shall be from Tables 23-IV-V-1 and 23-IV-V-2 only.

DESIGN CRITERIA:
Strength — Live load of 30 psf (1.44 kN/m²) plus dead load of 20 psf (0.96 kN/m²) determines the required bending design value.
Deflection — For 30 psf (1.44 kN/m²) live load.
Limited to span in inches (mm) divided by 180.

Rafter Size (in) ×25.4 for mm	Spacing (in)	Bending Design Value, F_b (psi) ×0.00689 for N/mm²														
		200	300	400	500	600	700	800	900	1000	1100	1200	1300	1400	1500	1600
2×4	12.0	2-10	3-6	4-0	4-6	4-11	5-4	5-9	6-1	6-5	6-8	7-0	7-3	7-7	7-10	8-1
	16.0	2-6	3-0	3-6	3-11	4-3	4-8	4-11	5-3	5-6	5-10	6-1	6-4	6-7	6-9	7-0
	19.2	2-3	2-9	3-2	3-7	3-11	4-3	4-6	4-10	5-1	5-4	5-6	5-9	6-0	6-2	6-5
	24.0	2-0	2-6	2-10	3-2	3-6	3-9	4-0	4-3	4-6	4-9	4-11	5-2	5-4	5-6	5-9
2×6	12.0	4-6	5-6	6-4	7-1	7-9	8-5	9-0	9-6	10-0	10-6	11-0	11-5	11-11	12-4	12-8
	16.0	3-11	4-9	5-6	6-2	6-9	7-3	7-9	8-3	8-8	9-1	9-6	9-11	10-3	10-8	11-0
	19.2	3-7	4-4	5-0	5-7	6-2	6-8	7-1	7-6	7-11	8-4	8-8	9-1	9-5	9-9	10-0
	24.0	3-2	3-11	4-6	5-0	5-6	5-11	6-4	6-9	7-1	7-5	7-9	8-1	8-5	8-8	9-0
2×8	12.0	5-11	7-3	8-4	9-4	10-3	11-1	11-10	12-7	13-3	13-11	14-6	15-1	15-8	16-3	16-9
	16.0	5-2	6-3	7-3	8-1	8-11	9-7	10-3	10-10	11-6	12-0	12-7	13-1	13-7	14-0	14-6
	19.2	4-8	5-9	6-7	7-5	8-1	8-9	9-4	9-11	10-6	11-0	11-6	11-11	12-5	12-10	13-3
	24.0	4-2	5-2	5-11	6-7	7-3	7-10	8-4	8-11	9-4	9-10	10-3	10-8	11-1	11-6	11-10
2×10	12.0	7-7	9-3	10-8	11-11	13-1	14-2	15-1	16-0	16-11	17-9	18-6	19-3	20-0	20-8	21-4
	16.0	6-6	8-0	9-3	10-4	11-4	12-3	13-1	13-10	14-8	15-4	16-0	16-8	17-4	17-11	18-6
	19.2	6-0	7-4	8-5	9-5	10-4	11-2	11-11	12-8	13-4	14-0	14-8	15-3	15-10	16-4	16-11
	24.0	5-4	6-6	7-7	8-5	9-3	10-0	10-8	11-4	11-11	12-6	13-1	13-7	14-2	14-8	15-1
E	12.0	0.04	0.08	0.12	0.17	0.23	0.29	0.35	0.42	0.49	0.57	0.65	0.73	0.82	0.91	1.00
	16.0	0.04	0.07	0.11	0.15	0.20	0.25	0.31	0.36	0.43	0.49	0.56	0.63	0.71	0.78	0.86
	19.2	0.03	0.06	0.10	0.14	0.18	0.23	0.28	0.33	0.39	0.45	0.51	0.58	0.65	0.72	0.79
	24.0	0.03	0.06	0.09	0.12	0.16	0.20	0.25	0.30	0.35	0.40	0.46	0.52	0.58	0.64	0.71

Rafter Size (in) ×25.4 for mm	Spacing (in)	Bending Design Value, F_b (psi) ×0.00689 for N/mm²													
		1700	1800	1900	2000	2100	2200	2300	2400	2500	2600	2700	2800	2900	3000
2×4	12.0	8-4	8-7	8-10	9-0	9-3	9-6	9-8	9-11	10-1	10-4	10-6	10-8	10-11	11-1
	16.0	7-3	7-5	7-8	7-10	8-0	8-2	8-5	8-7	8-9	8-11	9-1	9-3	9-5	9-7
	19.2	6-7	6-9	7-0	7-2	7-4	7-6	7-8	7-10	8-0	8-2	8-4	8-5	8-7	8-9
	24.0	5-11	6-1	6-3	6-5	6-7	6-8	6-10	7-0	7-2	7-3	7-5	7-7	7-8	7-10
2×6	12.0	13-1	13-6	13-10	14-2	14-7	14-11	15-3	15-7	15-11	16-2	16-6	16-10	17-1	17-5
	16.0	11-4	11-8	12-0	12-4	12-7	12-11	13-2	13-6	13-9	14-0	14-3	14-7	14-10	15-1
	19.2	10-4	10-8	10-11	11-3	11-6	11-9	12-0	12-4	12-7	12-10	13-1	13-3	13-6	13-9
	24.0	9-3	9-6	9-9	10-0	10-3	10-6	10-9	11-0	11-3	11-5	11-8	11-11	12-1	12-4
2×8	12.0	17-3	17-9	18-3	18-9	19-2	19-8	20-1	20-6	20-11	21-4	21-9	22-2	22-6	22-11
	16.0	14-11	15-5	15-10	16-3	16-7	17-0	17-5	17-9	18-1	18-6	18-10	19-2	19-6	19-10
	19.2	13-8	14-0	14-5	14-10	15-2	15-6	15-10	16-3	16-7	16-10	17-2	17-6	17-10	18-1
	24.0	12-2	12-7	12-11	13-3	13-7	13-11	14-2	14-6	14-10	15-1	15-5	15-8	15-11	16-3
2×10	12.0	22-0	22-8	23-3	23-11	24-6	25-1	25-7							
	16.0	19-1	19-7	20-2	20-8	21-2	21-8	22-2	22-8	23-1	23-7	24-0	24-6	24-11	25-4
	19.2	17-5	17-11	18-5	18-11	19-4	19-10	20-3	20-8	21-1	21-6	21-11	22-4	22-9	23-1
	24.0	15-7	16-0	16-6	16-11	17-4	17-9	18-1	18-6	18-11	19-3	19-7	20-0	20-4	20-8
E	12.0	1.09	1.19	1.29	1.39	1.50	1.61	1.72	1.83	1.95	2.07	2.19	2.31	2.43	2.56
	16.0	0.95	1.03	1.12	1.21	1.30	1.39	1.49	1.59	1.69	1.79	1.89	2.00	2.11	2.22
	19.2	0.86	0.94	1.02	1.10	1.19	1.27	1.36	1.45	1.54	1.63	1.73	1.83	1.92	2.03
	24.0	0.77	0.84	0.91	0.99	1.06	1.14	1.22	1.30	1.38	1.46	1.55	1.63	1.72	1.81

NOTE: The required modulus of elasticity, E, in 1,000,000 pounds per square inch (psi) (× 0.00689 for N/mm²) is shown at the bottom of this table, is limited to 2.6 million psi (17 914 N/mm²) and less, and is applicable to all lumber sizes shown. Spans are shown in feet-inches (1 foot = 304.8 mm, 1 inch = 25.4 mm) and are limited to 26 feet (7925 mm) and less.

TABLE 23-IV-V-1—VALUES FOR JOISTS AND RAFTERS—VISUALLY GRADED LUMBER
For Use in Tables 23-IV-J-1 through 23-IV-R-12 and Chapter 23, Division VII only.

These "F_b" values are for use where repetitive members are spaced not more than 24 inches (610 mm). For wider spacing, the "F_b" values shall be reduced 13 percent.

Values for surfaced dry or surfaced green lumber apply at 19 percent maximum moisture content in use.

SPECIES AND GRADE	SIZE (inches) × 25.4 for mm	DESIGN VALUE IN BENDING "F_b" psi			MODULUS OF ELASTICITY "E" psi	GRADING RULES AGENCY
		Normal Duration	Snow Loading	7-day Loading		
		× 0.00689 for N/mm²				
ASPEN						
Select Structural		1,510	1,735	1,885	1,100,000	
No. 1		1,080	1,240	1,350	1,100,000	
No. 2		1,035	1,190	1,295	1,000,000	
No. 3	2 × 4	605	695	755	900,000	
Stud		600	690	750	900,000	
Construction		805	925	1,005	900,000	
Standard		430	495	540	900,000	
Utility		200	230	250	800,000	
Select Structural		1,310	1,505	1,635	1,100,000	
No. 1		935	1,075	1,170	1,100,000	
No. 2	2 × 6	895	1,030	1,120	1,000,000	
No. 3		525	600	655	900,000	NELMA
Stud		545	630	685	900,000	NSLB
Select Structural		1,210	1,390	1,510	1,100,000	WWPA
No. 1		865	990	1,080	1,100,000	
No. 2	2 × 8	830	950	1,035	1,000,000	
No. 3		485	555	605	900,000	
Select Structural		1,105	1,275	1,385	1,100,000	
No. 1		790	910	990	1,100,000	
No. 2	2 × 10	760	875	950	1,000,000	
No. 3		445	510	555	900,000	
Select Structural		1,005	1,155	1,260	1,100,000	
No. 1		720	825	900	1,100,000	
No. 2	2 × 12	690	795	865	1,000,000	
No. 3		405	465	505	900,000	
BEECH-BIRCH-HICKORY						
Select Structural		2,500	2,875	3,125	1,700,000	
No. 1		1,810	2,085	2,265	1,600,000	
No. 2		1,725	1,985	2,155	1,500,000	
No. 3	2 × 4	990	1,140	1,240	1,300,000	
Stud		980	1,125	1,225	1,300,000	
Construction		1,325	1,520	1,655	1,400,000	
Standard		750	860	935	1,300,000	
Utility		345	395	430	1,200,000	
Select Structural		2,170	2,495	2,710	1,700,000	
No. 1		1,570	1,805	1,960	1,600,000	
No. 2	2 × 6	1,495	1,720	1,870	1,500,000	
No. 3		860	990	1,075	1,300,000	
Stud		890	1,025	1,115	1,300,000	NELMA
Select Structural		2,000	2,300	2,500	1,700,000	
No. 1		1,450	1,665	1,810	1,600,000	
No. 2	2 × 8	1,380	1,585	1,725	1,500,000	
No. 3		795	915	990	1,300,000	
Select Structural		1,835	2,110	2,295	1,700,000	
No. 1		1,330	1,525	1,660	1,600,000	
No. 2	2 × 10	1,265	1,455	1,580	1,500,000	
No. 3		725	835	910	1,300,000	
Select Structural		1,670	1,920	2,085	1,700,000	
No. 1		1,210	1,390	1,510	1,600,000	
No. 2	2 × 12	1,150	1,325	1,440	1,500,000	
No. 3		660	760	825	1,300,000	

(Continued)

TABLE 23-IV-V-1—VALUES FOR JOISTS AND RAFTERS—VISUALLY GRADED LUMBER—(Continued)

SPECIES AND GRADE	SIZE (inches) × 25.4 for mm	DESIGN VALUE IN BENDING "F_b" psi			MODULUS OF ELASTICITY "E" psi	GRADING RULES AGENCY
		Normal Duration	Snow Loading	7-day Loading		
		× 0.00689 for N/mm²				
COTTONWOOD						
Select Structural	2 × 4	1,510	1,735	1,885	1,200,000	
No. 1		1,080	1,240	1,350	1,200,000	
No. 2		1,080	1,240	1,350	1,100,000	
No. 3		605	695	755	1,000,000	
Stud		600	690	750	1,000,000	
Construction		805	925	1,005	1,000,000	
Standard		460	530	575	900,000	
Utility		200	230	250	900,000	
Select Structural	2 × 6	1,310	1,505	1,635	1,200,000	
No. 1		935	1,075	1,170	1,200,000	
No. 2		935	1,075	1,170	1,100,000	
No. 3		525	600	655	1,000,000	
Stud		545	630	685	1,000,000	NSLB
Select Structural	2 × 8	1,210	1,390	1,510	1,200,000	
No. 1		865	990	1,080	1,200,000	
No. 2		865	990	1,080	1,100,000	
No. 3		485	555	605	1,000,000	
Select Structural	2 × 10	1,105	1,275	1,385	1,200,000	
No. 1		790	910	990	1,200,000	
No. 2		790	910	990	1,100,000	
No. 3		445	510	555	1,000,000	
Select Structural	2 × 12	1,005	1,155	1,260	1,200,000	
No. 1		720	825	900	1,200,000	
No. 2		720	825	900	1,100,000	
No. 3		405	465	505	1,000,000	
DOUGLAS FIR-LARCH						
Select Structural	2 × 4	2,500	2,875	3,125	1,900,000	
No. 1 and better		1,985	2,280	2,480	1,800,000	
No. 1		1,725	1,985	2,155	1,700,000	
No. 2		1,510	1,735	1,885	1,600,000	
No. 3		865	990	1,080	1,400,000	
Stud		855	980	1,065	1,400,000	
Construction		1,150	1,325	1,440	1,500,000	
Standard		635	725	790	1,400,000	
Utility		315	365	395	1,300,000	
Select Structural	2 × 6	2,170	2,495	2,710	1,900,000	
No. 1 and better		1,720	1,975	2,150	1,800,000	
No. 1		1,495	1,720	1,870	1,700,000	
No. 2		1,310	1,505	1,635	1,600,000	
No. 3		750	860	935	1,400,000	
Stud		775	895	970	1,400,000	WCLIB WWPA
Select Structural	2 × 8	2,000	2,300	2,500	1,900,000	
No. 1 and better		1,585	1,825	1,985	1,800,000	
No. 1		1,380	1,585	1,725	1,700,000	
No. 2		1,210	1,390	1,510	1,600,000	
No. 3		690	795	865	1,400,000	
Select Structural	2 × 10	1,835	2,110	2,295	1,900,000	
No. 1 and better		1,455	1,675	1,820	1,800,000	
No. 1		1,265	1,455	1,580	1,700,000	
No. 2		1,105	1,275	1,385	1,600,000	
No. 3		635	725	790	1,400,000	
Select Structural	2 × 12	1,670	1,920	2,085	1,900,000	
No. 1 and better		1,325	1,520	1,655	1,800,000	
No. 1		1,150	1,325	1,440	1,700,000	
No. 2		1,005	1,155	1,260	1,600,000	
No. 3		575	660	720	1,400,000	

(Continued)

TABLE 23-IV-V-1—VALUES FOR JOISTS AND RAFTERS—VISUALLY GRADED LUMBER—(Continued)

SPECIES AND GRADE	SIZE (inches) × 25.4 for mm	Normal Duration	Snow Loading	7-day Loading	MODULUS OF ELASTICITY "E" psi	GRADING RULES AGENCY
DOUGLAS FIR-LARCH (North)			× 0.00689 for N/mm²			
Select Structural	2 × 4	2,245	2,580	2,805	1,900,000	NLGA
No. 1/No. 2		1,425	1,635	1,780	1,600,000	
No. 3		820	940	1,025	1,400,000	
Stud		820	945	1,030	1,400,000	
Construction		1,095	1,255	1,365	1,500,000	
Standard		605	695	755	1,400,000	
Utility		290	330	360	1,300,000	
Select Structural	2 × 6	1,945	2,235	2,430	1,900,000	
No. 1/No. 2		1,235	1,420	1,540	1,600,000	
No. 3		710	815	890	1,400,000	
Stud		750	860	935	1,400,000	
Select Structural	2 × 8	1,795	2,065	2,245	1,900,000	
No. 1/No. 2		1,140	1,310	1,425	1,600,000	
No. 3		655	755	820	1,400,000	
Select Structural	2 × 10	1,645	1,890	2,055	1,900,000	
No. 1/No. 2		1,045	1,200	1,305	1,600,000	
No. 3		600	690	750	1,400,000	
Select Structural	2 × 12	1,495	1,720	1,870	1,900,000	
No. 1/No. 2		950	1,090	1,185	1,600,000	
No. 3		545	630	685	1,400,000	
DOUGLAS FIR (South)						WWPA
Select Structural	2 × 4	2,245	2,580	2,805	1,400,000	
No. 1		1,555	1,785	1,940	1,300,000	
No. 2		1,425	1,635	1,780	1,200,000	
No. 3		820	940	1,025	1,100,000	
Stud		820	945	1,030	1,100,000	
Construction		1,065	1,225	1,330	1,200,000	
Standard		605	695	755	1,100,000	
Utility		290	330	360	1,000,000	
Select Structural	2 × 6	1,945	2,235	2,430	1,400,000	
No. 1		1,345	1,545	1,680	1,300,000	
No. 2		1,235	1,420	1,540	1,200,000	
No. 3		710	815	890	1,100,000	
Stud		750	860	935	1,100,000	
Select Structural	2 × 8	1,795	2,065	2,245	1,400,000	
No. 1		1,240	1,430	1,555	1,300,000	
No. 2		1,140	1,310	1,425	1,200,000	
No. 3		655	755	820	1,100,000	
Select Structural	2 × 10	1,645	1,890	2,055	1,400,000	
No. 1		1,140	1,310	1,425	1,300,000	
No. 2		1,045	1,200	1,305	1,200,000	
No. 3		600	690	750	1,100,000	
Select Structural	2 × 12	1,495	1,720	1,870	1,400,000	
No. 1		1,035	1,190	1,295	1,300,000	
No. 2		950	1,090	1,185	1,200,000	
No. 3		545	630	685	1,100,000	

(Continued)

TABLE 23-IV-V-1—VALUES FOR JOISTS AND RAFTERS—VISUALLY GRADED LUMBER—(Continued)

SPECIES AND GRADE	SIZE (inches) × 25.4 for mm	DESIGN VALUE IN BENDING "F_b" psi			MODULUS OF ELASTICITY "E" psi	GRADING RULES AGENCY
		Normal Duration	Snow Loading	7-day Loading		
		× 0.00689 for N/mm²				
EASTERN HEMLOCK—TAMARACK						
Select Structural	2 × 4	2,155	2,480	2,695	1,200,000	
No. 1		1,335	1,535	1,670	1,100,000	
No. 2		990	1,140	1,240	1,100,000	
No. 3		605	695	755	900,000	
Stud		570	655	710	900,000	
Construction		775	895	970	1,000,000	
Standard		430	495	540	900,000	
Utility		200	230	250	800,000	
Select Structural	2 × 6	1,870	2,150	2,335	1,200,000	
No. 1		1,160	1,330	1,450	1,100,000	
No. 2		860	990	1,075	1,100,000	
No. 3		525	600	655	900,000	
Stud		520	595	645	900,000	NELMA NSLB
Select Structural	2 × 8	1,725	1,985	2,155	1,200,000	
No. 1		1,070	1,230	1,335	1,100,000	
No. 2		795	915	990	1,100,000	
No. 3		485	555	605	900,000	
Select Structural	2 × 10	1,580	1,820	1,975	1,200,000	
No. 1		980	1,125	1,225	1,100,000	
No. 2		725	835	910	1,100,000	
No. 3		445	510	555	900,000	
Select Structural	2 × 12	1,440	1,655	1,795	1,200,000	
No. 1		890	1,025	1,115	1,100,000	
No. 2		660	760	825	1,100,000	
No. 3		405	465	505	900,000	
EASTERN SOFTWOODS						
Select Structural	2 × 4	2,155	2,480	2,695	1,200,000	
No. 1		1,335	1,535	1,670	1,100,000	
No. 2		990	1,140	1,240	1,100,000	
No. 3		605	695	755	900,000	
Stud		570	655	710	900,000	
Construction		775	895	970	1,000,000	
Standard		430	495	540	900,000	
Utility		200	230	250	800,000	
Select Structural	2 × 6	1,870	2,150	2,335	1,200,000	
No. 1		1,160	1,330	1,450	1,100,000	
No. 2		860	990	1,075	1,100,000	
No. 3		525	600	655	900,000	
Stud		520	595	645	900,000	NELMA NSLB
Select Structural	2 × 8	1,725	1,985	2,155	1,200,000	
No. 1		1,070	1,230	1,335	1,100,000	
No. 2		795	915	990	1,100,000	
No. 3		485	555	605	900,000	
Select Structural	2 × 10	1,580	1,820	1,975	1,200,000	
No. 1		980	1,125	1,225	1,100,000	
No. 2		725	835	910	1,100,000	
No. 3		445	510	555	900,000	
Select Structural	2 × 12	1,440	1,655	1,795	1,200,000	
No. 1		890	1,025	1,115	1,100,000	
No. 2		660	760	825	1,100,000	
No. 3		405	465	505	900,000	

(Continued)

TABLE 23-IV-V-1—VALUES FOR JOISTS AND RAFTERS—VISUALLY GRADED LUMBER—(Continued)

SPECIES AND GRADE	SIZE (inches) × 25.4 for mm	DESIGN VALUE IN BENDING "F_b" psi			MODULUS OF ELASTICITY "E" psi	GRADING RULES AGENCY
		Normal Duration	Snow Loading	7-day Loading		
		× 0.00689 for N/mm²				
EASTERN WHITE PINE						
Select Structural	2 × 4	2,155	2,480	2,695	1,200,000	
No. 1		1,335	1,535	1,670	1,100,000	
No. 2		990	1,140	1,240	1,100,000	
No. 3		605	695	755	900,000	
Stud		570	655	710	900,000	
Construction		775	895	970	1,000,000	
Standard		430	495	540	900,000	
Utility		200	230	250	800,000	
Select Structural	2 × 6	1,870	2,150	2,335	1,200,000	
No. 1		1,160	1,330	1,450	1,100,000	
No. 2		860	990	1,075	1,100,000	
No. 3		525	600	655	900,000	
Stud		520	595	645	900,000	NELMA NSLB
Select Structural	2 × 8	1,725	1,985	2,155	1,200,000	
No. 1		1,070	1,230	1,335	1,100,000	
No. 2		795	915	990	1,100,000	
No. 3		485	555	605	900,000	
Select Structural	2 × 10	1,580	1,820	1,975	1,200,000	
No. 1		980	1,125	1,225	1,100,000	
No. 2		725	835	910	1,100,000	
No. 3		445	510	555	900,000	
Select Structural	2 × 12	1,440	1,655	1,795	1,200,000	
No. 1		890	1,025	1,115	1,100,000	
No. 2		660	760	825	1,100,000	
No. 3		405	465	505	900,000	
HEM-FIR						
Select Structural	2 × 4	2,415	2,775	3,020	1,600,000	
No. 1 and better		1,810	2,085	2,265	1,500,000	
No. 1		1,640	1,885	2,050	1,500,000	
No. 2		1,465	1,685	1,835	1,300,000	
No. 3		865	990	1,080	1,200,000	
Stud		855	980	1,065	1,200,000	
Construction		1,120	1,290	1,400	1,300,000	
Standard		635	725	790	1,200,000	
Utility		290	330	360	1,100,000	
Select Structural	2 × 6	2,095	2,405	2,615	1,600,000	
No. 1 and better		1,570	1,805	1,960	1,500,000	
No. 1		1,420	1,635	1,775	1,500,000	
No. 2		1,270	1,460	1,590	1,300,000	
No. 3		750	860	935	1,200,000	
Stud		775	895	970	1,200,000	
Select Structural	2 × 8	1,930	2,220	2,415	1,600,000	WCLIB WWPA
No. 1 and better		1,450	1,665	1,810	1,500,000	
No. 1		1,310	1,510	1,640	1,500,000	
No. 2		1,175	1,350	1,465	1,300,000	
No. 3		690	795	865	1,200,000	
Select Structural	2 × 10	1,770	2,035	2,215	1,600,000	
No. 1 and better		1,330	1,525	1,660	1,500,000	
No. 1		1,200	1,380	1,500	1,500,000	
No. 2		1,075	1,235	1,345	1,300,000	
No. 3		635	725	790	1,200,000	
Select Structural	2 × 12	1,610	1,850	2,015	1,600,000	
No. 1 and better		1,210	1,390	1,510	1,500,000	
No. 1		1,095	1,255	1,365	1,500,000	
No. 2		980	1,125	1,220	1,300,000	
No. 3		575	660	720	1,200,000	

(Continued)

TABLE 23-IV-V-1—VALUES FOR JOISTS AND RAFTERS—VISUALLY GRADED LUMBER—(Continued)

SPECIES AND GRADE	SIZE (inches)	DESIGN VALUE IN BENDING "F_b" psi			MODULUS OF ELASTICITY "E" psi	GRADING RULES AGENCY
		Normal Duration	Snow Loading	7-day Loading		
	× 25.4 for mm	× 0.00689 for N/mm²				
HEM-FIR (North)						
Select Structural		2,245	2,580	2,805	1,700,000	
No. 1/No. 2		1,725	1,985	2,155	1,600,000	
No. 3		990	1,140	1,240	1,400,000	
Stud	2 × 4	980	1,125	1,225	1,400,000	
Construction		1,325	1,520	1,655	1,500,000	
Standard		720	825	900	1,400,000	
Utility		345	395	430	1,300,000	
Select Structural		1,945	2,235	2,430	1,700,000	
No. 1/No. 2	2 × 6	1,495	1,720	1,870	1,600,000	
No. 3		860	990	1,075	1,400,000	
Stud		890	1,025	1,115	1,400,000	NLGA
Select Structural		1,795	2,065	2,245	1,700,000	
No. 1/No. 2	2 × 8	1,380	1,585	1,725	1,600,000	
No. 3		795	915	990	1,400,000	
Select Structural		1,645	1,890	2,055	1,700,000	
No. 1/No. 2	2 × 10	1,265	1,455	1,580	1,600,000	
No. 3		725	835	910	1,400,000	
Select Structural		1,495	1,720	1,870	1,700,000	
No. 1/No. 2	2 × 12	1,150	1,325	1,440	1,600,000	
No. 3		660	760	825	1,400,000	
MIXED MAPLE						
Select Structural		1,725	1,985	2,155	1,300,000	
No. 1		1,250	1,440	1,565	1,200,000	
No. 2		1,210	1,390	1,510	1,100,000	
No. 3		690	795	865	1,000,000	
Stud	2 × 4	695	800	870	1,000,000	
Construction		920	1,060	1,150	1,100,000	
Standard		520	595	645	1,000,000	
Utility		260	300	325	900,000	
Select Structural		1,495	1,720	1,870	1,300,000	
No. 1		1,085	1,245	1,355	1,200,000	
No. 2	2 × 6	1,045	1,205	1,310	1,100,000	
No. 3		600	690	750	1,000,000	
Stud		635	725	790	1,000,000	NELMA
Select Structural		1,380	1,585	1,725	1,300,000	
No. 1	2 × 8	1,000	1,150	1,250	1,200,000	
No. 2		965	1,110	1,210	1,100,000	
No. 3		550	635	690	1,000,000	
Select Structural		1,265	1,455	1,580	1,300,000	
No. 1	2 × 10	915	1,055	1,145	1,200,000	
No. 2		885	1,020	1,105	1,100,000	
No. 3		505	580	635	1,000,000	
Select Structural		1,150	1,325	1,440	1,300,000	
No. 1	2 × 12	835	960	1,040	1,200,000	
No. 2		805	925	1,005	1,100,000	
No. 3		460	530	575	1,000,000	
MIXED OAK						
Select Structural		1,985	2,280	2,480	1,100,000	
No. 1		1,425	1,635	1,780	1,000,000	
No. 2		1,380	1,585	1,725	900,000	
No. 3		820	940	1,025	800,000	
Stud	2 × 4	790	910	990	800,000	NELMA
Construction		1,065	1,225	1,330	900,000	
Standard		605	695	755	800,000	
Utility		290	330	360	800,000	

(Continued)

TABLE 23-IV-V-1—VALUES FOR JOISTS AND RAFTERS—VISUALLY GRADED LUMBER—(Continued)

SPECIES AND GRADE	SIZE (inches)	DESIGN VALUE IN BENDING "F_b" psi			MODULUS OF ELASTICITY "E" psi	GRADING RULES AGENCY
		Normal Duration	Snow Loading	7-day Loading		
	× 25.4 for mm	× 0.00689 for N/mm²				
MIXED OAK—(continued)						
Select Structural		1,720	1,975	2,150	1,100,000	
No. 1		1,235	1,420	1,540	1,000,000	
No. 2	2 × 6	1,195	1,375	1,495	900,000	
No. 3		710	815	890	800,000	
Stud		720	825	900	800,000	
Select Structural		1,585	1,825	1,985	1,100,000	
No. 1		1,140	1,310	1,425	1,000,000	
No. 2	2 × 8	1,105	1,270	1,380	900,000	
No. 3		655	755	820	800,000	NELMA
Select Structural		1,455	1,675	1,820	1,100,000	
No. 1		1,045	1,200	1,305	1,000,000	
No. 2	2 × 10	1,010	1,165	1,265	900,000	
No. 3		600	690	750	800,000	
Select Structural		1,325	1,520	1,655	1,100,000	
No. 1		950	1,090	1,185	1,000,000	
No. 2	2 × 12	920	1,060	1,150	900,000	
No. 3		545	630	685	800,000	
MIXED SOUTHERN PINE						
Select Structural		2,360	2,710	2,950	1,600,000	
No. 1		1,670	1,920	2,080	1,500,000	
No. 2		1,500	1,720	1,870	1,400,000	
No. 3		865	990	1,080	1,200,000	
Stud	2 × 4	890	1,020	1,110	1,200,000	
Construction		1,150	1,320	1,440	1,300,000	
Standard		635	725	790	1,200,000	
Utility		315	365	395	1,100,000	
Select Structural		2,130	2,450	2,660	1,600,000	
No. 1		1,490	1,720	1,870	1,500,000	
No. 2	2 × 6	1,320	1,520	1,650	1,400,000	
No. 3		775	895	970	1,200,000	
Stud		775	895	970	1,200,000	SPIB
Select Structural		2,010	2,310	2,520	1,600,000	
No. 1		1,380	1,590	1,720	1,500,000	
No. 2	2 × 8	1,210	1,390	1,510	1,400,000	
No. 3		720	825	900	1,200,000	
Select Structural		1,730	1,980	2,160	1,600,000	
No. 1		1,210	1,390	1,510	1,500,000	
No. 2	2 × 10	1,060	1,220	1,330	1,400,000	
No. 3		605	695	755	1,200,000	
Select Structural		1,610	1,850	2,010	1,600,000	
No. 1		1,120	1,290	1,400	1,500,000	
No. 2	2 × 12	1,010	1,160	1,260	1,400,000	
No. 3		575	660	720	1,200,000	

(Continued)

TABLE 23-IV-V-1—VALUES FOR JOISTS AND RAFTERS—VISUALLY GRADED LUMBER—(Continued)

SPECIES AND GRADE	SIZE (inches)	DESIGN VALUE IN BENDING "F_b" psi			MODULUS OF ELASTICITY "E" psi	GRADING RULES AGENCY
	× 25.4 for mm	Normal Duration	Snow Loading	7-day Loading		
		× 0.00689 for N/mm²				
NORTHERN RED OAK						
Select Structural	2 × 4	2,415	2,775	3,020	1,400,000	
No. 1		1,725	1,985	2,155	1,400,000	
No. 2		1,680	1,935	2,100	1,300,000	
No. 3		950	1,090	1,185	1,200,000	
Stud		950	1,090	1,185	1,200,000	
Construction		1,265	1,455	1,580	1,200,000	
Standard		720	825	900	1,100,000	
Utility		345	395	430	1,000,000	
Select Structural	2 × 6	2,095	2,405	2,615	1,400,000	
No. 1		1,495	1,720	1,870	1,400,000	
No. 2		1,460	1,675	1,820	1,300,000	
No. 3		820	945	1,030	1,200,000	
Stud		865	990	1,080	1,200,000	NELMA
Select Structural	2 × 8	1,930	2,220	2,415	1,400,000	
No. 1		1,380	1,585	1,725	1,400,000	
No. 2		1,345	1,545	1,680	1,300,000	
No. 3		760	875	950	1,200,000	
Select Structural	2 × 10	1,770	2,035	2,215	1,400,000	
No. 1		1,265	1,455	1,580	1,400,000	
No. 2		1,235	1,420	1,540	1,300,000	
No. 3		695	800	870	1,200,000	
Select Structural	2 × 12	1,610	1,850	2,015	1,400,000	
No. 1		1,150	1,325	1,440	1,400,000	
No. 2		1,120	1,290	1,400	1,300,000	
No. 3		635	725	790	1,200,000	
NORTHERN SPECIES						
Select Structural	2 × 4	1,640	1,885	2,050	1,100,000	
No. 1/No. 2		990	1,140	1,240	1,100,000	
No. 3		605	695	755	1,000,000	
Stud		570	655	710	1,000,000	
Construction		775	895	970	1,000,000	
Standard		430	495	540	900,000	
Utility		200	230	250	900,000	
Select Structural	2 × 6	1,420	1,635	1,775	1,100,000	
No. 1/No. 2		860	990	1,075	1,100,000	
No. 3		525	600	655	1,000,000	
Stud		520	595	645	1,000,000	
Select Structural	2 × 8	1,310	1,510	1,640	1,100,000	NLGA
No. 1/No. 2		795	915	990	1,100,000	
No. 3		485	555	605	1,000,000	
Select Structural	2 × 10	1,200	1,380	1,500	1,100,000	
No. 1/No. 2		725	835	910	1,100,000	
No. 3		445	510	555	1,000,000	
Select Structural	2 × 12	1,095	1,255	1,365	1,100,000	
No. 1/No. 2		660	760	825	1,100,000	
No. 3		405	465	505	1,000,000	

(Continued)

TABLE 23-IV-V-1—VALUES FOR JOISTS AND RAFTERS—VISUALLY GRADED LUMBER—(Continued)

SPECIES AND GRADE	SIZE (inches) × 25.4 for mm	DESIGN VALUE IN BENDING "F_b" psi			MODULUS OF ELASTICITY "E" psi	GRADING RULES AGENCY
		Normal Duration	Snow Loading	7-day Loading		
		× 0.00689 for N/mm²				
NORTHERN WHITE CEDAR						
Select Structural	2 × 4	1,335	1,535	1,670	800,000	
No. 1		990	1,140	1,240	700,000	
No. 2		950	1,090	1,185	700,000	
No. 3		560	645	700	600,000	
Stud		540	620	670	600,000	
Construction		720	825	900	700,000	
Standard		405	465	505	600,000	
Utility		200	230	250	600,000	
Select Structural	2 × 6	1,160	1,330	1,450	800,000	
No. 1		860	990	1,075	700,000	
No. 2		820	945	1,030	700,000	
No. 3		485	560	605	600,000	
Stud		490	560	610	600,000	NELMA
Select Structural	2 × 8	1,070	1,230	1,335	800,000	
No. 1		795	915	990	700,000	
No. 2		760	875	950	700,000	
No. 3		450	515	560	600,000	
Select Structural	2 × 10	980	1,125	1,225	800,000	
No. 1		725	835	910	700,000	
No. 2		695	800	870	700,000	
No. 3		410	475	515	600,000	
Select Structural	2 × 12	890	1,025	1,115	800,000	
No. 1		660	760	825	700,000	
No. 2		635	725	790	700,000	
No. 3		375	430	465	600,000	
RED MAPLE						
Select Structural	2 × 4	2,245	2,580	2,805	1,700,000	
No. 1		1,595	1,835	1,995	1,600,000	
No. 2		1,555	1,785	1,940	1,500,000	
No. 3		905	1,040	1,130	1,300,000	
Stud		885	1,020	1,105	1,300,000	
Construction		1,210	1,390	1,510	1,400,000	
Standard		660	760	825	1,300,000	
Utility		315	365	395	1,200,000	
Select Structural	2 × 6	1,945	2,235	2,430	1,700,000	
No. 1		1,385	1,590	1,730	1,600,000	
No. 2		1,345	1,545	1,680	1,500,000	
No. 3		785	905	980	1,300,000	
Stud		805	925	1,005	1,300,000	NELMA
Select Structural	2 × 8	1,795	2,065	2,245	1,700,000	
No. 1		1,275	1,470	1,595	1,600,000	
No. 2		1,240	1,430	1,555	1,500,000	
No. 3		725	835	905	1,300,000	
Select Structural	2 × 10	1,645	1,890	2,055	1,700,000	
No. 1		1,170	1,345	1,465	1,600,000	
No. 2		1,140	1,310	1,425	1,500,000	
No. 3		665	765	830	1,300,000	
Select Structural	2 × 12	1,495	1,720	1,870	1,700,000	
No. 1		1,065	1,225	1,330	1,600,000	
No. 2		1,035	1,190	1,295	1,500,000	
No. 3		605	695	755	1,300,000	

(Continued)

TABLE 23-IV-V-1—VALUES FOR JOISTS AND RAFTERS—VISUALLY GRADED LUMBER—(Continued)

SPECIES AND GRADE	SIZE (inches) × 25.4 for mm	DESIGN VALUE IN BENDING "F_b" psi			MODULUS OF ELASTICITY "E" psi	GRADING RULES AGENCY
		Normal Duration	Snow Loading	7-day Loading		
		× 0.00689 for N/mm²				
RED OAK						
Select Structural	2 × 4	1,985	2,280	2,480	1,400,000	
No. 1		1,425	1,635	1,780	1,300,000	
No. 2		1,380	1,585	1,725	1,200,000	
No. 3		820	940	1,025	1,100,000	
Stud		790	910	990	1,100,006	
Construction		1,065	1,225	1,330	1,200,000	
Standard		605	695	755	1,100,000	
Utility		290	330	360	1,000,000	
Select Structural	2 × 6	1,720	1,975	2,150	1,400,000	
No. 1		1,235	1,420	1,540	1,300,000	
No. 2		1,195	1,375	1,495	1,200,000	
No. 3		710	815	890	1,100,000	
Stud		720	825	900	1,100,000	NELMA
Select Structural	2 × 8	1,585	1,825	1,985	1,400,000	
No. 1		1,140	1,310	1,425	1,300,000	
No. 2		1,105	1,270	1,380	1,200,000	
No. 3		655	755	820	1,100,000	
Select Structural	2 × 10	1,455	1,675	1,820	1,400,000	
No. 1		1,045	1,200	1,305	1,300,000	
No. 2		1,010	1,165	1,265	1,200,000	
No. 3		600	690	750	1,100,000	
Select Structural	2 × 12	1,325	1,520	1,655	1,400,000	
No. 1		950	1,090	1,185	1,300,000	
No. 2		920	1,060	1,150	1,200,000	
No. 3		545	630	685	1,100,000	
REDWOOD						
Clear Structural	2 × 4	3,020	3,470	3,775	1,400,000	
Select Structural		2,330	2,680	2,910	1,400,000	
Select Structural, open grain		1,900	2,180	2,370	1,100,000	
No. 1		1,680	1,935	2,100	1,300,000	
No. 1, open grain		1,335	1,535	1,670	1,100,000	
No. 2		1,595	1,835	1,995	1,200,000	
No. 2, open grain		1,250	1,440	1,565	1,000,000	
No. 3		905	1,040	1,130	1,100,000	
No. 3, open grain		735	845	915	900,000	
Stud		725	835	910	900,000	
Construction		950	1,090	1,185	900,000	
Standard		520	595	645	900,000	RIS
Utility		260	300	325	800,000	
Clear Structural	2 × 6	2,615	3,010	3,270	1,400,000	
Select Structural		2,020	2,320	2,525	1,400,000	
Select Structural, open grain		1,645	1,890	2,055	1,100,000	
No. 1		1,460	1,675	1,820	1,300,000	
No. 1, open grain		1,160	1,330	1,450	1,100,000	
No. 2		1,385	1,590	1,730	1,200,000	
No. 2, open grain		1,085	1,245	1,355	1,000,000	
No. 3		785	905	980	1,100,000	
No. 3, open grain		635	730	795	900,000	
Stud		660	760	825	900,000	

(Continued)

TABLE 23-IV-V-1—VALUES FOR JOISTS AND RAFTERS—VISUALLY GRADED LUMBER—(Continued)

SPECIES AND GRADE	SIZE (inches) × 25.4 for mm	DESIGN VALUE IN BENDING "F_b" psi			MODULUS OF ELASTICITY "E" psi	GRADING RULES AGENCY
		Normal Duration	Snow Loading	7-day Loading		
		× 0.00689 for N/mm²				
REDWOOD—(continued)						
Clear Structural	2 × 8	2,415	2,775	3,020	1,400,000	
Select Structural		1,865	2,140	2,330	1,400,000	
Select Structural, open grain		1,520	1,745	1,900	1,100,000	
No. 1		1,345	1,545	1,680	1,300,000	
No. 1, open grain		1,070	1,230	1,335	1,100,000	
No. 2		1,275	1,470	1,595	1,200,000	
No. 2, open grain		1,000	1,150	1,250	1,000,000	
No. 3		725	835	905	1,100,000	
No. 3, open grain		585	675	735	900,000	
Clear Structural	2 × 10	2,215	2,545	2,765	1,400,000	
Select Structural		1,710	1,965	2,135	1,400,000	
Select Structural, open grain		1,390	1,600	1,740	1,100,000	
No. 1		1,235	1,420	1,540	1,300,000	RIS
No. 1, open grain		980	1,125	1,225	1,100,000	
No. 2		1,170	1,345	1,465	1,200,000	
No. 2, open grain		915	1,055	1,145	1,000,000	
No. 3		665	765	830	1,100,000	
No. 3, open grain		540	620	670	900,000	
Clear Structural	2 × 12	2,015	2,315	2,515	1,400,000	
Select Structural		1,555	1,785	1,940	1,400,000	
Select Structural, open grain		1,265	1,455	1,580	1,100,000	
No. 1		1,120	1,290	1,400	1,300,000	
No. 1, open grain		890	1,025	1,115	1,100,000	
No. 2		1,065	1,225	1,330	1,200,000	
No. 2, open grain		835	960	1,040	1,000,000	
No. 3		605	695	755	1,100,000	
No. 3, open grain		490	560	610	900,000	
SOUTHERN PINE						
Dense Select Structural	2 × 4	3,510	4,030	4,380	1,900,000	
Select Structural		3,280	3,770	4,100	1,800,000	
Non-Dense Select Structural		3,050	3,500	3,810	1,700,000	
No. 1 Dense		2,300	2,650	2,880	1,800,000	
No. 1		2,130	2,450	2,660	1,700,000	
No. 1 Non-Dense		1,950	2,250	2,440	1,600,000	
No. 2 Dense		1,960	2,250	2,440	1,700,000	
No. 2		1,720	1,980	2,160	1,600,000	SPIB
No. 2 Non-Dense		1,550	1,790	1,940	1,400,000	
No. 3		980	1,120	1,220	1,400,000	
Stud		1,010	1,160	1,260	1,400,000	
Construction		1,270	1,450	1,580	1,500,000	
Standard		720	825	900	1,300,000	
Utility		345	395	430	1,300,000	

(Continued)

TABLE 23-IV-V-1—VALUES FOR JOISTS AND RAFTERS—VISUALLY GRADED LUMBER—(Continued)

SPECIES AND GRADE	SIZE (inches)	DESIGN VALUE IN BENDING "F_b" psi			MODULUS OF ELASTICITY "E" psi	GRADING RULES AGENCY
		Normal Duration	Snow Loading	7-day Loading		
	× 25.4 for mm	× 0.00689 for N/mm²				
SOUTHERN PINE—(continued)						
Dense Select Structural		3,100	3,570	3,880	1,900,000	
Select Structural		2,930	3,370	3,670	1,800,000	
Non-Dense Select Structural		2,700	3,110	3,380	1,700,000	
No. 1 Dense		2,010	2,310	2,520	1,800,000	
No. 1		1,900	2,180	2,370	1,700,000	
No. 1 Non-Dense	2 × 6	1,720	1,980	2,160	1,600,000	
No. 2 Dense		1,670	1,920	2,080	1,700,000	
No. 2		1,440	1,650	1,800	1,600,000	
No. 2 Non-Dense		1,320	1,520	1,650	1,400,000	
No. 3		865	990	1,080	1,400,000	
Stud		890	1,020	1,110	1,400,000	
Dense Select Structural		2,820	3,240	3,520	1,900,000	
Select Structural		2,650	3,040	3,310	1,800,000	
Non-Dense Select Structural		2,420	2,780	3,020	1,700,000	
No. 1 Dense		1,900	2,180	2,370	1,800,000	
No. 1	2 × 8	1,730	1,980	2,160	1,700,000	
No. 1 Non-Dense		1,550	1,790	1,940	1,600,000	
No. 2 Dense		1,610	1,850	2,010	1,700,000	
No. 2		1,380	1,590	1,720	1,600,000	
No. 2 Non-Dense		1,260	1,450	1,580	1,400,000	
No. 3		805	925	1,010	1,400,000	
Dense Select Structural		2,470	2,840	3,090	1,900,000	SPIB
Select Structural		2,360	2,710	2,9500²	1,800,000	
Non-Dense Select Structural		2,130	2,450	2,660	1,700,000	
No. 1 Dense		1,670	1,920	2,080	1,800,000	
No. 1	2 × 10	1,500	1,720	1,870	1,700,000	
No. 1 Non-Dense		1,380	1,590	1,730	1,600,000	
No. 2 Dense		1,380	1,590	1,730	1,700,000	
No. 2		1,210	1,390	1,510	1,600,000	
No. 2 Non-Dense		1,090	1,260	1,370	1,400,000	
No. 3		690	795	865	1,400,000	
Dense Select Structural		2,360	2,710	2,950	1,900,000	
Select Structural		2,190	2,510	2,730	1,800,000	
Non-Dense Select Structural		2,010	2,310	2,520	1,700,000	
No. 1 Dense		1,550	1,790	1,940	1,800,000	
No. 1	2 × 12	1,440	1,650	1,800	1,700,000	
No. 1 Non-Dense		1,320	1,520	1,650	1,600,000	
No. 2 Dense		1,320	1,520	1,650	1,700,000	
No. 2		1,120	1,290	1,400	1,600,000	
No. 2 Non-Dense		1,040	1,190	1,290	1,400,000	
No. 3		660	760	825	1,400,000	

(Continued)

TABLE 23-IV-V-1—VALUES FOR JOISTS AND RAFTERS—VISUALLY GRADED LUMBER—(Continued)

SPECIES AND GRADE	SIZE (inches) × 25.4 for mm	DESIGN VALUE IN BENDING "F_b" psi			MODULUS OF ELASTICITY "E" psi	GRADING RULES AGENCY
		Normal Duration	Snow Loading	7-day Loading		
		× 0.00689 for N/mm²				
SPRUCE-PINE-FIR						
Select Structural	2 × 4	2,155	2,480	2,695	1,500,000	
No. 1/No. 2		1,510	1,735	1,885	1,400,000	
No. 3		865	990	1,080	1,200,000	
Stud		855	980	1,065	1,200,000	
Construction		1,120	1,290	1,400	1,300,000	
Standard		635	725	790	1,200,000	
Utility		290	330	360	1,100,000	
Select Structural	2 × 6	1,870	2,150	2,335	1,500,000	
No. 1/No. 2		1,310	1,505	1,635	1,400,000	
No. 3		750	860	935	1,200,000	NLGA
Stud		775	895	970	1,200,000	
Select Structural	2 × 8	1,725	1,985	2,155	1,500,000	
No. 1/No. 2		1,210	1,390	1,510	1,400,000	
No. 3		690	795	865	1,200,000	
Select Structural	2 × 10	1,580	1,820	1,975	1,500,000	
No. 1/No. 2		1,105	1,275	1,385	1,400,000	
No. 3		635	725	790	1,200,000	
Select Structural	2 × 12	1,440	1,655	1,795	1,500,000	
No. 1/No. 2		1,005	1,155	1,260	1,400,000	
No. 3		575	660	720	1,200,000	
SPRUCE-PINE-FIR (South)						
Select Structural	2 × 4	2,245	2,580	2,805	1,300,000	
No. 1		1,465	1,685	1,835	1,200,000	
No. 2		1,295	1,490	1,615	1,100,000	
No. 3		735	845	915	1,000,000	
Stud		725	835	910	1,000,000	
Construction		980	1,125	1,220	1,000,000	
Standard		545	630	685	900,000	
Utility		260	300	325	900,000	
Select Structural	2 × 6	1,945	2,235	2,430	1,300,000	
No. 1		1,270	1,460	1,590	1,200,000	
No. 2		1,120	1,290	1,400	1,100,000	
No. 3		635	730	795	1,000,000	NELMA NSLB WCLIB WWPA
Stud		660	760	825	1,000,000	
Select Structural	2 × 8	1,795	2,065	2,245	1,300,000	
No. 1		1,175	1,350	1,465	1,200,000	
No. 2		1,035	1,190	1,295	1,100,000	
No. 3		585	675	735	1,000,000	
Select Structural	2 × 10	1,645	1,890	2,055	1,300,000	
No. 1		1,075	1,235	1,345	1,200,000	
No. 2		950	1,090	1,185	1,100,000	
No. 3		540	620	670	1,000,000	
Select Structural	2 × 12	1,495	1,720	1,870	1,300,000	
No. 1		980	1,125	1,220	1,200,000	
No. 2		865	990	1,080	1,100,000	
No. 3		490	560	610	1,000,000	
WESTERN CEDARS						
Select Structural	2 × 4	1,725	1,985	2,155	1,100,000	
No. 1		1,250	1,440	1,565	1,000,000	
No. 2		1,210	1,390	1,510	1,000,000	
No. 3		690	795	865	900,000	WCLIB WWPA
Stud		695	800	870	900,000	
Construction		920	1,060	1,150	900,000	
Standard		520	595	645	800,000	
Utility		260	300	325	800,000	

(Continued)

TABLE 23-IV-V-1—VALUES FOR JOISTS AND RAFTERS—VISUALLY GRADED LUMBER—(Continued)

SPECIES AND GRADE	SIZE (inches) × 25.4 for mm	DESIGN VALUE IN BENDING "F_b" psi			MODULUS OF ELASTICITY "E" psi	GRADING RULES AGENCY
		Normal Duration	Snow Loading	7-day Loading		
			× 0.00689 for N/mm²			
WESTERN CEDARS—(continued)						
Select Structural	2 × 6	1,495	1,720	1,870	1,100,000	
No. 1		1,085	1,245	1,355	1,000,000	
No. 2		1,045	1,205	1,310	1,000,000	
No. 3		600	690	750	900,000	
Stud		635	725	790	900,000	
Select Structural	2 × 8	1,380	1,585	1,725	1,100,000	
No. 1		1,000	1,150	1,250	1,000,000	
No. 2		965	1,110	1,210	1,000,000	WCLIB WWPA
No. 3		550	635	690	900,000	
Select Structural	2 × 10	1,265	1,455	1,580	1,100,000	
No. 1		915	1,055	1,145	1,000,000	
No. 2		885	1,020	1,105	1,000,000	
No. 3		505	580	635	900,000	
Select Structural	2 × 12	1,150	1,325	1,440	1,100,000	
No. 1		835	960	1,040	1,000,000	
No. 2		805	925	1,005	1,000,000	
No. 3		460	530	575	900,000	
WESTERN WOODS						
Select Structural	2 × 4	1,510	1,735	1,885	1,200,000	
No. 1		1,120	1,290	1,400	1,100,000	
No. 2		1,120	1,290	1,400	1,000,000	
No. 3		645	745	810	900,000	
Stud		635	725	790	900,000	
Construction		835	960	1,040	1,000,000	
Standard		460	530	575	900,000	
Utility		230	265	290	800,000	
Select Structural	2 × 6	1,310	1,505	1,635	1,200,000	
No. 1		970	1,120	1,215	1,100,000	
No. 2		970	1,120	1,215	1,000,000	
No. 3		560	645	700	900,000	
Stud		575	660	720	900,000	WCLIB WWPA
Select Structural	2 × 8	1,210	1,390	1,510	1,200,000	
No. 1		895	1,030	1,120	1,100,000	
No. 2		895	1,030	1,120	1,000,000	
No. 3		520	595	645	900,000	
Select Structural	2 × 10	1,105	1,275	1,385	1,200,000	
No. 1		820	945	1,030	1,100,000	
No. 2		820	945	1,030	1,000,000	
No. 3		475	545	595	900,000	
Select Structural	2 × 12	1,005	1,155	1,260	1,200,000	
No. 1		750	860	935	1,100,000	
No. 2		750	860	935	1,000,000	
No. 3		430	495	540	900,000	
WHITE OAK						
Select Structural	2 × 4	2,070	2,380	2,590	1,100,000	
No. 1		1,510	1,735	1,885	1,000,000	
No. 2		1,465	1,685	1,835	900,000	
No. 3		820	940	1,025	800,000	
Stud		820	945	1,030	800,000	NELMA
Construction		1,095	1,255	1,365	900,000	
Standard		605	695	755	800,000	
Utility		290	330	360	800,000	

(Continued)

TABLE 23-IV-V-1—VALUES FOR JOISTS AND RAFTERS—VISUALLY GRADED LUMBER—(Continued)

SPECIES AND GRADE	SIZE (inches) × 25.4 for mm	DESIGN VALUE IN BENDING "F_b" psi			MODULUS OF ELASTICITY "E" psi	GRADING RULES AGENCY
		Normal Duration	Snow Loading	7-day Loading		
		× 0.00689 for N/mm²				
WHITE OAK—(continued)						
Select Structural		1,795	2,065	2,245	1,100,000	
No. 1		1,310	1,505	1,635	1,000,000	
No. 2	2 × 6	1,270	1,460	1,590	900,000	
No. 3		710	815	890	800,000	
Stud		750	860	935	800,000	
Select Structural		1,655	1,905	2,070	1,100,000	
No. 1	2 × 8	1,210	1,390	1,510	1,000,000	
No. 2		1,175	1,350	1,465	900,000	NELMA
No. 3		655	755	820	800,000	
Select Structural		1,520	1,745	1,900	1,100,000	
No. 1	2 × 10	1,105	1,275	1,385	1,000,000	
No. 2		1,075	1,235	1,345	900,000	
No. 3		600	690	750	800,000	
Select Structural		1,380	1,585	1,725	1,100,000	
No. 1	2 × 12	1,005	1,155	1,260	1,000,000	
No. 2		980	1,125	1,220	900,000	
No. 3		545	630	685	800,000	
YELLOW POPLAR						
Select Structural		1,725	1,985	2,155	1,500,000	
No. 1		1,250	1,440	1,565	1,400,000	
No. 2		1,210	1,390	1,510	1,300,000	
No. 3		690	795	865	1,200,000	
Stud	2 × 4	695	800	870	1,200,000	
Construction		920	1,060	1,150	1,300,000	
Standard		520	595	645	1,100,000	
Utility		230	265	290	1,100,000	
Select Structural		1,495	1,720	1,870	1,500,000	
No. 1		1,085	1,245	1,355	1,400,000	
No. 2	2 × 6	1,045	1,205	1,310	1,300,000	
No. 3		600	690	750	1,200,000	
Stud		635	725	790	1,200,000	NSLB
Select Structural		1,380	1,585	1,725	1,500,000	
No. 1	2 × 8	1,000	1,150	1,250	1,400,000	
No. 2		965	1,110	1,210	1,300,000	
No. 3		550	635	690	1,200,000	
Select Structural		1,265	1,455	1,580	1,500,000	
No. 1	2 × 10	915	1,055	1,145	1,400,000	
No. 2		885	1,020	1,105	1,300,000	
No. 3		505	580	635	1,200,000	
Select Structural		1,150	1,325	1,440	1,500,000	
No. 1	2 × 12	835	960	1,040	1,400,000	
No. 2		805	925	1,005	1,300,000	
No. 3		460	530	575	1,200,000	

TABLE 23-IV-V-2—VALUES FOR JOISTS AND RAFTERS—MECHANICALLY GRADED LUMBER
For use in Tables 23-V-J-1 through 23-V-R-12 and Division V only.

GRADE DESIGNATION	SIZE (inches) × 25.4 for mm	DESIGN VALUE IN BENDING "F_b" psi			MODULUS OF ELASTICITY "E" psi	GRADING RULES AGENCIES
		Normal Duration	Snow Loading	7-day Loading		
		× 0.00689 for N/mm²				
MACHINE STRESS RATED (MSR) LUMBER						
900f-1.0E	2 × 4 and wider	1,040	1,190	1,290	1,000,000	WCLIB,WWPA
1200f-1.2E		1,380	1,590	1,730	1,200,000	NLGA,SPIB,WCLIB,WWPA
1350f-1.3E		1,550	1,790	1,940	1,300,000	SPIB,WCLIB,WWPA
1450f-1.3E		1,670	1,920	2,080	1,300,000	NLGA,WCLIB,WWPA
1500f-1.3E		1,730	1,980	2,160	1,300,000	SPIB
1500f-1.4E		1,730	1,980	2,160	1,400,000	NLGA,SPIB,WCLIB,WWPA
1650f-1.4E		1,900	2,180	2,370	1,400,000	SPIB
1650f-1.5E		1,900	2,180	2,370	1,500,000	NLGA,SPIB,WCLIB,WWPA
1800f-1.6E		2,070	2,380	2,590	1,600,000	NLGA,SPIB,WCLIB,WWPA
1950f-1.5E		2,240	2,580	2,800	1,500,000	SPIB
1950f-1.7E		2,240	2,580	2,800	1,700,000	NLGA,SPIB,WWPA
2100f-1.8E		2,420	2,780	3,020	1,800,000	NLGA,SPIB,WCLIB,WWPA
2250f-1.6E		2,590	2,980	3,230	1,600,000	SPIB
2250f-1.9E		2,590	2,980	3,230	1,900,000	NLGA,SPIB,WWPA
2400f-1.7E		2,760	3,170	3,450	1,700,000	SPIB
2400f-2.0E		2,760	3,170	3,450	2,000,000	NLGA,SPIB,WCLIB,WWPA
2550f-2.1E		2,930	3,370	3,670	2,100,000	NLGA,SPIB,WWPA
2700f-2.2E		3,110	3,570	3,880	2,200,000	NLGA,SPIB,WCLIB,WWPA
2850f-2.3E		3,280	3,770	4,100	2,300,000	SPIB,WWPA
3000f-2.4E		3,450	3,970	4,310	2,400,000	NLGA,SPIB
3150f-2.5E		3,620	4,170	4,530	2,500,000	SPIB
3300f-2.6E		3,800	4,360	4,740	2,600,000	SPIB
900f-1.2E	2 × 6 and wider	1,040	1,190	1,290	1,200,000	NLGA,WCLIB
1200f-1.5E		1,380	1,590	1,730	1,500,000	NLGA,WCLIB
1350f-1.8E		1,550	1,790	1,940	1,800,000	NLGA
1500f-1.8E		1,730	1,980	2,160	1,800,000	WCLIB
1800f-2.1E		2,070	2,380	2,590	2,100,000	NLGA,WCLIB
MACHINE EVALUATED LUMBER (MEL)						
M-10	2 × 4 and wider	1,610	1,850	2,010	1,200,000	SPIB
M-11		1,780	2,050	2,230	1,500,000	
M-12		1,840	2,120	2,300	1,600,000	
M-13		1,840	2,120	2,300	1,400,000	
M-14		2,070	2,380	2,590	1,700,000	
M-15		2,070	2,380	2,590	1,500,000	
M-16		2,070	2,380	2,590	1,500,000	
M-17		2,240	2,580	2,800	1,700,000	
M-18		2,300	2,650	2,880	1,800,000	
M-19		2,300	2,650	2,880	1,600,000	
M-20		2,300	2,650	2,880	1,900,000	
M-21		2,650	3,040	3,310	1,900,000	
M-22		2,700	3,110	3,380	1,700,000	
M-23		2,760	3,170	3,450	1,800,000	
M-24		3,110	3,570	3,880	1,900,000	
M-25		3,160	3,640	3,950	2,200,000	
M-26	2 × 4 and wider	3,220	3,700	4,030	2,000,000	SPIB
M-27		3,450	3,970	4,310	2,100,000	

The note to Table 23-IV-V-1 applies also to mechanically graded lumber.

Chapter 24
GLASS AND GLAZING

SECTION 2401 — SCOPE

2401.1 General. The provisions of this chapter apply to:

1. Exterior glass and glazing in all occupancies.

   EXCEPTION: Groups R and U Occupancies not over three stories in height and located in areas with a minimum basic wind speed less than 80 miles per hour (129 km/h).

2. Interior and exterior glass and glazing in all occupancies subject to human impact as specified in Section 2406 and hinged shower doors in all occupancies as specified in Section 2407.

3. Interior glass and glazing shall comply with Section 2404.1.

   EXCEPTION: Groups R and U Occupancies.

4. Skylights and sloped glazing.

2401.2 Standards. Standards for material shall be as specified in this chapter and UBC Standard 24-1.

Standards for glazing subject to human impact (hazardous location) as specified in Section 2406 shall be as specified in UBC Standard 24-2.

2401.3 Other Provisions. See Chapter 6 of this code for additional glass requirements where openings are required to be fire protected, and Section 2603.4 for openings glazed with plastics.

2401.4 Standards of Quality. The standards listed below labeled a "UBC Standard" are also listed in Chapter 35, Part II, and are part of this code.

1. UBC Standard 24-1, Flat Glass

2. UBC Standard 24-2, Safety Glazing

SECTION 2402 — IDENTIFICATION

Each light shall bear the manufacturer's label designating the type and thickness of glass. When approved by the building official, labels may be omitted, provided an affidavit is furnished by the glazing contractor certifying that each light is glazed in accordance with approved plans and specifications. Identification of glazing in hazardous locations shall be in accordance with Section 2406.

SECTION 2403 — AREA LIMITATIONS

Glass in windows, curtain and window walls, skylights, doors, and other exterior applications shall be chosen to withstand the loads for cladding as set forth in Chapter 16, Division III.

The area of individual lights shall not be more than as set forth in Graph 24-1, as adjusted by Table 24-A. Glass sizing for skylight applications shall be adjusted per Section 2409.5.

Graph 24-1 is applicable to rectangular glass firmly supported on all four edges.

When approved by the building official, alternate means for selecting glass may be used in place of Graph 24-1 and Table 24-A.

Glass and glazing subject to ice or snow loads shall be designed in accordance with Chapter 16.

SECTION 2404 — GLAZING SUPPORT AND FRAMING

2404.1 Support. Glass shall be firmly supported on all four edges.

   EXCEPTION: The building official may allow the use of glass that is not firmly supported on all four edges when justified by an approved design.

2404.2 Framing. The framing members for each individual glass pane shall be designed so the deflection perpendicular to the glass plane shall not exceed $1/175$ of the glass edge length or $3/4$ inch (19 mm), whichever is less, when subjected to the larger of the positive or negative load when loads are combined as specified in Section 1612.3.

SECTION 2405 — LOUVERED WINDOWS AND JALOUSIES

Regular float, wired and patterned glass in jalousies and louvered windows shall be no thinner than nominal $3/16$ inch (4.76 mm) and no longer than 48 inches (1219 mm). Exposed glass edges shall be smooth.

Wired glass with wire exposed on longitudinal edges shall not be used in jalousies or louvered windows.

SECTION 2406 — SAFETY GLAZING

2406.1 General. Glazing subject to human impact shall comply with this section.

2406.2 Identification. Each light of safety glazing material installed in hazardous locations as defined in Section 2406.4 shall be identified by a permanent label that specifies the labeler, whether the manufacturer or installer, and state that safety glazing material has been utilized in such installation. For additional identification requirements and for limitation on size and use by category classification, see UBC Standard 24-2, Part I.

Each unit of tempered glass shall be permanently identified by the manufacturer. The identification shall be etched or ceramic fired on the glass and be visible when the unit is glazed. Tempered spandrel glass is exempted from permanent labeling but such glass shall be identified by the manufacturer with a removable paper label.

2406.3 Human Impact Loads. Individual glazed areas in hazardous locations such as those indicated in Section 2406.4, including glazing used in fire assemblies in accordance with Section 713, shall pass the test requirements of UBC Standard 24-2, Part I.

   EXCEPTIONS: 1. Louvered windows and jalousies complying with Section 2405 need not comply with Section 2406.3.

   2. Polished wired glass complying with UBC Standard 24-2, Part II, may be used in smoke and draft control and fire assemblies.

Plastic glazing used in exterior applications also shall comply with the weathering requirements in UBC Standard 24-2, Part II.

2406.4 Hazardous Locations. The following shall be considered specific hazardous locations for the purposes of glazing:

1. Glazing in ingress and egress doors except jalousies.

2. Glazing in fixed and sliding panels of sliding door assemblies and panels in swinging doors other than wardrobe doors.

3. Glazing in storm doors.

4. Glazing in all unframed swinging doors.

5. Glazing in doors and enclosures for hot tubs, whirlpools, saunas, steam rooms, bathtubs and showers. Glazing in any portion of a building wall enclosing these compartments where the bottom exposed edge of the glazing is less than 60 inches (1525 mm) above a standing surface and drain inlet.

6. Glazing in fixed or operable panels adjacent to a door where the nearest exposed edge of the glazing is within a 24-inch (610 mm) arc of either vertical edge of the door in a closed position and where the bottom exposed edge of the glazing is less than 60 inches (1525 mm) above the walking surface.

7. Glazing in an individual fixed or operable panel, other than those locations described in Items 5 and 6, that meets all of the following conditions:

7.1 Exposed area of an individual pane greater than 9 square feet (0.84 m^2).

7.2 Exposed bottom edge less than 18 inches (457 mm) above the floor.

7.3 Exposed top edge greater than 36 inches (914 mm) above the floor.

7.4 One or more walking surfaces within 36 inches (914 mm) horizontally of the plane of the glazing.

8. Glazing in railings regardless of height above a walking surface. Included are structural baluster panels and nonstructural in-fill panels.

EXCEPTION: The following products and applications are exempt from the requirements for hazardous locations as listed in Items 1 through 8:

1. Glazing in Item 6 when there is an intervening wall or other permanent barrier between the door and the glazing.

2. Glazing in Item 7 when a protective bar is installed on the accessible sides of the glazing 34 inches (864 mm) to 38 inches (965 mm) above the floor. The bar shall be capable of withstanding a horizontal load of 50 pounds per linear foot (729 N/m) without contacting the glass and be a minimum of 1^1/$_2$ inches (38 mm) in height.

3. Outboard pane in insulating glass units and in other multiple glazed panels in Item 7 when the bottom exposed edge of the glass is 25 feet (7620 mm) or more above any grade, roof, walking surface, or other horizontal or sloped (within 45 degrees of horizontal) surface adjacent to the glass exterior.

4. Openings in door through which a 3-inch-diameter (76.2 mm) sphere will not pass.

5. Assemblies of leaded, faceted or carved glass in Items 1, 2, 6 and 7 when used for decorative purposes.

6. Curved panels in revolving door assemblies.

7. Doors in commercial refrigerated cabinets.

8. Glass block panels complying with Section 2110.

9. Glazing in walls and fences used as the barrier for indoor and outdoor swimming pools and spas when all of the following conditions are present:

9.1 The bottom edge of the glazing is less than 60 inches (1525 mm) above the pool side of the glazing.

9.2 The glazing is within 5 feet (1525 mm) of a swimming pool or spa water's edge.

10. Glazing in walls enclosing stairway landings or within 5 feet (1525 mm) of the bottom and top of stairways where the bottom edge of the glass is less than 60 inches (1525 mm) above a walking surface.

2406.5 Wardrobe Doors. Glazing in wardrobe doors shall meet the impact test requirements for safety glazing as set forth in UBC Standard 24-2, Part II. Laminated glass must also meet the boil test requirements of UBC Standard 24-2, Part II.

EXCEPTION: The impact test shall be modified so that if no breakage occurs when the impacting object is dropped from the height of 18 inches (457 mm), the test shall progress in height increments of 6 inches (152.5 mm) until the maximum of 48 inches (1219 mm) is reached.

2406.6 Glass Railings. Glass used as structural balustrade panels in railings shall be one of the following types:

1. Single fully tempered glass.

2. Laminated fully tempered glass.

3. Laminated heat-strengthened glass.

The panels and their support system shall be designed to withstand the load specified in Table 16-B. A safety factor of 4 shall be used.

Each handrail or guardrail section shall be supported by a minimum of three glass balusters or otherwise supported so that it remains in place should one baluster panel fail.

Glass balusters shall not be installed without a handrail or guardrail attached.

For all glazing types, the minimum nominal thickness shall be ^{1}/$_4$ inch (6.35 mm).

Glazing materials shall not be installed in railings in parking garages except for those locations where the railing is not exposed to impact from vehicles.

SECTION 2407 — HINGED SHOWER DOORS

Hinged shower doors shall open outward.

SECTION 2408 — RACQUETBALL AND SQUASH COURTS

2408.1 Test Method. Each panel of glass (including doors) in an actual installation or test mockup shall be impacted from the playing side at a point 59 inches (1499 mm) from the playing surface and its horizontal midpoint. The impactor and test procedure shall be as described in UBC Standard 24-2, Part I, Category II, using a drop height of 48 inches (1219 mm). Results from a test mockup shall apply only to actual installations in which the glass is no greater in either dimension and is at least as thick. Fittings and attachments for a mockup shall be identical to those used in actual installations. The conditions of Section 2408.2 shall be met.

2408.2 End Point Conditions. The following conditions shall be met when the glass is impacted as described in Section 2408.1:

1. The glass shall not break.

2. Deflection at the point of impact shall not exceed 1^1/$_2$ inches (38 mm).

3. Door hardware shall remain intact and operable.

4. The deflection of the door edges shall be no greater than the following for the listed drop heights. The impactor and procedures shall be as indicated in Section 2408.1.

Drop Height Deflection, inches (mm)

Drop Height	Deflection, inches (mm)
24 (610)	Thickness of adjacent glass + ^{1}/$_8$ (+ 3.2)
36 (914)	Thickness of adjacent glass + ^{1}/$_4$ (+ 6.4)
48 (1219)	Thickness of adjacent glass + ^{1}/$_2$ (+ 12.7)

SECTION 2409 — SLOPED GLAZING AND SKYLIGHTS

2409.1 Scope. This section applies to the installation of glass or other transparent, translucent or opaque glazing material installed at a slope of 15 degrees or more from the vertical plane, including glazing materials in skylights, roofs and sloped walls.

2409.2 Allowable Glazing Materials. Sloped glazing shall be any of the following materials, subject to the limitations in this section:

1. Laminated glass with a minimum 0.015-inch (0.38 mm) polyvinyl butyral interlayer for glass panes 16 square feet (1.5 m²) or less in area and with the highest point of the glass no more than 12 feet (3658 mm) above a walking surface; for larger or higher panes, the minimum interlayer thickness shall be 0.030 inch (0.76 mm).

2. Fully tempered glass.

3. Heat-strengthened glass.

4. Wired glass.

5. Approved rigid plastics meeting the requirements of Section 2603.7.

For multiple-layer glazing systems, each light or layer shall consist of any of the glazing materials specified above.

Annealed glass may be used as specified within Exceptions 2 and 3 of Section 2409.3.

2409.3 Screening. Heat-strengthened glass and fully tempered glass, when used in single-layer glazing systems, shall have screens installed below glazing. The screens shall be capable of supporting the weight of the glass and shall be substantially supported below and installed within 4 inches (102 mm) of the glass. They shall be constructed of a noncombustible material not thinner than 0.08 inch (2.03 mm) with a mesh not larger than 1 inch by 1 inch (25 mm by 25 mm). In a corrosive atmosphere, structurally equivalent noncorrosive screening materials shall be used. Heat-strengthened glass, fully tempered glass and wired glass, when used in multiple-layer glazing systems as the bottom glass layer over the walking surface, shall be equipped with screening that complies with the requirements for monolithic glazing systems.

EXCEPTIONS: 1. Fully tempered glass may be installed without required protective screens when located between intervening floors at a slope of 30 degrees or less from the vertical plane if the highest point of the glass is 10 feet (3048 mm) or less above the walking surface.

2. Allowable glazing material, including annealed glass, may be installed without required screens if the walking surface or any other accessible area below the glazing material is permanently protected from falling glass for a minimum horizontal distance equal to twice the height.

3. Allowable glazing material, including annealed glass, may be installed without screens in the sloped glazing systems of commercial or detached greenhouses used exclusively for growing plants and not intended for use by the public, provided the height of the greenhouse at the ridge does not exceed 20 feet (6096 mm) above grade.

4. Screens need not be provided within individual dwelling units when fully tempered glass is used as single glazing or in both panes of an insulating glass unit when all the following conditions are met:

4.1 The area of each pane (single glass) or unit (insulating glass) shall not exceed 16 square feet (1.49 m²).

4.2 The highest point of the glass shall not be more than 12 feet (3658 mm) above any walking surface or other accessible area.

4.3 The nominal thickness of each pane shall not exceed $^3/_{16}$ inch (4.76 mm).

2409.4 Framing. In Types I and II construction, skylight frames shall be constructed of noncombustible materials.

EXCEPTION: In foundries or buildings where acid fumes deleterious to metal are incidental to the use of the buildings, approved pressure-treated woods or other approved noncorrosive materials may be used for sash and frames.

Skylights set at an angle of less than 45 degrees from the horizontal plane shall be mounted at least 4 inches (102 mm) above the plane of the roof on a curb constructed of materials as required for the frame. Skylights may be installed in the plane of the roof when the roof slope is 45 degrees or greater from horizontal.

2409.5 Design Loads. Sloped glazing and skylights shall be designed to withstand the tributary loads specified in Section 1605. Sizing limitations specified within Graph 24-1 and Table 24-A may be utilized for glazing materials set forth in Section 2409.2, provided the design loads are increased by a factor of 2.67.

2409.6 Floors and Sidewalks. Glass used for the transmission of light, if placed in floors or sidewalks, shall be supported by metal or reinforced concrete frames, and such glass shall not be less than $^1/_2$ inch (12.7 mm) in thickness. Any such glass over 16 square inches (0.1 m²) in area shall have wire mesh embedded in the same or shall be provided a wire screen underneath, as specified for skylights in this section. All portions of the floor lights or sidewalk lights shall be of the same strength as is required by this code for floor or sidewalk construction, except in cases where the floor is surrounded by a railing not less than 3 feet 6 inches (1067 mm) in height, in which case the construction shall be calculated for not less than roof loads.

TABLE 24-A
GRAPH 24-1

1997 UNIFORM BUILDING CODE

TABLE 24-A—ADJUSTMENT FACTORS—RELATIVE RESISTANCE TO WIND LOADS

GLASS TYPE	ADJUSTMENT FACTOR[1]
Laminated[2]	0.75
Fully tempered	4.00
Heat strengthened	2.00
Wired	0.50
Insulating glass[3]—2 panes —3 panes	1.70 2.55
Patterned[4]	1.00
Regular (annealed)	1.00
Sandblasted	0.40[5]

[1]Loads determined from Chapter 16, Division III, shall be divided by this adjustment factor for use with Graph 24-1.
[2]Applies when two plies are identical in thickness and type; use total glass thickness, not thickness of one ply.
[3]Applies when each glass panel is the same thickness and type; use thickness of one panel.
[4]Use minimum glass thickness, i.e., measured at the thinnest part of the pattern; if necessary, interpolation of curves in Graph 24-1 may be required.
[5]Factor varies depending on depth and severity of sandblasting; value shown is minimum.

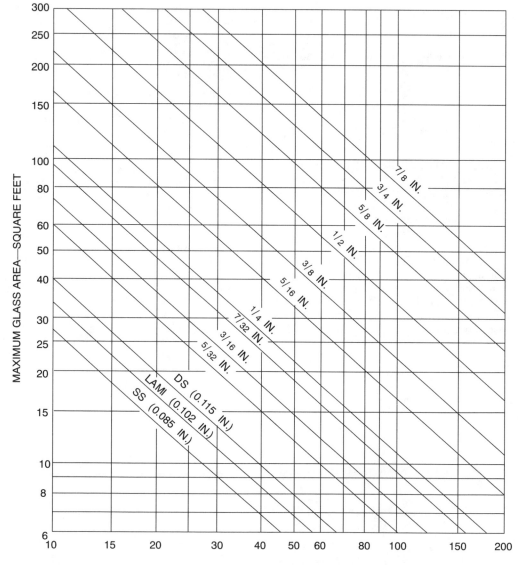

DESIGN WIND PRESSURE FROM CHAPTER 16, DIVISION III—POUNDS PER SQUARE FOOT

For **SI:** 1 inch = 25.4 mm, 1 square foot = 0.0929 m^2, 1 pound per square foot = 0.479 kN/m^2.

GRAPH 24-1—MAXIMUM ALLOWABLE AREA OF GLASS[1]

[1]Applicable for ratios of width to length of 1:1 to 5:1. Design safety factor = 2.5.

Chapter 25
GYPSUM BOARD AND PLASTER

SECTION 2501 — SCOPE

2501.1 General. The installation of lath, plaster and gypsum board shall be done in a manner and with materials as specified in this chapter and, when required for fire-resistive construction, also shall conform to the provisions of Chapter 7.

Other approved wall or ceiling coverings may be installed in accordance with the recommendations of the manufacturer and the conditions of approval.

2501.2 Inspection. No lath or gypsum board or their attachments shall be covered or finished until it has been inspected and approved by the building official in accordance with Section 108.5.

2501.3 Tests. The building official may require tests to be made in accordance with approved standards to determine compliance with the provisions of this chapter, provided the permit holder has been notified 24 hours in advance of the time of making such tests.

2501.4 Definitions. For purposes of this chapter, certain terms are defined as follows:

CEMENT PLASTER is a mixture of portland cement, portland cement and lime, masonry cement, or plastic cement and aggregate and other approved materials as specified in the code.

CORNER BEAD is a rigid formed unit or shape used at projecting or external angles to define and reinforce the corners of interior surfaces.

CORNERITE is a shaped reinforcing unit of expanded metal or wire fabric used for angle reinforcing and having minimum outstanding legs of not less than 2 inches (51 mm).

CORROSION-RESISTANT MATERIALS are materials that are inherently rust resistant or materials to which an approved rust-resistive coating has been applied either before or after forming or fabrication.

EXTERIOR SURFACES are weather-exposed surfaces as defined in Section 224.

EXTERNAL CORNER REINFORCEMENT is a shaped reinforcing unit for external corner reinforcement of cement plaster formed to ensure mechanical bond and a solid plaster corner.

INTERIOR SURFACES are surfaces other than weather-exposed surfaces.

MOIST CURING is any method employed to retain sufficient moisture for hydration of portland cement plaster.

PORTLAND CEMENT PLASTER is a mixture of portland cement or portland cement and lime and aggregate and other approved materials as specified in this code.

STEEL STUDS, LOAD-BEARING AND NONLOAD-BEARING, are prefabricated channel shapes, welded wire, or combination wire and steel angle types, galvanized or coated with rust-resistive material.

STRIPPING is flat reinforcing units of expanded metal or wire fabric or other materials not less than 3 inches (76 mm) wide to be installed as required over joints of gypsum lath.

TIE WIRE is wire for securing together metal framing or supports, for tying metal and wire fabric lath and gypsum lath and wallboard together, and for securing accessories.

WIRE BACKING is horizontal strands of tautened wire attached to surfaces of vertical wood supports that, when covered with building paper, provide a backing of cement plaster.

SECTION 2502 — MATERIALS

Lathing, plastering, wallboard materials and ceiling suspension systems shall conform to the applicable standards listed in Chapter 35.

The standards listed below labeled a "UBC Standard" are also listed in Chapter 35, Part II, and are part of this code. The other standards listed below are recognized standards (see Sections 3503 and 3504).

1. UBC Standard 21-11, Cement, Masonry

2. ASTM C 150, Portland Cement

3. UBC Standard 25-1, Plastic Cement

4. UBC Standard 25-2, Metal Suspension Systems for Acoustical Tile and for Lay-in Panel Ceilings

5. United States Government Military Specification MIL-B-19235 (Docks), Plaster Bonding Agents

6. ASTM C 557, Adhesives for Fastening Gypsum Wallboard to Wood Framing

7. ASTM C 35, Perlite, Vermiculite and Sand Aggregates for Gypsum Plaster

8. ASTM C 1002, Drill Screws

9. ASTM C 475 and C 474, Gypsum Wallboard Tape and Joint Compound

10. ASTM C 442, Gypsum Backing Board

11. ASTM C 37, Gypsum Lath

12. ASTM C 28, Gypsum Plasters

13. ASTM C 79, Gypsum Sheathing Board

14. ASTM C 36, Gypsum Wallboard

15. ASTM C 61, Keene's Cement

16. ASTM C 630, Water-resistant Gypsum Backing Board

17. ASTM C 588 and C 587, Gypsum Base for Veneer Plaster and Gypsum Veneer Plaster

18. ASTM C 6 and C 206, Lime

19. ASTM C 144 and C 897, Aggregate for Masonry Mortar and Aggregate for Job-mixed Portland Cement-based Plaster

20. ASTM C 22, C 472 and C 473, Testing Gypsum and Gypsum Products

21. ASTM C 843 and C 844, Application of Gypsum Base for Veneer Plaster and Gypsum Veneer Plaster

22. ASTM C 514, Nails for the Application of Gypsum Wallboard, Gypsum Backing Board and Gypsum Veneer Base

23. ASTM C 931, Exterior Gypsum Soffit Board

24. ANSI A42.4-1955 and Specification 2.6.73 of the California Lathing and Plastering Contractors Association, Metal Lath, Wire Lath, Wire Fabric Lath and Metal Accessories

SECTION 2503 — VERTICAL ASSEMBLIES

2503.1 General. In addition to the requirements of this section, vertical assemblies of plaster or gypsum board shall be designed to resist the loads specified in Chapter 16 of this code. For wood framing, see Chapter 23. For metal framing, see Chapter 22.

> **EXCEPTION:** Wood-framed assemblies meeting the requirements of Section 2320 need not be designed.

2503.2 Wood Framing. Wood supports for lath or gypsum board shall not be less than 2 inches (51 mm) nominal in least dimension. Wood stripping or furring shall not be less than 2 inches (51 mm) nominal thickness in the least dimension except that furring strips not less than 1-inch-by-2-inch (25 mm by 51 mm) nominal dimension may be used over solid backing.

2503.3 Studless Partitions. The minimum thickness of vertically erected studless solid plaster partitions of $^3/_8$-inch (9.5 mm) and $^3/_4$-inch (19.1 mm) rib metal lath or $^1/_2$-inch-thick (12.7 mm) long-length gypsum lath and gypsum board partitions shall be 2 inches (51 mm).

SECTION 2504 — HORIZONTAL ASSEMBLIES

2504.1 General. In addition to the requirements of this section, supports for horizontal assemblies of plaster or gypsum board shall be designed to support all loads as specified in Chapter 16 of this code.

> **EXCEPTION:** Wood-framed assemblies meeting the requirements of Section 2320 need not be designed.

2504.2 Wood Framing. Wood stripping or suspended wood systems, where used, shall not be less than 2 inches (51 mm) nominal thickness in the least dimension, except that furring strips not less than 1-inch-by-2-inch (25 mm by 51 mm) nominal dimension may be used over solid backing.

2504.3 Hangers. Hangers for suspended ceilings shall not be less than the sizes set forth in Table 25-A, fastened to or embedded in the structural framing, masonry or concrete.

Hangers shall be saddle-tied around main runners to develop the full strength of the hangers. Lower ends of flat hangers shall be bolted with $^3/_8$-inch (9.5 mm) bolts to runner channels or bent tightly around runners and bolted to the main part of the hanger.

2504.4 Runners and Furring. The main runner and cross-furring shall not be less than the sizes set forth in Table 25-A, except that other steel sections of equivalent strength may be substituted for those set forth in this table. Cross-furring shall be securely attached to the main runner by saddle-tying with not less than one strand of 0.051-inch (1.30 mm) (No. 16 A.W. gage) or two strands of 0.040-inch (1.02 mm) (No. 18 A.W. gage) tie wire or approved equivalent attachments.

SECTION 2505 — INTERIOR LATH

2505.1 General. Gypsum lath shall not be installed until weather protection for the installation is provided. Where wood-frame walls and partitions are covered on the interior with cement plaster or tile of similar material and are subject to water splash, the framing shall be protected with an approved moisture barrier.

Showers and public toilet walls shall conform to Section 807.1.

2505.2 Application of Gypsum Lath. The thickness, spacing of supports and the method of attachment of gypsum lath shall be as set forth in Tables 25-B and 25-C. Approved wire and sheet metal attachment clips may be used.

Gypsum lath shall be applied with the long dimension perpendicular to supports and with end joints staggered in successive courses. End joints may occur on one support when stripping is applied the full length of the joints.

Where electrical radiant heat cables are installed on ceilings, the stripping, if conductive, may be omitted a distance not to exceed 12 inches (305 mm) from the walls.

Where lath edges are not in moderate contact and have joint gaps exceeding $^3/_8$ inch (9.5 mm), the joint gaps shall be covered with stripping or cornerite. Stripping or cornerite may be omitted when the entire surface is reinforced with not less than 1-inch (25 mm) 0.035-inch (0.89 mm) (No. 20 B.W. gage) woven wire. When lath is secured to horizontal or vertical supports not used as structural diaphragms, end joints may occur between supports when lath ends are secured together with approved fasteners. Vertical assemblies also shall conform to Section 1611.5.

Cornerite shall be installed so as to retain position during plastering at all internal corners. Cornerite may be omitted when plaster is not continuous from one plane to an adjacent plane.

2505.3 Application of Metal Plaster Bases. The type and weight of metal lath, and the gage and spacing of wire in welded or woven lath, the spacing of supports, and the methods of attachment to wood supports shall be as set forth in Tables 25-B and 25-C.

Metal lath shall be attached to metal supports with not less than 0.049-inch (1.2 mm) (No. 18 B.W. gage) tie wire spaced not more than 6 inches (152 mm) apart or with approved equivalent attachments.

Metal lath or wire fabric lath shall be applied with the long dimension of the sheets perpendicular to supports.

Metal lath shall be lapped not less than $^1/_2$ inch (12.7 mm) at sides and 1 inch (25 mm) at ends. Wire fabric lath shall be lapped not less than one mesh at sides and ends, but not less than 1 inch (25 mm). Rib metal lath with edge ribs greater than $^1/_8$ inch (3.2 mm) shall be lapped at sides by nesting outside ribs. When edge ribs are $^1/_8$ inch (3.2 mm) or less, rib metal lath may be lapped $^1/_2$ inch (12.7 mm) at sides, or outside ribs may be nested. Where end laps of sheets do not occur over supports, they shall be securely tied together with not less than 0.049-inch (1.2 mm) (No. 18 B.W. gage) wire.

Cornerite shall be installed in all internal corners to retain position during plastering. Cornerite may be omitted when lath is continuous or when plaster is not continuous from one plane to an adjacent plane.

SECTION 2506 — EXTERIOR LATH

2506.1 General. Exterior surfaces are weather-exposed surfaces as defined in Section 224. For eave overhangs required to be fire resistive, see Section 705.

2506.2 Corrosion Resistance. All lath and lath attachments shall be of corrosion-resistant material. See Section 2501.4.

2506.3 Backing. Backing or a lath shall provide sufficient rigidity to permit plaster application.

Where lath on vertical surfaces extends between rafters or other similar projecting members, solid backing shall be installed to provide support for lath and attachments.

Gypsum lath or gypsum board shall not be used, except that on horizontal supports of ceilings or roof soffits it may be used as backing for metal lath or wire fabric lath and cement plaster.

Backing is not required under metal lath or paperbacked wire fabric lath.

2506.4 Weather-resistive Barriers. Weather-resistive barriers shall be installed as required in Section 1402.1 and, when applied over wood base sheathing, shall include two layers of Grade D paper.

2506.5 Application of Metal Plaster Bases. The application of metal lath or wire fabric lath shall be as specified in Section 2505.3 and they shall be furred out from vertical supports or backing not less than $1/4$ inch (6.4 mm) except as set forth in Table 25-B, Footnote 2.

Where no external corner reinforcement is used, lath shall be furred out and carried around corners at least one support on frame construction.

A minimum 0.019-inch (0.48 mm) (No. 26 galvanized sheet gage) corrosion-resistant weep screed with a minimum vertical attachment flange of $3^1/2$ inches (89 mm) shall be provided at or below the foundation plate line on all exterior stud walls. The screed shall be placed a minimum of 4 inches (102 mm) above the earth or 2 inches (51 mm) above paved areas and shall be of a type that will allow trapped water to drain to the exterior of the building. The weather-resistive barrier shall lap the attachment flange, and the exterior lath shall cover and terminate on the attachment flange of the screed.

SECTION 2507 — INTERIOR PLASTER

2507.1 General. Plastering with gypsum plaster or cement plaster shall not be less than three coats when applied over metal lath or wire fabric lath and shall not be less than two coats when applied over other bases permitted by this chapter. Showers and public toilet walls shall conform to Section 807.1.

Plaster shall not be applied directly to fiber insulation board. Cement plaster shall not be applied directly to gypsum lath, gypsum masonry or gypsum plaster except as specified in Section 2506.3.

When installed, grounds shall ensure the minimum thickness of plaster as set forth in Table 25-D. Plaster thickness shall be measured from the face of lath and other bases.

2507.2 Base Coat Proportions. Proportions of aggregate to cementitious materials shall not exceed the volume set forth in Table 25-E for gypsum plaster and Table 25-F for cement plaster.

2507.3 Base Coat Application.

2507.3.1 General. Base coats shall be applied with sufficient material and pressure to form a complete key or bond.

2507.3.2 Gypsum plaster. For two-coat work, the first coat shall be brought out to grounds and straightened to a true surface, leaving the surface rough to receive the finish coat. For three-coat work, the surface of the first coat shall be scored sufficiently to provide adequate bond for the second coat and shall be permitted to harden and set before the second coat is applied. The second coat shall be brought out to grounds and straightened to a true surface, leaving the surface rough to receive the finish coat.

2507.3.3 Cement plaster. The first two coats shall be as required for the first coats of exterior plaster, except that the moist-curing time period between the first and second coats shall not be less than 24 hours and the thickness shall be as set forth in Table 25-D. Moist curing shall not be required where job and weather conditions are favorable to the retention of moisture in the cement plaster for the required time period.

2507.4 Finish Coat Application. Finish coats shall be applied with sufficient material and pressure to form a complete bond.

Finish coats shall be proportioned and mixed in an approved manner. Gypsum and lime and other interior finish coats shall be applied over gypsum base coats that have hardened and set. Thicknesses shall not be less than $1/16$ inch (1.6 mm).

Cement plaster finish coats may be applied over interior cement plaster base coats that have been in place not less than 24 hours.

Approved acoustical finish plaster may be applied over any base coat plaster, over clean masonry or concrete, or other approved surfaces.

2507.5 Interior Masonry or Concrete. Condition of surfaces shall be as specified in Section 2508.8. Approved specially prepared gypsum plaster designed for application to concrete surfaces or approved acoustical plaster may be used. The total thickness of base coat plaster applied to concrete ceilings shall be as set forth in Table 25-D. Should ceiling surfaces require more than the maximum thickness permitted in Table 25-D, metal lath or wire fabric lath shall be installed on such surfaces before plastering.

SECTION 2508 — EXTERIOR PLASTER

2508.1 General. Plastering with cement plaster shall not be less than three coats when applied over metal lath or wire fabric lath and shall not be less than two coats when applied over masonry, concrete or gypsum backing as specified in Section 2506.3. If plaster surface is completely covered by veneer or other facing material, or is completely concealed by another wall, plaster application need be only two coats, provided the total thickness is as set forth in Table 25-F.

On wood-frame or metal stud construction with an on-grade concrete floor slab system, exterior plaster shall be applied in such a manner as to cover, but not extend below, lath and paper. See Section 2506.5 for the application of paper and lath, and flashing or weep screeds.

Only approved plasticity agents and approved amounts thereof may be added to portland cement. When plastic cement is used, no additional lime or plasticizers shall be added. Hydrated lime or the equivalent amount of lime putty used as a plasticizer may be added to cement plaster or cement and lime plaster in an amount not to exceed that set forth in Table 25-F.

Gypsum plaster shall not be used on exterior surfaces. See Section 224.

2508.2 Base Coat Proportions. The proportion of aggregate to cementitious materials shall be as set forth in Table 25-F.

2508.3 Base Coat Application. The first coat shall be applied with sufficient material and pressure to fill solidly all openings in the lath. The surface shall be scored horizontally sufficiently rough to provide adequate bond to receive the second coat.

The second coat shall be brought out to proper thickness, rodded and floated sufficiently rough to provide adequate bond for the finish coat. The second coat shall have no variation greater than $1/4$ inch (6.4 mm) in any direction under a 5-foot (1524 mm) straight edge.

2508.4 Environmental Conditions. Portland cement-based plaster shall not be applied to frozen base or those bases containing frost. Plaster mixes shall not contain frozen ingredients. Plaster coats shall be protected from freezing for a period of not less than 24 hours after set has occurred.

2508.5 Curing and Interval. First and second coats of plaster shall be applied and moist cured as set forth in Table 25-F.

When applied over gypsum backing as specified in Section 2506.3 or directly to unit masonry surfaces, the second coat may be applied as soon as the first coat has attained sufficient hardness.

2508.6 Alternate Method of Application. As an alternate method of application, the second coat may be applied as soon as the first coat has attained sufficient rigidity to receive the second coat.

When using this method of application, calcium aluminate cement up to 15 percent of the weight of the portland cement may be added to the mix.

Curing of the first coat may be omitted and the second coat shall be cured as set forth in Table 25-F.

2508.7 Finish Coats. Finish coats shall be proportioned and mixed in an approved manner and in accordance with Table 25-F.

Cement plaster finish coats shall be applied over base coats that have been in place for the time periods set forth in Table 25-F. The third or finish coat shall be applied with sufficient material and pressure to bond to and to cover the brown coat and shall be of sufficient thickness to conceal the brown coat.

2508.8 Preparation of Masonry and Concrete. Surfaces shall be clean, free from efflorescence, sufficiently damp and rough to ensure proper bond. If surface is insufficiently rough, approved bonding agents or a portland cement dash bond coat mixed in proportions of one and one half parts volume of sand to one part volume of portland cement or plastic cement shall be applied. Dash bond coat shall be left undisturbed and shall be moist cured not less than 24 hours. When dash bond is applied, first coat of base coat plaster may be omitted. See Table 25-D for thickness.

SECTION 2509 — EXPOSED AGGREGATE PLASTER

2509.1 General. Exposed natural or integrally colored aggregate may be partially embedded in a natural or colored bedding coat of cement plaster or gypsum plaster, subject to the provisions of this section.

2509.2 Aggregate. The aggregate may be applied manually or mechanically and shall consist of marble chips, pebbles or similar durable, nonreactive materials, moderately hard (three or more on the Mohs scale).

2509.3 Bedding Coat Proportions. The exterior bedding coat shall be composed of one part portland cement, one part Type S lime, and a maximum three parts of graded white or natural sand by volume. The interior bedding coat shall be composed of 100 pounds (45.4 kg) neat gypsum plaster and a maximum 200 pounds (90.7 kg) of graded white sand, or exterior or interior may be a factory-prepared bedding coat. The exterior bedding coat shall have a minimum compressive strength of 1,000 pounds per square inch (6894.8 kPa).

2509.4 Application. The bedding coat may be applied directly over the first (scratch) coat of plaster, provided the ultimate overall thickness is a minimum of $^7/_8$ inch (22.2 mm), including lath. Over concrete or masonry surfaces, the overall thickness shall be a minimum of $^1/_2$ inch (12.7 mm).

2509.5 Bases. Exposed aggregate plaster may be applied over concrete, masonry, cement plaster base coats or gypsum plaster base coats.

2509.6 Preparation of Masonry and Concrete. Masonry and concrete surfaces shall be prepared in accordance with the provisions of Section 2508.8.

2509.7 Curing. Cement plaster base coats shall be cured in accordance with Table 25-F. Cement plaster bedding coat shall retain sufficient moisture for hydration (hardening) for 24 hours

minimum or, where necessary, shall be kept damp for 24 hours by light water spraying.

SECTION 2510 — PNEUMATICALLY PLACED PLASTER (GUNITE)

Pneumatically placed portland cement plaster shall be a mixture of portland cement and sand, mixed dry, conveyed by air through a pipe or flexible tube, hydrated at the nozzle at the end of the conveyor, and deposited by air pressure in its final position.

Rebound material may be screened and reused as sand in an amount not greater than 25 percent of the total sand in any batch.

Pneumatically placed portland cement plaster shall consist of a mixture of one part cement to not more than five parts sand. Plasticity agents may be used as specified in Section 2508.1. Except when applied to concrete or masonry, such plaster shall be applied in not less than two coats to a minimum total thickness of $^7/_8$ inch (22.2 mm). The first coat shall be rodded as specified in Section 2508.3 for the second coat. The curing period and time interval shall be as set forth in Table 25-F.

SECTION 2511 — GYPSUM WALLBOARD

2511.1 General. Gypsum wallboard shall not be installed on exterior surfaces. See Section 224. For use as backing under stucco, see Section 2506.3.

Gypsum wallboard shall not be installed until weather protection for the installation is provided.

2511.2 Supports. Supports shall be spaced not to exceed the spacing set forth in Table 25-G for single-ply application and Table 25-H for two-ply application. Vertical assemblies shall comply with Section 2503. Horizontal assemblies shall comply with Section 2504.

2511.3 Single-ply Application. All edges and ends of gypsum wallboard shall occur on the framing members, except those edges and ends that are perpendicular to the framing members. All edges and ends of gypsum wallboard shall be in moderate contact except in concealed spaces where fire-resistive construction or diaphragm action is not required.

The size and spacing of fasteners shall comply with Table 25-G except where modified by fire-resistive construction meeting the requirements of Section 703.2. Fasteners shall not be spaced less than $^3/_8$ inch (9.5 mm) from edges and ends of gypsum wallboard. Fasteners at the top and bottom plates of vertical assemblies, or the edges and ends of horizontal assemblies perpendicular to supports, and at the wall line may be omitted except on shear-resisting elements or fire-resistive assemblies. Fasteners shall be applied in such a manner as not to fracture the face paper with the fastener head.

Gypsum wallboard may be applied to wood-framing members with an approved adhesive. A continuous bead of the adhesive shall be applied to the face of all framing members, except top and bottom plates, of sufficient size as to spread to an average width of 1 inch (25 mm) and thickness of $^1/_{16}$ inch (1.6 mm) when the gypsum wallboard is applied. Where the edges or ends of two pieces of gypsum wallboard occur on the same framing member, two continuous parallel beads of adhesive shall be applied to the framing member. Fasteners shall be used with adhesive application in accordance with Table 25-G.

2511.4 Two-ply Application. The base of gypsum wallboard shall be applied with fasteners of the type and size as required for the nonadhesive application of single-ply gypsum wallboard. Fastener spacings shall be in accordance with Table 25-H except where modified by fire-resistive construction meeting the requirements of Section 703.2.

The face ply of gypsum wallboard may be applied with gypsum wallboard joint compound or approved adhesive furnishing full coverage between the plies or with fasteners in accordance with Table 25-H. When the face ply is installed with joint compound or adhesive, the joints of the face ply need not occur on supports. Temporary nails or shoring shall be used to hold face ply in position until the joint compound or adhesive develops adequate bond.

2511.5 Joint Treatment. Gypsum wallboard single-layer fire-rated assemblies shall have joints treated.

> **EXCEPTION:** Joint treatment need not be provided when any of the following conditions occur:
>
> 1. Where the wallboard is to receive a decorative finish such as wood paneling, battens, acoustical finishes or any similar application that would be equivalent to joint treatment.
>
> 2. Joints occur over wood-framing members.
>
> 3. Assemblies tested without joint treatment.

SECTION 2512 — USE OF GYPSUM IN SHOWERS AND WATER CLOSETS

When gypsum is used as a base for tile or wall panels for tub, shower or water closet compartment walls (see Sections 807.1.2 and 807.1.3), water-resistant gypsum backing board shall be used. Regular gypsum wallboard is permitted under tile or wall panels in other wall and ceiling areas when installed in accordance with Table 25-G. Water-resistant gypsum board shall not be used in the following locations:

1. Over a vapor retarder.

2. In areas subject to continuous high humidity, such as saunas, steam rooms or gang shower rooms.

3. On ceilings where frame spacing exceeds 12 inches (305 mm) on center.

SECTION 2513 — SHEAR-RESISTING CONSTRUCTION WITH WOOD FRAME

2513.1 General. Cement plaster, gypsum lath and plaster, gypsum veneer base, gypsum sheathing board, and gypsum wallboard may be used on wood studs for vertical diaphragms if applied in accordance with this section. Shear-resisting values shall not exceed those set forth in Table 25-I. The effects of overturning on vertical diaphragms shall be investigated in accordance with Section 1605.2.2.

The shear values tabulated shall not be cumulative with the shear value of other materials applied to the same wall. The shear values may be additive when the identical materials applied as specified in this section are applied to both sides of the wall.

2513.2 Masonry and Concrete Construction. Cement plaster, gypsum lath and plaster, gypsum veneer base, gypsum sheathing board, and gypsum wallboard shall not be used in vertical diaphragms to resist forces imposed by masonry or concrete construction.

2513.3 Wall Framing. Framing for vertical diaphragms shall comply with Section 2320.11 for bearing walls, and studs shall not be spaced farther apart than 16 inches (406 mm) center to center. Sills, plates and marginal studs shall be adequately connected to framing elements located above and below to resist all design forces.

2513.4 Height-to-length Ratio. The maximum allowable height-to-length ratio for the construction in this section shall be 2 to 1. Wall sections having height-to-length ratios in excess of $1^1/_2$ to 1 shall be blocked.

2513.5 Application. End joints of adjacent courses of gypsum lath, gypsum veneer base, gypsum sheathing board or gypsum wallboard sheets shall not occur over the same stud.

Where required in Table 25-I, blocking having the same cross-sectional dimensions as the studs shall be provided at all joints that are perpendicular to the studs.

The size and spacing of nails shall be as set forth in Table 25-I. Nails shall not be spaced less than $^3/_8$ inch (9.5 mm) from edges and ends of gypsum lath, gypsum veneer base, gypsum sheathing board and gypsum wallboard, or from sides of studs, blocking, and top and bottom plates.

2513.5.1 Gypsum lath. Gypsum lath shall be applied perpendicular to the studs. Maximum allowable shear values shall be as set forth in Table 25-I.

2513.5.2 Gypsum sheathing board. Four-foot-wide (1219 mm) pieces may be applied parallel or perpendicular to studs. Two-foot-wide (610 mm) pieces shall be applied perpendicular to the studs. Maximum allowable shear values shall be as set forth in Table 25-I.

2513.5.3 Gypsum wallboard or veneer base. Gypsum wallboard or veneer base may be applied parallel or perpendicular to studs. Maximum allowable shear values shall be as set forth in Table 25-I.

TABLE 25-A

1997 UNIFORM BUILDING CODE

TABLE 25-A—SUSPENDED AND FURRED CEILINGS[1]
[For support of ceilings weighing not more than 10 pounds per square foot (4.89 kg/m²)]

MINIMUM SIZES FOR WIRE AND RIGID HANGERS			
Size and Type		Maximum Area Supported (square feet) × 0.09 for m²	Size × 25.4 for mm
Hangers for suspended ceilings		12.5	0.148-inch (3.76 mm) (No. 9 B.W. gage) wire
		16	0.145-inch (4.19 mm) (No. 8 B.W. gage) wire
		18	$^3/_{16}''$ diameter, mild steel rod[2]
		20	$^7/_{32}''$ diameter, mild steel rod[2]
		22.5	$^1/_4''$ diameter, mild steel rod[2]
		22.0	$1'' \times ^3/_{16}''$ mild steel flats[3]
Hangers for attaching runners and furring directly to beams and joists	For supporting runners	Single hangers between beams[4]	8	0.109-inch (2.77 mm) (No. 12 B.W. gage) wire
		12	0.134-inch (3.40 mm) (No. 10 B.W. gage) wire
		16	0.165-inch (4.19 mm) (No. 8 B.W. gage) wire
		Double wire loops at beams or joists[3]	8	0.083-inch (2.11 mm) (No. 14 B.W. gage) wire
		12	0.109-inch (2.77 mm) (No. 12 B.W. gage) wire
		16	0.120-inch (3.05 mm) (No. 11 B.W. gage) wire
	For supporting furring without runners[4] (wire loops at supports)	Type of support: Concrete / Steel / Wood	8	0.083-inch (2.11 mm) (No. 14 B.W. gage) wire / 0.065-inch (1.65 mm) (No. 16 B.W. gage) wire (2 loops)[5] / 0.065-inch (1.65 mm) (No. 16 B.W. gage) wire (2 loops)[5]

MINIMUM SIZES AND MAXIMUM SPANS FOR MAIN RUNNERS[6,7]		
Size and Type × 25.4 for mm × 1.49 for kg/m	Maximum Spacing of Hangers or Supports (Along Runners) × 304.8 for mm	Maximum Spacing of Runners (Transverse)
$^3/_4''$ — 0.3 pound per foot, cold- or hot-rolled channel	2'	3'
$1^1/_2''$ — 0.475 pound per foot, cold-rolled channel	3'	4'
$1^1/_2''$ — 0.475 pound per foot, cold-rolled channel	3.5'	3.5'
$1^1/_2''$ — 0.475 pound per foot, cold-rolled channel	4'	3'
$1^1/_2''$ — 1.12 pounds per foot, hot-rolled channel	4'	5'
$2''$ — 1.26 pounds per foot, hot-rolled channel	5'	5'
$2''$ — 0.59 pounds per foot, cold-rolled channel	5'	3.5'
$1^1/_2'' \times 1^1/_2'' \times ^3/_{16}''$ angle	5'	3.5'

MINIMUM SIZES AND MAXIMUM SPANS FOR CROSS FURRING[6,7]		
Size and Type of Cross-furring × 25.4 for mm × 1.49 for kg/m	Maximum Spacing of Runners or Supports × 304.8 for mm	Maximum Spacing of Cross-furring Members (Transverse) × 25.4 for mm
$^1/_4''$ diameter pencil rods	2'	12''
$^3/_8''$ diameter pencil rods	2'	19''
$^3/_8''$ diameter pencil rods	2.5'	12''
$^3/_4''$ — 0.3 pound per foot, cold- or hot-rolled channel	3'	24''
	3.5'	16''
	4'	12''
$1''$ — 0.410 pound per foot, hot-rolled channel	4'	24''
	4.5'	19''
	5'	12''

[1]Metal suspension systems for acoustical tile and lay-in panel ceiling systems weighing not more than 4 pounds per square foot (19.5 kg/m²), including light fixtures and all ceiling-supported equipment and conforming to UBC Standard 25-2, are exempt from Table 25-A.

[2]All rod hangers shall be protected with a zinc or cadmium coating or with a rust-inhibitive paint.

[3]All flat hangers shall be protected with a zinc or cadmium coating or with a rust-inhibitive paint.

[4]Inserts, special clips or other devices of equal strength may be substituted for those specified.

[5]Two loops of 0.049-inch (1.24 mm) (No. 18 B.W. gage) wire may be substituted for each loop of 0.065-inch (1.65 mm) (No. 16 B.W. gage) wire for attaching furring to steel or wood joists.

[6]Spans are based on webs of channels being erected vertically.

[7]Other sections of hot- or cold-rolled members of equivalent strength may be substituted for those specified.

TABLE 25-B[1]—TYPES OF LATH—MAXIMUM SPACING OF SUPPORTS

TYPE OF LATH[2]		MINIMUM WEIGHT (per square yard) ($\times$ 0.38 for kg/m^2) GAGE AND MESH SIZE ($\times$ 25.4 for mm)	VERTICAL (inches) $\times$ 25.4 for mm			HORIZONTAL (inches) $\times$ 25.4 for mm	
			Wood	Metal		Wood or Concrete	Metal
				Solid Plaster Partitions	Other		
1. Expanded metal lath (diamond mesh)		2.5	16[3]	16[3]	12	12	12
		3.4	16[3]	16[3]	16	16	16
2. Flat rib expanded metal lath		2.75	16	16	16	16	16
		3.4	19	24	19	19	19
3. Stucco mesh expanded metal lath		1.8 and 3.6	16[4]	—	—	—	—
4. $^3/_8''$ (9.5 mm) rib expanded metal lath		3.4	24	24[5]	24	24	24
		4.0	24	24[5]	24	24	24
5. Sheet lath		4.5	24	5	24	24	24
6. Wire fabric lath	Welded	1.95 pounds, 0.120 inch (No. 11 B.W. gage), 2″ x 2″	24	24	24	24	24
		1.16 pounds, 0.065 inch (No. 16 B.W. gage), 2″ x 2″	16	16	16	16	16
		1.4 pounds, 0.049 inch (No. 18 B.W. gage), 1″ x 1″[6]	16[4]	—	—	—	—
	Woven	1.1 pounds, 0.049 inch (No. 18 B.W. gage), $1^1/_2''$ hexagonal[6]	24	16	16	24	16
		1.4 pounds, 0.058 inch (No. 17 B.W. gage), $1^1/_2''$ hexagonal[6]	24	16	16	24	16
		1.4 pounds, 0.049 inch (No. 18 B.W. gage), 1″ hexagonal[6]	24	16	16	24	16
7. $^3/_8''$ (9.5 mm) gypsum lath (plain)			16	—	16[7]	16	16
8. $^1/_2''$ (12.7 mm) gypsum lath (plain)			24	—	24	24	24

[1]For fire-resistive construction, see Tables 7-A, 7-B and 7-C. For shear-resisting elements, see Table 25-I.

[2]Metal lath and wire fabric lath used as reinforcement for cement plaster shall be furred out away from vertical supports at least $^1/_4$ inch (6.4 mm). Self-furring lath meets furring requirements.

EXCEPTION: Furring of expanded metal lath is not required on supports having a bearing surface width of $1^5/_8$ inches (41 mm) or less.

[3]Span may be increased to 24 inches (610 mm) with self-furred metal lath over solid sheathing assemblies approved for this use.

[4]Wire backing required on open vertical frame construction except under expanded metal lath and paperbacked wire fabric lath.

[5]May be used for studless solid partitions.

[6]Woven wire or welded wire fabric lath not to be used as base for gypsum plaster without absorbent paperbacking or slot-perforated separator.

[7]Span may be increased to 24 inches (610 mm) on vertical screw or approved nailable assemblies.

TABLE 25-C
TABLE 25-D

1997 UNIFORM BUILDING CODE

TABLE 25-C—TYPES OF LATH—ATTACHMENT TO WOOD AND METAL[1] SUPPORTS

TYPE OF LATH	NAILS[2,3] Type and Size	NAILS Max. Spacing[6] Vertical	NAILS Max. Spacing[6] Horizontal	SCREWS[3,4] Max. Spacing[6] Vertical	SCREWS Max. Spacing[6] Horizontal	STAPLES[3,5] Wire Gage No.	STAPLES Crown	STAPLES Leg	STAPLES Max. Spacing[6] Vertical	STAPLES Max. Spacing[6] Horizontal
		(inches) $\times$ 25.4 for mm		(inches)			(inches) $\times$ 25.4 for mm			
1. Diamond mesh expanded metal lath and flat rib metal lath	4d blued smooth box $1^1/_2$"[7] No. 14 gage $^7/_{32}$" head (clinched)[8] 1" No. 11 gage $^7/_{16}$" head, barbed $1^1/_2$" No. 11 gage $^7/_{16}$" head, barbed	6 6 6	— — 6	6	6	16	$^3/_4$	$^7/_8$	6	6
2. $^3/_8$" (9.5 mm) rib metal lath and sheet lath	$1^1/_2$" No. 11 gage $^7/_{16}$" head, barbed	6	6	6	6	16	$^3/_4$	$1^1/_4$	At ribs	At ribs
3. $^3/_4$" (19.1 mm) rib metal lath	4d common $1^1/_2$" No. $12^1/_2$ gage $^1/_4$" head 2" No. 11 gage $^7/_{16}$" head, barbed	At ribs	— At ribs	At ribs	At ribs	16	$^3/_4$	$1^5/_8$	At ribs	At ribs
4. Wire fabric lath[9]	4d blued smooth box (clinched)[8] 1" No. 11 gage $^7/_{16}$" head, barbed	6 6	— —	6	6	16	$^3/_4$	$^7/_8$	6	6
	$1^1/_2$" No. 11 gage $^7/_{16}$" head, barbed $1^1/_4$" No. 12 gage $^3/_8$" head, furring 1" No. 12 gage $^3/_8$" head	6 6 6	6 6			16	$^7/_{16}$[9]	$^7/_8$	6	6
5. $^3/_8$" (9.5 mm) gypsum lath	$1^1/_8$" No. 13 gage $^{19}/_{64}$" head, blued	8[10]	8[10]	8[10]	8[10]	16	$^3/_4$	$^7/_8$[11]	8[10]	8[10]
6. $^1/_2$" (12.7 mm) gypsum lath	$1^1/_4$" No. 13 gage $^{19}/_{64}$" head, blued	8	8[10] 6[7]	8[10]	8[10] 6[7]	16	$^3/_4$	$1^1/_8$[11]	8[10]	8[10] 6[7]

[1]Metal lath, wire lath, wire fabric lath and metal accessories shall conform to approved standards.

[2]For nailable nonload-bearing metal supports, use annular threaded nails or approved staples.

[3]For fire-resistive construction, see Tables 7-B and 7-C. For shear-resisting elements, see Table 25-I. Approved wire and sheet metal attachment clips may be used.

[4]Screws shall be an approved type long enough to penetrate into wood framing not less than $^5/_8$ inch (15.9 mm) and through metal supports adaptable for screw attachment not less than $^1/_4$ inch (6.4 mm).

[5]With chisel or divergent points.

[6]Maximum spacing of attachments from longitudinal edges shall not exceed 2 inches (51 mm).

[7]Supports spaced 24 inches (610 mm) on center. Four attachments per 16-inch-wide (406 mm) lath per bearing. Five attachments per 24-inch-wide (610 mm) lath per bearing.

[8]For interiors only.

[9]Attach self-furring wire fabric lath to supports at furring device.

[10]Three attachments per 16-inch-wide (406.4 mm) lath per bearing. Four attachments per 24-inch-wide (610 mm) lath per bearing.

[11]When lath and stripping are stapled simultaneously, increase leg length of staple $^1/_8$ inch (3.2 mm).

TABLE 25-D—THICKNESS OF PLASTER[1]

PLASTER BASE	FINISHED THICKNESS OF PLASTER FROM FACE OF LATH, MASONRY, CONCRETE $\times$ 25.4 for mm — Gypsum Plaster	Portland Cement Plaster
1. Expanded metal lath	$^5/_8$" minimum[2]	$^5/_8$" minimum[2]
2. Wire fabric lath	$^5/_8$" minimum[2]	$^3/_4$" minimum (interior)[3] $^7/_8$" minimum (exterior)[3]
3. Gypsum lath	$^1/_2$" minimum	
4. Masonry walls[4]	$^1/_2$" minimum	$^1/_2$" minimum
5. Monolithic concrete walls[4,5]	$^5/_8$" maximum[6]	$^7/_8$" maximum[6]
6. Monolithic concrete ceilings[4,5]	$^3/_8$" maximum[6,7,8]	$^1/_2$" maximum[7,8]

[1]For fire-resistive construction, see Tables 7-A, 7-B and 7-C.

[2]When measured from back plane of expanded metal lath, exclusive of ribs, or self-furring lath, plaster thickness shall be $^3/_4$ inch (19 mm) minimum.

[3]When measured from face of support or backing.

[4]Because masonry and concrete surfaces may vary in plane, thickness of plaster need not be uniform.

[5]When applied over a liquid bonding agent, finish coat may be applied directly to concrete surface.

[6]An approved skim-coat plaster $^1/_{16}$ inch (1.6 mm) thick may be applied directly to concrete.

[7]On concrete ceilings, where the base coat plaster thickness exceeds the maximum thickness shown, metal lath or wire fabric lath shall be attached to the concrete.

[8]Approved acoustical plaster may be applied directly to concrete, or over base coat plaster, beyond the maximum plaster thickness shown.

TABLE 25-E—GYPSUM PLASTER PROPORTIONS[1]

NUMBER	COAT	PLASTER BASE OR LATH	MAXIMUM VOLUME AGGREGATE PER 100 POUNDS (45.4 kg) NEAT PLASTER[2,3] (cubic feet) × 0.028 for m³	
			Damp Loose Sand[4]	Perlite or Vermiculite[4]
1. Two-coat work	Base coat	Gypsum lath	$2^1/_2$	2
	Base coat	Masonry	3	3
2. Three-coat work	First coat	Lath	2^5	2
	Second coat	Lath	3^5	2^6
	First and second coats	Masonry	3	3

[1]Wood-fibered gypsum plaster may be mixed in the proportions of 100 pounds (45.4 kg) of gypsum to not more than 1 cubic foot (0.028 m³) of sand where applied on masonry or concrete.
[2]For fire-resistive construction, see Tables 7-A, 7-B and 7-C.
[3]When determining the amount of aggregate in set plaster, a tolerance of 10 percent shall be allowed.
[4]Combinations of sand and lightweight aggregate may be used, provided the volume and weight relationship of the combined aggregate to gypsum plaster is maintained.
[5]If used for both first and second coats, the volume of aggregate may be $2^1/_2$ cubic feet (0.07 m³).
[6]Where plaster is 1 inch (25 mm) or more in total thickness, the proportions for the second coat may be increased to 3 cubic feet (0.08 m³).

TABLE 25-F—CEMENT PLASTERS[1]

PORTLAND CEMENT PLASTER							

Coat	Volume Cement	Maximum Weight (or Volume) Lime per Volume Cement	Maximum Volume Sand per Combined Volumes Cement and Lime[2]	Approximate Minimum Thickness[3]	× 25.4 for mm	Minimum Period Moist Curing	Minimum Interval between Coats
First	1	20 lbs. (9.07 kg)	4		$3/_8″$[4]	48 hours[5]	48 hours[6]
Second	1	20 lbs. (9.07 kg)	5	1st and 2nd coats total $3/_4″$		48 hours	7 days[7]
Finish	1	1^8	3	1st, 2nd and finish coats $7/_8″$		—	7

PORTLAND CEMENT-LIME PLASTER[9]							

Coat	Volume Cement	Maximum Volume Lime per Volume Cement	Maximum Volume Sand per Combined Volumes Cement and Lime[3]	Approximate Minimum Thickness[3]	× 25.4 for mm	Minimum Period Moist Curing	Minimum Interval between Coats
First	1	1	4		$3/_8″$[4]	48 hours[5]	48 hours[6]
Second	1	1	$4^1/_2$	1st and 2nd coats total $3/_4″$		48 hours	7 days[7]
Finish	1	1^8	3	1st, 2nd and finish coats $7/_8″$		—	7

PLASTIC CEMENT PLASTER[9]							

Coat	Volume Cement	Maximum Weight (or Volume) Lime per Volume Cement	Maximum Volume Sand per Volume Cement[2]	Approximate Minimum Thickness[3]	× 25.4 for mm	Minimum Period Moist Curing	Minimum Interval between Coats
First	1	—	4		$3/_8″$[4]	48 hours[5]	48 hours[6]
Second	1	—	5	1st and 2nd coats total $3/_4″$		48 hours	7 days[7]
Finish	1	—	3	1st, 2nd and finish coats $7/_8″$		—	7

[1]Exposed aggregate plaster shall be applied in accordance with Section 2509. Minimum overall thickness shall be $3/_4$ inch (19 mm).
[2]When determining the amount of sand in set plaster, a tolerance of 10 percent may be allowed.
[3]See Table 25-D.
[4]Measured from face of support or backing to crest of scored plaster.
[5]See Section 2507.3.3.
[6]Twenty-four-hour minimum interval between coats of interior cement plaster. For alternate method of application, see Section 2508.6.
[7]Finish coat plaster may be applied to interior portland cement base coats after a 48-hour period.
[8]For finish coat plaster, up to an equal part of dry hydrated lime by weight (or an equivalent volume of lime putty) may be added to Types I, II and III standard portland cement.
[9]No additions of plasticizing agents shall be made.

TABLE 25-G
TABLE 25-H

1997 UNIFORM BUILDING CODE

TABLE 25-G—SINGLE-PLY GYPSUM WALLBOARD APPLIED PARALLEL (||) OR PERPENDICULAR (⊥) TO FRAMING MEMBERS

THICKNESS OF GYPSUM WALLBOARD (inch) × 25.4 for mm	PLANE OF FRAMING SURFACE	MAXIMUM SPACING OF FRAMING MEMBER[1] (Center to Center) (inches) × 25.4 for mm	LONG DIMENSION OF GYPSUM WALLBOARD SHEETS IN RELATION TO DIRECTION OF FRAMING MEMBERS ||	LONG DIMENSION OF GYPSUM WALLBOARD SHEETS IN RELATION TO DIRECTION OF FRAMING MEMBERS ⊥	MAXIMUM SPACING OF FASTENERS[1] (Center to Center) (inches) × 25.4 for mm Nails[3]	MAXIMUM SPACING OF FASTENERS[1] (Center to Center) (inches) × 25.4 for mm Screws[4]	NAILS[2]—TO WOOD × 25.4 for mm
$1/2$	Horizontal	16	P	P	7	12	No. 13 gage, $1^3/8''$ long, $^{19}/_{64}''$ head; 0.098″ diameter, $1^1/4''$ long, annular ringed; 5d, cooler (0.086″ dia., $1^5/8''$ long, $^{15}/_{64}''$ head) or wallboard (0.086″ dia., $1^5/8''$ long, $^9/_{32}''$ head) nail.
$1/2$	Horizontal	24	NP	P	7	12	
$1/2$	Vertical	16	P	P	8	16	
$1/2$	Vertical	24	P	P	8	12	
$5/8$	Horizontal	16	P	P	7	12	No. 13 gage, $1^5/8''$ long, $^{19}/_{64}''$ head; 0.098″ diameter, $1^3/8''$ long, annular ringed; 6d, cooler (0.092″ dia., $1^7/8''$ long, $1/4''$ head) or wallboard (0.0915″ dia., $1^7/8''$ long, $^{19}/_{64}''$ head) nail.
$5/8$	Horizontal	24	NP	P	7	12	
$5/8$	Vertical	16	P	P	8	16	
$5/8$	Vertical	24	P	P	8	12	

Nail or Screw Fastenings with Adhesives (Maximum Center to Center in Inches)

× 25.4 for mm

(Column headings as above)					End	Edges	Field	
$1/2$ or $5/8$	Horizontal	16	P	P	16	16	24	As required for $1/2''$ and $5/8''$ gypsum wallboard, see above.
$1/2$ or $5/8$	Horizontal	24	NP	P	16	24	24	
$1/2$ or $5/8$	Vertical	24	P	P	16	24	NR	

NOTES: Horizontal refers to applications such as ceilings. Vertical refers to applications such as walls.

|| denotes parallel.

⊥ denotes perpendicular.   P—Permitted.   NP—Not permitted.   NR—Not required.

[1] A combination of fasteners consisting of nails along the perimeter and screws in the field of the gypsum board may be used with the spacing of the fasteners shown in the table.

For fire-resistive construction, see Tables 7-B and 7-C. For shear-resisting elements, see Table 25-I.

[2] Where the metal framing has a clinching design formed to receive the nails by two edges of metal, the nails shall not be less than $5/8$ inch (15.9 mm) longer than the wallboard thickness, and shall have ringed shanks. Where the metal framing has a nailing groove formed to receive the nails, the nails shall have barbed shanks or be 5d, No. $13^1/2$ gage, $1^5/8$ inches (41 mm) long, $^{15}/_{64}$-inch (6.0 mm) head for $1/2$-inch (12.7 mm) gypsum wallboard; 6d, No. 13 gage, $1^7/8$ (48 mm) inches long, $^{15}/_{64}$-inch (6.0 mm) head for $5/8$-inch (15.9 mm) gypsum wallboard.

[3] Two nails spaced 2 inches to $2^1/2$ inches (51 mm to 64 mm) apart may be used where the pairs are spaced 12 inches (305 mm) on center except around the perimeter of the sheets.

[4] Screws shall be long enough to penetrate into wood framing not less than $5/8$ inch (15.9 mm) and through metal framing not less than $1/4$ inch (6.4 mm).

TABLE 25-H—APPLICATION OF TWO-PLY GYPSUM WALLBOARD[1]

				FASTENERS ONLY				

Thickness of Gypsum Wallboard (Each Ply) (inch) × 25.4 for mm	Plane of Framing Surface	Long Dimension of Gypsum Wallboard Sheets	Maximum Spacing of Framing Members (Center to Center) (inches) × 25.4 for mm	Maximum Spacing of Fasteners (Center to Center) (inches) × 25.4 for mm Base Ply Nails[2]	Maximum Spacing of Fasteners (Center to Center) (inches) × 25.4 for mm Base Ply Screws[3]	Maximum Spacing of Fasteners (Center to Center) (inches) × 25.4 for mm Base Ply Staples[4]	Maximum Spacing of Fasteners (Center to Center) (inches) × 25.4 for mm Face Ply Nails[2]	Maximum Spacing of Fasteners (Center to Center) (inches) × 25.4 for mm Face Ply Screws[3]
$3/8$	Horizontal	Perpendicular only	16	16	24	16	7	12
$3/8$	Vertical	Either direction	16	16	24	16	8	12
$1/2$	Horizontal	Perpendicular only	24	16	24	16	7	12
$1/2$	Vertical	Either direction	24	16	24	16	8	12
$5/8$	Horizontal	Perpendicular only	24	16	24	16	7	12
$5/8$	Vertical	Either direction	24	16	24	16	8	12

				FASTENERS AND ADHESIVES				
$3/8$ Base ply	Horizontal	Perpendicular only	16	7			5	Temporary nailing or shoring to comply with Section 2511.4
$3/8$ Base ply	Vertical	Either direction	24	8			7	
$1/2$ Base ply	Horizontal	Perpendicular only	24	7	12		5	
$1/2$ Base ply	Vertical	Either direction	24	8	12		7	
$5/8$ Base ply	Horizontal	Perpendicular only	24	7			5	
$5/8$ Base ply	Vertical	Either direction	24	8			7	

[1] For fire-resistive construction, see Tables 7-B and 7-C. For shear-resisting elements, see Table 25-I.

[2] Nails for wood framing shall be long enough to penetrate into wood members not less than $3/4$ inch (19.1 mm), and the sizes shall comply with the provisions of Table 25-G. For nails not included in Table 25-G, use the appropriate size cooler or wallboard nails. Nails for metal framing shall comply with the provisions of Table 25-G.

[3] Screws shall comply with the provisions of Table 25-G.

[4] Staples shall not be less than No. 16 gage by $3/4$-inch (19.1 mm) crown width with leg length of $7/8$ inch (22.2 mm), $1^1/8$ inches (28.6 mm) and $1^3/8$ inches (34.9 mm) for gypsum wallboard thicknesses of $3/8$ inch (9.5 mm), $1/2$ inch (12.7 mm) and $5/8$ inch (15.9 mm), respectively.

TABLE 25-I—ALLOWABLE SHEAR FOR WIND OR SEISMIC FORCES IN POUNDS PER FOOT FOR VERTICAL DIAPHRAGMS OF LATH AND PLASTER OR GYPSUM BOARD FRAME WALL ASSEMBLIES[1]

TYPE OF MATERIAL	THICKNESS OF MATERIAL × 25.4 for mm × 304.8 for mm	WALL CONSTRUCTION	NAIL SPACING[2] MAXIMUM (inches) × 25.4 for mm	SHEAR VALUE × 14.6 for N/m	MINIMUM NAIL SIZE[3] × 25.4 for mm
1. Expanded metal, or woven wire lath and portland cement plaster	$7/8''$	Unblocked	6	180	No. 11 gage, $1^1/_2''$ long, $7/_{16}''$ head No. 16 gage staple, $7/_8''$ legs
2. Gypsum lath	$3/_8''$ lath and $1/_2''$ plaster	Unblocked	5	100	No. 13 gage, $1^1/_8''$ long, $19/_{64}''$ head, plasterboard blued nail
3. Gypsum sheathing board	$1/_2'' \times 2' \times 8'$	Unblocked	4	75	No. 11 gage, $1^3/_4''$ long, $7/_{16}''$ head, diamond-point, galvanized
	$1/_2'' \times 4'$	Blocked	4	175	
	$1/_2'' \times 4'$	Unblocked	7	100	
4. Gypsum wallboard or veneer base	$1/_2''$	Unblocked	7	100	5d cooler (0.086″ dia., $1^5/_8''$ long, $15/_{64}''$ head) or wallboard (0.086″ dia., $1^5/_8''$ long, $9/_{32}''$ head)
			4	125	
		Blocked	7	125	
			4	150	
	$5/_8''$	Unblocked	7	115	6d cooler (0.092″ dia., $1^7/_8''$ long, $1/_4''$ head) or wallboard (0.0915″ dia., $1^7/_8''$ long, $19/_{64}''$ head)
			4	145	
		Blocked	7	145	
			4	175	
		Blocked Two ply	Base ply: 9 Face ply: 7	250	Base ply—6d cooler (0.092″ dia., $1^7/_8''$ long, $1/_4''$ head) or wallboard (0.0915″ dia., $1^7/_8''$ long, $19/_{64}''$ head) Face ply—8d cooler (0.113″ dia., $2^3/_8''$ long, $9/_{32}''$ head) or wallboard (0.113″ dia., $2^3/_8''$ long, $3/_8''$ head)

[1]These vertical diaphragms shall not be used to resist loads imposed by masonry or concrete construction. See Section 2513.2. Values shown are for short-term loading due to wind or due to seismic loading. Values shown must be reduced 25 percent for normal loading. The values shown in Items 2, 3 and 4 shall be reduced 50 percent for loading due to earthquake in Seismic Zones 3 and 4.

[2]Applies to nailing at all studs, top and bottom plates, and blocking.

[3]Alternate nails may be used if their dimensions are not less than the specified dimensions.

Chapter 26

PLASTIC

SECTION 2601 — SCOPE

Foam plastics, light-transmitting plastics and plastic veneers shall comply with this chapter.

See Section 1404 for requirements for vinyl siding.

SECTION 2602 — FOAM PLASTIC INSULATION

2602.1 General. The provisions of this section shall govern the requirements and uses of foam plastic insulation in buildings and structures. For trim, see Section 601.5.5.

2602.2 Labeling and Identification. Packages and containers of foam plastic insulation and foam plastic insulation components delivered to the jobsite shall bear the label of an approved agency showing the manufacturer's name, the product listing, product identification and information to show that the end use will comply with the code requirements.

2602.3 Surface-burning Characteristics. Foam plastic insulation used in building construction shall have a flame-spread rating of not more than 75 and a smoke-developed rating of not more than 450 when tested in accordance with UBC Standard 8-1 in the maximum thickness intended for use.

> **EXCEPTION:** Foam plastic insulation when tested in a minimum thickness of 4 inches (102 mm) may be used in a greater thickness in cold-storage buildings, ice plants, food-processing rooms and similar areas. For rooms within a building, the foam plastic insulation shall be protected by a thermal barrier on both sides having an index of 15.

2602.4 Thermal Barrier. The interior of the building shall be separated from the foam plastic insulation by an approved thermal barrier having an index of 15 when tested in accordance with UBC Standard 26-2. The thermal barrier shall be installed in such a manner that it will remain in place for the time of its index classification based on approved diversified tests.

> **EXCEPTION:** The thermal barrier is not required:
>
> 1. For siding backer board, provided the foam plastic insulation is not of more than 2,000 Btu per square foot (22.7 MJ/m^2) as determined by UBC Standard 26-1 and when it is separated from the interior of the building by not less than 2 inches (51 mm) of mineral fiber insulation or equivalent, or applied as re-siding over existing wall construction.
>
> 2. For walk-in coolers and freezer units having an aggregate floor area less than 400 square feet (37.2 m^2).
>
> 3. In a masonry or concrete wall, floor or roof system when the foam plastic insulation is covered by a minimum of 1-inch (25 mm) thickness of masonry or concrete. Loose-fill-type foam plastic insulation shall be tested as board stock for flame spread and smoke development as described above.
>
> 4. Within an attic or crawl space where entry is made only for service of utilities, and when foam plastic insulation is covered with a material such as $1^1/_2$-inch-thick (38 mm) mineral fiber insulation; $^1/_4$-inch-thick (6.4 mm) plywood, hardboard or gypsum wallboard; corrosion-resistant sheet metal having a base metal thickness not less than 0.0160 inch (0.4 mm) at any point; or other approved material installed in such a manner that the foam plastic insulation is not exposed.
>
> 5. In cooler and freezer walls when:
>
> > 5.1 The foam plastic insulation has a flame-spread rating of 25 or less when tested in a minimum 4-inch (102 mm) thickness;
> >
> > 5.2 Has flash and self-ignition temperatures of not less than 600°F and 800°F (316°C and 427°C), respectively;

> > 5.3 Is covered by not less than 0.032-inch (0.8 mm) aluminum or corrosion-resistant steel having a base metal thickness not less than 0.0160 inch (0.4 mm) at any point; and
> >
> > 5.4 Is protected by an automatic sprinkler system. When the cooler or freezer is within a building, both the cooler or freezer and that part of the building in which it is located shall be sprinklered.
>
> 6. Exterior garage doors in Group U, Division 1 Occupancies.

2602.5 Special Provisions.

2602.5.1 General. Foam plastic insulation may be used in the applications set forth in this section.

2602.5.2 Noncombustible exterior walls.

2602.5.2.1 One-story buildings. Foam plastic insulation may be used in exterior walls of one-story buildings where exterior walls are required to be of noncombustible construction subject to the following:

1. The building is protected throughout with automatic sprinklers.

2. Foam plastic insulation, tested in the maximum thickness and density intended for use, has a flame-spread rating of 25 or less and a smoke-developed rating of 450 or less in accordance with UBC Standard 8-1.

3. The foam plastic insulation has a maximum 4-inch (102 mm) thickness.

4. The thermal barrier may be omitted when the foam plastic insulation is covered by not less than 0.032-inch-thick (0.8 mm) aluminum or corrosion-resistant sheet steel, having a base metal thickness of 0.0160 (0.4 mm) inch.

5. When the wall is required to have a fire-resistive rating, data based on tests conducted in accordance with UBC Standard 7-1 are provided to substantiate that the required fire-resistive rating is maintained.

2602.5.2.2 Buildings of any height. Except for foam plastic insulation in masonry or concrete construction complying with Section 2602.4, Exception 3, assemblies employing foam plastic insulation in or on exterior walls of buildings where the exterior walls are required to be of noncombustible construction shall comply with the following:

1. When the wall is required to have a fire-resistive rating, data based on tests conducted in accordance with UBC Standard 7-1 are provided to substantiate that the fire-resistive rating is maintained.

2. The foam plastic insulation is separated from the interior of the building by a thermal barrier having an index of 15 unless specifically approved under Section 2602.6.

3. Combustible content of foam plastic insulation in any portion of the wall or panel does not exceed 6,000 Btu per square foot (68.2 MJ/m^2) of wall area as determined by tests in accordance with UBC Standard 26-1.

4. Foam plastic insulation, exterior coatings and facings tested separately shall each have a flame-spread rating of 25 or less and a smoke-developed rating of 450 or less in accordance with UBC Standard 8-1. The foam plastic shall be tested in the thickness intended for use.

5. The wall assembly passes the conditions of acceptance of UBC Standard 26-4 or 26-9.

6. Foam plastic insulation is listed and the edge or face of each piece is labeled with the following information:

6.1 Inspection agency name.

6.2 Product for which the insulation is listed.

6.3 Identification of the insulation manufacturer.

6.4 Flame-spread and smoke-development classifications.

2602.5.3 Roofing. Foam plastic insulation meeting the requirements of Sections 2602.2, 2602.3 and 2602.4 may be used as part of a roof-covering assembly, provided the assembly with the foam plastic insulation is a Class A, B or C roofing assembly when tested in accordance with UBC Standard 15-2. Foam plastic insulation, which is a part of a Class A, B or C roof-covering assembly, need not meet the requirements of Sections 2602.2, 2602.3 and 2602.4, provided the assembly with the foam plastic insulation satisfactorily passes a test for insulated roof decks.

Any roofing assembly or roof covering installed in accordance with this code and the manufacturer's instructions may be applied over foam plastic insulation when the foam is separated from the interior of the building by wood structural panel sheathing not less than $^1/_2$ inch (12.7 mm) in thickness bonded with exterior glue, with edges supported by blocking, tongue-and-groove joints or other approved type of edge support, or an equivalent material. The thermal barrier requirement is waived.

For all roof applications, the smoke-developed rating shall not be limited.

2602.5.4 Doors. Where pivoted or side-hinged swinging doors are permitted without a fire-resistive rating, foam plastic insulation having a flame-spread rating of 75 or less may be used as a core material when the door facing is metal having a minimum thickness of 0.032-inch (0.8 mm) aluminum or steel having a base metal thickness not less than 0.0160 inch (0.4 mm) at any point. The thermal barrier is not required for this condition.

2602.5.5 Garage doors. In other than Group U Occupancies and where garage doors are permitted without a fire-resistive rating, foam plastic insulation may be used without thermal barrier in such garage doors subject to the following:

1. The foam plastic insulation shall have a flame-spread index of 75 or less and smoke-developed rating of 450 or less, and

2. The facing on the side of the garage door located inside a building shall be minimum 0.010-inch (0.3 mm) steel or minimum $^1/_8$-inch (3.2 mm) wood or the garage door shall be tested in accordance with, and shall meet the acceptance criteria of, UBC Standard 26-8.

2602.6 Specific Approval. Foam plastic insulation or assemblies using foam plastic insulation may be used based on approved tests such as, but not limited to, tunnel tests in accordance with UBC Standard 8-1, fire tests related to actual end use such as UBC Standard 26-3 and an ignition temperature test establishing a minimum self-ignition temperature of 650°F (343°C). In lieu of testing, the specific approval may be based on the end use, quantity, location and similar considerations where such tests would not be applicable or practical.

Foam plastic insulation in a thickness greater than 4 inches (102 mm) may be used if it has been tested for flame spread and smoke development at a minimum thickness of 4 inches (102 mm) provided the end use has been specifically approved in accordance with this section with the thickness and density intended for use.

SECTION 2603 — LIGHT-TRANSMITTING PLASTICS

2603.1 General.

2603.1.1 Scope. The provisions of this section shall govern the quality and methods of application of plastics for use as light-transmitting materials in buildings and structures. For foam plastics, see Sections 601.5.5 and 2602. Light-transmitting plastic materials that meet the other code requirements for walls and roofs may be used in accordance with the other applicable chapters of the code.

2603.1.2 Approval for use. The building official shall require that sufficient technical data be submitted to substantiate the proposed use of any light-transmitting material and, if it is determined that the evidence submitted is satisfactory for the use intended, the building official may approve its use subject to the requirements of this section.

2603.1.3 Identification. Each unit or package of plastic shall be identified with a mark or decal satisfactory to the building official, which includes identification as to the material classification.

2603.1.4 Combination of glazing and exterior wall panels. Combinations of plastic glazing and plastic exterior wall panels shall be subject to the area, height, percentage and separation requirements applicable to the class of plastics as prescribed for wall panel installation.

2603.1.5 Combination of roof panels and skylights. Combinations of plastic roof panels and plastic skylights shall be subject to the area percentage and separation requirements applicable to roof panel installation.

2603.1.6 Standards of quality. The standards listed below labeled a "UBC Standard" are also listed in Chapter 35, Part II, and are part of this code.

1. UBC Standard 15-2, Test Standard for Determining the Fire Retardancy of Roof Assemblies

2. UBC Standard 26-5, Chamber Method of Test for Measuring the Density of Smoke from the Burning or Decomposition of Plastic Materials

3. UBC Standard 26-6, Ignition Properties of Plastics

4. UBC Standard 26-7, Method of Test for Determining Classification of Approved Light-transmitting Plastics

2603.2 Definitions. For the purpose of this section, certain terms are defined as follows:

EXTERIOR WALL PANELS are materials that are not classified as plastic glazing and are used as light-transmitting media in exterior walls.

GLASS FIBER REINFORCED PLASTIC is plastic reinforced with glass fiber having not less than 20 percent of glass fibers by weight.

GLAZING is material that has all edges set in a frame or sash and is not held by mechanical fasteners that pass through the material.

LIGHT-DIFFUSING SYSTEM is construction consisting in whole or in part of lenses, panels, grids or baffles made with approved plastics positioned below independently mounted electrical light sources. Lenses, panels, grids and baffles that are part of an electrical fixture shall not be considered as a light-diffusing system.

PLASTIC MATERIALS, APPROVED. See Chapter 2.

ROOF PANELS are structural panels other than skylights that are fastened to structural members or structural panels or sheathing and are used as light-transmitting media in the plane of the roof.

THERMOPLASTIC MATERIAL is a plastic material capable of being repeatedly softened by increase of temperature and hardened by decrease of temperature.

THERMOSETTING MATERIAL is a plastic material capable of being changed into a substantially nonreformable product when cured.

2603.3 Design and Installation.

2603.3.1 Structural requirements. Plastic materials in their assembly shall be of adequate strength and durability to withstand the design loads as prescribed elsewhere in this code. Technical data shall be submitted to establish stresses, maximum unsupported spans and such other information for the various thicknesses and forms used as may be deemed necessary by the building official.

2603.3.2 Fastening. Fastening shall be adequate to withstand design loads as prescribed elsewhere in this code. Proper allowance shall be made for expansion and contraction of plastic materials in accordance with accepted data on coefficient of expansion of the material and other material in conjunction with which it is employed.

2603.4 Glazing of Unprotected Openings. In Type V-N construction, doors, sash and framed openings not required to be fire protected may be glazed or equipped with approved plastic material.

In types of construction other than Type V-N, openings not required to be fire protected may be glazed or equipped with approved plastic, subject to the following requirements:

1. The aggregate area of plastic glazing shall not exceed 25 percent of the area of any wall face of the story in which it is installed. The area of a single pane of glazing installed above the first story shall not exceed 16 square feet (1.5 m^2) and the vertical dimension of a single pane shall not exceed 4 feet (1219 mm).

> EXCEPTION: When an approved automatic sprinkler system is provided throughout, the area of glazing may be increased to a maximum of 50 percent of the wall face of the story in which it is installed with no limit on the maximum dimension or area of a single pane of glazing.

2. Approved flame barriers extending 30 inches (762 mm) beyond the exterior wall in the plane of the floor, or vertical panels not less than 4 feet (1219 mm) in height, shall be installed between glazed units located in adjacent stories.

3. Plastics shall not be installed more than 65 feet (19 812 mm) above grade level.

2603.5 Light-transmitting Exterior Wall Panels. In Type V-N construction, approved plastics may be installed in exterior walls provided the walls are not required to have a fire-resistive rating.

In types of construction other than Type V-N, approved plastics may be installed in exterior walls, provided the walls are not required to have a fire-resistive rating, subject to the following requirements:

1. Approved exterior wall panels shall not be installed more than 40 feet (12 192 mm) above grade level.

2. Approved exterior wall panels shall not be installed in exterior walls located less than 10 feet (3048 mm) from the property line determined in accordance with Section 503.

3. The area and size shall be limited to that set forth in Table 26-A.

> EXCEPTIONS: 1. In structures provided with approved flame barriers extending 30 inches (762 mm) beyond the exterior wall in the plane of the floor, there need be no vertical separation at the floor ex-

cept that provided by the vertical thickness of the flame-barrier projection.

> 2. When an approved automatic sprinkler system is provided throughout the building, the maximum percentage area of plastic panels in the exterior wall and the maximum square feet of any individual panel may be increased 50 percent above that set forth in Table 26-A, and the separation requirements, both vertical and horizontal, as set forth in Table 26-A may be reduced by 50 percent.

2603.6 Roof Panels. Approved plastic roof panels may be installed in roofs of buildings not required to have a fire-resistive rating, subject to the following limitations:

1. Individual roof panels or units shall be separated from each other by distances of not less than 4 feet (1219 mm) measured in a horizontal plane.

2. Roof panels or units shall not be installed within that portion of a roof located within a distance to property line or public way where openings in exterior walls are prohibited or required to be protected, whichever is most restrictive.

3. Roof panels of Class CC1 plastics shall be limited to a maximum individual panel area of 150 square feet (13.9 m^2), and the total maximum aggregate area of all panels shall not exceed $33^1/_3$ percent of the floor area of the room or space sheltered. Roof panels of Class CC2 plastics shall be limited to a maximum individual panel area of 100 square feet (9.3 m^2), and the total maximum aggregate area of all panels shall not exceed 25 percent of the floor area of the room or space sheltered.

> EXCEPTION: Swimming pool shelters are exempt from the area limitations of Section 2603.6, provided such shelters do not exceed 5,000 square feet (464.5 m^2) in area and are not closer than 10 feet (3048 mm) to the property line or adjacent building.

2603.7 Skylights.

2603.7.1 General. Skylight assemblies may be glazed with approved plastic materials in accordance with the following provisions:

1. The plastics shall be mounted at least 4 inches (102 mm) above the plane of the roof by a curb constructed consistent with the requirements for the type of construction classification.

> EXCEPTION: Curbs may be omitted on roofs of Group R, Division 3 Occupancies with a minimum slope of 3 units vertical in 12 units horizontal (25% slope) when self-flashing skylights are used.

2. Flat or corrugated plastic skylights shall slope at least 4 units vertical in 12 units horizontal (33.3% slope). Dome-shaped skylights shall rise above the mounting flange a minimum distance equal to 10 percent of the maximum span of the dome but not less than 5 inches (127 mm).

> EXCEPTION: Skylights that pass the Class B Burning Brand Test specified in UBC Standard 15-2.

3. The edges of the plastic lights or domes shall be protected by metal or other noncombustible materials or shall be tested to show that equivalent fire protection is provided.

> EXCEPTION: The metal or noncombustible edge is not required where nonrated roof coverings are permitted.

4. Each skylight unit may have a maximum area within the curb of 100 square feet (9.3 m^2) for CC2 material and 200 square feet (18.6 m^2) for CC1 material.

> EXCEPTIONS: 1. The maximum area within the curb need not be limited if the building on which the skylights are located is not more than one story in height, the building has an exterior separation from other buildings of at least 30 feet (9144 mm), and the room or space sheltered by the roof is not classified in a Group I, Division 1.1, 1.2 or 3 Occupancy or as a required means of egress.

> 2. Except for Groups A, Divisions 1 and 2; I, Divisions 1.1, 1.2 and 2; and H, Divisions 1 and 2 Occupancies, the maximum area within the curb need not be limited where skylights are:

2.1 Serving as a fire venting system complying with this code, or

2.2 Used in a building completely equipped with an approved automatic sprinkler system complying with UBC Standard 9-1 or 9-3.

5. The aggregate area of skylights installed in the roof shall not exceed 33$^1/_3$ percent of the floor area of the room or space sheltered by the roof when CC1 materials are used and 25 percent when CC2 materials are used.

6. Skylight units shall be separated from each other by a distance of not less than 4 feet (1219 mm) measured in a horizontal plane.

EXCEPTIONS: 1. Except for Groups A, Divisions 1 and 2; I, Divisions 1.1, 1.2 and 2; and H, Divisions 1 and 2 Occupancies, the separation is not required where the skylights are:

1.1 Serving as a fire venting system complying with this code, or

1.2 Used in a building completely equipped with an approved automatic sprinkler system complying with UBC Standard 9-1 or 9-3.

2. Multiple skylights located above the same room or space with a combined area not exceeding the limits set forth in Section 2603.7.1, Item 4.

7. Skylights shall not be installed within that portion of a roof located within a distance to property line or public way where openings in exterior walls are prohibited or required to be protected, whichever is most restrictive.

2603.7.2 Plastics over stair shafts. Approved plastic materials that will not automatically vent but are able to be vented may be used over stairways and shafts, provided the installation conforms to the requirements of Section 2603.7.1.

2603.8 Light-diffusing Systems.

2603.8.1 General. Plastic diffusers in light-diffusing systems shall be supported directly or indirectly by the use of noncombustible hangers.

Light-transmitting plastic materials in light-diffusing systems shall comply with Chapter 8 unless the approved plastic used in the light-diffusing system meets the following requirements:

1. Diffusers shall fall from their mounting at an ambient temperature of at least 200°F (93°C) below the ignition temperature of the plastic material.

2. Diffusers shall remain in place at an ambient room temperature of 175°F (79°C) for a period of not less than 15 minutes.

3. The maximum length of any single plastic panel shall not exceed 10 feet (3048 mm), and the maximum area of any single plastic panel shall not exceed 30 square feet (2.8 m^2).

4. The area of approved plastic materials when used in required means of egress as defined in Chapter 10 shall not exceed 30 percent of the aggregate area of the ceiling in which they are installed.

EXCEPTION: The aggregate area need not be limited in a building equipped with an approved automatic sprinkler system complying with UBC Standard 9-1 or 9-3.

2603.8.2 Limitations. A plastic light-diffusing system shall not be installed in the areas to be equipped with automatic sprinklers unless appropriate tests have shown that the system does not prevent effective operation of the sprinklers or unless sprinklers are located both above and below the light-diffusing system to give effective sprinkler protection.

2603.9 Diffusers in Electrical Fixtures. Use of approved plastics as light-diffuser panels installed in approved electrical lighting fixtures in or on walls or ceilings shall comply with Chapter 8 unless the plastic panels meet the requirements of Section 2603.8.1.

2603.10 Partitions. Approved light-transmitting plastics may be used in or as partitions, in accordance with the requirements of this code.

2603.11 Awnings and Patio Covers. Approved plastics may be used in awnings and patio covers. All such awnings shall be constructed in accordance with provisions specified in Section 3206 for projections and appendages. For patio covers, see Appendix Chapter 31.

2603.12 Greenhouses. Approved plastics may be used in lieu of plain glass in greenhouses.

2603.13 Canopies. Plastic panels constructed of approved plastic materials may be installed in canopies erected over motor vehicle fuel-dispensing station fuel dispensers, provided the panels are located at least 10 feet (3048 mm) from any building on the same property and face yards or streets not less than 40 feet (12 192 mm) in width on the other sides. The aggregate area of plastics shall not exceed 1,000 square feet (92.9 m^2). The maximum area of any individual panel shall not exceed 100 square feet (9.3 m^2).

2603.14 Solar Collectors. Solar collectors having noncombustible sides and bottoms may be equipped with plastic covers on buildings not over three stories in height or 9,000 square feet (836.1 m^2) in total floor area, provided the plastic cover when exceeding a thickness of 0.010 inch (0.3 mm) shall be of approved plastic and the total area shall not exceed 33$^1/_3$ percent of the roof area for CC1 materials or 25 percent of the roof area for CC2 materials.

EXCEPTION: Plastic covers having a thickness of 0.010 inch (0.3 mm) or less may be of any plastic, provided the total area of the collectors does not exceed 33$^1/_3$ percent of the roof area.

SECTION 2604 — PLASTIC VENEER

2604.1 Interior Use. When used within a building, plastic veneer shall comply with the interior finish requirements of Chapter 8.

2604.2 Exterior Use. Exterior plastic veneer may be installed on the exterior walls of buildings of any type of construction in accordance with the following requirements:

1. Plastic veneer shall be of approved plastics materials as defined in Chapter 2.

2. Plastic veneer shall not be attached to any exterior wall to a height greater than 50 feet (15 240 mm) above grade.

3. Sections of plastic veneer shall not exceed 300 square feet (27.9 m^2) in area and shall be separated by a minimum of 4 feet (1219 mm) vertically.

EXCEPTION: The area and separation requirements and the smoke-density limitation are not applicable to plastic veneer applied to Type V-N buildings, provided the walls are not required to have a fire-resistive rating.

TABLE 26-A—AREA LIMITATION AND SEPARATION REQUIREMENTS FOR EXTERIOR WALL PANELS[1]

CLASS OF PLASTIC	MAXIMUM PERCENT AREA OF EXTERIOR WALLS IN PLASTIC PANELS	MAXIMUM SQUARE FEET SINGLE INDIVIDUAL PANELS × 0.093 for m²	MAXIMUM PANEL HEIGHT (feet) × 304.8 for mm	MINIMUM SEPARATION OF PANELS (feet) × 304.8 for mm	
				Vertical	Horizontal
CC1	25	100	16	6	4
CC2	15	75	8	8	4

[1]The maximum percent area of exterior walls limitation shall be based on the individual story wall area.

Chapter 27
ELECTRICAL SYSTEMS

SECTION 2701 — ELECTRICAL CODE

Electrical systems shall be in accordance with the Electrical Code.
(See Section 206, Electrical Code.)

Chapter 28
MECHANICAL SYSTEMS

SECTION 2801 — MECHANICAL CODE

The installation and maintenance of heating, ventilating, product removal, cooling and refrigerating systems shall be in accordance with the Mechanical Code.

SECTION 2802 — REFRIGERATION SYSTEM MACHINERY ROOM

Refrigeration systems shall comply with the Mechanical Code. When a refrigeration machinery room is required, it shall be separated from the remainder of the building or located on the property as required for a Group H, Division 7 Occupancy, regardless of area. A horizontal occupancy separation may be limited to the actual floor area of the machinery room. Structural supporting elements shall be protected only for the type of construction and not the occupancy separation. Means of egress from the machinery room shall comply with Section 1007.7.2. Nothing contained herein shall be used to limit the height or area of the building or the machinery room. The refrigeration system, its refrigerant and its safety devices shall be maintained in accordance with the Fire Code.

Chapter 29
PLUMBING SYSTEMS

SECTION 2901 — PLUMBING CODE

Plumbing systems shall comply with the Plumbing Code.

SECTION 2902 — NUMBER OF FIXTURES

2902.1 General. The number of plumbing fixtures within a building shall not be less than set forth in Section 2902. Fixtures located within unisex toilet and bathing rooms shall be included in determining the number of fixtures provided in an occupancy.

2902.2 Group A Occupancies. In Group A Occupancies, at least one lavatory for each two water closets for each sex shall be provided at an approved location. At least one drinking fountain shall be provided at each floor level in an approved location.

> **EXCEPTION:** A drinking fountain need not be provided in a drinking or dining establishment.

For other requirements on water closets, see Sections 807 and 2903. See Chapter 11 for access to water closets and drinking fountains.

2902.3 Groups B, F, H, M and S Occupancies. In Groups B, F, H, M and S Occupancies, buildings or portions thereof where persons are employed shall be provided with at least one water closet. Separate facilities shall be provided for each sex when the number of employees exceeds four. Such toilet facilities shall be located either in such building or conveniently in a building adjacent thereto on the same property.

Such water closet rooms in connection with food establishments where food is prepared, stored or served shall have a nonabsorbent interior finish as specified in Section 807.1, shall have hand-washing facilities therein or adjacent thereto, and shall be separated from food preparation or storage rooms as specified in Section 302.6.

For other requirements on water closets, see Section 2903.

2902.4 Group E Occupancies. Water closets shall be provided on the basis of the following ratio of water closets to the number of students:

	Boys	Girls
Elementary Schools	1:100	1:35
Secondary Schools	1:100	1:45

In addition, urinals shall be provided for boys on the basis of 1:30 in elementary and secondary schools.

There shall be provided at least one lavatory for each two water closets or urinals, and at least one drinking fountain on each floor for elementary and secondary schools.

For other requirements on water closets, see Sections 807 and 2903.

2902.5 Group I Occupancies. In Group I Occupancies, sanitation facilities for employees shall be provided as specified in Section 2902.3. Additional sanitation facilities shall be provided for other occupants when the facilities for employees are not accessible to such other occupants.

For other requirements on water closets, see Sections 807 and 2903.

2902.6 Group R Occupancies. Buildings classified as Group R Occupancies shall be provided with at least one water closet. Hotels or subdivisions thereof where both sexes are accommodated shall contain at least two separate toilet facilities that are conspicuously identified for male or female use, each of which contains at least one water closet.

> **EXCEPTION:** Hotel guest rooms may have one unidentified toilet facility.

Additional water closets shall be provided on each floor for each sex at the rate of one for every additional 10 guests, or fractional part thereof, in excess of 10.

Dwelling units shall be provided with a kitchen equipped with a kitchen sink. Dwelling units, congregate residences and lodging houses shall be provided with a bathroom equipped with facilities consisting of a water closet, lavatory, and either a bathtub or shower. Each sink, lavatory, and either a bathtub or shower shall be equipped with hot and cold running water necessary for its normal operation.

For other requirements on water closets, see Sections 807 and 2903.

SECTION 2903 — ALTERNATE NUMBER OF FIXTURES

As an alternate to the minimum number of plumbing fixtures required by this chapter, see Appendix Chapter 29. When adopted, as set forth in Section 101.3, it will take precedence over the requirements of this chapter.

SECTION 2904 — ACCESS TO WATER CLOSET STOOL

The water closet stool in all occupancies shall be located in a clear space not less than 30 inches (762 mm) in width. The clear space in front of the water closet stool shall not be less than 24 inches (610 mm).

See Chapter 11 for requirements for water closets on floors required to be accessible to persons with disabilities.

Chapter 30
ELEVATORS, DUMBWAITERS, ESCALATORS
AND MOVING WALKS

SECTION 3001 — SCOPE

The provisions of this chapter shall apply to the design, construction, installation, operation, alteration and repair of elevators, dumbwaiters, escalators and moving walks and their hoistways.

SECTION 3002 — ELEVATOR AND ELEVATOR LOBBY ENCLOSURES

Walls and partitions enclosing elevator and dumbwaiter hoistway shafts and escalator shafts shall not be of less than the fire-resistive construction required under Types of Construction in Chapter 6 of this code.

Elevator hoistway shaft enclosure walls not required to have a fire-resistive rating may be constructed with glass. Such glass shall be laminated glass that passes the test requirements of UBC Standard 24-2, Part I.

Elevator lobbies shall have at least one means of egress. The use of exit or exit-access doors shall not require keys, tools, or special knowledge or effort.

SECTION 3003 — SPECIAL PROVISIONS

3003.1 Number of Cars in Hoistway. When there are three or fewer elevator cars in a building, they may be located within the same hoistway enclosure. When there are four elevator cars, they shall be divided in such a manner that at least two separate hoistway enclosures are provided. When there are more than four elevators, not more than four elevator cars may be located within a single hoistway enclosure.

3003.2 Smoke-detection Recall. When the elevator vertical travel is 25 feet (7620 mm) or more, each associated elevator lobby or entrance area and associated machine rooms shall be provided with an approved, listed smoke detector for elevator recall purposes only. The detector may serve to close the elevator lobby door and additional doors at the hoistway opening allowed in Section 3006.

When the lobby or entrance area smoke detector, or machine room smoke detector, is activated, elevator doors shall be prevented from opening and all cars serving that lobby or entrance area, or served by equipment in that machine room, shall return to the main floor where they shall be under manual control only. If the main floor or transfer floor lobby or entrance-area smoke detector is activated, all cars serving the main floor or transfer floor shall return to a location approved by the chief of the fire department and building official where they shall be under manual control only.

3003.3 Standby Power. Standby power when required by Section 403 shall be provided to at least one elevator in each bank. Standby power shall be manually transferable to all elevators in each bank. Standby power shall be provided by an approved self-contained generator set to operate automatically whenever there is a loss of electrical power to the building. The generator set shall be located in a separate room enclosed by at least a one-hour fire-resistive occupancy separation. The generator shall have a fuel supply adequate to operate the equipment connected to it for a minimum of two hours.

EXCEPTIONS: 1. Where a single elevator serves all floor levels in the building and is located so that all areas of the building can be reached within a travel distance of 300 feet (91 440 mm) from the elevator, then only that elevator need be provided with standby power.

2. Standby power shall be capable of operating one elevator at a time in any bank or group of banks having a common lobby.

NOTE: A bank of elevators is a group of elevators or a single elevator controlled by a common operating system; that is, all those elevators that respond to a single call button constitute a bank of elevators. There is no limit on the number of cars that may be in a bank or group, but there may not be more than four cars within a common hoistway.

3003.4 Size of Cab and Control Locations.

3003.4.1 General. In buildings three or more stories in height served by an elevator or a building served by an elevator required by Chapter 11, at least one elevator serving all floors shall accommodate a wheelchair, in accordance with this section.

3003.4.2 Operation and leveling. The elevator shall be automatic and be provided with a self-leveling feature that will automatically bring the car to the floor landings within a tolerance of plus or minus $^1/_2$ inch (12.7 mm) under normal loading and unloading conditions. This self-leveling shall, within its zone, be entirely automatic and independent of the operating device and shall correct the overtravel or undertravel. The car shall also be maintained approximately level with the landing, irrespective of load.

3003.4.3 Door operation. Power-operated horizontally sliding car and hoistway doors opened and closed by automatic means shall be provided.

3003.4.4 Door size. Minimum clear width for elevator doors shall be 36 inches (914 mm).

EXCEPTION: When approved by the building official, the minimum door width may be reduced to 32 inches (813 mm) for a car with dimensions as permitted by the exception to Section 3003.4.7.

3003.4.5 Door protective and reopening device. Doors closed by automatic means shall be provided with a door-reopening device that will function to stop and reopen a car door and adjacent hoistway door in case the car door is obstructed while closing. This reopening device shall also be capable of sensing an object or person in the path of a closing door without requiring contact for activation at a nominal 5 inches and 29 inches (127 mm and 737 mm) above the floor.

Door-reopening devices shall remain effective for a period of not less than 20 seconds.

3003.4.6 Door delay (passenger service time).

3003.4.6.1 Hall call. The minimum acceptable time from notification that a car is answering a call (lantern and audible signal) until the doors of that car start to close shall be as indicated in the following table:

DISTANCE, feet (mm)	TIME (seconds)
0 to 5 (0 to 1524)	4
10 (3048)	7
15 (4572)	10
20 (6096)	13

The distance shall be established from a point in the center of the corridor or lobby [maximum 5 feet (1524 mm)] directly opposite the farthest hall button to the center line of the hoistway entrance.

3003.4.6.2 Car call. The minimum acceptable time for doors to remain fully open shall not be less than three seconds.

3003.4.7 Car inside. The car inside shall allow for the turning of a wheelchair. The minimum clear distance between walls or between wall and door, excluding return panels, shall not be less than 68 inches by 54 inches (1727 mm by 1372 mm). Minimum distance from wall to return panel shall not be less than 51 inches (1295 mm).

> **EXCEPTION:** When approved by the building official, elevators provided in schools, institutions or other buildings may have a minimum clear distance between walls or between wall and door, excluding return panels, of not less than 54 inches by 54 inches (1372 mm by 1372 mm). Minimum distance from wall to return panel shall not be less than 51 inches (1295 mm).

3003.4.8 Car controls. Controls shall be readily accessible from a wheelchair upon entering an elevator.

The center line of the alarm button and emergency stop switch shall be at a nominal 35 inches (889 mm), and the highest floor button no higher than 54 inches (1372 mm) from the floor. Floor registration buttons, exclusive of border, shall be a minimum $3/4$ inch (19.1 mm) in size, raised, flush or recessed. Visual indication shall be provided to show each call registered and extinguished when call is answered. Depth of flush or recessed buttons when operated shall not exceed $3/8$ inch (9.5 mm).

Markings shall be adjacent to the controls on a contrasting color background to the left of the controls. Letters or numbers shall be a minimum of $5/8$ inch (15.9 mm) high and raised or recessed 0.030 inch (0.8 mm).

Applied plates permanently attached shall be acceptable.

Emergency controls shall be grouped together at the bottom of the control panel.

Controls not essential to the automatic operation of the elevator may be located as convenient.

3003.4.9 Car position indicator and signal. A car position indicator shall be provided above the car operating panel or over the opening of each car to show the position of the car in the hoistway by illumination of the indication corresponding to the landing at which the car is stopped or passing.

Indications shall be on a contrasting color background and a minimum of $1/2$ inch (12.7 mm) in height.

In addition, an audible signal shall sound to tell a passenger that the car is stopping or passing a floor served by the elevator.

A special button located with emergency controls may be provided. Operation of the button will activate an audible signal only for the desired trip.

3003.4.10 Telephone or intercommunicating system. A means of two-way communication shall be provided between the elevator and a point outside the hoistway.

If a telephone is provided, it shall be located a maximum of 54 inches (1372 mm) from the floor with a minimum cord length of 29 inches (737 mm). Markings or the international symbol for telephones shall be adjacent to the control on a contrasting color background. Letters or numbers shall be a minimum of $5/8$ inch (15.9 mm) high and raised or recessed 0.030 inch (0.8 mm).

Applied plates permanently attached shall be acceptable.

3003.4.11 Floor covering. Floor covering shall have a nonslip hard surface that permits easy movement of wheelchairs.

If carpeting is used, it shall be securely attached, heavy duty, with a tight weave and low pile, installed without padding.

3003.4.12 Handrails. A handrail shall be provided on one wall of the car, preferably the rear. The rails shall be smooth and the inside surface at least $1^1/_2$ inches (38 mm) clear of the walls at a nominal height of 32 inches (813 mm) from the floor.

Nominal = ± 1 inch (25 mm).

> **NOTE:** Thirty-two inches (813 mm) required to reduce interference with car controls where lowest button is centered at 35 inches (889 mm) above floor.

3003.4.13 Minimum illumination. The minimum illumination at the car controls and the landing when the car and landing doors are open shall not be less than 5 footcandles (54 lx).

3003.4.14 Hall buttons. The center line of the hall call buttons shall be a nominal 42 inches (1067 mm) above the floor.

Direction buttons, exclusive of border, shall be a minimum of $3/4$ inch (19.1 mm) in size, raised, flush or recessed. Visual indication shall be provided to show each call registered and extinguished when the call is answered. Depth of flush or recessed button when operated shall not exceed $3/8$ inch (9.5 mm).

3003.4.15 Hall lantern. A visual and audible signal shall be provided at each hoistway entrance indicating to the prospective passenger the car answering the call and its direction of travel.

The visual signal for each direction shall be a minimum of $2^1/2$ inches (64 mm) in size and visible from the proximity of the hall call button.

The audible signal shall sound once for the up direction and twice for the down direction.

The center line of the fixture shall be located a minimum of 6 feet (1829 mm) from the floor.

The use of in-car lanterns conforming to the above and located in jamb shall be acceptable.

3003.4.16 Doorjamb marking. The floor designation shall be provided at each hoistway entrance on both sides of jamb visible from within the car and the elevator lobby at a height of 60 inches (1524 mm) above the floor. Designations shall be on a contrasting background 2 inches (51 mm) high and raised 0.030 inch (0.8 mm).

Applied plates permanently attached shall be acceptable.

3003.5 Stretcher Requirements. In all structures four or more stories in height, at least one elevator shall be provided with a minimum clear distance between walls or between walls and door excluding return panels, of not less than 80 inches by 54 inches (2032 mm by 1372 mm), and a minimum distance from wall to return panel of not less than 51 inches (1295 mm) with a 42-inch (1067 mm) side slide door, unless otherwise designed to accommodate an ambulance-type stretcher 76 inches (1930 mm) by 24 inches (610 mm) in the horizontal position.

In buildings where one elevator does not serve all floors, two or more elevators may be used. The elevators shall be identified by the international symbol for emergency medical services (Star of Life). The symbol shall not be less than 3 inches (76 mm) and placed inside on both sides of the hoistway door frame. The symbol shall be placed no lower than 78 inches (1981 mm) from the floor level or higher than 84 inches (2134 mm) from floor level.

3003.6 Emergency Signs. Except at the main entrance level, an approved pictorial sign of a standardized design shall be posted adjacent to each elevator call station to indicate that, in case of fire, the elevator will not operate and that exit stairways should be used.

3003.7 Restricted or Limited-use Elevators. The building official may waive the requirements of this section for any elevator designed for limited or restricted use serving only specific floors or a specific function.

SECTION 3004 — HOISTWAY VENTING

Shafts (hoistways) housing elevators extending through more than two floor levels shall be vented to the outside. The area of the vent shall not be less than $3^1/_2$ percent of the area of the elevator shaft, provided a minimum of 3 square feet (0.279 m^2) per elevator is provided. Vents shall be capable of manual operation only.

The venting of each individual hoistway shall be independent from any other hoistway venting, and the interconnection of separate hoistways for the purpose of venting is prohibited.

SECTION 3005 — ELEVATOR MACHINE ROOM

3005.1 Operation of Solid-state Equipment. When solid-state equipment is used to operate the elevators, the elevator equipment room shall be provided with an independent ventilation or air-conditioning system to prevent overheating of the electrical equipment. The operating temperature shall be established by the elevator equipment manufacturer's specification. When standby power is connected to elevators, the machine room ventilation or air conditioning shall be connected to standby.

3005.2 Detection. The elevator machine room serving a pressurized elevator hoistway shall be pressurized upon activation of a heat or smoke detector located in the elevator machine room. See Section 905, Smoke Control.

SECTION 3006 — CHANGE IN USE

Any change in use of an elevator, freight to passenger, passenger to freight, or from one freight class to another, shall not be made without the approval of the building official. Said approval shall be granted only after it is demonstrated that the installation conforms to the requirements in the Elevator Code.

SECTION 3007 — ADDITIONAL DOORS

Doors other than the hoistway door and the elevator car door shall be prohibited at the point of access to an elevator car.

> **EXCEPTION:** Doors that are readily openable from the car side without a key, tool, or special knowledge or effort.

Chapter 31
SPECIAL CONSTRUCTION

SECTION 3101 — SCOPE

The provisions of this chapter shall apply to special construction described herein.

SECTION 3102 — CHIMNEYS, FIREPLACES AND BARBECUES

3102.1 Scope. Chimneys, flues, fireplaces and barbecues, and their connections, carrying products of combustion shall conform to the requirements of this section.

3102.2 Definitions.

BARBECUE is a stationary open hearth or brazier, either fuel fired or electric, used for food preparation.

CHIMNEY is a hollow shaft containing one or more passageways, vertical or nearly so, for conveying products of combustion to the outside atmosphere.

CHIMNEY CLASSIFICATIONS:

Chimney, High-heat Industrial Appliance-type, is a factory-built, masonry or metal chimney suitable for removing the products of combustion from fuel-burning high-heat appliances producing combustion gases in excess of 2,000°F (1093°C) measured at the appliance flue outlet.

Chimney, Low-heat Industrial Appliance-type, is a factory-built, masonry or metal chimney suitable for removing the products of combustion from fuel-burning low-heat appliances producing combustion gases not in excess of 1,000°F (538°C) under normal operating conditions but capable of producing combustion gases of 1,400°F (760°C) during intermittent forced firing for periods up to one hour. All temperatures are measured at the appliance flue outlet.

Chimney, Medium-heat Industrial Appliance-type, is a factory-built, masonry or metal chimney suitable for removing the products of combustion from fuel-burning medium-heat appliances producing combustion gases not in excess of 2,000°F (1093°C) measured at the appliance flue outlet.

Chimney, Residential Appliance-type, is a factory-built or masonry chimney suitable for removing products of combustion from residential-type appliances producing combustion gases not in excess of 1,000°F (538°C) measured at the appliance flue outlet.

CHIMNEY CONNECTOR is the pipe or breeching that connects a fuel-burning appliance to a chimney. (See Mechanical Code, Chapter 9.)

CHIMNEY, FACTORY-BUILT, is a chimney manufactured at a location other than the building site and composed of listed factory-built components assembled in accordance with the terms of the listing to form the completed chimney.

CHIMNEY LINER is a lining material of fireclay or approved refractory brick. For recognized standards on fireclay refractory brick see Sections 3503 and 3504; ASTM C 27, Fireclay and High-Alumina Refractory Brick; or ASTM C 1261, Firebox Brick for Residential Fireplaces.

FIREBRICK is a refractory brick.

FIREPLACE is a hearth and fire chamber or similar prepared place in which a fire may be made and which is built in conjunction with a chimney.

Factory-built Fireplace is a listed assembly of a fire chamber, its chimney and related factory-made parts designed for unit assembly without requiring field construction. Factory-built fireplaces are not dependent on mortar-filled joints for continued safe use.

Masonry Fireplace is a hearth and fire chamber of solid masonry units such as bricks, stones, masonry units or reinforced concrete provided with a suitable chimney.

MASONRY CHIMNEY is a chimney of masonry units, bricks, stones or listed masonry chimney units lined with approved flue liners. For the purpose of this chapter, masonry chimneys shall include reinforced concrete chimneys.

3102.3 Chimneys, General.

3102.3.1 Chimney support. Chimneys shall be designed, anchored, supported and reinforced as required in this chapter and applicable provisions of Chapters 16, 18, 19, 21 and 22 of this code. A chimney shall not support any structural load other than its own weight unless designed as a supporting member.

3102.3.2 Construction. Each chimney shall be so constructed as to safely convey flue gases not exceeding the maximum temperatures for the type of construction as set forth in Table 31-B and shall be capable of producing a draft at the appliance not less than that required for safe operation.

3102.3.3 Clearance. Clearance to combustible material shall be as required by Table 31-B.

3102.3.4 Lining. When required by Table 31-B, chimneys shall be lined with clay flue tile, firebrick, molded refractory units or other approved lining not less than $5/8$ inch (15.9 mm) thick as set forth in Table 31-B. Chimney liners shall be carefully bedded in approved medium-duty refractory mortar with close-fitting joints left smooth on the inside. Medium-duty fractory motor shall be in accordance with Sections 3503, 3504 and ASTM C 199.

3102.3.5 Area. The minimum net cross-sectional area of the chimney flue for fireplaces shall be determined in accordance with Figure 31-1. The minimum cross-sectional area shown or a flue size providing equivalent net cross-sectional area shall be used. The height of the chimney shall be measured from the firebox floor to the top of the last chimney flue tile. Chimney passageways for low-heat chimneys and incinerators shall not be smaller in area than the vent connection on the appliance attached thereto or not less than that set forth in Table 31-A.

> **EXCEPTION:** Chimney passageways designed by engineering methods approved by the building official.

3102.3.6 Height and termination. Every chimney shall extend above the roof and the highest elevation of any part of a building as shown in Table 31-B. For altitudes over 2,000 feet (610 m), the building official shall be consulted in determining the height of the chimney.

3102.3.7 Cleanouts. Cleanout openings shall be provided within 6 inches (152 mm) of the base of every masonry chimney.

3102.3.8 Spark arrester. Where determined necessary by the building official due to local climatic conditions or where sparks escaping from the chimney would create a hazard, chimneys attached to any appliance or fireplace that burns solid fuel shall be equipped with an approved spark arrester. The net free area of the spark arrester shall not be less than four times the net free area of the outlet of the chimney. The spark arrester screen shall have heat and corrosion resistance equivalent to 0.109-inch (2.77 mm) (No. 12 B.W. gage) wire, 0.042-inch (1.07 mm) (No. 19 B.W. gage) galvanized wire or 0.022-inch (0.56 mm) (No. 24 B.W. gage) stainless steel. Openings shall not permit the passage of spheres having a diameter larger than $^1/_2$ inch (12.7 mm) and shall not block the passage of spheres having a diameter of less than $^3/_8$ inch (9.5 mm).

Chimneys used with fireplaces or heating appliances in which solid or liquid fuel is used shall be provided with a spark arrester as required in the Fire Code.

> **EXCEPTION:** Chimneys that are located more than 200 feet (60 960 mm) from any mountainous, brush-covered or forest-covered land or land covered with flammable material and that are not attached to a structure having less than a Class C roof covering, as set forth in Chapter 15.

3102.4 Masonry Chimneys.

3102.4.1 Design. Masonry chimneys shall be designed and constructed to comply with Sections 3102.3.2 and 3102.4.2.

3102.4.2 Walls. Walls of masonry chimneys shall be constructed as set forth in Table 31-B.

3102.4.3 Reinforcing and seismic anchorage. Unless a specific design is provided, every masonry or concrete chimney in Seismic Zones 2, 3 and 4 shall be reinforced with not less than four No. 4 steel reinforcing bars conforming to the provisions of Chapter 19 or 21 of this code. The bars shall extend the full height of the chimney and shall be spliced in accordance with the applicable requirements of Chapter 19 or 21. In masonry chimneys, the vertical bars shall have a minimum cover of $^1/_2$ inch (12.7 mm) of grout or mortar tempered to a pouring consistency. The bars shall be tied horizontally at 18-inch (457 mm) intervals with not less than $^1/_4$-inch-diameter (6.4 mm) steel ties. The slope of the inclined portion of the offset in vertical bars shall not exceed 2 units vertical in 1 unit horizontal (200% slope). Two ties shall also be placed at each bend in vertical bars. Where the width of the chimney exceeds 40 inches (1016 mm), two additional No. 4 vertical bars shall be provided for each additional flue incorporated in the chimney or for each additional 40 inches (1016 mm) in width or fraction thereof.

In Seismic Zones 2, 3 and 4, all masonry and concrete chimneys shall be anchored at each floor or ceiling line more than 6 feet (1829 mm) above grade, except when constructed completely within the exterior walls of the building. Anchorage shall consist of two $^3/_{16}$-inch-by-1-inch (4.8 mm by 25 mm) steel straps cast at least 12 inches (305 mm) into the chimney with a 180-degree bend with a 6-inch (152 mm) extension around the vertical reinforcing bars in the outer face of the chimney.

Each strap shall be fastened to the structural framework of the building with two $^1/_2$-inch-diameter (12.7 mm) bolts per strap. Where the joists do not head into the chimney, the anchor strap shall be connected to 2-inch-by-4-inch (51 mm by 102 mm) ties crossing a minimum of four joists. The ties shall be connected to each joist with two 16d nails. As an alternative to the 2-inch-by-4-inch (51 mm by 102 mm) ties, each anchor strap shall be connected to the structural framework by two $^1/_2$-inch-diameter (12.7 mm) bolts in an approved manner.

3102.4.4 Chimney offset. Masonry chimneys may be offset at a slope of not more than 4 units vertical in 24 units horizontal (16.7% slope), but not more than one third of the dimension of the chimney, in the direction of the offset. The slope of the transition from the fireplace to the chimney shall not exceed 2 units vertical in 1 unit horizontal (200% slope).

3102.4.5 Change in size or shape. Masonry chimneys shall not change in size or shape within 6 inches (152 mm) above or below any combustible floor, ceiling or roof component penetrated by the chimney.

3102.4.6 Separation of masonry chimney passageways. Two or more flues in a chimney shall be separated by masonry not less than 4 inches (102 mm) thick bonded into the masonry wall of the chimney.

3102.4.7 Inlets. Every inlet to any masonry chimney shall enter the side thereof and shall not be of less than $^1/_8$-inch-thick (3.2 mm) metal or $^5/_8$-inch-thick (15.9 mm) refractory material. Where there is no other opening below the inlet other than the cleanout, a masonry plug shall be constructed in the chimney not more than 16 inches (406 mm) below the inlet and the cleanout shall be located where it is accessible above the plug. If the plug is located less than 6 inches (152 mm) below the inlet, the inlet may serve as the cleanout.

3102.5 Factory-built Chimneys and Fireplaces.

3102.5.1 General. Factory-built chimneys and factory-built fireplaces shall be listed and shall be installed in accordance with the terms of their listings and the manufacturer's instructions as specified in the Mechanical Code.

3102.5.2 Hearth extensions. Hearth extensions of listed factory-built fireplaces shall conform to the conditions of listing and the manufacturer's installation instructions.

3102.5.3 Multiple venting in vertical shafts. Factory-built chimneys utilized with listed factory-built fireplaces may be used in a common vertical shaft having the required fire-resistance rating.

3102.6 Metal Chimneys. Metal chimneys shall be constructed and installed to meet the requirements of the Mechanical Code.

Metal chimneys shall be anchored at each floor and roof with two $1^1/_2$-inch-by-$^1/_8$-inch (38 mm by 3.2 mm) metal straps looped around the outside of the chimney installation and nailed with not less than six 8d nails per strap at each joist.

3102.7 Masonry and Concrete Fireplaces and Barbecues.

3102.7.1 General. Masonry fireplaces, barbecues, smoke chambers and fireplace chimneys shall be of masonry or reinforced concrete and shall conform to the requirements of this section.

3102.7.2 Support. Masonry fireplaces shall be supported on foundations designed as specified in Chapters 16, 18 and 21.

When an approved design is not provided, foundations for masonry and concrete fireplaces shall not be less than 12 inches (305 mm) thick, extend not less than 6 inches (152 mm) outside the fireplace wall and project below the natural ground surface in accordance with the depth of foundations set forth in Table 18-I-C.

3102.7.3 Fireplace walls. Masonry walls of fireplaces shall not be less than 8 inches (203 mm) in thickness. Walls of fireboxes shall not be less than 10 inches (254 mm) in thickness, except that where a lining of firebrick is used, such walls shall not be less than a total of 8 inches (203 mm) in thickness. The firebox shall not be less than 20 inches (508 mm) in depth. Joints in firebrick shall not exceed $^1/_4$ inch (6.4 mm).

> **EXCEPTION:** For Rumford fireplaces, the depth may be reduced to 12 inches (305 mm) when
>
> 1. The depth is at least one third the width of the fireplace opening.

2. The throat is at least 12 inches (305 mm) above the lintel and is at least $^1/_{20}$ of the cross-sectional area of the fireplace opening.

3102.7.4 Hoods. Metal hoods used as part of a fireplace or barbecue shall not be less than 0.036-inch (0.92 mm) (No. 19 carbon sheet steel gage) copper, galvanized steel or other equivalent corrosion-resistant ferrous metal with all seams and connections of smokeproof unsoldered constructions. The hoods shall be sloped at an angle of 45 degrees or less from the vertical and shall extend horizontally at least 6 inches (152 mm) beyond the limits of the firebox. Metal hoods shall be kept a minimum of 18 inches (457 mm) from combustible materials unless approved for reduced clearances.

3102.7.5 Metal heat circulators. Approved metal heat circulators may be installed in fireplaces.

3102.7.6 Smoke chamber. Front and side walls shall not be less than 8 inches (203 mm) in thickness. Smoke chamber back walls shall not be less than 6 inches (152 mm) in thickness. A minimum $^5/_8$-inch-thick (16 mm) clay flue lining, complying with Sections 3503, 3504 and ASTM C 315, shall be permitted to form the inside surface of the 8-inch (203 mm) and 6-inch (152 mm) smoke chamber walls.

3102.7.7 Chimneys. Chimneys for fireplaces shall be constructed as specified in Sections 3102.3, 3102.4 and 3102.5 for residential-type appliances.

3102.7.8 Clearance to combustible material. Combustible materials shall not be placed within 2 inches (51 mm) of fireplace, smoke chamber or chimney walls. Combustible material shall not be placed within 6 inches (152 mm) of the fireplace opening. No such combustible material within 12 inches (305 mm) of the fireplace opening shall project more than $^1/_8$ inch (3.2 mm) for each 1-inch (25 mm) clearance from such opening.

No part of metal hoods used as part of a fireplace or barbecue shall be less than 18 inches (457 mm) from combustible material. This clearance may be reduced to the minimum requirements specified in the Mechanical Code.

3102.7.9 Areas of flues, throats and dampers. The throat shall be at least 8 inches (203 mm) above the fireplace opening and shall be at least 4 inches (102 mm) in depth. The net cross-sectional area of the flue and of the throat between the firebox and the smoke chamber of a fireplace shall not be less than that set forth in Figure 31-1 or Table 31-A. Metal dampers equivalent to not less than 0.097-inch (2.46 mm) (No. 12 carbon sheet metal gage) steel shall be installed. When fully opened, damper openings shall not be less than 90 percent of the required flue area.

3102.7.10 Lintel. Masonry over the fireplace opening shall be supported by a noncombustible lintel unless the masonry is self-supporting.

3102.7.11 Hearth. Masonry fireplaces shall be provided with a brick, concrete, stone or other approved noncombustible hearth slab. This slab shall not be less than 4 inches (102 mm) thick and shall be supported by noncombustible materials or reinforced to carry its own weight and all imposed loads. Combustible forms and centering shall be removed.

3102.7.12 Hearth extensions. Hearths shall extend at least 16 inches (406 mm) from the front of, and at least 8 inches (203 mm) beyond each side of, the fireplace opening. Where the fireplace opening is 6 square feet (0.56 m^2) or larger, the hearth extension shall extend at least 20 inches (508 mm) in front of, and at least 12 inches (305 mm) beyond each side of, the fireplace opening.

Except for fireplaces that open to the exterior of the building, the hearth slab shall be readily distinguishable from the surrounding or adjacent floor.

3102.7.13 Fire blocking. Fire blocking between chimneys and combustible construction shall meet the requirements specified in Section 708.

SECTION 3103 — TEMPORARY BUILDINGS OR STRUCTURES

Temporary buildings or structures such as reviewing stands and other miscellaneous structures, sheds, canopies or fences used for the protection of the public around and in conjunction with construction work may be erected by special permit from the building official for a limited period of time. Such buildings or structures need not comply with the type of construction or fire-resistive time periods required by this code. Temporary buildings or structures shall be completely removed upon the expiration of the time limit stated in the permit.

TABLE 31-A   **1997 UNIFORM BUILDING CODE**

TABLE 31-A—MINIMUM PASSAGEWAY AREAS FOR MASONRY CHIMNEYS[1]

	MINIMUM CROSS-SECTIONAL AREA		
	× 645 for mm²		
Type of Masonry Chimney	**Tile Lined**		**Lined with Firebrick or Unlined**
	Round	**Square or Rectangle**	
1. Residential (other than fireplaces)	50 square inches	50 square inches	85 square inches
2. Fireplace	See Figure 31-1	See Figure 31-1	$^{1}/_{8}$ of opening minimum 100 square inches
3. Low heat	50 square inches	57 square inches	135 square inches

4. Incinerator
Apartment type
  1 opening
  2 to 6 openings
  7 to 14 openings
  15 or more openings
	196 square inches
	324 square inches
	484 square inches
	484 square inches plus 10 square inches for each additional opening
			Not applicable

NOTE: For altitudes over 2,000 feet (610 m) above sea level, the building official shall be consulted in determining the area of the passageway.

[1]Areas for medium- and high-heat chimneys shall be determined using accepted engineering methods and as approved by the building official.

TABLE 31-B—CONSTRUCTION, CLEARANCE AND TERMINATION REQUIREMENTS FOR MASONRY AND CONCRETE CHIMNEYS

CHIMNEYS SERVING	THICKNESS (min. inches) × 25.4 for mm		HEIGHT ABOVE ROOF OPENING (feet) × 304.8 for mm	HEIGHT ABOVE ANY PART OF BUILDING WITHIN (feet) × 304.8 for mm			CLEARANCE TO COMBUSTIBLE CONSTRUCTION (inches) × 25.4 for mm	
	Walls	Lining		10	25	50	Int. Inst.	Ext. Inst.
1. **RESIDENTIAL-TYPE APPLIANCES**[1,2] (Low Btu input) Clay, shale or concrete brick Reinforced concrete Hollow masonry units Stone	4[3] 4[3] 4[4] 12	5/8 fire-clay tile or 2 firebrick	2	2			2	1 or 1/2 gypsum[5]
Unburned clay units	8	4 1/2 firebrick						
2. **BUILDING HEATING AND INDUSTRIAL-TYPE LOW-HEAT APPLIANCES**[1,2] [1,000°F (538°C) operating temp.—1,400°F (760°C) maximum] Clay, shale or concrete brick Hollow masonry units Reinforced concrete Stone	8 8[4] 8 12	5/8 fire-clay tile or 2 firebrick	3	2			2	2
3. **MEDIUM-HEAT INDUSTRIAL-TYPE APPLIANCES**[1,6] [2,000°F (1093°C) maximum] Clay, shale or concrete brick Hollow masonry units (Grouted solid) Reinforced concrete Stone	8 8 8 12	4 1/2 medium-duty firebrick	10	10			4	4
4. **HIGH-HEAT INDUSTRIAL-TYPE APPLIANCES**[1,6] [Over 2,000°F (1093°C)] Clay, shale or concrete brick Hollow masonry units (Grouted solid) Reinforced concrete	16[7] 16[7] 16[7]	4 1/2 high-duty firebrick	20			20	8	8
5. **RESIDENTIAL-TYPE INCINERATORS**	Same as for residential-type appliances as shown above.							
6. **CHUTE-FED AND FLUE-FED INCINERATORS WITH COMBINED HEARTH AND GRATE AREA 7 SQ. FT. (0.65 m²) OR LESS** Clay, shale or concrete brick or hollow units Portion extending to 10 ft. (3048 mm) above combustion chamber roof Portion more than 10 ft. (3048 mm) above combustion chamber roof	4 8	4 1/2 medium-duty firebrick 5/8 fire-clay tile liner	3	2			2	2
7. **CHUTE-FED AND FLUE-FED INCINERATORS—COMBINED HEARTH AND GRATE AREAS LARGER THAN 7 SQ. FT. (0.65 m²)** Clay, shale or concrete brick or hollow units grouted solid or reinforced concrete Portion extending to 40 ft. (12 192 mm) above combustion chamber roof Portion more than 40 ft. (12 192 mm) above combustion chamber roof Reinforced concrete	4 8 8	4 1/2 medium-duty firebrick 5/8 fire-clay tile liner 4 1/2 medium-duty firebrick laid in medium-duty refract mortar		10			2	2
8. **COMMERCIAL OR INDUSTRIAL-TYPE INCINERATORS**[2] Clay or shale solid brick Reinforced concrete	8 8	4 1/2 medium-duty firebrick laid in medium-duty refract mortar		10			4	4

[1]See Table 8-B of the Mechanical Code for types of appliances allowed with each type of chimney.
[2]Lining shall extend from bottom to top of chimney.
[3]Chimneys having walls 8 inches (203 mm) or more in thickness may be unlined.
[4]Equivalent thickness including grouted cells when grouted solid. The equivalent thickness may also include the grout thickness between the liner and masonry unit.
[5]Chimneys for residential-type appliances installed entirely on the exterior of the building. For fireplace and barbecue chimneys, see Section 3102.7.8.
[6]Lining to extend from 24 inches (610 mm) below connector to 25 feet (7620 mm) above.
[7]Two 8-inch (203 mm) walls with 2-inch (51 mm) airspace between walls. Outer and inner walls may be of solid masonry units or reinforced concrete or any combination thereof.
[8]Clearance shall be approved by the building official and shall be such that the temperature of combustible materials will not exceed 160°F (710°C).

FIGURE 31-1   1997 UNIFORM BUILDING CODE

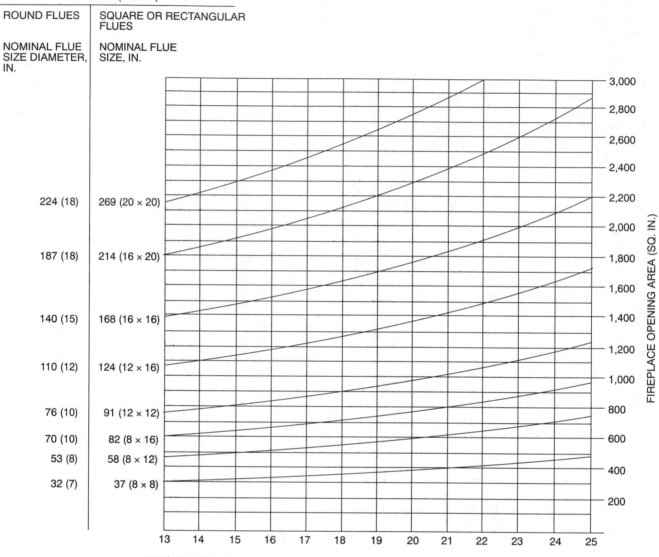

MINIMUM CROSS-SECTIONAL
FLUE AREA (SQ. IN.)

ROUND FLUES	SQUARE OR RECTANGULAR FLUES
NOMINAL FLUE SIZE DIAMETER, IN.	NOMINAL FLUE SIZE, IN.
224 (18)	269 (20 × 20)
187 (18)	214 (16 × 20)
140 (15)	168 (16 × 16)
110 (12)	124 (12 × 16)
76 (10)	91 (12 × 12)
70 (10)	82 (8 × 16)
53 (8)	58 (8 × 12)
32 (7)	37 (8 × 8)

CHIMNEY HEIGHT, MEASURED FROM FLOOR OF COMBUSTION CHAMBER TO TOP OF FLUE (FT.)

For **SI:** 1 inch = 25.4 mm, 1 square inch = 654.16 mm^2, 1 foot = 304.8 mm.

FIGURE 31-1—FLUE SIZES FOR MASONRY CHIMNEYS[1]

[1]The smaller flue area shall be utilized where the fireplace opening area and the chimney height selected intersect between flue area curves.

Chapter 32
CONSTRUCTION IN THE PUBLIC RIGHT OF WAY

SECTION 3201 — GENERAL

No part of any structure or any appendage thereto, except signs, shall project beyond the property line of the building site, except as specified in this chapter.

Structures or appendages regulated by this code shall be constructed of materials as specified in Section 705.

The projection of any structure or appendage shall be the distance measured horizontally from the property line to the outermost point of the projection.

Nothing in this code shall prohibit the construction and use of a structure between buildings and over or under a public way, provided the structure complies with all requirements of this code.

No provisions of this chapter shall be construed to permit the violation of other laws or ordinances regulating the use and occupancy of public property.

SECTION 3202 — PROJECTION INTO ALLEYS

No part of any structure or any appendage thereto shall project into any alley.

> **EXCEPTIONS:** 1. A curb or buffer block may project not more than 9 inches (229 mm) and not exceed a height of 9 inches (229 mm) above grade.
>
> 2. Footings located at least 8 feet (2438 mm) below grade may project not more than 12 inches (305 mm).

SECTION 3203 — SPACE BELOW SIDEWALK

The space adjoining a building below a sidewalk on public property may be used and occupied in connection with the building for any purpose not inconsistent with this code or other laws or ordinances regulating the use and occupancy of such spaces on condition that the right to so use and occupy may be revoked by the city at any time and that the owner of the building will construct the necessary walls and footings to separate such space from the building and pay all costs and expenses attendant therewith.

Footings located at least 8 feet (2438 mm) below grade may project not more than 12 inches (305 mm).

SECTION 3204 — BALCONIES, SUN-CONTROL DEVICES AND APPENDAGES

Oriel windows, balconies, sun-control devices, unroofed porches, cornices, belt courses and appendages such as water tables, sills, capitals, bases and architectural projections may project over the public property of the building site a distance as determined by the clearance of the lowest point of the projection above the grade immediately below, as follows:

Clearance above grade less than 8 feet (2438 mm)—no projection is permitted.

Clearance above grade over 8 feet (2438 mm)—1 inch (25 mm) of projection is permitted for each additional inch of clearance, provided that such projection shall exceed a distance of 4 feet (1219 mm).

SECTION 3205 — MARQUEES

3205.1 General. For the purpose of this section, a marquee shall include any object or decoration attached to or a part of said marquee.

3205.2 Projection and Clearance. The horizontal clearance between a marquee and the curb line shall not be less than 2 feet (610 mm).

A marquee projecting more than two thirds of the distance from the property line to the curb line shall not be less than 12 feet (3658 mm) above the ground or pavement below.

A marquee projecting less than two thirds of the distance from the property line to the curb line shall not be less than 8 feet (2438 mm) above the ground or pavement below.

3205.3 Length. A marquee projecting more than two thirds of the distance from the property line to the curb line shall not exceed 25 feet (7620 mm) in length along the direction of the street.

3205.4 Thickness. The maximum height or thickness of a marquee measured vertically from its lowest to its highest point shall not exceed 3 feet (914 mm) when the marquee projects more than two thirds of the distance from the property line to the curb line and shall not exceed 9 feet (2743 mm) when the marquee is less than two thirds of the distance from the property line to the curb line.

3205.5 Construction. A marquee shall be supported entirely by the building and constructed of noncombustible material or, when supported by a building of Type V construction, may be of one-hour fire-resistive construction.

3205.6 Roof Construction. The roof or any part thereof may be a skylight, provided glass skylights are of laminated or wired glass complying with Section 2409. Plastic skylights shall comply with Section 2603.7.

Every roof and skylight of a marquee shall be sloped to downspouts that shall conduct any drainage from the marquee under the sidewalk to the curb.

3205.7 Location Prohibited. Every marquee shall be so located as not to interfere with the operation of any exterior standpipe or to obstruct the clear passage of a means of egress from the building or the installation or maintenance of electroliers.

SECTION 3206 — AWNINGS

3206.1 Definition. For the purpose of this section:

AWNING is a shelter supported entirely from the exterior wall of a building.

3206.2 Construction. Awnings shall have noncombustible frames but may have combustible coverings. Awnings shall be either fixed, retractable, folding or collapsible. Awnings in any configuration shall not obstruct the use of a required means of egress.

3206.3 Projection. Awnings may extend over public property not more than 7 feet (2134 mm) from the face of a supporting building, but no portion shall extend nearer than 2 feet (610 mm) to the face of the nearest curb line measured horizontally. In no case shall the awning extend over public property greater than two thirds of the distance from the property line to the nearest curb in front of the building site.

3206.4 Clearances. All portions of any awning shall be at least 8 feet (2438 mm) above any public walkway.

> **EXCEPTION:** Any valance attached to an awning shall not project above the roof of the awning at the point of attachment and shall not extend more than 12 inches (305 mm) below the roof of the awning at the point of attachment, but in no case shall any portion of a valance be less than 7 feet (2134 mm) in height above a public way.

SECTION 3207 — DOORS

Power-operated doors and their guide rails shall not project over public property. Other doors, either fully opened or when opening, shall not project more than 1 foot (305 mm) beyond the property line, except that in alleys no projection beyond the property line is permitted.

Chapter 33
SITE WORK, DEMOLITION AND CONSTRUCTION

SECTION 3301 — EXCAVATIONS AND FILLS

3301.1 General. Excavation or fills for buildings or structures shall be so constructed or protected that they do not endanger life or property.

Slopes for permanent fills shall not be steeper than 1 unit vertical in 2 units horizontal (50% slope). Cut slopes for permanent excavations shall not be steeper than 1 unit vertical in 2 units horizontal (50% slope) unless substantiating data justifying steeper cut slopes are submitted. Deviation from the foregoing limitations for cut slopes shall be permitted only upon the presentation of a soil investigation report acceptable to the building official.

No fill or other surcharge loads shall be placed adjacent to any building or structure unless such building or structure is capable of withstanding the additional loads caused by the fill or surcharge.

Existing footings or foundations that may be affected by any excavation shall be underpinned adequately or otherwise protected against settlement and shall be protected against lateral movement.

For footings on adjacent slopes, see Section 1806.5.

Fills to be used to support the foundations of any building or structure shall be placed in accordance with accepted engineering practice. A soil investigation report and a report of satisfactory placement of fill, both acceptable to the building official, shall be submitted.

Where applicable (see Section 101.3), see Appendix Chapter 33 for excavation and grading.

3301.2 Protection of Adjoining Property. The requirements for protection of adjacent property and depth to which protection is required shall be as defined by prevailing law. Where not defined by law, the following shall apply: Any person making or causing an excavation to be made to a depth of 12 feet (3658 mm) or less below the grade shall protect the excavation so that the soil of adjoining property will not cave in or settle, but shall not be liable for the expense of underpinning or extending the foundation of buildings on adjoining properties when the excavation is not in excess of 12 feet (3658 mm) in depth. Before commencing the excavation, the person making or causing the excavation to be made shall notify in writing the owners of adjoining buildings not less than 10 days before such excavation is to be made that the excavation is to be made and that the adjoining buildings should be protected.

The owners of the adjoining properties shall be given access to the excavation for the purpose of protecting such adjoining buildings.

Any person making or causing an excavation to be made exceeding 12 feet (3658 mm) in depth below the grade shall protect the excavation so that the adjoining soil will not cave in or settle and shall extend the foundation of any adjoining buildings below the depth of 12 feet (3658 mm) below grade at the expense of the person causing or making the excavation. The owner of the adjoining buildings shall extend the foundation of these buildings to a depth of 12 feet (3658 mm) below grade at such owner's expense, as provided in the preceding paragraph.

SECTION 3302 — PREPARATION OF BUILDING SITE

All stumps and roots shall be removed from the soil to a depth of at least 12 inches (305 mm) below the surface of the ground in the area to be occupied by the building.

All wood forms that have been used in placing concrete, if within the ground or between foundation sills and the ground, shall be removed before a building is occupied or used for any purpose. Before completion, loose or casual wood shall be removed from direct contact with the ground under the building.

SECTION 3303 — PROTECTION OF PEDESTRIANS DURING CONSTRUCTION OR DEMOLITION

3303.1 General. No person shall use or occupy a street, alley or public sidewalk for the performance of work under a building permit except in accordance with the provisions of this chapter.

No person shall perform any work on any building or structure adjacent to a public way in general use by the public for pedestrian travel unless the pedestrians are protected as specified in this chapter.

Any material or structure temporarily occupying public property, including fences and walkways, shall be adequately lighted between sunset and sunrise.

For additional requirements for temporary buildings or structures, see Section 3103.

3303.2 Temporary Use of Streets and Alleys. The use of public property shall meet the requirements of the public agency having jurisdiction. Whenever requested, plot plans and construction details shall be submitted for review by the agencies concerned.

3303.3 Storage on Public Property. Material and equipment necessary for work to be done under a permit shall not be placed or stored on public property so as to obstruct free and convenient approach to and use of any fire hydrant, fire or police alarm box, utility box, catch basin, or manhole or so as to interfere with the free flow of water in any street or alley gutter.

3303.4 Mixing Mortar on Public Property. The mixing or handling of mortar, concrete or other material on public property shall be done in a manner that will not deface public property or create a nuisance.

3303.5 Protection of Utilities. A substantial protective frame and boarding shall be built around and over every street lamp, utility box, fire or police alarm box, fire hydrant, catch basin, and manhole that may be damaged by any work being done under the permit. This protection shall be maintained while such work is being done and shall not obstruct the normal functioning of the device.

3303.6 Walkway. A walkway not less than 4 feet (1219 mm) wide shall be maintained on the sidewalk in front of the building site during construction, alteration or demolition unless the public agency having jurisdiction authorizes the sidewalk to be fenced and closed. Adequate signs and railings shall be provided to direct pedestrian traffic. Railings shall be provided when required by Section 3303.7.

The walkway shall be capable of supporting a uniform live load of 150 pounds per square foot (psf) (7.18 kN/m^2). A durable wearing surface shall be provided.

3303.7 Pedestrian Protection.

3303.7.1 Protection required. Pedestrian traffic shall be protected by a railing on the street side when the walkway extends into the roadway, by a railing adjacent to excavations and by such other protection as set forth in Table 33-A. The construction of such protective devices shall be in accordance with the provisions of this chapter.

3303.7.2 Railings. Railings shall be substantially built and, when of wood, shall be constructed of new material having a nominal size of at least 2 inches by 4 inches (51 mm by 102 mm). Railings shall be at least 3 feet 6 inches (1067 mm) in height and, when adjacent to excavations, shall be provided with a midrail.

3303.7.3 Fences. Fences shall be solid and substantially built, be not less than 8 feet (2438 mm) in height above grade and be placed on the side of the walkway nearest to the building site. Fences shall extend the entire length of the building site and each end shall be returned to the building line.

Openings in such fences shall be protected by doors that are normally kept closed.

All fences shall be provided with 2-inch-by-4-inch (51 mm by 102 mm) plates, top and bottom, and shall be well braced. The fence material shall be a minimum of $^3/_4$-inch (19.1 mm) boards or $^1/_4$-inch (6.4 mm) plywood. Plywood fences shall conform to the following requirements:

1. Plywood panels shall be bonded with an adhesive identical to that for exterior plywood.

2. Plywood $^1/_4$ inch (6.4 mm) or $^5/_{16}$ inch (7.9 mm) in thickness shall have studs spaced not more than 2 feet (610 mm) on center.

3. Plywood $^3/_8$ inch (9.5 mm) or $^1/_2$ inch (12.7 mm) in thickness shall have studs spaced not more than 4 feet (1219 mm) on center, provided a 2-inch-by-4-inch (51 mm by 102 mm) stiffener is placed horizontally at the midheight when the stud spacing exceeds 2 feet (610 mm) on center.

4. Plywood $^5/_8$ inch (15.9 mm) or thicker shall not span over 8 feet (2438 mm).

3303.7.4 Canopies. The protective canopy shall have a clear height of 8 feet (2438 mm) above the walkway. The roof shall be tightly sheathed. The sheathing shall be 2-inch (51 mm) nominal wood planking or equal. Every canopy shall have a solid fence built along its entire length on the construction side.

If materials are stored or work is done on the roof of the canopy, the street sides and ends of the canopy roof shall be protected by a tight curb board not less than 1 foot (305 mm) high and a railing not less than 3 feet 6 inches (1067 mm) high.

The entire structure shall be designed to carry the loads to be imposed on it, provided the live load shall not be less than 150 psf (7.18 kN/m^2). In lieu of such design, a protection canopy supporting not more than 150 psf (7.18 kN/m^2) may be constructed as follows:

1. Footings shall be continuous 2-inch-by-6-inch (51 mm by 152 mm) members with scabbed joints.

2. Posts not less than 4 inches by 6 inches (102 mm by 152 mm) in size shall be provided on both sides of the canopy and spaced not more than 12 feet (3658 mm), center to center.

3. Stringers not less than 4 inches by 12 inches (102 mm by 305 mm) in size shall be placed on edge upon the posts.

4. Joists resting upon the stringers shall be at least 2 inches by 8 inches (51 mm by 305 mm) in size and shall be spaced not more than 2 feet (610 m), center to center.

5. The deck shall be of planks at least 2 inches (51 mm) thick nailed to the joists.

6. Each post shall be knee-braced to joists and stringers by members 4 feet (1219 mm) long, not less than 2 inches by 4 inches (51 mm by 102 mm) in size.

7. A curb not less than 2 inches by 12 inches (51 mm by 305 mm) in size shall be set on edge along the outside edge of the deck.

EXCEPTION: Protection canopies for new, light-frame construction not exceeding two stories in height may be designed for a live load of 75 psf (3.59 kN/m^2) or the loads to be imposed on it, whichever is the greater.

3303.8 Maintenance and Removal of Protective Devices.

3303.8.1 Maintenance. Pedestrian protection required by Section 3303.7 shall be maintained in place and kept in good order for the entire length of time pedestrians may be endangered.

3303.8.2 Removal. Every protection fence or canopy shall be removed within 30 days after such protection is no longer required by this chapter for protection of pedestrians.

3303.9 Demolition. The work of demolishing any building shall not commence until the required pedestrian protection structures are in place.

The building official may require the permittee to submit plans and a complete schedule for demolition. Where such are required, no work shall be done until such plans or schedule, or both, are approved by the building official.

TABLE 33-A—TYPE OF PROTECTION REQUIRED FOR PEDESTRIANS

HEIGHT OF CONSTRUCTION	DISTANCE FROM CONSTRUCTION	PROTECTION REQUIRED
	× 304.8 for mm	
8 feet or less	Less than 6 feet 6 feet or more	Railing None
More than 8 feet	Less than 6 feet	Fence and canopy
	6 feet or more, but not more than one fourth the height of construction	Fence and canopy
	6 feet or more, but between one fourth to one half the height of construction	Fence
	6 feet or more, but exceeding one half the construction height	None

Chapter 34
EXISTING STRUCTURES

SECTION 3401 — GENERAL

Buildings in existence at the time of the adoption of this code may have their existing use or occupancy continued, if such use or occupancy was legal at the time of the adoption of this code, provided such continued use is not dangerous to life.

Any change in the use or occupancy of any existing building or structure shall comply with the provisions of Sections 109 and 3405 of this code.

For existing buildings, see Appendix Chapter 34. See also Section 101.3.

For a comprehensive code and guidelines on the treatment of existing buildings, see the *Uniform Code for Building Conservation.*

SECTION 3402 — MAINTENANCE

All buildings and structures, both existing and new, and all parts thereof, shall be maintained in a safe and sanitary condition. All devices or safeguards required by this code shall be maintained in conformance with the code edition under which installed. The owner or the owner's designated agent shall be responsible for the maintenance of buildings and structures. To determine compliance with this subsection, the building official may cause a structure to be reinspected.

SECTION 3403 — ADDITIONS, ALTERATIONS OR REPAIRS

3403.1 General. Buildings and structures to which additions, alterations or repairs are made shall comply with all the requirements of this code for new facilities except as specifically provided in this section. See Section 310.9 for provisions requiring installation of smoke detectors in existing Group R, Division 3 Occupancies.

3403.2 When Allowed. Additions, alterations or repairs may be made to any building or structure without requiring the existing building or structure to comply with all the requirements of this code, provided the addition, alteration or repair conforms to that required for a new building or structure.

Additions or alterations shall not be made to an existing building or structure that will cause the existing building or structure to be in violation of any of the provisions of this code and such additions or alterations shall not cause the existing building or structure to become unsafe. An unsafe condition shall be deemed to have been created if an addition or alteration will cause the existing building or structure to become structurally unsafe or overloaded, will not provide adequate egress in compliance with the provisions of this code or will obstruct existing exits, will create a fire hazard, will reduce required fire resistance, or will otherwise create conditions dangerous to human life. Any building so altered, which involves a change in use or occupancy, shall not exceed the height, number of stories and area permitted for new buildings. Any building plus new additions shall not exceed the height, number of stories and area specified for new buildings.

Additions or alterations shall not be made to an existing building or structure when such existing building or structure is not in full compliance with the provisions of this code except when such addition or alteration will result in the existing building or structure being no more hazardous based on life safety, firesafety and sanitation, than before such additions or alterations are undertaken. (See also Section 307.11.3 for Group H, Division 6 Occupancies.)

EXCEPTION: Alterations of existing structural elements, or additions of new structural elements, which are not required by Section 3401 and are initiated for the purpose of increasing the lateral-force-resisting strength or stiffness of an existing structure, need not be designed for forces conforming to these regulations provided that an engineering analysis is submitted to show that

1. The capacity of existing structural elements required to resist forces is not reduced,

2. The lateral loading to required existing structural elements is not increased beyond their capacity,

3. New structural elements are detailed and connected to the existing structural elements as required by these regulations,

4. New or relocated nonstructural elements are detailed and connected to existing or new structural elements as required by these regulations, and

5. An unsafe condition as defined above is not created.

3403.3 Nonstructural. Alterations or repairs to an existing building or structure that are nonstructural and do not adversely affect any structural member or any part of the building or structure having required fire resistance may be made with the same materials of which the building or structure is constructed.

3403.4 Glass Replacement. The installation or replacement of glass shall be as required for new installations.

3403.5 Historic Buildings. Repairs, alterations and additions necessary for the preservation, restoration, rehabilitation or continued use of a building or structure may be made without conformance to all the requirements of this code when authorized by the building official, provided

1. The building or structure has been designated by official action of the legally constituted authority of this jurisdiction as having special historical or architectural significance.

2. Any unsafe conditions as described in this code are corrected.

3. The restored building or structure will be no more hazardous based on life safety, firesafety and sanitation than the existing building.

SECTION 3404 — MOVED BUILDINGS

Buildings or structures moved into or within the jurisdiction shall comply with the provisions of this code for new buildings or structures.

SECTION 3405 — CHANGE IN USE

No change shall be made in the character of occupancies or use of any building that would place the building in a different division of the same group of occupancy or in a different group of occupancies, unless such building is made to comply with the requirements of this code for such division or group of occupancy.

EXCEPTION: The character of the occupancy of existing buildings may be changed subject to the approval of the building official, and the building may be occupied for purposes in other groups without conforming to all the requirements of this code for those groups, provided the new or proposed use is less hazardous, based on life and fire risk, than the existing use.

No change in the character of occupancy of a building shall be made without a certificate of occupancy, as required in Section 109 of this code. The building official may issue a certificate of occupancy pursuant to the intent of the above exception without certifying that the building complies with all provisions of this code.

Chapter 35
UNIFORM BUILDING CODE STANDARDS

Part I—General

SECTION 3501 — UBC STANDARDS

The Uniform Building Code standards referred to in various parts of this code, which are also listed in Part II of this chapter, are hereby declared to be part of this code and are referred to in this code as a "UBC standard."

SECTION 3502 — ADOPTED STANDARDS

The standards referred to in various parts of the code, which are listed in Part III of this chapter, are hereby declared to be part of this code.

SECTION 3503 — STANDARD OF DUTY

The standard of duty established for the recognized standards listed in Part IV of this chapter is that the design, construction and quality of materials of buildings and structures be reasonably safe for life, limb, health, property and public welfare.

SECTION 3504 — RECOGNIZED STANDARDS

The standards listed in Part IV of this chapter are recognized standards. Compliance with these recognized standards shall be prima facie evidence of compliance with the standard of duty set forth in Section 3503.

Part II—UBC Standards

UBC STD. AND SEC.	TITLE AND SOURCE

CHAPTER 2

2-1; 201.2, 215
Noncombustible Material—Tests. Standard Method of Test E 136-79 of the ASTM.*

CHAPTER 4

4-1; 303.8, 405.1.1, 405.3.4
Proscenium Firesafety Curtains. Installation Standard of the International Conference of Building Officials.

CHAPTER 7

7-1; 405.1.1, 601.3, 703.2, 703.4, 706, 709.3.2.2, 709.5, 709.6, 710.2, Table 7-A, 2602.5.2
Fire Tests of Building Construction and Materials. Standard Methods E 119-83 of the ASTM.

7-2; 302.4, 703.4, 713.5, 713.9, 1004.3.4.3.2, 1005.3.3.5
Fire Tests of Door Assemblies. Standard 10B-1988 of Underwriters Laboratories Inc. and International Conference of Building Officials Test Standard for Smoke- and Draft-Control Door Assemblies.

7-3; 703.4, 713.5
Tinclad Fire Doors. Specification of the American National Standards Institute/Underwriters Laboratories Inc. 10A-1979 (R 1985).

7-4; 703.4, 713.5, 713.9
Fire Tests of Window Assemblies. Standard Methods E 163-76 of the ASTM.

7-5; 703.4, 714
Fire Tests of Through-penetration Fire Stops. Standard Method E 814-83 of the ASTM.

7-6; 703.4; 704.6; 1701.4; 1701.5, Item 10
Thickness and Density Determination for Spray-applied Fireproofing. Test Standard of the International Conference of Building Officials.

7-7; 703.3, 703.4
Methods for Calculating Fire Resistance of Steel, Concrete, Wood, Concrete Masonry and Clay Masonry Construction. Standard of the International Conference of Building Officials.

7-8; 308.2.2.1, 1001.2, 1003.3.1.2
Horizontal Sliding Fire Doors Used in an Exit. Test Standard of the International Conference of Building Officials.

CHAPTER 8

8-1; 201.2; 207; 215; 217; 405.1.1; 601.3; 707.2; 707.3; 801.2; 802.1, Item 1; 802.2; 2602.3; 2602.5.2; 2602.6
Test Method for Surface-burning Characteristics of Building Materials. Standard Test Method E 84-84 of the ASTM.

8-2; 801.2, 805
Standard Test Method for Evaluating Room Fire Growth Contribution of Textile Wall Covering. Test Method of the International Conference of Building Officials.

CHAPTER 9

9-1; 307.11.3, 321.1, 403.2, 404.3.1, 405.1.1, 804.1, 902, 904.1.2, 904.1.3, 904.2.6.3, 904.2.7, 904.3.2, 2603.7.1, 2603.8.1
Installation of Sprinkler Systems. Standard for the Installation of Sprinkler Systems, NFPA 13-1991, National Fire Protection Association.

9-2; 902, 904.1.2, 904.5.1
Standpipe Systems. The Standard for Installation of Standpipe Systems and Hose Systems, NFPA 14-1993, National Fire Protection Association.

9-3; 804.1, 805, 902, 904.1.2, 904.1.3, 2603.7.1, 2603.8.1
Installation of Sprinkler Systems in Group R Occupancies Four Stories or Less. Standard for the Installation of Sprinkler Systems in Residential Occupancies up to Four Stories in Height, NFPA 13R-1989, National Fire Protection Association.

CHAPTER 10

10-1; 1001.2, 1003.3.1.2
Power-operated Exit Doors. Test Standard of the International Conference of Building Officials.

10-2; 1001.2, 1003.3.3.13
Stairway Identification. Specification Standard of the International Conference of Building Officials.

10-3; Appendix 3407.1
Exit Ladder Device. Test Standard of the International Conference of Building Officials.

10-4; 1001.2, 1003.3.1.9
Panic Hardware. Standard 305, July 30, 1979, of Underwriters Laboratories Inc.

---

*ASTM refers to the American Society for Testing and Materials.

1997 UNIFORM BUILDING CODE

CHAPTER 14

14-1; 601.3, 1401.2, 1402.1
Kraft Waterproof Building Paper. Federal Specification UU-B-790a (February 5, 1968).

14-2; 1401.2, 1404
Vinyl Siding. Standard Specification D 3679-91 for Rigid Polyvinyl Chloride (PVC) of the ASTM.

CHAPTER 15

15-1; 1501.2, Table 15-E
Roofing Aggregates. Material Standard of the International Conference of Building Officials.

15-2; 601.3, 1501.2, 1502, Table 15-A, 2602.5.3
Test Standard for Determining the Fire Retardancy of Roof-covering Materials. Standard Specification 790 (October 5, 1983) of Underwriters Laboratories Inc.

15-3; 1501.2, 1502, 1507.2, 1507.12
Wood Shakes. Part I—Wood Shakes (nonpreservative treated). Grading and Packing Rules for Red Cedar Shakes. Grading Rules of the Red Cedar Shingle & Handsplit Shake Bureau, 1975. Part II—Wood Shake Hip and Ridge Units (nonpreservative). Shake and Shingle Council and Material Standard of the International Conference of Building Officials. Part III—Wood Shakes (preservative treated). Grading and Packing Rules for Treated Southern Pine and Red Pine, Black Gum/Sweet Gum Taper-sawn Shakes. Grading Rules of the Red Cedar Shingle & Handsplit Shake Bureau, 1982, and Material Standard of the International Conference of Building Officials. Part IV—Southern Yellow Pine, Red Pine, Black Gum/Sweet Gum Taper-sawn Shake Hip and Ridge Units. Material Standard of the International Conference of Building Officials.

15-4; 1501.2, 1502, 1507.2, 1507.13
Wood Shingles. Standard of the Red Cedar Shingle & Handsplit Shake Bureau and Material Standard of the International Conference of Building Officials.

15-5; 1501.2, 1502, 1507.7
Roof Tile. Test Standard of the International Conference of Building Officials.

15-6; 1501.2, 1502
Modified Bitumen, Thermoplastic and Thermoset Membranes Used for Roof Coverings. Standard Specifications D 412-87, D 471-79, D 570-81, D 624-86, D 638-84, D 751-79, D 816-82, D 1004-66 (1981), D 1204-84, D 2136-84 and D 2137-83 of the ASTM.

15-7; 906.1, 906.4, 1501.2
Automatic Smoke and Heat Vents. Material Standard of the International Conference of Building Officials.

CHAPTER 18

18-1; 1801.2, 1803.1
Soils Classification. Standard Method D 2487-69 of the ASTM.

18-2; 1801.2, 1803.2
Expansion Index Test. Recommendation of the Los Angeles Section of the ASCE Soil Committee.

CHAPTER 19

19-1; 1903.5.2, 1912.14.3
Welding Reinforcing Steel, Metal Inserts and Connections in Reinforced Concrete Construction. Structural Welding Code—Reinforcing Steel ANSI/AWS D1.4-92 of the American Welding Society, Inc.

19-2; 1903.9, 1925.1, 1925.3
Mill-Mixed Gypsum Concrete and Poured Gypsum Roof Diaphragms. Standard Specification C 317-70 of the ASTM. Poured Gypsum Roof Diaphragm, based on reports of test programs by S. B. Barnes and Associates, dated February 1955, November 1956, January 1958 and February 1962.

CHAPTER 21

21-1; 2102.2, Item 4
Building Brick, Facing Brick and Hollow Brick. (Made from Clay or Shale.) Standard Specifications C 62-92c, C 216-94a and C 652-94a of the ASTM.

21-2; 2102.2, Item 6
Calcium Silicate Face Brick (Sand-lime Brick). Standard Specification C 73-95 of the ASTM.

21-3; 2102.2, Item 5
Concrete Building Brick. Standard Specification C 55-95 of the ASTM.

21-4; 2102.2, Item 5
Hollow and Solid Load-bearing Concrete Masonry Units. Standard Specification C 90-95 of the ASTM.

21-5; 2102.2, Item 5
Nonload-bearing Concrete Masonry Units. Standard Specification C 129-95 of the ASTM.

21-6; See *Uniform Code for Building Conservation.*
In-Place Masonry Shear Tests. Test Standard of the International Conference of Building Officials.

21-7; See *Uniform Code for Building Conservation.*
Tests of Anchors in Unreinforced Masonry Walls. Test Standard of the International Conference of Building Officials.

21-8; See *Uniform Code for Building Conservation.*
Pointing of Unreinforced Masonry Walls. Construction Specification of the International Conference of Building Officials.

21-9; 2102.2, Item 6
Unburned Clay Masonry Units and Standard Methods of Sampling and Testing Unburned Clay Masonry Units. Test Standard of the International Conference of Building Officials.

21-10; 2102.2, 2104.8
Part I—Joint Reinforcement for Masonry. Specification Standard of the International Conference of Building Officials. Part II—Cold-drawn Steel Wire for Concrete Reinforcement. Standard Specification A 82-90a of the ASTM.

21-11; 2102.2, Item 2; Table 21-A
Cement, Masonry. Standard Specification C 91-93a of the ASTM.

21-12; 2102.2, Item 3
Quicklime for Structural Purposes. Standard Specification C 5-79 (Reapproved 1992) of the ASTM.

21-13; 2102.2, Item 3
Hydrated Lime for Masonry Purposes. Standard Specification C 207-91 (Reapproved 1992) of the ASTM.

21-14; 2102.2, Item 2; Table 21-A
Mortar Cement. Test Standard of the International Conference of Building Officials.

21-15; 2102.2, Item 8
Mortar for Unit Masonry and Reinforced Masonry Other Than Gypsum. Standard Specification C 270-95T of the ASTM.

21-16; 2102.2, Item 8
Field Tests Specimens for Mortar. Test Standard of the International Conference of Building Officials.

21-17; 2102.2, Item 6; 2105.3.2, 2105.3.3
Test Method for Compressive Strength of Masonry Prisms. Standard Test Method E 447-80 of the ASTM.

21-18; 2102.2, Item 9; Table 21-B
Method of Sampling and Testing Grout. Standard Method C 1019-89a (93) of the ASTM.

21-19; 2102.2, Item 9
Grout for Masonry. Standard Specification C 476-91 of the ASTM.

21-20; 2102.2, Item 8
Standard Test Method for Flexural Bond Strength of Mortar Cement. Test Standard of the International Conference of Building Officials.

CHAPTER 22

22-1; 1808.6.1, 1808.7.1, 2202.2
Material Specifications for Structural Steel. Standard Specifications A 27, A 36, A 48, A 53, A 148, A 242, A 252, A 283, A 307, A 325, A 336, A 441, A 446, A 449, A 490, A 500, A 501, A 514, A 529, A 563, A 569, A 570, A 572, A 588, A 606, A 607, A 611, A 618, A 666, A 668, A 690, A 715 and A 852 of the ASTM.

CHAPTER 23

23-1; 2302.1, 2303
Classification, Definition and Methods of Grading for All Species of Lumber. Standard Methods D 245-88 and D 2555-88 of the ASTM, Handbook No. 72 of the United States Department of Agriculture, American Softwood Lumber Standard PS20-70 and National Grading Rule for Dimension Lumber of the National Grading Rule Committee.

23-2; 2302.1, 2303, 2304.2
Construction and Industrial Plywood. Product Standard PS 1-95 of the United States Department of Commerce, and National Bureau of Standards Calculation Diaphragm Action, an Engineering Standard of the International Conference of Building Officials.

23-3; 2302.1, 2303, 2304.2
Wood-Based Structural-Use Panels. Product Standard PS 2-92 of the United States Department of Commerce and the American Plywood Association.

23-4; 201.2, 207, 2303
Fire-retardant-treated Wood Tests on Durability and Hygroscopic Properties. Standard Test Methods D 2898-81 and D 3201-79 of the ASTM and Standards C 20-83 and C 27-83 of the American Wood Preservers Association.

23-5; 2303,
Fire-retardant-treated Wood. Design Values for Fire-retardant-treated Lumber.

CHAPTER 24

24-1; 2401.2, 2401.4
Flat Glass. Standard Specification C 1036-85 of the ASTM.

24-2; 2401.2, 2401.4, 2406.2, 2406.3, 2406.5, 2408.1
Safety Glazing. Safety Standard for Architectural Glazing Materials (16 C.F.R., Part 1201) of the United States Consumer Product Safety Commission and Performance Specifications and Methods of Test for Transparent Safety Glazing Material Used in Buildings. ANSI Z97.1-1975 of the American National Standards Institute, Inc.

CHAPTER 25

25-1; 2502
Plastic Cement. Test Standard of the International Conference of Building Officials.

25-2; 2502, Table 25-A
Metal Suspension Systems for Acoustical Tile and for Lay-in Panel Ceilings. Standard Specification C 635-69 and Standard Recommended Practice C 636-69 of the ASTM.

CHAPTER 26

26-1; 601.3, 2602.4, 2602.5.2
Test Method to Determine Potential Heat of Building Materials. Test Standard of the International Conference of Building Officials.

26-2; 601.3, 2602.4
Test Method for the Evaluation of Thermal Barriers. Standard of the International Conference of Building Officials.

26-3; 601.3, 2602.6
Room Fire Test Standard for Interior of Foam Plastic Systems. Test Standard of the International Conference of Building Officials.

26-4; 601.3, 2602.5.2
Method of Test for the Evaluation of Flammability Characteristics of Exterior, Nonload-bearing Wall Panel Assemblies Using Foam Plastic Insulation. Test Standard of the International Conference of Building Officials.

26-5; 201.2, 217, 2603.1.6
Chamber Method of Test for Measuring the Density of Smoke from the Burning or Decomposition of Plastic Materials. Standard Test Method D 2843-70 of the ASTM.

26-6; 201.2, 217, 601.3, 2603.1.6
Test Method for Ignition Properties of Plastics. Standard Test Method D 1929-68 (1975) of the ASTM.

26-7; 201.2, 217, 2603.1.6
Method of Test for Determining Classification of Approved Light-transmitting Plastics. Standard Test Method D 635-74 of the ASTM.

26-8; 601.3, 2602.5.5
Room Fire Test for Garage Doors Using Foam Plastic Insulation, Test Standard of the International Conference of Building Officials.

26-9; 601.3, 2602.5.2
Method of Test for the Evaluation of Flammability Characteristics of Exterior Nonload-Bearing Wall Assemblies Containing Combustible Components Using the Intermediate-Scale, Multistory Test Apparatus. Test Standard of the International Conference of Building Officials.

CHAPTER 31

31-1; Appendix 3112.2
Flame-retardant Membranes. Test Standard of the International Conference of Building Officials.

APPENDIX CHAPTER 10

10-5; See Section 1009 and *Uniform Building Security Code*
Tests for Doors and Locking Hardware Used for Security. Standard of the International Conference of Building Officials.

10-6; See Section 1009 and *Uniform Building Security Code*
Tests for Window Assemblies. Standard of the International Conference of Building Officials.

Part III—Standards Adopted by Reference

TITLE AND SOURCE	SECTION REFERENCE
CHAPTER 7	
Lightweight Aggregates for Structural Concrete ASTM C 330-89	702, 703.4
CPSC 16 CFR, Part 1209 Interim Safety Standard for Cellulose Insulation and Part 1404 Cellulose Insulation	707.3
Fire-Resistance Design Manual, Fourteenth Edition, April 1994 Gypsum Association	Tables 7-A, 7-B and 7-C

Appendix

Appendix Chapter 3
USE OR OCCUPANCY
Division I—DETENTION AND CORRECTIONAL FACILITIES

SECTION 313 — SCOPE

The provisions of this chapter apply to the design and construction of Group I, Division 3 Occupancies housing detention or correctional facilities (prisons, jails and reformatories).

SECTION 314 — APPLICATION

This appendix chapter may be used as alternative provisions to requirements found in Chapter 3 of this code. If this appendix chapter is used for design or construction purposes, all requirements in this appendix chapter shall be used. Chapter 3 provisions may be used if not specifically noted in this appendix chapter.

SECTION 315 — DEFINITIONS

For the purpose of this chapter, certain terms are defined as follows:

CELL is a housing unit in a detention or correctional facility for the confinement of not more than two inmates or prisoners.

CELL COMPLEX is a cluster or group of cells in a jail, prison or similar detention facility, together with rooms used for accessory purposes, all of which open into the cell complex, and are used for functions such as dining, counseling, exercise, classrooms, sick call, visiting, storage, staff offices, control rooms or similar functions, and interconnecting corridors all within the cell complex.

CELL, MULTIPLE-OCCUPANCY, is a housing area in a detention or correctional facility designed to house no less than three or no more than 16 inmates.

CELL TIER are cells located one level above the other, not exceeding two levels per floor.

DAY ROOM is a room adjacent to a cell, cell complex or cell tier and is used as a dining, exercise or other activity room for inmates.

SECTION 316 — CONSTRUCTION, REQUIREMENT EXCEPTIONS

316.1 General. Except as provided in this appendix chapter, buildings shall be constructed in accordance with the provisions of this code.

316.2 Exceptions to Table 6-A. Regardless of the provisions of Table 6-A, nonbearing cell walls within cell complexes may be of nonfire-rated, noncombustible construction, provided the cell complex is separated from all other areas of the building, including corridors that connect to the cell complex by construction and opening protection as required for corridors.

The open space in front of a cell tier not exceeding two tiers in height in detention or correctional facilities shall not be considered a vertical shaft whether extending from the floor to ceiling above or from floor to underside of roof.

SECTION 317 — COMPARTMENTATION

Every story having an occupant load of more than 50 inmates in a detention or correctional facility shall be divided into not less than two approximately equal compartments by a smoke-stop partition, constructed pursuant to the provisions of Section 308.2.

> **EXCEPTIONS:** 1. Protection may be accomplished with horizontal exits. (See Section 1005.3.5.)
>
> 2. In restraint areas, there are no restrictions on the total area of glazed openings in a smoke barrier, provided vision panels are of glazing material as specified in Section 713.9.

SECTION 318 — OCCUPANCY SEPARATIONS

Regardless of the provisions of Table 3-B, a three-hour fire-resistive occupancy separation as set forth in Section 302.3 may be used between a Group I, Division 3 Occupancy and a Group S, Division 3 Occupancy used only for the parking of vehicles used to transport inmates or prisoners provided no repair work or fueling is performed.

> **EXCEPTION:** Such occupancy separations need not be provided unless the Group S, Division 3 Occupancy area is enclosed with both surrounding walls and a solid roof.

SECTION 319 — GLAZING

In restraint areas of fully sprinklered detention and correctional facilities, the area of glazing in one-hour corridor walls is not restricted, provided:

1. All glazing is approved $1/4$-inch-thick (6.4 mm) wired glass or other approved and fire-tested glazing material set in steel frames.

2. In lieu of the sizes set forth in Section 1004.3.4.3.2, the size and area of wired glass assemblies shall conform to Sections 713.7 and 713.8. Other glazing material shall not exceed the sizes and areas as specified in the fire test.

SECTION 320 — ELECTRICAL

Approved special electrical systems, exit illumination, power installations and alternate on-site electrical supplies shall be provided for every building or portion of a building housing 10 or more inmates in a detention or correctional facility.

SECTION 321 — AUTOMATIC SPRINKLER AND STANDPIPE SYSTEMS

321.1 General. Every building, or portion thereof, housing more than six inmates in a detention or correctional facility or similar occupancy shall be protected by an automatic sprinkler system conforming to the provisions of UBC Standard 9-1. The main sprinkler control valve or valves and all other control valves in the system shall be electrically supervised so that at least a local alarm will sound at a constantly attended location when valves are closed.

> **EXCEPTION:** The sprinkler and piping serving single cells may be imbedded in the concrete construction. Protection for sprinklers and piping shall meet the provisions of UBC Standard 9-1.

When a complete approved automatic sprinkler system conforming to this section is installed in a building or buildings of a detention or correctional facility, pressurized enclosures need not be provided. However all required stairways shall be pressurized to a minimum of 0.15 inch of water column (37.3 Pa) upon actuation of the smoke-detection system.

321.2 Wet Standpipes. Every building in a detention or correctional facility, housing 50 or more inmates, shall be provided with Class II standpipes with hoses, conforming to the provisions of Chapter 9. Wet standpipes shall be located in cell complexes and in other cell areas of the building. In addition, Class II standpipes shall be located so that it will not be necessary to extend hose lines through interlocking security doors or any exit doors in smoke-stop partitions or horizontal exit walls.

321.3 Dry Standpipes. Regardless of the height of the building or number of stories, every detention or correctional facility shall be provided with a Class I standpipe.

> **EXCEPTION:** In lieu of dry standpipes, combined systems meeting the provisions of UBC Standard 9-2 may be used.

When acceptable to the fire authority having jurisdiction, fire department connections may be located inside all security walls or fences on the property.

Standpipes shall be located in accordance with Chapter 9 and, when located in cell complexes, may be placed in secured pipe chases.

SECTION 322 — FIRE ALARM SYSTEMS

Fire alarm systems shall be provided in accordance with the Fire Code.

SECTION 323 — SMOKE MANAGEMENT

323.1 Smoke Management System. A mechanically operated smoke management system or systems shall be provided in every detention or correctional facility.

323.2 Design and Installation. Every smoke management system shall be designed with zones that shall not exceed one smoke compartment per zone, except cell zones. Upon activation, the system shall operate at 100 percent exhaust from any zone of smoke generation and at 100 percent supply to all floors with returns closed in all zones adjacent to zone of smoke generation at not less than eight air changes per hour.

323.3 Automatic Initiation. Operation of the smoke-management system shall be initiated automatically upon the actuation of appropriately zoned automatic sprinkler flow indicators or smoke detectors, or both. Smoke detectors shall be installed in accordance with Section 608 of the Mechanical Code and their listing.

323.4 Manual Controls. Zone operation status indicators and manual controls capable of overriding the automatic controls shall be provided in a location approved by the fire department.

323.5 Location of Intakes. Exhaust discharges and fresh air supply intakes shall be so located as to prevent the reintroduction of smoke into the building.

323.6 Plans. The location of required fire dampers or combination smoke-fire dampers shall be clearly indicated on plans.

323.7 Omission of Fire Dampers. Fire dampers required by other provisions of this code are not required if such dampers interfere with the operation of the smoke management system.

> **EXCEPTION:** Those required to maintain the integrity of a floor-ceiling assembly.

323.8 Duct Materials. Duct materials shall be capable of safely conveying heat, smoke and toxic gases, to withstand both positive and negative pressures that may be imposed during the smoke-control mode, and to maintain their structural integrity under fire exposure conditions.

SECTION 324 — MEANS OF EGRESS

324.1 Number of Means of Egress. Multiple-occupancy rooms and day rooms in buildings or portions thereof in correctional or detention facilities constructed of not less than one-hour fire-resistive construction shall be provided with a minimum of two means of egress when the occupant load is more than 20.

The occupant load of any restraint area shall be determined by Table 10-A and classified as to the occupancy group it most nearly resembles, and means of egress shall be provided as required by Section 1003.1.

A minimum of two means of egress shall be provided in all areas of restraint (cells, day rooms, cell tiers and cell complexes) within a detention or correctional facility when the occupant load is more than 20.

324.2 Adjoining or Accessory Areas. Means of egress from a room may open into an adjoining or intervening room or area, provided such adjoining room is accessory to the area served and provides a direct means of egress to a corridor, exit or exterior exit balcony.

> **EXCEPTIONS:** 1. Means of egress are not to pass through kitchens, storerooms, restrooms, closets or spaces used for similar purposes.
>
> 2. The space in front of cells normally called a day room and used as means of egress in a detention or correctional facility shall not be considered an adjoining or accessory area if individual cells open directly into the space.

324.3 Cell Door Width. Cell doors shall not be less than 2 feet (610 mm) in width and 6 feet (1829 mm) in height.

324.4 Sliding Doors in Detention or Correctional Facilities. Electrically controlled and operated sliding doors may be used as exit doors regardless of occupant load served. Electrically controlled doors shall be designed to allow for manual operation by staff in the event of power failure.

324.5 Dead-end Balconies. Exit-access balconies serving cell tiers shall not extend more than 50 feet (15 240 mm) beyond an exit stairway.

> **NOTE:** For number of means of egress, see Section 1004.2.3.

324.6 Electrically Operable Doors. All doors (except those opening directly to the exterior of the building) and doors from cells and holding rooms in detention and correctional occupancies shall be electrically operable from the facility control center. Electric operation shall override any manual device.

SECTION 325 — FENCED ENCLOSURES

Exterior fenced enclosures into which a means of egress from a building or buildings terminate shall be provided with a safe dispersal area located not less than 50 feet (15 240 mm) from any building. Dispersal areas shall be based on an area of not less than 3 square feet (0.28 m^2) per occupant. A gate shall be provided from the safe dispersal area to allow for necessary relocation of occupants.

Exterior fenced enclosures used for exit discharge and which do not provide a safe dispersal area shall have not less than two means of egress.

Fenced enclosures located on roofs of buildings one or more stories in height shall be provided with not less than two means of egress regardless of occupant load.

Fenced enclosures used for recreational or activity purposes only shall be provided with exits in accordance with Chapter 10.

Division II—AGRICULTURAL BUILDINGS

SECTION 326 — SCOPE

The provisions of this appendix shall apply exclusively to agricultural buildings. Such buildings shall be classified as Group U, Division 3 Occupancies and shall include the following uses:

1. Storage, livestock and poultry.

2. Milking barns.

3. Shade structures.

4. Horticultural structures (greenhouse and crop protection).

SECTION 327 — CONSTRUCTION, HEIGHT AND ALLOWABLE AREA

327.1 General. Buildings classed as Group U, Division 3 Occupancies shall be of one of the types of construction specified in this code and shall not exceed the area or height limits specified in Sections 504, 505 and 506 and Table A-3-A.

327.2 Special Provisions. The area of a Group U, Division 3 Occupancy in a one-story building shall not be limited if the building is entirely surrounded and adjoined by public ways or yards not less than 60 feet (18 288 mm) in width, regardless of the type of construction.

The area of a two-story Group U, Division 3 Occupancy shall not be limited if the building is entirely surrounded and adjoined by public ways or yards not less than 60 feet (18 288 mm) in width and is provided with an approved automatic sprinkler system throughout conforming to UBC Standard 9-1.

Buildings using plastics shall comply with Type V-N construction. Plastics shall be approved plastics as defined in Chapter 2 and regulated by Chapter 26. For foam plastic, see Section 2602.

EXCEPTIONS: 1. When used as skylights or roofs, the areas of plastic skylights shall not be limited.

2. Except where designs must consider snow loads, plastics less than 20 mil (0.51 mm) thick may be used without regard to structural considerations. The structural frame of the building, however, shall comply.

SECTION 328 — OCCUPANCY SEPARATIONS

Occupancy separations shall be as specified in Section 302 and Table A-3-B.

SECTION 329 — EXTERIOR WALLS AND OPENINGS

Except where Table 6-A requires greater protection, exterior walls of agricultural buildings shall not be less than one-hour fire-resistive construction when less than 20 feet (6096 mm) from property line.

Openings in exterior walls of agricultural buildings that are less than 20 feet (6096 mm) from property lines shall be protected by fire assemblies having a fire-protection rating of not less than three-fourths hour.

SECTION 330 — MEANS OF EGRESS

Means of egress shall be as specified in Chapter 10.

EXCEPTIONS: 1. The maximum travel distance shall not exceed 300 feet (91 440 mm).

2. One means of egress is required for each 15,000 square feet (1394 m^2) of floor area and fraction thereof.

3. Exit and exit-access openings shall not be less than 2 feet 6 inches by 6 feet 8 inches (762 mm by 2032 mm).

TABLE A-3-A—BASIC ALLOWABLE AREA FOR A GROUP U, DIVISION 3 OCCUPANCY, ONE STORY IN HEIGHT AND MAXIMUM HEIGHT OF SUCH OCCUPANCY

I	II			III and IV		V	
	F.R.	One-hour	N	One-hour or Type IV	N	One-hour	N
ALLOWABLE AREA[1] (square feet)							
× 0.093 for m²							
Unlimited	60,000	27,100	18,000	27,100	18,000	21,100	12,000[1]
MAXIMUM HEIGHT IN STORIES[2]							
Unlimited	12	4	2	4	2	3	2

[1]See Section 327 for unlimited area under certain conditions.
[2]For maximum height in feet, see Chapter 5, Table 5-B.

TABLE A-3-B—REQUIRED SEPARATIONS BETWEEN GROUP U, DIVISION 3 AND OTHER OCCUPANCIES (In Hours)

Occupancy	A	E	I	H[1]	S-3	B	S-1, 2, 4 and 5	F and M	R-1	R-3	U
Rating	4	4	4	4	4	1	1	1	1	1	N

[1]See Chapter 3 for Group H, Division 1 Occupancies.

Division III—REQUIREMENTS FOR GROUP R, DIVISION 3 OCCUPANCIES

SECTION 331 — GENERAL

331.1 Purpose. The purpose of this division is to provide minimum standards for the protection of life, limb, health, property and environment and for the safety and welfare of the consumer, general public, and the owners and occupants of Group R, Division 3 Occupancies regulated by this code.

331.2 Scope. The provisions of this division apply to the construction, prefabrication, alteration, repair, use, occupancy and maintenance of detached one- or two-family dwellings not more than three stories in height and their accessory structures.

SECTION 332 — ONE AND TWO FAMILY DWELLING CODE ADOPTED

Buildings regulated by this division shall be designed and constructed to comply with the requirements of the Council of American Building Officials *One and Two Family Dwelling Code,* 1995 edition (as it applies to detached one- and two-family dwellings), promulgated jointly by the International Conference of Building Officials, Building Officials and Code Administrators International, and the Southern Building Code Congress International.

Division IV—REQUIREMENTS FOR GROUP R, DIVISION 4 OCCUPANCIES

SECTION 333 — GENERAL

333.1 Purpose. The purpose of this division is to provide minimum standards of safety for group care facilities.

333.2 Scope.

333.2.1 General. The provisions of this division shall apply to buildings or portions thereof that are to be used for Group R, Division 4 Occupancies.

333.2.2 Applicability of other provisions. Except as specifically required by this division, Group R, Division 4 Occupancies shall meet all applicable provisions of this code. Group R, Division 4 Occupancies need not be accessible to persons with disabilities.

333.3 Definitions. For the purpose of this division, certain terms are defined as follows:

AMBULATORY PERSONS are those capable of achieving mobility sufficient to exit without the assistance of another person.

GROUP R, DIVISION 4 OCCUPANCIES shall be residential group care facilities for ambulatory, nonrestrained persons who may have a mental or physical impairment (each accommodating more than five and not more than 16 clients or residents, excluding staff).

SECTION 334 — CONSTRUCTION, HEIGHT AND ALLOWABLE AREA

334.1 General. Buildings or portions of buildings classified as Group R, Division 4 may be constructed of any materials allowed by this code, shall not exceed two stories in height or be located above the second story in any building, and shall not exceed 3,000 square feet (278.7 m^2) in floor area per story except as provided in Sections 504, 505 and 506.

334.2 Special Provisions. Group R, Division 4 Occupancies having more than 3,000 square feet (278.7 m^2) of floor area above the first story shall not be of less than one-hour fire-resistive construction throughout.

334.3 Mixed Occupancies. Group R, Division 4 Occupancies shall be separated from Group H Occupancies by a four-hour fire-resistive occupancy separation and shall be separated from all other occupancies by a one-hour fire-resistive occupancy separation.

> **EXCEPTIONS:** 1. An occupancy separation need not be provided between a Group R, Division 4 Occupancy and a carport having no enclosed uses above, provided the carport is entirely open on two or more sides.
>
> 2. In the one-hour occupancy separation between a Group R, Division 4 and Group U, Division 1 Occupancy, the separation may be limited to the installation of materials approved for one-hour fire-resistive construction on the garage side, and a self-closing, tightfitting, solid-wood door 1^3/$_8$ inches (35 mm) in thickness will be permitted in lieu of a one-hour fire assembly. Fire dampers need not be installed in air ducts passing through the wall, floor or ceiling separating a Group R, Division 4 Occupancy from a Group U, Division 1 Occupancy, provided such ducts within the Group U Occupancy are constructed of steel having a thickness not less than 0.019 inch (0.48 mm) (No. 26 galvanized sheet gage) and have no openings into the Group U Occupancy.

SECTION 335 — LOCATION ON PROPERTY

Exterior walls located less than 3 feet (914 mm) from property lines shall be of one-hour fire-resistive construction. Openings shall not be permitted in exterior walls located less than 3 feet (914 mm) from property lines. For other requirements, see Section 503 and Chapter 6.

SECTION 336 — MEANS OF EGRESS AND EMERGENCY ESCAPES

336.1 General. Group R, Division 4 Occupancies shall be provided with means of egress as required by this section and Chapter 10 of this code.

336.2 Exits Required.

336.2.1 Number of exits. Every story, basement or portion thereof housing a Group R, Division 4 Occupancy shall not have less than two exit or exit-access doors.

> **EXCEPTIONS:** 1. Basements used exclusively for the service of the building may have one exit or exit-access door. For the purpose of this exception, storage rooms, laundry rooms, maintenance offices and similar uses shall not be considered as providing service to the building.
>
> 2. Storage rooms, laundry rooms and maintenance offices not exceeding 300 square feet (27.9 m^2) in floor area may be provided with only one exit or exit-access door.

336.2.2 Distance to exits. The maximum travel distance specified in Chapter 10 shall be reduced by 50 percent.

336.3 Corridor Width. Corridors shall not be less than 36 inches (914 mm) in width.

336.4 Stairways. Stairways shall be constructed as required by Section 1003.3.3 of this code.

> **EXCEPTION:** In buildings that are converted to a Group R, Division 4 Occupancy, existing stairways may have an 8-inch-maximum (203 mm) rise, 9-inch-minimum (229 mm) run and may be 30 inches (762 mm) in width.

336.5 Emergency Means of Egress Illumination. In the event of power failure, means of egress illumination shall be automatically provided from an emergency system. Emergency systems shall be supplied from storage batteries or an on-site generator set and the system shall be installed in accordance with the requirements of the Electrical Code.

336.6 Emergency Escape. Every sleeping room shall be provided with emergency escape or rescue facilities as required by Section 310.4 of this code.

SECTION 337 — LIGHT, VENTILATION AND SANITATION

Light and ventilation shall be as specified in Section 1203.

Sanitation shall be as specified in Section 2902.6.

SECTION 338 — YARDS AND COURTS

Yards and courts shall be as specified in Section 1203.4.

SECTION 339 — ROOM DIMENSIONS

Room dimensions shall be as specified in Section 310.6.

SECTION 340 — SHAFT ENCLOSURES

Exits shall be enclosed as specified in Chapter 10.

Elevator shafts, vent shafts, dumbwaiter shafts, clothes chutes and other vertical openings shall be enclosed and the enclosure shall be as specified in Section 711.

SECTION 341 — FIRE ALARM SYSTEMS

An approved automatic and manual fire alarm system shall be provided in Group R, Division 4 Occupancies.

SECTION 342 — HEATING

All habitable rooms shall be provided with heating facilities capable of maintaining a room temperature of 70°F (21°C) at a point 3 feet (914 mm) above the floor.

SECTION 343 — SPECIAL HAZARDS

343.1 Heating Equipment. All heating equipment shall be permanently installed. Chimneys and heating apparatus shall conform to the requirements of Chapter 31 of this code and the Mechanical Code.

343.2 Flammable Liquids. The storage and handling of gasoline, fuel oil or other flammable liquids shall be in accordance with the Fire Code.

Appendix Chapter 4
SPECIAL USE AND OCCUPANCY
Division I— BARRIERS FOR SWIMMING POOLS, SPAS AND HOT TUBS

SECTION 419 — GENERAL

419.1 Scope. The provisions of this section apply to the design and construction of barriers for swimming pools located on the premises of Group R, Division 3 Occupancies.

419.2 Standards of Quality. In addition to the other requirements of this code, safety covers for pools and spas shall meet the requirements for pool and spa safety covers as listed below. The standard listed below is a recognized standard. (See Section 3504.)

1. ASTM F 1346, Standard Performance Specification for Safety Covers and Labeling Requirement for All Covers for Swimming Pools, Spas and Hot Tubs

SECTION 420 — DEFINITIONS

For the purpose of this section, certain terms, words and phrases are defined as follows:

ABOVEGROUND/ON-GROUND POOL. See definition of "swimming pool."

BARRIER is a fence, wall, building wall or combination thereof that completely surrounds the swimming pool and obstructs access to the swimming pool.

GRADE is the underlying surface, such as earth or a walking surface.

HOT TUB. See definition of "spa, nonself-contained" and "spa, self-contained."

IN-GROUND POOL. See definition of "swimming pool."

SEPARATION FENCE is a barrier that separates all doors of a dwelling unit with direct access to a swimming pool from the swimming pool.

SPA, NONSELF-CONTAINED, is a hydromassage pool or tub for recreational or therapeutic use, not located in health-care facilities, designed for immersion of users and usually having a filter, heater and motor-driven blower. It may be installed indoors or outdoors, on the ground or on a supporting structure, or in the ground or in a supporting structure. A nonself-contained spa is intended for recreational bathing and contains water over 24 inches (610 mm) deep.

SPA, SELF-CONTAINED, is a continuous-duty appliance in which all control, water-heating and water-circulating equipment is an integral part of the product, located entirely under the spa skirt. A self-contained spa is intended for recreational bathing and contains water over 24 inches (610 mm) deep.

SWIMMING POOL is any structure intended for swimming or recreational bathing that contains water over 24 inches (610 mm) deep. This includes in-ground, aboveground and on-ground swimming pools, and fixed-in-place wading pools.

SWIMMING POOL, INDOOR, is a swimming pool that is totally contained within a residential structure and surrounded on all four sides by walls of said structure.

SWIMMING POOL, OUTDOOR, is any swimming pool that is not an indoor pool.

SECTION 421 — REQUIREMENTS

421.1 Outdoor Swimming Pool. An outdoor swimming pool shall be provided with a barrier that shall be installed, inspected and approved prior to plastering or filling with water. The barrier shall comply with the following:

1. The top of the barrier shall be at least 48 inches (1219 mm) above grade measured on the side of the barrier that faces away from the swimming pool. The maximum vertical clearance between grade and the bottom of the barrier shall be 2 inches (51 mm) measured on the side of the barrier that faces away from the swimming pool. The maximum vertical clearance at the bottom of the barrier may be increased to 4 inches (102 mm) when grade is a solid surface such as a concrete deck, or when the barrier is mounted on the top of the aboveground pool structure. When barriers have horizontal members spaced less than 45 inches (1143 mm) apart, the horizontal members shall be placed on the pool side of the barrier. Any decorative design work on the side away from the swimming pool, such as protrusions, indentations or cutouts, which render the barrier easily climbable, is prohibited.

2. Openings in the barrier shall not allow passage of a $1^3/_4$-inch-diameter (44.5 mm) sphere.

> **EXCEPTIONS:** 1. When vertical spacing between such openings is 45 inches (1143 mm) or more, the opening size may be increased such that the passage of a 4-inch-diameter (102 mm) sphere is not allowed.
>
> 2. For fencing composed of vertical and horizontal members, the spacing between vertical members may be increased up to 4 inches (102 mm) when the distance between the tops of horizontal members is 45 inches (1143 mm) or more.

3. Chain link fences used as the barrier shall not be less than 11 gage.

4. Access gates shall comply with the requirements of Items 1 through 3. Pedestrian access gates shall be self-closing and have a self-latching device. Where the release mechanism of the self-latching device is located less than 54 inches (1372 mm) from the bottom of the gate, (1) the release mechanism shall be located on the pool side of the barrier at least 3 inches (76 mm) below the top of the gate, and (2) the gate and barrier shall have no opening greater than $1/_2$ inch (12.7 mm) within 18 inches (457 mm) of the release mechanism. Pedestrian gates shall swing away from the pool. Any gates other than pedestrian access gates shall be equipped with lockable hardware or padlocks and shall remain locked at all times when not in use.

5. Where a wall of a Group R, Division 3 Occupancy dwelling unit serves as part of the barrier and contains door openings between the dwelling unit and the outdoor swimming pool that provide direct access to the pool, a separation fence meeting the requirements of Items 1, 2, 3 and 4 of Section 421.1 shall be provided.

> **EXCEPTION:** When approved by the building official, one of the following may be used:
>
> 1. Self-closing and self-latching devices installed on all doors with direct access to the pool with the release mechanism located a minimum of 54 inches (1372 mm) above the floor.
>
> 2. An alarm installed on all doors with direct access to the pool. The alarm shall sound continuously for a minimum of 30 seconds within seven seconds after the door and its screen, if present, are opened, and be capable of providing a sound pressure level of not less than 85 dBA when measured indoors at 10 feet (3048 mm). The alarm shall automatically reset under all condi-

tions. The alarm system shall be equipped with a manual means, such as a touchpad or switch, to temporarily deactivate the alarm for a single opening. Such deactivation shall last no longer than 15 seconds. The deactivation switch shall be located at least 54 inches (1372 mm) above the threshold of the door.

  3. Other means of protection may be acceptable so long as the degree of protection afforded is not less than that afforded by any of the devices described above.

  6. Where an aboveground pool structure is used as a barrier or where the barrier is mounted on top of the pool structure, and the means of access is a ladder or steps, then (1) the ladder or steps shall be capable of being secured, locked or removed to prevent access or (2) the ladder or steps shall be surrounded by a barrier that meets the requirements of Items 1 through 5. When the ladder

or steps are secured, locked or removed, any opening created shall be protected by a barrier complying with Items 1 through 5.

421.2 Indoor Swimming Pool. For an indoor swimming pool, protection shall comply with the requirements of Section 421.1, Item 5.

421.3 Spas and Hot Tubs. For a nonself-contained and self-contained spa or hot tub, protection shall comply with the requirements of Section 421.1.

    EXCEPTION: A self-contained spa or hot tub equipped with a listed safety cover shall be exempt from the requirements of Section 421.1.

Division II—AVIATION CONTROL TOWERS

SECTION 422 — GENERAL

The provisions of this appendix apply exclusively to aviation control towers not exceeding 1,500 square feet (139.35 m²) per floor. Such buildings shall be classified as Group B Occupancies and shall be used only for the following uses:

1. Airport traffic control cab.
2. Electrical and mechanical equipment rooms.
3. Airport terminal radar and electronics rooms.
4. Office spaces incidental to the tower operation.
5. Lounges for employees, including sanitary facilities.

SECTION 423 — CONSTRUCTION, HEIGHT AND ALLOWABLE AREA

Buildings or portions of buildings constructed under the provisions of this chapter shall be either Type I-F.R., Type II-F.R., Type II One-hour, Type II-N or Type III One-hour construction. The height of the building or parts thereof shall not exceed the limitations specified in Table A-4-A and the area of such buildings shall not exceed 1,500 square feet (139.35 m²) on any floor.

SECTION 424 — MEANS OF EGRESS

A single stairway may be used for means of egress in towers of any height, provided the occupant load per floor does not exceed 15. Access to the stairway and the elevator shall be separated from each other a distance apart equal to no less than one half of the length of the maximum overall diagonal dimension of the area served measured in a straight line. The exit stairway and elevator hoistway may be located in a common shaft enclosure, provided they are separated from each other by a four-hour separation having no openings. Such stairway shall be constructed to comply with the requirements for pressurized enclosures as specified in Section 1005.3.3. Stairways, however, need not extend to the roof as specified in Section 1003.3.3.11. The provisions of Section 403 do not apply.

SECTION 425 — FIRE ALARMS

Smoke detectors shall be installed in all occupied levels. These devices shall be part of an approved fire alarm system having audible alarms mounted in all occupied levels.

SECTION 426 — ACCESSIBILITY

Aviation control towers need not be accessible to the handicapped as specified in the provisions of Chapter 11 and Section 2903.

SECTION 427 — STANDBY POWER AND EMERGENCY GENERATION SYSTEMS

A standby power-generation system conforming to the Electrical Code shall be installed in aviation control towers over 65 feet (19 812 mm) in height and shall provide power to the following equipment:

1. Pressurized enclosure, mechanical equipment and lighting.
2. Elevator operational power.
3. Smoke-detection systems.

TABLE A-4-A—MAXIMUM HEIGHT OF AVIATION CONTROL TOWERS (feet)

TYPES OF CONSTRUCTION				
I-F.R.	II-F.R.	II One-hour	III One-hour	II-N
× 0.3048 for m				
Unlimited	240	100	65	85

Appendix Chapter 9
BASEMENT PIPE INLETS

SECTION 907 — BASEMENT PIPE INLETS

907.1 General. All basement pipe inlets shall be installed in accordance with the requirements of this section.

907.2 Where Required. Basement pipe inlets shall be installed in the first floor of every store, warehouse or factory having basements.

> **EXCEPTIONS:** 1. Where the basement is equipped with an automatic sprinkler system as specified in Section 904.2.
>
> 2. Where the basement is used for the storage of permanent archives or valuables such as safe deposit vaults or similar uses adversely affected by water.

907.3 Location. The location of basement pipe inlets shall be as required by the fire department.

907.4 Detailed Requirements. All basement pipe inlets shall be of cast iron, steel, brass or bronze with lids of cast brass or bronze.

The basement pipe inlet shall consist of a sleeve not less than 8 inches (203 mm) inside diameter extending through the floor and terminating flush with or through the basement ceiling and shall have a top flange recessed with an inside shoulder to receive the lid. The top flange shall be installed flush with finish floor surface. The lid shall be a solid casting and have a lift recessed in the top. This lid shall be provided with a cast-in sign reading: FIRE DEPARTMENT ONLY, DO NOT COVER. The lid shall be installed in such a manner to permit its easy removal from the flange shoulder.

Appendix Chapter 10
BUILDING SECURITY

SECTION 1010 — BUILDING SECURITY

Building security shall be in accordance with the *Uniform Building Security Code.*

Appendix Chapter 11
ACCESSIBILITY
Division I—SITE ACCESSIBILITY

SECTION 1107 — ACCESSIBLE EXTERIOR ROUTES

1107.1 General. Accessible exterior routes shall be provided from public transportation stops, accessible parking and accessible passenger loading zones and public sidewalks to the accessible building entrance they serve.

When more than one building or facility is located on a site, at least one accessible route shall connect accessible elements, facilities and buildings that are on the same site. The accessible route between accessible parking and accessible building entrances shall be the most practical direct route.

1107.2 Definition. CABO/ANSI A117.1 is American National Standard A117.1-1992 published by the Council of American Building Officials.

1107.3 Design and Construction. When accessibility is required by this section, it shall be designed and constructed in accordance with CABO/ANSI A117.1.

SECTION 1108 — PARKING FACILITIES

1108.1 Accessible Parking Required. When parking lots or garage facilities are provided, accessible parking spaces shall be provided in accordance with Table A-11-A except for the following occupancies:

1. For Group I, Divisions 1.1 and 2 medical care occupancies specializing in the treatment of persons with mobility impairments, 20 percent of the parking spaces provided shall be accessible.

2. For Group I, Divisions 1.1 and 1.2 and Group B Occupancies providing outpatient medical care facilities, 10 percent of the parking spaces provided shall be accessible.

3. For Group R, Division 1 apartment building containing accessible or adaptable dwelling units where parking is provided, 2 percent of the parking spaces shall be accessible. Where parking is provided within or beneath a building, accessible parking spaces shall also be provided within or beneath the building.

One van accessible parking space shall be provided for every eight accessible parking spaces, or fraction thereof.

Accessible parking spaces shall be located on the shortest possible accessible route from adjacent parking to an accessible building entrance. In facilities with multiple accessible building entrances with adjacent parking, accessible parking spaces shall be dispersed and located near the accessible entrances.

> **EXCEPTION:** In multilevel parking structures, accessible van parking spaces may be located on one level.

Where a parking facility is not accessory to a particular building, accessible parking spaces shall be located on the shortest accessible route to an accessible pedestrian entrance to the parking facility.

1108.2 Design and Construction. When accessible and van accessible parking spaces are required by this section, they shall be designed and constructed in accordance with CABO/ANSI A117.1.

1108.3 Signs. Accessible parking spaces required by this section shall be identified by a sign complying with CABO/ANSI A117.1.

> **EXCEPTION:** Accessible parking space signs need not be provided in parking garages or parking facilities that have five or less total parking spaces.

SECTION 1109 — PASSENGER LOADING ZONES

1109.1 Location. When provided, passenger loading zones shall be located on an accessible route.

1109.2 Design and Construction. Passenger loading zones shall be designed and constructed in accordance with CABO/ANSI A117.1.

TABLE A-11-A—NUMBER OF ACCESSIBLE PARKING SPACES

TOTAL PARKING SPACES IN LOT OR GARAGE	MINIMUM REQUIRED NUMBER OF ACCESSIBLE SPACES
1-25	1
26-50	2
51-75	3
76-100	4
101-150	5
151-200	6
201-300	7
301-400	8
401-500	9
501-1,000	2% of total spaces
Over 1,000	20 spaces plus 1 space for every 100 spaces, or fraction thereof, over 1,000

Division II—ACCESSIBILITY FOR EXISTING BUILDINGS

SECTION 1110 — SCOPE

The provisions of this division apply to renovations, alterations and additions to existing buildings, including those identified as historic buildings. This division includes minimum standards for removing architectural barriers, and providing and maintaining access to existing buildings and facilities for persons with disabilities.

SECTION 1111 — DEFINITIONS

For the purpose of this division, certain terms are defined as follows:

ALTERATION is any change, addition or modification in construction or occupancy.

TECHNICALLY INFEASIBLE is an alteration of a building or facility that has little likelihood of being accomplished because existing structural conditions would require removing or altering a load-bearing member that is an essential part of the structural frame, or because existing physical or site constraints prohibit modification or addition of elements, spaces or features that are in full and strict compliance with the minimum requirements for new construction and which are necessary to provide accessibility.

SECTION 1112 — ALTERATIONS

1112.1 General.

1112.1.1 Compliance. Alterations to existing buildings or facilities shall comply with this section. Alterations shall not reduce or have the effect of reducing accessibility or usability of a building, portion of a building, or facility. If compliance with this section is technically infeasible, the alteration shall provide access to the maximum extent technically feasible.

> **EXCEPTION:** Alterations to Group R, Division 1 apartment occupancies need not comply with this section.

1112.1.2 Existing elements. If existing elements, spaces, essential features or common areas are altered, each such altered element, space, feature or area shall comply with the applicable provisions in Division I of this appendix chapter and CABO/ANSI A117.1.

> **EXCEPTION:** Accessible means of egress required by Section 1104 need not be provided in alterations of existing buildings and facilities.

When an alteration is to an area of primary function, the accessible route to the altered area shall be made accessible. The accessible route to the primary function area shall include toilet facilities or drinking fountains serving the area of primary function.

> **EXCEPTIONS:** 1. The costs of providing the accessible route need not exceed 20 percent of the costs of the alterations affecting the area of primary function.
>
> 2. Alterations to windows, hardware, operating controls, electrical outlets and signs.
>
> 3. Alterations to mechanical systems or electrical systems, installation or alteration of fire-protection systems, and abatement of hazardous materials.
>
> 4. Alterations undertaken for the primary purpose of increasing the accessibility of an existing building, facility or element.

1112.2 Modifications.

1112.2.1 General. Modifications set forth in this section may be used for compliance when the required standard is technically infeasible.

1112.2.2 Hotel guest rooms. When guest rooms of a hotel are being altered, at least one of every 25 guest rooms being altered shall be accessible, and at least one additional guest room for every 25 guest rooms being altered shall be provided with visible and audible alarm-indicating appliances for persons with hearing impairments. The total number of accessible guest rooms and guest rooms accessible to persons with hearing impairments need not exceed the number required by Section 1103.1.9.2.

1112.2.3 Performance areas. When it is technically infeasible to alter performance areas to be on an accessible route, at least one of each type of performance area shall be made accessible.

1112.2.4 Platform lifts. Platform lifts may be used when installation of an elevator is technically infeasible.

1112.2.5 Toilet rooms. The addition of one accessible unisex toilet facility accessible to occupants on the floor may be provided in lieu of making existing toilet facilities accessible when it is technically infeasible to alter existing toilet and bathing facilities to be accessible. The unisex facility shall be located on the same floor and in the same area as the existing toilet facilities. Each unisex toilet facility shall contain one accessible water closet and lavatory, and the door shall be lockable from within the room.

When existing toilet facilities are being altered and are not made accessible, directional signs shall be provided indicating the location of the nearest accessible toilet or bathing facility within the building.

1112.2.6 Assembly areas. Seating shall adjoin an accessible route that also serves as a means of egress. When it is technically infeasible to disperse accessible seating throughout an altered assembly area, accessible seating areas may be clustered. Each accessible seating area shall have provisions for companion seating.

1112.2.7 Dressing rooms. When it is technically infeasible to provide accessible dressing rooms in each group of rooms, one dressing room for each sex, or a unisex dressing room, on each level shall be accessible.

SECTION 1113 — CHANGE OF OCCUPANCY

Requirements for new construction provided in Chapter 11 shall apply to existing buildings that undergo a change of occupancy group, unless technically infeasible.

SECTION 1114 — HISTORIC PRESERVATION

Accessibility provisions of this division shall be applied to historic buildings and facilities as defined in Section 3403.5 of this code.

The building official, after consulting with the appropriate historic preservation officer, shall determine whether provisions required by this division for accessible routes, ramps, entrances, toilets, parking or signage would threaten or destroy the historic significance of the building or facility.

If it is determined that any of the accessibility requirements listed above would threaten or destroy the historic significance of a building or facility, the modifications of Section 1112.2 for that feature may be utilized.

Appendix Chapter 12
INTERIOR ENVIRONMENT
Division I—VENTILATION

SECTION 1206 — SCOPE

Buildings and structures enclosing spaces intended for human occupancy shall be provided with ventilation in accordance with this appendix chapter.

SECTION 1207 — VENTILATION

1207.1 General. Enclosed portions of buildings and structures in occupancies, other than the locations specified in Sections 1207.3 through 1207.7, shall be provided with natural ventilation by means of openable exterior openings with an area of not less than $1/20$ of the total floor area of such portions, or shall be provided with a mechanically operated ventilating system. The mechanically operated ventilating system shall be capable of supplying ventilation air in accordance with Table A-12-A during such time as the building or space is occupied.

1207.2 Register Velocity. In assembly, educational and institutional occupancies when the velocity of the air at the register exceeds 10 feet per second (3.048 m/s), the register shall be placed more than 8 feet (2438 mm) above the floor directly beneath.

1207.3 Toilet Rooms. Toilet rooms shall be provided with a fully openable exterior window at least 3 square feet (0.27 m^2) in area; a vertical duct not less than 100 square inches (0.064 516 m^2) in area for the first toilet facility, with 50 additional square inches (0.032 m^2) for each additional facility; or a mechanically operated exhaust system capable of exhausting 50 cubic feet of air per minute (23.6 L/s) for each water closet or urinal installed in the toilet room. Such systems shall be connected directly to the outside, and the point of discharge shall be at least 3 feet (914 mm) from any openable window.

1207.4 Ventilation in Hazardous Locations. Rooms, areas or spaces in which explosive, corrosive, combustible, flammable or highly toxic dusts, mists, fumes, vapors or gases are or may be emitted due to the processing, use, handling or storage of materials shall be mechanically ventilated as required by the Fire Code and the Mechanical Code.

Emissions generated at work stations shall be confined to the area in which they are generated as specified in the Fire Code and the Mechanical Code.

Supply and exhaust openings shall be in accordance with the Mechanical Code. Exhaust air contaminated by highly toxic material shall be treated in accordance with the Fire Code.

A manual shutoff control for ventilation equipment shall be provided outside the room adjacent to the principal access door to the room. The switch shall be of the break-glass type and shall be labeled "Ventilation System Emergency Shutoff."

1207.5 Groups B, F, M and S Occupancies. In Groups B, F, M and S Occupancies, or portions thereof, where Class I, II or III-A liquids are used, mechanical exhaust shall be provided sufficient to produce six air changes per hour. Such mechanical exhaust shall be taken from a point at or near the floor level.

1207.6 Group S Parking Garages. In parking garages, other than open parking garages as defined in Section 405.2, used for storing or handling of automobiles operating under their own power and on loading platforms in bus terminals, ventilation shall be provided capable of exhausting a minimum of 1.5 cubic feet per minute (cfm) per square foot (0.761 L/s/m^2) of gross floor area. The building official may approve an alternate ventilation system designed to exhaust a minimum of 14,000 cfm (6608 L/s) for each operating vehicle. Such system shall be based on the anticipated instantaneous movement rate of vehicles, but not less than 2.5 percent (or one vehicle) of the garage capacity. Automatic carbon monoxide-sensing devices may be employed to modulate the ventilation system to maintain a maximum average concentration of carbon monoxide of 50 parts per million during any eight-hour period, with a maximum concentration not greater than 200 parts per million for a period not exceeding one hour.

> **EXCEPTION:** In Group S, Division 3 repair garages and motor vehicle fuel-dispensing stations without lubrication pits; storage garages; and in Group S, Division 5 aircraft hangars, such ventilating system may be omitted when, in the building official's opinion, the building is supplied with unobstructed openings to the outer air that are sufficient to provide the necessary ventilation.

Connecting offices, waiting rooms, ticket booths and similar uses shall be supplied with conditioned air under positive pressure.

1207.7 Group H, Division 4 Occupancies. In buildings used for the repair or handling of motor vehicles operating under their own power, mechanical ventilation shall be provided capable of exhausting a minimum of 1.5 cfm per square foot (7.62 L/s/m^2) of floor area. Each engine repair stall shall be equipped with an exhaust pipe extension duct, extending to the outside of the building, that, if over 10 feet (3048 mm) in length, shall mechanically exhaust 300 cfm (141.6 L/s). Connecting offices and waiting rooms shall be supplied with conditioned air under positive pressure.

> **EXCEPTION:** In repair garages and aircraft hangars, the building official may authorize the omission of such ventilating equipment when, in his or her opinion, the building is supplied with unobstructed openings to the outer air that are well distributed and sufficient in size to provide the necessary ventilation.

TABLE A-12-A—OUTDOOR AIR REQUIREMENTS FOR VENTILATION

OCCUPANCY [1]	OUTDOOR VENTILATION AIR (cfm per square foot of area unless noted) [2] × 0.472 for L/s per m²
Group A Occupancies	
Applications similar to:	
Food and Beverage Service	
Bars, cocktail lounges	3.00
Cafeteria, fast food	2.00
Dining rooms	1.40
Kitchens (cooking) [3]	0.30
Sports and Amusement	
Assembly rooms	1.80
Ballrooms and discos	2.50
Bowling alleys (seating areas)	1.75
Conference rooms	1.00
Gambling casinos	3.60
Game rooms	1.75
Ice arenas	0.50 (playing areas)
Playing floors (gymnasium)	0.60
Spectator areas	2.25
Swimming pools (pool and deck area)	0.50
Theaters	
Auditorium	2.25
Lobbies	3.00
Stages, studios	1.05
Ticket booths	1.20
Transportation	
Platforms	1.50
Waiting rooms	1.50
Group B Occupancies	
Applications similar to:	
Offices	
Bank vaults	0.08
Conference rooms	1.00
Corridors and utilities	0.05
Darkrooms	0.50
Duplicating, printing areas	0.50
Elevators	1.00[4]
Locker and dressing rooms	0.50
Meat-processing areas	0.15
Office spaces	0.14
Pharmacies	0.30
Photo studios	0.15
Public restrooms (per water closet or urinal)	50 cfm/water closet or urinal[4]
Reception areas	0.90
Smoking lounges	4.20[4]
Telecommunication centers and data entry spaces	1.20
Group E Occupancies	
Applications similar to:	
Education	
Auditoriums	2.25
Classrooms	0.75
Corridors	0.00
Laboratories	0.60
Libraries	0.30
Locker rooms	0.50
Music rooms	0.75
Smoking lounges	4.20[4]
Training shop	0.60

TABLE A-12-A—OUTDOOR AIR REQUIREMENTS FOR VENTILATION—(Continued)

OCCUPANCY [1]	OUTDOOR VENTILATION AIR (cfm per square foot of area unless noted) [2]
	× 0.472 for L/s per m^2
Group F Occupancies	
Applications similar to:	
Dry Cleaners, Laundries	
Coin-operated dry cleaners	0.30
Coin-operated laundries	0.30
Commercial dry cleaners	0.90
Commercial laundries	0.25
Storage, pick-up areas	1.05
Group I Occupancies	
Applications similar to:	
Hospitals, Nursing and Convalescent Homes	
Autopsy rooms	0.50[4]
Medical procedure rooms	0.30
Operating rooms	0.60
Patient rooms	0.25
Physical therapy rooms	0.30
Recovery and ICU rooms	0.30
Correctional facilities	
Cells	0.40
Dining halls	1.50
Guard stations	0.60
Public restrooms	50 cfm/water closet or urinal[4]
Group M Occupancies	
Applications similar to:	
Stores, Sales Floors and Showroom Floors	
Basement and street levels	0.30
Dressing rooms	0.20
Malls and arcades	0.20
Shipping and receiving areas	0.15
Smoking lounges	4.20[4]
Storage rooms	0.15
Upper levels	0.20
Warehouse	0.05
Specialty Shops	
Barber shops	0.38
Beauty shops	0.63
Clothiers	0.30
Drug stores	0.12
Fabric stores	0.12
Florists	0.12
Food stores	0.12
Furniture stores	0.30
Hardware stores	0.12
Pet shops	1.00
Reducing salons	0.30

(continued)

TABLE A-12-A—OUTDOOR AIR REQUIREMENTS FOR VENTILATION—(Continued)

OCCUPANCY [1]	OUTDOOR VENTILATION AIR (cfm per square foot of area unless noted) [2]
	× 0.472 for L/s per m^2
Group R Occupancies	
Division 1	
Hotels, motels, resorts, dormitories	
Assembly rooms	1.80
Bedrooms	30 cfm/room[5]
Conference rooms	1.00
Dormitory sleeping rooms	0.30
Living rooms	30 cfm/room[5]
Lobbies	0.45
Private bathrooms (intermittent exhaust)	35 cfm/room[5]
Division 1 Apartment Houses and Division 3 Dwellings and Lodging Houses	
Individual Dwelling Units, Lodging Houses	
Bathrooms (intermittent exhaust) or (continuous exhaust)	50 cfm/room[4,5] 20 cfm/room[4,5]
Kitchens (intermittent exhaust) or (continuous exhaust)	100 cfm/room[4,5] 25 cfm/room[4,5]
Living areas	0.35 ACH[6]
Group S Occupancies	
Applications similar to:	
Division 3	
Enclosed parking garages	1.50

[1]Applications may not be unique to a single occupancy group. Where specific use is not listed, judgment as to similarity shall be by the building official.

[2]Based on net occupiable space. The minimum amount of outdoor air supplied during occupancy shall be permitted to be based on the rate per square foot (m^2) of floor area indicated in Table A-12-A or cubic feet per minute (L/s) per person in accordance with nationally recognized standards. See Chapter 35. Controls shall be permitted to adjust outdoor air ventilation rates to provide equivalent rates per person under different conditions of occupancy.

[3]The sum of the outdoor and transfer air from adjacent spaces shall be sufficient to provide an exhaust rate of not less than 1.5 cubic feet per minute per square foot (0.708 L/s per m^2).

[4]Normally supplied by transfer air with local mechanical exhaust with no recirculation.

[5]Independent of room size.

[6]Air changes per hour, but not less than 15 cubic feet per minute (7.08 L/s) per person. Occupancy shall be based on the number of bedrooms; first bedroom, two persons each additional bedroom, one person.

Division II—SOUND TRANSMISSION CONTROL

SECTION 1208 — SOUND TRANSMISSION CONTROL

1208.1 General. In Group R Occupancies, wall and floor-ceiling assemblies separating dwelling units or guest rooms from each other and from public space such as interior corridors and service areas shall provide airborne sound insulation for walls, and both airborne and impact sound insulation for floor-ceiling assemblies.

The standards listed below are recognized standards (see Sections 3503 and 3504).

1. ASTM E 90 and E 413, Laboratory Determination of Airborne Sound Transmission Class (STC)

2. ASTM E 492, Impact Sound Insulation

3. ASTM E 336, Airborne Sound Insulation Field Test

1208.2 Airborne Sound Insulation. All such separating walls and floor-ceiling assemblies shall provide an airborne sound insulation equal to that required to meet a sound transmission class (STC) of 50 (45 if field tested).

Penetrations or openings in construction assemblies for piping; electrical devices; recessed cabinets; bathtubs; soffits; or heating, ventilating or exhaust ducts shall be sealed, lined, insulated or otherwise treated to maintain the required ratings.

Entrance doors from interior corridors together with their perimeter seals shall have a laboratory-tested STC rating of not less than 26 and such perimeter seals shall be maintained in good operating condition.

1208.3 Impact Sound Insulation. All separating floor-ceiling assemblies between separate units or guest rooms shall provide impact sound insulation equal to that required to meet an impact insulation class (IIC) of 50 (45 if field tested). Floor coverings may be included in the assembly to obtain the required ratings and must be retained as a permanent part of the assembly and may be replaced only by other floor covering that provides the same sound insulation required above.

1208.4 Tested Assemblies. Field or laboratory tested wall or floor-ceiling designs having an STC or IIC of 50 or more may be used without additional field testing when, in the opinion of the building official, the tested design has not been compromised by flanking paths. Tests may be required by the building official when evidence of compromised separations is noted.

1208.5 Field Testing and Certification. Field testing, when required, shall be done under the supervision of a professional acoustician who shall be experienced in the field of acoustical testing and engineering and who shall forward certified test results to the building official that minimum sound insulation requirements stated above have been met.

1208.6 Airborne Sound Insulation Field Tests. When required, airborne sound insulation shall be determined according to the applicable Field Airborne Sound Transmission Loss Test procedures. All sound transmitted from the source room to the receiving room shall be considered to be transmitted through the test partition.

1208.7 Impact Sound Insulation Field Test. When required, impact sound insulation shall be determined.

SECTION 1209 — SOUND TRANSMISSION CONTROL SYSTEMS

Generic systems as listed in the *Fire Resistance Design Manual,* Fourteenth Edition, dated April 1994, as published by the Gypsum Association, may be accepted where a laboratory test indicates that the requirements of Section 1208 are met by the system.

Appendix Chapter 13
ENERGY CONSERVATION IN NEW BUILDING CONSTRUCTION

SECTION 1302 — GENERAL

1302.1 Purpose. The purpose of this appendix is to regulate the design and construction of the exterior envelopes and selection of heating, ventilating and air-conditioning, service water heating, electrical distribution, and illuminating systems and equipment required for the purpose of effective conservation of energy within a building or structure governed by this code.

1302.2 *Model Energy Code* **Adopted.** To comply with the purpose of this appendix, buildings shall be designed to comply with the requirements of the *Model Energy Code* promulgated jointly by the International Conference of Building Officials, the Southern Building Code Congress International, Building Officials and Code Administrators International, and the National Conference of States on Building Codes and Standards, dated 1995.

Appendix Chapter 15
REROOFING

NOTE: This appendix chapter has been revised in its entirety.

SECTION 1514 — GENERAL

All reroofing shall conform to the applicable provisions of Chapter 15 of this code and as otherwise required in this chapter.

Roofing materials and methods of application shall comply with the UBC standards or shall follow manufacturer's installation requirements when approved by the building official.

SECTION 1515 — INSPECTION AND WRITTEN APPROVAL

1515.1 Written Approval Required. New roofing shall not be applied without first obtaining written approval from the building official.

The building official may allow existing roof coverings to remain when inspection or other evidence reveals all of the following:

1. The roof structure is sufficient to sustain the weight of the additional dead load of the new roofing.

2. The roof deck is structurally sound.

3. Roof drains and drainage are sufficient to prevent extensive accumulation of water.

4. The existing roofing is securely attached to the deck.

5. Existing insulation is not water soaked.

6. Fire-retardant requirements are maintained.

1515.2 Required Inspections.

1515.2.1 Preroofing inspection. Inspection prior to the installation of new roofing must be obtained from the building official to verify the existing roofing meets all the conditions in Section 1515.1. The building official may accept an inspection report of above-listed conditions prepared by a special inspector.

1515.2.2 Final inspection. A final inspection and approval shall be obtained from the building official when the reroofing is complete.

SECTION 1516 — REROOFING OVERLAYS ALLOWED

1516.1 General. No roof shall have in any combination more than that allowed in Table A-15-A. Roofing conforming to Section 1503 overlaid on existing roofing shall comply with the provisions of this section and manufacturer's installation requirements as an overlay when approved by the building official.

1516.2 Overlay on Existing Built-up Roofs. The building official may allow reroofing over existing built-up roofing when the conditions specified in Section 1515.1 have been met. When an existing built-up roof has been removed and prior to application of new roofing on a nailable deck that has residual bitumen, rosin-sized or other dry sheet shall be installed. Prior to the application of any reroofing, the existing surface shall be prepared as follows:

1. **Gravel-surfaced roofing.** Not more than one overlay shall be approved over an existing built-up roof. The existing built-up roof shall be cleaned of all loose gravel and debris. All blisters, buckles and other irregularities shall be cut and made smooth and secure. On nonnailable decks, minimum $^3/_8$-inch (9.5 mm) insula-

tion board shall be securely cemented to the existing roofing with hot bitumen after the existing surface has been adequately primed. On nailable decks, a rosin-sized or other dry sheet shall be installed and a base sheet shall be mechanically fastened in place.

2. **Smooth or cap-sheet surface.** Not more than one overlay shall be applied over an existing built-up roof. All blisters, buckles and other irregularities of existing built-up roof shall be cut and made smooth and secure. On nonnailable decks, a base sheet shall be spot cemented to the existing roof. On nailable decks, a base sheet shall be mechanically fastened in place and where residual materials on the existing surface may cause the new base sheet to adhere to the old roof, a rosin-sized or other dry sheet shall be installed under the new base sheet.

3. **Intersecting walls.** All concrete and masonry walls shall be completely cleaned and primed to receive new flashing. All other walls shall have the surface finish material removed to a minimum height of 6 inches (152 mm) above the new roof deck surface to receive new roofing and flashing. All rotted wood shall be replaced with new material. Surface finish material shall be replaced or reinstalled.

4. **Parapets.** Parapets of area separation walls shall have noncombustible faces, including counterflashing and coping materials.

> **EXCEPTION:** Combustible roofing may extend 7 inches (178 mm) above the roof surface.

5. **Cant strips.** Where space permits, cant strips shall be installed at all angles. All angles shall be flashed with at least two more layers than in the new roof with an exposed finish layer of inorganic felt or mineral surfaced cap sheet.

6. **Asphalt and wood shingle application.** Not more than one overlay of asphalt shingles shall be applied over one existing built-up roof on structures with a slope of 2 units vertical in 12 units horizontal (16.7% slope) or greater. Not more than one overlay of wood shingles shall be applied over one existing built-up roof on structures with a slope of 3 units vertical in 12 units horizontal (25% slope) or greater. The existing built-up roof shall have all surfaces cleaned of gravel and debris, all blisters and irregularities cut and made smooth and secure, and an underlayment of not less than Type 30 nonperforated felt shall be installed prior to reroofing.

7. **Spray-applied polyurethane foam application.** Spray-applied polyurethane foam may be applied directly to existing built-up roofing systems when the completed assembly is a Class A, B or C fire-retardant roofing assembly and complies with Section 2602.5.3. When applied on a fire-resistive roof-ceiling assembly, the completed assembly shall also comply with Section 710.1.

Base sheets or dry sheets are not required over existing surfaces when applying spray polyurethane foam roofing systems.

Miscellaneous materials such as adhesives, elastomeric caulking compounds, metal, vents and drains shall be a composite part of the roof system.

1516.3 Overlay on Existing Wood Roofs or Asphalt Shingle Roofs. The building official may allow reroofing over existing wood shingle roofing or asphalt shingle roofing. Only fire-retardant roofing assemblies or noncombustible roof covering may be applied over existing wood shake roofs in accordance with the listing or manufacturer's installation requirements when approved by the building official.

When the application of new roofing over existing wood shingle or wood shake roofs creates a combustible concealed space, the entire existing surface shall be covered with gypsum board, mineral fiber, glass fiber or other approved materials securely fastened in place.

Hip and ridge cover on existing shake or shingle roofing shall be removed prior to reroofing application. Roofing overlays may be installed in accordance with the following:

1. **Asphalt shingles.** Not more than two overlays of asphalt shingles shall be applied over an existing asphalt or wood shingle roof. Asphalt shingles applied over wood shingles shall not have less than Type 30 nonperforated felt underlayment installed prior to reroofing.

2. **Wood shakes.** Not more than one overlay of wood shakes shall be applied over an existing asphalt shingle or wood shingle roofing on structures with a slope of 4 units vertical in 12 units horizontal (33% slope) or greater. One layer of 18-inch (457 mm), Type 30 nonperforated felt shall be shingled between each course in such a manner that no felt is exposed to the weather below the shake butts.

3. **Wood shingles.** Not more than one overlay of wood shingles shall be applied over existing wood or asphalt shingles. Wood shingles applied over asphalt shingles shall not have less than Type 30 nonperforated felt underlayment installed prior to reroofing.

SECTION 1517 — TILE

Tile may be applied to roofs with a slope of 4 units vertical in 12 units horizontal (33% slope) or greater over existing roof coverings in accordance with Table A-15-A. Such installations shall be substantiated by a report prepared by an engineer or architect licensed by the state to practice as such, indicating that the existing or modified framing system is adequate to support the additional tile roof covering.

Tile shall be applied in accordance with the original manufacturer's specifications or when the original manufacturer's specifications are no longer available, in accordance with Section 1507.7.

Tile may be repaired to match the prior installation except that clay and terra-cotta hips and ridge tile shall be reinstalled with portland cement mortar.

SECTION 1518 — METAL ROOF COVERING

Metal roof covering may be applied over existing roofing in accordance with Table A-15-A. Reroofing with metal roof covering shall be in accordance with the original manufacturer's specifications or when the original manufacturer's specifications are no longer available as required by Section 1507.8.

SECTION 1519 — OTHER ROOFING

Reroofing with systems not covered elsewhere in Chapter 15 or this appendix, such as, but not limited to, those that are fluid applied or applied as nonasphaltic sheets, shall be done with materials and procedures approved by the building official.

SECTION 1520 — FLASHING AND EDGING

Missing, rusted or damaged flashing and counterflashing, vent caps, and metal edging shall be installed or replaced with new materials. When existing built-up roofs remain, vent flashing, metal edging, drain outlets, metal counterflashing and collars shall be removed and cleaned. All metal allowed to be reinstalled shall be primed prior to reroofing installation. Collars and flanges shall be flashed per the roofing manufacturer's instructions.

TABLE A-15-A—ALLOWABLE REROOFS OVER EXISTING ROOFING
(Inspection and Written Approval Required Prior to Application)

EXISTING ROOFING	NEW OVERLAY ROOFING							
	Built Up	Wood Shake	Wood Shingle	Asphalt Shingle	Tile Roof	Metal Roof	Modified Bitumen	Spray Polyurethane Foam
Built Up	Yes	NP	Yes (3:12)	Yes (2:12)	Yes (2.5:12)	Yes	Yes	Yes
Wood Shake[1]	NP	NP	NP	NP	Yes[2]	Yes[2]	NP	NP
Wood Shingle[1]	NP	Yes[3] (4:12)	Yes[4]	Yes[4]	Yes[2]	Yes[2]	NP	NP
Asphalt Shingle[1]	NP	Yes[3] (4:12)	Yes[4] (3:12)	Yes	Yes (2.5:12)	Yes	Yes	NP
Asphalt over Wood	NP	NP	NP	Yes	Yes[2]	Yes[2]	Yes	NP
Asphalt over Asphalt	NP	NP	NP	Yes	Yes	Yes	Yes	NP
Tile Roof	NP	NP	NP	NP	NP	NP	NP	NP
Metal Roof	NP	NP	NP	NP	NP	Yes	NP	NP
Modified Bitumen	Yes	NP	Yes (3:12)	Yes	Yes (2.5:12)	Yes	Yes	NP

NP = Not Permitted.

Note: (Minimum Roof Slope)

[1]See Section 1515.2 for specific requirements.

[2]Board and batten leveling system must be firestopped in accordance with Section 1516.3.

[3]One layer 18-inch (457 mm) Type 30 nonperforated felt interlaced between shake courses required.

[4]Type 30 nonperforated felt underlayment required for reroofing.

**Appendix Chapter 16 is printed in its entirety in Volume 2 of the *Uniform Building Code*.
Excerpts from Appendix Chapter 16 are reprinted herein.**

Excerpts from Appendix Chapter 16
STRUCTURAL FORCES
Division II—EARTHQUAKE RECORDING INSTRUMENTATION

SECTION 1649 — GENERAL

In Seismic Zones 3 and 4 every building over six stories in height with an aggregate floor area of 60,000 square feet (5574 m^2) or more, and every building over 10 stories in height regardless of floor area, shall be provided with not less than three approved recording accelerographs.

The accelerographs shall be interconnected for common start and common timing.

SECTION 1650 — LOCATION

The instruments shall be located in the basement, midportion, and near the top of the building. Each instrument shall be located so that access is maintained at all times and is unobstructed by room contents. A sign stating MAINTAIN CLEAR ACCESS TO THIS INSTRUMENT shall be posted in a conspicuous location.

SECTION 1651 — MAINTENANCE

Maintenance and service of the instruments shall be provided by the owner of the building, subject to the approval of the building official. Data produced by the instruments shall be made available to the building official on request.

SECTION 1652 — INSTRUMENTATION OF EXISTING BUILDINGS

All owners of existing structures selected by the jurisdiction authorities shall provide accessible space for the installation of appropriate earthquake-recording instruments. Location of said instruments shall be determined by the jurisdiction authorities. The jurisdiction authorities shall make arrangements to provide, maintain and service the instruments. Data shall be the property of the jurisdiction, but copies of individual records shall be made available to the public on request and the payment of an appropriate fee.

Division III—SEISMIC ZONE TABULATION

NOTE: This division has been revised in its entirety.

SECTION 1653 — FOR AREAS OUTSIDE THE UNITED STATES

Location	Seismic Zone	Location	Seismic Zone
AFRICA		Mali	
Algeria		Bamako	0
Alger	3	Mauritania	
Oran	3	Nouakchott	0
Angola		Mauritius	
Luanda	0	Port Louis	0
Benin		Morocco	
Cotonou	0	Casablanca	2A
Botswana		Port Lyautcy	1
Gaborone	0	Rabat	2A
Burundi		Tangier	3
Bujumbura	3	Mozambique	
Cameroon		Maputo	2A
Douala	0	Niger	
Yaounde	0	Niamey	0
Cape Verde		Nigeria	
Praia	0	Ibadan	0
Central African Republic		Kaduna	0
Bangui	0	Lagos	0
Chad		Republic of Rwanda	
Ndjamena	0	Kigali	3
Congo		Senegal	
Brazzaville	0	Dakar	0
Djibouti	3	Seychelles	
Egypt		Victoria	0
Alexandria	2A	Sierra Leone	
Cairo	2A	Freetown	0
Port Said	2A	Somalia	
Equatorial Guinea		Mogadishu	0
Malabo	0	South Africa	
Ethiopia		Cape Town	3
Addis Ababa	3	Durban	2A
Asmara	3	Johannesburg	2A
Gabon		Natal	1
Libreville	0	Pretoria	2A
Gambia		Swaziland	
Banjul	0	Mbabane	2A
Ghana		Tanzania	
Accra	3	Dar es Salaam	2A
Guinea		Zanzibar	2A
Bissau	1	Togo	
Conakry	0	Lome	1
Ivory Coast		Tunisia	
Abidjan	0	Tunis	3
Kenya		Uganda	
Nairobi	2A	Kampala	2A
Lesotho		Upper Volta	
Maseru	2A	Ougadougou	0
Liberia		Zaire	
Monrovia	1	Bukavu	3
Libya		Kinshasa	0
Tripoli	2A	Lubumbashi	2A
Wheelus AFB	2A	Zambia	
Malagasy Republic		Lukasa	2A
Tananarive	0	Zimbabwe	
Malawi		Harare (Salisbury)	3
Blantyre	3	ASIA	
Lilongwe	3	Afghanistan	
Zomba	3	Kabul	4

Location	Seismic Zone	Location	Seismic Zone
Bahrain		**Laos**	
Manama	0	Vientiane	1
Bangladesh		**Lebanon**	
Dacca	3	Beirut	3
Brunei		**Malaysia**	
Bandar Seri Begawan	1	Kuala Lumpur	1
Burma		**Nepal**	
Mandalay	3	Kathmandu	4
Rangoon	3	**Oman**	
China		Muscat	2A
Beijing	4	**Pakistan**	
Chengdu	3	Islamabad	4
Guangzhou	2A	Karachi	4
Nanjing	2A	Lahore	2A
Qingdao	3	Peshawar	4
Shanghai	2A	**Qatar**	
Shengyang	4	Doha	0
Taiwan		**Saudi Arabia**	
All	4	Al Batin	1
Tihwa	4	Dharan	1
Wuhan	2A	Jiddah	2A
Xianggang	2A	Khamis Mushayf	1
Cyprus		Riyadh	0
Nicosia	3	**Singapore**	
India		All	1
Bombay	3	**South Yemen**	
Calcutta	2A	Aden City	3
Madras	1	**Sri Lanka**	
New Delhi	3	Colombo	0
Indonesia		**Syria**	
Bandung	4	Aleppo	3
Jakarta	4	Damascus	3
Medan	3	**Thailand**	
Surabaya	4	Bangkok	1
Iran		Chiang Mai	2A
Isfahan	3	Songkhla	0
Shiraz	3	Udorn	1
Tabriz	4	**Turkey**	
Tehran	4	Adana	2A
Iraq		Ankara	2A
Baghdad	3	Ismir	4
Basra	1	Istanbul	4
Israel		Karamursel	3
Haifa	3	**United Arab Emirates**	
Jerusalem	3	Abu Dhabi	0
Tel Aviv	3	Dubai	0
Japan		**Viet Nam**	
Fukuoka	3	Ho Chi Minh (Saigon)	0
Itazuke AFB	3	**Yemen Aran Republic**	
Misawa AFB	3	Sanaa	3
Naha, Okinawa	4	**ATLANTIC OCEAN AREA**	
Osaka/Kobe	4	**Azores**	
Sapporo	3	All	2A
Tokyo	4	**Bermuda**	
Wakkami	3	All	1
Yokohama	4	**CARIBBEAN SEA**	
Yokota	4	**Bahama Islands**	
Jordan		All	1
Amman	3	**Cuba**	
Korea		All	2A
Kimhae	1	**Dominican Republic**	
Kwangju	1	Santo Domingo	3
Pusan	1	**French West Indies**	
Seoul	0	Martinique	3
Kuwait		**Grenada**	
Kuwait	1	Saint Georges	3

Location	Seismic Zone	Location	Seismic Zone
Haiti		Paris	0
Port au Prince	3	Strasbourg	2A
Jamaica		Germany, Federal Republic	
Kingston	3	Berlin	0
Leeward Islands		Bonn	2A
All	3	Bremen	0
Trinidad & Tobago		Dusseldorf	1
All	3	Frankfurt	2A
CENTRAL AMERICA		Hamburg	0
Belize		Munich	1
Belmopan	2A	Stuttgart	2A
Canal Zone		Vaihigen	2A
All	2A	Greece	
Costa Rica		Athens	3
San Jose	3	Kavalla	4
El Salvador		Makri	4
San Salvador	4	Rhodes	3
Guatemala		Sauda Bay	4
Guatemala	4	Thessaloniki	4
Honduras		Hungary	
Tegucigalpa	3	Budapest	2A
Mexico		Iceland	
Ciudad Juarez	2A	Keflavick	3
Guadalajara	3	Reykjavik	4
Hermosillo	3	Ireland	
Matamoros	0	Dublin	0
Mazatlan	2A	Italy	
Merida	0	Aviano AFB	3
Mexico City	3	Brindisi	0
Monterrey	0	Florence	3
Nuevo Laredo	0	Genoa	3
Tijuana	3	Milan	2A
Nicaragua		Naples	3
Managua	4	Palermo	3
Panama		Rome	2A
Colon	3	Sicily	3
Galeta	2B	Trieste	3
Panama	3	Turin	2A
EUROPE		Luxembourg	
Albania		Luxembourg	1
Tirana	3	Malta	
Austria		Valleta	2A
Salzburg	2A	Netherlands	
Vienna	2A	All	0
Belgium		Norway	
Antwerp	1	Oslo	2A
Brussels	2A	Poland	
Bosnia-Herzegovina		Krakow	2A
Belgrade	2A	Poznan	1
Bulgaria		Warszawa	1
Sofia	3	Portugal	
Croatia		Lisbon	4
Zagreb	3	Opporto	3
Czechoslovakia		Romania	
Bratislava	2A	Bucharest	3
Prague	1	Russia	
Denmark		Moscow	0
Copenhagen	1	St. Petersburg	0
Finland		Spain	
Helsinki	1	Barcelona	2A
France		Bilbao	2A
Bordeaux	2A	Madrid	0
Lyon	1	Rota	2A
Marseille	3	Seville	2A
Nice	3	Sweden	
		Goteborg	2A
		Stockholm	1

Location	Seismic Zone	Location	Seismic Zone
Switzerland		Peru	
Bern	2A	Lima	4
Geneva	1	Piura	4
Zurich	2A	Uruguay	
Ukraine		Montevideo	0
Kiev	0	Venezuela	
United Kingdom		Caracas	4
Belfast	0	Maracaibo	2A
Edinburgh	1	PACIFIC OCEAN AREA	
Edzell	1	Australia	
Glasgow/Renfrew	1	Brisbane	1
Hamilton	1	Canberra	1
Liverpool	1	Melbourne	1
London	2A	Perth	1
Londonderry	1	Sydney	1
Thurso	1	Caroline Islands	
NORTH AMERICA		Koror, Palau Is.	2A
Greenland		Ponape	0
All	1	Fiji	
Canada		Suva	3
Argentia NAS	2A	Johnson Island	
Calgary, Alb	1	All	1
Churchill, Man	0	Mariana Islands	
Cold Lake, Alb	1	Guam	3
Edmonton, Alb	1	Saipan	3
E. Harmon AFB	2A	Tinian	3
Fort Williams, Ont	0	Marshall Islands	
Frobisher N.W. Ter.	0	All	1
Goose Airport	1	New Zealand	
Halifax	1	Auckland	3
Montreal, Quebec	3	Wellington	4
Ottawa, Ont	2A	Papau New Guinea	
St. John's Nfd	3	Port Moresby	3
Toronto, Ont	1	Phillipine Islands	
Vancouver	3	Baguio	3
Winnepeg, Man	1	Cebu	4
SOUTH AMERICA		Manila	4
Argentina		Samoa	
Buenos Aires	0	All	3
Bolivia		Wake Island	
La Paz	3	All	0
Santa Cruz	1		
Brazil			
Belem	0		
Belo Horizonte	0		
Brasilia	0		
Manaus	0		
Porto Allegre	0		
Recife	0		
Rio de Janeiro	0		
Salvador	0		
Sao Paulo	1		
Chile			
Santiago	4		
Valparaiso	4		
Colombia			
Bogota	3		
Ecuador			
Guayaquil	3		
Quito	4		
Paraguay			
Asuncion	0		

The above compilation is a partial listing of seismic zones for cities and countries outside of the United States. It has been provided in this code primarily as a source of information, and may not, in all cases, reflect local ordinances or current scientific information.

When an authority having jurisdiction requires seismic design forces that are higher than would be indicated by the above zones, the local requirements shall govern. When an authority having jurisdiction requires seismic design forces that are lower than would be indicated by the above zones, and these forces have been developed with consideration of regional tectonics and up-to-date geologic and seismologic information, the local requirements may be used.

When no local seismic design requirements exist, properly determined information on site-specific ground motions may be used to justify a lower seismic zone. Such site-specific ground motions shall have been developed with proper consideration of regional tectonics and local geologic and seismologic information, and shall have no more than a 10 percent chance of being exceeded in a 50-year period.

**Appendix Chapter 18 is printed in its entirety in Volume 2 of the *Uniform Building Code*.
Excerpts from Appendix Chapter 18 are reprinted herein.**

Excerpts from Appendix Chapter 18
WATERPROOFING AND DAMPPROOFING FOUNDATIONS

SECTION 1820 — SCOPE

Walls, or portions thereof, retaining earth and enclosing interior spaces and floors below grade shall be waterproofed or dampproofed according to this appendix chapter.

> **EXCEPTION:** Walls enclosing crawl spaces.

SECTION 1821 — GROUNDWATER TABLE INVESTIGATION

A subsurface soils investigation shall be made in accordance with Section 1804.3, Item 3, to determine the possibility of the groundwater table rising above the proposed elevation of the floor or floors below grade. The building official may require that this determination be made by an engineer or architect licensed by the state to practice as such.

> **EXCEPTIONS:** 1. When foundation waterproofing is provided.
>
> 2. When dampproofing is provided and the building official finds that there is satisfactory data from adjacent areas to demonstrate that groundwater has not been a problem.

SECTION 1822 — DAMPPROOFING REQUIRED

Where the groundwater investigation required by Section 1821 indicates that a hydrostatic pressure caused by the water table will not occur, floors and walls shall be dampproofed and a subsoil drainage system shall be installed in accordance with this appendix chapter.

> **EXCEPTION:** Wood foundation systems shall be constructed in accordance with Chapter 18, Division II.

SECTION 1823 — FLOOR DAMPPROOFING

1823.1 General. Dampproofing materials shall be installed between the floor and base materials required by Section 1825.2.

> **EXCEPTION:** Where a separate floor is provided above a concrete slab, the dampproofing may be installed on top of the slab.

1823.2 Dampproofing Materials. Dampproofing installed beneath the slab shall consist of not less than 6-mil (0.152 mm) polyethylene, or other approved methods or materials. When permitted to be installed on top of the slab, dampproofing shall consist of not less than 4-mil (0.1 mm) polyethylene, mopped-on bitumen or other approved methods or materials. Joints in membranes shall be lapped and sealed in an approved manner.

SECTION 1824 — WALL DAMPPROOFING

1824.1 General. Dampproofing materials shall be installed on the exterior surface of walls, and shall extend from a point 6 inches (152 mm) above grade, down to the top of the spread portion of the footing.

1824.2 Surface Preparation. Prior to application of dampproofing materials on concrete walls, fins or sharp projections that may pierce the membrane shall be removed and all holes and recesses resulting from the removal of form ties shall be sealed with a dry-pack mortar, bituminous material, or other approved methods or materials.

1824.3 Dampproofing Materials. Wall dampproofing shall consist of a bituminous material, acrylic modified cement base coating, any of the materials permitted for waterproofing in Section 1828.4, or other approved methods or materials. When such materials are not approved for direct application to unit masonry, the wall shall be parged on the exterior surface below grade with not less than $^3/_8$ inch (9.5 mm) of portland cement mortar.

SECTION 1825 — OTHER DAMPPROOFING REQUIREMENTS

1825.1 Subsoil Drainage System. When dampproofing is required, a base material shall be installed under the floor and a drain shall be installed around the foundation perimeter in accordance with this subsection.

> **EXCEPTION:** When the finished ground level is below the floor level for more than 25 percent of the perimeter of the building, the base material required by Section 1825.2 need not be provided and the foundation drain required by Section 1825.3 need be provided only around that portion of the building where the ground level is above the floor level.

1825.2 Base Material. Floors shall be placed over base material not less than 4 inches (102 mm) in thickness consisting of gravel or crushed stone containing not more than 10 percent material that passes a No. 4 sieve (4.75 mm).

1825.3 Foundation Drain. The drain shall consist of gravel, crushed stone or drain tile.

Gravel or crushed stone drains shall contain not more than 10 percent material that passes a No. 4 sieve (4.75 mm). The drain shall extend a minimum of 12 inches (305 mm) beyond the outside edge of the footing. The depth shall be such that the bottom of the drain is not higher than the bottom of the base material under the floor, and the top of the drain is not less than 6 inches (152 mm) above the spread portion of the footing. The top of the drain shall be covered with an approved filter membrane material.

When drain tile or perforated pipe is used, the invert of the pipe or tile shall be not higher than the floor elevation. The top of joints or the top of perforations shall be protected with an approved filter membrane material. The pipe or tile shall be placed on not less than 2 inches (51 mm) of gravel or crushed stone complying with this section and covered with not less than 6 inches (152 mm) of the same material.

1825.4 Drainage Disposal. The floor base and foundation perimeter drain shall discharge by gravity or mechanical means into an approved drainage system.

> **EXCEPTION:** Where a site is located in well-drained gravel or sand-gravel mixture soils, a dedicated drainage system need not be provided.

SECTION 1826 — WATERPROOFING REQUIRED

Where the groundwater investigation required by Section 1821 indicates that a hydrostatic pressure caused by the water table does exist, walls and floors shall be waterproofed in accordance with this appendix chapter.

> **EXCEPTIONS:** 1. When the groundwater table can be lowered and maintained at an elevation not less than 6 inches (152 mm) below

the bottom of the lowest floor, dampproofing provisions in accordance with Section 1822 may be used in lieu of waterproofing.

The design of the system to lower the groundwater table shall be based on accepted principles of engineering which shall consider, but not necessarily be limited to, the permeability of the soil, the rate at which water enters the drainage system, the rated capacity of pumps, the head against which pumps are to pump, and the rated capacity of the disposal area of the system.

2. Wood foundation systems constructed in accordance with Chapter 18, Division II, are to be provided with additional moisture-control measures as specified in Section 1812.

SECTION 1827 — FLOOR WATERPROOFING

1827.1 General. Floors required to be waterproofed shall be of concrete designed to withstand anticipated hydrostatic pressure.

1827.2 Waterproofing Materials. Waterproofing of floors shall be accomplished by placing under the slab a membrane of rubberized asphalt, polymer-modified asphalt, butyl rubber, neoprene, or not less than 6-mil (0.15 mm) polyvinyl chloride or polyethylene, or other approved materials capable of bridging nonstructural cracks. Joints in the membrane shall be lapped not less than 6 inches (152 mm) and sealed in an approved manner.

SECTION 1828 — WALL WATERPROOFING

1828.1 General. Walls required to be waterproofed shall be of concrete or masonry designed to withstand the anticipated hydrostatic pressure and other lateral loads.

1828.2 Wall Preparation. Prior to the application of waterproofing materials on concrete or masonry walls, the wall surfaces shall be prepared in accordance with Section 1824.2.

1828.3 Where Required. Waterproofing shall be applied from a point 12 inches (305 mm) above the maximum elevation of the groundwater table down to the top of the spread portion of the

footing. The remainder of the wall located below grade shall be dampproofed with materials in accordance with Section 1824.3.

1828.4 Waterproofing Materials. Waterproofing shall consist of rubberized asphalt, polymer-modified asphalt, butyl rubber, or other approved materials capable of bridging nonstructural cracks. Joints in the membrane shall be lapped and sealed in an approved manner.

1828.5 Joints. Joints in walls and floors, and between the wall and floor, and penetrations of the wall and floor shall be made watertight using approved methods and materials.

SECTION 1829 — OTHER DAMPPROOFING AND WATERPROOFING REQUIREMENTS

1829.1 Placement of Backfill. The excavation outside the foundation shall be backfilled with soil which is free of organic material, construction debris and large rocks. The backfill shall be placed in lifts and compacted in a manner which does not damage the waterproofing or dampproofing material or structurally damage the wall.

1829.2 Site Grading. The ground immediately adjacent to the foundation shall be sloped away from the building at not less than 1 unit vertical in 12 units horizontal (8.3% slope) for a minimum distance of 6 feet (1829 mm) measured perpendicular to the face of the wall or an approved alternate method of diverting water away from the foundation shall be used. Consideration shall be given to possible additional settlement of the backfill when establishing final ground level adjacent to the foundation.

1829.3 Erosion Protection. Where water impacts the ground from the edge of the roof, downspout, scupper, valley, or other rainwater collection or diversion device, provisions shall be used to prevent soil erosion and direct the water away from the foundation.

**Appendix Chapter 19 is printed in its entirety in Volume 2 of the *Uniform Building Code.*
Excerpts from Appendix Chapter 19 are reprinted herein.**

Excerpts from Appendix Chapter 19
PROTECTION OF RESIDENTIAL CONCRETE EXPOSED TO FREEZING AND THAWING

SECTION 1928 — GENERAL

1928.1 Purpose. The purpose of this appendix is to provide minimum standards for the protection of residential concrete exposed to freezing and thawing conditions.

1928.2 Scope. The provisions of this appendix apply to concrete used in buildings of Groups R and U Occupancies that are three stories or less in height.

1928.3 Special Provisions. Normal-weight aggregate concrete used in buildings of Groups R and U Occupancies three stories or less in height which are subject to de-icer chemicals or freezing and thawing conditions as determined from Figure A-19-1 shall comply with the requirements of Table A-19-A.

TABLE A-19-A—MINIMUM SPECIFIED COMPRESSIVE STRENGTH OF CONCRETE[1]

TYPE OR LOCATION OF CONCRETE CONSTRUCTION	MINIMUM SPECIFIED COMPRESSIVE STRENGTH[2] (f'_c)		
	× 6.89 for kPa		
	Weathering Potential[3]		
	Negligible	Moderate	Severe
Basement walls and foundations not exposed to the weather	2,500	2,500	2,500[4]
Basement slabs and interior slabs on grade, except garage floor slabs	2,500	2,500	2,500[4]
Basement walls, foundation walls, exterior walls and other vertical concrete work exposed to the weather	2,500	3,000[5]	3,000[5]
Porches, carport slabs and steps exposed to the weather, and garage floor slabs	2,500	3,000[5]	3,500[5]

[1]Increases in compressive strength above those used in the design shall not cause implementation of the special inspection provisions of Section 1701.5, Item 1.

[2]At 28 days, pounds per square inch (kPa).

[3]See Figure A-19-1 for weathering potential.

[4]Concrete in these locations which may be subject to freezing and thawing during construction shall be air-entrained concrete in accordance with Footnote 5.

[5]Concrete shall be air entrained. Total air content (percentage by volume of concrete) shall not be less than 5 percent or more than 7 percent.

SEVERE

MODERATE

MILD

<u>WEATHERING REGIONS (WEATHERING INDEX)</u>

FIGURE A-19-1—WEATHERING REGIONS FOR RESIDENTIAL CONCRETE

NOTES:

[1]The three exposures are:

A. Severe—Outdoor exposure in a cold climate where concrete may be exposed to the use of de-icing salts or where there may be a continuous presence of moisture during frequent cycles of freezing and thawing. Examples are pavements, driveways, walks, curbs, steps, porches and slabs in unheated garages. Destructive action from de-icing salts may occur either from direct application or from being carried onto an unsalted area from a salted area, such as on the undercarriage of a car traveling on a salted street but parked on an unsalted driveway or garage slab.

B. Moderate—Outdoor exposure in a climate where concrete will not be exposed to the application of de-icing salts but will occasionally be exposed to freezing and thawing.

C. Mild—Any exposure where freezing and thawing in the presence of moisture is rare or totally absent.

[2]Data needed to determine the weathering index for any locality may be found or estimated from the tables of Local Climatological Data, published by the Weather Bureau, U.S. Department of Commerce.

[3]The weathering regions map provides the location of severe, moderate and mild winter weathering areas as they occur in the United States (Alaska and Hawaii are classified as severe and mild, respectively). The map cannot be precise. This is especially true in mountainous areas where conditions change dramatically within very short distances. It is intended to classify as severe any area in which weathering conditions may cause de-icing salt to be used, either by individuals or for street or highway maintenance. These conditions are significant snowfall combined with extended periods during which there is little or no natural thawing. If there is any doubt about which of two regions is applicable, the more severe exposure should be selected.

[4]The Weathering Index:

    Severe—As a guideline, the number of days during which the temperature does not rise above 32°F (0°C) is multiplied by the inches of snowfall. An index of 150 or more is classified as severe. Cold, humid climates may be more severe than cold, dry climates for a given index.

    Moderate, Mild—Multiply the inches of precipitation times the number of days the temperature registers below 32°F (0°C) Use the occurrence between the first day in the fall and the last day in the spring that the temperature registers below 32°F (0°C) An index above 200 is moderate. An index below 200 is mild.

Appendix Chapter 21 is printed in its entirety in Volume 2 of the *Uniform Building Code.*
Excerpts from Appendix Chapter 21 are reprinted herein.

Excerpts from Appendix Chapter 21
PRESCRIPTIVE MASONRY CONSTRUCTION IN HIGH-WIND AREAS

SECTION 2112 — GENERAL

2112.1 Purpose. The provisions of this chapter are intended to promote public safety and welfare by reducing the risk of wind-induced damages to masonry construction.

2112.2 Scope. The requirements of this chapter shall apply to masonry construction in buildings when all of the following conditions are met:

1. The building is located in an area with a basic wind speed from 80 through 110 miles per hour (mph) (129 km/h through 177 km/h).

2. The building is located in Seismic Zone 0, 1 or 2.

3. The building does not exceed two stories.

4. Floor and roof joists shall be wood or steel or of precast hollowcore concrete planks with a maximum span of 32 feet (9754 mm) between bearing walls. Masonry walls shall be provided for the support of steel joists or concrete planks.

5. The building is of regular shape.

2112.3 General. The requirements of Chapter 21 are applicable except as specifically modified by this chapter. Other methods may be used provided a satisfactory design is submitted showing compliance with the provisions of Chapter 16, Part II, and other applicable provisions of this code.

Wood floor, roof and interior walls shall be constructed as specified in Appendix Chapter 23 and as further regulated in this section.

In areas where the wind speed exceeds 110 mph (177 km/h), masonry buildings shall be designed in accordance with Chapter 16, Part II, and other applicable provisions of this code.

Buildings of unusual shape or size, or split-level construction, shall be designed in accordance with Chapter 16, Part II, and other applicable provisions of this code.

In addition to the other provisions of this chapter, foundations for buildings in areas subject to wave action or tidal surge shall be designed in accordance with approved national standards.

All metal connectors and fasteners used in exposed locations or in areas otherwise subject to corrosion shall be of corrosion-resistant or noncorrosive material. When the terms "corrosion resistant" or "noncorrosive" are used in this chapter, they shall mean having a corrosion resistance equal to or greater than a hot-dipped galvanized coating of 1.5 ounces of zinc per square foot (3.95 g/m^2) of surface area. When an element is required to be corrosion resistant or noncorrosive, all of its parts, such as screws, nails, wire, dowels, bolts, nuts, washers, shims, anchors, ties and attachments, shall be corrosion resistant.

2112.4 Materials.

2112.4.1 General. All masonry materials shall comply with Section 2102.2 as applicable for standards of quality.

2112.4.2 Hollow-unit masonry.

1. Exterior concrete block shall be a minimum of Grade N-II with a compressive strength of not less than 1,900 pounds per square inch (psi) (13 091 kPa) on the net area.

2. Interior concrete block shall be a minimum of Grade S-II with a compressive strength of not less than 700 psi (4823 kPa) on the gross area.

3. Exterior clay or shale hollow brick shall have a compressive strength of not less than 2,500 psi (17 225 kPa) on the net area. Such hollow brick shall be at least Grade MW except that where subject to severe freezing it shall be Grade SW.

4. Interior clay or shale hollow brick shall be Grade MW with a compressive strength of 2,000 psi (13 780 kPa) on the net area.

2112.4.3 Solid masonry.

1. Exterior clay or shale bricks shall have a compressive strength of not less than 2,500 psi (17 225 kPa) on the net area.

2. Exterior clay or shale bricks shall be Grade MW, except that where subject to severe freezing they shall be Grade SW.

3. Interior clay or shale bricks shall have a compressive strength of not less than 2,000 psi (13 780 kPa).

2112.4.4 Grout. Grout shall achieve a compressive strength of not less than 2,000 psi (13 780 kPa).

2112.4.5 Mortar. Mortar for exterior walls and for interior shear walls shall be Type M or Type S.

2112.5 Construction Requirements. Grouted cavity wall and block wall construction shall comply with Section 2104.

Unburned clay masonry and stone masonry shall not be used.

2112.6 Foundations. Footings shall have a thickness of not less than 8 inches (203 mm) and shall comply with Tables A-21-A-1 and A-21-A-2 for width. See Figure A-21-1 for other applicable details.

Footings shall extend 18 inches (457 mm) below the undisturbed ground surface or the frost depth, whichever is deeper.

Foundation stem walls shall be as wide as the wall they support. They shall be reinforced with reinforcing bar sizes and spacing to match the reinforcement of the walls they support.

Basement and other below-grade walls shall comply with Table A-21-B.

2112.7 Drainage. Basement walls and other walls or portions thereof retaining more than 3 feet (914 mm) of earth and enclosing interior spaces or floors below grade shall have a minimum 4-inch-diameter (102 mm) footing drain as illustrated in Table A-21-B and Figure A-21-3.

The finish elevations around the building shall be graded to provide a slope away from the building of not less than $^1/_4$ unit vertical in 12 units horizontal (2% slope).

2112.8 Wall Construction.

2112.8.1 Minimum thickness. Reinforced exterior bearing walls shall have a minimum 8-inch (203 mm) nominal thickness.

Interior masonry nonbearing walls shall have a minimum 6-inch (152 mm) nominal thickness. Unreinforced grouted brick walls shall have a minimum 10-inch (254 mm) thickness. Unreinforced hollow-unit and solid masonry shall have a minimum 8-inch (203 mm) nominal thickness.

> **EXCEPTION:** In buildings not more than two stories or 26 feet (7924.8 mm) in height, masonry walls may be of 8-inch (203 mm) nominal thickness. Solid masonry walls in one-story buildings may be of 6-inch (152 mm) nominal thickness when not over 9 feet (2743 mm) in height, provided that when gable construction is used an additional 6 feet (1829 mm) are permitted to the peak of the gable.

2112.8.2 Lateral support and height. All walls shall be laterally supported at the top and bottom. The maximum unsupported height of bearing walls or other masonry walls shall be 12 feet (3658 mm). Gable-end walls may be 15 feet (4572 mm) at their peak.

Wood-framed gable-end walls on buildings shall comply with Table A-21-I and Figure A-21-17 or A-21-18.

2112.8.3 Walls in Seismic Zone 2 and use of stack bond. In Seismic Zone 2, walls shall comply with Figure A-21-2 as a minimum. Walls with stack bond shall be designed.

2112.8.4 Lintels. The span of lintels over openings shall not exceed 12 feet (3658 mm), and lintels shall be reinforced. The reinforcement bars shall extend not less than 2 feet (610 mm) beyond the edge of opening and into lintel supports.

Lintel reinforcement shall be within fully grouted cells in accordance with Table A-21-E.

2112.8.5 Reinforcement. Walls shall be reinforced as shown in Tables A-21-C-1 through A-21-C-5 and Figure A-21-2.

2112.8.6 Anchorage of walls to floors and roofs. Anchors between walls and floors or roofs shall be embedded in grouted cells or cavities and shall conform to Section 2112.9.

2112.9 Floor and Roof Systems. The anchorage of wood roof systems which are supported by masonry walls shall comply with Appendix Sections 2337.5.1 and 2337.5.8, Table A-21-D and Figure A-21-7.

Wood roof and floor systems which are supported by ledgers at the inside face of masonry walls shall comply with Table A-21-D, Part I.

The ends of joist girders shall extend a distance of not less than 6 inches (152 mm) over masonry or concrete supports and be attached to a steel bearing plate. This plate is to be located not more than $1/2$ inch (12.7 mm) from the face of the wall and is to be not less than 9 inches (229 mm) wide perpendicular to the length of the joist girder. Ends of joist girders resting on steel bearing plates on masonry or structural concrete shall be attached thereto with a minimum of two $1/4$-inch (6.4 mm) fillet welds 2 inches (51 mm) long, or with two $3/4$-inch (19 mm) bolts.

Ends of joist girders resting on steel supports shall be connected thereto with a minimum of two $1/4$-inch (6.4 mm) fillet welds 2 inches (51 mm) long, or with two $3/4$-inch (19 mm) bolts. In steel frames, joist girders at column lines shall be field bolted to the columns to provide lateral stability during construction.

Steel joist roof and floor systems shall be anchored in accordance with Table A-21-H.

Wall ties spaced as shown in Table A-21-D, Part II, shall connect to framing or blocking at roofs and walls. Wall ties shall enter grouted cells or cavities and shall be $1^1/8$-inch (29 mm) minimum

width by 0.036 inch (0.91 mm) (No. 20 galvanized sheet gage) sheet steel.

Roof and floor hollow-core precast plank systems shall be anchored in accordance with Table A-21-G.

Roof uplift anchorage shall enter a grouted bond beam reinforced with horizontal bars as shown in Tables A-21-C-1 through A-21-C-5 and Figure A-21-7.

2112.10 Lateral Force Resistance.

2112.10.1 Complete load path and uplift resistance. Strapping, approved framing anchors, and mechanical fasteners, bond beams, and vertical reinforcement shall be installed to provide a continuous tie from the roof to the foundation system. (See Figure A-21-8.) In addition, roof and floor systems, masonry shear walls, or masonry or wood cross walls shall provide lateral stability.

2112.10.2 Floor and roof diaphragms. Floor and roof diaphragms shall be connected to masonry walls as shown in Table A-21-F, Part II.

Gabled and sloped roof members not supported at the ridge shall be tied by ceiling joists or equivalent lateral ties located as close to where the roof member bears on the wall as is practically possible, at not more than 48 inches (1219 mm) on center. Collar ties shall not be used for these lateral ties. (See Figure A-21-17 and Table A-21-I.)

2112.10.3 Walls. Masonry walls shall be provided around all sides of floor and roof systems in accordance with Figure A-21-9 and Table A-21-F.

The cumulative length of exterior masonry walls along each side of the floor or roof systems shall be at least 20 percent of the parallel dimension. Required elements shall be without openings and shall not be less than 48 inches (1219 mm) in width.

Interior cross walls (nonbearing) at right angles to bearing walls shall be provided when the length of the building perpendicular to the span of the floor or roof framing exceeds twice the distance between shear walls or 32 feet (9754 mm), whichever is greater. Cross walls, when required, shall conform to Section 2112.10.4.

2112.10.4 Interior cross walls. When required by Table A-21-F, Part I, masonry walls shall be at least 6 feet (1829 mm) long and reinforced with 9 gage wire joint reinforcement spaced not more than 16 inches (406 mm) on center. Cross walls shall comply with Footnote 3 of Table A-21-F, Part I.

Interior wood stud walls may be used to resist the wind load from one-story masonry buildings in areas where the basic wind speed is 100 mph (161 km/h), Exposure C or less, and 110 mph (177 km/h), Exposure B. When wood stud walls are so used, they shall:

1. Be perpendicular to exterior masonry walls at 15 feet (4572 mm) or less on center.

2. Be at least 8 feet (2438 mm) long without openings and be sheathed on at least one side with $15/32$-inch (12 mm) wood structural panel nailed with 8d common or galvanized box nails at 6 inches (152 mm) on center edge and field nailing. All unsupported edges of wood structural panels shall be blocked.

3. Be connected to wood blocking or wood joists below with two 16d nails at 16 inches (406 mm) on center through their sill plates. They shall be connected to footings with $1/2$-inch-diameter (12.7 mm) bolts at 3 feet 6 inches (1067 mm) on center.

4. Connect to wood roof systems as outlined in Table A-21-F, Part II, as a cross wall. Wood structural panel roof sheathing shall have all unsupported edges blocked.

TABLE A-21-A-1—EXTERIOR FOUNDATION REQUIREMENTS FOR MASONRY BUILDINGS WITH 6- AND 8-INCH-THICK WALLS
(Wood or Steel Framing)
(Width of Footings in Inches)[1,2,3]
See Figure A-21-1 for typical details.

WALL HEIGHT (feet)	SPAN TO BEARING WALLS (feet)	ONE-STORY BUILDINGS Roof Live Load[4] × 0.0479 for kN/m² 20 psf (inches)	30 psf (inches)	40 psf (inches)	TWO-STORY BUILDINGS Roof Live Load[4] (psf) × 0.0479 for kN/m² — Plus Floor Live Load[5] (psf) × 0.0479 for kN/m² — Minimum Width of Footing (inches) — 20 / 50	20 / 100	30 / 50	30 / 100	40 / 50	40 / 100
× 304.8 for mm		× 25.4 for mm			× 25.4 for mm					
8	8	12			12	12	12	12	12	12
	16				12	14	12	14	12	14
	24				14	18	14	18	16	18
	32				16	20	18	20	18	20
10	8	12			12	12	12	12	12	12
	16				14	16	14	16	14	16
	24				16	20	16	18	16	20
	32				20	24	20	22	20	24
12	8	12	12	12	12	14	12	14	12	14
	16	12	12	12	16	18	16	16	14	16
	24	12	12	14	18	20	18	20	18	20
	32	12	14	16	20	22	22	22	22	24

[1]For buildings with under-floor space or basements, footing thickness is to be a minimum of 12 inches (305 mm). It shall be reinforced with No. 4 bars at 24 inches (610 mm) on center when its width is required to be 18 inches (457 mm) or larger and it supports more than the roof and one floor.

[2]Soil to be at least Class 4 as shown in Table 18-I-A.

[3]Footings are a minimum of 10 inches (254 mm) thick for a one-story building and 12 inches (305 mm) thick for a two-story building. Bottom of footing to be 18 inches (457 mm) below grade or the frost depth, whichever is deeper. Footing to be reinforced with No. 4 bars at 24 inches (610 mm) on center when supporting more than the roof and one floor.

[4]From Table 21-C or local snow load tables. For areas without snow loads use 20 pounds per square foot (0.96 kN/m²).

[5]From Table 21-A. For intermediate floor loads go to next higher value.

TABLE A-21-A-2—INTERIOR FOUNDATION REQUIREMENTS FOR MASONRY BUILDINGS WITH 6- AND 8-INCH-THICK WALLS
(Wood or Steel Framing)
(Width of Footings in Inches)[1,2,3,4]
See Figure A-21-1 for typical details.

WALL HEIGHT (feet)	SPAN TO BEARING WALLS (feet)	ONE-STORY BUILDINGS Roof Live Load[5] × 0.0479 for kN/m²			TWO-STORY BUILDINGS Roof Live Load[5] (psf) × 0.0479 for kN/m²					
					20		30		40	
					Plus Floor Live Load[6] (psf) × 0.0479 for kN/m²					
		20 psf (inches)	30 psf (inches)	40 psf (inches)	50	100	50	100	50	100
× 304.8 for mm		Minimum Width of Footing (inches) × 25.4 for mm								
8	8	12	12	12	12	14	12	14	12	14
	16	12	12	12	16	20	18	20	18	22
	24	12	12	14	20	26	22	28	22	28
	32	14	14	16	24	28	26	32	28	34
10	8	12	12	12	14	16	14	16	14	16
	16	12	12	12	20	24	20	22	20	22
	24	12	14	14	22	28	22	28	22	28
	32	14	14	16	26	34	26	32	28	34
12	8	12	12	12	14	16	16	18	16	18
	16	12	14	16	20	24	20	22	20	22
	24	14	14	16	24	28	22	28	24	28
	32	16	16	18	28	30	28	32	28	34

[1]For buildings with under-floor space or basements, footing thickness is to be a minimum of 12 inches (305 mm). It shall be reinforced with No. 4 bars at 24 inches (610 mm) on center when its width is required to be 18 inches (457 mm) or larger and it supports more than the roof and one floor.

[2]Soil to be at least Class 4 as shown in Table 18-I-A.

[3]Footings are 10 inches (254 mm) thick for up to 24 inches (610 mm) wide and 12 inches (305 mm) thick for up to 34 inches (864 mm) wide. Footings shall be reinforced with No. 4 bars at 24 inches (610 mm) on center when supporting more than the roof and one floor.

[4]These interior footings support roof-ceiling or floors or both for a distance on each side equal to the span length shown. A tributary width equal to the span length may be used.

[5]From Table 16-C or local snow load tables. For areas without snow loads use 20 pounds per square foot (0.96 kN/m²).

[6]From Table 16-A. For intermediate floor loads go to next higher value.

TABLE A-21-B—VERTICAL REINFORCEMENT AND TOP RESTRAINT FOR VARIOUS HEIGHTS OF BASEMENT AND OTHER BELOW-GRADE WALLS

DESIGN ASSUMPTIONS
A. Materials:
  1. **Concrete Masonry Units**—Grade hollow load-bearing units conforming to Section 2112.4.2 for strength of units should not be less than that required for applicable f'_m.
  2. **Mortar**—Type M, 2,500 psi (17 240 kPa) strength.
  3. **Corefill**—Fine or coarse grout (UBC Standard 21-19) with an ultimate strength (28 days) of at least 2,500 psi. (17 240 kPa)
  4. **Reinforcement**—Deformed billet-steel bars.
  5. 1,500 psf (71.8 kPa) soil bearing required.[1]
B. Allowable stresses in accordance with Section 2106 and Table 21-M.

Soil Equiv.-fluid wt. = 30 pcf[1] (4.71 kN/m³)			Vertical Reinforcement with Axial Compressive Load (P) Equal to or Less than 5,000 lb./lin. ft. (72.92 kN/m)		
	Floor Connection[2,3]		f'_m = 1,500 psi (10 335 kPa)		
Wall Depth below Grade h (feet)	Wood Floor		Spacing of Reinforcement (inches)[4]		
× 304.8 for mm	Bolt and Spacing	Angle Clip Spacing	× 25.4 for mm		
8-Inch Walls			No. 3	No. 4	No. 5
× 25.4 for mm					
4	1/2″ at 60″	48″ o.c.	24	40	56
5	1/2″ at 40″	32″ o.c.	16	24	40
6	5/8″ at 32″	20″ o.c.	—	16	24
			Spacing of Reinforcement (inches)		
10-Inch Walls			× 25.4 for mm		
× 25.4 for mm			No. 4	No. 5	No. 6	No. 7
6	5/8″ at 32″	20″ o.c.	40	56	64	72
7	5/8″ at 24″	16″ o.c.	24	40	48	56
9	3/4″ at 20″	2 at 24″ o.c.	16	24	32	40
			Spacing of Reinforcement (inches)		
12-Inch Walls			× 25.4 for mm		
× 25.4 for mm			No. 4	No. 5	No. 6	No. 7
7	5/8″ at 24″	16″ o.c.	40	56	80	80
8	3/4″ at 20″	2 at 24″ o.c.	32	48	56	64
9	7/8″ at 18″	2 at 18″ o.c.	24	40	48	48
10	1″ at 16″	2 at 16″ o.c.	16	32	40	40

[1]Soil type is at least Class 4 as shown in Table 18-I-A.
[2]There shall be no backfill placed until after the wall is anchored to the floor and seven days have passed after grouting.
[3]For Figure A-21-4 only.
[4]See Figure A-21-5 for placement of reinforcement.

TABLE A-21-C-1—VERTICAL REINFORCING STEEL REQUIREMENTS FOR 6-INCH-THICK (153 mm) MASONRY WALLS[1] IN AREAS WHERE BASIC WIND SPEEDS ARE 80 MILES PER HOUR (129 km/h) OR GREATER[2,3,4,5]
(Wood or Steel Roof and Floor Framing)

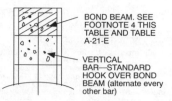

BOND BEAM. SEE FOOTNOTE 4 THIS TABLE AND TABLE A-21-E

VERTICAL BAR—STANDARD HOOK OVER BOND BEAM (alternate every other bar)

Criteria: Roof Live Load = 20 psf to 40 psf (0.96 kN/m^2 to 1.9 kN/m^2); Floor Live Load = 50 psf (2.4 kN/m^2); enclosed building[6]

EXPO-SURE	STORIES	UNSUP-PORTED HEIGHT (feet) × 304.8 for mm	80 MPH 8	16	24	32	90 MPH 8	16	24	32	100 MPH 8	16	24	32	110 MPH 8	16	24	32
B	One-story building	8	NR*								No. 4 80	No. 4 80	No. 4 80	No. 4 80	No. 4 64	No. 4 64	No. 4 72	No. 4 88
		10	No. 4 80	No. 4 88	No. 4 96	No. 4 96	No. 4 64	No. 4 64	No. 4 72	No. 4 80	No. 4 48	No. 4 48	No. 4 48	No. 4 56	No. 4 40	No. 4 40	No. 4 40	No. 4 48
		12	No. 4 48	No. 4 48	No. 4 56	No. 4 64	No. 4 40	No. 4 40	No. 4 48	No. 4 48	No. 5 48	No. 5 48	No. 5 56	No. 5 56	No. 5 40	No. 5 40	No. 5 40	No. 5 40
	Two-story building		Design required or use 8-inch or larger units for two-story condition.															
C	One-story building	8	No. 4 72	No. 4 72	No. 4 72	No. 4 96	No. 4 56	No. 4 56	No. 4 56	No. 4 56	No. 4 40	No. 4 40	No. 4 48	No. 4 48	No. 4 32	No. 4 32	No. 4 32	No. 4 40
		10	No. 4 40	No. 4 40	No. 4 40	No. 4 48	No. 4 32	No. 4 32	No. 4 32	No. 4 32	No. 5 40	No. 5 40	No. 5 40	No. 5 48	No. 5 32	No. 5 32	No. 5 32	No. 5 40
		12	No. 5 40	No. 5 48	No. 5 48	No. 5 48	No. 5 32	No. 5 32	No. 5 32	No. 5 40	Use 8-inch or larger units							
	Two-story building		Design required or use 8-inch or larger units for two-story condition.															
D	One-story building	8	No. 4 56	No. 4 56	No. 4 64	No. 4 80	No. 4 48	No. 4 48	No. 4 48	No. 4 48	No. 4 32	No. 4 40	No. 4 40	No. 4 40	No. 4 32	No. 4 32	No. 4 32	No. 4 32
		10	No. 4 32	No. 4 32	No. 4 32	No. 4 40	No. 5 40	No. 5 40	No. 5 48	No. 5 48	No. 5 48	No. 5 32	No. 5 32	No. 5 40				
		12	No. 5 32	No. 5 40	No. 5 40	No. 5 40	Use 8-inch or larger units											
	Two-story building		Design required or use 8-inch or larger units for two-story condition.															

Column headers note: × 1.61 for km/h; Span between Bearing Walls (feet) × 304.8 for mm; Size of Rebar and Spacing (inches) × 25.4 for mm

*NR — No vertical reinforcement required. However, see Table A-21-F for shear wall reinforcement.

[1] These values are for walls with running bond. For stack bond see Section 2112.8.3.

[2] The figure on top of the listed data is the bar size; the figure below it is the maximum spacing in inches (mm). Reinforcing bar strength shall be A 615 Grade 60. The vertical bars are centered in the middle of the wall.

[3] Roof load is assumed to be concentrically loaded on the wall. For roofs which hang on ledgers, a design is required.

[4] Minimum horizontal reinforcement shall be one No. 4 at the ledger and foundation. Also, see Table A-21-E for lintels and Table A-21-F for shear wall reinforcing where applicable.

[5] Hook vertical bars over bond beam bars as shown. Extend bars into footing using lap splices where necessary.

[6] Design required for open buildings of 6-inch-thick (153 mm) masonry.

To use this table, check criteria by the following method:

[6.1] Choose proper roof live load from Table 16-C or snow load criteria for the locality in which the building is located.

[6.2] Check if building is enclosed or partially enclosed by the procedure in Chapter 16, Part III.

[6.3] Choose proper floor load from Table 16-A. [For loads less than 50 pounds per square foot (psf) (2.4 kN/m^2), use 50 psf (2.4 kN/m^2), and for loads between 50 psf (2.4 kN/m^2) and 100 psf (4.8 kN/m^2), use 100 psf (4.8 kN/m^2).]

[6.4] Find proper wind speed and exposure for the site—see Figure 16-1, Chapter 16, Sections 1619 and 1620.

[6.5] Within the proper vertical column, choose appropriate span-to-bearing wall and appropriate height and story.

[6.6] Read proper size and spacing of reinforcement for the thickness of the wall mentioned in the title of the table. (Equivalent area of steel, taking spacing into account, may be substituted.)

[6.7] For buildings in Seismic Zone 2 (see Figure 16-2 in Chapter 16), use minimum reinforcement in Figure A-21-2 if it is more restrictive than the table values.

TABLE A-21-C-2—VERTICAL REINFORCING STEEL REQUIREMENTS FOR 8-INCH-THICK (203 mm) MASONRY WALLS[1] IN AREAS WHERE BASIC WIND SPEEDS ARE 80 MILES PER HOUR (129 km/h) OR GREATER[2,3,4,5] (Wood or Steel Roof and Floor Framing)

BOND BEAM. SEE FOOTNOTE 4 THIS TABLE AND TABLE A-21-E

VERTICAL BAR—STANDARD HOOK OVER BOND BEAM (alternate every other bar)

Criteria: Roof Live Load = 20 psf to 40 psf (0.96 kN/m^2 to 1.9 kN/m^2); Floor Live Load = 50 psf (2.4 kN/m^2); enclosed building

			80 MPH				90 MPH				100 MPH				110 MPH			
			\(\times\) 1.61 for km/h															
			Span between Bearing Walls (feet)															
			\(\times\) 304.8 for mm															
		UNSUPPORTED HEIGHT (feet)	8	16	24	32	8	16	24	32	8	16	24	32	8	16	24	32
			Size of Rebar and Spacing (inches)															
EXPOSURE	STORIES	\(\times\) 304.8 for mm	\(\times\) 25.4 for mm															
B	One-story building or top story of two-story building	8													No. 3 / 56	No. 3 / 56	No. 3 / 64	No. 3 / 64
		10	NR*				No. 4 / 80	No. 4 / 80	No. 4 / 88	No. 4 / 88	No. 4 / 64	No. 4 / 64	No. 4 / 64	No. 4 / 72	No. 4 / 48	No. 4 / 48	No. 4 / 56	No. 4 / 56
		12	No. 4 / 64	No. 4 / 72	No. 4 / 72	No. 4 / 72	No. 4 / 56	No. 4 / 56	No. 4 / 56	No. 4 / 56	No. 4 / 40	No. 4 / 40	No. 5 / 64	No. 5 / 64	No. 5 / 56	No. 5 / 56	No. 5 / 56	No. 5 / 56
	First story of a two-story building	8	No. 3 / 96	No. 3 / 96	No. 3 / 96	No. 3 / 96	No. 3 / 96	No. 3 / 96	No. 3 / 96	No. 3 / 96	No. 3 / 96	No. 3 / 88	No. 3 / 80	No. 3 / 72	No. 3 / 72	No. 3 / 72	No. 3 / 64	No. 3 / 64
		10	No. 3 / 88	No. 3 / 80	No. 3 / 72	No. 3 / 64	No. 3 / 64	No. 3 / 64	No. 3 / 56	No. 3 / 56	No. 4 / 72	No. 4 / 72	No. 4 / 64	No. 4 / 64	No. 4 / 64	No. 4 / 56	No. 4 / 56	No. 4 / 56
		12	No. 4 / 80	No. 4 / 72	No. 4 / 64	No. 4 / 64	No. 4 / 64	No. 4 / 56	No. 4 / 56	No. 4 / 48	No. 4 / 48	No. 4 / 48	No. 5 / 64	No. 5 / 56	No. 5 / 56	No. 5 / 56	No. 5 / 48	No. 5 / 48
C	One-story building or top story of two-story building	8	NR*				No. 3 / 48	No. 3 / 48	No. 3 / 48	No. 3 / 56	No. 4 / 64	No. 4 / 64	No. 4 / 72	No. 4 / 72	No. 4 / 48	No. 4 / 56	No. 4 / 56	No. 4 / 56
		10	No. 4 / 56	No. 4 / 56	No. 4 / 64	No. 4 / 64	No. 4 / 48	No. 4 / 48	No. 4 / 48	No. 4 / 48	No. 5 / 56	No. 5 / 56	No. 5 / 56	No. 5 / 56	No. 5 / 48	No. 5 / 48	No. 5 / 48	No. 5 / 48
		12	No. 5 / 56	No. 5 / 64	No. 5 / 64	No. 5 / 64	No. 5 / 48	No. 5 / 48	No. 5 / 48	No. 5 / 48	No. 6 / 56	No. 6 / 56	No. 6 / 56	No. 6 / 56	No. 6 / 40	No. 6 / 40	No. 6 / 40	No. 6 / 48
	First story of a two-story building	8	No. 3 / 80	No. 3 / 80	No. 3 / 56	No. 3 / 72	No. 3 / 56	No. 3 / 56	No. 3 / 56	No. 3 / 56	No. 4 / 72	No. 4 / 72	No. 4 / 72	No. 4 / 64	No. 4 / 56	No. 4 / 56	No. 4 / 56	No. 4 / 56
		10	No. 4 / 72	No. 4 / 64	No. 4 / 64	No. 4 / 56	No. 4 / 56	No. 4 / 48	No. 4 / 48	No. 4 / 48	No. 5 / 64	No. 5 / 56	No. 5 / 56	No. 5 / 56	No. 5 / 48	No. 5 / 48	No. 5 / 48	No. 5 / 48
		12	No. 5 / 64	No. 5 / 64	No. 5 / 56	No. 5 / 56	No. 5 / 48	No. 5 / 48	No. 5 / 48	No. 5 / 48	No. 6 / 56	No. 6 / 56	No. 6 / 48	No. 6 / 48	No. 6 / 48	No. 6 / 40	No. 6 / 40	No. 6 / 40
D	One-story building or top story of two-story building	8	No. 3 / 48	No. 3 / 48	No. 3 / 56	No. 3 / 56	No. 4 / 64	No. 4 / 72	No. 4 / 72	No. 4 / 80	No. 4 / 56	No. 4 / 56	No. 4 / 56	No. 4 / 56	No. 4 / 40	No. 4 / 48	No. 4 / 48	No. 4 / 48
		10	No. 4 / 48	No. 4 / 48	No. 4 / 48	No. 4 / 56	No. 4 / 56	No. 4 / 64	No. 4 / 64	No. 4 / 64	No. 5 / 48	No. 5 / 48	No. 5 / 48	No. 5 / 48	No. 5 / 40	No. 5 / 40	No. 5 / 40	No. 5 / 40
		12	No. 5 / 48	No. 5 / 48	No. 5 / 56	No. 5 / 56	No. 6 / 56	No. 6 / 56	No. 6 / 56	No. 6 / 56	No. 6 / 48	No. 6 / 48	No. 6 / 48	No. 6 / 48	No. 6 / 40	No. 6 / 40	No. 6 / 40	No. 6 / 40
	First story of a two-story building	8	No. 3 / 64	No. 3 / 64	No. 3 / 64	No. 3 / 56	No. 4 / 80	No. 4 / 80	No. 4 / 72	No. 4 / 72	No. 4 / 64	No. 4 / 56	No. 4 / 56	No. 4 / 56	No. 4 / 48	No. 4 / 48	No. 4 / 48	No. 4 / 48
		10	No. 4 / 56	No. 4 / 56	No. 4 / 56	No. 4 / 48	No. 4 / 48	No. 5 / 64	No. 5 / 64	No. 5 / 56	No. 5 / 48	No. 5 / 48	No. 5 / 48	No. 5 / 48	No. 5 / 40	No. 5 / 40	No. 5 / 40	No. 5 / 32
		12	No. 5 / 56	No. 5 / 56	No. 5 / 48	No. 5 / 48	No. 6 / 56	No. 6 / 56	No. 6 / 56	No. 6 / 48	No. 6 / 48	No. 6 / 48	No. 6 / 40	No. 6 / 40	No. 6 / 40	No. 6 / 32	No. 6 / 32	No. 6 / 32

*NR — No vertical reinforcement required. However, see Table A-21-F for shear wall reinforcement.

[1]These values are for walls with running bond. For stack bond see Section 2112.8.3.

[2]The figure on top of the listed data is the bar size; the figure below it is the maximum spacing in inches (mm). Reinforcing bar strength shall be A 615 Grade 60.

[3]Roof load is assumed to be concentrically loaded on the wall. For roofs which hang on ledgers, a design is required.

[4]Minimum horizontal reinforcement shall be one No. 4 at the ledger and foundation. Also, see Table A-21-E for lintels and Table A-21-F for shear wall reinforcing where applicable.

[5]Hook vertical bars over bond beam as shown. Extend bars into footing using lap splices where necessary. Where second-story bar spacing does not match those on the first story, hook bars around floor bond beam also.

To use this table, check criteria by the following method:

[5.1]Choose proper roof live load from Table 16-C or snow load criteria for the locality in which the building is located.

[5.2]Check if building is enclosed or partially enclosed by the procedure in Chapter 16, Part III.

[5.3]Choose proper floor load from Table 16-A. [For loads less that 50 psf (2.4 kN/m^2), use 50 psf (2.4 kN/m^2), and for loads between 50 psf (2.4 kN/m^2) and 100 psf (4.8 kN/m^2), use 100 psf (4.8 kN/m^2).]

[5.4]Find proper wind speed and exposure for the site—see Figure 16-1, Chapter 16, Sections 1619 and 1620.

[5.5]Within the proper vertical column, choose appropriate span-to-bearing wall and appropriate height and story.

[5.6]Read proper size and spacing of reinforcement for the thickness of the wall mentioned in the title of the table. (Equivalent area of steel, taking spacing into account, may be substituted.)

[5.7]For buildings in Seismic Zone 2 (see Figure 16-2 in Chapter 16), use minimum reinforcement in Figure A-21-2 if it is more restrictive than the table values.

TABLE A-21-C-3—VERTICAL REINFORCING STEEL REQUIREMENTS FOR 8-INCH-THICK (203 mm) MASONRY WALLS[1] IN AREAS WHERE BASIC WIND SPEEDS ARE 80 MILES PER HOUR (129 km/h) OR GREATER[2,3,4,5]
(Wood or Steel Roof and Floor Framing)

BOND BEAM. SEE FOOTNOTE 4 THIS TABLE AND TABLE A-21-E

VERTICAL BAR—STANDARD HOOK OVER BOND BEAM (alternate every other bar)

Criteria: Roof Live Load = 20 psf to 40 psf (0.96 kN/m² to 1.9 kN/m²); Floor Live Load = 100 psf (4.8 kN/m²); enclosed building

			80 MPH				90 MPH				100 MPH				110 MPH			
			× 1.61 for km/h — Span between Bearing Walls (feet) × 304.8 for mm															
		UNSUPPORTED HEIGHT (feet) × 304.8 for mm	8	16	24	32	8	16	24	32	8	16	24	32	8	16	24	32
EXPOSURE	STORIES		Size of Rebar and Spacing (inches) × 25.4 for mm															
B	One-story building or top story of two-story building	8	NR*	NR*	NR*	NR*	NR*	NR*	NR*	NR*	NR*	NR*	NR*	NR*	No.3 56	No.3 56	No.3 64	No.3 64
		10	NR*	NR*	NR*	NR*	No.4 64	No.4 64	No.4 64	No.4 72	No.4 64	No.4 64	No.4 64	No.4 72	No.4 48	No.4 48	No.4 56	No.4 56
		12	No.4 64	No.4 72	No.4 72	No.4 72	No.4 56	No.4 56	No.4 56	No.4 56	No.4 40	No.4 40	No.4 40	No.4 40	No.5 56	No.5 56	No.5 56	No.5 56
	First story of a two-story building	8	No.3 96	No.3 96	No.3 80	No.3 64	No.3 96	No.3 88	No.3 72	No.3 56	No.3 80	No.3 64	No.3 56	No.3 48	No.3 64	No.3 56	No.3 48	No.4 64
		10	No.3 72	No.3 64	No.3 56	No.3 48	No.3 56	No.3 48	No.4 64	No.4 56	No.4 72	No.4 64	No.4 56	No.4 48	No.4 56	No.4 48	No.4 48	No.5 56
		12	No.4 72	No.4 64	No.4 56	No.4 48	No.4 56	No.4 48	No.4 48	No.4 40	No.4 48	No.4 40	No.5 48	No.5 48	No.5 56	No.5 48	No.5 48	No.5 40
C	One-story building or top story of two-story building	8	NR*	NR*	NR*	NR*	No.3 48	No.3 48	No.3 48	No.3 56	No.4 64	No.4 64	No.4 72	No.4 72	No.4 48	No.4 56	No.4 56	No.4 56
		10	No.4 56	No.4 56	No.4 64	No.4 64	No.4 48	No.4 48	No.4 48	No.4 48	No.5 56	No.5 56	No.5 56	No.5 56	No.5 48	No.5 48	No.5 48	No.5 48
		12	No.4 40	No.5 64	No.5 64	No.5 64	No.5 48	No.5 48	No.5 48	No.5 48	No.5 40	No.6 56	No.6 56	No.6 56	No.6 40	No.6 40	No.6 40	No.5 32
	First story of a two-story building	8	No.3 72	No.3 64	No.3 56	No.3 48	No.3 56	No.3 48	No.3 64	No.3 56	No.4 72	No.4 64	No.4 56	No.4 48	No.4 56	No.4 48	No.4 48	No.5 56
		10	No.4 64	No.4 56	No.4 48	No.4 48	No.4 48	No.4 48	No.5 56	No.5 56	No.5 56	No.5 56	No.5 48	No.5 48	No.5 48	No.5 48	No.5 40	No.5 40
		12	No.4 40	No.5 56	No.5 48	No.5 48	No.5 48	No.5 40	No.5 40	No.6 48	No.5 40	No.6 48	No.6 48	No.6 40	No.6 40	No.6 40	No.6 40	No.6 32
D	One-story building or top story of two-story building	8	No.3 48	No.3 48	No.3 56	No.3 56	No.3 64	No.4 72	No.4 72	No.4 80	No.4 56	No.4 56	No.4 56	No.4 56	No.4 40	No.4 48	No.4 48	No.4 48
		10	No.4 48	No.4 48	No.4 48	No.4 56	No.5 56	No.5 64	No.5 64	No.5 64	No.5 48	No.5 48	No.5 48	No.5 48	No.5 40	No.5 40	No.5 40	No.5 40
		12	No.5 48	No.5 48	No.5 56	No.5 56	No.5 40	No.5 40	No.6 56	No.6 56	No.6 48	No.6 48	No.6 48	No.5 32	No.6 40	No.6 40	No.6 40	No.6 40
	First story of a two-story building	8	No.3 56	No.3 56	No.3 48	No.3 64	No.3 48	No.4 64	No.4 56	No.4 56	No.4 56	No.4 56	No.4 48	No.5 64	No.4 48	No.5 64	No.5 56	No.5 56
		10	No.4 56	No.4 48	No.4 48	No.5 56	No.5 64	No.5 56	No.5 48	No.5 48	No.5 48	No.5 48	No.5 40	No.5 40	No.5 40	No.5 40	No.6 48	No.6 48
		12	No.5 48	No.5 48	No.5 40	No.5 40	No.5 40	No.5 40	No.6 48	No.6 40	No.6 48	No.6 40	No.6 40	No.6 32	No.6 32	No.6 32	No.6 32	No.6 24

*NR — No vertical reinforcement required. However, see Table A-21-F for shear wall reinforcement.

[1]These values are for walls with running bond. For stack bond see Section 2112.8.3.

[2]The figure on top of the listed data is the bar size; the figure below it is the maximum spacing in inches (mm). Reinforcing bar strength shall be A 615 Grade 60.

[3]Roof load is assumed to be concentrically loaded on the wall. For roofs which hang on ledgers, a design is required.

[4]Minimum horizontal reinforcement shall be one No. 4 at the ledger and foundation. Also, see Table A-21-E for lintels and Table A-21-F for shear wall reinforcing where applicable.

[5]Hook vertical bars over bond beam as shown. Extend bars into footing using lap splices where necessary. Where second-story bar spacing does not match those on the first story, hook bars around floor bond beam also.

To use this table, check criteria by the following method:

[5.1]Choose proper roof live load from Table 16-C or snow load criteria for the locality in which the building is located.

[5.2]Check if building is enclosed or partially enclosed by the procedure in Chapter 16, Part III.

[5.3]Choose proper floor load from Table 16-A. [For loads less than 50 psf (2.4 kN/m²), use 50 psf (2.4 kN/m²), and for loads between 50 psf (2.4 kN/m²) and 100 psf (4.8 kN/m²), use 100 psf (4.8 kN/m²).]

[5.4]Find proper wind speed and exposure for the site—see Figure 16-1, Chapter 16, Sections 1619 and 1620.

[5.5]Within the proper vertical column, choose appropriate span-to-bearing wall and appropriate height and story.

[5.6]Read proper size and spacing of reinforcement for the thickness of the wall mentioned in the title of the table. (Equivalent area of steel, taking spacing into account, may be substituted.)

[5.7]For buildings in Seismic Zone 2 (see Figure 16-2 in Chapter 16), use minimum reinforcement in Figure A-21-2 if it is more restrictive than the table values.

TABLE A-21-C-4—VERTICAL REINFORCING STEEL REQUIREMENTS FOR 8-INCH-THICK (203 mm) MASONRY WALLS[1] IN AREAS WHERE BASIC WIND SPEEDS ARE 80 MILES PER HOUR (129 km/h) OR GREATER[2,3,4,5]
(Wood or Steel Roof and Floor Framing)

BOND BEAM. SEE FOOTNOTE 4 THIS TABLE AND TABLE A-21-E

VERTICAL BAR—STANDARD HOOK OVER BOND BEAM (alternate every other bar)

Criteria: Roof Live Load = 20 psf to 40 psf (0.96 kN/m² to 1.9 kN/m²);
Floor Live Load = 50 psf (2.4 kN/m²); partially enclosed building

EXPO-SURE	STORIES	UNSUP-PORTED HEIGHT (feet) × 304.8 for mm	80 MPH 8	16	24	32	90 MPH 8	16	24	32	100 MPH 8	16	24	32	110 MPH 8	16	24	32
							× 304.8 for mm Span between Bearing Walls (feet) — Size of Rebar and Spacing (inches) × 25.4 for mm											
B	One-story building or top story of two-story building	8	No. 4 96	No. 4 96	No. 3 80	No. 3 88	No. 3 56	No. 3 56	No. 3 64	No. 3 64	No. 3 40	No. 3 48	No. 3 48	No. 3 48	No. 4 64	No. 4 64	No. 4 64	No. 4 72
		10	No. 4 64	No. 4 64	No. 4 72	No. 4 72	No. 4 48	No. 4 56	No. 4 56	No. 4 56	No. 4 40	No. 4 40	No. 4 40	No. 4 40	No. 5 48	No. 5 56	No. 5 56	No. 5 56
		12	No. 4 40	No. 4 48	No. 4 48	No. 4 48	No. 5 56	No. 5 56	No. 5 56	No. 5 56	No. 5 40	No. 5 40	No. 5 40	No. 5 40	No. 6 48	No. 6 48	No. 6 48	No. 6 48
	First story of a two-story building	8	No. 3 96	No. 3 96	No. 3 88	No. 3 80	No. 3 72	No. 3 72	No. 3 64	No. 3 64	No. 3 48	No. 3 48	No. 3 48	No. 3 48	No. 4 64	No. 4 64	No. 4 64	No. 4 64
		10	No. 3 48	No. 3 48	No. 3 48	No. 3 48	No. 4 64	No. 4 56	No. 4 56	No. 4 56	No. 4 48	No. 4 48	No. 4 48	No. 4 40	No.4 40	No. 5 56	No. 5 56	No. 5 48
		12	No. 4 48	No. 4 48	No. 4 48	No. 4 40	No. 4 40	No. 5 56	No. 5 56	No. 5 48	No. 5 48	No. 5 40	No. 5 40	No. 5 40	No. 6 48	No. 6 48	No. 6 48	No. 6 48
C	One-story building or top story of two-story building	8	No. 3 40	No. 3 40	No. 3 40	No. 3 40	No. 4 56	No. 4 56	No. 4 56	No. 4 56	No. 4 40	No. 4 40	No. 4 40	No. 4 48	No. 5 56	No. 5 56	No. 5 56	No. 5 56
		10	No. 4 40	No. 4 40	No. 4 40	No. 4 40	No. 5 48	No. 5 48	No. 5 48	No. 5 48	No. 5 32	No. 5 32	No. 5 40	No. 5 40	No. 6 40	No. 6 40	No. 6 40	No. 6 40
		12	No. 5 40	No. 5 40	No. 5 40	No. 5 40	No. 6 40	No. 6 48	No. 6 48	No. 6 48	No. 6 32	No. 6 32	No. 6 32	No. 6 32	Use 10-inch or larger units			
	First story of a two-story building	8	No. 3 48	No. 3 48	No. 3 48	No. 3 48	No. 4 56	No. 4 56	No. 4 56	No. 4 56	No. 4 48	No. 4 48	No. 4 40	No. 4 40	No. 5 56	No. 5 56	No. 5 56	No. 5 48
		10	No. 4 40	No. 4 40	No. 4 40	No. 4 40	No. 5 48	No. 5 48	No. 5 48	No. 5 48	No. 5 40	No. 5 40	No. 5 40	No. 5 48	No. 6 48	No. 6 48	No. 6 40	No. 6 40
		12	No. 5 40	No. 5 40	No. 5 40	No. 6 48	No. 6 48	No. 6 48	No. 6 40	No. 6 40	No. 6 32	No. 6 32	No. 6 32	No. 6 32	Use 10-inch or larger units			
D	One-story building or top story of two-story building	8	No. 4 56	No. 4 56	No. 4 56	No. 4 64	No. 4 40	No. 4 48	No. 4 48	No. 4 48	No. 5 56	No. 5 56	No. 5 56	No. 5 56	No. 5 48	No. 5 48	No. 5 48	No. 5 48
		10	No. 5 48	No. 5 48	No. 5 48	No. 5 56	No. 5 40	No. 5 40	No. 5 40	No. 5 40	No. 6 56	No. 6 48	No. 6 48	No. 6 48	No. 6 32	No. 6 32	No. 6 40	No. 6 40
		12	No. 6 48	No. 6 48	No. 6 48	No. 6 48	No. 6 40	No. 6 40	No. 6 40	No. 6 40	No. 6 24	No. 6 32	No. 6 32	No. 6 32	Use 10-inch or larger units			
	First story of a two-story building	8	No. 4 64	No. 4 64	No. 4 64	No. 4 56	No. 4 48	No. 4 48	No. 4 48	No. 4 48	No. 4 40	No. 4 40	No. 5 56	No. 5 56	No. 5 48	No. 5 48	No. 5 48	No. 5 40
		10	No. 5 56	No. 5 56	No. 5 48	No. 5 48	No. 5 40	No. 5 40	No. 5 40	No. 5 40	No. 6 48	No. 6 48	No. 6 48	No. 6 40	No. 6 40	No. 6 40	No. 6 32	No. 6 32
		12	No. 6 48	No. 6 48	No. 6 48	No. 6 48	No. 6 40	No. 6 40	Use 10-inch or larger units									

*NR — No vertical reinforcement required. However, see Table A-21-F for shear wall reinforcement.

[1]These values are for walls with running bond. For stack bond see Section 2112.8.3.
[2]The figure on top of the listed data is the bar size; the figure below it is the maximum spacing in inches (mm). Reinforcing bar strength shall be A 615 Grade 60.
[3]Roof load is assumed to be concentrically loaded on the wall. For roofs which hang on ledgers, a design is required.
[4]Minimum horizontal reinforcement shall be one No. 4 at the ledger and foundation. Also, see Table A-21-E for lintels and Table A-21-F for shear wall reinforcing where applicable.
[5]Hook vertical bars over bond beam as shown. Extend bars into footing using lap splices where necessary.
    To use this table, check criteria by the following method:
    5.1 Choose proper roof live load from Table 16-C or snow load criteria for the locality in which the building is located.
    5.2 Check if building is enclosed or partially enclosed by the procedure in Chapter 16, Part III.
    5.3 Choose proper floor load from Table 16-A. [For loads less than 50 psf (2.4 kN/m²), use 50 psf (2.4 kN/m²), and for loads between 50 psf (2.4 kN/m²) and 100 psf (4.8 kN/m²) , use 100 psf (4.8 kN/m²).]
    5.4 Find proper wind speed and exposure for the site—see Figure 16-1, Chapter 16, Sections 1619 and 1620.
    5.5 Within the proper vertical column, choose appropriate span-to-bearing wall and appropriate height and story.
    5.6 Read proper size and spacing of reinforcement for the thickness of the wall mentioned in the title of the table. (Equivalent area of steel, taking spacing into account, may be substituted.)
    5.7 For buildings in Seismic Zone 2 (see Figure 16-2 in Chapter 16), use minimum reinforcement in Figure A-21-2 if it is more restrictive than the table values.

TABLE A-21-C-5—VERTICAL REINFORCING STEEL REQUIREMENTS FOR 8-INCH-THICK (203 mm) MASONRY WALLS[1] IN AREAS WHERE BASIC WIND SPEEDS ARE 80 MILES PER HOUR (129 km/h) OR GREATER[2,3,4,5]
(Wood or Steel Roof and Floor Framing)

Criteria: Roof Live Load = 20 psf to 40 psf (0.96 kN/m^2 to 1.9 kN/m^2);
Floor Live Load = 100 psf (4.8 kN/m^2); partially enclosed building

BOND BEAM. SEE FOOTNOTE 4 THIS TABLE AND TABLE A-21-E

VERTICAL BAR—STANDARD HOOK OVER BOND BEAM (alternate every other bar)

EXPO-SURE	STORIES	UNSUPPORTED HEIGHT (feet) × 304.8 for mm	80 MPH				90 MPH				100 MPH				110 MPH			
			span Span between Bearing Walls (feet) × 304.8 for mm — Size of Rebar and Spacing (inches) × 25.4 for mm															
			8	16	24	32	8	16	24	32	8	16	24	32	8	16	24	32
B	One-story building or top story of two-story building	8	No. 3 72	No. 4 96	No. 3 80	No. 3 88	No. 3 56	No. 3 56	No. 3 64	No. 3 64	No. 4 80	No. 4 80	No. 4 80	No. 4 88	No. 4 64	No. 4 64	No. 4 64	No. 4 72
		10	No. 4 64	No. 4 64	No. 4 72	No. 4 72	No. 4 48	No. 4 56	No. 4 56	No. 4 56	No. 4 40	No. 4 40	No. 4 40	No. 4 40	No. 5 48	No. 5 56	No. 5 56	No. 5 56
		12	No. 4 40	No. 4 48	No. 4 48	No. 4 48	No. 5 56	No. 5 56	No. 5 56	No. 5 56	No. 5 40	No. 5 40	No. 5 40	No. 5 40	No. 6 48	No. 6 48	No. 6 48	No. 6 48
	First story of a two-story building	8	No. 3 88	No. 3 96	No. 3 56	No. 4 72	No. 3 64	No. 3 56	No. 3 64	No. 4 64	No. 4 80	No. 4 72	No. 4 64	No. 4 56	No. 4 64	No. 4 56	No. 4 48	No. 4 48
		10	No. 4 72	No. 4 64	No. 4 56	No. 4 48	No. 4 56	No. 4 48	No. 4 48	No. 4 40	No. 4 48	No. 4 40	No. 5 56	No. 5 48	No. 5 56	No. 5 48	No. 5 48	No. 5 40
		12	No. 4 48	No. 4 40	No. 4 40	No. 5 48	No. 5 56	No. 5 48	No. 5 48	No. 5 40	No. 5 40	No. 5 40	No. 6 48	No. 6 48	No. 6 48	No. 6 48	No. 6 40	No. 6 40
C	One-story building or top story of two-story building	8	No. 4 64	No. 4 64	No. 4 72	No. 4 72	No. 4 56	No. 4 56	No. 4 56	No. 4 56	No. 4 40	No. 4 40	No. 4 40	No. 4 48	No. 5 56	No. 5 56	No. 5 56	No. 5 56
		10	No. 5 56	No. 5 56	No. 4 40	No. 4 40	No. 5 48	No. 5 48	No. 5 48	No. 5 48	No. 5 32	No. 6 56	No. 6 56	No. 5 40	No. 6 40	No. 6 40	No. 6 40	No. 6 40
		12	No. 5 40	No. 5 40	No. 5 40	No. 5 40	No. 6 40	No. 6 48	No. 6 48	No. 6 48	No. 6 32	No. 6 32	No. 6 32	No. 6 32	Use 10-inch or larger units			
	First story of a two-story building	8	No. 4 72	No. 4 64	No. 4 56	No. 4 48	No. 4 56	No. 4 48	No. 4 48	No. 4 40	No. 4 40	No. 4 40	No. 4 40	No. 5 48	No. 5 56	No. 5 48	No. 5 48	No. 5 40
		10	No. 5 64	No. 5 56	No. 5 48	No. 5 48	No. 5 48	No. 5 48	No. 5 40	No. 5 40	No. 5 40	No. 6 48	No. 6 48	No. 6 40	No. 6 40	No. 6 40	No. 6 40	No. 6 32
		12	No. 5 40	No. 6 48	No. 6 48	No. 6 40	No. 6 40	No. 6 40	No. 6 40	No. 6 32	No. 6 32	No. 6 32	No. 6 32	Use 10-inch or larger units				
D	One-story building or top story of two-story building	8	No. 4 56	No. 4 56	No. 4 56	No. 4 64	No. 4 40	No. 5 72	No. 5 72	No. 5 72	No. 5 56	No. 5 56	No. 5 56	No. 5 56	No. 5 48	No. 5 48	No. 5 48	No. 5 48
		10	No. 5 48	No. 5 48	No. 5 48	No. 5 56	No. 5 40	No. 5 40	No. 5 40	No. 5 40	No. 6 56	No. 6 48	No. 6 48	No. 6 48	No. 6 32	No. 6 32	No. 6 40	No. 6 40
		12	No. 6 48	No. 6 48	No. 6 48	No. 6 48	No. 6 40	No. 6 40	No. 6 40	No. 6 40	No. 6 24	No. 6 32	No. 6 32	No. 6 32	Use 10-inch or larger units			
	First story of a two-story building	8	No. 4 64	No. 4 56	No. 4 48	No. 4 48	No. 4 48	No. 5 64	No. 5 56	No. 5 56	No. 5 56	No. 5 48	No. 5 48	No. 5 48	No. 5 48	No. 5 40	No. 5 40	No. 5 40
		10	No. 5 56	No. 5 48	No. 5 48	No. 5 40	No. 5 40	No. 5 40	No. 6 48	No. 6 48	No. 6 48	No. 6 40	No. 6 40	No. 6 40	No. 6 40	No. 6 32	No. 6 32	No. 6 32
		12	No. 6 48	No. 6 48	No. 6 40	No. 6 40	No. 6 40	No. 6 32	No. 6 32	No. 6 32	Use 10-inch or larger units							

*NR — No vertical reinforcement required. However, see Table A-21-F for shear wall reinforcement.

[1]These values are for walls with running bond. For stack bond see Section 2112.8.3.

[2]The figure on top of the listed data is the bar size; the figure below it is the maximum spacing in inches (mm). Reinforcing bar strength shall be A 615 Grade 60.

[3]Roof load is assumed to be concentrically loaded on the wall. For roofs which hang on ledgers, a design is required.

[4]Minimum horizontal reinforcement shall be one No. 4 at the ledger and foundation. Also, see Table A-21-E for lintels and Table A-21-F for shear wall reinforcing where applicable.

[5]Hook vertical bars over bond beam as shown. Extend bars into footing using lap splices where necessary.

To use this table, check criteria by the following method:

5.1 Choose proper roof live load from Table 16-C or snow load criteria for the locality in which the building is located.

5.2 Check if building is enclosed or partially enclosed by the procedure in Chapter 16, Part III.

5.3 Choose proper floor load from Table 16-A. [For loads less than 50 psf (2.4 kN/m^2), use 50 psf (2.4 kN/m^2), and for loads between 50 psf (2.4 kN/m^2) and 100 psf (4.8 kN/m^2), use 100 psf (4.8 kN/m^2).]

5.4 Find proper wind speed and exposure for the site—see Figure 16-1, Chapter 16, Sections 1619 and 1620.

5.5 Within the proper vertical column, choose appropriate span-to-bearing wall and appropriate height and story.

5.6 Read proper size and spacing of reinforcement for the thickness of the wall mentioned in the title of the table. (Equivalent area of steel, taking spacing into account, may be substituted.)

5.7 For buildings in Seismic Zone 2 (see Figure 16-2 in Chapter 16), use minimum reinforcement in Figure A-21-2 if it is more restrictive than the table values.

TABLE A-21-D—ANCHORAGE OF WOOD MEMBERS TO EXTERIOR WALLS FOR VERTICAL AND UPLIFT FORCES
[In areas where basic wind speeds are 80 miles per hour (129 km/h) or greater]
See Figure A-21-7 for details

Part I—Anchor bolt size and spacing [in inches (mm)][1,2,3] on wood ledgers carrying vertical loads from roofs and floors[4,5]
Douglas fir-larch, California redwood (close grain) and southern pine[6,7]

TYPE OF LOADING	LIVE LOAD[8,9] psf (× 0.0479 for kN/m²)	2-INCH (51 mm) × LEDGER				3-INCH (76 mm) × LEDGER				4-INCH (102 mm) × LEDGER			
		8	16	24	32	8	16	24	32	8	16	24	32
Roof	20	$1/2$ 32	$(2)1/2$ 16	$5/8$ 16	$7/8$ 16	$1/2$ 32	$1/2$ 16	$(2)1/2$ 32	$7/8$ 16	—	$5/8$ 32	$7/8$ 32	$(2)5/8$ 32
	30	$(2)1/2$ 32	$1/2$ 16	$3/4$ 16	$7/8$ 16	$1/2$ 16	$(2)7/8$ 32	$7/8$ 16	$7/8$ 16	—	$(2)1/2$ 32	$5/8$ 16	$3/4$ 16
	40	$1/2$ 16	$5/8$ 16	$3/4$ 8	—	$5/8$ 16	$(2)5/8$ 32	$7/8$ 16	1 16	$5/8$ 32	$5/8$ 16	$3/4$ 16	$7/8$ 16
Floor[10]	50	$1/2$ 16	1 12	—	—	$5/8$ 24	$3/4$ 32	$3/4$ 12	$1 1/4$ 12	$5/8$ 24	$7/8$ 24	$7/8$ 16	$7/8$ 12
	100	1 16	$(2)3/4$ 12	—	—	$5/8$ 16	1 12	$(2)3/4$ 12	$(2)1$ 12	$7/8$ 16	$3/4$ 12	1 12	$(2)3/4$ 12

[1]Closer spacing may be used.
[2]Use two bolts, one above the other, at splices and locate them away from the splice end by $3^1/2$ inches (89 mm) for $1/2$-inch (13 mm) diameter, $4^1/2$ inches (114.3 mm) for $5/8$-inch (15.9 mm) diameter, $5^1/4$ inches (133 mm) for $3/4$-inch (19 mm) diameter, $6^1/4$ inches (158 mm) for $7/8$-inch (22.2 mm) diameter and 7 inches (178 mm) for 1-inch (25.4 mm) diameter.
[3]See Table A-21-F for lateral force requirements (when applicable).
[4]Tabulated values are based on short-term loading due to roof loads (25 percent) or snow loads (15 percent), whichever controls. No increase is allowed for floor loads.
[5]See details in Figure A-21-7 for location relative to other construction. Note that roofs are concentrically loaded.
[6]See Chapter 23, Division III, Part I, for other species. Adjust spacing in direct proportion to the perpendicular-to-grain values for the applicable ledger and bolt sizes shown using the procedure described in Chapter 23, Division III, Part I. No increase is allowed for special inspection.
[7]Values on top are bolt sizes and underneath are spacing. Multiple bolts are shown in parenthesis: example (2) = two.
[8]See Table 16-C or Appendix Chapter 16, Division I, for values.
[9]Joist spacing is limited to 30 inches (762 mm) on center maximum.
[10]Where two bolts are required they shall be staggered at half the spacing shown or be placed one above the other.

Part II—Uplift anchors[1] for wood roof members [number of common nails in a 0.036 inch (0.91 mm) (No. 20 galvanized sheet gage) by $1^1/8$-inch (28.6 mm) tie strap embedded 5 inches (127 mm) into a masonry bond beam[2]]

ENCLOSURE[3]	EXPOSURE[4]	80 MPH				90 MPH				100 MPH				110 MPH			
		8	16	24	32	8	16	24	32	8	16	24	32	8	16	24	32
Enclosed	B	NR	NR	NR	NR	NR	NR	NR	NR	NR	NR	NR	NR	NR	NR	2-8d	2-8d
	C	NR	NR	NR	NR	NR	2-8d	3-8d	4-8d	2-8d	4-8d	5-10d	5-10d	2-10d	4-10d	3-10d 24"	4-10d 24"
	D	NR	2-8d	3-8d	4-8d	2-8d	4-8d	4-10d	5-10d	3-8d	5-8d	5-10d	4-10d 24"	3-10d	5-10d	4-10d 24"	5-10d 24"
Open	B	NR	NR	NR	NR	NR	NR	2-8d	2-8d	NR	2-8d	4-8d	5-10d	2-8d	4-8d	5-8d	6-10d
	C	2-8d	4-8d	5-8d	5-10d	3-8d	5-8d	3-10d 24"	4-10d 24"	3-10d	5-10d	5-10d 24"	5-10d 16"	5-8d	4-10d 24"	5-10d 16"	6-10d 16"
	D	2-8d	5-8d	5-10d	5-10d 24"	4-8d	5-10d	4-10d 24"	5-10d 24"	5-8d	4-10d 24"	6-10d 24"	6-10d 16"	4-8d	5-10d 24"	6-10d 16"	6-10d 12"

NR — No requirements; use Table 23-II-B-1 minimum.

[1]Tie straps are at 48 inches (1219 mm) on center unless otherwise stated. See Figure A-21-7 for illustration of tie straps.
[2]Bond beam to be at least 48 inches (1219 mm) deep nominal and shall be reinforced as shown in Table A-21-E for lintels, or Tables A-21-C-1 through A-21-C-5 for walls in general where they are more restrictive.
[3]See Chapter 21, Part II, for definitions.
[4]See Section 1616 for definitions.
[5]For flat roofs connected to interior walls, the span shall be one half the larger distance on either side of the wall.

TABLE A-21-E—LINTEL REINFORCEMENT OVER EXTERIOR OPENINGS[1,2]—WOOD AND STEEL FRAMING[3]
[Lintels larger than 12 feet 0 inch (3658 mm) shall be designed.][4]
8-INCH (203 mm) MASONRY UNITS[5]

Part I—Roof Loads[5]

ANY WALL HEIGHT (feet)	SPAN TO BEARING WALLS (feet)[9]	SECOND STORY OF A TWO-STORY OR ONE-STORY BUILDINGS ROOF LIVE LOAD[6,7,8]					
		20-30 psf			40 psf		
		× 0.0479 for kN/m²					
		Width of Opening[9] (feet)					
		× 304.8 for mm					
		4	8	12	4	8	12
		Lintel depth (inches) number and size of rebar					
× 304.8 for mm		× 25.4 for mm					
Any (up to 12′)	8	8 1 No. 3	8 1 No. 3	16 1 No. 4 (B)	8 1 No. 3	8 1 No. 4	16 1 No. 4 (B)
	16	8 1 No. 3	8 1 No. 3	16 1 No. 4 (B)	8 1 No. 3	8 2 No. 4 (A)	16 2 No. 5 (B)
	24	8 1 No. 3	8 1 No. 4	16 1 No. 4 (B)	8 1 No. 3	16 1 No. 4 (B)	24 2 No. 5 (B)
	32	8 1 No. 3	16 1 No. 4 (B)	16 1 No. 5 (B)	8 1 No. 3	16 1 No. 5 (B)	24 2 No. 5 (C)

Part II—Floor and Roof Loads[5]

WALL HEIGHT	SPAN TO BEARING[9,11] WALLS (feet)	FIRST STORY OF TWO-STORY BUILDINGS FLOOR LIVE LOAD[10]					
		50 psf			100 psf		
		× 0.0479 for kN/m²					
		Width of Opening[9] (feet)					
		× 304.8 for mm					
		4	8	12	4	8	12
		Lintel depth (inches) number and size of rebar					
× 304.8 for mm		× 25.4 for mm					
Any (up to 12′)	8	8 1 No. 3	8 1 No. 4	16 1 No. 4 (B)	8 1 No. 3	8 2 No. 4 (A)	16 2 No. 5 (B)
	16	8 1 No. 3	8 2 No. 4 (A)	16 2 No. 5 (B)	8 1 No. 3	16 1 No. 4 (B)	24 2 No. 4 (C)
	24	8 1 No. 3	16 1 No. 4 (B)	24 2 No. 5 (B)	8 1 No. 3 (A)	16 1 No. 5 (B)	24 3 No. 5 (C)
	32	8 1 No. 3	16 1 No. 5 (B)	24 2 No. 5 (C)	8 1 No. 4 (B)	24 2 No. 5 (C)	Design Required

[1]The values shown are number and size of A 615, 60 grade steel reinforcement bars: Example—2 No. 4 is two $1/2$-inch-diameter (13 mm) deformed reinforcing bars. See also Figure A-21-8 for continuous load path.

[2]Stirrup spacing requirements: A = No. 3 at 8 inches (203 mm) on center, B = No. 3 at 4 inches (102 mm) on center, C = No. 4 at 8 inches (203 mm) on center. None are required unless specifically mentioned in the table.

[3]Design required for lintels supporting precast planks.

[4]Lintels are 8-inch (203 mm) nominal depth where supporting roof loads only and 16-inch (406 mm) nominal depth where supporting floor and roofs unless otherwise stated. All lintels are solidly grouted.

[5]Wall weight is included.

[6]The stirrup size and spacing, where required, as indicated in parenthesis below the reinforcing bar requirements.

[7]All exposure categories are included for wind uplift on the lintel. See Footnote 4 of Tables A-21-C-1 through A-21-C-5 as a minimum bond beam. Table A-21-F may also control.

[8]Two No. 5 vertical bars minimum are required on each side of the lintel for 100 and 110 miles per hour (161 and 177 km/h), Exposure D. Bar to extend 25 inches (635 mm) beyond opening or hook over top bars.

[9]For spans between the figures shown, go to next higher span width.

[10]From Table 21-A. For other floor loads go to next higher value. Where required floor load exceeds 100 pounds per square foot (4.8 kN/m²), a design is required.

[11]When interior walls support floors from each side, these values may be used if the spans on each side are less than 16 feet 0 inch (4877 mm) each. Enter the table with the total of both span widths.

TABLE A-21-F—MASONRY SHEAR WALL[1,2,3] AND DIAPHRAGM REQUIREMENTS IN HIGH-WIND AREAS[4]

Part I—Minimum wall length and horizontal bar reinforcement required for exterior shear walls and cross walls[5] (all wall heights). [Design criteria: 20 psf to 40 psf (0.96 kN/m² to 1.9 kN/m²) roof load; 50 psf or 100 psf (2.4 kN/m² or 4.8 kN/m²) floor load; open or enclosed buildings.]

Wind Speed	Exposure	Distance between Shear Resisting Walls[7] "L" or "b" (feet)	One-story Building or Second Story of a Two-story Building	First Story of a Two-story Building
× 1.61 for km/h		× 304.8 for mm	inch × 25.4 for mm foot × 304.8 mm	
80 mph	B	32	NSR	9'-4"
		48	NSR	5'-4" DBL (D)
		64	10'-0"	7'-6" DBL (C)
	C	32	NSR	5'-4" DBL (C)
		48	11'-0"	8'-8" DBL (C)
		64	13'-4"	15'-0" (D)
	D	32	8'-8"	7'-0" (C)
		48	9'-4" (C)	10'-8" (D)
		64	10'-0" (D)	13'-8" (D)
90 mph	B	32	NSR	7'-8" DBL (C)
		48	NSR	8'-0" (D)
		64	12'-8"	12'-0" (D)
	C	32	NSR	14'-8"
		48	13'-8"	10'-0" (D)
		64	10'-8" (C)	15'-6" DBL (B)
	D	32	7'-8" (C)	11'-8" (D)
		48	12'-0" (C)	12'-8" DBL (B)
		64	11'-8" (D)	18'-4" DBL (C)
100 mph	B	32	NSR	5'-4" DBL (C)
		48	10'-0"	10'-0" (D)
		64	15'-4"	64'-8" DBL (C)

(Continued)

**TABLE A-21-F—MASONRY SHEAR WALL[1,2,3] AND DIAPHRAGM
REQUIREMENTS IN HIGH-WIND AREAS[4]—(Part I Continued)**

Wind Speed	Exposure	Distance between Shear Resisting Walls[7] "L" or "b" (feet)	One-story Building or Second Story of a Two-story Building	First Story of a Two-story Building
× 1.61 for km/h		× 304.8 for mm	inch × 25.4 for mm foot × 304.8 for mm	
100 mph (cont.)	C	32	5'-4" (D)	11'-8" (D)
		48	12'-8" (C)	12'-8" DBL (C)
		64	12'-4" (D)	19'-8" DBL (C)
	D	32	5'-4" DBL (B)	9'-4" DBL (C)
		48	9'-4" (D)	14'-8" DBL (C)
		64	17'-4" (D)	21'-0" DBL (C)
110 mph	B	32	NSR	6'-0" DBL (C)
		48	12'-0"	10'-0" DBL (C)
		64	12'-8" (C)	14'-0" (D)
	C	32	5'-4" DBL (B)	9'-8" (D)
		48	12'-0" (C)	15'-4" (D)
		64	16'-8" (C)	18'-8" DBL (C)
	D	32	8'-8" (C)	11'-4" (D)
		48	12'-4" (C)	18'-0" (D)
		64	18'-8" (C)	20'-8" DBL (C)

*NSR—No special horizontal reinforcement required for shear resistance if 5 feet 4 inches (1626 mm) long minimum.

[1]Cumulative shear wall length is to be at least as long as is shown in this table. However, see Figure A-21-9. The top figure is the minimum length. When required, the figure below it in parenthesis is the spacing of steel reinforcing wire installed as shown in Figure A-21-10, below. (A) = two 0.148 inch (3.76 mm) (No. 9 B.W. gage) at 16 inches (406 mm) on center, (B) = two $^3/_{16}$ inch (4.76 mm) at 16 inches (406 mm) on center, (C) = two 0.148 inches (3.76 mm) (No. 9 B.W. gage) at 8 inches (203 mm) on center, (D) = two $^3/_{16}$ inch (4.76 mm) at 8 inches (203 mm) on center. The symbol DBL means double these amounts. Equivalent areas of reinforcing bars spaced not over 4 feet 0 inch (1219 mm) on center may be used.

[2]All bearing and shear walls are to be in-plane with vertical reinforcement, when required, extending from one floor to the other as dictated in Tables A-21-C-1 through A-21-C-5.

[3]Minimum bond beam shall be 100 miles per hour (mph) (161 km/h), Exposure B; 90 mph (145 km/h), Exposure B, and 80 mph (129 km/h), Exposures B and C, one No. 4; 100 mph (161 km/h), Exposure C; 80 and 90 mph (129 and 145 km/h); Exposures C and D, two No. 4; all others two No. 5.

[4]Table is adjusted to include provisions for Seismic Zones 0, 1 and 2.

[5]Cross walls are to be at least twice as long as shown in the table for shear walls. The tributary width (L/2) shall be the distance used in the third column above to find minimum reinforcement and length.

[6]For walls which width is equal to or less than half its height, add an extra No. 5 vertical bar at each end.

[7]Use 32-foot (9753 mm) requirements for distances less than 32 feet (9754 mm). Also use it for bearing walls used as shear walls.

(Continued)

TABLE A-21-F—MASONRY SHEAR WALL[1,2,3] AND DIAPHRAGM REQUIREMENTS IN HIGH-WIND AREAS[4]—(Continued)

Part II—Wood floor and roof diaphragms and connections[8,9]
[All wall heights 8 feet to 12 feet (2438 mm to 3657 mm).]

Wind Speed		Distance between Shear Walls[10] "L" or "b" (feet)	Minimum Wood Structural Panel/Particleboard Size[9] and Nailing[11,12]		
	Exposure		Thickness (inches)	Common Nail Size (penny)	Nail Spacing (inches)
× 1.61 for km/h		× 304.8 for mm	× 25.4 for mm		× 25.4 for mm
80 mph	B	16	$5/16$	6	6 o.c.
		32	$3/8$	6	6 o.c.
		48	$3/8$	8	6 o.c.
		64	$3/8$	8	6 o.c.
	C	16	$3/8$	8	6 o.c.
		32	$1/2$ or $15/32$	8	6 o.c.
		48	$1/2$ or $15/32$	10	6 o.c.
		64	$5/8$ or $19/32$	10	6 o.c.
	D	16	$1/2$ or $15/32$	8	6 o.c.
		32	$5/8$ or $19/32$	10	6 o.c.
		48	$1/2$ or $15/32$ blocked	8	4/6 o.c.
		64	$1/2$ or $15/32$ blocked	8	4/6 o.c.
90 mph	B	16	$5/16$	6	6 o.c.
		32	$3/8$	8	6 o.c.
		48	$3/8$	8	6 o.c.
		64	$3/8$	8	6 o.c.
	C	16	$1/2$ or $15/32$	10	6 o.c.
		32	$3/8$ blocked	8	4/6 o.c.
		48	$3/8$ blocked	8	4/6 o.c.
		64	$5/8$ or $19/32$ blocked	10	6 o.c.
	D	16	$5/8$ or $19/32$	10	6 o.c.
		32	$1/2$ or $15/32$ blocked	10	4/6 o.c.
		48	$1/2$ or $15/32$ blocked	10	4/6 o.c.
		64	Design required or provide extra cross walls		
100 mph	B	16	$3/8$	8	6 o.c.
		32	$1/2$ or $15/32$	8	6 o.c.
		48	$1/2$ or $15/32$	8	6 o.c.
		64	$5/8$ or $19/32$	10	6 o.c.

(Continued)

TABLE A-21-F—MASONRY SHEAR WALL[1,2,3] AND DIAPHRAGM REQUIREMENTS IN HIGH-WIND AREAS[4]—(Part II Continued)

Wind Speed × 1.61 for km/h	Exposure	Distance between Shear Walls[10] "L" or "b" (feet) × 304.8 for mm	Minimum Wood Structural Panel/Particleboard Size[9] and Nailing[11,12]		
			Thickness (inches) × 25.4 for mm	Common Nail Size (penny)	Nail Spacing (inches) × 25.4 for mm
100 mph (cont.)	C	16	$3/8$ blocked	8	4/6 o.c.
		32	$5/8$ or $19/32$ blocked	10	4/6 o.c.
		48	$5/8$ or $19/32$ blocked	10	4/6 o.c.
		64	Design required or provide extra cross walls		
	D	16	$1/2$ or $15/32$ blocked	10	4/6 o.c.
		32	$5/8$ or $19/32$ blocked	10	4/6 o.c.
		48	Design required or provide extra cross walls		
		64	Design required or provide extra cross walls		
110 mph	B	16	$1/2$ or $15/32$	8	6 o.c.
		32	$1/2$ or $15/32$	10	6 o.c.
		48	$5/8$ or $19/32$	10	6 o.c.
		64	$1/2$ or $15/32$ blocked	8	4/6 o.c.
	C	16	$1/2$ or $15/32$ blocked	8	4/6 o.c.
		32	$5/8$ or $19/32$ blocked	10	4/6 o.c.
		48	Design required or provide extra cross walls		
		64	Design required or provide extra cross walls		
	D	16	$5/8$ or $19/32$ blocked	10	4/6 o.c.
		32	Design required or provide extra cross walls		
		48	Design required or provide extra cross walls		
		64	Design required or provide extra cross walls		

The table heading spans: FRAMING OF DOUGLAS FIR-LARCH OR SOUTHERN PINE

[8]These requirements represent the maximum values for a diaphragm which is within a maximum 32-foot-by-64-foot (9.75 m by 19.5 m) module surrounded by shear walls, cross walls or bearing walls. (See Figure A-21-9.)

[9]See Tables 23-II-E-1 and 23-II-E-2 for minimum sizes depending on span between joists.

[10]See Figure A-21-9 for "L" and "b."

[11]The wood structural panel/particleboard (all grades) thickness is given first. The nailing size and boundary/supported edge spacing is shown next.. Blocking of unsupported edges is stated where required. Twelve-inch (305 mm) spacing required in the field of the roof/floor. Boundary nailing is required over interior walls [see Figure A-21-12 (b)].

[12]Use Case 1 for unblocked diaphragms and any case for blocked diaphragms.

TABLE A-21-G—MINIMUM WALL CONNECTION REQUIREMENTS IN HIGH-WIND AREAS
Precast Hollow-core Plank Floors and Roofs

Spacing of No. 4 bent reinforcing bar in block or brick walls connected to precast concrete planks[1,2]

WIND SPEED AND EXPOSURE	EXTERIOR WALLS	INTERIOR WALLS
	× 25.4 for mm	
90 mph (145 km/h) Exposure C and less 100 mph (161 km/h) Exposure B	32″ o.c.	16″ o.c.
90 mph (145 km/h) Exposure D 100 mph (161 km/h) Exposure C 110 mph (177 km/h) Exposure B	24″ o.c.	12″ o.c.
100 mph (161 km/h) Exposure D 110 mph (177 km/h) Exposures C and D	16″ o.c.	12″ o.c.

[1]This table assumes maximum wall height of 12 feet (3.7 m) and a width-to-length ratio of diaphragm between shear walls of 3:1 or less.
[2]The precast planks shall be designed as shall the walls and footings supporting them.

TABLE A-21-H—MINIMUM HOLD-DOWN REQUIREMENTS IN HIGH-WIND AREAS
Steel Floors and Roofs

WIND SPEED AND EXPOSURE	MAXIMUM SPACING OF ROOF JOISTS WITH CONNECTION SHOWN[1,2,3]
	× 25.4 for mm
100 mph (161 km/h) Exposure B 90 mph (145 km/h) Exposures B and C 80 mph (129 km/h) Exposures B, C and D	48″
110 mph (177 km/h) Exposure B 100 mph (161 km/h) Exposure C	30″
110 mph (177 km/h) Exposures C and D 100 mph (161 km/h) Exposure D	Design required

[1]Maximum span is 32 feet (9.75 m) to bearing walls.
[2]Joists and decking to be designed.
[3]Bottom chord of joists to be braced for reversal of stresses caused by wind uplift.

TABLE A-21-I—DIAGONAL BRACING REQUIREMENTS
FOR GABLE-END WALL[1,2] ROOF PITCH 3:12 to 5:12

EXPOSURE	BASIC WIND SPEED (mph)							
	× 1.61 for km/h							
	80		90		100		110	
	3:12 (25%)	4:12 (33%) and 5:12 (42%)	3:12 (25%)	4:12 (33%) and 5:12 (42%)	3:12 (25%)	4:12 (33%) and 5:12 (42%)	3:12 (25%)	4:12 (33%) and 5:12 (42%)
	× 25.4 for mm							
B	I at 48″ o.c.	III at 48″ o.c.	I at 48″ o.c.	III at 48″ o.c.	I at 24″ o.c.	III at 24″ o.c.	I at 24″ o.c.	III at 24″ o.c.
C	I at 24″ o.c.	III at 48″ o.c.	I at 24″ o.c.	III at 24″ o.c.	II at 24″ o.c.	IV at 24″ o.c.	II at 24″ o.c.	IV at 24″ o.c.
D	I at 24″ o.c.	III at 48″ o.c.	II at 24″ o.c.	IV at 24″ o.c.	II at 24″ o.c.	IV at 24″ o.c.	Two-II at 24″ o.c.	Two-III at 24″ o.c.

[1] I = 2-inch-by-4-inch brace, one clip angle (51 mm × 102 mm).

  II = 2-inch-by-4-inch brace, two clip angles (one each side) (51 mm × 102 mm).

  III = 3-inch-by-4-inch brace, one clip angle (76 mm × 102 mm).

  IV = 3-inch-by-4-inch brace, two clip angles (one each side) (76 mm × 102 mm).

The spacing requirements of the brace are shown below the symbol.

[2]See Figures A-21-17 and A-21-18 for details and size of clip angles.

NOTE: Horizontal and vertical reinforcement to be determined
by Tables A-21-C-1 through A-21-C-5 and A-21-F.

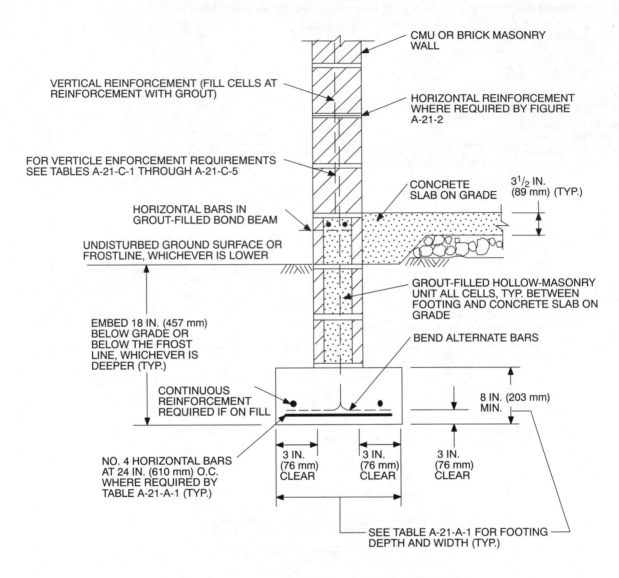

HOLLOW-MASONRY UNIT EXTERIOR FOUNDATION WALL

FIGURE A-21-1—VARIOUS DETAILS OF FOOTINGS
(See Tables A-21-A-1 and A-21-A-2 for widths.)

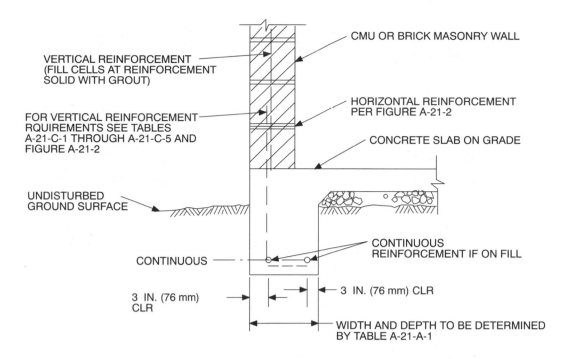

VERTICAL REINFORCEMENT
(FILL CELLS AT REINFORCEMENT
SOLID WITH GROUT)

FOR VERTICAL REINFORCEMENT
RQUIREMENTS SEE TABLES
A-21-C-1 THROUGH A-21-C-5 AND
FIGURE A-21-2

UNDISTURBED
GROUND SURFACE

CMU OR BRICK MASONRY WALL

HORIZONTAL REINFORCEMENT
PER FIGURE A-21-2

CONCRETE SLAB ON GRADE

CONTINUOUS
REINFORCEMENT IF ON FILL

CONTINUOUS

3 IN. (76 mm)
CLR

3 IN. (76 mm) CLR

WIDTH AND DEPTH TO BE DETERMINED
BY TABLE A-21-A-1

GRADE BEAM OR CONTINUOUS CONCRETE SLAB—TURN DOWN

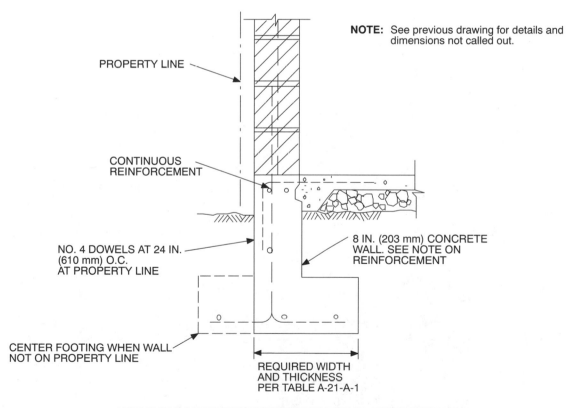

PROPERTY LINE

NOTE: See previous drawing for details and
dimensions not called out.

CONTINUOUS
REINFORCEMENT

NO. 4 DOWELS AT 24 IN.
(610 mm) O.C.
AT PROPERTY LINE

8 IN. (203 mm) CONCRETE
WALL. SEE NOTE ON
REINFORCEMENT

CENTER FOOTING WHEN WALL
NOT ON PROPERTY LINE

REQUIRED WIDTH
AND THICKNESS
PER TABLE A-21-A-1

HOLLOW-MASONRY UNIT CONCRETE EXTERIOR FOUNDATION WALL

FIGURE A-21-1—VARIOUS DETAILS OF FOOTINGS—(Continued)
(See Tables A-21-A-1 and A-21-A-2 for widths.)

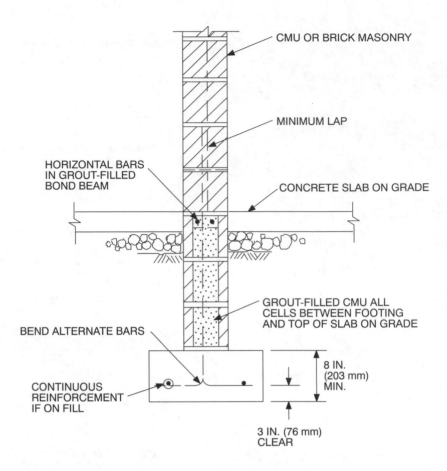

HOLLOW–MASONRY UNIT INTERIOR FOUNDATION WALL

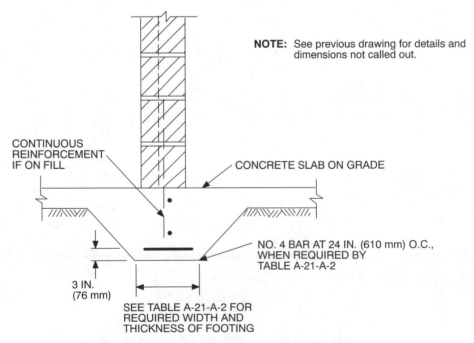

CONCRETE INTERIOR NONBEARING WALL FOOTING

FIGURE A-21-1—VARIOUS DETAILS OF FOOTINGS—(Continued)
(See Tables A-21-A-1 and A-21-A-2 for widths.)

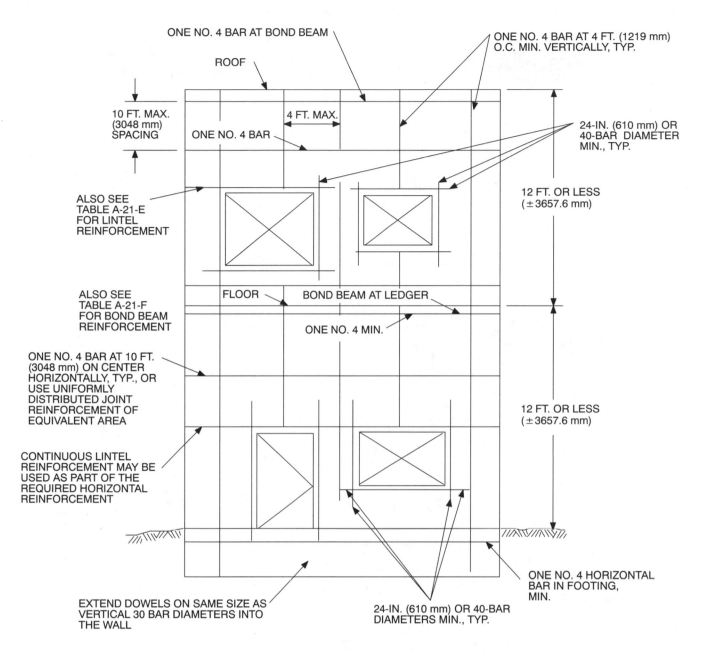

FIGURE A-21-2—MINIMUM MASONRY WALL REQUIREMENTS IN SEISMIC ZONE 2

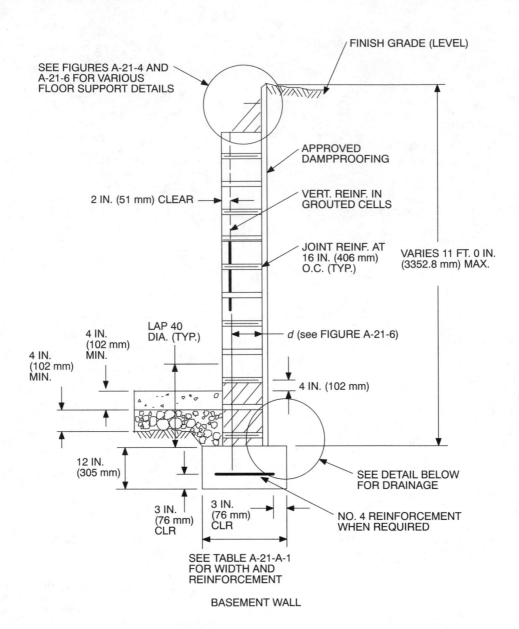

SEE FIGURES A-21-4 AND A-21-6 FOR VARIOUS FLOOR SUPPORT DETAILS

FINISH GRADE (LEVEL)

APPROVED DAMPPROOFING

VERT. REINF. IN GROUTED CELLS

2 IN. (51 mm) CLEAR

JOINT REINF. AT 16 IN. (406 mm) O.C. (TYP.)

VARIES 11 FT. 0 IN. (3352.8 mm) MAX.

4 IN. (102 mm) MIN.

4 IN. (102 mm) MIN.

4 IN. (102 mm) MIN.

LAP 40 DIA. (TYP.)

d (see FIGURE A-21-6)

4 IN. (102 mm)

12 IN. (305 mm)

3 IN. (76 mm) CLR

3 IN. (76 mm) CLR

SEE DETAIL BELOW FOR DRAINAGE

NO. 4 REINFORCEMENT WHEN REQUIRED

SEE TABLE A-21-A-1 FOR WIDTH AND REINFORCEMENT

BASEMENT WALL

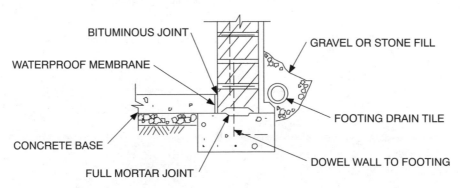

BITUMINOUS JOINT

WATERPROOF MEMBRANE

GRAVEL OR STONE FILL

FOOTING DRAIN TILE

CONCRETE BASE

FULL MORTAR JOINT

DOWEL WALL TO FOOTING

FIGURE A-21-3—BELOW-GRADE WALL AND DRAINAGE DETAILS

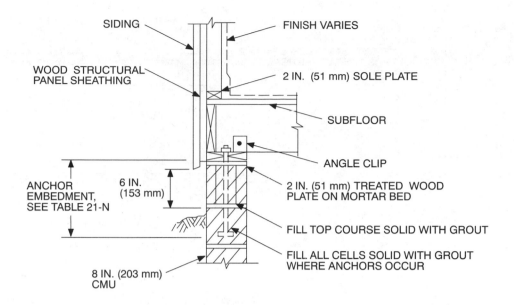

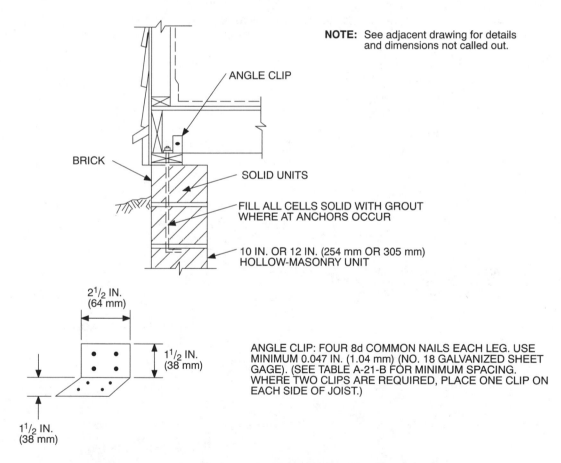

NOTE: See adjacent drawing for details and dimensions not called out.

ANGLE CLIP: FOUR 8d COMMON NAILS EACH LEG. USE MINIMUM 0.047 IN. (1.04 mm) (NO. 18 GALVANIZED SHEET GAGE). (SEE TABLE A-21-B FOR MINIMUM SPACING. WHERE TWO CLIPS ARE REQUIRED, PLACE ONE CLIP ON EACH SIDE OF JOIST.)

FIGURE A-21-4—HOLLOW-MASONRY UNIT FOUNDATION WALL—WOOD FLOOR

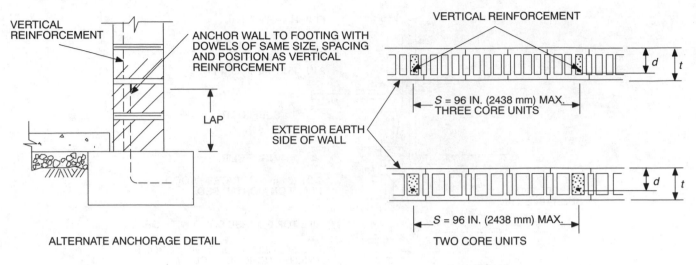

8 IN. (203 mm) WALLS: $t = 7^5/_8$ IN. (194 mm) $d = 5$ IN. (127 mm)
10 IN. (254 mm) WALLS: $t = 9^5/_8$ IN. (245 mm) $d = 7$ IN. (153 mm)
12 IN. (305 mm) WALLS: $t = 11^5/_8$ IN. (295 mm) $d = 8^3/_4$ IN. (225 mm)

FIGURE A-21-5—PLACEMENT OF REINFORCEMENT

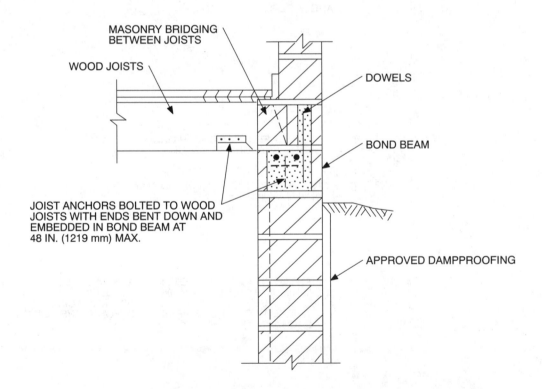

(A) HOLLOW–MASONRY UNIT WALL—WOOD FLOOR

FIGURE A-21-6—VARIOUS CONNECTIONS OF FLOORS TO BASEMENT WALLS

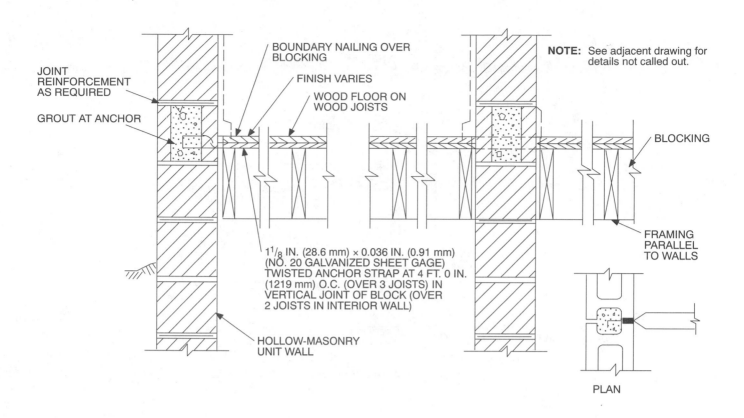

JOINT REINFORCEMENT AS REQUIRED

GROUT AT ANCHOR

BOUNDARY NAILING OVER BLOCKING

FINISH VARIES

WOOD FLOOR ON WOOD JOISTS

NOTE: See adjacent drawing for details not called out.

BLOCKING

FRAMING PARALLEL TO WALLS

$1\frac{1}{8}$ IN. (28.6 mm) × 0.036 IN. (0.91 mm) (NO. 20 GALVANIZED SHEET GAGE) TWISTED ANCHOR STRAP AT 4 FT. 0 IN. (1219 mm) O.C. (OVER 3 JOISTS) IN VERTICAL JOINT OF BLOCK (OVER 2 JOISTS IN INTERIOR WALL)

HOLLOW-MASONRY UNIT WALL

PLAN

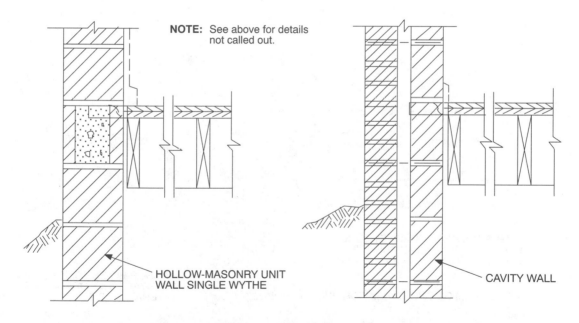

NOTE: See above for details not called out.

HOLLOW-MASONRY UNIT WALL SINGLE WYTHE

CAVITY WALL

(B) WOOD FLOOR, JOISTS PARALLEL TO WALL

FIGURE A-21-6—VARIOUS CONNECTIONS OF FLOORS TO BASEMENT WALLS—(Continued)

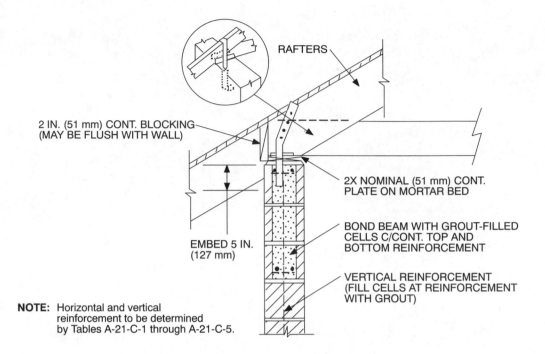

RAFTERS

2 IN. (51 mm) CONT. BLOCKING
(MAY BE FLUSH WITH WALL)

EMBED 5 IN.
(127 mm)

2X NOMINAL (51 mm) CONT.
PLATE ON MORTAR BED

BOND BEAM WITH GROUT-FILLED
CELLS C/CONT. TOP AND
BOTTOM REINFORCEMENT

VERTICAL REINFORCEMENT
(FILL CELLS AT REINFORCEMENT
WITH GROUT)

NOTE: Horizontal and vertical
reinforcement to be determined
by Tables A-21-C-1 through A-21-C-5.

ROOF WITH OVERHANG

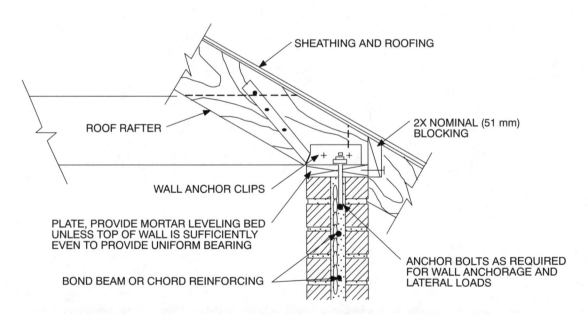

SHEATHING AND ROOFING

ROOF RAFTER

2X NOMINAL (51 mm)
BLOCKING

WALL ANCHOR CLIPS

PLATE, PROVIDE MORTAR LEVELING BED
UNLESS TOP OF WALL IS SUFFICIENTLY
EVEN TO PROVIDE UNIFORM BEARING

ANCHOR BOLTS AS REQUIRED
FOR WALL ANCHORAGE AND
LATERAL LOADS

BOND BEAM OR CHORD REINFORCING

FIGURE A-21-7—VARIOUS DETAILS ASSOCIATED WITH TABLE A-21-D (Uplift Resistance)

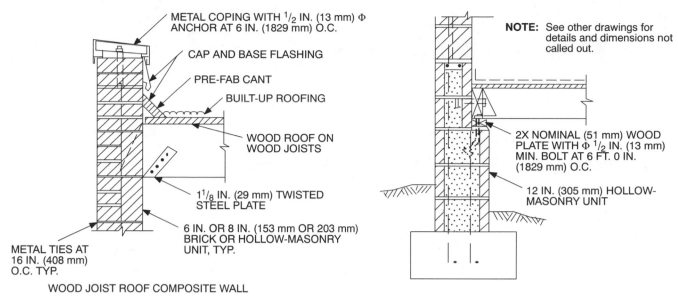

METAL COPING WITH $1/2$ IN. (13 mm) Φ ANCHOR AT 6 IN. (1829 mm) O.C.

CAP AND BASE FLASHING

PRE-FAB CANT

BUILT-UP ROOFING

WOOD ROOF ON WOOD JOISTS

$1^1/_8$ IN. (29 mm) TWISTED STEEL PLATE

6 IN. OR 8 IN. (153 mm OR 203 mm) BRICK OR HOLLOW-MASONRY UNIT, TYP.

METAL TIES AT 16 IN. (408 mm) O.C. TYP.

WOOD JOIST ROOF COMPOSITE WALL

NOTE: See other drawings for details and dimensions not called out.

2X NOMINAL (51 mm) WOOD PLATE WITH Φ $1/2$ IN. (13 mm) MIN. BOLT AT 6 FT. 0 IN. (1829 mm) O.C.

12 IN. (305 mm) HOLLOW-MASONRY UNIT

HOLLOW-MASONRY UNIT FOUNDATION WALL—JOIST PERPENDICULAR

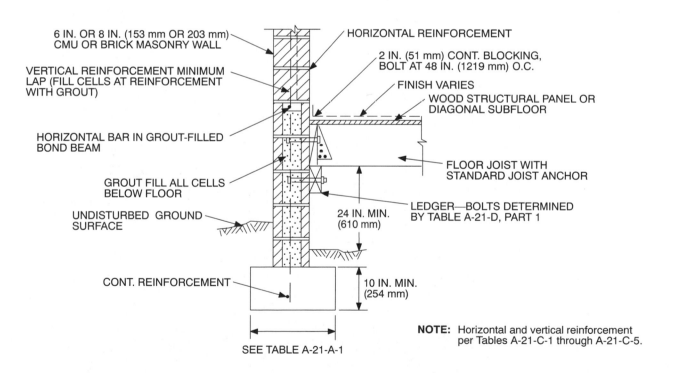

6 IN. OR 8 IN. (153 mm OR 203 mm) CMU OR BRICK MASONRY WALL

VERTICAL REINFORCEMENT MINIMUM LAP (FILL CELLS AT REINFORCEMENT WITH GROUT)

HORIZONTAL BAR IN GROUT-FILLED BOND BEAM

GROUT FILL ALL CELLS BELOW FLOOR

UNDISTURBED GROUND SURFACE

CONT. REINFORCEMENT

HORIZONTAL REINFORCEMENT

2 IN. (51 mm) CONT. BLOCKING, BOLT AT 48 IN. (1219 mm) O.C.

FINISH VARIES

WOOD STRUCTURAL PANEL OR DIAGONAL SUBFLOOR

FLOOR JOIST WITH STANDARD JOIST ANCHOR

LEDGER—BOLTS DETERMINED BY TABLE A-21-D, PART 1

24 IN. MIN. (610 mm)

10 IN. MIN. (254 mm)

SEE TABLE A-21-A-1

NOTE: Horizontal and vertical reinforcement per Tables A-21-C-1 through A-21-C-5.

HOLLOW-MASONRY UNIT FOUNDATION WALL—JOIST PERPENDICULAR—(Continued)

FIGURE A-21-7—VARIOUS DETAILS ASSOCIATED WITH TABLE A-21-D (Uplift Resistance)—(Continued)

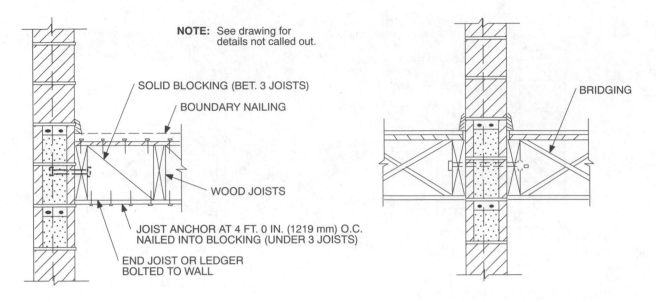

NOTE: See drawing for details not called out.

SOLID BLOCKING (BET. 3 JOISTS)

BOUNDARY NAILING

WOOD JOISTS

JOIST ANCHOR AT 4 FT. 0 IN. (1219 mm) O.C. NAILED INTO BLOCKING (UNDER 3 JOISTS)

END JOIST OR LEDGER BOLTED TO WALL

EXTERIOR WALL—JOIST PARALLEL

BRIDGING

INTERIOR WALL—JOIST PARALLEL

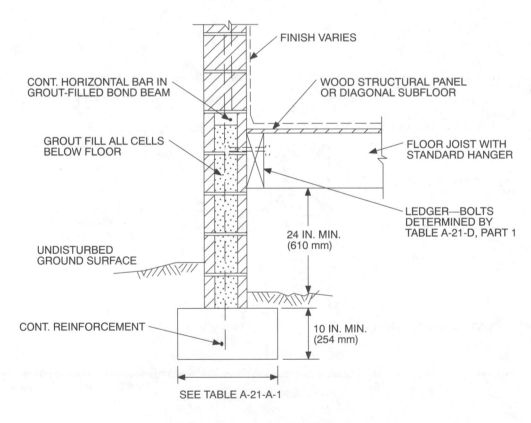

FINISH VARIES

CONT. HORIZONTAL BAR IN GROUT-FILLED BOND BEAM

WOOD STRUCTURAL PANEL OR DIAGONAL SUBFLOOR

GROUT FILL ALL CELLS BELOW FLOOR

FLOOR JOIST WITH STANDARD HANGER

LEDGER—BOLTS DETERMINED BY TABLE A-21-D, PART 1

24 IN. MIN. (610 mm)

UNDISTURBED GROUND SURFACE

CONT. REINFORCEMENT

10 IN. MIN. (254 mm)

SEE TABLE A-21-A-1

HOLLOW-MASONRY UNIT FOUNDATION WALL—JOIST PERPENDICULAR

FIGURE A-21-7—VARIOUS DETAILS ASSOCIATED WITH TABLE A-21-D (Uplift Resistance)—(Continued)

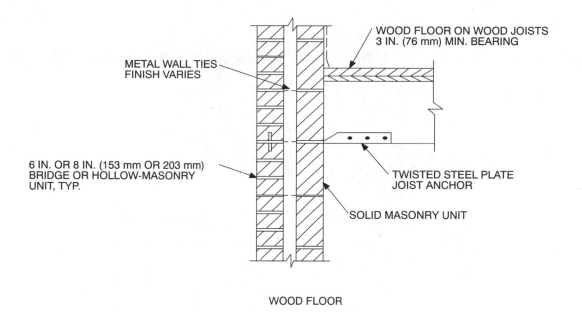

METAL WALL TIES
FINISH VARIES

WOOD FLOOR ON WOOD JOISTS
3 IN. (76 mm) MIN. BEARING

6 IN. OR 8 IN. (153 mm OR 203 mm)
BRIDGE OR HOLLOW-MASONRY
UNIT, TYP.

TWISTED STEEL PLATE
JOIST ANCHOR

SOLID MASONRY UNIT

WOOD FLOOR

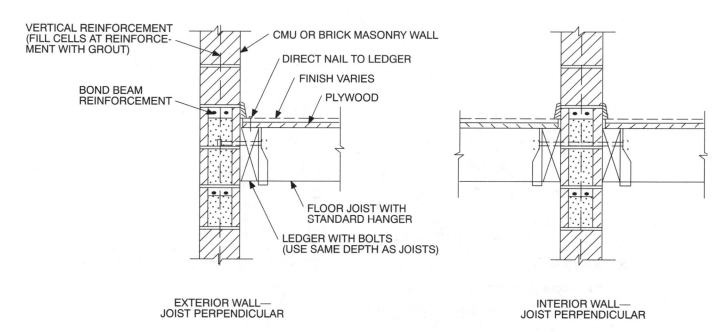

VERTICAL REINFORCEMENT
(FILL CELLS AT REINFORCE-
MENT WITH GROUT)

CMU OR BRICK MASONRY WALL

DIRECT NAIL TO LEDGER

FINISH VARIES

PLYWOOD

BOND BEAM
REINFORCEMENT

FLOOR JOIST WITH
STANDARD HANGER

LEDGER WITH BOLTS
(USE SAME DEPTH AS JOISTS)

EXTERIOR WALL—
JOIST PERPENDICULAR

INTERIOR WALL—
JOIST PERPENDICULAR

FIGURE A-21-7—VARIOUS DETAILS ASSOCIATED WITH TABLE A-21-D (Uplift Resistance)—(Continued)

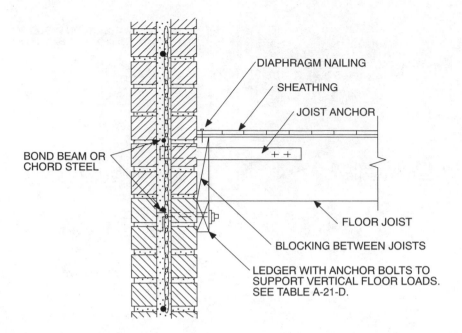

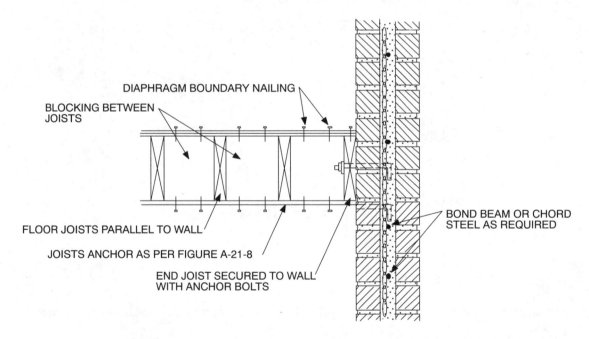

FIGURE A-21-7—VARIOUS DETAILS ASSOCIATED WITH TABLE A-21-D (Uplift Resistance)—(Continued)

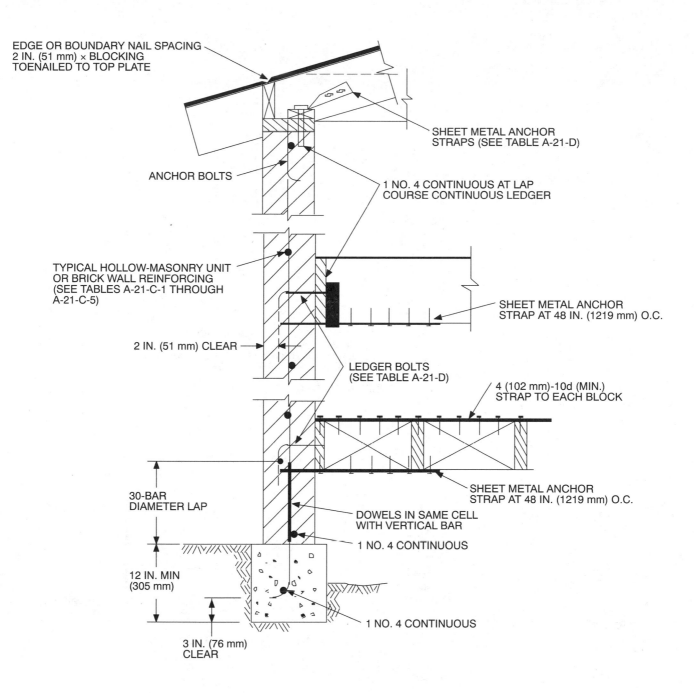

EDGE OR BOUNDARY NAIL SPACING
2 IN. (51 mm) × BLOCKING
TOENAILED TO TOP PLATE

SHEET METAL ANCHOR
STRAPS (SEE TABLE A-21-D)

ANCHOR BOLTS

1 NO. 4 CONTINUOUS AT LAP
COURSE CONTINUOUS LEDGER

TYPICAL HOLLOW-MASONRY UNIT
OR BRICK WALL REINFORCING
(SEE TABLES A-21-C-1 THROUGH
A-21-C-5)

SHEET METAL ANCHOR
STRAP AT 48 IN. (1219 mm) O.C.

2 IN. (51 mm) CLEAR

LEDGER BOLTS
(SEE TABLE A-21-D)

4 (102 mm)-10d (MIN.)
STRAP TO EACH BLOCK

30-BAR
DIAMETER LAP

SHEET METAL ANCHOR
STRAP AT 48 IN. (1219 mm) O.C.

DOWELS IN SAME CELL
WITH VERTICAL BAR

1 NO. 4 CONTINUOUS

12 IN. MIN
(305 mm)

1 NO. 4 CONTINUOUS

3 IN. (76 mm)
CLEAR

FIGURE A-21-8—CONTINUOUS LOAD PATH

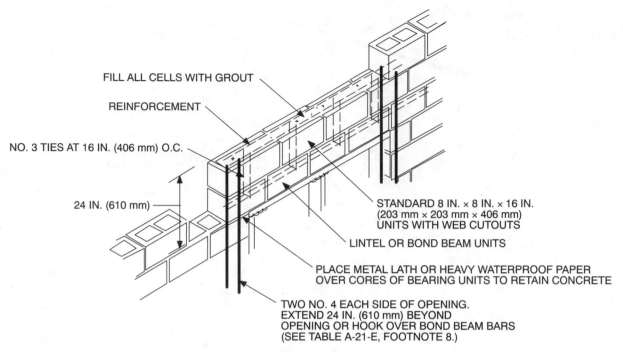

FILL ALL CELLS WITH GROUT

REINFORCEMENT

NO. 3 TIES AT 16 IN. (406 mm) O.C.

24 IN. (610 mm)

STANDARD 8 IN. × 8 IN. × 16 IN.
(203 mm × 203 mm × 406 mm)
UNITS WITH WEB CUTOUTS

LINTEL OR BOND BEAM UNITS

PLACE METAL LATH OR HEAVY WATERPROOF PAPER
OVER CORES OF BEARING UNITS TO RETAIN CONCRETE

TWO NO. 4 EACH SIDE OF OPENING.
EXTEND 24 IN. (610 mm) BEYOND
OPENING OR HOOK OVER BOND BEAM BARS
(SEE TABLE A-21-E, FOOTNOTE 8.)

REINFORCING DETAILS

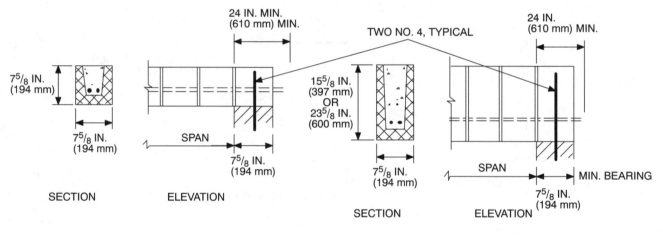

WITHOUT STIRRUPS

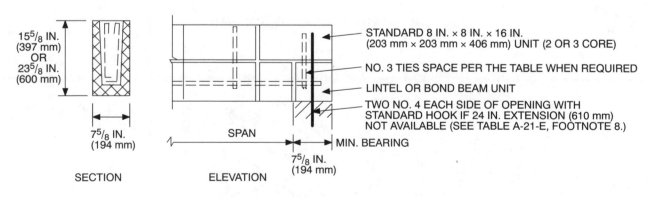

STANDARD 8 IN. × 8 IN. × 16 IN.
(203 mm × 203 mm × 406 mm) UNIT (2 OR 3 CORE)

NO. 3 TIES SPACE PER THE TABLE WHEN REQUIRED

LINTEL OR BOND BEAM UNIT

TWO NO. 4 EACH SIDE OF OPENING WITH
STANDARD HOOK IF 24 IN. EXTENSION (610 mm)
NOT AVAILABLE (SEE TABLE A-21-E, FOOTNOTE 8.)

WITH STIRRUPS

FIGURE A-21-8—CONTINUOUS LOAD PATH—(Continued)

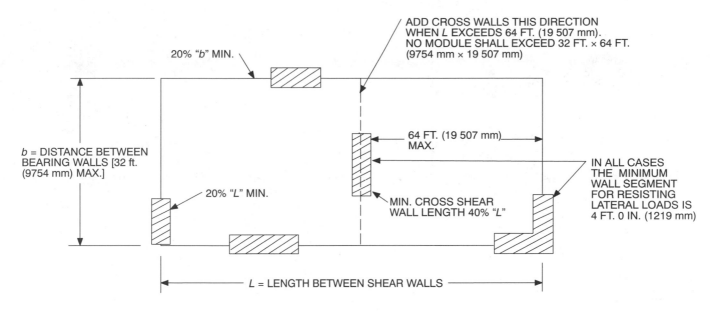

FIGURE A-21-9—SPACING AND LENGTHS OF SHEAR WALLS

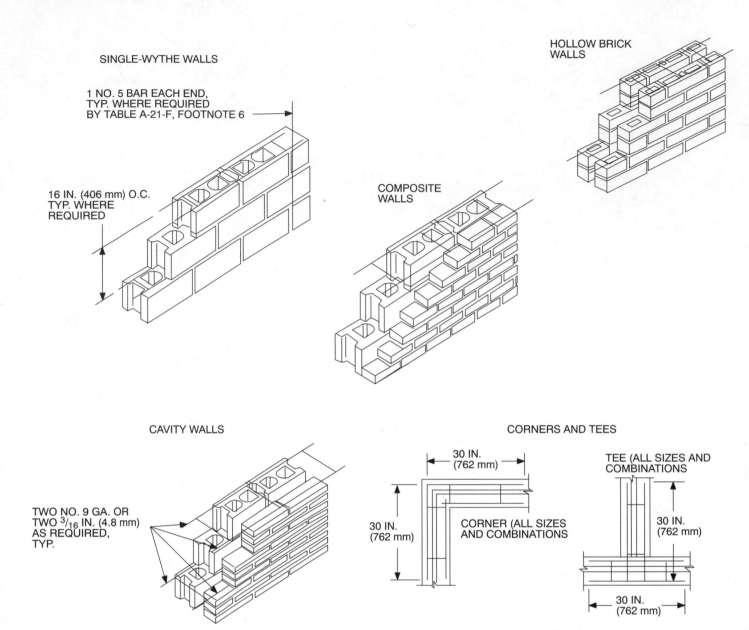

SINGLE-WYTHE WALLS

1 NO. 5 BAR EACH END, TYP. WHERE REQUIRED BY TABLE A-21-F, FOOTNOTE 6

16 IN. (406 mm) O.C. TYP. WHERE REQUIRED

HOLLOW BRICK WALLS

COMPOSITE WALLS

CAVITY WALLS

TWO NO. 9 GA. OR TWO $^3/_{16}$ IN. (4.8 mm) AS REQUIRED, TYP.

CORNERS AND TEES

30 IN. (762 mm)

30 IN. (762 mm)

CORNER (ALL SIZES AND COMBINATIONS

TEE (ALL SIZES AND COMBINATIONS

30 IN. (762 mm)

30 IN. (762 mm)

FIGURE A-21-10—SPACING OF STEEL REINFORCING WIRE

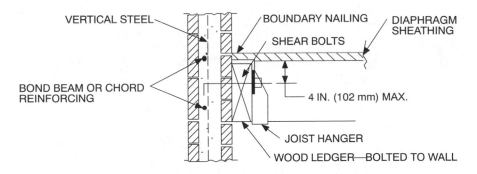

(a) FLOOR JOISTS PERPENDICULAR TO WALL JOIST HANGER SUPPORTS

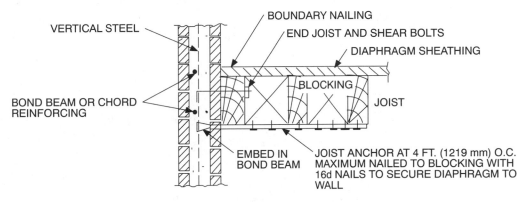

(b) FLOOR JOISTS PARALLEL TO WALL

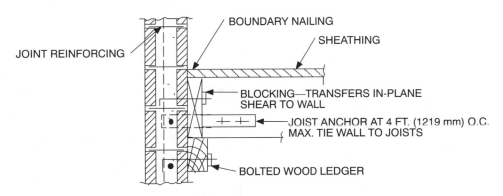

(c) WOOD LEDGER FLOOR JOIST SUPPORT

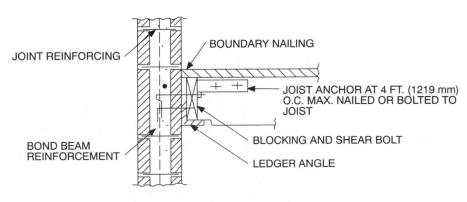

(d) STEEL LEDGER FLOOR JOIST SUPPORT

FIGURE A-21-11—FLOOR-TO-WALL CONNECTION DETAILS

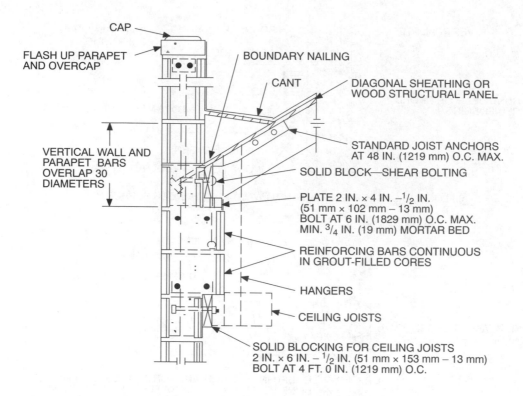

CAP

FLASH UP PARAPET
AND OVERCAP

BOUNDARY NAILING

CANT

DIAGONAL SHEATHING OR
WOOD STRUCTURAL PANEL

VERTICAL WALL AND
PARAPET BARS
OVERLAP 30
DIAMETERS

STANDARD JOIST ANCHORS
AT 48 IN. (1219 mm) O.C. MAX.

SOLID BLOCK—SHEAR BOLTING

PLATE 2 IN. × 4 IN. −$\frac{1}{2}$ IN.
(51 mm × 102 mm − 13 mm)
BOLT AT 6 IN. (1829 mm) O.C. MAX.
MIN. $\frac{3}{4}$ IN. (19 mm) MORTAR BED

REINFORCING BARS CONTINUOUS
IN GROUT-FILLED CORES

HANGERS

CEILING JOISTS

SOLID BLOCKING FOR CEILING JOISTS
2 IN. × 6 IN. − $\frac{1}{2}$ IN. (51 mm × 153 mm − 13 mm)
BOLT AT 4 FT. 0 IN. (1219 mm) O.C.

(a) EXTERIOR WALL SUPPORT

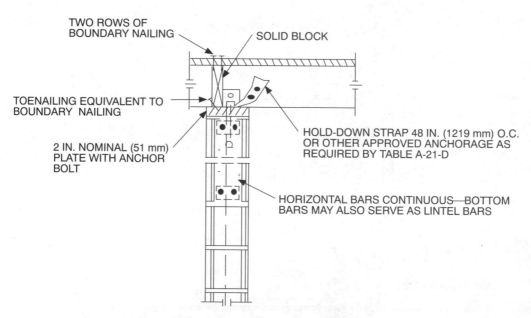

TWO ROWS OF
BOUNDARY NAILING

SOLID BLOCK

TOENAILING EQUIVALENT TO
BOUNDARY NAILING

2 IN. NOMINAL (51 mm)
PLATE WITH ANCHOR
BOLT

HOLD-DOWN STRAP 48 IN. (1219 mm) O.C.
OR OTHER APPROVED ANCHORAGE AS
REQUIRED BY TABLE A-21-D

HORIZONTAL BARS CONTINUOUS—BOTTOM
BARS MAY ALSO SERVE AS LINTEL BARS

(b) INTERIOR WALL SUPPORT BOND—BEAM SUPPORTS

FIGURE A-21-12—ROOF-TO-WALL CONNECTION DETAILS

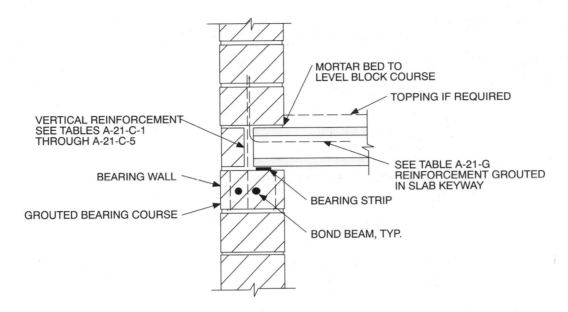

(a) SLAB PERPENDICULAR TO WALL

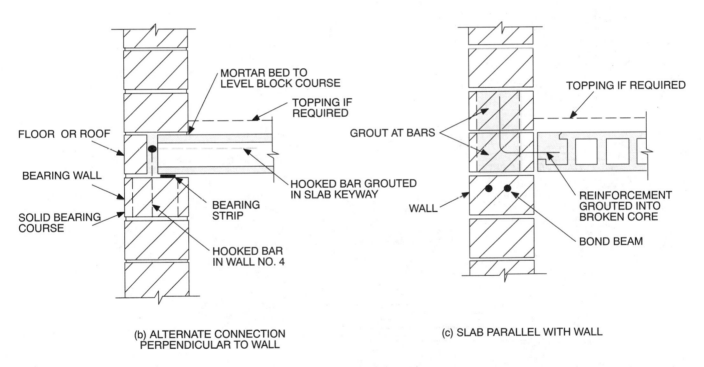

(b) ALTERNATE CONNECTION
PERPENDICULAR TO WALL

(c) SLAB PARALLEL WITH WALL

FIGURE A-21-13—VARIOUS TYPES OF WALL CONNECTIONS

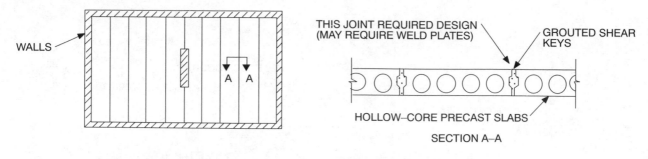

WALLS

THIS JOINT REQUIRED DESIGN
(MAY REQUIRE WELD PLATES)

GROUTED SHEAR
KEYS

HOLLOW–CORE PRECAST SLABS

SECTION A–A

(d) PLAN VIEW OF FLOOR OR ROOF AND CROSS SECTION THROUGH PLANKS

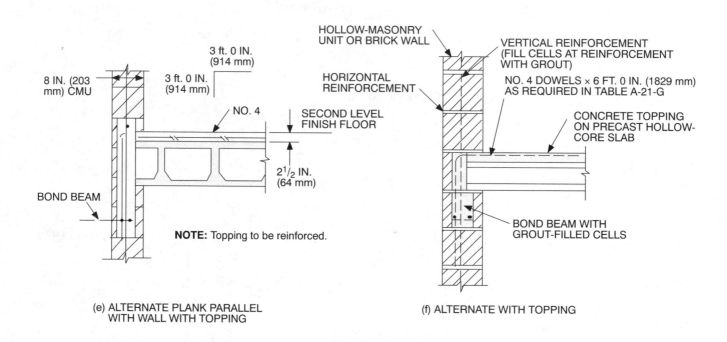

8 IN. (203 mm) CMU

3 ft. 0 IN. (914 mm)

3 ft. 0 IN. (914 mm)

NO. 4

SECOND LEVEL FINISH FLOOR

$2^1/_2$ IN. (64 mm)

BOND BEAM

NOTE: Topping to be reinforced.

HOLLOW-MASONRY UNIT OR BRICK WALL

HORIZONTAL REINFORCEMENT

VERTICAL REINFORCEMENT (FILL CELLS AT REINFORCEMENT WITH GROUT)

NO. 4 DOWELS × 6 FT. 0 IN. (1829 mm) AS REQUIRED IN TABLE A-21-G

CONCRETE TOPPING ON PRECAST HOLLOW-CORE SLAB

BOND BEAM WITH GROUT-FILLED CELLS

(e) ALTERNATE PLANK PARALLEL WITH WALL WITH TOPPING

(f) ALTERNATE WITH TOPPING

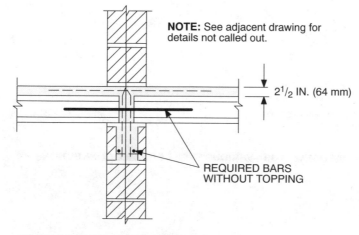

NOTE: See adjacent drawing for details not called out.

$2^1/_2$ IN. (64 mm)

REQUIRED BARS WITHOUT TOPPING

(g) INTERIOR WALL MINIMUM CONNECTION

FIGURE A-21-13—VARIOUS TYPES OF WALL CONNECTIONS—(Continued)

BENT P̱ 0.114 IN. (2.90 mm) (NO. 11 GALVANIZED SHEET GAGE) × ⌐ 4 IN. (102 mm) / 4 IN. (102 mm)

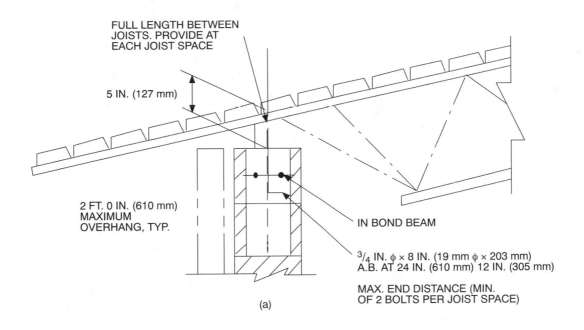

FULL LENGTH BETWEEN JOISTS. PROVIDE AT EACH JOIST SPACE

5 IN. (127 mm)

2 FT. 0 IN. (610 mm) MAXIMUM OVERHANG, TYP.

IN BOND BEAM

³/₄ IN. φ × 8 IN. (19 mm φ × 203 mm) A.B. AT 24 IN. (610 mm) 12 IN. (305 mm)

MAX. END DISTANCE (MIN. OF 2 BOLTS PER JOIST SPACE)

(a)

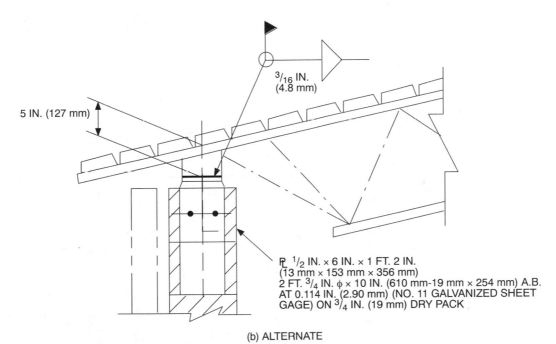

³/₁₆ IN. (4.8 mm)

5 IN. (127 mm)

P̱ ¹/₂ IN. × 6 IN. × 1 FT. 2 IN. (13 mm × 153 mm × 356 mm) 2 FT. ³/₄ IN. φ × 10 IN. (610 mm-19 mm × 254 mm) A.B. AT 0.114 IN. (2.90 mm) (NO. 11 GALVANIZED SHEET GAGE) ON ³/₄ IN. (19 mm) DRY PACK

(b) ALTERNATE

FIGURE A-21-14—EXTERIOR WALL DETAILS

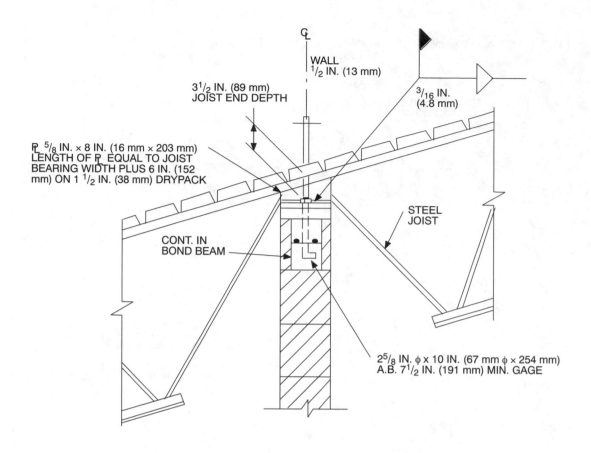

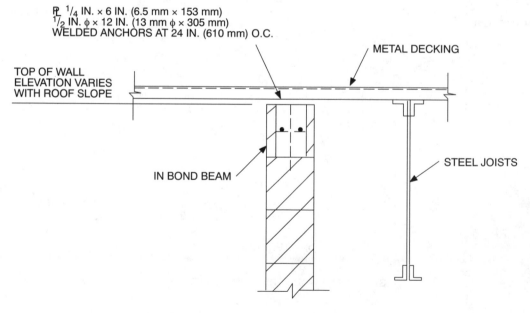

FIGURE A-21-15—INTERIOR WALL DETAILS

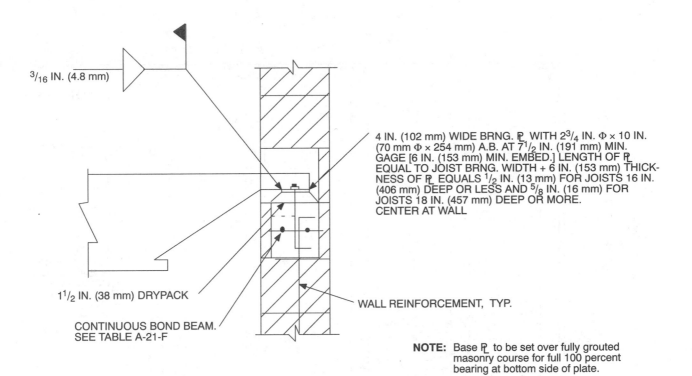

3/16 IN. (4.8 mm)

4 IN. (102 mm) WIDE BRNG. PL WITH 2³/₄ IN. Φ × 10 IN. (70 mm Φ × 254 mm) A.B. AT 7¹/₂ IN. (191 mm) MIN. GAGE [6 IN. (153 mm) MIN. EMBED.] LENGTH OF PL EQUAL TO JOIST BRNG. WIDTH + 6 IN. (153 mm) THICK-NESS OF PL EQUALS ¹/₂ IN. (13 mm) FOR JOISTS 16 IN. (406 mm) DEEP OR LESS AND ⁵/₈ IN. (16 mm) FOR JOISTS 18 IN. (457 mm) DEEP OR MORE. CENTER AT WALL

1¹/₂ IN. (38 mm) DRYPACK

CONTINUOUS BOND BEAM. SEE TABLE A-21-F

WALL REINFORCEMENT, TYP.

NOTE: Base PL to be set over fully grouted masonry course for full 100 percent bearing at bottom side of plate.

FIGURE A-21-16—FLOOR DETAILS
(Design Required for Joists and Wall)

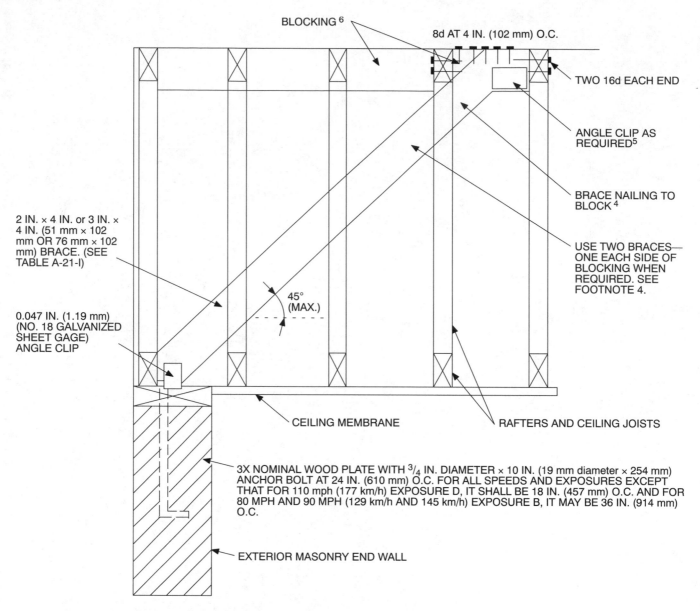

BLOCKING [6]

8d AT 4 IN. (102 mm) O.C.

TWO 16d EACH END

ANGLE CLIP AS REQUIRED[5]

BRACE NAILING TO BLOCK [4]

USE TWO BRACES— ONE EACH SIDE OF BLOCKING WHEN REQUIRED. SEE FOOTNOTE 4.

2 IN. × 4 IN. or 3 IN. × 4 IN. (51 mm × 102 mm OR 76 mm × 102 mm) BRACE. (SEE TABLE A-21-I)

45° (MAX.)

0.047 IN. (1.19 mm) (NO. 18 GALVANIZED SHEET GAGE) ANGLE CLIP

CEILING MEMBRANE

RAFTERS AND CEILING JOISTS

3X NOMINAL WOOD PLATE WITH $3/4$ IN. DIAMETER × 10 IN. (19 mm diameter × 254 mm) ANCHOR BOLT AT 24 IN. (610 mm) O.C. FOR ALL SPEEDS AND EXPOSURES EXCEPT THAT FOR 110 mph (177 km/h) EXPOSURE D, IT SHALL BE 18 IN. (457 mm) O.C. AND FOR 80 MPH AND 90 MPH (129 km/h AND 145 km/h) EXPOSURE B, IT MAY BE 36 IN. (914 mm) O.C.

EXTERIOR MASONRY END WALL

[1] For roof slopes up to 5 units vertical in 12 units horizontal (42%); see Table A-21-I.
[2] See Detail 2, Table A-21-B, for size of angle clip.
[3] Angle clip one side or both sides as required by Table A-21-I.
[4] Use six 16d nails to fasten brace to block, except use two braces and six 16d nails each for 110 miles per hour (mph) (177 km/h), Exposure D. Place on brace on each side block.
[5] Add angle clip each end of block for 90 mph (145 km/h), Exposure D, and 100 and 110 mph (161 and 177 km/h) for Exposures C and D.
[6] Use 2 in. × 6 in. (51 mm × 153 mm) block with 2 in. (51 mm) × brace, 2 in. × 8 in. (51 mm × 203 mm) block with 3 in. (76 mm) × brace.

FIGURE A-21-17—DIAGONAL BRACING OF GABLE-END WALL[1]

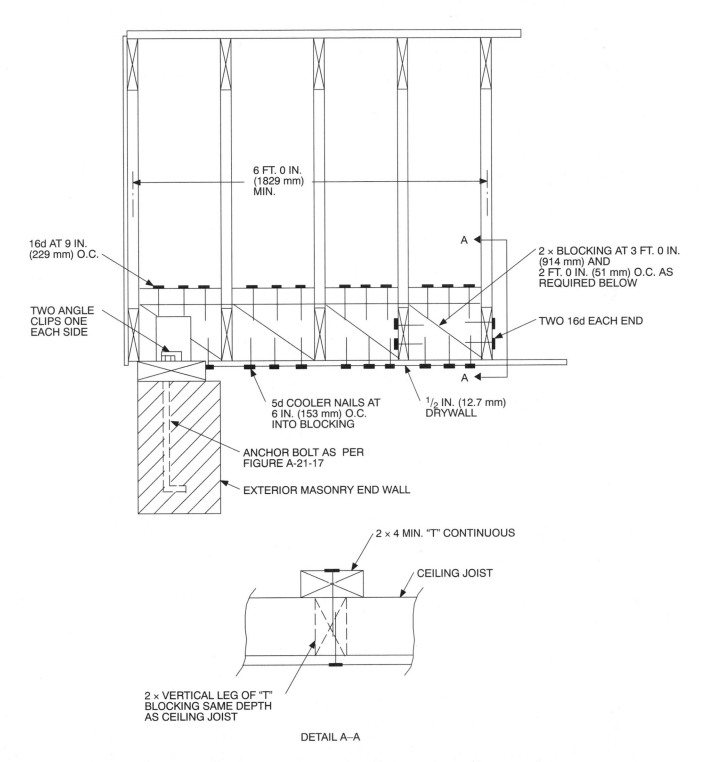

6 FT. 0 IN.
(1829 mm)
MIN.

A

2 × BLOCKING AT 3 FT. 0 IN.
(914 mm) AND
2 FT. 0 IN. (51 mm) O.C. AS
REQUIRED BELOW

16d AT 9 IN.
(229 mm) O.C.

TWO 16d EACH END

TWO ANGLE
CLIPS ONE
EACH SIDE

A

5d COOLER NAILS AT
6 IN. (153 mm) O.C.
INTO BLOCKING

1/2 IN. (12.7 mm)
DRYWALL

ANCHOR BOLT AS PER
FIGURE A-21-17

EXTERIOR MASONRY END WALL

2 × 4 MIN. "T" CONTINUOUS

CEILING JOIST

2 × VERTICAL LEG OF "T"
BLOCKING SAME DEPTH
AS CEILING JOIST

DETAIL A–A

NOTE: This detail may be used for flat roofs also, except use full height blocking connected to roof sheathing in lieu of "T."
2 × 4 "T" at 36 in. (914 mm) on center—90 miles per hour (mph) (145 km/h) Exposure C and less, and 100 mph and
110 mph (161 km/h and 177 km/h), Exposure B.
2 × 4 "T" at 24 in. (610 mm) on center—required for 90 mph (145 mm) exposure.
See Figure A-21-4 for details of clip angle and connections.

FIGURE A-21-18—ALTERNATE HORIZONTAL BRACING OF GABLE-END WALL

**Appendix Chapter 23 is printed in its entirety in Volume 2 of the *Uniform Building Code*.
Excerpts from Appendix Chapter 23 are reprinted herein.**

Excerpts from Appendix Chapter 23
CONVENTIONAL LIGHT-FRAME CONSTRUCTION IN HIGH-WIND AREAS

SECTION 2337 — GENERAL

2337.1 Purpose. The provisions of this chapter are intended to promote public safety and welfare by reducing the risk of wind-induced damages to conventional light-frame construction.

2337.2 Scope. This chapter applies to regular-shaped buildings which have roof structural members spanning 32 feet (9.75 m) or less, are not more than three stories in height, are of conventional light-frame construction and are located in areas with a basic wind speed from 80 through 110 miles per hour (mph) (129 km/h through 177 km/h).

> **EXCEPTION:** Detached carports and garages not exceeding 600 square feet (55.7 m²) and accessory to Group R, Division 3 Occupancies need only comply with the roof-member-to-wall-tie requirements of Section 2337.5.8.

2337.3 Definitions. For the purpose of this chapter, certain terms are defined as follows:

CORROSION RESISTANT or **NONCORROSIVE** is material having a corrosion resistance equal to or greater than a hot-dipped galvanized coating of 1.5 ounces of zinc per square foot (4 g/m²) of surface area.

2337.4 General. The requirements of Section 2320 are applicable except as specifically modified by this chapter. Other methods may be used, provided a satisfactory design is submitted showing compliance with the provisions of Section 1611.4 and other applicable portions of this code.

In addition to the other provisions of this chapter, foundations for buildings in areas subject to wave action or tidal surge shall be designed in accordance with approved national standards.

When an element is required to be corrosion resistant or noncorrosive, all of its parts, such as screws, nails, wire, dowels, bolts, nuts, washers, shims, anchors, ties and attachments, shall also be corrosion resistant or noncorrosive.

2337.5 Complete Load Path and Uplift Ties.

2337.5.1 General. Blocking, bridging, straps, approved framing anchors or mechanical fasteners shall be installed to provide continuous ties from the roof to the foundation system. (See Figure A-23-1.)

Tie straps shall be $1^1/_8$-inch (28.6 mm) by 0.036-inch (0.91 mm) (No. 20 gage) sheet steel and shall be corrosion resistant as herein specified. All metal connectors and fasteners used in exposed locations or in areas otherwise subject to corrosion shall be of corrosion-resistant or noncorrosive material.

2337.5.2 Walls-to-foundation tie. Exterior walls shall be tied to a continuous foundation, or an elevated foundation system in accordance with Section 2337.10.

2337.5.3 Sills and foundation tie. Foundation plates resting on concrete or masonry foundations shall be bolted to the foundation with not less than $1/_2$-inch-diameter (13 mm) anchor bolts with 7-inch-minimum (178 mm) embedment into the foundation. In areas where the basic wind speed is 90 mph (145 km/h) or greater, the maximum spacing of anchor bolts shall be 4 feet (1219 mm) on center. Structures located where the basic wind speed is less than 90 mph (145 km/h) may have anchor bolts spaced not more than 6 feet (1829 mm) on center.

2337.5.4 Floor-to-foundation tie. The lowest-level exterior wall studs shall be connected to the foundation sill plate or an approved elevated foundation system with bent tie straps spaced not more than 48 inches (1219 mm) on center. Tie straps shall be nailed and installed in accordance with Table A-23-B and Figure A-23-1.

2337.5.5 Wall framing details. The spacing of 2-inch-by-4-inch studs (51 mm by 102 mm) in exterior walls shall not exceed 16 inches (406 mm) on center for areas with a basic wind speed of 90 mph (145 km/h) or greater.

Mechanical fasteners complying with this chapter shall be installed as required to connect studs to the sole plates, foundation sill plate and top plates of the wall.

Interior main cross-stud partitions shall be installed approximately perpendicular to the exterior wall when the length of the structure exceeds the width. The maximum distance between these partitions shall not exceed the width of the structure. Interior main cross-stud partition walls shall be securely fastened to exterior walls at the point of intersection with fasteners as required by Table 23-II-B-1. The main cross-stud partitions shall be covered on both sides by materials as described in Section 2337.5.6.

2337.5.6 Wall sheathing. All exterior walls and required interior main cross-stud partitions shall be sheathed in accordance with Table A-23-A. The total width of sheathed wall elements shall not be less than 50 percent of the exterior wall length or 60 percent of the width of the building for required interior main cross-stud partitions. The exterior wall sheathing or covering shall extend from the foundation sill plate or girder to the top plates at the roof level and shall be adequately attached thereto.

A sheathed wall element not less than 4 feet (1219 mm) in width shall be installed at each corner or as near thereto as possible. There shall not be less than one 4-foot (1219 mm) sheathed wall element for every 20 feet (6096 mm) or fraction thereof of wall length. The height-to-length ratio of required sheathed wall elements shall not exceed 3 for wood structural panel or particleboard and $1^1/_2$ (38 mm) for other sheathing materials listed in Table A-23-A.

2337.5.7 Floor-to-floor tie. Upper-level exterior wall studs shall be aligned and connected to the wall studs below with a tie strap as required by Table A-23-B.

2337.5.8 Roof-members-to-wall tie. Tie straps shall be provided from the side of the roof-framing member to the exterior studs, posts or other supporting members below the roof. The wall studs to which the roof-framing members are tied shall be aligned with the roof-framing member and be connected in accordance with Table A-23-B.

The eave overhang shall not exceed 3 feet (914 mm) unless an analysis is provided showing that the required resistance is provided to prevent uplift.

Where openings exceed 6 feet (1829 mm) in width, the required tie straps shall be doubled at each edge of the opening and connected to a doubled full-height wall stud. When openings exceed

12 feet (3658 mm) in width, ties designed to prevent uplift shall be provided.

> **EXCEPTION:** The opening width may be increased to 16 feet (4877 mm) for garages and carports accessory to Group R, Division 3 Occupancies when constructed in accordance with the following:
>
> 1. Approved column bases shall be a minimum $^3/_{16}$-inch (4.8 mm) steel plate embedded not less than 8 inches (203 mm) into the concrete footing and connected to a minimum 4-inch-by-4-inch (102 mm by 102 mm) wood post with two $^5/_8$-inch-diameter (15.9 mm) through bolts.
> 2. Beams over openings shall be connected to minimum 4-inch by 4-inch (102 mm by 102 mm) wood posts below with an approved $^3/_{16}$-inch (4.8 mm) steel post cap with two $^5/_8$-inch-diameter (15.9 mm) through bolts to the posts and to the beams.

2337.5.9 Ridge ties. Opposing rafters shall be aligned at the ridge and be connected at the rafters with a tie strap in accordance with Table A-23-C.

2337.6 Masonry Veneer. Anchor ties shall be spaced so as to support not more than $1^1/_3$ square feet (860 mm²) of wall area but not more than 12 inches (305 mm) on center vertically. The materials and connection details shall comply with Chapter 14.

2337.7 Roof Sheathing. Solid roof sheathing shall be applied and shall consist of a minimum 1-inch-thick (25 mm) nominal lumber applied diagonally or a minimum $^{15}/_{32}$-inch-thick (11.9 mm) wood structural panel or particleboard or other approved sheathing applied with the long dimension perpendicular to supporting rafters. Sheathing shall be nailed to roof framing in an approved manner. The end joints of wood structural panels or particleboard shall be staggered and shall occur over blocking, rafters or other supports.

2337.8 Gable-end Walls. The roof overhang at gabled ends shall not exceed 2 feet (610 mm) unless an analysis showing that the required resistance to prevent uplift is provided.

Gable-end wall studs shall be continuous between points of lateral support which are perpendicular to the plane of the wall.

Gable-end wall studs shall be attached with approved mechanical fasteners at the top and bottom.

2337.9 Roof Covering. Roof coverings shall be approved and shall be installed and fastened in accordance with Chapter 15 and with the manufacturer's instructions. In areas with basic wind speeds of 90 mph (145 km/h) or greater strip asphalt shingles shall be fastened with a minimum of six fasteners and hand sealed.

2337.10 Elevated Foundation.

2337.10.1 General. When approved, elevated foundations supporting not more than one story and meeting the provisions of this section may be used. A foundation investigation may be required by the building official.

2337.10.2 Material. All exposed wood-framing members shall be treated wood. All metal connectors and fasteners used in exposed locations shall be corrosion-resistant or noncorrosive steel.

2337.10.3 Wood piles. The spacing of wood piles shall not exceed 8 feet (2438 mm) on center. Square piles shall not be less than 10 inches (254 mm) and tapered piles shall have a tip of not less than 8 inches (203 mm). Ten-inch-square (64 516 mm²) piles shall have a minimum embedment length of 10 feet (3048 mm) and shall project not more than 8 feet (2438 mm) above undisturbed ground surface. Eight-inch (203 mm) taper piles shall have a minimum embedment length of 14 feet (4267 mm) and shall project not more than 7 feet (2134 mm) above undisturbed ground surface.

2337.10.4 Girders. Floor girders shall be solid sawn timber, built-up 2-inch-thick (51 mm) lumber or trusses. Splices shall occur over wood piles. The floor girders shall span in the direction parallel to the potential floodwater and wave action.

2337.10.5 Connections. Wood piles may be notched to provide a shelf for supporting the floor girders. The total notching shall not exceed 50 percent of the pile cross section. Approved bolted connections with $^1/_4$-inch (6.4 mm) corrosion-resistant or noncorrosive steel plates and $^3/_4$-inch-diameter (19 mm) bolts shall be provided. Each end of the girder shall be connected to the piles using a minimum of two $^3/_4$-inch-diameter (19 mm) bolts.

**TABLE A-23-A—WALL SHEATHING AT EXTERIOR WALLS AND
INTERIOR MAIN CROSS-STUD PARTITIONS[1]**

BASIC WIND SPEED (mph) × 1.61 for km/h	STORIES	LEVEL[2]	EXPOSURE B	C	D
80	1		A	A	B
	2	2	A	A	B
		1	C	D	D
	3	3	A	A	B
		2	C	D	D
		1	C	D	E
90	1		A	B	B
	2	2	A	B	B
		1	C	D	D
	3	3	A	B	Not permitted
		2	C	D	
		1	D	E	
100	1		A	C	C
	2	2	A	C	C
		1	C	D	E
	3	3	A	Not permitted	Not permitted
		2	C		
		1	D		
110	1		B	C	C
	2	2	B	C	C
		1	D	E	E

[1]Sheathing types; exterior walls with sheathing at one face, interior main cross-stud partitions with sheathing at each face. The values for sheathing are listed in order of increased capacity. Sheathing with a capacity greater than required may be substituted for the sheathing listed. Particleboard sheathing in accordance with Table 23-IV-D-2 may be substituted for sheathing Types A and B.

A. One-half-inch (12.7 mm) gypsum board or gypsum sheathing with 5d cooler nails at 7 inches (178 mm) or $^3/_8$-inch (9.5 mm) gypsum lath and $^1/_2$-inch (12.7 mm) plaster.

B. One-half-inch (12.7 mm) gypsum board or gypsum sheathing with 5d cooler nails at 4 inches (102 mm).

C. Expanded metal lath and $^7/_8$-inch (22 mm) portland cement plaster.

D. Three-eighths-inch (9.5 mm) wood structural panel or particleboard sheathing with 8d nails at 6 inches (153 mm) all edges and 12 inches (305 mm) intermediate.

E. Three-eighths-inch (9.5 mm) plywood or particleboard sheathing with 8d nails at 4 inches (102 mm) all edges and 12 inches (305 mm) intermediate.
   The application of these sheathing materials shall comply with Section 2513.5 and Table 25-I for Types A, B and C, and Section 2315.1 and Table 23-II-I-1 or 23-II-I-2 for Types D and E. All panel edges of Types D and E shall be backed with 2-inch (51 mm) nominal or wider framing.

[2]Level refers to the space between the upper surface of any floor and upper surface of floor next above. The topmost level shall be the space between upper surface of the topmost floor and the ceiling or roof above. Wall sheathing at useable or unused under-floor space shall be provided as required for the level directly above.

TABLE A-23-B—ROOF AND FLOOR ANCHORAGE AT EXTERIOR WALLS

BASIC WIND SPEED (mph) × 1.61 for km/h	LOCATION[1]	NUMBER OF NAILS[2] Exposure B	C	D
80	roof to wall	6-8d	8-8d	8-10d
	floor to floor	—	4-10d	6-10d
	floor to foundation	—	4-10d	4-10d
90	roof to wall	8-8d	8-10d	10-10d
	floor to floor	—	6-10d	8-10d
	floor to foundation	—	4-10d	6-10d
100	roof to wall	8-10d	10-10d	12-10d
	floor to floor	6-10d	8-10d	10-10d
	floor to foundation	4-10d	6-10d	8-10d
110	roof to wall	10-10d	12-10d	12-10d
	floor to floor	8-10d	10-10d	10-10d
	floor to foundation	6-10d	8-10d	8-10d

[1]For floor-to-foundation anchorage, see Section 2337.5.4.

[2]Number of common nails listed is total required for each tie strap. The tie straps shall be spaced at 48 inches (1219 mm) on center along the length of the wall. The number of nails on each side of the roof or floor plate joints shall be equal. Nails shall be spaced to avoid splitting of the wood. See Figure A-23-1 for illustration of these tie straps.

TABLE A-23-C—RIDGE TIE-STRAP NAILING[1]

BASIC WIND SPEED (mph)	NUMBER OF NAILS[1]		
	Exposure		
× 1.61 for km/h	B	C	D
80	6-10d	8-10d	10-10d
90	8-10d	10-10d	12-10d
100	10-10d	12-10d	14-10d
110	12-10d	14-10d	16-10d

[1]Number of common nails listed is total required for each tie strap. The tie straps shall be spaced at 48 inches (1219 mm) on center along the length of the roof. The number of nails on each side of the rafter/ridge joint shall be equal. Nails shall be spaced to avoid splitting of the wood. See Figure A-23-1 for illustration of these tie straps.

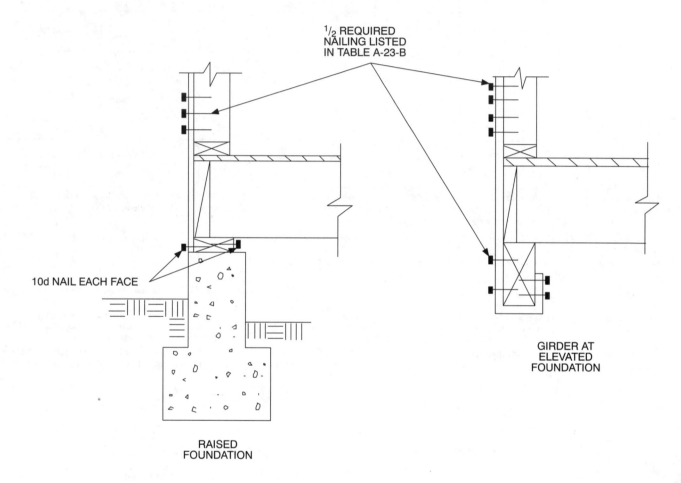

FIGURE A-23-1—COMPLETE LOAD PATH DETAILS

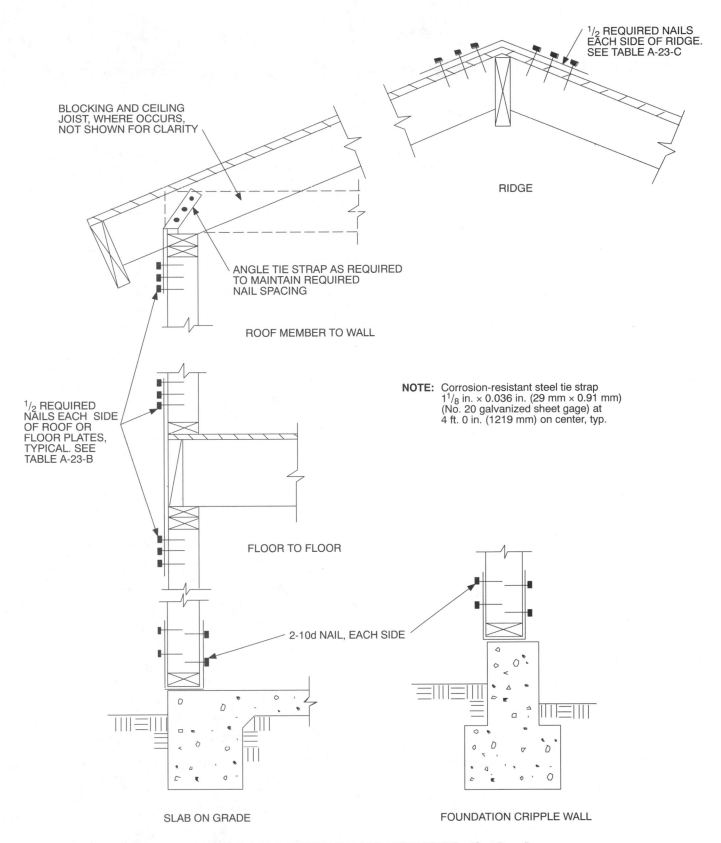

BLOCKING AND CEILING JOIST, WHERE OCCURS, NOT SHOWN FOR CLARITY

ANGLE TIE STRAP AS REQUIRED TO MAINTAIN REQUIRED NAIL SPACING

ROOF MEMBER TO WALL

$^1/_2$ REQUIRED NAILS EACH SIDE OF RIDGE. SEE TABLE A-23-C

RIDGE

NOTE: Corrosion-resistant steel tie strap $1^1/_8$ in. × 0.036 in. (29 mm × 0.91 mm) (No. 20 galvanized sheet gage) at 4 ft. 0 in. (1219 mm) on center, typ.

$^1/_2$ REQUIRED NAILS EACH SIDE OF ROOF OR FLOOR PLATES, TYPICAL. SEE TABLE A-23-B

FLOOR TO FLOOR

2-10d NAIL, EACH SIDE

SLAB ON GRADE

FOUNDATION CRIPPLE WALL

FIGURE A-23-1—COMPLETE LOAD PATH DETAILS—(Continued)

Appendix Chapter 29
MINIMUM PLUMBING FIXTURES

SECTION 2905 — GENERAL

Each building shall be provided with sanitary facilities, including provisions for accessibility in accordance with Chapter 11. Plumbing fixtures shall be provided for the type of building occupancy with the minimum numbers as shown in Table A-29-A. The number of fixtures are the minimum required as shown in Table A-29-A and are assumed to be based on 50 percent male and 50 percent female. The occupant load factors shall be as shown in Table A-29-A.

EXCEPTION: Where circumstances dictate that a different ratio is needed, the adjustment shall be approved by the building official.

TABLE A-29-A—MINIMUM PLUMBING FIXTURES[1,2,3]

TYPE OF BUILDING OR OCCUPANCY[4]	WATER CLOSETS[5] (fixtures per person) MALE	FEMALE	LAVATORIES[6] (fixtures per person) MALE	FEMALE	BATHTUB OR SHOWER (fixtures per person)
For the occupancies listed below, use 30 square feet (2.78 m²) per occupant for the minimum number of plumbing fixtures.					
Group A Conference rooms, dining rooms, drinking establishments, exhibit rooms, gymnasiums, lounges, stages and similar uses including restaurants classified as Group B Occupancies	1:1-25 2:26-75 3:76-125 4:126-200 5:201-300 6:301-400 Over 400, add one fixture for each additional 200 males or 150 females.	1:1-25 2:26-75 3:76-125 4:126-200 5:201-300 6:301-400	one for each water closet up to four; then one for each two additional water closets		
For the assembly occupancies listed below, use the number of fixed seating or, where no fixed seating is provided, use 15 square feet (1.39 m²) per occupant for the minimum number of plumbing fixtures.					
Assembly places— Auditoriums, convention halls, dance floors, lodge rooms, stadiums and casinos	1:1-50 2:51-100 3:101-150 4:151-300 Over 300 males, add one fixture for each additional 200, and over 400 females add one for each 125.	3:1-50 4:51-100 6:101-200 8:201-400	1:1-200 2:201-400 3:401-750 Over 750, add one fixture for each additional 500 persons.	1:1-200 2:201-400 3:401-750	
For the assembly occupancies listed below, use the number of fixed seating or, where no fixed seating is provided, use 30 square feet (2.29 m²) per occupant for the minimum number of plumbing fixtures.					
Worship places Principal assembly area	one per 150	one per 75	one per two water closets		
Worship places Educational and activity unit	one per 125	one per 75	one per two water closets		
For the occupancies listed below, use 200 square feet (18.58 m²) per occupant for the minimum number of plumbing fixtures.					
Group B Offices or public buildings	1:1-15 2:16-35 3:36-55 Over 55, add one for each 50 persons.	1:1-15 2:16-35 3:36-55	one per two water closets		
For the occupancies listed below, use 50 square feet (4.65 m²) per occupant for the minimum number of plumbing fixtures.					
Group E Schools—for staff use All schools	1:1-15 2:16-35 3:36-55 Over 55, add one fixture for each additional 40 persons.	1:1-15 2:16-35 3:36-55	one per 40	one per 40	
Schools—for student use Day care	1:1-20 2:21-50 Over 50, add one fixture for each additional 50 persons.	1:1-20 2:21-50	1:1-25 2:26-50 Over 50, add one fixture for each additional 50 persons.	1:1-25 2:26-50	
Elementary	one per 30	one per 25	one per 35	one per 35	
Secondary	one per 40	one per 30	one per 40	one per 40	
For the occupancies listed below, use 50 square feet (4.65 m²) per occupant for the minimum number of plumbing fixtures.					
Education Facilities other than Group E Others (colleges, universities, adult centers, etc.)	one per 40	one per 30	one per 40	one per 40	

(Continued)

TABLE A-29-A—MINIMUM PLUMBING FIXTURES[1,2,3]—(Continued)

TYPE OF BUILDING OR OCCUPANCY[4]	WATER CLOSETS[5] (fixtures per person)		LAVATORIES[6] (fixtures per person)		BATHTUB OR SHOWER (fixtures per person)
	MALE	FEMALE	MALE	FEMALE	
For the occupancies listed below, use 2,000 square feet (185.8 m²) per occupant for the minimum number of plumbing fixtures.					
Group F Workshop, foundries and similar establishments, and Group H Occupancies	1:1-10 2:11-25 3:26-50 4:51-75 5:76-100 Over 100, add one fixture for each additional 300 persons.	1:1-10 2:11-25 3:26-50 4:51-75 5:76-100	one for each two water closets		one shower for each 15 persons exposed to excessive heat or to skin contamination with irritating materials
For the occupancies listed below, use the designated application and 200 square feet (18.58 m²) per occupant of the general use area for the minimum number of plumbing fixtures.					
Group I Hospital waiting rooms	one per room (usable by either sex)		one per room		
Hospital general use areas	1:1-15 2:16-35 3:36-55 Over 55, add one fixture for each additional 40 persons.	1:1-15 3:16-35 4:36-55	one per each two water closets		
Hospitals Patient room Ward room	one per room one per eight patients		one per room one per 10 patients		one per room one per 20 patients
Jails and reformatories Cell Exercise room	one per cell one per exercise room		one per cell one per exercise room		
Other institutions (on each occupied floor)	one per 25	one per 25	one per 10	one per 10	one per eight
For the occupancies listed below, use 200 square feet (18.58 m²) per occupant for the minimum number of plumbing fixtures.					
Group M Retail or wholesale stores	1:1-50 2:51-100 3:101-400 Over 400, add one fixture for each additional 500 males and one for each 150 females.	1:1-50 2:51-100 3:101-200 4:201-300 5:301-400	one for each two water closets		
For Group R Occupancies, dwelling units and hotel guest rooms, use the chart. For congregate residences, use 200 square feet (18.58 m²) for Group R, Division 1 Occupancies and 300 square feet (27.87 m²) for Group R, Division 3 Occupancies for the minimum plumbing fixtures.					
Group R Dwelling units Hotel guest rooms	one per dwelling unit one per guest room		one per dwelling unit one per guest room		one per dwelling unit one per guest room
Congregate residences	one per 10 Add one fixture for each additional 25 males and one for each additional 20 females.	one per 8	one per 12 Over 12, add one fixture for each additional 20 males and one for each additional 15 females	one per 12	one per eight For females, add one bathtub per 30. Over 150, add one per 20.
For the occupancies listed below, use 5,000 square feet (464.5 m²) per occupant for the minimum number of plumbing fixtures.					
Group S Warehouses	1:1-10 2:11-25 3:26-50 4:51-75 5:76-100 Over 100, add one for each 300 males and females.	1:1-10 2:11-25 3:26-50 4:51-75 5:76-100	one per 40 occupants of each sex		one shower for each 15 persons exposed to excessive heat or to skin contamination with poisonous, infectious or irritating materials

NOTE: Occupant loads over 30 shall have one drinking fountain for each 150 occupants.

[1]The figures shown are based on one fixture being the minimum required for the number of persons indicated or any fraction thereof.

[2]Drinking fountains shall not be installed in toilet rooms.

[3]When the design occupant load is less than 10 persons, a facility usable by either sex may be approved by the building official.

[4]Any category not mentioned specifically or about which there are any questions shall be classified by the building official and included in the category which it most nearly resembles, based on the expected use of the plumbing facilities.

[5]Where urinals are provided, one water closet less than the number specified may be provided for each urinal installed, except the number of water closets in such cases shall not be reduced to less than one half of the minimum specified.

[6]Twenty-four inches (610 mm) of wash sink or 18 inches (457 mm) of a circular basin, when provided with water outlets for such space, shall be considered equivalent to one lavatory.

Appendix Chapter 30
ELEVATORS, DUMBWAITERS, ESCALATORS AND MOVING WALKS

SECTION 3008 — PURPOSE

The purpose of this appendix is to safeguard life, limb, property and public welfare by establishing minimum requirements regulating the design, construction, alteration, operation and maintenance of elevators, dumbwaiters, escalators and moving walks and by establishing procedures by which these requirements may be enforced.

SECTION 3009 — SCOPE

This appendix shall apply to new and existing installations of elevators, dumbwaiters, escalators and moving walks, requiring permits therefore and providing for the inspection and maintenance of such conveyances.

SECTION 3010 — DEFINITIONS

For purposes of this appendix, certain terms are defined as follows:

ANSI CODE is the ASME/ANSI A17.1-1987 with Supplements A17.1a-1988 and A17.1b-1989, Safety Code for Elevators and Escalators, an American National Standard published by the American Society of Mechanical Engineers.

SECTION 3011 — PERMITS—CERTIFICATES OF INSPECTION

3011.1 Permits Required. It shall be unlawful to hereafter install any new elevator, moving walk, escalator or dumbwaiter, or to make major alterations to any existing elevator, dumbwaiter, escalator or moving walk as defined in Part XII of the ANSI code, without having first obtained a permit for such installation from the building official. Permits shall not be required for maintenance or minor alterations.

3011.2 Certificates of Inspection Required. It shall be unlawful to operate any elevator, dumbwaiter, escalator or moving walk without a current certificate of inspection issued by the building official. Such certificate shall be issued upon payment of prescribed fees and the presentation of a valid inspection report indicating that the conveyance is safe and that the inspections and tests have been performed in accordance with Part X of the ANSI code. Certificates shall not be issued when the conveyance is posted as unsafe pursuant to Section 3015.

> **EXCEPTION:** Certificates of inspection shall not be required for conveyances within a dwelling unit.

3011.3 Application for Permits. Application for a permit to install shall be made on forms provided by the building official, and the permit shall be issued to an owner upon payment of the permit fees specified in this section.

3011.4 Application for Certificates of Inspection. Application for a certificate of inspection shall be made by the owner of an elevator, dumbwaiter, escalator or moving walk. Applications shall be accompanied by an inspection report as described in Section 3014. Fees for certificates of inspection shall be as specified in this section.

3011.5 Fees. A fee for each permit or certificate of inspection shall be paid to the building official as follows:

New Installations:

Passenger or freight elevator, escalator, moving walk:

Up to and including $40,000 of valuation—$89.00

Over $40,000 of valuation—$89.00 plus $1.65 for each $1,000 or fraction thereof over $40,000

Dumbwaiter or Private Residence Elevator:

Up to and including $10,000 of valuation—$25.00

Over $10,000 of valuation—$25.00 plus $1.65 for each $1,000 or fraction thereof over $10,000

Major Alterations:

Fees for major alterations shall be as set forth in Table 1-A.

Installation fees include charges for the first year's annual inspection fee and charges for electrical equipment on the conveyance side of the disconnect switch.

Annual certificates of inspection:

For each elevator	$41.50
For each escalator or moving walk	$24.65
For each commercial dumbwaiter	$16.75

(Each escalator or moving walk unit powered by one motor shall be considered as a separate escalator or moving walk.)

SECTION 3012 — ANSI CODE ADOPTED

New elevators, dumbwaiters, escalators and moving walks and major alterations to such conveyances and the installation thereof shall conform to the requirements of the American National Standards Institute ASME/ANSI A17.1-1987, Safety Code for Elevators and Escalators, including Supplements A17.1a-1988 and A17.1b-1989, published by the American Society of Mechanical Engineers. Existing elevators and escalators shall conform with ASME/ANSI A17.3-1986, Safety Code for Existing Elevators and Escalators, including Supplement A17.3a-1989, published by the American Society of Mechanical Engineers.

SECTION 3013 — DESIGN

For detailed design, construction and installation requirements, see Chapter 16 and the appropriate requirements of the ANSI code.

In Seismic Zones 3 and 4, elevators shall conform to Appendix F of the ANSI code.

SECTION 3014 — REQUIREMENTS FOR OPERATION AND MAINTENANCE

3014.1 General. The owner shall be responsible for the safe operation and maintenance of each elevator, dumbwaiter, escalator or moving walk installation and shall cause periodic inspections, tests and maintenance to be made on such conveyances as required in this section.

3014.2 Periodic Inspections and Tests. Routine and periodic inspections and tests shall be made as required by Part X of the ANSI code.

3014.3 Alterations, Repairs and Maintenance. Alterations, repairs and maintenance shall be made as required by Part XII of the ANSI code.

3014.4 Inspection Costs. All costs of such inspections and tests shall be paid by the owner.

3014.5 Inspection Reports. After each required inspection, a full and correct report of such inspection shall be filed with the building official.

SECTION 3015 — UNSAFE CONDITIONS

When an inspection reveals an unsafe condition, the inspector shall immediately file with the owner and the building official a full and true report of such inspection and such unsafe condition. If the building official finds that the unsafe condition endangers human life, the building official shall cause to be placed on such elevator, escalator or moving walk, in a conspicuous place, a notice stating that such conveyance is unsafe. The owner shall see to it that such notice of unsafe condition is legibly maintained where placed by the building official. The building official shall also issue an order in writing to the owner requiring the repairs or alterations to be made to such conveyance that are necessary to render it safe and may order the operation thereof discontinued until the repairs or alterations are made or the unsafe conditions are removed. A posted notice of unsafe conditions shall be removed only by the building official when satisfied that the unsafe conditions have been corrected.

Appendix Chapter 31
SPECIAL CONSTRUCTION
Division I—FLOOD-RESISTANT CONSTRUCTION

SECTION 3104 — GENERAL

3104.1 Purpose. The provisions of this division are intended to promote public safety and welfare by reducing the risk of flood damage in areas prone to flooding.

3104.2 Scope. Buildings and structures erected in areas prone to flooding shall be constructed as required by the provisions of this division. The base flood elevation shown on the approved flood hazard map is the minimum elevation used to define areas prone to flooding, unless records indicate a higher elevation is to be used. The flood-prone areas are defined in the jurisdiction's floodplain management ordinance.

3104.3 Definitions. For the purpose of this division, certain terms are defined as follows:

BASE FLOOD ELEVATION is the depth or peak elevation of flooding, including wave height, having 1 percent chance of being equaled or exceeded in any given year.

FLOOD HAZARD MAP is a map published by an approved agency that defines the flood boundaries, elevations and insurance risk zones as determined by a detailed flood insurance study.

HAZARD ZONES are areas that have been determined to be prone to flooding and are classified as either flood hazard zones, A zones, or coastal high-hazard zones, V zones, in accordance with Sections 3107.1 and 3108.1.

SECTION 3105 — MANUFACTURED STRUCTURES

New or replacement manufactured structures located in any flood hazard zone shall be located in accordance with the applicable elevation requirements of Sections 3107.2 and 3108.2, and the anchor and tie-down requirements of Section 3110.1.

SECTION 3106 — PROTECTION OF MECHANICAL AND ELECTRICAL SYSTEMS

New or replacement electrical equipment and heating, ventilating, air conditioning and other service facilities shall be either placed above the base flood elevation or protected to prevent water from entering or accumulating within the system components during floods up to the base flood elevation. Installation of electrical wiring and outlets, switches, junction boxes and panels below the base flood elevation shall conform to the provisions of the Electrical Code for such items in wet locations.

SECTION 3107 — FLOOD HAZARD ZONES—A ZONES

3107.1 General. Areas that have been determined as prone to flooding but not subject to wave heights of more than 3 feet (914 mm) are designated as flood hazard zones. Buildings or structures erected in flood hazard zones shall be designed and constructed in accordance with this section.

3107.2 Elevation. Buildings or structures erected within a flood hazard zone shall have the lowest floor, including basement floors, located at or above the base flood elevation.

> **EXCEPTIONS:** 1. Except for Group R Occupancies, any occupancy may have floors below the base flood elevation in accordance with Section 3107.4.

> 2. Floors of buildings or structures that are used only for building access, means of egress, foyers, storage and parking garages may be below the base flood elevation in accordance with Section 3107.3.

3107.3 Enclosures below Base Flood Elevation. Enclosed spaces below the base flood elevation shall not be used with the exception of building access, means of egress, foyers, storage and parking garages. Enclosed spaces shall be provided with vents, valves or other openings that will automatically equalize the lateral pressure of waters acting on the exterior wall surfaces. The bottom of the openings shall not be higher than 12 inches (305 mm) above finish grade. A minimum of two openings per building, or one opening for each enclosure below the base flood elevation, whichever is greater, shall be provided. The total net area of such openings shall not be less than 4 square feet (0.37 m²) or 1 square inch for every square foot (0.007 m² for every 1 m²) of enclosed area, whichever is greater.

3107.4 Flood-resistant Construction. Buildings or structures of any occupancy other than Group R may, in lieu of meeting the elevation provisions of Section 3107.2, be erected with floors usable for human occupancy below the base flood elevation, provided the following conditions are met:

1. Space below the base flood elevation shall be constructed with exterior walls and floors that are impermeable to the passage of water.

2. Structural components subject to hydrostatic and hydrodynamic loads during the occurrence of flooding to the base flood elevation shall be capable of resisting such forces, including the effect of buoyancy.

3. Openings below the base flood elevation shall be provided with watertight closures and shall have adequate structural capacity to support flood loads acting upon closure surfaces.

4. Floor and wall penetrations for plumbing, mechanical and electrical systems shall be made watertight to prevent flood water seepage through spaces between penetration and wall construction materials. Sanitary sewer and storm drainage systems that have openings below the base flood elevation shall be provided with closure devices to prevent backwater flow during conditions of flooding.

3107.5 Plan Requirements for Flood-resistant Construction. When buildings or structures are to be constructed in accordance with Section 3107.4, an architect or engineer licensed by the state to practice as such shall prepare plans showing details of the floor wall and foundation support components. Calculations and approved technical data used to comply with the conditions of Section 3107.4 shall also be provided.

SECTION 3108 — COASTAL HIGH HAZARD ZONES—V ZONES

3108.1 General. Areas that have been determined to be subject to wave heights in excess of 3 feet (914 mm) or subject to high-velocity wave run-up or wave-induced erosion are designated as coastal high-hazard zones. Buildings or structures erected in coastal high-hazard zones shall be designed and constructed in accordance with this section.

3108.2 Elevation. Buildings or structures erected within a coastal high-hazard zone shall be elevated so that the lowest portion of horizontal structural members, with the exception of footings, mat or raft foundations, piles, pile caps, columns, grade beams, and bracing, shall be located at or above the base flood elevation.

3108.3 Enclosures below Base Flood Elevation. Spaces below the base flood elevation in a coastal high-hazard zone shall be free of obstruction.

> **EXCEPTIONS:** 1. Footings, mat or raft foundations, piles, pile caps, columns, grade beams, and bracing that provide structural stability for the building.
>
> 2. Structural systems of entrances and required exits.
>
> 3. Storage of portable or mobile items that can be moved in the event of a storm.
>
> 4. Walls or partitions may be used to enclose all or part of the space, provided they are not part of the structural support of the building and are designed to break away under high tides or wave action without causing damage to the structural system of the building (see Section 3110.6.) Screening, lattice-type arrangements or other materials that allow the passage of water may also be used.

3108.4 Foundations. Buildings or structures erected in coastal high-hazard zones shall be supported on piles or columns. When piles are used, they shall have soil penetration to resist the combined wave and wind loads to which they may be subject during a flood equal to the base flood elevation. Pile design shall include consideration of decreased resistance capacity caused by scour of the soil strata surrounding the piles. Pile system design and installation shall be made in accordance with the provisions of this code. When mat or raft foundations are used, they shall be located at a depth to provide protection from erosion or scour.

3108.5 Plan Requirements for Coastal High-hazard Construction. When buildings or structures are to be constructed in accordance with Section 3108, an architect or engineer licensed by the state to practice as such shall submit plans showing details of the foundation support and connection components to comply with the requirements of Section 3108.4. When solid walls or partitions are proposed below the base flood elevation, wall, framing and connection details of such walls in accordance with Section 3108.3 shall be provided.

SECTION 3109 — ELEVATION CERTIFICATION

A land surveyor, architect or engineer licensed by the state to practice as such shall certify that the actual elevation in relation to mean sea level of the lowest floor, if in a flood hazard zone, or the bottom of the lowest horizontal structural member, if in a coastal high-hazard zone, are at or above the minimum elevation when required by the provisions of Sections 3107.2 and 3108.2.

SECTION 3110 — DESIGN REQUIREMENTS

3110.1 Structural Systems. Structural systems of buildings or structures shall be constructed, connected and anchored to resist flotation, collapse or permanent lateral movement due to loads from flooding equal to the base flood elevation.

3110.2 Design Loads. The structural system shall be designed in accordance with well-established engineering principles and with consideration of hydrodynamic and hydrostatic loads. The required loading shall be established by site-specific criteria or approved national standards. Impact loads shall be considered in the analysis of the structural system.

3110.3 Load Combinations. Loading combinations shall be subject to approval by the building official. The structural system shall be designed to resist each combination of loading acting simultaneously. In lieu of site-specific loading requirements, load combinations from an approved national standard may be used.

3110.4 Stress Increases. Allowable stresses may be increased one third for flood loads in combination with dead load or dead and live load combinations. When strength design is used, flood loads may be considered as dead loads when considering dead and live load conditions. Flood loads may be considered as wind loads in other load combinations.

3110.5 Overturning. Buildings and structures and parts or elements shall be designed to resist sliding or overturning by at least 1.5 times the lateral force or overturning moment caused by wind and flood loads acting simultaneously. For the purpose of providing stability, only the dead load shall be considered effective in resisting overturning.

3110.6 Breakaway Walls. When walls or partitions located below the base flood elevation are required to break away in accordance with Section 3108.3, such walls shall be designed for not less than 10 pounds per square foot (psf) (0.48 kN/m^2) or more than 20 psf (0.96 kN/m^2) on the vertical projected area.

Division II—MEMBRANE STRUCTURES

SECTION 3111 — GENERAL

3111.1 Purpose. The purpose of this appendix is to establish minimum standards of safety for the construction and use of air-supported, air-inflated and membrane-covered cable or frame structures, collectively known as membrane structures.

3111.2 Scope. The provisions of this appendix shall apply to membrane structures erected for a period of 180 days or longer. Those erected for a shorter period of time shall comply with applicable provisions of the Fire Code.

> **EXCEPTION:** Water storage facilities, water clarifiers, water treatment plants, sewer plants, aquaculture pond covers, residential and agricultural greenhouses, and similar facilities not used for human occupancy need meet only the requirements of Section 3112.2 and Section 3115.

3111.3 Definitions. For the purpose of this appendix, certain terms are defined as follows:

AIR-INFLATED STRUCTURE is a building where the shape of the structure is maintained by air pressurization of cells or tubes to form a barrel vault over the usable area. Occupants of such a structure do not occupy the pressurized area used to support the structure.

AIR-SUPPORTED STRUCTURE is a building wherein the shape of the structure is attained by air pressure and occupants of the structure are within the elevated pressure area. Air-supported structures are of two basic types:

1. **Single skin**—Where there is only the single outer skin and the air pressure is directly against that skin.

2. **Double skin**—Similar to a single skin, but with an attached liner that is separated from the outer skin and provides an air space that serves for insulation, acoustic, aesthetic or similar purposes.

A cable-restrained air-supported structure is one in which the uplift is resisted by cables or webbing that are anchored to either foundations or deadmen. Reinforcing cable or webbing may be attached by various methods to the membrane or may be an integral part of the membrane. This is not a cable-supported structure.

CABLE STRUCTURE is a nonpressurized structure in which a mast and cable system provide support and tension to the membrane weather barrier and the membrane imparts structural stability to the structure.

FRAME-COVERED STRUCTURE is a nonpressurized building wherein the structure is composed of a rigid framework to support tensioned membrane that provides the weather barrier.

MEMBRANE is a thin, flexible, impervious material capable of being supported by an air pressure of 1.5 inches of water column (373 Pa).

NONCOMBUSTIBLE MEMBRANE STRUCTURE is a membrane structure in which the membrane and all component parts of the structure are noncombustible as defined by Section 215.

TENT is any structure, enclosure or shelter constructed of canvas or pliable material supported by any manner except by air or the contents it protects.

SECTION 3112 — TYPE OF CONSTRUCTION AND GENERAL REQUIREMENTS

3112.1 General. Membrane structures shall be classified as Type V-N construction, except that noncombustible membrane structures may be classified as Type II-N construction.

> **EXCEPTION:** A noncombustible membrane structure used exclusively as a roof and located more than 25 feet (7620 mm) above any floor, balcony or gallery is deemed to comply with the roof construction requirements for Type I and Type II fire-resistant construction, provided that such a structure complies with the requirements of this section.

3112.2 Membrane Material. Membranes shall be either noncombustible as defined by Section 215, or flame retardant conforming to UBC Standard 31-1, which is a part of this code (see Chapter 35).

> **EXCEPTION:** Plastic less than 20-mil (0.51 mm) thickness used in greenhouses and for aquaculture pond covers need not be flame retardant.

3112.3 Applicability of Other Provisions. Except as otherwise specifically required by this section, membrane structures shall meet all applicable provisions of this code. Roof coverings shall be fire retardant.

> **EXCEPTION:** Roof coverings for Group U, Division 1 Occupancies not exceeding 1,000 square feet (93 m^2) in area need not be fire retardant.

3112.4 Allowable Floor Areas. The area of a membrane structure shall not exceed the limits set forth in Table 5-B, except as provided in Section 505.

3112.5 Maximum Height. Membrane structures shall not exceed one story and shall not exceed the height limits in feet (mm) set forth in Table 5-B.

> **EXCEPTION:** Noncombustible membrane structures serving as roofs only.

SECTION 3113 — INFLATION SYSTEMS

3113.1 General. Air-supported and air-inflated structures shall be provided with primary and auxiliary inflation systems to meet the minimum requirements of this section.

3113.2 Equipment Requirements. The inflation system shall consist of one or more blowers and shall include provisions for automatic control to maintain the required inflation pressures. The system shall be so designed as to prevent overpressurization of the system.

In addition to the primary inflation system, in buildings exceeding 1,500 square feet (139.4 m^2) in area, there shall be provided an auxiliary inflation system with sufficient capacity to maintain the inflation of the structure in case of primary system failure.

The auxiliary inflation system shall operate automatically if there is a loss of internal pressure or should the primary blower system become inoperative.

Blower equipment shall meet the following requirements:

1. Blowers shall be powered by continuous rated motors at the maximum power required for any flow condition as required by the structural design.

2. Blowers shall be provided with inlet screens, belt guards and other protective devices as may be required by the building official to provide protection from injury.

3. Blowers shall be housed within a weather-protecting structure.

4. Blowers shall be equipped with back draft check dampers to minimize air loss when inoperative.

5. Blower inlets shall be located to provide protection from air contamination. Location of inlets shall be approved by the building official.

3113.3 Emergency Power. Whenever an auxiliary inflation system is required, an approved standby power-generating system shall be provided. The system shall be equipped with a suitable means for automatically starting the generator set upon failure of the normal electrical service and for automatic transfer and operation of all the required electrical functions at full power within 60 seconds of such normal service failure. Standby power shall be capable of operating independently for a minimum of four hours.

SECTION 3114 — SECTION PROVISIONS

A system capable of supporting the membrane in the event of deflation shall be provided in all air-supported and air-inflated structures having an occupant load of more than 50 or when covering a swimming pool regardless of occupant load. Such system shall maintain the membrane at least 7 feet (2134 mm) above the floor, seating area or surface of the water.

> **EXCEPTION:** Membrane structures used as a roof for Type I or Type II fire-resistant construction must be maintained not less than 25 feet (7620 mm) above floor or seating areas.

SECTION 3115 — ENGINEERING DESIGN

All membrane structures shall be structurally designed in accordance with criteria approved by the building official and developed by an engineer or architect licensed by the state to practice as such.

Division III—PATIO COVERS

SECTION 3116 — PATIO COVERS DEFINED

Patio covers are one-story structures not exceeding 12 feet (3657 mm) in height. Enclosure walls may have any configuration, provided the open area of the longer wall and one additional wall is equal to at least 65 percent of the area below a minimum of 6 feet 8 inches (2032 mm) of each wall, measured from the floor. Openings may be enclosed with insect screening or plastic that is readily removable translucent or transparent plastic not more than 0.125 inch (3.2 mm) in thickness.

Patio covers may be detached or attached to other buildings as accessories to Group U; Group R, Division 3; or single dwelling units in Group R, Division 1 Occupancies. Patio covers shall be used only for recreational, outdoor living purposes and not as carports, garages, storage rooms or habitable rooms.

SECTION 3117 — DESIGN LOADS

Patio covers shall be designed and constructed to sustain the loads required by Chapter 16, combined in accordance with Section 1612.2 for load and resistance factor design or Section 1612.3 for allowable stress design, except that the live load, L, shall not be taken as less than 10 pounds per square foot (0.48 kN/m^2) and,

where less than 12 feet (3658 mm) high, the horizontal wind load shall be as indicated in Table A-31-A. In addition, they shall be designed to support a minimum wind uplift equal to the horizontal wind load acting vertical upward normal to the roof surface, except that for structures not more than 10 feet (3048 mm) above grade the uplift may be three fourths of the horizontal wind load. When enclosed with insect screening or plastic that is readily removable translucent or transparent plastic not more than 0.125 inch (3.2 mm) in thickness, wind loads shall be applied to the structure, assuming it is fully enclosed.

SECTION 3118 — LIGHT AND VENTILATION

Exterior openings required for light and ventilation may open into a patio structure conforming to Section 3116.

SECTION 3119 — FOOTINGS

A patio cover may be supported on a concrete slab on grade without footings, provided the slab is not less than $3^{1}/_{2}$ inches (89 mm) thick and further provided that the columns do not support live and dead loads in excess of 750 pounds (3.34 kN) per column.

TABLE A-31-A—DESIGN WIND PRESSURES FOR PATIO COVERS[1]

HEIGHT ZONE IN FEET	WIND SPEED—MAP AREAS (miles per hour)						
	× 1.61 for km/h						
	70	80	90	100	110	120	130
× 304.8 for mm	× 0.048 for kN/m^2 (psf)						
Less than 12	10	13	15	19	23	27	32

[1]See Chapter 16, Figure 16-1, for basic wind speeds.

Appendix Chapter 33
EXCAVATION AND GRADING

SECTION 3304 — PURPOSE

The purpose of this appendix is to safeguard life, limb, property and the public welfare by regulating grading on private property.

SECTION 3305 — SCOPE

This appendix sets forth rules and regulations to control excavation, grading and earthwork construction, including fills and embankments; establishes the administrative procedure for issuance of permits; and provides for approval of plans and inspection of grading construction.

The standards listed below are recognized standards (see Sections 3503 and 3504).

1. **Testing.**

   1.1 ASTM D 1557, Moisture-density Relations of Soils and Soil Aggregate Mixtures

   1.2 ASTM D 1556, In Place Density of Soils by the Sand-Cone Method

   1.3 ASTM D 2167, In Place Density of Soils by the Rubber-Balloon Method

   1.4 ASTM D 2937, In Place Density of Soils by the Drive-Cylinder Method

   1.5 ASTM D 2922 and D 3017, In Place Moisture Contact and Density of Soils by Nuclear Methods

SECTION 3306 — PERMITS REQUIRED

3306.1 Permits Required. Except as specified in Section 3306.2 of this section, no person shall do any grading without first having obtained a grading permit from the building official.

3306.2 Exempted Work. A grading permit is not required for the following:

1. When approved by the building official, grading in an isolated, self-contained area if there is no danger to private or public property.

2. An excavation below finished grade for basements and footings of a building, retaining wall or other structure authorized by a valid building permit. This shall not exempt any fill made with the material from such excavation or exempt any excavation having an unsupported height greater than 5 feet (1524 mm) after the completion of such structure.

3. Cemetery graves.

4. Refuse disposal sites controlled by other regulations.

5. Excavations for wells or tunnels or utilities.

6. Mining, quarrying, excavating, processing or stockpiling of rock, sand, gravel, aggregate or clay where established and provided for by law, provided such operations do not affect the lateral support or increase the stresses in or pressure upon any adjacent or contiguous property.

7. Exploratory excavations under the direction of soil engineers or engineering geologists.

8. An excavation that (1) is less than 2 feet (610 mm) in depth or (2) does not create a cut slope greater than 5 feet (1524 mm) in height and steeper than 1 unit vertical in $1^1/_2$ units horizontal (66.7% slope).

9. A fill less than 1 foot (305 mm) in depth and placed on natural terrain with a slope flatter than 1 unit vertical in 5 units horizontal (20% slope), or less than 3 feet (914 mm) in depth, not intended to support structures, that does not exceed 50 cubic yards (38.3 m^3) on any one lot and does not obstruct a drainage course.

Exemption from the permit requirements of this chapter shall not be deemed to grant authorization for any work to be done in any manner in violation of the provisions of this chapter or any other laws or ordinances of this jurisdiction.

SECTION 3307 — HAZARDS

Whenever the building official determines that any existing excavation or embankment or fill on private property has become a hazard to life and limb, or endangers property, or adversely affects the safety, use or stability of a public way or drainage channel, the owner of the property upon which the excavation or fill is located, or other person or agent in control of said property, upon receipt of notice in writing from the building official, shall within the period specified therein repair or eliminate such excavation or embankment to eliminate the hazard and to be in conformance with the requirements of this code.

SECTION 3308 — DEFINITIONS

For the purposes of this appendix, the definitions listed hereunder shall be construed as specified in this section.

APPROVAL shall mean that the proposed work or completed work conforms to this chapter in the opinion of the building official.

AS-GRADED is the extent of surface conditions on completion of grading.

BEDROCK is in-place solid rock.

BENCH is a relatively level step excavated into earth material on which fill is to be placed.

BORROW is earth material acquired from an off-site location for use in grading on a site.

CIVIL ENGINEER is a professional engineer registered in the state to practice in the field of civil works.

CIVIL ENGINEERING is the application of the knowledge of the forces of nature, principles of mechanics and the properties of materials to the evaluation, design and construction of civil works.

COMPACTION is the densification of a fill by mechanical means.

EARTH MATERIAL is any rock, natural soil or fill or any combination thereof.

ENGINEERING GEOLOGIST is a geologist experienced and knowledgeable in engineering geology.

ENGINEERING GEOLOGY is the application of geologic knowledge and principles in the investigation and evaluation of naturally occurring rock and soil for use in the design of civil works.

EROSION is the wearing away of the ground surface as a result of the movement of wind, water or ice.

EXCAVATION is the mechanical removal of earth material.

FILL is a deposit of earth material placed by artificial means.

GEOTECHNICAL ENGINEER. See "soils engineer."

GRADE is the vertical location of the ground surface.

Existing Grade is the grade prior to grading.

Finish Grade is the final grade of the site that conforms to the approved plan.

Rough Grade is the stage at which the grade approximately conforms to the approved plan.

GRADING is any excavating or filling or combination thereof.

KEY is a designed compacted fill placed in a trench excavated in earth material beneath the toe of a proposed fill slope.

PROFESSIONAL INSPECTION is the inspection required by this code to be performed by the civil engineer, soils engineer or engineering geologist. Such inspections include that performed by persons supervised by such engineers or geologists and shall be sufficient to form an opinion relating to the conduct of the work.

SITE is any lot or parcel of land or contiguous combination thereof, under the same ownership, where grading is performed or permitted.

SLOPE is an inclined ground surface the inclination of which is expressed as a ratio of horizontal distance to vertical distance.

SOIL is naturally occurring superficial deposits overlying bedrock.

SOILS ENGINEER (GEOTECHNICAL ENGINEER) is an engineer experienced and knowledgeable in the practice of soils engineering (geotechnical) engineering.

SOILS ENGINEERING (GEOTECHNICAL ENGINEERING) is the application of the principles of soils mechanics in the investigation, evaluation and design of civil works involving the use of earth materials and the inspection or testing of the construction thereof.

TERRACE is a relatively level step constructed in the face of a graded slope surface for drainage and maintenance purposes.

SECTION 3309 — GRADING PERMIT REQUIREMENTS

3309.1 Permits Required. Except as exempted in Section 3306 of this code, no person shall do any grading without first obtaining a grading permit from the building official. A separate permit shall be obtained for each site, and may cover both excavations and fills.

3309.2 Application. The provisions of Section 106.3.1 are applicable to grading. Additionally, the application shall state the estimated quantities of work involved.

3309.3 Grading Designation. Grading in excess of 5,000 cubic yards (3825 m³) shall be performed in accordance with the approved grading plan prepared by a civil engineer, and shall be designated as "engineered grading." Grading involving less than 5,000 cubic yards (3825 m³) shall be designated "regular grading" unless the permittee chooses to have the grading performed as engineered grading, or the building official determines that special conditions or unusual hazards exist, in which case grading shall conform to the requirements for engineered grading.

3309.4 Engineered Grading Requirements. Application for a grading permit shall be accompanied by two sets of plans and specifications, and supporting data consisting of a soils engineering report and engineering geology report. The plans and specifications shall be prepared and signed by an individual licensed by the state to prepare such plans or specifications when required by the building official.

Specifications shall contain information covering construction and material requirements.

Plans shall be drawn to scale upon substantial paper or cloth and shall be of sufficient clarity to indicate the nature and extent of the work proposed and show in detail that they will conform to the provisions of this code and all relevant laws, ordinances, rules and regulations. The first sheet of each set of plans shall give location of the work, the name and address of the owner, and the person by whom they were prepared.

The plans shall include the following information:

1. General vicinity of the proposed site.

2. Property limits and accurate contours of existing ground and details of terrain and area drainage.

3. Limiting dimensions, elevations or finish contours to be achieved by the grading, and proposed drainage channels and related construction.

4. Detailed plans of all surface and subsurface drainage devices, walls, cribbing, dams and other protective devices to be constructed with, or as a part of, the proposed work, together with a map showing the drainage area and the estimated runoff of the area served by any drains.

5. Location of any buildings or structures on the property where the work is to be performed and the location of any buildings or structures on land of adjacent owners that are within 15 feet (4572 mm) of the property or that may be affected by the proposed grading operations.

6. Recommendations included in the soils engineering report and the engineering geology report shall be incorporated in the grading plans or specifications. When approved by the building official, specific recommendations contained in the soils engineering report and the engineering geology report, which are applicable to grading, may be included by reference.

7. The dates of the soils engineering and engineering geology reports together with the names, addresses and phone numbers of the firms or individuals who prepared the reports.

3309.5 Soils Engineering Report. The soils engineering report required by Section 3309.4 shall include data regarding the nature, distribution and strength of existing soils, conclusions and recommendations for grading procedures and design criteria for corrective measures, including buttress fills, when necessary, and opinion on adequacy for the intended use of sites to be developed by the proposed grading as affected by soils engineering factors, including the stability of slopes.

3309.6 Engineering Geology Report. The engineering geology report required by Section 3309.4 shall include an adequate description of the geology of the site, conclusions and recommendations regarding the effect of geologic conditions on the proposed development, and opinion on the adequacy for the intended use of sites to be developed by the proposed grading, as affected by geologic factors.

3309.7 Liquefaction Study. The building official may require a geotechnical investigation in accordance with Sections 1804.2 and 1804.5 when, during the course of an investigation, all of the following conditions are discovered, the report shall address the potential for liquefaction:

1. Shallow ground water, 50 feet (15 240 mm) or less.

2. Unconsolidated sandy alluvium.

3. Seismic Zones 3 and 4.

3309.8 Regular Grading Requirements. Each application for a grading permit shall be accompanied by a plan in sufficient clarity to indicate the nature and extent of the work. The plans shall give the location of the work, the name of the owner and the name of the person who prepared the plan. The plan shall include the following information:

1. General vicinity of the proposed site.

2. Limiting dimensions and depth of cut and fill.

3. Location of any buildings or structures where work is to be performed, and the location of any buildings or structures within 15 feet (4572 mm) of the proposed grading.

3309.9 Issuance. The provisions of Section 106.4 are applicable to grading permits. The building official may require that grading operations and project designs be modified if delays occur which incur weather-generated problems not considered at the time the permit was issued.

The building official may require professional inspection and testing by the soils engineer. When the building official has cause to believe that geologic factors may be involved, the grading will be required to conform to engineered grading.

SECTION 3310 — GRADING FEES

3310.1 General. Fees shall be assessed in accordance with the provisions of this section or shall be as set forth in the fee schedule adopted by the jurisdiction.

3310.2 Plan Review Fees. When a plan or other data are required to be submitted, a plan review fee shall be paid at the time of submitting plans and specifications for review. Said plan review fee shall be as set forth in Table A-33-A. Separate plan review fees shall apply to retaining walls or major drainage structures as required elsewhere in this code. For excavation and fill on the same site, the fee shall be based on the volume of excavation or fill, whichever is greater.

3310.3 Grading Permit Fees. A fee for each grading permit shall be paid to the building official as set forth in Table A-33-B. Separate permits and fees shall apply to retaining walls or major drainage structures as required elsewhere in this code. There shall be no separate charge for standard terrace drains and similar facilities.

TABLE A-33-A—GRADING PLAN REVIEW FEES

50 cubic yards (38.2 m^3) or less	No fee
51 to 100 cubic yards (40 m^3 to 76.5 m^3)	$23.50
101 to 1,000 cubic yards (77.2 m^3 to 764.6 m^3)	37.00
1,001 to 10,000 cubic yards (765.3 m^3 to 7645.5 m^3)	49.25
10,001 to 100,000 cubic yards (7646.3 m^3 to 76 455 m^3)—$49.25 for the first 10,000 cubic yards (7645.5 m^3), plus $24.50 for each additional 10,000 yards (7645.5 m^3) or fraction thereof.	
100,001 to 200,000 cubic yards (76 456 m^3 to 152 911 m^3)—$269.75 for the first 100,000 cubic yards (76 455 m^3), plus $13.25 for each additional 10,000 cubic yards (7645.5 m^3) or fraction thereof.	
200,001 cubic yards (152 912 m^3) or more—$402.25 for the first 200,000 cubic yards (152 911 m^3), plus $7.25 for each additional 10,000 cubic yards (7645.5 m^3) or fraction thereof.	
Other Fees: Additional plan review required by changes, additions or revisions to approved plans . . . . $50.50 per hour* 　(minimum charge—one-half hour)	

*Or the total hourly cost to the jurisdiction, whichever is the greatest. This cost shall include supervision, overhead, equipment, hourly wages and fringe benefits of the employees involved.

TABLE A-33-B—GRADING PERMIT FEES[1]

50 cubic yards (38.2 m^3) or less	$23.50
51 to 100 cubic yards (40 m^3 to 76.5 m^3)	37.00
101 to 1,000 cubic yards (77.2 m^3 to 764.6 m^3)—$37.00 for the first 100 cubic yards (76.5 m^3) plus $17.50 for each additional 100 cubic yards (76.5 m^3) or fraction thereof.	
1,001 to 10,000 cubic yards (765.3 m^3 to 7645.5 m^3)—$194.50 for the first 1,000 cubic yards (764.6 m^3), plus $14.50 for each additional 1,000 cubic yards (764.6 m^3) or fraction thereof.	
10,001 to 100,000 cubic yards (7646.3 m^3 to 76 455 m^3)—$325.00 for the first 10,000 cubic yards (7645.5 m^3), plus $66.00 for each additional 10,000 cubic yards (7645.5 m^3) or fraction thereof.	
100,001 cubic yards (76 456 m^3) or more—$919.00 for the first 100,000 cubic yards (76 455 m^3), plus $36.50 for each additional 10,000 cubic yards (7645.5 m^3) or fraction thereof.	
Other Inspections and Fees: 1. Inspections outside of normal business 　hours . . . . $50.50 per hour[2] 　(minimum charge—two hours) 2. Reinspection fees assessed under provisions of 　Section 108.8 . . . . $50.50 per hour[2] 3. Inspections for which no fee is specifically 　indicated . . . . $50.50 per hour[2] 　(minimum charge—one-half hour)	

[1]The fee for a grading permit authorizing additional work to that under a valid permit shall be the difference between the fee paid for the original permit and the fee shown for the entire project.

[2]Or the total hourly cost to the jurisdiction, whichever is the greatest. This cost shall include supervision, overhead, equipment, hourly wages and fringe benefits of the employees involved.

SECTION 3311 — BONDS

The building official may require bonds in such form and amounts as may be deemed necessary to ensure that the work, if not completed in accordance with the approved plans and specifications, will be corrected to eliminate hazardous conditions.

In lieu of a surety bond the applicant may file a cash bond or instrument of credit with the building official in an amount equal to that which would be required in the surety bond.

SECTION 3312 — CUTS

3312.1 General. Unless otherwise recommended in the approved soils engineering or engineering geology report, cuts shall conform to the provisions of this section.

In the absence of an approved soils engineering report, these provisions may be waived for minor cuts not intended to support structures.

3312.2 Slope. The slope of cut surfaces shall be no steeper than is safe for the intended use and shall be no steeper than 1 unit vertical in 2 units horizontal (50% slope) unless the permittee furnishes a soils engineering or an engineering geology report, or both, stating that the site has been investigated and giving an opinion that a cut at a steeper slope will be stable and not create a hazard to public or private property.

SECTION 3313 — FILLS

3313.1 General. Unless otherwise recommended in the approved soils engineering report, fills shall conform to the provisions of this section.

In the absence of an approved soils engineering report, these provisions may be waived for minor fills not intended to support structures.

3313.2 Preparation of Ground. Fill slopes shall not be constructed on natural slopes steeper than 1 unit vertical in 2 units horizontal (50% slope). The ground surface shall be prepared to receive fill by removing vegetation, noncomplying fill, topsoil and other unsuitable materials scarifying to provide a bond with the new fill and, where slopes are steeper than 1 unit vertical in 5 units horizontal (20% slope) and the height is greater than 5 feet (1524 mm), by benching into sound bedrock or other competent material as determined by the soils engineer. The bench under the toe of a fill on a slope steeper than 1 unit vertical in 5 units horizontal (20% slope) shall be at least 10 feet (3048 mm) wide. The area beyond the toe of fill shall be sloped for sheet overflow or a paved drain shall be provided. When fill is to be placed over a cut, the bench under the toe of fill shall be at least 10 feet (3048 mm) wide but the cut shall be made before placing the fill and acceptance by the soils engineer or engineering geologist or both as a suitable foundation for fill.

3313.3 Fill Material. Detrimental amounts of organic material shall not be permitted in fills. Except as permitted by the building official, no rock or similar irreducible material with a maximum dimension greater than 12 inches (305 mm) shall be buried or placed in fills.

> **EXCEPTION:** The building official may permit placement of larger rock when the soils engineer properly devises a method of placement, and continuously inspects its placement and approves the fill stability. The following conditions shall also apply:
>
> 1. Prior to issuance of the grading permit, potential rock disposal areas shall be delineated on the grading plan.

> 2. Rock sizes greater than 12 inches (305 mm) in maximum dimension shall be 10 feet (3048 mm) or more below grade, measured vertically.
>
> 3. Rocks shall be placed so as to assure filling of all voids with well-graded soil.

3313.4 Compaction. All fills shall be compacted to a minimum of 90 percent of maximum density.

3313.5 Slope. The slope of fill surfaces shall be no steeper than is safe for the intended use. Fill slopes shall be no steeper than 1 unit vertical in 2 units horizontal (50% slope).

SECTION 3314 — SETBACKS

3314.1 General. Cut and fill slopes shall be set back from site boundaries in accordance with this section. Setback dimensions shall be horizontal distances measured perpendicular to the site boundary. Setback dimensions shall be as shown in Figure A-33-1.

3314.2 Top of Cut Slope. The top of cut slopes shall not be made nearer to a site boundary line than one fifth of the vertical height of cut with a minimum of 2 feet (610 mm) and a maximum of 10 feet (3048 mm). The setback may need to be increased for any required interceptor drains.

3314.3 Toe of Fill Slope. The toe of fill slope shall be made not nearer to the site boundary line than one half the height of the slope with a minimum of 2 feet (610 mm) and a maximum of 20 feet (6096 mm). Where a fill slope is to be located near the site boundary and the adjacent off-site property is developed, special precautions shall be incorporated in the work as the building official deems necessary to protect the adjoining property from damage as a result of such grading. These precautions may include but are not limited to:

1. Additional setbacks.

2. Provision for retaining or slough walls.

3. Mechanical or chemical treatment of the fill slope surface to minimize erosion.

4. Provisions for the control of surface waters.

3314.4 Modification of Slope Location. The building official may approve alternate setbacks. The building official may require an investigation and recommendation by a qualified engineer or engineering geologist to demonstrate that the intent of this section has been satisfied.

SECTION 3315 — DRAINAGE AND TERRACING

3315.1 General. Unless otherwise indicated on the approved grading plan, drainage facilities and terracing shall conform to the provisions of this section for cut or fill slopes steeper than 1 unit vertical in 3 units horizontal (33.3% slope).

3315.2 Terrace. Terraces at least 6 feet (1829 mm) in width shall be established at not more than 30-foot (9144 mm) vertical intervals on all cut or fill slopes to control surface drainage and debris except that where only one terrace is required, it shall be at midheight. For cut or fill slopes greater than 60 feet (18 288 mm) and up to 120 feet (36 576 mm) in vertical height, one terrace at approximately midheight shall be 12 feet (3658 mm) in width. Terrace widths and spacing for cut and fill slopes greater than 120 feet (36 576 mm) in height shall be designed by the civil engineer and approved by the building official. Suitable access shall be provided to permit proper cleaning and maintenance.

Swales or ditches on terraces shall have a minimum gradient of 5 percent and must be paved with reinforced concrete not less than

3 inches (76 mm) in thickness or an approved equal paving. They shall have a minimum depth at the deepest point of 1 foot (305 mm) and a minimum paved width of 5 feet (1524 mm).

A single run of swale or ditch shall not collect runoff from a tributary area exceeding 13,500 square feet (1254.2 m^2) (projected) without discharging into a down drain.

3315.3 Subsurface Drainage. Cut and fill slopes shall be provided with subsurface drainage as necessary for stability.

3315.4 Disposal. All drainage facilities shall be designed to carry waters to the nearest practicable drainage way approved by the building official or other appropriate jurisdiction as a safe place to deposit such waters. Erosion of ground in the area of discharge shall be prevented by installation of nonerosive downdrains or other devices.

Building pads shall have a drainage gradient of 2 percent toward approved drainage facilities, unless waived by the building official.

> **EXCEPTION:** The gradient from the building pad may be 1 percent if all of the following conditions exist throughout the permit area:
>
> 1. No proposed fills are greater than 10 feet (3048 mm) in maximum depth.
>
> 2. No proposed finish cut or fill slope faces have a vertical height in excess of 10 feet (3048 mm).
>
> 3. No existing slope faces steeper than 1 unit vertical in 10 units horizontal (10% slope) have a vertical height in excess of 10 feet (3048 mm).

3315.5 Interceptor Drains. Paved interceptor drains shall be installed along the top of all cut slopes where the tributary drainage area above slopes toward the cut and has a drainage path greater than 40 feet (12 192 mm) measured horizontally. Interceptor drains shall be paved with a minimum of 3 inches (76 mm) of concrete or gunite and reinforced. They shall have a minimum depth of 12 inches (305 mm) and a minimum paved width of 30 inches (762 mm) measured horizontally across the drain. The slope of drain shall be approved by the building official.

SECTION 3316 — EROSION CONTROL

3316.1 Slopes. The faces of cut and fill slopes shall be prepared and maintained to control against erosion. This control may consist of effective planting. The protection for the slopes shall be installed as soon as practicable and prior to calling for final approval. Where cut slopes are not subject to erosion due to the erosion-resistant character of the materials, such protection may be omitted.

3316.2 Other Devices. Where necessary, check dams, cribbing, riprap or other devices or methods shall be employed to control erosion and provide safety.

SECTION 3317 — GRADING INSPECTION

3317.1 General. Grading operations for which a permit is required shall be subject to inspection by the building official. Professional inspection of grading operations shall be provided by the civil engineer, soils engineer and the engineering geologist retained to provide such services in accordance with Section 3317.5 for engineered grading and as required by the building official for regular grading.

3317.2 Civil Engineer. The civil engineer shall provide professional inspection within such engineer's area of technical specialty, which shall consist of observation and review as to the establishment of line, grade and surface drainage of the develop-

ment area. If revised plans are required during the course of the work they shall be prepared by the civil engineer.

3317.3 Soils Engineer. The soils engineer shall provide professional inspection within such engineer's area of technical specialty, which shall include observation during grading and testing for required compaction. The soils engineer shall provide sufficient observation during the preparation of the natural ground and placement and compaction of the fill to verify that such work is being performed in accordance with the conditions of the approved plan and the appropriate requirements of this chapter. Revised recommendations relating to conditions differing from the approved soils engineering and engineering geology reports shall be submitted to the permittee, the building official and the civil engineer.

3317.4 Engineering Geologist. The engineering geologist shall provide professional inspection within such engineer's area of technical specialty, which shall include professional inspection of the bedrock excavation to determine if conditions encountered are in conformance with the approved report. Revised recommendations relating to conditions differing from the approved engineering geology report shall be submitted to the soils engineer.

3317.5 Permittee. The permittee shall be responsible for the work to be performed in accordance with the approved plans and specifications and in conformance with the provisions of this code, and the permittee shall engage consultants, if required, to provide professional inspections on a timely basis. The permittee shall act as a coordinator between the consultants, the contractor and the building official. In the event of changed conditions, the permittee shall be responsible for informing the building official of such change and shall provide revised plans for approval.

3317.6 Building Official. The building official shall inspect the project at the various stages of work requiring approval to determine that adequate control is being exercised by the professional consultants.

3317.7 Notification of Noncompliance. If, in the course of fulfilling their respective duties under this chapter, the civil engineer, the soils engineer or the engineering geologist finds that the work is not being done in conformance with this chapter or the approved grading plans, the discrepancies shall be reported immediately in writing to the permittee and to the building official.

3317.8 Transfer of Responsibility. If the civil engineer, the soils engineer, or the engineering geologist of record is changed during grading, the work shall be stopped until the replacement has agreed in writing to accept their responsibility within the area of technical competence for approval upon completion of the work. It shall be the duty of the permittee to notify the building official in writing of such change prior to the recommencement of such grading.

SECTION 3318 — COMPLETION OF WORK

3318.1 Final Reports. Upon completion of the rough grading work and at the final completion of the work, the following reports and drawings and supplements thereto are required for engineered grading or when professional inspection is performed for regular grading, as applicable.

1. An as-built grading plan prepared by the civil engineer retained to provide such services in accordance with Section 3317.5 showing original ground surface elevations, as-graded ground surface elevations, lot drainage patterns, and the locations and elevations of surface drainage facilities and of the outlets of subsurface drains. As-constructed locations, elevations and details of subsurface drains shall be shown as reported by the soils engineer.

Civil engineers shall state that to the best of their knowledge the work within their area of responsibility was done in accordance with the final approved grading plan.

2. A report prepared by the soils engineer retained to provide such services in accordance with Section 3317.3, including locations and elevations of field density tests, summaries of field and laboratory tests, other substantiating data, and comments on any changes made during grading and their effect on the recommendations made in the approved soils engineering investigation report. Soils engineers shall submit a statement that, to the best of their knowledge, the work within their area of responsibilities is in accordance with the approved soils engineering report and applicable provisions of this chapter.

3. A report prepared by the engineering geologist retained to provide such services in accordance with Section 3317.5, including a final description of the geology of the site and any new infor-

mation disclosed during the grading and the effect of same on recommendations incorporated in the approved grading plan. Engineering geologists shall submit a statement that, to the best of their knowledge, the work within their area of responsibility is in accordance with the approved engineering geologist report and applicable provisions of this chapter.

4. The grading contractor shall submit in a form prescribed by the building official a statement of conformance to said as-built plan and the specifications.

3318.2 Notification of Completion. The permittee shall notify the building official when the grading operation is ready for final inspection. Final approval shall not be given until all work, including installation of all drainage facilities and their protective devices, and all erosion-control measures have been completed in accordance with the final approved grading plan, and the required reports have been submitted.

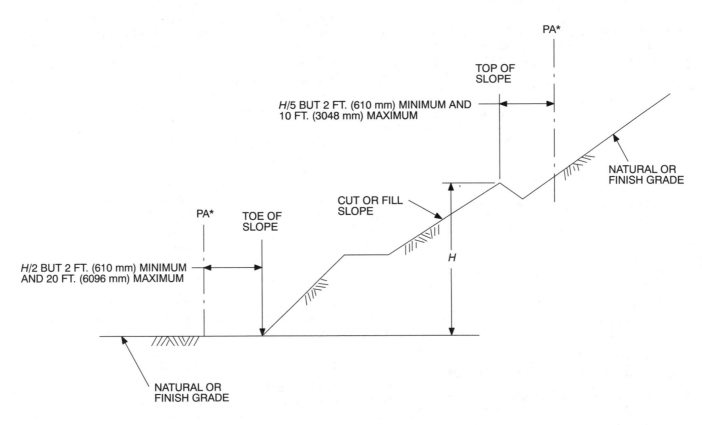

* PERMIT AREA BOUNDARY

FIGURE A-33-1—SETBACK DIMENSIONS

Appendix Chapter 34
EXISTING STRUCTURES
Division I—LIFE-SAFETY REQUIREMENTS FOR EXISTING BUILDINGS
OTHER THAN HIGH-RISE BUILDINGS

SECTION 3406 — GENERAL

3406.1 Purpose. The purpose of this division is to provide a reasonable degree of safety to persons occupying existing buildings by providing for alterations to such existing buildings that do not conform with the minimum requirements of this code.

> **EXCEPTION:** Group U Occupancies regulated by Appendix Chapter 34, Division II, and Group R, Division 3 Occupancies, except that Group R, Division 3 Occupancies shall comply with Section 3411.

3406.2 Effective Date. Within 18 months of the effective date of this division, plans for compliance shall be submitted and approved, and within 18 months thereafter, the work shall be completed or the building shall be vacated until made to conform.

SECTION 3407 — EXITS

3407.1 Number of Means of Egress. Every floor above the first story used for human occupancy shall have at least two means of egress, one of which may be an exterior fire escape complying with Section 3407.4. Subject to the approval of the building official, an approved ladder device may be used in lieu of a fire escape when the construction feature or location of the building on the property makes the installation of a fire escape impracticable.

> **EXCEPTION:** In all occupancies, second stories with an occupant load of 10 or less may have one means of egress.

An exit ladder device when used in lieu of a fire escape shall conform with UBC Standard 10-3, which is a part of this code (see Chapter 35), and the following:

1. Serves an occupant load of 10 or less or a single dwelling unit or guest room.

2. The building does not exceed three stories in height.

3. The access is adjacent to an opening as specified for emergency egress or rescue or from a balcony.

4. Shall not pass in front of any building opening below the unit being served.

5. The availability of activating the device for the ladder is accessible only from the opening or balcony served.

6. Installed so that it will not cause a person using it to be within 6 feet (1829 mm) of exposed electrical wiring.

3407.2 Stair Construction. All required stairs shall have a minimum run of 9 inches (229 mm) and a maximum rise of 8 inches (203 mm) and shall have a minimum width of 30 inches (762 mm) exclusive of handrails. Every stairway shall have at least one handrail. A landing having a minimum 30-inch (762 mm) run in the direction of travel shall be provided at each point of access to the stairway.

> **EXCEPTION:** Fire escapes as provided for in this section.

Exterior stairs shall be of noncombustible construction.

> **EXCEPTION:** On buildings of Types III, IV and V construction, provided the exterior stairs are constructed of wood not less than 2-inch (51 mm) nominal thickness.

3407.3 Corridors. Corridors of Groups A, B, E, F, H, I, M and R, Division 1, and S Occupancies serving an occupant load of 30 or more, shall have walls and ceilings of not less than one-hour fire-resistive construction as required by this code. Existing walls surfaced with wood lath and plaster in good condition or $^1/_2$-inch (12.7 mm) gypsum wallboard or openings with fixed wired glass set in steel frames are permitted for corridor walls and ceilings and occupancy separations when approved. Doors opening into such corridors shall be protected by 20-minute fire assemblies or solid wood doors not less than $1^3/_4$ inches (45 mm) thick. Where the existing frame will not accommodate the $1^3/_4$-inch-thick (45 mm) door, a $1^3/_8$-inch-thick (35 mm) solid bonded wood-core door or equivalent insulated steel door shall be permitted. Doors shall be self-closing or automatic closing by smoke detection. Transoms and openings other than doors from corridors to rooms shall comply with Section 1004.3.4.3.2 of this code or shall be covered with a minimum of $^3/_4$-inch (19.1 mm) plywood or $^1/_2$-inch (12.7 mm) gypsum wallboard or equivalent material on the room side.

> **EXCEPTION:** Existing corridor walls, ceilings and opening protection not in compliance with the above may be continued when such buildings are protected with an approved automatic sprinkler system throughout. Such sprinkler system may be supplied from the domestic water system if it is of adequate volume and pressure.

3407.4 Fire Escapes.

1. Existing fire escapes that, in the opinion of the building official, comply with the intent of this section may be used as one of the required exits. The location and anchorage of fire escapes shall be of approved design and construction.

2. Fire escapes shall comply with the following:

Access from a corridor shall not be through an intervening room.

All openings within 10 feet (3048 mm) shall be protected by three-fourths-hour fire assemblies. When located within a recess or vestibule, adjacent enclosure walls shall not be of less than one-hour fire-resistive construction.

Egress from the building shall be by a clear opening having a minimum dimension of not less than 29 inches (737 mm). Such openings shall be openable from the inside without the use of a key or special knowledge or effort. The sill of an opening giving access shall not be more than 30 inches (762 mm) above the floor of the building or balcony.

Fire escape stairways and balconies shall support the dead load plus a live load of not less than 100 pounds per square foot (4.79 kN/m^2) and shall be provided with a top and intermediate handrail on each side. The pitch of the stairway shall not exceed 60 degrees with a minimum width of 18 inches (457 mm). Treads shall not be less than 4 inches (102 mm) in width and the rise between treads shall not exceed 10 inches (254 mm). All stair and balcony railings shall support a horizontal force of not less than 50 pounds per lineal foot (729.5 N/m) of railing.

Balconies shall not be less than 44 inches (1118 mm) in width with no floor opening other than the stairway opening greater than $^5/_8$ inch (16 mm) in width. Stairway openings in such balconies shall not be less than 22 inches by 44 inches (599 mm by 1118 mm). The balustrade of each balcony shall not be less than 36 inches (914 mm) high with not more than 9 inches (229 mm) between balusters.

Fire escapes shall extend to the roof or provide an approved gooseneck ladder between the top floor landing and the roof when serving buildings four or more stories in height having roofs with a slope of less than 4 units vertical in 12 units horizontal (33.3%

slope). Fire escape ladders shall be designed and connected to the building to withstand a horizontal force of 100 pounds per lineal foot (1459 N/m); each rung shall support a concentrated load of 500 pounds (2224 N) placed anywhere on the rung. All ladders shall be at least 15 inches (381 mm) wide, located within 12 inches (305 mm) of the building and shall be placed flatwise relative to the face of the building. Ladder rungs shall be $^3/_4$ inch (19 mm) in diameter and shall be located 12 inches (305 mm) on center. Openings for roof access ladders through cornices and similar projections shall have minimum dimensions of 30 inches by 33 inches (762 mm by 838 mm).

The lowest balcony shall not be more than 18 feet (5486 mm) from the ground. Fire escapes shall extend to the ground or be provided with counterbalanced stairs reaching to the ground.

Fire escapes shall not take the place of stairways required by the codes under which the building was constructed.

Fire escapes shall be kept clear and unobstructed at all times and maintained in good working order.

3407.5 Exit and Fire Escape Signs. Exit signs shall be provided as required by this code.

> **EXCEPTION:** The use of existing exit signs may be continued when approved by the building official.

All doors or windows providing access to a fire escape shall be provided with fire escape signs.

SECTION 3408 — ENCLOSURE OF VERTICAL SHAFTS

Interior vertical shafts, including but not limited to stairways, elevator hoistways, service and utility shafts, shall be enclosed by a minimum of one-hour fire-resistive construction. All openings into such shafts shall be protected with one-hour fire assemblies that shall be maintained self-closing or be automatic closing by smoke detection. All other openings shall be fire protected in an approved manner. Existing fusible link-type automatic door-closing devices may be permitted if the fusible link rating does not exceed 135°F (57.2°C).

> **EXCEPTIONS:** 1. In other than Group I Occupancies, an enclosure will not be required for openings serving only one adjacent floor.
>
> 2. Stairways need not be enclosed in a continuous vertical shaft if each story is separated from other stories by one-hour fire-resistive construction or approved wired glass set in steel frames. In addition, all exit corridors shall be sprinklered and the openings between the corridor and occupant space shall have at least one sprinkler head above the openings on the tenant side. The sprinkler system may be supplied from the domestic water supply if of adequate volume and pressure.
>
> 3. Vertical openings need not be protected if the building is protected by an approved automatic sprinkler system.

SECTION 3409 — BASEMENT ACCESS OR SPRINKLER PROTECTION

An approved automatic sprinkler system shall be provided in basements or stories exceeding 1,500 square feet (139.3 m^2) in

area and not having a minimum of 20 square feet (1.86 m^2) of opening entirely above the adjoining ground level in each 50 lineal feet (15 240 mm) or fraction thereof of exterior wall on at least one side of the building. Openings shall have a minimum clear dimension of 30 inches (762 mm).

If any portion of a basement is located more than 75 feet (22 860 mm) from required openings, the basement shall be provided with an approved automatic sprinkler system throughout.

SECTION 3410 — STANDPIPES

Any buildings over four stories in height shall be provided with an approved Class I or Class III standpipe system.

SECTION 3411 — SMOKE DETECTORS

3411.1 General. Dwelling units and hotel or lodging house guest rooms that are used for sleeping purposes shall be provided with smoke detectors. Detectors shall be installed in accordance with the approved manufacturer's instructions.

3411.2 Power Source. Smoke detectors may be battery operated or may receive their primary power from the building wiring when such wiring is served from a commercial source. Wiring shall be permanent and without disconnecting switches other than those required for overcurrent protection.

3411.3 Location within Dwelling Units. In dwelling units, detectors shall be mounted on the ceiling or wall at a point centrally located in the corridor or area giving access to each separate sleeping area. Where sleeping rooms are on an upper level, the detector shall be placed at the center of the ceiling directly above the stairway. Detectors shall also be installed in the basements of dwelling units having stairways that open from the basement into the dwelling. Detectors shall sound an alarm audible in all sleeping areas of the dwelling unit in which they are located.

3411.4 Location in Efficiency Dwelling Units and Hotels. In efficiency dwelling units, hotel suites and in hotel sleeping rooms, detectors shall be located on the ceiling or wall of the main room or hotel sleeping room. When sleeping rooms within an efficiency dwelling unit or hotel suite are on an upper level, the detector shall be placed at the center of the ceiling directly above the stairway. When actuated, the detector shall sound an alarm audible within the sleeping area of the dwelling unit, hotel suite or sleeping room in which it is located.

SECTION 3412 — SEPARATION OF OCCUPANCIES

Occupancy separations shall be provided as specified in Section 302 of this code. When approved by the building official, existing wood lath and plaster in good condition or $^1/_2$-inch (12.7 mm) gypsum wallboard may be acceptable where one-hour occupancy separations are required.

Division II—LIFE-SAFETY REQUIREMENTS FOR EXISTING HIGH-RISE BUILDINGS

SECTION 3413 — SCOPE

These provisions apply to existing high-rise buildings constructed prior to the adoption of this division and which house Group B offices or Group R, Division 1 Occupancies, each having floors used for human occupancy located more than 75 feet (22 860 mm) above the lowest level of fire department vehicle access.

SECTION 3414 — GENERAL

Existing high-rise buildings as specified in Section 3413 shall be modified to conform with not less than the minimum provisions specified in Table A-34-A and as further enumerated within this division.

The provisions of this division shall not be construed to allow the elimination of fire-protection systems or a reduction in the level of firesafety provided in buildings constructed in conformance with previously adopted codes.

SECTION 3415 — COMPLIANCE DATA

After adoption of this division, the building official shall duly notify the owners whose buildings are subject to the provisions of this division. Upon receipt of such notice, the owner shall, subject to the following time limits, take necessary actions to comply with the provisions of this division.

Plans and specifications for the necessary alterations shall be filed with the building official within the time period established by the local jurisdiction after the date of owner notification. Work on the required alterations to the building shall commence within 30 months of the date of owner notification and such work shall be completed within five years from the date of owner notification.

The building official shall grant necessary extensions of time when it can be shown that the specified time periods are not physically practical or pose an undue hardship. The granting of an extension of time for compliance shall be based on the showing of good cause and subject to the filing of an acceptable systematic progressive plan of correction with the building official.

SECTION 3416 — AUTHORITY OF THE BUILDING OFFICIAL

For the purpose of applying the provisions of this division, the building official shall have the authority to consider alternative approaches and grant necessary deviations from this division as follows:

1. Allow alternate materials or methods of compliance if such alternate materials or methods of compliance will provide levels of fire and life safety equal to or greater than those specifically set forth in this division.

2. Waive specific individual requirements if it can be shown that such requirements are not physically possible or practical and that a practical alternative cannot be provided.

SECTION 3417 — APPEALS BOARD

Appeals of the determinations of the building official in applying the provisions of this code may be made by an appeal directed to the board of appeals as established by Section 105 of this code.

SECTION 3418 — SPECIFIC PROVISIONS AND ALTERNATES

3418.1 Specific Provisions. The following provisions shall apply when required by Table A-34-A.

3418.1.1 Type of construction. Buildings classified as Type II-N, III-N or V-N construction shall be equipped with an approved automatic sprinkler system installed in accordance with UBC Standard 9-1, which is a part of this code (see Chapter 35).

> **EXCEPTION:** Installation of meters or backflow preventers for the connection to the water works system need not be provided unless required by other regulations of the authority having jurisdiction.

3418.1.2 Automatic sprinklers. All required corridors, stairwells, elevator lobbies, public assembly areas occupied by 100 or more persons and commercial kitchens shall be protected by an approved automatic sprinkler system meeting the design criteria of UBC Standard 9-1, which is a part of this code (see Chapter 35). A minimum of one sprinkler shall be provided on the room side of every corridor opening.

> **EXCEPTION:** Sprinklers may be omitted in stairwells of noncombustible construction.

3418.1.3 Fire department communication system. When it is determined by test that the portable fire department communication equipment is ineffective, a communication system acceptable to the fire department shall be installed within the existing high-rise building to permit emergency communication between fire-suppression personnel.

3418.1.4 Single-station smoke detectors. Single-station smoke detectors shall be installed within all dwelling units or guest rooms in accordance with the manufacturer's installation instructions. In dwelling units, the detector shall be mounted on the ceiling or wall at a point centrally located in the corridor or area giving access to each separate sleeping area. When sleeping rooms are located on an upper level, the detector shall be installed at the center of the ceiling directly above the stairway within the unit. In efficiency dwelling units, hotel suites and in hotel guest rooms, detectors shall be located on the ceiling or wall of the main room or hotel sleeping room. When actuated, the detector shall provide an audible alarm in the sleeping area of the dwelling unit, hotel suite or guest room in which it is located.

Such detectors may be battery operated.

3418.1.5 Manual fire alarm system. An approved manual fire alarm system connected to a central, proprietary or remote station service, or an approved manual fire alarm system that will provide an audible signal at a constantly attended location, shall be provided.

3418.1.6 Occupant voice notification system. An approved occupant voice notification system shall be provided. Such system shall provide communication from a location acceptable to the fire department and shall permit voice notification to at least all normally occupied areas of the building.

The occupant voice notification system may be combined with a fire alarm system, provided the combined system has been approved and listed for such use. The sounding of a fire alarm signal in any given area or floor shall not prohibit voice communication to other areas or floors. Combination systems shall be designed to permit voice transmission to override the fire alarm signal, but the fire alarm shall not terminate in less than three minutes.

3418.1.7 Vertical shaft enclosures. Openings through two or more floors, except mezzanine floors, that contain a stairway or elevator, shall be provided with vertical shaft enclosure protection as specified herein. Such floor openings, when not enclosed by existing shaft enclosure construction, shall be protected by one-hour fire-resistive-rated shaft enclosure construction. For floor openings that are enclosed by existing shaft enclosure construction having fire-resistive capabilities similar to wood lath and plaster in good condition, $^1/_2$-inch (12.7 mm) gypsum wallboard or approved $^1/_4$-inch-thick (6.4 mm) wired glass is acceptable. Wired glass set in a steel frame may be installed in existing shaft enclosure walls but shall be rendered inoperative and be fixed in a closed position.

Openings through two or more floors for other than stairways or elevators, such as openings provided for piping, ducts, gas vents, dumbwaiters, and rubbish and linen chutes, shall be provided with vertical shaft enclosure protection as specified for stairways and elevators.

EXCEPTION: Openings for piping, ducts, gas vents, dumbwaiters, and rubbish and linen chutes of copper or ferrous construction are permitted without a shaft enclosure, provided the floor openings are effectively firestopped at each floor level.

3418.1.8 Shaft enclosure opening protection. Openings other than those provided for elevator doors in new vertical shaft enclosures constructed of one-hour fire-resistive construction shall be equipped with approved fire assemblies having a fire-protection rating of not less than one hour. Openings other than those provided for elevator doors in existing vertical shaft enclosures shall be equipped with approved 20-minute-rated fire assemblies, $1^3/_4$-inch (44 mm) solid wood doors or the equivalent thereto. Doors shall be either self-closing or automatic closing and automatic latching.

All elevators on all floors shall open into elevator lobbies that are separated from the remainder of the building as is required for corridor construction in the Building Code, unless the building is protected throughout by a sprinkler system.

3418.1.9 Manual shutoff of heating, ventilating and air-conditioning (HVAC) systems. Heating, ventilating and air-conditioning systems shall be equipped with manual shutoff controls installed at an approved location when required by the fire department.

3418.1.10 Automatic elevator recall system. Elevators shall be equipped with an approved automatic recall system as required by Section 403.7, Item 2.

3418.1.11 Unlocked stairway doors. Exit doors into exit stairway enclosures shall be maintained unlocked from the stairway side on at least every fifth floor level. All unlocked doors shall bear a sign stating ACCESS ONTO FLOOR THIS LEVEL.

Stairway doors may be locked, subject to the following conditions:

1. Stairway doors that are to be locked from the stairway side shall have the capability of being unlocked simultaneously without unlatching upon a signal from an approved location.

2. A telephone or other two-way communications system connected to an approved emergency service that operates continuously shall be provided at not less than every fifth floor in each required stairway.

3418.1.12 Stair shaft ventilation. Stair shaft enclosures that extend to the roof shall be provided with an approved manually openable hatch to the exterior having an area not less than 16 square feet (1.486 m^2) with a minimum dimension of 2 feet (610 mm).

EXCEPTIONS: 1. Stair shaft enclosures complying with the requirements for pressurized enclosures.

2. Stair shaft enclosures pressurized as required for mechanically operated pressurized enclosures to a minimum of 0.15-inch water column (37 Pa) and a maximum of 0.50-inch water column (124 Pa).

3418.1.13 Elevator shaft ventilation. Elevator shaft enclosures that extend to the roof shall be vented to the outside with vents whose area shall be not less than $3^1/_2$ percent of the area of the elevator shaft, with a minimum of 3 square feet (0.28 m^2) per elevator.

EXCEPTION: Where energy conservation or hoistway pressurization requires that the vents be normally closed, automatic venting by actuation of an elevator lobby detector or power failure may be accepted.

3418.1.14 Posting of elevators. A permanent sign shall be installed in each elevator cab adjacent to the floor status indicator and at each elevator call station on each floor reading IN FIRE EMERGENCY, DO NOT USE ELEVATOR—USE EXIT STAIRS, or similar verbiage approved by the building official.

EXCEPTION: Sign may be omitted at the main entrance floor-level call station.

3418.1.15 Exit stairways. All buildings shall have a minimum of two approved exit stairways.

3418.1.16 Corridor construction. Corridors serving an occupant load of 30 or more shall have walls and ceilings of not less than one-hour fire-resistive construction as required by this code. Existing walls may be surfaced with wood lath and plaster in good condition or $^1/_2$-inch (12.7 mm) gypsum wallboard for corridor walls and ceilings and occupancy separations when approved.

3418.1.17 Corridor openings. Openings in corridor walls and ceilings shall be protected by not less than $1^3/_8$-inch (35 mm) solid-bonded wood-core doors; approved $^1/_4$-inch-thick (6.4 mm) wired glass; approved fire dampers conforming to UBC Standard 7-7, which is a part of this code; or by equivalent protection in lieu of any of these items (see Chapter 35). Transoms shall be fixed closed and covered with $^1/_2$-inch (12.7 mm) Type X gypsum wallboard or equivalent material installed on both sides of the opening.

3418.1.18 Corridor door closers. Exit-access doors into corridors shall be equipped with self-closing devices or shall be automatic closing by actuation of a smoke detector. When spring hinges are used as the closing device, not less than two such hinges shall be installed on each door leaf.

3418.1.19 Corridor dead ends. The length of dead-end corridors serving an occupant load of more than 30 shall not exceed 20 feet (6096 mm).

3418.1.20 Interior finish. The interior finish in corridors, exit stairways and extensions thereof shall conform to the provisions of Chapter 8 of this code.

3418.1.21 Exit stairway illumination. When the building is occupied, exit stairways shall be illuminated with lights having an intensity of not less than 1 footcandle (10.8 lx) at the floor level. Such lighting shall be equipped with an independent alternate source of power such as a battery pack or on-site generator.

3418.1.22 Corridor illumination. When the building is occupied, corridors shall be illuminated with lights having an intensity of not less than 1 footcandle (10.8 lx) at the floor level. Such lighting shall be equipped with an independent alternate source of power such as a battery pack or on-site generator.

3418.1.23 Exit stairway exit signs. The location of exit stairways shall be clearly indicated by illuminated exit signs. Such exit signs shall be equipped with an independent alternate source of power such as a battery pack or on-site generator or shall be of an approved self-illuminating type.

3418.1.24 Exit signs. Illuminated exit signs shall be provided in all means of egress and located in such a manner as to clearly indicate the direction of egress. Such exit signs shall be equipped with an independent alternate source of power such as a battery pack or on-site generator or shall be of an approved self-illuminating type.

3418.1.25 Emergency plan. The management for all buildings shall establish and maintain a written fire- and life-safety emergency plan that has been approved by the chief. The chief shall develop written criteria and guidelines on which all plans shall be based.

3418.1.26 Posting of emergency plan and exit plans. Copies of the emergency plan and exiting plans (including elevator and stairway placarding) shall be posted in locations approved by the chief.

3418.1.27 Fire drills. The management of all buildings shall conduct fire drills for their staff and employees at least every 120 days. The fire department must be advised of such drills at least 24 hours in advance. A written record of each drill shall be maintained in the building management office and made available to the fire department for review.

3418.2 Sprinkler Alternatives. The requirements of Table A-34-A may be modified as specified by the following for existing high-rise buildings of Type I, II-F.R., II One-hour, III One-hour, IV or V One-hour construction when an approved automatic sprinkler system is installed throughout the building in accordance with UBC Standard 9-1:

Item 5—Manual fire alarm system shall not be required.

Item 6—Occupant voice notification system shall not be required; however, if the building is equipped with a public address system, the public address system shall be available for use as an occupant voice notification system.

Item 7—Vertical shaft enclosures may be of nonrated construction for required exit stairway enclosures. Vertical shaft enclosures of openings in floors provided for elevators, escalators and supplemental stairways shall not be required, provided such openings are protected by an approved curtain board and water curtain sprinkler system.

Item 8—Protection of openings in vertical shaft enclosures may be nonrated but shall not be less than a $1^3/_4$-inch (44 mm) solid-wood door or the equivalent thereto. Closing and latching hardware shall be provided.

Item 10—An automatic elevator recall system shall not be required.

Item 12—Stair shaft ventilation shall not be required.

Item 16—Existing corridor construction need not be altered.

Item 17—Door openings into corridors may be protected by assemblies other than those specified in Section 3418.1, provided an effective smoke barrier is maintained. Closing and latching hardware shall be provided. Protection of duct penetrations is not required.

Item 19—The length of existing corridor dead ends shall not be limited.

Item 20—Interior finish in means of egress may be reduced by one classification but shall not be less than Class III.

Installation of meters or backflow preventers for the connection to the water works system need not be provided unless required by other regulations of the authority having jurisdiction.

TABLE A-34-A—OCCUPANCY CLASSIFICATION AND USE[1]

ITEMS REQUIRED	GROUP R, DIVISION 1						GROUP B		
	Apartment			Hotel			Office		
	Height Zones[2]								
	1	2	3	1	2	3	1	2	3
1. Automatic sprinklers in buildings of Type II-N, III-N or V-N construction. See Section 3418.1.1.	R	R	—	R	R	—	R	R	—
2. Automatic sprinklers in corridors, stairways, elevator lobbies, public assembly areas, kitchens and at doors opening to corridors. See Section 3418.1.2.	R	R	R	R	R	R	R	R	R
3. Fire department communication system or radios. See Section 3418.1.3.	R	R	R	R	R	R	R	R	R
4. Single-station smoke detectors. See Section 3418.1.4.	R	R	R	R	R	R	NR	NR	NR
5. Manual fire alarm system. See Section 3418.1.5.	R	R	R	R	R	R	R	R	R
6. Occupant voice notification system. See Section 3418.1.6.	NR	R	R	NR	R	R	NR	NR	NR
7. Vertical shaft enclosure walls of one-hour fire resistance. See Section 3418.1.7.	R	R	R	R	R	R	R	R	R
8. Protection of openings in vertical shaft enclosures by 20-minute-rated assemblies. See Section 3418.1.8.	R	R	R	R	R	R	R	R	R
9. Manual shutoff of HVAC systems. See Section 3418.1.9.	R	R	R	R	R	R	R	R	R
10. Automatic elevator recall system. See Section 3418.1.10.	R	R	R	R	R	R	R	R	R
11. Unlocked stairway doors every fifth floor. See Section 3418.1.11.	R	R	R	R	R	R	NR	R	R
12. Stair shaft ventilation. See Section 3418.1.12.	R	R	R	R	R	R	R	R	R
13. Elevator shaft ventilaton. See Section 3418.1.13.	R	R	R	R	R	R	R	R	R
14. Posting of elevators as not intended for exiting purposes. See Section 3418.1.14.	R	R	R	R	R	R	R	R	R
15. Minimum of two exit stairways. See Section 3418.1.15.	R	R	R	R	R	R	R	R	R
16. Corridor wall construction. See Section 3418.1.16.	R	R	R	R	R	R	R	R	R
17. Protected corridor openings with 20-minute-rated assemblies or $1^3/_4$-inch (44 mm) solid-wood door. See Section 3418.1.17.	R	R	R	R	R	R	NR	NR	NR
18. Corridor doors equipped with self-closing devices. See Section 3418.1.18.	R	R	R	R	R	R	NR	NR	NR
19. Corridor dead ends limited to 20 feet (6096 mm) maximum. See Section 3418.1.19.	R	R	R	R	R	R	NR	NR	NR
20. Interior finish controlled in corridors, exit stairways and extensions thereof. See Section 3418.1.20.	R	R	R	R	R	R	R	R	R
21. Exit stairway illumination. See Section 3418.1.21.	R	R	R	R	R	R	R	R	R
22. Corridor illumination. See Section 3418.1.22.	R	R	R	R	R	R	NR	NR	NR
23. Exit stairway exit signs. See Section 3418.1.23.	R	R	R	R	R	R	R	R	R
24. Exit signs. See Section 3418.1.24.	R	R	R	R	R	R	R	R	R
25. Emergency planning. See Section 3418.1.25.	R	R	R	R	R	R	R	R	R
26. Posting of emergency instructions. See Section 3418.1.26.	R	R	R	R	R	R	R	R	R
27. Fire drills. See Section 3418.1.27.	NR	NR	NR	R	R	R	NR	NR	NR

[1]R—Provisions are required.
NR—Provisions are not required.
[2]Height zones are established based on a building having a floor as measured to the top of the floor surface used for human occupancy located within the ranges of heights above the lowest level of the fire department vehicle access in accordance with the following:
    Height Zone 1: More than 75 feet (22 860 mm), but not in excess of 149 feet (45 415 mm).
    Height Zone 2: More than 149 feet (45 415 mm), but not in excess of 399 feet (121.6 m).
    Height Zone 3: More than 399 feet (121.6 m).

Division III—REPAIRS TO BUILDINGS AND STRUCTURES DAMAGED BY THE OCCURRENCE OF A NATURAL DISASTER

NOTE: This is a new division.

SECTION 3419 — PURPOSE

The purpose of this division is to provide a defined level of repair for buildings damaged by a natural disaster in jurisdictions where a formal state of emergency has been proclaimed.

SECTION 3420 — GENERAL

Required repair levels shall be based on the ratio of the estimated value of the repairs required to restore the structural members to their pre-event condition to the estimated replacement value of the building or structure.

SECTION 3421 — STRUCTURAL REPAIRS

When the damage ratio does not exceed 0.10 (10 percent), buildings and structures, except essential service facilities included as Category I buildings and structures in Table 16-K, shall at a minimum be restored to their pre-event condition.

When the damage ratio is greater than 0.10 (10 percent) but less than 0.5 (50 percent), buildings and structures, except essential service facilities included as Category I buildings and structures in Table 16-K, shall have the damaged structural members including all critical ties and connections associated with the damaged structural members, all structural members supported by the damaged member, and all structural members supporting the damaged members repaired and strengthened to bring them into com-

pliance with the force levels and connection requirements of the Building Code. This criteria shall apply to essential service facilities when the damage ratio is less than 0.3 (30 percent).

> **EXCEPTION:** For buildings with rigid diaphragms where the above-required repair and strengthening increases the rigidity of the resisting members, the entire lateral-force-resisting system of the building shall be investigated. When, in the opinion of the building official, an unsafe or adverse condition has been created as a result of the increase in rigidity, the condition shall be corrected.

When the damage ratio is greater than 0.5 (50 percent), buildings and structures, except essential service facilities included as Category I buildings and structures in Table 16-K, shall at a minimum have the entire building or structure strengthened to comply with the force levels and connection requirements of the Building Code. This criteria shall apply to essential service facilities when the damage ratio is greater than or equal to 0.3 (30 percent).

SECTION 3422 — NONSTRUCTURAL REPAIRS TO LIGHT FIXTURES AND SUSPENDED CEILINGS

Under all damage ratios, when light fixtures and the suspension system of suspended ceiling are damaged, the damaged light fixtures and suspension systems shall be repaired to fully comply with the requirements of this code and UBC Standard 25-2. Undamaged light fixtures and suspension systems shall have the additional support and bracing, provided that is required in UBC Standard 25-2.

UNIT CONVERSION TABLES

SI SYMBOLS AND PREFIXES

BASE UNITS		
Quantity	Unit	Symbol
Length	Meter	m
Mass	Kilogram	kg
Time	Second	s
Electric current	Ampere	A
Thermodynamic temperature	Kelvin	K
Amount of substance	Mole	mol
Luminous intensity	Candela	cd

SI SUPPLEMENTARY UNITS		
Quantity	Unit	Symbol
Plane angle	Radian	rad
Solid angle	Steradian	sr

SI PREFIXES		
Multiplication Factor	Prefix	Symbol
$1\ 000\ 000\ 000\ 000\ 000\ 000 = 10^{18}$	exa	E
$1\ 000\ 000\ 000\ 000\ 000 = 10^{15}$	peta	P
$1\ 000\ 000\ 000\ 000 = 10^{12}$	tera	T
$1\ 000\ 000\ 000 = 10^{9}$	giga	G
$1\ 000\ 000 = 10^{6}$	mega	M
$1\ 000 = 10^{3}$	kilo	k
$100 = 10^{2}$	hecto	h
$10 = 10^{1}$	deka	da
$0.1 = 10^{-1}$	deci	d
$0.01 = 10^{-2}$	centi	c
$0.001 = 10^{-3}$	milli	m
$0.000\ 001 = 10^{-6}$	micro	μ
$0.000\ 000\ 001 = 10^{-9}$	nano	n
$0.000\ 000\ 000\ 001 = 10^{-12}$	pico	p
$0.000\ 000\ 000\ 000\ 001 = 10^{-15}$	femto	f
$0.000\ 000\ 000\ 000\ 000\ 001 = 10^{-18}$	atto	a

SI DERIVED UNIT WITH SPECIAL NAMES			
Quantity	Unit	Symbol	Formula
Frequency (of a periodic phenomenon)	hertz	Hz	1/s
Force	newton	N	$kg \cdot m/s^2$
Pressure, stress	pascal	Pa	N/m^2
Energy, work, quantity of heat	joule	J	$N \cdot m$
Power, radiant flux	watt	W	J/s
Quantity of electricity, electric charge	coulomb	C	$A \cdot s$
Electric potential, potential difference, electromotive force	volt	V	W/A
Capacitance	farad	F	C/V
Electric resistance	ohm	Ω	V/A
Conductance	siemens	S	A/V
Magnetic flux	weber	Wb	$V \cdot s$
Magnetic flux density	tesla	T	Wb/m^2
Inductance	henry	H	Wb/A
Luminous flux	lumen	lm	$cd \cdot sr$
Illuminance	lux	lx	lm/m^2
Activity (of radionuclides)	becquerel	Bq	1/s
Absorbed dose	gray	Gy	J/kg

CONVERSION FACTORS

To convert	to	multiply by
LENGTH		
1 mile (U.S. statute)	km	1.609 344
1 yd	m	0.9144
1 ft	m	0.3048
	mm	304.8
1 in	mm	25.4
AREA		
1 mile^2 (U.S. statute)	km^2	2.589 998
1 acre (U.S. survey)	ha	0.404 6873
	m^2	4046.873
1 yd^2	m^2	0.836 1274
1 ft^2	m^2	0.092 903 04
1 in^2	mm^2	645.16
VOLUME, MODULUS OF SECTION		
l acre ft	m^3	1233.489
1 yd^3	m^3	0.764 5549
100 board ft	m^3	0.235 9737
1 ft^3	m^3	0.028 316 85
	L(dm^3)	28.3168
1 in^3	mm^3	16 387.06
	mL (cm^3)	16.3871
1 barrel (42 U.S. gallons)	m^3	0.158 9873
(FLUID) CAPACITY		
1 gal (U.S. liquid)*	L**	3.785 412
1 qt (U.S. liquid)	mL	946.3529
1 pt (U.S. liquid)	mL	473.1765
1 fl oz (U.S.)	mL	29.5735
1 gal (U.S. liquid)	m^3	0.003 785 412
*1 gallon (UK) approx. 1.2 gal (U.S.)	**1 liter approx. 0.001 cubic meter	
SECOND MOMENT OF AREA		
1 in^4	mm^4	416 231 4
	m^4	416 231 4 $\times$ 10^{-7}
PLANE ANGLE		
1° (degree)	rad	0.017 453 29
	mrad	17.453 29
1' (minute)	urad	290.8882
1" (second)	urad	4.848 137
VELOCITY, SPEED		
1 ft/s	m/s	0.3048
1 mile/h	km/h	1.609 344
	m/s	0.447 04
VOLUME RATE OF FLOW		
1 ft^3/s	m^3/s	0.028 316 85
1 ft^3/min	L/s	0.471 9474
1 gal/min	L/s	0.063 0902
1 gal/min	m^3/min	0.0038
1 gal/h	mL/s	1.051 50
1 million gal/d	L/s	43.8126
1 acre ft/s	m^3/s	1233.49
TEMPERATURE INTERVAL		
1°F	°C or K	0.555 556 $^5/_9$°C = $^5/_9$K
EQUIVALENT TEMPERATURE ($t_{°C}$ = T_K − 273.15)		
$t_{°F}$	$t_{°C}$	$t_{°F}$ = $^9/_5 t_{°C}$ + 32

(Continued)

CONVERSION FACTORS—(Continued)

To convert	to	multiply by
MASS		
1 ton (short ***)	metric ton	0.907 185
	kg	907.1847
1 lb	kg	0.453 5924
1 oz	g	28.349 52
***1 long ton (2,240 lb)	kg	1016.047
MASS PER UNIT AREA		
1 lb/ft^2	kg/m^2	4.882 428
1 oz/yd^2	g/m^2	33.905 75
1 oz/ft^2	g/m^2	305.1517
DENSITY (MASS PER UNIT VOLUME)		
1 lb/ft^3	kg/m^3	16.01846
1 lb/yd^3	kg/m^3	0.593 2764
1 ton/yd^3	t/m^3	1.186 553
FORCE		
1 tonf (ton-force)	kN	8.896 44
1 kip (1,000 lbf)	kN	4.448 22
1 lbf (pound-force)	N	4.448 22
MOMENT OF FORCE, TORQUE		
1 lbf·ft	N·m	1.355 818
1 lbf·in	N·m	0.112 9848
1 tonf·ft	kN·m	2.711 64
1 kip·ft	kN·m	1.355 82
FORCE PER UNIT LENGTH		
1 lbf/ft	N/m	14.5939
1 lbf/in	N/m	175.1268
1 tonf/ft	kN/m	29.1878
PRESSURE, STRESS, MODULUS OF ELASTICITY (FORCE PER UNIT AREA) (1 Pa = 1 N/m^2)		
1 tonf/in^2	MPa	13.7895
1 tonf/ft^2	kPa	95.7605
1 kip/in^2	MPa	6.894 757
1 lbf/in^2	kPa	6.894 757
1 lbf/ft^2	Pa	47.8803
Atmosphere	kPa	101.3250
1 inch mercury	kPa	3.376 85
1 foot (water column at 32°F)	kPa	2.988 98
WORK, ENERGY, HEAT(1J = 1N·m = 1W·s)		
1 kWh (550 ft·lbf/s)	MJ	3.6
1 Btu (Int. Table)	kJ	1.055 056
	J	1055.056
1 ft·lbf	J	1.355 818
COEFFICIENT OF HEAT TRANSFER		
1 Btu/(ft^2·h·°F)	W/(m^2·K)	5.678 263
THERMAL CONDUCTIVITY		
1 Btu/(ft·h·°F)	W/(m·K)	1.730 735
ILLUMINANCE		
1 lm/ft^2 (footcandle)	lx (lux)	10.763 91
LUMINANCE		
1 cd/ft^2	cd/m^2	10.7639
1 foot lambert	cd/m^2	3.426 259
1 lambert	kcd/m^2	3.183 099

GAGE CONVERSION TABLE

APPROXIMATE MINIMUM THICKNESS (inch/mm) FOR CARBON SHEET STEEL CORRESPONDING TO MANUFACTURER'S STANDARD GAGE AND GALVANIZED SHEET GAGE NUMBERS

Manufacturer's Standard Gage No.	CARBON SHEET STEEL				Galvanized Sheet Gage No.	GALVANIZED SHEET			
	Decimal and Nominal Thickness Equivalent		Recommended Minimum Thickness Equivalent[1]			Decimal and Nominal Thickness Equivalent		Recommended Minimum Thickness Equivalent[1]	
	(inch)	(mm)[2]	(inch)	(mm)[2]		(inch)	(mm)[2]	(inch)	(mm)[2]
8	0.1644	4.17	0.156	3.46	8	0.1681	4.27	0.159	4.04
9	0.1495	3.80	0.142	3.61	9	0.1532	3.89	0.144	3.66
10	0.1345	3.42	0.127	3.23	10	0.1382	3.51	0.129	3.23
11	0.1196	3.04	0.112	2.84	11	0.1233	3.13	0.114	2.90
12	0.1046	2.66	0.097	2.46	12	0.1084	2.75	0.099	2.51
13	0.0897	2.28	0.083	2.11	13	0.0934	2.37	0.084	2.13
14	0.0747	1.90	0.068	1.73	14	0.0785	1.97	0.070	1.78
15	0.0673	1.71	0.062	1.57	15	0.0710	1.80	0.065	1.65
16	0.0598	1.52	0.055	1.40	16	0.0635	1.61	0.058	1.47
17	0.0538	1.37	0.050	1.27	17	0.0575	1.46	0.053	1.35
18	0.0478	1.21	0.044	1.12	18	0.0516	1.31	0.047	1.19
19	0.0418	1.06	0.038	0.97	19	0.0456	1.16	0.041	1.04
20	0.0359	0.91	0.033	0.84	20	0.0396	1.01	0.036	0.91
21	0.0329	0.84	0.030	0.76	21	0.0366	0.93	0.033	0.84
22	0.0299	0.76	0.027	0.69	22	0.0336	0.85	0.030	0.76
23	0.0269	0.68	0.024	0.61	23	0.0306	0.78	0.027	0.69
24	0.0239	0.61	0.021	0.53	24	0.0276	0.70	0.024	0.61
25	0.0209	0.53	0.018	0.46	25	0.0247	0.63	0.021	0.53
26	0.0179	0.45	0.016	0.41	26	0.0217	0.55	0.019	0.48
27	0.0164	0.42	0.014	0.36	27	0.0202	0.51	0.017	0.43
28	0.0149	0.38	0.013	0.33	28	0.0187	0.47	0.016	0.41
					29	0.0172	0.44	0.014	0.36
					30	0.0157	0.40	0.013	0.33

[1]The thickness of the sheets set forth in the code correspond to the thickness shown under these columns. They are the approximate minimum thicknesses and are based on the following references:

Carbon sheet steel—Thickness 0.071 inch and over:
  ASTM A 568-74, Table 3, Thickness Tolerances of Hot-Rolled Sheet
  (Carbon Steel).

Carbon sheet steel—Thickness less than 0.071 inch:
  ASTM A 568-74, Table 23, Thickness Tolerances of Cold-Rolled Sheet
  (Carbon and High Strength Low Alloy).

Galvanized sheet steel—All thicknesses:
  ASTM A 525-79, Table 4, Thickness Tolerances of Hot-Dip Galvanized Sheet.

Minimum thickness is the difference between the thickness equivalent of each gage and the maximum negative tolerance for the widest rolled width.

[2]The SI equivalents are calculated and rounded to two significant figures following the decimal point.

INDEX

Index is not all inclusive of code items.